New Trends in Probability and Statistics

NEW TRENDS IN PROBABILITY AND STATISTICS

Volume 1

Proceedings of the Bakuriani Colloquium in Honour of Yu.V. Prohorov, Bakuriani, Georgia, USSR, 24 February - 4 March, 1990

Editors
V.V. Sazonov and T. Shervashidze

MOKSLAS
Vilnius, Lithuania

///VSP///
Utrecht, The Netherlands
Tokyo, Japan

VSP BV
P.O. Box 346
3700 AH Zeist
The Netherlands

MOKSLAS
Žvaigždžiu 23
Vilnius
Lithuania

First published in 1991
ISBN 90-6764-133-2

CIP-DATA KONINKLIJKE BIBLIOTHEEK, DEN HAAG

New trends in probability and statistics: Proceedings of the Bakuriani Colloquium
in honour of Yu.V. Prohorov / eds.: V.V. Sazonov and T. Shervashidze.
- Utrecht: VSP / Vilnius: Mokslas
Vol. 1
ISBN 90-6764-133-2 bound
SISO 517 UDC 519.2
Subject headings: probability theory / mathematical statistics

Typeset by JV DIALOGUE Center STAT-DIALOGUE, Moscow
Publishing Service Group TEV, Vilnius
Printed in Lithuania by Spindulys, Kaunas

CONTENTS

Preface xiii

Bakuriani Colloquia on Probability Theory and Mathematical Statistics xv

I. LIMIT THEOREMS: FINITE – DIMENSIONAL CASE

On the Bounds for the Probabilities of Large Deviations for Random Variables and Vectors
Sh.S. Ebralidze 3

The Influence of Extremes Summands on the Law of Iterated Logarithm
M.U. Gafurov and I.M. Hamdamov 11

Generalized Marcinkiewicz's Theorem and Asymptotic Expansions in the Central Limit Theorem
L.B. Klebanov and J.A. Melamed 30

On Integrability of $\sup |S_{n_k}/n_k|$
O.I. Klesov 38

Asymptotic Expansions in Large Deviation Zones for the Distribution Density of Sums of Independent Random Variables
L. Saulis 43

Strong Law of Large Numbers for Operator Normalized Sums of Independent Random Vectors
S. Solntsev 57

II. LIMIT THEOREMS IN GENERAL SPACES

On the Distribution of the Sup–Norm for a Stable Motion and the Rate of Convergence
M. Bloznelis 73

Semi–Invariant Conditions of Weak Convergence of Random Processes in the Space of Continuous Functions
V. Buldygin 78

Almost Sure Permutational Convergence of Vector Random Series and Kolmogorov's Problem
S.A. Chobanyan and G.J. Giorgobiani 93

Limit Theorems for Stable Laws in Banach Spaces
A.N. Chuprunov 106

An Estimate of the Rate of Convergence in the CLT for Dependent Hilbert Space Valued Random Variables
P. Gudynas 111

An Improved Estimate of the Accuracy of Gaussian Approximation in Hilbert Space
V.V. Sazonov and V.V. Ulyanov 123

On Large Deviations for L–Estimates
R. Zitikis 137

III. LIMIT THEOREMS FOR STOCHASTIC PROCESSES

Rates of Convergence in the Invariance Principle for Empirical Processes
V.I. Kolchinskii 167

Ergodic Theorem for Semimartingale–Helix
R.Sh. Liptser 183

Ergodic Theorems for Discrete Time Random Processes
S.V. Nagaev 190

On Functional Principle of Large Deviations
A. Puhalskii 198

IV. PROBABILITIES IN GENERAL SPACES

Some Properties of Gaussian Measures and Potentials in Martingale Spaces with Mixed Norm
A.D. Bendikov 221

Unexpected Limit Behaviour of the Gaussian Tail Probabilities
M.A. Lifshits 233

Topologies on the Space of σ–Additive Cylindrical Probabilities
D.H. Mushtari 241

Two–Sided Exponential Inequalities for the Distribution of the Norm of Banach Space Valued Random Variables
E.I. Ostrovskii 246

Nonuniform Estimates of the Density of the Squared Norm of a Gaussian Vector in Hilbert Space
V.V. Ulyanov 255

On Sequential Conditions for a Cylindrical Measure
to be Countably Additive
Yu.N. Vladimirskii 264

A Family of Measures Admitting Consistent Estimates
Z.S. Zerakidze 276

V. STOCHASTIC ANALYSIS AND CONTROL

Asymptotic Behaviour of Solutions of Stochastic Equations
V. Buldygin and V. Koval 289

On the Decomposition of a Maximum of Semimartingales
and Itô's Generalized Formula
R.J. Chitashvili and M.G. Mania 301

Optimal Transmission of Gaussian Signals Involving Cost
for Transmission and Feedback
O.A. Glonti and L.G. Jamburia 351

Stationary and Periodic Solutions of Stochastic Difference
and Differential Equations in Banach Space
A.Ya. Dorogovtsev 375

Controlled Random Fields on Graphs and Stochastic Models
of Economic Equilibrium
I.V. Evstigneev 391

Asymptotic Expansions for Singulary Perturbed Stochastic
Differential Equations
Yu.M. Kabanov, S.M. Pergamenshchikov, and J.M. Stoyanov 413

On Uniformly Optimal Coding of Gaussian Messages
in White Gaussian Channels with Feedback
P.K. Katyshev 436

On the Structure of a Random Graph with Nonuniform
Distribution
V.I. Khokhlov and V.F. Kolchin 445

Identities for Ladder Functionals of Homogeneous Random
Processes and Walks with Independent Increments
V.S. Lugavov and B.A. Rogozin 457

On Convergence of Solutions of Stochastic Equations
S. Makhno 474

Stochastic Integrals and Stochastic Differential Equations
on the Plane Involving Strong Semimartingales
Yu.S. Mishura 485

Itô–Ventsel' Formula For Anticipative Processes
O. Purtukhia 503

Connectivity Property and Optimality Almost Surely and in Probability
V.I. Rotar' 528

Absolute Continuity of Smooth Measures Generated by Nonlinear Equations
G. Sokhadze 540

VI. MATHEMATICAL STATISTICS

A New Method of Asymptotic Analysis of Simple Linear Rank Statistics
D.M. Chibisov 553

On Stable M-Estimators in the Partial Likelihood Scheme
N.L. Lazrieva and T.A. Toronjadze 567

Asymptotic Estimation of Integrals of Certain Random Fields
A.V. Nagaev 597

On Minimax Bound for Parameter Estimation in Ball (Bias Accounting)
A.V. Nazin 612

Limit Theorems for Some Random Processes Based on Frequencies
R.M. Mnatsakanov 617

From Asymptotic Efficiency in Minimax Sense to Bahadur Efficiency
M. Radavičius 629

On Estimating Parameters of a Distribution Concentrated in a Circle
A.E. Shemyakin 636

On the Approximation of the Likelihood Function by a Vector Analogue of the Whittle Statistics
K.A. Tsertsvadze 644

VII. TOWARDS NATURAL SCIENCES

On Condensable Point Processes
R.V. Ambartzumian 655

The Ground State of a Random Stationary Medium in the Mean Field Approximation
A. Astrauskas and S.A. Molchanov 668

Scaling Behavior of Random Walks with Topological Constraints
S.K. Nechaev and Ya.G. Sinai 683

Convergence of the Stochastic Quantization Method for Lattice R–Gauge Theories
V. Sidoravičius 694

List of Contributors 701

Organizing Committee 702

PREFACE

In the picturesque Georgian village of Bakuriani situated in the Caucasus mountains, the colloquium on Probability Theory and Mathematical Statistics was held to honor one of the most eminent probabilists of our time – Academician Yu.V. Prokhorov.

The scientific thinking of Yu.V. Prokhorov is characterized by his ability to find and clearly state natural problems of Probability and Statistics and to search for new approaches and methods for solving them.

The contribution of Yu.V. Prokhorov to Probability and Statistics is widely known and covers many different topics. We mention here only a few of them which we believe are the most important:

– He solved a difficult classical problem by finding necessary and sufficient conditions for the validity of the strong law of large numbers for sums of independent random variables.

– He made a deep contribution to the area of local limit theorems for both lattice and continuously distributed random variables.

– He wrote important papers on the problem of approximation of convolutions of probability distributions by different classes of laws (such as Gaussian, Poisson and general infinitely divisible laws).

– Especially important is his fundamental paper of 1956 dealing with functional methods in the theory of stochastic processes. This paper had a great influence on the future development of a large part of modern Probability Theory.

– There is also his brilliant work on asymptotic methods in queuing theory.

Among the participants in the colloquium there were many of Prokhorov's former students and the students of his students who continued to develop his ideas.

Unfortunately Yu.V. Prokhorov could not take part in the colloquium in person. He sent, however, the following greeting to the participants:

"Dear Colleagues! Let me express my cordial gratitude to the organizers of this colloquium for their decision to connect it with my recent 60th birthday.

The rapid development of Probability Theory and Mathematical Statistics in the modern world is easily noticeable, e.g., new journals are being founded often and widely distributed to promote different directions of these sciences.

Together with you, dear participants, I would like to devote the rest of my life to the development of 'our' science. In this I see the source of genuine happiness and real stability in these quickly changing circumstances of life."

There were around 70 participants in the colloquium who represented almost all major probability and statistics centers of the Soviet Union. The main topics of the colloquium were limit theorems, probabilities on general spaces, statistics and control of stochastic processes.

This colloquium by far was not the first probability and statistics colloquium in Bakuriani: in fact, it was the 23rd. In the previous ones a number of prominent specialists took part, e.g., in several of them A.N. Kolmogorov participated with great enthusiasm. Up to now the Bakuriani colloquia were important and influential events in the life of the probability and statistics community of the Soviet Union.

In the present proceedings, we included the papers actually delivered at the colloquium as well as some papers whose authors could not participate in person but wanted to make a contribution to the Festschrift for Yu.V. Prokhorov.

V.V. Sazonov *T. Shervashidze*

BAKURIANI COLLOQUIA ON PROBABILITY THEORY AND MATHEMATICAL STATISTICS

The colloquia were organized almost every year since 1966 by several Georgian mathematical institutions jointly with the Steklov Mathematical Institute.

Since 1979, reports on the colloquia were published in the journal "Theory of Probability and Its Applications". Since 1980, the abstracts of papers presented at the colloquia, and sometimes proceedings, have appeared in Russian in separate Tbilisi publications.

The average number of participants at Bakuriani colloquia has been 77 (with the maximum 103 and the minimum 52).

This year the colloquium was organized by the Razmadze Mathematical Institute and the Muskhelishvili Institute of Computational Mathematics of the Georgian Academy of Sciences, the Tbilisi State University, and the Steklov Mathematical Institute.

Part One

LIMIT THEOREMS: FINITE–DIMENSIONAL CASE

New Trends in Probab. and Statist., pp. 3 - 10
V. Sazonov and T. Shervashidze (Eds.)
1991 VSP/Mokslas

ON THE BOUNDS FOR THE PROBABILITIES OF LARGE DEVIATIONS FOR RANDOM VARIABLES AND VECTORS

Sh.S. EBRALIDZE
Laboratory of Probabilistic and Statistical Methods, I. Javakhishvili State University, David Agmashenebeli Ave. 150[a], 380012 Tbilisi , Georgia, USSR.

ABSTRACT

Inequalities for the probabilities of large deviations for sums of independent random variables and vectors are studied. The obtained estimates use the closeness of distributions of summands to the normal one.

1. BOUNDS IN TERMS OF GENERALIZED MOMENTS

This section is devoted to the bounds of $\mathbf{P}(S_n > x)$, where S_n is a sum of independent random variables, in terms of exponential moments (or generalized pseudomoments). First bounds in such characteristics were given in (Nagaev and Sakoyan, 1976). In (Nagaev, 1979) the significance of these bounds is discussed in detail. In (Shorgin, 1982) the rate of convergence in the central limit theorem in the same terms was obtained. Let $X_1, ..., X_n$ be independent random variables with continuous strictly increasing distribution functions $F_i(x) = \mathbf{P}(X_i < x)$. For any $\delta \geqslant 2$ define a class of functions $G(\delta)$ with continuous nonincreasing derivatives on $(0, \infty)$ satisfying the conditions:

(a) $g'(u) \to 0$ as $u \to \infty$,

(b) $g'(u) > \delta/u$.

For a random variable x denote by $\Phi_X(x)$ the normal distribution function with the same first and second moments as X, $\Phi_{(x)}$ denotes the standart normal distribution function.
Define

$$b_{gi} = \int_{-\infty}^{+\infty} e^{g(|u|)} |F_i(u) - \Phi_{X_i}(u)| du, \quad B_{gn} = \sum_{i=1}^{n} b_{gi}, \quad i = 1, ..., n.$$

Let us put

$$\sigma_i^2 = \operatorname{var} X_i, \quad B_n^2 = \sum_{i=1}^{n} \sigma_i^2, \quad S_n = \sum_{i=1}^{n} X_i,$$

and suppose

$$\mathbf{E}X_i = 0, \quad \mathbf{E}\exp g(|X_i|) < \infty, \quad g \in G(\delta), \quad i = 1, ..., n. \tag{1}$$

THEOREM 1. *Let independent random variables $X_1, ..., X_n$ in addition to* (1) *have zero medians, then, for any α, $0 < \alpha < 1$, and $x > 0$,*

$$\mathbf{P}(|S_n| > x) \leqslant 2 - 2\Phi\left(\frac{(1-\alpha)x}{B_n}\right) + \left(\frac{1+\delta}{\alpha x} Bgn\right)^{\alpha\delta/(1+\delta)} \exp\left\{-\frac{\alpha\delta}{1+\delta} g(x)\right\}.$$

For the proof we shall need an inequality proved in (Ebralidze, 1971a).
Let X be a random variable with a distribution function $F(x)$ and a zero median. It is know that if Z is a random variable uniformly distributed on [0,1] then $F^{-1}(Z)$ is a random variable distributed as X (F^{-1} is the inverse function of F). Put

$$\widetilde{X} = F^{-1}(Z), \quad Y = \Phi_X^{-1}(Z).$$

LEMMA. *Let $h > 0$, then*

$$\mathbf{E}e^{\pm h(\tilde{X} - Y)} \leqslant 1 + \int_{-\infty}^{+\infty} |F(u) - \Phi(u)| h(e^{hu} - 1) du. \tag{2}$$

Proof of Theorem 1. Let $Z_1, ..., Z_n$ be independent random variables uniformly distributed on [0,1]. Define the random variables

$$\widetilde{X}_i = F^{-1}(Z_i), \quad Y_i = \Phi_{X_i}^{-1}(Z_i).$$

As it has already been noted, $\widetilde{X}_i$ are distributed as X_i, and Y_i are normally distributed with distribution functions $\Phi_{X_i}(x)$. Obviously, the random variables $\widetilde{X}_1, ..., \widetilde{X}_n$ and $Y_1, ..., Y_n$ are independent.
For any x and any α, $0 < \alpha < 1$,

$$\mathbf{P}\Big(|S_n| > x\Big) = \mathbf{P}\Big(|\widetilde{X}_1 + ... + \widetilde{X}_n| > x\Big) \leqslant$$

$$\leqslant \mathbf{P}\Big(|\widetilde{X}_1 - Y_1 + ... + \widetilde{X}_n - Y_n| > \alpha x\Big) + \mathbf{P}\Big(|Y_1 + ... + Y_n| > (1-\alpha)x\Big). \tag{3}$$

For any $h > 0$

$$\mathbf{P}\Big(|\widetilde{S}_n| > \alpha x\Big) \leqslant e^{-\alpha h x} \mathbf{E}e^{h|\widetilde{S}_n|}, \tag{4}$$

where $\widetilde{S}_n = \sum_{i=1}^n (\widetilde{X}_i - Y_i)$. The independence of $\widetilde{X}_1 - Y_1, ..., \widetilde{X}_n - Y_n$ implies

$$\mathbf{E}e^{h|\widetilde{S}_n|} = \prod_{i=1}^n \mathbf{E}e^{h|\tilde{X}_i - Y_i|} \leqslant \exp\left[\sum_{i=1}^n \mathbf{E}e^{h|\tilde{X}_i - Y_i|} - 1\right].$$

Using (2), we have

$$\mathbf{E}e^{h|\widetilde{S}_n|} \leqslant \exp\Big(\sum_{i=1}^{n}\int_{-\infty}^{\infty}|F_i(x)-\Phi_{X_i}(x)|h(e^{h|x|}-1)dx\Big).$$

Taking this into account, we have from (4)

$$\mathbf{P}\Big(|\widetilde{S}_n|>\alpha x\Big) \leqslant \exp\Big(-\alpha hx+\sum_{i=1}^{n}\int_{-\infty}^{\infty}|F_i(u)-\Phi_{X_i}(u)|h(e^{h|u|}-1)du\Big). \quad (5)$$

Consider

$$\int_{-\infty}^{\infty} e^{h|u|}|F_i(u)-\Phi_{X_i}(u)|du =$$

$$= \int_{-\infty}^{\infty} e^{h|u|-g(u)}e^{g(|u|)}|F_i(u)-\Phi_{X_i}(u)|du \leqslant \sup_u e^{h|u|-g(|u|)}b_{gi}.$$

Introduce $f(u) = hu - g(u)$, $u > 0$, $f'(u) = h - g'(u)$. If $h = g'(x)$, then $f'(x) = 0$, i.e.,

$$\sup_u e^{h|u|-g(|u|)} = e^{hx-g(x)}, \quad x > 0.$$

Using this, we have from (5),

$$\mathbf{P}\Big(|\widetilde{S}_n|>\alpha x\Big) \leqslant \exp\Big[-\alpha hx + hB_{gn}e^{hx-g(x)}\Big]. \quad (6)$$

Examine the function $S(h) = -\alpha hx + hB_{gn}e^{hx-g(x)}$ for $h > 0$. Evidently,

$$S(0) = 0, \quad S'(h) = -\alpha hx + B_{gn}e^{hx-g(x)} +$$

$$+B_{gn}hxe^{hx-g(x)}, \quad S''(h) > 0$$

Hence, the function $S(h)$ has a unique minimum at some point h_0.

$$.S'(h_0) = -\alpha x + B_{gn}e^{h_0x-g(x)} + B_{gn}h_0e^{h_0x-g(x)} = 0,$$

$$S(h_0) = -h_0^2xB_{gn}e^{hx-g(x)}.$$

It is not easy to find h_0, but we need the upper bound for $\min_{h>0} S(h)$ rather than an exact value of h_0.
We have

$$-\alpha x + B_{gn}(1+h_0x)e^{h_0x-g(x)} = 0,$$

$$\alpha x = B_{gn}(1+h_0x)e^{h_0x-g(x)}.$$

From condition (b) for the function $g(u)$, and keeping in mind that

$$h = g'(x), \quad hx \geqslant \delta > 2,$$

we obtain

$$h_0 < \frac{1}{x} \log \left[\frac{\alpha x}{(1+\delta)B_{gn}} e^{g(x)} \right].$$

Since $S(h)$ has a minimum at h_0 for any other value $h < h_0$,

$$S(h_0) < S(h) = -\alpha \log \left[\frac{\alpha x}{(1+\delta)B_{gn}} e^{g(x)} \right] +$$

$$+ B_{gn} \frac{1}{x} \log \left[\frac{\alpha x}{(1+\delta)B_{gn}} e^{g(x)} \right] \exp \left\{ \log \frac{\alpha x}{(1+\delta)B_{gn}} \right\} =$$

$$= -\frac{\alpha\delta}{1+\delta} \log \left[\frac{\alpha x}{(1+\delta)B_{gn}} e^{g(x)} \right].$$

From (6) we obtain

$$\mathbf{P}\Big(|\widetilde{S}_n| > \alpha x \Big) \leqslant \exp \left\{ -\frac{\alpha\delta}{1+\delta} \log \left[\frac{\alpha x}{(1+\delta)B_{gn}} e^{g(x)} \right] \right\}.$$

Substituting the latter relation into (3) and taking into account that the random variable $Y_1 + ... + Y_n$ is normally distributed with parameters $(0, B_n)$, we obtain the assertion of the theorem.

2. THE MULTIVARIATE CASE

This section is devoted to estimates of probabilities for sums and maximums of sums of random vectors. The estimates are obtained in terms of conditional difference pseudomoments. The obtained estimates are of Nagaev–Fuk's inequality type (Fuk and Nagaev, 1971).

Let $X_1, ..., X_n$ be independent random vectors with values in the k–dimensional Euclidean space $\mathbf{R}^k$ satisfying the conditions:

$$\mathbf{E}X_i = 0, \quad \mathbf{E}|X_i|^2 = \Lambda_i, \quad \mathbf{E}|X_i|^3 < \infty, \quad i = 1, ..., n, \tag{7}$$

where $|X|$ is the length of the vector X.

Let $F_{ij}(x_j \mid x_1, ..., x_{j-1}) = \mathbf{P}(X_{ij} < x_j \mid X_{i1} = x_1, ..., X_{i,j-1} = x_{j-1})$ be the conditional distribution function of the component X_{ij} of the vector $X_i = (X_{i1}, ..., X_{ik})$ with respect to the first $j-1$ components. And $F_{i1}(x_1)$ will denote the marginal distribution function of the first component X_{i1} of the vector X_i.

Suppose that all distribution functions $F_{i1}(x_1), F_{i2}(x_2 \mid x_1), ..., F_{ik}(x_k \mid x_1, ..., x_{k-1})$ are continuous and strictly increasing, thus the inverse functions F_{ij}^{-1} exist, they are defined by the relation

$$F_{ij}\Big(F_{ij}^{-1}(y \mid x_1, ..., x_{j-1}) \mid x_1, ..., x_{j-1} \Big) = y.$$

Let the functions $\Phi_{ik}(x_k \mid x_1, ..., x_{k-1})$ be similarly defined conditional distribution functions for some normal distribution, the first and second moments of which coincide with the first and second moments of the vector X_i.

Suppose $Z_{i1},...,Z_{ik}$, $i=1,2,...,n$, are independent sets of random variables uniformly distributed on [0,1]. A sequence of such independent uniformly distributed on [0,1] random variables is produced if we define the random variables Z_{ij} in the following way

$$Z_{ij}=F_{ij}(X_{ij}\mid X_{i1},...,X_{i,j-1}),$$

which are uniformly distributed and independent. Indeed,

$$\mathbf{P}\Big(Z_{ij}<t\mid Z_{i1}=z_1,...,Z_{i,j-1}=z_{j-1}\Big)=$$

$$=\mathbf{P}\Big(F_{ij}(X_{ij}\mid X_{i1},...,X_{i,j-1})<t\mid X_{i1}=F_{i1}^{-1}(z_1),...,X_{i,j-1}=$$

$$=F_{i,j-1}^{-1}(z_{j-1}\mid z_1,...,z_{j-1})\Big)=$$

$$=\mathbf{P}\Big(X_{ij}<F_{ij}^{-1}(t\mid...)\mid...\Big)=F_{ij}\Big(F_{ij}^{-1}(t\mid...)\mid...\Big)=t.$$

Consequently, Z_{ij} are independent since the conditional probability of Z_{ij} under the condition $X_{ij},...,X_{i,j-1}$ is independent of the condition.
Define the vectors $\widetilde{X}_i=(\widetilde{X}_{i1},...,\widetilde{X}_{ik})$, $Y_i=(Y_{i1},...,Y_{ik})$ by the relations

$$\widetilde{X}_{ij}=F_{ij}^{-1}\Big(Z_{ij}\mid X_{i1},...,X_{i,j-1}\Big),\quad Y_{ij}=\Phi_{ij}^{-1}\Big(Z_{ij}\mid Y_{i1},...,Y_{i,j-1}\Big),\quad j=1,...,k,$$

F_{ij}^{-1} and Φ_{ij}^{-1} are inverse functions of the introduced distribution functions.
The vectors $\widetilde{X}_i$ are distributed as X_i, and Y_i are normally distributed vectors

$$\mathbf{P}\Big(\widetilde{X}_{ij}<x_j\mid X_{i1}=x_1,...,X_{i,j-1}=x_{j-1}\Big)=$$

$$=\mathbf{P}\Big(F_{ij}^{-1}(Z_{ij}\mid X_{i1},..,X_{i,j-1})<x_j\mid X_{i1}=x_1,...,X_{i,j-1}=x_{j-1}\Big)=$$

$$=\mathbf{P}\Big(Z_{ij}<F_{ij}(x_j\mid X_{i1},...,X_{i,j-1})\mid X_{i1}=x_1,...,X_{i,j-1}=x_{j-1}\Big)=$$

$$=F_{ij}\Big(x_j\mid x_1,...,x_{j-1}\Big).$$

Obviously, for any t, $t>0$,

$$\mathbf{P}\Big(|X_1+...+X_n|>t\Big)=\mathbf{P}\Big(|\widetilde{X}_1+...+\widetilde{X}_n|>t\Big)\leqslant$$

$$\leqslant\mathbf{P}\Big(|\widetilde{X}_1-Y_1+...+\widetilde{X}_n-Y_n|>\alpha t\Big)+\mathbf{P}\Big(|Y_1+...+Y_n|>(1-\alpha)t\Big),\qquad(8)$$

where α is some constant, $0<\alpha<1$.
Estimating the first summand in (8) by means of a multivariate analogue of Nagaev–Fuk's estimate (Fuk and Nagaev, 1971), we obtain

$$\mathbf{P}\Big(|\widetilde{X}_1-Y_1+...+\widetilde{X}_n-Y_n|>\alpha t\Big)<4\exp\Big(-C_1\frac{\alpha^2t^2}{\bar{\Lambda}_n}\Big)+C_2\frac{M_n}{\alpha^3t^3},$$

where C_1 and C_2 are constants and

$$\widetilde{\Lambda}_n = \sum_{i=1}^{n} \mathbf{E}|\widetilde{X}_i - Y_i|^2, \quad M_n = \sum_{i=1}^{n} \mathbf{E}|\widetilde{X}_i - Y_i|^3.$$

We can estimate $\widetilde{\Lambda}_n$ from above as

$$\widetilde{\Lambda}_n = \sum_{i=1}^{n} \mathbf{E}|\widetilde{X}_i - Y_i|^2 \leqslant 2\sum_{i=1}^{n} \Lambda_i.$$

Now we can estimate the third order moment

$$\mathbf{E}|\widetilde{X}_i - Y_i|^3 = \mathbf{E}\Big(\sum_{j=1}^{k}(\widetilde{X}_{ij} - Y_{ij})^2\Big)^{3/2} \leqslant k^{1/2}\sum_{j=1}^{k}\mathbf{E}|\widetilde{X}_{ij} - Y_{ij}|^3.$$

Keeping in mind the definition of the random variables $\widetilde{X}_{ij}$ and Y_{ij}, we have

$$\mathbf{E}|\widetilde{X}_{ij} - Y_{ij}|^3 = \mathbf{E}\Big|F_{ij}^{-1}\Big(Z_{ij} \mid X_{i1}, \dots, X_{i,j-1}\Big) - \Phi_{ij}\Big(Z_{ij} \mid Y_{i1}, \dots, Y_{i,j-1}\Big)\Big|^3.$$

We can apply an elementary inequality which is valid for all a and b : $|a - b|^3 \leqslant 2^2|a^3 - b^3|$, and repeat the scheme of the proof from (Ebralidze, 1971b)

$$\mathbf{E}|\widetilde{X}_{ij} - Y_{ij}|^3 \leqslant 4\mathbf{E}\int_0^1 \Big|[F_{ij}^{-1}(z \mid \dots)]^3 - [\Phi_{ij}^{-1}(z \mid \dots)]^3\Big| dz.$$

The function $z = \widetilde{F}_{ij}(u \mid \dots) = F_{ij}(u^{1/3} \mid \dots)$ is a conditional distribution function and its inverse function is $u = [F_{ij}^{-1}(z \mid \dots)]^3$.
Similarly, for $\widetilde{\Phi}_{ij}(u \mid \dots)$,

$$\int_0^1 \Big|[F_{ij}^{-1}(z \mid \dots)]^3 - [\Phi_{ij}^{-1}(z \mid \dots)]^3\Big| dz =$$

$$= \int_0^1 \Big|\widetilde{F}_{ij}(z \mid \dots) - \widetilde{\Phi}_{ij}(z \mid \dots)\Big| dz =$$

$$= \int_{-\infty}^{\infty} \Big|\widetilde{F}_{ij}^{-1}(u \mid \dots) - \widetilde{\Phi}_{ij}^{-1}(u \mid \dots)\Big| du =$$

$$= \int_{-\infty}^{\infty} \Big|F_{ij}(u^{1/3} \mid \dots) - \Phi_{ij}(u^{1/3} \mid \dots)\Big| du =$$

$$= 3 \int_{-\infty}^{\infty} u^2 \Big| F_{ij}(u|...) - \Phi_{ij}(u|...) \Big| du.$$

Define

$$\nu_{ij}^{(3)} = 3\mathbf{E} \int_{-\infty}^{\infty} \Big| F_{ij}(u \mid X_{i1}, ..., X_{i,j-1}) - \Phi_{ij}(u \mid Y_{i1}, ..., Y_{i,j-1}) \Big| u^2 du.$$

Summing up stated above, we have

$$\mathbf{E}|\widetilde{X}_{ij} - Y_{ij}|^3 \leqslant 4k^{1/2} \nu_{ij}^{(3)}.$$

We have proved the following theorem.

THEOREM 2. *Let $X_1, ..., X_n$ be independent random vectors in $\mathbf{R}^k$ satisfying conditions (7). Then, for any constant $t > 0$ and any α, $0 < \alpha < 1$,*

$$\mathbf{P}\Big(|X_1 + ... + X_n| > t\Big) < \mathbf{P}\Big(|Y_1 + ... + Y_n| > (1-\alpha)t\Big) +$$

$$+4 \exp\left(- C_1 \frac{\alpha^2 t^2}{\Lambda_n} \right) + C_2 k^{1/2} \frac{\nu_n^{(3)}}{\alpha^3 t^3},$$

where C_1 and C_2 are absolute constants,

$$\Lambda_n = \sum_{i=1}^{n} \Lambda, \quad \nu_n^{(3)} = \sum_{i=1}^{n} \sum_{j=1}^{k} \nu_{ij}^{(3)}.$$

Using the results of (Borovkov, 1972) one can easily see that a similar estimate is also valid for

$$\mathbf{P}\left(\max_{1 \leqslant m \leqslant n} |X_1, ..., X_m| > t \right).$$

THEOREM 3. *Let $X_1, ..., X_n$ be independent random vectors in R^k satisfying conditions (7). Then there exist constants N_1, N_2, N_3 such that, for any $t > 0$ and any α, $0 < \alpha < 1$,*

$$\mathbf{P}\left(\max_{1 \leqslant m \leqslant n} |X_1 + ... + X_m| > t \right) \leqslant \mathbf{P}\left(\max_{1 \leqslant m \leqslant n} |Y_1 + ... + Y_m| > (1-\alpha)t \right) +$$

$$+N_1 \exp\left(- N_2 \frac{\alpha^2 t^2}{\Lambda_n} \right) + N_3 k^{1/2} \frac{\nu_n^{(3)}}{\alpha^3 t^3}.$$

REFERENCES

Borovkov, A.A. (1972). Notes on inequalities for sums of independent variables. *Theory Probab. Appl.* **17**, 556–557.

Ebralidze, Sh.S. (1971a). Inequalities for the probabilities of large deviations in terms of pseudomoments. *Theory Probab. Appl.* **16**, 761–765.

Ebralidze, Sh.S. (1971 b). Inequalities for the probabilities of large deviations in the multi-dimensional case. *Theory Probab. Appl.* **16**, 765–769.

Fuk, D.X. and Nagaev, S.V. (1971). Probability inequalities for sums of independent random variables. *Theory Probab. Appl.* **16**, 643–660.

Nagaev, S.V. (1979). Large deviations of sums of independent random variables. *Ann. Probab.* **7**, 745–789.

Nagaev, S.V. and Sakoyan, S.K. (1976). On a bound for the probability of large deviations., In: *Limit Theorems and Mathematical Statistics*, FAN, Tashkent, pp. 132–140 (in Russian).

Shorgin, A.Ya. (1982). Nonclassical estimates of the rate of convergence in the central limit theorem taking into account large deviations. *Theory Probab. Appl.* **27**, 324–337.

New Trends in Probab. and Statist., pp. 11 - 29
V. Sazonov and T. Shervashidze (Eds.)
1991 VSP/Mokslas

THE INFLUENCE OF EXTREMES SUMMANDS ON THE LAW OF ITERATED LOGARITHM

M.U. GAFUROV and I.M. HAMDAMOV
Tashkent Motor Highway Institute, Department of High Mathematics,
K. Marx st. 32, 700000 Tashkent,
Institute of Mathematics of the Uzbek Academy of Sciences,
F. Khodjaev str., 29, 700143 Tashkent, USSR

ABSTRACT

Let $X_1, \dots, X_n$ be i.i.d. random variables with a common symmetric distribution, and let $X_n^{(1)} \geqslant \dots \geqslant X_n^{(n)}$ denote the corresponding order statistics. Define $S_n^{(r)} = = \sum_{i=1}^n X_i - \sum_{j=1}^r X_n^{(j)}$, $n > r \geqslant 1$. In (Feller, 1968) LIL for $S_n^{(1)}$ was proved under some conditions. In this paper we show that Feller's conditions are not sufficient and establish that LIL holds for $S_n^{(r)}$, $r = \text{const} \geqslant 1$, under some stronger conditions.

1. INTRODUCTION, THE MAIN RESULT AND DISCUSSION

Let $X_1, X_2, \dots$ be a sequence of independent and identically distributed (i.i.d.) random variables (r.v.) with common symmetric distribution function $F(x)$. For any $n \geqslant 1$, let $X_n^{(1)} \geqslant \dots \geqslant X_n^{(n)}$ be the order statistics based on $X_1, \dots, X_n$.
Put $S_n^{(0)} = S_n = \sum_{i=1}^n X_i$, and if $r \geqslant 1$ is a fixed integer and $n > r$, then let $S_n^{(r)} = S_n - \sum_{i=1}^r X_n^{(i)}$ be the (lightly) trimmed sum.
Suppose first $\mathbf{E}X_1^2 < \infty$. This is equivalent to the convergence of the series $\sum_{n=1}^\infty \mathbf{P}\{X_1 \geqslant \varepsilon\sqrt{n}\}$ for any $\varepsilon > 0$. From this and by the Borel–Cantelli lemma it follows that the maximal term in the sum is negligible, i.e., $X_n^{(1)} = o(\sqrt{n})$, a.s., $n \to \infty$, and, hence, also $X_n^{(k)} = o(\sqrt{n})$ a.s., $k \geqslant 2$. Thus, in the case of finite variance,

$$\frac{S_n}{\sqrt{n}} = \frac{S_n^{(r)}}{\sqrt{n}} + o(1) \quad \text{a.s.} \tag{1.1}$$

as $n \to \infty$ and for both sums holds the law of the iterated logarithm (LIL).
An entirely different situation is observed when $\mathbf{E}X_1^2 = \infty$. For an illustration let us consider an example of Feller, who was one of the first to observe this phenomenon.
Let the distribution $F(x)$ be concentrated on $|x| > 1$ and has density $|x|^{-3}$.
Then obviously $\mathbf{E}X_1^2 = \infty$, however the central limit theorem holds here with the normalization $\sqrt{(n \log n)/2}$. On the other hand it can be shown that the series

$\sum_{n=1}^{\infty} \mathbf{P}\{X_1 \geqslant \varepsilon\sqrt{n \log n}\}$ and even the series $\sum_{n=1}^{\infty} \mathbf{P}\{X_1 \geqslant \varepsilon\sqrt{n \log n \log\log n}\}$ diverges and then

$$\limsup_{n\to\infty} \frac{S_n}{\sqrt{n \log n \log\log n}} = \infty \quad \text{a.s.},$$

i.e., the LIL does not hold here.

If the maximal term of the sum S_n is excluded then it is shown in (Feller, 1968) that

$$\limsup_{n\to\infty} \frac{S_n^{(1)}}{\sqrt{n \log n \log\log n}} = 1 \quad \text{a.s.};$$

this happens due to the fact that

$$\limsup_{n\to\infty} \frac{X_n^{(1)}}{\sqrt{n \log n \log\log n}} = \infty \quad \text{a.s.}$$

Furthermore, Feller obtained the following result for a class of the distribution functions $F(x)$. Assuming that

$$B(x) = \int\limits_{|t|<x} t^2 dF(t), \quad B(2x)/B(x) < C, \quad \text{for all} \quad x > x_0,$$

$$I(\gamma) = \int\limits_3^{\infty} \frac{x^2 dF(x)}{B(x)(\log\log x)^{1+\gamma}} < \infty, \quad \text{for some} \quad \gamma \in (0; 1/4), \tag{1.2}$$

and $I(0) = \infty$, he showed that

$$\limsup_{n\to\infty} \frac{S_n^{(1)}}{\rho_n\sqrt{2\log\log\rho_n}} = 1 \quad \text{a.s.},$$

but

$$\limsup_{n\to\infty} \frac{S_n}{\rho_n\sqrt{2\log\log\rho_n}} = \infty \quad \text{a.s.},$$

where ρ_n is the largest root of the equation

$$\rho_n^2 = nB(\rho_n). \tag{1.3}$$

Note that, under the conditions of Feller,

$$\limsup_{n\to\infty} \frac{X_n^{(1)}}{\rho_n\sqrt{2\log\log\rho_n}} = \infty \quad \text{a.s.},$$

but the extremum terms of the sum $S_n^{(1)}$ do not play the essential role, i.e., $X_n^{(r)} = o(\rho_n\sqrt{\log\log\rho_n})$ a.s, $n \to \infty$, for any $r \geqslant 2$.

A natural question arises: What are the conditions on $F(x)$, under which $X_n^{(1)}, \ldots, X_n^{(r)}$ $(r \geqslant 2)$ play the same essential role in the sum S_n, as $X_n^{(1)}$ in

S_n and is it possible at all? The answer to this question is the main aim of this paper.
Let $\mathbf{E}X_1^2 = \infty$, $P(x) = 1 - F(x) = F(-x)$ and

$$I_r(\gamma) = \int_3^\infty \frac{x^{2r}P^{r-1}(x)\,dF(x)}{B^r(x)(\log\log x)^{r+\gamma}}, \quad \gamma \geqslant 0.$$

We introduce the condition

$$2^\alpha \leqslant \liminf_{x\to\infty} \frac{P(x)}{P(2x)} \leqslant \limsup_{x\to\infty} \frac{P(x)}{P(2x)} < \infty, \tag{1.4}$$

where $0 < \alpha \leqslant 2$. It is easy to see that this condition holds, for example, if $P(x) \sim x^{-\beta}l(x)$, $x \to \infty$ ($\beta \geqslant \alpha$, $l(x)$ is slowly varying function (s.v.f.)). Note also that (1.4) implies $B(2x)/B(x) < C$.
The next theorem is the main result of this paper.

THEOREM. *Assume (1.4), and let $I_r(0) = \infty$* and *$I_r(\gamma) < \infty$ for some $\gamma \in (0; r\gamma_0)$. Then*

$$\limsup_{n\to\infty} \frac{S_n^{(r)}}{g_n\sqrt{2\log\log g_n}} = 1 \quad \text{a.s.},$$

but

$$\limsup_{n\to\infty} \frac{S_n^{(r-1)}}{g_n\sqrt{2\log\log g_n}} = \infty \quad \text{a.s.},$$

where

$$g_n = \begin{cases} \sqrt{nB(d_n)} & \text{if } \ 0 < \alpha \leqslant 2(r+1)/(r+2), \\ \rho_n & \text{if } \ 2(r+1)/(r+2) < \alpha \leqslant 2; \end{cases}$$

$$\gamma_0 = \begin{cases} \alpha/\big(2(r+1) - \alpha\big) & \text{if } \ 0 < \alpha \leqslant 2(r+1)/(r+2), \\ \big(\alpha(r+2) - 2(r+1)\big)/\big(2(r+1) - \alpha\big) & \text{if } \ 2(r+1)/(r+2) < \alpha \leqslant 2; \end{cases}$$

and d_n is the largest root of the equation

$$d_n^2 = nB(d_n)\log\log d_n. \tag{1.5}$$

It should be noted that our theorem is not a generalization of the result of Feller. We have to impose on the original distribution condition (1.4) which is stronger than the analogous condition $B(2x)/B(x) < M$ in (Feller, 1968). Example 1 below shows that it is impossible to reject condition (1.4) completely.
However, there are situations where our theorem is a generalization of the Feller result. For example, if $P(x) = x^{-2}l(x)$ then condition (1.4) and condition $B(2x)/B(x) < M$ are equivalent and $\rho_n = g_n$. From the theorem it is easy to obtain the following

COROLLARY. *Under the conditions of the theorem there exists a sequence w_n such that $w_n/\sqrt{n}$ increases and*

$$\limsup_{n\to\infty} \frac{S_n^{(r)}}{w_n} = 0 \quad \text{a.s.},$$

but

$$\limsup_{n\to\infty} \frac{S_n^{(r-1)}}{w_n} = \infty \quad \text{a.s.} \tag{1.6}$$

Indeed, if

$$\limsup_{n\to\infty} \frac{S_n^{(r-1)}}{g_n\sqrt{\log\log g_n}} = \infty \quad \text{a.s.},$$

then we can choose w_n such that

$$\frac{w_n}{g_n\sqrt{\log\log g_n}} \longrightarrow \infty \quad \text{and} \quad \limsup_{n\to\infty} \frac{S_n^{(r-1)}}{w_n} = \infty \quad \text{a.s.}$$

and (1.6) is valid for such w_n. Note that to get (1.6), Mori in (Mori, 1977) required the increase of $w_n/n^{1/\delta}$, for $\delta < 2$. We extend the results of Mori to the case $\delta = 2$. In the next example all conditions of Theorem 2 in (Feller, 1968) are satisfied, but nevertheless the conclusion of the theorem is not true.

Example 1. We construct an example when, as required in Theorem 2 (Feller, 1968), $B(x)$ be a monotonously increasing function, $B(x) \uparrow \infty$ as $x \to \infty$ such that

$$B(2x)/B(x) < M,$$

condition (1.2) is satisfied, but

$$\limsup_{n\to\infty} \frac{S_n^{(1)}}{\rho_n\sqrt{\log\log \rho_n}} = \infty. \tag{1.7}$$

Let $y_k = 2^{2^k}$, $0 < \varepsilon < \delta < 1/4$, x_k is the root of the equation

$$2^{(\log\log x_k)^{1+\varepsilon}} = 2^{k^{1+\varepsilon}} k^{1+\delta}$$

(we write $\log x$ instead of $\log_2 x$) and k_0 is such that $x_{k_0} \leqslant y_{k_0+1}$. For $k > k_0$, we define $B(x) = C_0 G(x)$, where

$$G(x) = \begin{cases} 2^{k^{1+\varepsilon}} x/y_k & \text{if} \quad x \in (y_k, y_k k^{1+\delta}], \\ 2^{k^{1+\varepsilon}} k^{1+\delta} & \text{if} \quad x \in (y_k k^{1+\delta}, x_k], \\ 2^{(\log\log x)^{1+\varepsilon}} & \text{if} \quad x \in (x_k, y_{k+1}] \end{cases}$$

and C_0 is a normalized constant.

Let us check the fulfillment of the conditions of Theorem 2 in (Feller, 1968). It is obvious that $B(x)$ is monotonously increasing function and $B(2x)/B(x) \leqslant 2$. It is easy to see that, for all sufficiently large k,

$$x_k \geqslant 2^{2^k(1+\log k/(2k^{\varepsilon}))} \geqslant y_k 2^{2^{k/2}}. \tag{1.8}$$

First we show (1.2). Using the equation $B'(x) = 2x^2 f(x)$ and the definition of $B(x)$, it is not difficult to see that, for $k \geqslant k_0$, the density function is

$$f(x) = \begin{cases} C_0 2^{k^{1+\epsilon}}/(y_k x^2) & \text{if} \quad |x| \in (y_k; y_k k^{1+\delta}] \\ 0 & \text{if} \quad |x| \in (y_k k^{1+\delta}; x_k] \\ C_0 B(x)(\log\log x)^{\epsilon}/(x^3 \log x) & \text{if} \quad |x| \in (x_k; y_{k+1}]. \end{cases} \tag{1.9}$$

From here and the obvious relation $\log\log y_k \sim \log\log y_{k+1}$ as $k \to \infty$, we have

$$I_1(k) = \int_{y_k}^{y_k k^{1+\delta}} \frac{x^2 dF(x)}{B(x)(\log\log x)^{1+\gamma}} \sim \frac{C_0 \log k}{k^{1+\delta}},$$

$$I_2(k) = \int_{y_k k^{1+\delta}}^{x_k} \frac{x^2 dF(x)}{B(x)(\log\log x)^{1+\gamma}} = 0,$$

$$I_3(k) = \int_{x_k}^{y_{k+1}} \frac{x^2 dF(x)}{B(x)(\log\log x)^{1+\gamma}} \sim \int_{x_k}^{y_{k+1}} \frac{dx}{x \log x \, (\log\log x)^{1+\gamma-\epsilon}}.$$

Putting $A = y_{k_0}$, we get

$$\int_A^{\infty} \frac{x^2 dF(x)}{B(x)(\log\log x)^{1+\gamma}} = \sum_{k=k_0}^{\infty} \sum_{i=1}^{3} I_i(k) = \begin{cases} = \infty & \text{if} \quad \gamma = 0; \\ < \infty & \text{if} \quad \gamma > \epsilon. \end{cases}$$

Thus the validity of (1.2) is proved.
Let us prove now (1.7). Let b_n be the largest root of the equation

$$b_n^2 = nB(b_n)(\log\log b_n)^{1+\gamma}, \quad \gamma \geqslant 0. \tag{1.10}$$

It is not difficult to check that $b_{n+1}/b_n \to 1$ and $b_n \uparrow \infty$ as $n \to \infty$. Let us denote $d_n = b_n$ if $\gamma = 0$, and $a_n = b_n$ if $\gamma = \delta$. From the proof of Lemma 3, in particular, it follows that

$$\int_A^{\infty} \frac{x^2 dF(x)}{B(x)(\log\log x)^{1+\gamma}} < \infty$$

is equivalent to the convergence of the series $\sum_{n=1}^{\infty} P(b_n)$. From this and (1.2), by the Borel–Cantelli lemma, we obtain

$$\limsup_{n\to\infty} \frac{|X_n^{(1)}|}{d_n} = \infty \quad \text{a.s.,} \quad \text{but} \quad \limsup_{n\to\infty} \frac{|X_n^{(1)}|}{a_n} = 0 \quad \text{a.s.} \tag{1.11}$$

Further, we choose subsequences n_k and n'_k such that $a_{n_k-1} \leqslant y_k \leqslant a_{n_k}$ and $a_{n'_k-1} \leqslant y_k k^{1+\delta} \leqslant a_{n'_k}$, respectively. Obviously $a_{n_k} \sim y_k$ and $a_{n'_k} \sim y_k k^{1+\delta}$. From this and (1.10), it is easy to see that $\rho_{n'_k} \sim a_{n_k} \sim y_k$, $n'_k \sim y_{k+1}/2^{k^{1+\epsilon}}$ and $n_k \sim y_{k+1}/(2^{k^{1+\epsilon}} k^{1+\delta})$, $k \to \infty$.

We will show that, for the subsequence n'_k, relation (1.7) is fulfilled.
Let $X^{(1)}_{n_k,n'_k} \geqslant \ldots \geqslant X^{(n'_k-n_k)}_{n_k,n'_k}$ denote the order statistics based on $X_{n_k+1},\ldots,X_{n'_k}$,

$$S^{(1)}_{n_k,n'_k} = \sum_{i=n_k+1}^{n'_k} X_i - X^{(1)}_{n_k,n'_k}, \quad z_k = \rho_{n'_k}\sqrt{\log\log \rho_{n'_k}}.$$

Then, in order to show (1.7), it is sufficient to prove that

$$\limsup_{k\to\infty} \frac{S^{(1)}_{n_k,n'_k}}{z_k} = \infty \quad \text{a.s.,} \tag{1.12}$$

$$\limsup_{k\to\infty} \frac{|S_{n_k}|}{z_k} = 0 \quad \text{a.s.} \tag{1.13}$$

Since $\{S^{(1)}_{n_k,n'_k} > a\}_{k=k_0}^{\infty}$ is a sequence of independent events, for all $a \in (-\infty;\infty)$, by the Borel–Cantelli lemma, to prove (1.12) it is sufficient to show that

$$\sum_{k=k_0}^{\infty} \mathbf{P}\Big\{S^{(1)}_{n_k,n'_k} \geqslant \tau z_k\Big\} = \infty \quad \text{for any} \quad \tau > 0. \tag{1.14}$$

On the other hand, from (1.9) it is easy to see that, as $k \to \infty$,

$$P(a_{n_k}) \sim \int_{a_{n_k}}^{a_{n'_k}} f(x)dx \sim 2^{k^{1+\varepsilon}}/y_{k+1}, \quad k\to\infty,$$

$$P(\tau z_k) \sim \int_{\tau z_k}^{a_{n_k}} f(x)dx \sim 2^{k^{1+\varepsilon}}/(\tau y_{k+1}\sqrt{k}), \quad n'_k P(\tau z_k) \to 0, \tag{1.15}$$

for all $\tau > 0$. We have, for all sufficiently large k,

$$\mathbf{P}\Big\{S^{(1)}_{n_k,n'_k} \geqslant \tau z_k\Big\} \geqslant \mathbf{P}\Bigg\{\bigcup_{n_k\leqslant i,\ j\leqslant n'_k,\ i\neq j} \Big\{X_i \geqslant \tau z_k,\ X_j \geqslant \tau z_k,\ \sum_{m=n_k+1}^{n'_k} X_m \geqslant 0,$$

$$X_m < \tau z_k;\ n_k < m \leqslant n'_k,\ m \neq i,j\Big\}\Bigg\} \geqslant C\Big(n'_k P(\tau z_k)\Big)^2 \geqslant C/k.$$

This implies (1.14) and (1.12) is proved.
In order to prove (1.13), it is sufficient to show that

$$\sum_{k=k_0}^{\infty} \mathbf{P}\Big\{|S_{n_k}| \geqslant \tau z_k\Big\} < \infty \quad \text{for any} \quad \tau > 0. \tag{1.16}$$

We have

$$\sum_{k=k_0}^{\infty} \mathbf{P}\Big\{|S_{n_k}| \geqslant \tau z_k\Big\} \leqslant \Sigma_1 + \Sigma_2,$$

where

$$\Sigma_1 = \sum_{k=k_0}^{\infty} \mathbf{P}\Big\{|\overline{X}_{n_k}^{(1)}| \geqslant a_{n_k}\Big\}, \quad \Sigma_2 = \sum_{k=k_0}^{\infty} \mathbf{P}\Big\{|S_{n_k}| \geqslant \tau z_k, \ |\overline{X}_{n_k}^{(1)}| < a_{n_k}\Big\},$$

and $|\overline{X}_n^{(1)}| = \max\{|X_1|, \dots, |X_n|\}$. Since $\mathbf{P}\{|\overline{X}_n^{(1)}| \geqslant a_{n_k}\} \leqslant 2n_k P(a_{n_k})$, in view of (1.15), Σ_1 converges. Further, using the Chebyshev inequality, we obtain, for all sufficiently large k,

$$\mathbf{P}\Big\{|S_{n_k}| \geqslant \tau z_k, \ |\overline{X}_{n_k}^{(1)}| < a_{n_k}\Big\} \leqslant \mathbf{P}\Big\{\sum_{i=1}^{n_k} X_i I(|X_i| < a_{n_k}) \geqslant \tau z_k\Big\} \leqslant$$

$$\leqslant \frac{n_k B(a_{n_k})}{\tau^2 z_k^2} \leqslant \frac{C}{k^{2+\delta}} .$$

The convergence of the series Σ_2 follows from these estimates. Thus, relation (1.16) is proved.
The next example shows that there exists a distribution which satisfies the conditions of our theorem, i.e., the set of applicability of the theorem is not empty.

Example 2. Let $B(x) = C_0 L(x)$, where

$$L(x) = \exp\Big\{\int_a^x \Big((\log\log z)^{(r-1)/r} \Big/ z(\log z)^{1/r}\Big) dz\Big\}$$

and C_0 is a normalizing constant. Obviously, by the integral representation of the s.v. f. (see (Feller, 1971)), $B(x)$ is a s.v. f., $B(x) \uparrow \infty$ and

$$\lim_{x\to\infty} \sup B(2x)/B(x) = 1.$$

It is easy to see that the density function $f(x)$ has the form

$$f(x) = \frac{C_0 B(x)(\log\log x)^{(r-1)/r}}{x^3(\log x)^{1/r}}$$

and $\lim_{x\to\infty} P(x)/P(2x) = 4$, i.e., $\alpha = 2$.
It is easy to check that the integral

$$\int_a^{\infty} \frac{x^{2r} P^{r-1}(x) dF(x)}{B^r(x)(\log\log x)^{r+\gamma}}$$

diverges for $\gamma = 0$ and converges for any $\gamma > 0$. Furthermore, by definition, $g_n = \rho_n$. Consequently,

$$\limsup_{n\to\infty} \frac{S_n^{(r)}}{\rho_n\sqrt{2\log\log\rho_n}} = 1 \quad \text{a.s.,} \quad \text{but} \quad \limsup_{n\to\infty} \frac{S_n^{(r-1)}}{\rho_n\sqrt{2\log\log\rho_n}} = \infty \quad \text{a.s.}$$

2. PROOF OF THE THEOREM

Let $(\overline{X}_n^{(i)})$ be $(X_i)_{i=1}^n$ arranged in the decreasing order of the absolute value and put $\overline{S}_n^{(r)} = S_n - \sum_{i=1}^r \overline{X}_n^{(i)}$. We will call $\overline{S}_n^{(r)}$ an absolutely trimmed sum.
The proof of the theorem is contained in the following two lemmas. The first one shows the equivalence of the problem for the trimmed sum $S_n^{(r)}$ an the absolutely trimmed sum $\overline{S}_n^{(r)}$.

LEMMA 1. *Let $\{u_n\}$ be an increasing sequence such that $u_n \to \infty$ as $n \to \infty$. Then*

$$\limsup_{n\to\infty} \frac{\overline{S}_n^{(r)}}{u_n} = \limsup_{n\to\infty} \frac{S_n^{(r)}}{u_n} \quad \text{a.s.}$$

Proof. First, we prove that

$$\mathbf{P}\Big\{\bigcup_{n=k}^{\infty}\Big(\overline{S}_n^{(r)} > u_n,\ X_n^{(r)} = \overline{X}_n^{(r)}\Big)\Big\} \geqslant \frac{1}{2^r}\mathbf{P}\Big\{\bigcup_{n=k}^{\infty}\Big(\overline{S}_n^{(r)} > u_n\Big)\Big\}. \tag{2.1}$$

Let $\Lambda_r = \{i_1, \ldots, i_r\}$ be a subset of the set of natural numbers $\{1, 2, \ldots, n\}$, and

$$M_{\Lambda_r} = \Big\{\overline{X}_n^{(k)} \in \{X_j : j \in \Lambda_r\},\ k = 1, \ldots, r\Big\}.$$

It is easy to see that

$$M_{\Lambda_r} \cap \{X_n^{(r)} = \overline{X}_n^{(r)}\} = M_{\Lambda_r} \cap \{X_j \geqslant 0,\ j \in \Lambda_r\}.$$

Hence, taking into account the simmetry of r.v. $X_1, \ldots, X_n$, for any $m \leqslant n$, we have

$$\mathbf{P}\{\overline{S}_m^{(r)} \leqslant u_m, \ldots, \overline{S}_{n-1}^{(r)} \leqslant u_{n-1},\ \overline{S}_n^{(r)} > u_n,\ X_n^{(r)} = \overline{X}_n^{(r)},\ M_{\Lambda_r}\} \geqslant$$

$$\geqslant \frac{1}{2^r}\mathbf{P}\{\overline{S}_m^{(r)} \leqslant u_m, \ldots, \overline{S}_{n-1}^{(r)} \leqslant u_{n-1},\ \overline{S}_n^{(r)} > u_n,\ M_{\Lambda_r}\}.$$

Consequently,

$$\mathbf{P}\Big\{\overline{S}_m^{(r)} \leqslant u_m, \ldots, \overline{S}_{n-1}^{(r)} \leqslant u_{n-1},\ \overline{S}_n^{(r)} > u_n,\ X_n^{(r)} = \overline{X}_n^{(r)}\Big\} =$$

$$= \sum_{\Lambda_r}\mathbf{P}\Big\{\overline{S}_m^{(r)} \leqslant u_m, \ldots, \overline{S}_{n-1}^{(r)} \leqslant u_{n-1},\ \overline{S}_n^{(r)} > u_n,\ X_n^{(r)} = \overline{X}_n^{(r)},\ M_{\Lambda_r}\Big\} \geqslant$$

$$\geqslant \frac{1}{2^r}\sum_{\Lambda_r}\mathbf{P}\Big\{\overline{S}_m^{(r)} \leqslant u_m,\dots,\overline{S}_{n-1}^{(r)} \leqslant u_{n-1},\ \overline{S}_n^{(r)} > u_n,\ M_{\Lambda_r}\Big\} =$$

$$= \frac{1}{2^r}\mathbf{P}\Big\{\overline{S}_m^{(r)} \leqslant u_m,\dots,\overline{S}_{n-1}^{(r)} \leqslant u_{n-1},\ \overline{S}_n^{(r)} > u_n\Big\}. \tag{2.2}$$

On the other hand, for any sequences of events $\{A_n\}$ and $\{B_n\}$, obviously, we have

$$\bigcup_{n\geqslant k}(A_nB_n) = (A_kB_k)\cup(\overline{A_kB_k}A_{k+1}B_{k+1})\cup\ldots\supseteq$$

$$\supseteq (A_kB_k)\cup(\overline{A}_kA_{k+1}B_{k+1})\cup(\overline{A}_k\overline{A}_{k+1}A_{k+2}B_{k+2})\cup\ldots$$

Therefore, using (2.2), we obtain

$$\mathbf{P}\Big\{\bigcup_{n\geqslant k}(\overline{S}_n^{(r)} > u_n,\ X_n^{(r)} = \overline{X}_n^{(r)})\Big\} \geqslant$$

$$\geqslant \sum_{n\geqslant k}\mathbf{P}\Big\{\overline{S}_k^{(r)} \leqslant u_k,\dots,\overline{S}_{n-1}^{(r)} \leqslant u_{n-1},\ \overline{S}_n^{(r)} > u_n,\ X_n^{(r)} = \overline{X}_n^{(r)}\Big\} \geqslant$$

$$\geqslant \frac{1}{2^r}\sum_{n\geqslant k}\mathbf{P}\Big\{\overline{S}_k^{(r)} \leqslant u_k,\dots,\overline{S}_{n-1}^{(r)} \leqslant u_{n-1},\ \overline{S}_n^{(r)} > u_n\Big\} = \frac{1}{2^r}\mathbf{P}\Big\{\bigcup_{n\geqslant k}(\overline{S}_n^{(r)} \geqslant u_n)\Big\}$$

and (2.1) is proved.
Now, from (2.1), we have

$$\mathbf{P}\Big\{\bigcup_{n\geqslant k}(\overline{S}_n^{(r)} > u_n)\Big\} \geqslant \mathbf{P}\Big\{\bigcup_{n\geqslant k}(S_n^{(r)} > u_n)\Big\} \geqslant$$

$$\geqslant \Big\{\mathbf{P}\bigcup_{n\geqslant k}(S_n^{(r)} > u_n,\ \overline{X}_n^{(r)} = X_n^{(r)})\Big\} \geqslant \frac{1}{2^r}\mathbf{P}\Big\{\bigcup_{n\geqslant k}(\overline{S}_n^{(r)} > u_n)\Big\}.$$

This, together with Kolmogorov's zero–one law (see (Shiryayev, 1984)), imply the statement of Lemma 1.
The next lemma characterizes the transition from $\overline{S}_n^{(r-1)}$ to $\overline{S}_n^{(r)}$ in the LIL, i.e., essentially proves the theorem for absolutely trimmed sums.

LEMMA 2. *Let all the conditions of the theorem hold. Then*

$$\limsup_{n\to\infty}\frac{\overline{S}_n^{(r)}}{g_n\sqrt{2\log\log g_n}} = 1 \quad \text{a.s.,} \quad \textit{but} \quad \limsup_{n\to\infty}\frac{\overline{S}_n^{(r-1)}}{g_n\sqrt{2\log\log g_n}} = \infty \quad \text{a.s.}$$

Before the proof of Lemma 2, let us introduce some auxiliary notations and statements.
Let $a(x)$ be a positive increasing function on $[1;\infty)$ satisfying the condition

$$a^2(x)/x = A\big(a(x)\big), \tag{2.3}$$

for some increasing function $A(x)$.
Putting $a_n = a(n)$, we will assume that

$$a_{2n}/a_n < C, \quad \text{for all} \quad n \geqslant 1. \tag{2.4}$$

Further, let

$$J_r(a,b) = \int_a^b \frac{x^{2r} P^{r-1}(x)\, dF(x)}{A^r(x)}, \quad J_r = J_r(a_1, \infty).$$

Let $v(x) = A(x)/B(x)$ and assume that

$$v(x) \uparrow \infty \quad \text{as} \quad x \uparrow \infty. \tag{2.5}$$

For any $p > \eta > 0$, we put

$$b(x) = a(x) v^{-\eta}(a(x)), \quad c(x) = a(x) v^{-p}(a(x)).$$

Denoting $b_n = b(n)$, $c_n = c(n)$, define

$$X_i' = X_i I(|X_i| < \varepsilon b_i), \quad X_i'' = X_i I(c_i \leqslant |X_i| < \varepsilon b_i), \quad X_i^T = X_i I(|X_i| < c_i),$$

$$S_n' = \sum_{i=1}^n X_i', \quad S_n'' = \sum_{i=1}^n X_i'', \quad S_n^T = \sum_{i=1}^n X_i^T.$$

Then we have the following lemmas.

LEMMA 3. *Assume the conditions* (1.4), (2.3) *and* (2.4) *to be satisfied. Then*

$$1) \quad \limsup_{n\to\infty} \frac{X_n^{(r)}}{a_n} = \limsup_{n\to\infty} \frac{\overline{X}_n^{(r)}}{a_n} = \infty \quad \text{a.s.} \quad \textit{if} \quad J_r = \infty;$$

$$2) \quad \limsup_{n\to\infty} \frac{X_n^{(r)}}{a_n} = \limsup_{n\to\infty} \frac{\overline{X}_n^{(r)}}{a_n} = 0 \quad \text{a.s.} \quad \textit{if} \quad J_r < \infty.$$

LEMMA 4. *Assume the conditions* (1.4), (2.3)–(2.5) *hold and* $J_r < \infty$. *Then, for any* $\eta < 1/(2(r+1))$,

$$\text{a)} \quad \limsup_{n\to\infty} \frac{X_n^{(r+1)}}{b_n} = \limsup_{n\to\infty} \frac{\overline{X}_n^{(r+1)}}{b_n} = 0 \quad \text{a.s.};$$

$$\text{b)} \quad \limsup_{n\to\infty} b_n^{-1} \left|\overline{S}_n^{(r)} - S_n'\right| < r\varepsilon\left(2 + 2^\eta/(2^\eta - 1)\right) \quad \text{a.s.}$$

LEMMA 5. *Let the conditions* (1.4), (2.3)–(2.5) *be satisfied and* $J_r < \infty$. *Furthermore, let* $v(tx)/v(x) < Ct^\mu$, *for some* $\mu \in [0; 1/p)$ *and for all* $t \geqslant 1$. *Then, for any* $\delta < \eta$,

$$\limsup_{n\to\infty} \frac{S_n''}{a_n v^{-\delta}(a_n)} = 0 \quad \text{a.s.}$$

The next lemma essentially is the LIL for S_n^T.

LEMMA 6. *Let* $J_r < \infty$, $v(x) = (\log\log x)^{1+\gamma/r}$ *and let condition* (1.4) *be satisfied. Then*

$$\limsup_{n\to\infty} \frac{S_n^T}{g_n\sqrt{2\log\log g_n}} = 1 \quad \text{a.s.}$$

Proofs of Lemmas 3–6 are given in Section 3.

Proof of Lemma 2. Denote $v(x) = (\log\log x)^{1+\gamma/r}$. Then $J_r = I_r(\gamma)$ and all conditions of Lemmas 3–6 are satisfied. Also, by the definition of γ_0, and, from (1.3) and (1.5), it is not difficult to deduce that, for any $\alpha \in (0;2]$ and $\gamma \in (0; r\gamma_0)$, exist a number $\delta \in (0;\eta)$ such that

$$b_n \leqslant a_n v^{-\delta}(a_n) \leqslant \begin{cases} Cd_n & \text{if} \quad 0 < \alpha \leqslant 2(r+1)/(r+2); \\ C\rho_n\sqrt{\log\log \rho_n} & \text{if} \quad 2(r+1)/(r+2) < \alpha \leqslant 2. \end{cases}$$

From this and Lemmas 4–6, the first statement of Lemma 2 follows if we apply the following two obvious inequalities:

$$\limsup_{n\to\infty} \frac{\overline{S}_n^{(r)}}{g_n\sqrt{2\log\log g_n}} \geqslant \limsup_{n\to\infty} \frac{S_n^T}{g_n\sqrt{2\log\log g_n}} -$$

$$- \limsup_{n\to\infty} \frac{|\overline{S}_n^{(r)} - S'_n|}{g_n\sqrt{2\log\log g_n}} - \limsup_{n\to\infty} \frac{|S''_n|}{g_n\sqrt{2\log\log g_n}}$$

and

$$\limsup_{n\to\infty} \frac{\overline{S}_n^{(r)}}{g_n\sqrt{2\log\log g_n}} \leqslant \limsup_{n\to\infty} \frac{S_n^T}{g_n\sqrt{2\log\log g_n}} +$$

$$+ \limsup_{n\to\infty} \frac{|\overline{S}_n^{(r)} - S'_n|}{g_n\sqrt{2\log\log g_n}} + \limsup_{n\to\infty} \frac{|S''_n|}{g_n\sqrt{2\log\log g_n}}.$$

The second statement of Lemma 2 follows from the first relation and from Lemma 3 if we apply the inequality

$$\limsup_{n\to\infty} \frac{\overline{S}_n^{(r-1)}}{g_n\sqrt{2\log\log g_n}} \geqslant \limsup_{n\to\infty} \frac{\overline{X}_n^{(r)}}{2d_n} - \limsup_{n\to\infty} \frac{\overline{S}_n^{(r)}}{g_n\sqrt{2\log\log g_n}}.$$

Here it is necessary to use Lemma 3, putting $A(x) = B(x)\log\log x$. This gives $J_r = I_r(0)$ and $d_n \geqslant g_n\sqrt{\log\log g_n}$. Lemma 2 is proved.

3. PROOFS OF THE PRELIMINARY LEMMAS 3–6

Our proof of the lemmas is based on the following result which is proved in (Mori, 1977).

LEMMA 7. *Let $\{a_n\}_{n\geqslant 1}$ be a nondecreasing sequence of positive numbers. If $a_n \uparrow \infty$, then*

$$\mathbf{P}\Big\{ |\overline{X}_n^{(r)}| \geqslant a_n \text{ i.o. } \Big\} = 0 \text{ or } 1,$$

according to $\sum_{n=1}^{\infty} n^{r-1} P^r(a_n)$ converges or diverges.

To prove Lemma 3 using Lemma 7, it is sufficient to show the equivalence of the following statements

$$(\mathbf{A}) \quad \sum_{n=r}^{\infty} n^{r-1} P^r(a_n) < \infty; \quad (\mathbf{B}) \quad J_r < \infty.$$

First we prove that if (**A**) or (**B**) is valid then

$$nP(a_n) \to 0 \quad \text{as} \quad n \to \infty. \tag{3.1}$$

Since the reasonings are similar, we will only prove (3.1) if (**A**) is satisfied. Assume that (**A**) is valid, but (3.1) is not. Then there exists some sequence $\{n_i\}$ such that $n_i P(a_{n_i}) \geqslant C > 0$ (without loss of generality we can assume that $n_i \geqslant 2n_{i-1}$, $n_0 = 1$). From this, using standard arguments (see (Mori, 1976)), we obtain

$$\sum_{n=1}^{\infty} n^{r-1} P^r(a_n) \geqslant C \sum_{i=1}^{\infty} \big(n_i P(a_{n_i})\big)^r = \infty$$

which contradicts (**A**). This contradiction proves (3.1).
Now, using Abel's transformation and taking into consideration (1.4) and (3.1), we obtain

$$\sum_{n=r}^{\infty} n^{r-1} P^r(a_n) = \lim_{N\to\infty} \sum_{n=r}^{N} n^{r-1} P^r(a_n) =$$

$$= \sum_{n=r}^{\infty} \Big(\sum_{k=r}^{r} k^{r-1} \Big) \big(P^r(a_n) - P^r(a_{n+1})\big) \asymp \sum_{n=r}^{\infty} n^r P^{r-1}(a_n) \big(F(a_{n+1}) - F(a_n)\big).$$

It is easy to see, that the convergence of the above series is equivalent to the convergence of the integral $\int_r^{\infty} x^r P^{r-1}\big(a(x)\big)\, dF\big(a(x)\big)$. On the other hand, from (2.3), it follows that this integral is equal to

$$\int_r^{\infty} \frac{\big(xA(a(x))\big)^r P^{r-1}(a(x))\, dF\big(a(x)\big)}{A^r\big(a(x)\big)} = J_r - \int_{a_1}^{a_r} \frac{y^{2r} P^{r-1}(y)\, dF(y)}{A^r(y)}.$$

Consequently, (**A**) $\Leftrightarrow$ (**B**). Thus Lemma 3 is proved.
Let us prove now Lemma 4.
a) By condition (1.4), there exists a number $\lambda = \lambda(\alpha) > 0$ such that, for all sufficiently large x,

$$P(x/2) \geqslant (1 + \lambda) P(x).$$

From this it follows that $B(x) \geqslant \lambda x^2 P(x)/4$ or

$$x^2 P(x)/B(x) \leqslant 4/\lambda. \tag{3.2}$$

Now, by Lemma 3, it is sufficient to show that $\int_1^\infty x^{r+1}P^r\big(b(x)\big)\, dF\big(b(x)\big) < \infty$. We have

$$\int_1^\infty x^{r+1}P^r(b(x))\, dF(b(x)) = \int_1^\infty \frac{(xA(a(x))^{r+1}P^r(b(x))\, dF(b(x))}{A^{r+1}(x)} =$$

$$= \int_1^\infty \frac{a^{2(r+1)}(x)v^{-2\eta(r+1)}(a(x))P^r(b(x))\, dF(b(x))}{A^r(a(x))B(a(x))v^{1-2\eta(r+1)}(a(x))} \leqslant C\int_{b_1}^\infty \frac{y^{2r}P^r(y)\, dF(y)}{A^r(y)}\,.$$

b) Let $S_n^* = \sum_{i=1}^n X_i I(|X_i| < \varepsilon b_n)$. Then, from the statement **a)** of Lemma 4, it follows that there exist a sufficiently large N such that, for all $n > 2^N$,

$$\big|\overline{X}_n^{r+1}\big| < \varepsilon b_n \quad \text{and} \quad \big|S_n^* - \overline{S}_n^{(r)}\big| < r\varepsilon b_n. \tag{3.3}$$

Assume that $k \geqslant N+1$ and $2^k \leqslant n < 2^{k+1}$. Then

$$|S_n^* - S'_n| = \left|\sum_{i=1}^n X_i I(\varepsilon b_i \leqslant |X_i| < \varepsilon b_n)\right| \leqslant \Sigma_1 + \Sigma_2, \tag{3.4}$$

where

$$\Sigma_1 = \sum_{i=1}^{2^k} |X_i| I\big(\varepsilon b_i \leqslant |X_i| < \varepsilon b_{2^k}\big),$$

$$\Sigma_2 = \sum_{i=1}^{n} |X_i| I\big(\varepsilon b_{2^k} \leqslant |X_i| < \varepsilon b_n\big).$$

Let us estimate Σ_1 and Σ_2.
By the choice of N, it is not difficult to see that $\Sigma_2 < r\varepsilon b_n$. For Σ_1 we have

$$\Sigma_1 \leqslant \sum_{i=1}^{2^k} |X_i| \sum_{j=[\log i]}^{k-1} I\Big(\varepsilon b_{2^j} \leqslant |X_i| < \varepsilon b_{2^{j+1}}\Big).$$

Changing the order of the summation and taking into consideration the choice of N, we obtain

$$\Sigma_1 \leqslant \sum_{j=0}^{k-1} \sum_{i=1}^{2^{j+1}} |X_i| I\Big(\varepsilon b_{2^j} \leqslant |X_i| < \varepsilon b_{2^{j+1}}\Big) =$$

$$= \sum_{j=0}^{N-1} \sum_{i=1}^{2^{j+1}} + \sum_{j=N}^{k-1} \sum_{i=1}^{2^{j+1}} \leqslant 2^N \varepsilon b_{2^N} + r\varepsilon \sum_{j=N}^{k-1} b_{2^{j+1}}. \tag{3.5}$$

On the other hand, by the definition of $v(x)$,

$$\frac{v(a(2^{j+1}))}{v(a(2^j))} \leqslant \frac{A(a(2^{j+1}))}{A(a(2^j))} = \frac{1}{2}\left(\frac{a(2^{j+1})}{a(2^j)}\right)^2$$

and

$$\frac{b(2^j)}{b(2^{j+1})} = \frac{a(2^j)}{a(2^{j+1})}\left(\frac{v(a(2^{j+1}))}{v(a(2^j))}\right)^{\eta} \leqslant \frac{1}{2^{\eta}}\left(\frac{a(2^j)}{a(2^{j+1})}\right)^{1-2\eta} \leqslant \frac{1}{2^{\eta}}.$$

Hence,

$$\sum_{j=N}^{k-1} b_{2^{j+1}} = b_{2^k}\sum_{j=N+1}^{k}\frac{b(2^j)}{b(2^k)} \leqslant b_{2^k}\sum_{j=0}^{\infty}\left(\frac{1}{2^{\eta}}\right)^j = b_{2^k}\frac{2^{\eta}}{2^{\eta}-1}.$$

Thus, using (3.5), we obtain

$$\Sigma_1 \leqslant 2^N\varepsilon b_{2^N} + r\varepsilon b_n 2^{\eta}/(2^{\eta}-1).$$

Applying the obtained estimates of Σ_1 and Σ_2 in (3.4) and using (3.3), we obtain

$$\left|\overline{S}_n^{(r)} - S'_n\right| \leqslant \left|\overline{S}_n^{(r)} - S^*_n\right| + \left|S^*_n - S'_n\right| \leqslant$$

$$\leqslant 2^N\varepsilon b_{2^N} + r\varepsilon b_n\left(2 + 2^{\eta}/(2^{\eta}-1)\right).$$

Dividing both sides of this inequality by b_n and letting $n \to \infty$, we obtain the limiting inequality **b**) of Lemma 4.
The proof of Lemma 5 relies essentially on a result of Egorov which we present here for the sake of completeness.

LEMMA 8 (Egorov, 1970). *Let $\xi_1, \ldots, \xi_n$ be independent r.v. with $\mathbf{E}\xi_i = 0$ and $V_i^2 = \mathbf{E}\xi_i^2 < \infty$. Furthermore, let h_n be an increasing sequence such that $h_n \to \infty$ as $n \to \infty$. Then $\sum_{i=1}^n \xi_i/h_n \to \infty$ a.s., as $n \to \infty$ if, for some $n_0 \geqslant 2$,*

$$\sum_{n=n_0}^{\infty}\frac{V_n^2}{h_n^{2n_0}}\sum_{1\leqslant j_1\leqslant\ldots\leqslant j_{n_0-1}\leqslant n} V_{j_1}^2\ldots V_{j_{n_0-1}}^2 < \infty$$

and

$$\sum_{n=1}^{\infty}\mathbf{P}\left\{|\xi_n| \geqslant \varepsilon h_n\right\} < \infty \quad \text{for any} \quad \varepsilon > 0.$$

Proof of Lemma 5. By Lemma 8, it is enough to prove the convergence of the following series

$$\sum_{n=1}^{\infty}\mathbf{P}\left\{X''_n > \varepsilon a_n v^{-\delta}(a_n)\right\} \tag{3.6}$$

and

$$\Sigma_0 = \sum_{n=N^*}^{\infty}\frac{\sigma_n^2}{(a_n v^{-\delta}(a_n))^{2N^*}}\sum_{1\leqslant j_1\leqslant\ldots\leqslant j_{N^*-1}\leqslant n}\sigma_{j_1}^2\ldots\sigma_{j_{N^*-1}}^2,$$

where $N^* > 2$ is some natural number which we choose later and

$$\sigma_i^2 = \mathbf{E}(X'')^2 = B\left(\varepsilon a_i v^{-\eta}(a_i)\right) - B\left(a_i v^{-p}(a_i)\right).$$

Since $\delta < \eta$, the convergence of (3.6) is obvious.

Let us investigate the convergence of the series Σ_0. It is not difficult to check that

$$\Sigma_0 \leqslant \sum_{n=N^*}^{\infty} \frac{\sigma_n^2 v^{2N^*\delta}(a_n)}{a_n^{2N^*}} \sum_{1\leqslant j_1 \leqslant \ldots \leqslant j_{N^*-1} \leqslant n} \left(\sigma_{j_1}^{2(N^*-1)} + \ldots + \sigma_{j_{N^*-1}}^{2(N^*+1)}\right) \leqslant$$

$$\leqslant \sum_{n=1}^{\infty} \frac{\sigma_n^2 v^{2N^*\delta}(a_n)}{a_n^{2N^*}} (N^* - 1) n^{N^*-2} \sum_{j=1}^{n} \sigma_j^{2(N^*-1)} =$$

$$= (N^* - 1) \sum_{n=1}^{\infty} \frac{\sigma_n^2 v^{2N^*\delta}(a_n)}{n^2 A^{N^*}(a_n)} \sum_{j=1}^{n} \sigma_j^{2(N^*-1)}$$

Since $v(x) = A(x)/B(x)$ and $\delta < \eta < 1/2$, changing the order of the summation and putting $\nu = 2N^*\delta - N^*$, we obtain

$$\Sigma_0 \leqslant C \sum_{j=1}^{\infty} \sigma_j^{2(N^*-1)} \sum_{n=j}^{\infty} \frac{\sigma_n^2 v^{2N^*\delta}}{n^2 A^{N^*}(a_n)} \leqslant$$

$$\leqslant C \sum_{j=1}^{\infty} \sigma_j^{2(N^*-1)} \sum_{n=j}^{\infty} \frac{v^{\nu}(a_n)}{n^2 B^{N^*-1}(a_n)} \leqslant C \sum_{j=1}^{\infty} \frac{\sigma_j^{2(N^*-1)} v^{\nu}(a_j)}{B^{N^*-1}(a_j)} \sum_{n=j}^{\infty} \frac{1}{n^2} \leqslant$$

$$\leqslant C \sum_{j=1}^{\infty} \frac{\sigma_j^{2(N^*-1)} v^{\nu}(a_j)}{j B^{N^*-1}(a_j)}.$$

Let $N^* \geqslant r + 1$. Then $\sigma_j^{2(N^*-1)}/B^{N^*-1}(a_j) \leqslant \sigma_j^{2r}/B^r(a_j)$. Using this and again changing the order of summation, we have

$$\Sigma_0 \leqslant C \sum_{j=1}^{\infty} \frac{\sigma_j^{2(r-1)} v^{\nu}(a_j)}{j B^r(a_j)} \Big(B(a_j) - B(a_j v^{-p}(a_j))\Big) \leqslant$$

$$\leqslant C \sum_{j=1}^{\infty} \frac{\sigma_j^{2(r-1)} v^{\nu}(a_j)}{j B^r(a_j)} \sum_{\{k:\ a_j v^{-p}(a_j) < a_k \leqslant a_j\}} (B(a_k) - B(a_{k-1})) \leqslant$$

$$\leqslant C \sum_{k=k_0}^{\infty} (B(a_k) - B(a_{k-1})) \sum_{\{j:\ a_k \leqslant a_j < c a_k v^{p/(1-p\mu)}(a_k)\}} \frac{\sigma_j^{2(r-1)} v^{\nu}(a_j)}{j B^r(a_j)} \leqslant$$

$$\leqslant C \sum_{k=k_0}^{\infty} \frac{(B(a_k) - B(a_{k-1})) v^{\nu}(a_k)}{B^r(a_k)} \sum_{\{j:\ a_k \leqslant a_j < c a_k v^{p/(1-p\mu)}(a_k)\}} \frac{\sigma_j^{2(r-1)}}{j}. \qquad (3.7)$$

From condition (1.4), it follows that in the domain of summation by j,

$$\sigma_j^2 \leqslant 2 \int_{T_1}^{T_2} x^2 dF(x) \leqslant c a_k^2 P(a_k) (v(a_k))^{2p/(1-p\mu)+Mp}, \qquad (3.8)$$

where $T_1 = a_k v^{-p}(a_k)$, $T_2 = a_k(v(a_k))^{p/(1-p\mu)}$.
On the other hand, it is easy to see that

$$\sum_{\{j:\ a_k \leqslant a_j < c a_k v^{p/(1-p\mu)}(a_k)\}} \frac{1}{j} \leqslant C \log v(a_k).$$

Thus, from the above inequality and from (3.7), (3.8), it follows that

$$\Sigma_0 \leqslant C \exp_{v(a_k)} \left\{ 2N^*\delta + 2p(r-1)/(1-p\mu) + Mp + \delta' - N^* \right\} \times$$

$$\times \sum_{k=k_0}^{\infty} \frac{(B(a_k) - B(a_{k-1}))a_k^{2(r-1)} P^{r-1}(a_k)}{B^r(a_k)}$$

for any $\delta' > 0$, where $\exp_s\{t\} = s^t$.
Now, choosing N^* such that

$$2N^*\delta + 2p(r-1)/(1-p\mu) + Mp + \delta' + r - N^* < 0,$$

we obtain

$$\Sigma_0 \leqslant C \sum_{k=k_0}^{\infty} \frac{(B(a_k) - B(a_{k-1}))a_k^{2(r-1)} P^{r-1}(a_k)}{B^r(a_k) v^r(a_k)} \leqslant C \int_{a_{k_0}}^{\infty} \frac{x^{2r} P^{r-1}(x) dF(x)}{A^r(x)} < \infty.$$

Lemma 5 is proved.

Proof of Lemma 6. Since $v(x) = (\log\log x)^{1+\gamma'}$, where $\gamma' = \gamma/r$, it is easy to see that, as $x > y \to \infty$,

$$B(x)/B(y) < c(x/y)^{2-\alpha'} \quad \text{for any} \quad \alpha' < \alpha. \tag{3.9}$$

From this it easily follows that

$$\rho_n \geqslant a_n v^{-\tau}(a_n) \quad \text{for any} \quad \tau > 1/\alpha. \tag{3.10}$$

Put $(\sigma_n^T)^2 = \mathbf{E}(S_n^T)^2$. Then, for a sufficiently large n,

$$(\sigma_n^T)^2 = \sum_{i=1}^{n} \int_{|x|<c_i} x^2 dF(x) \geqslant (1-\varepsilon) n B(a_n v^{-q}(a_n)), \tag{3.11}$$

for any $q > p$ and $\varepsilon > 0$.
Thus, choosing p such that $p > \tau + 1/2$ and $\alpha p - 1 > 1/2$, from (2.10) and (3.11), we obtain

$$a_n v^{-p}(a_n) = o\left(\min \left\{ \frac{\rho_n}{\sqrt{\log\log \rho_n}}, \frac{\sigma_n^T}{\sqrt{\log\log \sigma_n^T}} \right\} \right), \quad n \to \infty. \tag{3.12}$$

It should be noted that with this p the relations $\log\log a_n \sim \log\log \sigma_n^T \sim \log\log \rho_n$ as $n \to \infty$, and

$$\sigma_n^T \leqslant \rho_n \leqslant g_n \tag{3.13}$$

hold.

The most important in the proof is the relation

$$\limsup_{n\to\infty} \frac{S_n^T}{\sigma_n^T \sqrt{2\log\log \sigma_n^T}} = 1 \quad \text{a.s.,} \tag{3.14}$$

which follows easily from the classical Kolmogorov LIL (see (Petrov, 1975)), if (3.12) is true. The rest of the proof consists in showing that $g_n \sim \sigma_n^T$, for a subsequence which is of the LIL type. If we assume that, for a sufficiently large n,

$$B\Big(a_n v^{-q}(a_n)\Big) > (1-\varkappa)B(a_n), \tag{3.15}$$

for any $\varkappa > 0$ and some $q > p$, then, from the obvious inequality $B(a_n v^{-q}(a_n)) < < B(a_n)$, it follows that

$$B(\rho_n) \sim B(a_n) \sim B\Big(a_n v^{-q}(a_n)\Big) \sim B(d_n), \quad n \to \infty.$$

This implies $g_n \sim \sigma_n^T$. However, (1.4) does not necessarily imply (3.15). We now consider the cases when (3.15) is not valid. Let the sequence $\{n_k\}$ be such that

$$B\big(a_{n_k} v^{-q}(a_{n_k})\big) < (1-\varkappa_0)B(a_{n_k}), \tag{3.16}$$

for some $q > p$ and $0 < \varkappa_0 < 1$.

Without loss of generality we can assume that $a_{n_k}/a_{n_{k-1}} \geqslant C > 1$. Then, for any $\nu > 0$, we have

$$\Sigma^* = \sum_{k=1}^{\infty} v^{-\nu}(a_{n_k}) J_r\big(a_{n_k} v^{-q}(a_{n_k}), a_{n_k}\big) \leqslant$$

$$\leqslant \sum_{k=1}^{\infty} v^{-\nu}(a_{n_k}) \sum_{\{j:\ a_{n_k} v^{-q}(a_{n_k}) \leqslant a_{n_j} \leqslant a_{n_k}\}} J_r(a_{n_{j-1}}, a_{n_j}).$$

Changing the order of summation, we obtain

$$\Sigma^* \leqslant C \sum_{j=1}^{\infty} J_r(a_{n_{j-1}}, a_{n_j}) v^{-\nu}(a_{n_j}) \sum_{\{k:\ a_{n_j} \leqslant a_{n_k} \leqslant c a_{n_j} v^{2q}(a_{n_j})\}} 1 \leqslant C J_r < \infty. \tag{3.17}$$

On the other hand, from (3.16), it is easy to see that

$$2P(a_{n_k} v^{-q}(a_{n_k})) = \int_{a_{n_k} v^{-q}(a_{n_k})}^{\infty} \frac{dB(x)}{x^2} \geqslant \varkappa_0 a_{n_k}^{-2} B(a_{n_k}).$$

From this and (1.4) we obtain

$$P(a_{n_k}) \geqslant C a_{n_k}^{-2} B(a_{n_k}) v^{-Mq}(a_{n_k}).$$

Then

$$\Sigma^* \geqslant C \sum_{k=1}^{\infty} \frac{a_{n_k}^{2(r-1)} v^{-2q(r-1)-\nu}(a_{n_k}) P^{r-1}(a_{n_k}) B(a_{n_k})}{A^r(a_{n_k})} \geqslant C \sum_{k=1}^{\infty} v^{-\Delta}(a_{n_k}),$$

where $\Delta = 2q(r-1) + \nu + Mq(r-1) + r$. From the above relation and (3.17) we deduce that

$$\sum_{k=1}^{\infty} v^{-\Delta}(a_{n_k}) < \infty. \tag{3.18}$$

Applying Corollary 2 in (Nagaev, 1979) to the probability

$$\mathbf{P}\left\{ S_{n_k}^T \geqslant \varepsilon \sigma_{n_k}^T \sqrt{\log\log \sigma_{n_k}^T} \right\},$$

and taking into account (3.18), we obtain

$$\sum_{k=1}^{\infty} \mathbf{P}\left\{ S_{n_k}^T \geqslant \varepsilon \sigma_{n_k}^T \sqrt{\log\log \sigma_{n_k}^T} \right\} < \infty \quad \text{for any} \quad \varepsilon > 0.$$

Together with the Borel–Cantelli lemma this implies

$$\limsup_{k\to\infty} \frac{S_{n_k}^T}{\sigma_{n_k}^T \sqrt{\log\log \sigma_{n_k}^T}} = 0 \quad \text{a.s.}, \tag{3.19}$$

which, by (3.13), is equivalent to

$$\limsup_{k\to\infty} \frac{S_{n_k}^T}{g_{n_k}^T \sqrt{\log\log g_{n_k}}} = 0 \quad \text{a.s.} \tag{3.20}$$

Therefore, only sequences $\{m_k\}$ for which $B(a_{m_k}) \sim B(\rho_{m_k}) \sim B(\sigma_{m_k}^T)$, $k \to \infty$, are of the LIL type.
Together with (3.19), (3.20) and (14) this implies the lemma.

REFERENCES

Egorov, V.A. (1970). On the strong law of large numbers and the law of the iterated logarithm for sequences of independent random variables. *Theory Probab. Appl.* **15**, 509–515.

Feller, W. (1968). An extention of the law of the iterated logarithm to variables without variance. *J. Math. Mech.* **18**, 343–355.

Feller, W. (1971). *An Introduction to Probability Theory and Its Applications.* Wiley, New York.

Mori, T. (1976). The strong law of large numbers when extreme terms are excluded from sums. *Z. Warsch. verw. Geb.* **36**, 189–194.

Mori, T. (1977). Stability for sums of i.i.d. variables when extreme terms are excluded. *Z. Wahrsch. verw. Geb.* **40**, 159–167.

Nagaev, S.V. (1979). Large deviations of sums of independent random variables. *Ann. Probab.* **7**, 745–789.

Petrov, V.V. (1987). *Limit Theorems for Sums of Independent Random Variables.* Nauka, Moscow. (in Russian) .

Shiryayev, A.N. (1984). *Probability.* Springer , Berlin etc.

New Trends in Probab. and Statist., pp. 30 - 37
V. Sazonov and T. Shervashidze (Eds.)
1991 VSP/Mokslas

GENERALIZED MARCINKIEWICZ'S THEOREM AND ASYMPTOTIC EXPANSIONS IN THE CENTRAL LIMIT THEOREM

L.B. KLEBANOV and J.A. MELAMED
Leningrad Civil Engineering Institute, Second Krasnoarmeiskaya str. 4, 198005 Leningrad, USSR
Razmadze Mathematical Institute of the Georgian Academy of Sciences, Z. Rukhadze str. 1, 380093 Tbilisi, Georgia, USSR

ABSTRACT

A generalization of the well-known Marcinkiewicz's theorem is obtained which has naturally originated in the study of sums of a random number of random variables. This generalized Marcinkiewicz's theorem can be used to analyze the coefficients of the asymptotic expansion in the central limit theorem. Namely, let F_n be the distribution function of the normalized sum of independent identically distributed random variables X_j satisfying the Cramér condition, and let $H_\ell(x)$ $(\ell = 1, 2, ...)$ be the Hermite polynomials. If, for all $n \geqslant 2$,

$$\int_{-\infty}^{\infty} (F_n(x) - \Phi(x))H_{\ell+1}(x)\, dx = \sum_{j=1}^{\ell-1} c_j/n^{j/2},$$

for all $\ell \notin \{\ell_k\}$, where $\{\ell_k\}$ is a sequence of indices satisfying the condition $k/\ell_k \to 0$ as $k \to \infty$, then X_1 has the Gaussian distribution.

Let ν_p be a random variable with non-negative integer values which depends on a parameter $p \in (0,1)$. We suppose that $\mathbf{E}\nu_p = 1/p$ and $\lim_{p\to 0} \mathbf{P}\{p\nu_p < x\} = A(x)$, where $A(x)$ is some distribution function. Let $X_1, ..., X_n, ...$ be independent identically distributed random variables independent of ν_p. When sums $\sum_{j=1}^{\nu_p} X_j$ are considered, then according to (Gnedenko, 1983) the role of the exponent is played by $\int_0^\infty \exp\{\tau x\}\, dA(x)$. If this integral converges for all complex τ, then the maximum of its modulus is attained at the real positive semi-axis.
In the characterization theory of probability distributions by properties of linear forms with a non-random number of terms the Marcinkiewicz theorem (Marcinkiewicz, 1938) plays an important role. According to it, the function

$$f(t) = \exp\{P_n(t)\},$$

where $P_n(t)$ is a polynomial of degree n, can be a characteristic function (ch.f.) of some probability distribution iff $f(t)$ is the ch.f. of the normal distribution and, in particular, $n \leqslant 2$. In order to obtain the corresponding characterizations in the case of sums of random numbers of random variables, we must have analogies and generalizations of the above mentioned assertion.

We can look at the Marcinkiewicz theorem from a somewhat different point of view. Namely, let $f(t)$ be a ch.f. which has derivatives of all orders at the point $t = 0$ (i.e., the corresponding probability distribution has moments of all orders). Denote by $\{\varkappa_j,\ j \geqslant 1\}$ the sequence of cumulants corresponding to the ch.f. $f(t)$. Remind that

$$\varkappa_j = \frac{d^j}{dt^j} \log f(t) \big|_{t=0}. \tag{1}$$

The Marcinkiewicz theorem states that if $\varkappa_j \neq 0$, for all $j \geqslant j_0$ ($j_0 \geqslant 2$ is some positive integer), then $f(t)$ is a ch.f. of the normal distribution. One of the authors suggested the hypothesis that if $f(t)$ is a ch.f. analytic in some strip

$$S_\rho = \{t\colon |\mathrm{Im}\, t| < \rho\} \qquad (\rho > 0)$$

and all cumulants whose numbers do not belong to some rare sequence vanish, then $f(t)$ is a ch.f. of the normal law. More precisely, if the ch.f. $f(t)$ is analytic in S_ρ and $\varkappa_j = 0$ for all $j \notin \{j_k,\ k \geqslant 1\}$, $\lim_{k\to\infty} k/j_k = 0$, then $f(t)$ is the ch.f. of a normal distribution.

This hypothesis was recently proved in (Fryntov, 1988, 1989) (earlier Ostrovskii informed the authors on one intermediate result). Below we replace cumulants by some other characteristics; this is necessary since we replace the logarithm in (1) by a function of a rather general character.

Now we will state our results. Let $h(z)$ be an entire function. As usually, we denote by $\mathbf{C}$ the set of all complex numbers. Put for a real number $R > 0$

$$\mathcal{H}_R = \{z\colon |h(z)| \leqslant R\}.$$

THEOREM 1. *Let $f(t)$ be a ch.f. analytic in the strip S_ρ for some $\rho > 0$. Assume that the entire function $h(z)$ satisfies the condition:*
for any $R > 0$ there exists a complex number $\alpha = \alpha(R)$ such that

$$\exp\{\alpha\mathcal{H}_R\} \neq \mathbf{C}\backslash\{0\}.$$

Assume also that

$$f(t) = h\Big(\sum_{j=0}^{\infty} a_{\lambda_j} t^{\lambda_j}\Big),$$

where $\lambda_j \geqslant 0$ are integers ($0 = \lambda_0 < \lambda_1 < \dots$) and $\sum_{j=1}^{\infty} 1/\lambda_j < \infty$. Then

$$f(t) = h(a_0 + a_1 t + a_2 t^2).$$

Thus, Theorem 1 means that if a ch.f. $f(t)$ is analytic in S_ρ and represents an entire function h (satisfying the above mentioned condition) of a gap series, then the condition $\sum_{j=1}^{\infty} 1/\lambda_j < \infty$ on the gaps guarantees the termination of the gap series, and this happens not further than at the second degree of t.

Proof of Theorem 1 we split into several steps.
1) $f(t)$ is an entire function. In fact, if the series

$$g(t) = \sum_{j=0}^{\infty} a_{\lambda_j} t^{\lambda_j}$$

converges in a circle of a radius $R < \infty$, then, according to the Fabri's theorem on the gaps (see, e.g.,(Bieberbach, 1967)), the circle $|t| = R$ is a natural boundary for $g(t)$ and, hence, $f(t)$ must have infinitely many singular points on $\{|t| = R\} \cap S_\rho$ in spite of the assumption that $f(t)$ is analytic in S_ρ. Thus, $R = \infty$.
2) Let us take now an arbitrary $\rho_1 > 0$ and estimate the growth of $g(t)$ in the strip $S_{\rho_1} = \{t\colon |\mathrm{Im}\, t| \leqslant \rho_1\}$. We have

$$|f(t)| \leqslant \max\big(|f(i\rho_1)|,\ |f(-i\rho_1)|\big)$$

for all $t \in S_{\rho_1}$. Denote by R_1 the right-hand side of the previous inequality. By the assumption of Theorem 1, there is an $\alpha = \alpha(R_1)$ such that for $t \in S_{\rho_1}$ the function $T(t) = \exp\{\alpha g(t)\}$ does not take at least one more value $\widetilde{T}$, except zero. The function $\zeta(t) = \mathrm{th}\big(\pi t/(4\rho_1)\big)$ maps conformally the strip S_{ρ_1} onto the unit circle $\{|\zeta| < 1\}$ $\big(\zeta(0) = 0\big)$. Therefore, the function $\xi(\zeta) = T\big(t(\zeta)\big)/\widetilde{T}$ is analytic in the unit circle and does not take in it the values 0 and 1. According to the Schottky theorem (see, e.g.,(Hayman, 1966))

$$M(r;\xi) \leqslant \exp\Big\{\frac{1}{1-r}\big[(1+r)\log\max\big(1, |\xi(0)|\big) + 2Cr\big]\Big\},$$

where C is an absolute constant, and $M(r;\xi) = \max_{|\zeta| \leqslant r} |\xi(\zeta)|$. Since $0 < r < 1$, the last inequality can be written in a somewhat weaker form

$$M(r;\xi) \leqslant \exp\big\{A/(1-r)\big\},$$

where A does not depend on r. Thus,

$$|T\big(t(\zeta)\big)| \leqslant |\widetilde{T}| \exp\big\{A/(1-r)\big\},$$

and, hence,

$$\mathrm{Re}\,\big\{\alpha g\big(t(\zeta)\big)\big\} \leqslant \log|\widetilde{T}| + A/(1-r) \tag{2}$$

for $|\zeta| \leqslant r$.
Denote by $\mathfrak{A}(r)$ the largest value of $\mathrm{Re}\,\big\{\alpha\, g\big(t(\zeta)\big)\big\}$ in the circle $\{|\zeta| \leqslant \sqrt{r}\}$. It follows from (2) that

$$\mathfrak{A}(r) \leqslant \log|\widetilde{T}| + 2A/(1-r). \tag{3}$$

According to the Carathéodory theorem (see, e.g., (Polya and Szegö, 1978, Part I, Section III, Chapter 6, §2))

$$M\Big(r; \alpha g\big(t(\zeta)\big)\Big) \leqslant \frac{8A}{(1-r)^2} + \frac{4\big(\log|\widetilde{T}| + 2|\alpha g(0)|\big)}{1-r} \leqslant \frac{B}{(1-r)^2},$$

where B does not depend on r.
But, for $t = x + iy \in S_{\rho_1}$, we have (see (Ostrovskii, 1983))

$$1 - |\zeta(t)| \geqslant (1 - |\zeta(t)|^2)/2 \geqslant \frac{1}{2} \exp\left\{ -\frac{\pi|x|}{2\rho_1} \right\} \cos \frac{\pi y}{2\rho_1}.$$

Therefore,

$$|\alpha g(t)| \leqslant 8B \exp\left\{ \frac{\pi|x|}{\rho_1} \right\} \tag{4}$$

for $t \in S_{\rho_1/2}$.
3) Let us extend estimate (4) to the whole plane. According to Theorem 4.4.10. in (Leont'ev, 1980), for any $\mu > 1$ there is a constant K such that, for all $t \in \mathbf{C}$,

$$|g(t)| \leqslant 8KB \exp\left\{ \frac{\pi\mu|x|}{\rho_1} \right\}, \quad t = x + iy.$$

Since $\rho_1 > 0$ is arbitrary,

$$\lim_{r\to\infty} (\log M(r; g))/r = 0.$$

4) The conclusion of Theorem 1 follows now from (Ostrovskii, 1963).
Under more restrictive assumptions on the function $h(z)$ than those of Theorem 1, the restrictions on the values of gaps can be weakened.
Let now $h(z)$ be an entire function satisfying the conditions

i) $|h(z)| \leqslant h(\operatorname{Re} z)$ for any z.

ii) $h(z)$ increases monotonically for $z \in [0, \infty) = \mathbf{R}^1_+$.

THEOREM 2. *Let $f(t)$ be a ch.f. analytic in the strip S_ρ ($\rho > 0$), and $h(z)$ be an entire function satisfying conditions* i), ii). *Assume that*

$$f(t) = h\left(\sum_{j=0}^{\infty} a_{\lambda_j} t^{\lambda_j} \right)$$

and suppose that there exists the limit

$$\lim_{j\to\infty} j/\lambda_j = \sigma < 1/2.$$

Then

a) *if $h(z)$ does not have exceptional values, then*

$$f(t) = h(a_0 + a_1 t + a_2 t^2); \tag{5}$$

b) *if $h_0 \neq 1$ is an exceptional value for $h(z)$, then either $f(t)$ has the form (5) or $f(t)$ takes the value h_0 at least at $k = [1/\sigma] - 1$ points of the circle $|t| = R$, and, moreover, $f(t)$ admits analytic continuation at least into the* *strip $\{t\colon |\mathrm{Im}\, t| < R\}$. These values are taken on each arc of length $2\pi R/k$ of the above mentioned circle.*

Note that if $\sigma = 0$, then $f(t)$ necessarily has the form (5). Remark also that the results of (Fryntov, 1988, 1989) are contained in Theorem 2 as a special case.

Proof of Theorem 2. Denote

$$g(t) = \sum_{j=0}^{\infty} a_{\lambda_j} t^{\lambda_j}.$$

If the series $g(t)$ converges in a circle of a radius $R < \infty$, then by the Polya–Fabry theorem (see, e.g., (Bieberbach, 1967)) $g(t)$ has at least $[1/\sigma] + 1$ singular points on the circle $|t| = R$, and also there is at least one singular point on each arc of length $2\pi R/k$. It is clear that the ch.f. $f(t)$ is analytic at least in the strip $\{t\colon |\mathrm{Im}\, t| < R\}$. It is easily seen that inside this strip the series $g(t)$ can have as its singular points only those points t for which $f(t) = h_0$. Therefore, $f(t)$ is either an entire function or $f(t)$ takes the value h_0 at least at $k = [1/\sigma] - 1$ points of the circle $|t| = R$. To complete the proof, it is sufficient to show that if $f(t)$ is an entire function, then it has the form (5). Let T be the angle $|\arg z| \leqslant \pi\sigma_1$, where $\sigma_1 > \sigma$. Since $f(t)$ is an entire function,

$$M(r; f) = \max_{|t|\leqslant r} |f(t)| = \max\big(|f(ir)|,\ |f(-ir)|\big).$$

Denote, for brevity, $M(r) = M(r; f)$. We have

$$\max_{|t|\leqslant r,\, t\in T} |f(t)| \leqslant M\big(r \sin(\pi\sigma_1)\big).$$

Therefore,

$$\max_{|z|\leqslant r,\, z\in T} \big|h\big(g(z)\big)\big| \leqslant M\big(r \sin(\pi\sigma_1)\big),$$

and by i), ii),

$$\max_{|z|\leqslant r,\, z\in T} \mathrm{Re}\, g(z) \leqslant h^{-1}\Big(M\big(r \sin(\pi\sigma_1)\big)\Big).$$

Applying the Carathéodory theorem, one can show that

$$\max_{|z|\leqslant r,\, z\in T_1} |g(z)| \leqslant C h^{-1}\Big(M\big(r \sin(\pi\sigma_1)\big)\Big),$$

where T_1 is the angle $|\arg z| \leqslant \pi\sigma_2$ $(\sigma_1 > \sigma_2 > \sigma)$ and C does not depend on r. Theorem 2.2.6 from (Leont'ev, 1980) shows that for any $q > 1$ there is $A(q)$ such that

$$\max_{|z|\leqslant r} |g(z)| \leqslant C A(q)\, h^{-1}\Big(M\big(qr \sin(\pi\sigma_1)\big)\Big).$$

We may assume that $q > 1$ is chosen so that

$$\lambda = q \sin(\pi\sigma_1) < 1.$$

Then we have

$$|f(t)| = |h(g(t))| \leqslant h(|g(t)|) \leqslant h\Big(CA(q)\, h^{-1}(M(\lambda r))\Big)$$

for $|t| \leqslant r$, i.e.,

$$M(r) \leqslant h\Big(CA(q)\, h^{-1}(M(\lambda r))\Big).$$

Thus,

$$M(r;g) = h^{-1}(M(r)) \leqslant CA(q)\, h^{-1}(M(\lambda r))$$

for all $r > 0$ and some $\lambda < 1$.
It is easily seen that the monotone function

$$\psi(r) = h^{-1}(M(r))$$

satisfying the inequality $\psi(r) \leqslant C_1\psi(\lambda r)$, $0 < \lambda < 1$, increases not faster than some polynomial. In particular,

$$\lim_{r\to\infty} (\log \psi(r))/r = 0.$$

Hence,

$$\lim_{r\to\infty} \big(\log M(r;g)/r\big) = 0$$

and the required assertion follows from (Ostrovskii, 1963).
Note that in general one cannot get rid of the assumption of Theorem 1 that $h(z)$ is an entire function. In fact, if

$$h(z) = z^{-2}, \quad g(t) = (1+t^2)^2,$$

then the function

$$f(t) = h\big(g(t)\big) = (1+t^2)^{-4}$$

is the ch.f. of a convolution of two Laplace distributions, however, the series $g(t)$ terminates not at the second, but at the fourth degree of t.
Now let us apply Theorem 2 to the analysis of coefficients of asymptotic expansion in the central limit theorem.
Consider a sequence of independent identically distributed random variables $\{X_n,\ n = 1,2,\ldots\}$. Let

$$V(x) = \mathbf{P}\{X_1 < x\}, \quad \mathbf{E}X_1 = 0, \quad \mathbf{E}X_1^2 = \sigma^2 > 0,$$

$$F_n(x) = \mathbf{P}\left\{\frac{1}{\sigma\sqrt{n}}\sum_{j=1}^{n} X_j < x\right\},$$

$$v(t) = \mathbf{E}e^{itX_1}, \quad \limsup_{t\to\infty}|v(t)| < 1. \tag{6}$$

Assume that $v(t)$ is analytic in some strip S_ρ. It is known (see, e.g., (Petrov, 1975)) that under the above assumptions the expansion

$$F_n(x) - \Phi(x) = \sum_{\nu=1}^{\infty} Q_\nu(x)/n^{\nu/2} \tag{7}$$

is valid, where $\Phi(x)$ is the distribution function of the standard normal law, and

$$Q_\nu(x) = -\frac{1}{\sqrt{2\pi}} e^{-x^2/2} \sum H_{\nu+2s-1}(x) \prod_{m=1}^{\nu} \frac{1}{k!} \left(\frac{\gamma_{m+2}}{(m+2)!\,\sigma^{m+2}} \right)^{k_m}. \tag{8}$$

The summation in (8) is carried out over all non-negative integer solutions of the equation $k_1 + 2k_2 + \ldots + \nu k_\nu = \nu$; $s = k_1 + k_2 + \ldots + k_\nu$; $H_\ell(x)$ is the Hermite polynomial of degree ℓ, and γ_m is the m-th cumulant of the random variable X_1. Multiplying both sides of (7) by $H_{\ell+1}(x)$ and integrating over the real line, by the orthogonality of the Hermite polynomials, we get

$$\int_{-\infty}^{\infty} (F_n(x) - \Phi(x)) H_{\ell+1}(x)\, dx = \sum_{j=1}^{\ell} c_j/n^{j/2}, \tag{9}$$

where c_j are some coefficients.

THEOREM 3. *Let $\{X_n,\ n = 1,2\ldots\}$ be a sequence of independent identically distributed random variables satisfying conditions (6) and $v(x)$ be analytic in some strip S_p. Assume that in (9) the coefficients $c_\ell = 0$ for all $\ell \notin \{\ell_k\}$, where ℓ_k is an increasing sequence of indices, satisfying the condition $\lim_{k\to\infty} k/\ell_k = 0$. Then $V(x)$ is a normal distribution function.*

Proof. A direct calculation shows that

$$c_\ell = -\frac{\gamma_{\ell+2}}{(\ell+2)!\,\sigma^{\ell+2}} \int_{-\infty}^{\infty} H_{\ell+1}^2(x) \frac{e^{-x^2/2}}{\sqrt{2\pi}}\, dx.$$

Hence, if $c_\ell = 0$, then $\gamma_{\ell+2} = 0$ for all $\ell \notin \{\ell_k\}$, and the desired result follows if we apply Theorem 2 to the case of exponential function $h(z)$.

THEOREM 4. *Suppose that conditions of Theorem 3 are fulfilled and, in addition, the ch.f. $v(t)$ of the random variable X_1 is an entire function without zeros. If in (9) the coefficients $c_\ell = 0$ for all $\ell \notin \{\ell_k\}$ and $\lim_{k\to\infty} k/\ell_k < 1/2$, then $V(x)$ is a normal distribution function.*

In order to prove the theorem, it is sufficient to observe that the only exceptional value for the exponential function is zero, and to apply Theorem 2 assuming that $h(x)$ is the exponential function.

REFERENCES

Bieberbach, L. (1955). *Analytische Fortsetzung*. Springer, Berlin.

Fryntov, A.E. (1988, 1989). Characterization of a Gaussian distribution by gaps in its sequence of cumulants. *Theory Probab. Appl.* **33**, 638–644; **34**, 613–614.

Gnedenko, B.V. (1983). On limit theorems for a random number of random variables. *Lect. Notes Math.* **1021**, 167–176.

Hayman, W.K. (1964). *Meromorphic Functions*. Clarendon Press, Oxford.

Leont'ev, A.F. (1980). *Sequences of Polynomials of Exponents*. Nauka, Moscow (in Russian).

Marcinkiewicz, J. (1938). Sur une propertiété de la loi de Gauss. *Math. Z.* **44**, 612–618.

Ostrovskii, I.V. (1963). On entire functions satisfying certain specific inegualities connected with the theory of characteristic functions of probability laws. In: *Notes Mech.-Math. Fak. Khar'kov Gos. Univ. and Khar'kov. Math. Soc.*, **29**, 145–168 (in Russian).

Ostrovskii, I.V. (1983). On the growth of entire characteristic functions *Lect. Notes Math.* **982**, 151–155.

Petrov, V.V. (1975). *Sums of Independent Random Variables*. Springer, Berlin etc.

Polya, G. and Szegö, G. (1964). *Aufgaben und Lehrsatze aus der Analysis*. **1**. Springer, Berlin etc.

New Trends in Probab. and Statist., pp. 38 - 42
V. Sazonov and T. Shervashidze (Eds.)
1991 VSP/Mokslas

ON INTEGRABILITY OF $\sup|S_{n_k}/n_k|$

O.I. KLESOV
Kiev State University, Vladimirskaya str. 64, 252017 Kiev, Ukraine, USSR

ABSTRACT

Let $\{S_n,\ n \geqslant 1\}$ denote the partial sums of i.i.d. random variables and let $\{n_k,\ k \geqslant 1\}$ be a strictly increasing subsequence of positive integers. We determine necessary and sufficient conditions for

$$\mathbf{E}\sup_{k\geqslant 1}|S_{n_k}/n_k| < \infty.$$

Let $\{X_n,\ n \geqslant 1\}$ be a sequence of i.i.d. random variables, $S_n = \sum_{k\leqslant n} X_k$, and let $\{n_k,\ k \geqslant 1\}$ denote a strictly increasing sequence of natural numbers. In this note necessary and sufficient conditions for existence of the first moment of the random variable $\sup_{k\geqslant 1}|S_{n_k}/n_k|$ is obtained.
In (Marcinkiewicz and Zygmund, 1937) the sufficiency part was proved for $n_k = k$. In (Burkholder, 1962) the necessity was proved for this case.
Put $\gamma_k = n_{k-1}/n_k$. Gut in (Gut, 1979) obtained similar results for the cases

a) $\liminf_{k\to\infty} \gamma_k > 0$;

b) $\limsup_{k\to\infty} \gamma_k < 1$.

For $L \subseteq N$ we denote card(L) the number of elements of L and denote

$$\Psi(\{m_k\},\ z) = \text{card}\ (k\colon\ m_k \leqslant z),$$

for every strictly increasing sequence $\{m_k,\ k \geqslant 1\}$. It is easy to see that

$$\Psi(\{m_k\},\ z) \leqslant \Psi(\{m_k\},\ z-1) + 1.$$

Gut in (Gut, 1979) proved that

$$\mathbf{E}\sup_{k\geqslant 1}|S_{n_k}/n_k| < \infty \tag{1}$$

if and only if

$$\mathbf{E}\,|X_1|\log^+|X_1| < \infty \quad \text{when} \quad \liminf_{k\to\infty}\gamma_k > 0; \tag{2}$$

$$\mathbf{E}\,|X_1|\,\Psi(\{n_k\},\ |X_1|) < \infty \quad \text{when} \quad \limsup_{k\to\infty}\gamma_k < 1, \tag{3}$$

where $\log^+ z = \log(1+z)$ for $z \geqslant 0$. In what follows $[z]$ denotes the largest integer $m \leqslant z$. For $c > 1$ consider the sequence $\{m_k,\ k \geqslant 1\}$ defined as follows:

$$m_1 = n_1, \quad m_k = \min(n_i\colon\ n_i \geqslant [cm_{k-1}]), \quad k \geqslant 2.$$

THEOREM. i) *Condition* (1) *is satisfied if, for some* $c > 1$, *we have*

$$\mathbf{E}\,|X_1|\,\Psi(\{m_k\},\ |X_1|) < \infty. \tag{4}$$

ii) *Condition* (4) *is satisfied for all* $c > 1$ *if condition* (1) *holds true.*

Remarks. 1. If condition **a)** is satisfied, then there exists $\varepsilon > 0$ for which $n_{k-1} \geqslant \varepsilon n_k$. Therefore, $\varepsilon m_{k+1} \leqslant c m_k$ and

$$\Psi(\{m_k\},\ z) \geqslant \text{const} \cdot \log^+ z.$$

On the other hand, $m_{k+1} \geqslant c m_k$ and

$$\Psi(\{m_k\},\ z) \leqslant \text{const} \cdot \log^+ z.$$

So, in this case conditions (4) and (2) are equivalent.
2. If condition **b)** is satisfied, then there exists $\varepsilon > 0$, for which $n_{k-1} \leqslant (1-\varepsilon)\,n_k$. Therefore $n_k = m_k$ for $c = 1/(1-\varepsilon)$ and conditions (3) and (4) are equivalent in this case.
3. Now consider the sequence $n_k = k!$ For this sequence,

$$\Psi(\{n_k\},\ z) \sim \log z / \log\log z$$

and condition (1) is equivalent to

$$\mathbf{E}\,|X_1|\,\log^+|X_1| \,/\, \log^+\log^+|X_1| < \infty.$$

It is clear that this result is true for the sequence $n_{2k} = k!$, $n_{2k+1} = k! + 1$, $k \geqslant 1$. But

$$\liminf_{k\to\infty} \gamma_k = 0, \qquad \limsup_{k\to\infty} \gamma_k = 1,$$

and this case is not covered by Gut's theorem.

COROLLARY. *If* $n_k = k$, *then conditions* (1) *and* (2) *are equivalent.*

This result has been obtained in (Marcinkiewicz and Zygmund, 1937) and (Burkholder, 1962).
For the proof of the theorem we need the following lemma. Let $\mu(\xi)$ denote the median of a random variable ξ.

LEMMA. *Let* $\{Y_n,\ n \geqslant 1\}$ *be independent symmetric random variables,* $T_n = \sum_{k\leqslant n} Y_k$. *Assume that* $b_n \geqslant 0$ *and let* $\{r_n\}$, $\{t_n\}$ *be two sequences of natural numbers, for which*

$$r_n \leqslant t_n \leqslant r_{n+1}.$$

If $x > 0$, then

$$\mathbf{P}\Big(\sup_{n\geqslant 1} b_n \max_{r_n\leqslant j\leqslant t_n} T_j \geqslant x\Big) \leqslant 2\mathbf{P}\Big(\sup_{n\geqslant 1} b_n T_{t_n} \geqslant x\Big), \tag{5}$$

$$\mathbf{P}\Big(\sup_{n\geqslant 1} b_n \max_{r_n\leqslant j\leqslant t_n} |T_j| \geqslant x\Big) \leqslant 2\mathbf{P}\Big(\sup_{n\geqslant 1} b_n |T_{t_n}| \geqslant x\Big). \tag{6}$$

This lemma is a generalization of the Lévy inequality which it reduces to, if we put $b_1 = 1$ and $b_n = 0$ for $n \geqslant 2$.

Proof of Lemma. First we shall prove inequality (5). Let $L_n = \{j\colon r_n \leqslant j \leqslant t_n\}$ and

$$V_{ni} = \max_{r_n\leqslant j\leqslant i} T_j, \quad n\geqslant 1, \quad i\in L_n;$$

$$W_n = \max_{j\leqslant n} b_j V_{jt_j}, \quad n \geqslant 1.$$

Define the sets

$$A_1 = \{\omega\colon\ b_1 V_{1t_1} \geqslant x\},$$

$$A_n = \{\omega\colon\ W_{n-1} < x,\ b_n V_{nt_n} \geqslant x\}, \quad n > 1;$$

$$B_1 = \{\omega\colon\ b_1 T_{t_1} \geqslant x\},$$

$$B_n = \{\omega\colon\ \max_{j<n} b_j T_{t_j} < x,\ b_n T_{t_n} \geqslant x\}, \quad n>1;$$

$$C_{ni} = \{\omega\colon\ T_{t_n} \geqslant T_i\}, \quad n\geqslant 1,\ i\in L_n;$$

$$D_{ni} = \begin{cases} \{\omega\colon\ b_1 V_{1r_1} \geqslant x\}, & n=1,\ i=r_1,\\ \{\omega\colon\ b_1 V_{1(i-1)} < x,\ b_1 V_{1i} \geqslant x\}, & n=1,\ r_1 < i \leqslant t_1,\\ \{\omega\colon\ W_{n-1} < x,\ b_n V_{nr_n} \geqslant x\}, & n>1,\ i=r_n,\\ \{\omega\colon\ W_{n-1} < x,\ b_n V_{n(i-1)} < x,\ b_n V_{ni} \geqslant x\}, & n>1,\ r_n < i \leqslant t_n. \end{cases}$$

Since the sets C_{ni} and D_{ni} are independent, and

$$E = \{\omega\colon\ \sup_{n\geqslant 1} b_n \max_{j\in L_n} T_j \geqslant x\} = \bigcup_{n\geqslant 1} A_n,$$

$$A_n \cap A_m = \varnothing, \quad n\neq m, \qquad A_n = \bigcup_{i\in L_n} D_{ni}, \quad n\geqslant 1,$$

$$D_{ni}\cap D_{nj} = \varnothing, \quad i\neq j, \qquad 2\mathbf{P}(C_{ni}) \geqslant 1, \quad n\geqslant 1,\ i\in L_n,$$

we obtain

$$\mathbf{P}(E) = \sum_{n\geqslant 1}\sum_{i\in L_n} \mathbf{P}(D_{ni}) \leqslant \sum_{n\geqslant 1}\sum_{i\in L_n} 2\mathbf{P}(C_{ni}\cap D_{ni}) \leqslant$$

$$\leqslant 2\sum_{n\geqslant 1} \mathbf{P}(B_n) = 2\mathbf{P}\Big(\sup_{n\geqslant 1} b_n T_{t_n} \geqslant x\Big).$$

These relations complete the proof of inequality (5).

Inequality (6) follows immediately from inequality (5).

Proof of Theorem. First observe that

$$\limsup_{k\to\infty} m_{k-1} / m_k \leqslant 1/c < 1$$

and, consequently, from Gut's theorem we have

$$\mathbf{E}|X_1|\,\Psi(\{m_k\}, |X_1|) < \infty \iff \mathbf{E}\sup_{k\geqslant 1}|S_{m_k} / m_k| < \infty. \tag{7}$$

Therefore, it is necessary to prove only part i) of Theorem.
We first assume that X_1 is symmetric. Let

$$t_1 = m_1, \qquad t_k = [cm_{k-1}], \quad k > 1;$$

$$l_0 = 0, \qquad l_1 = m_1, \qquad l_k = \max(n_i\colon\ n_i < m_k), \quad k \geqslant 2.$$

By Lemma, for all $x > 0$, we have

$$\mathbf{P}\Big(\sup_{k\geqslant 1}|S_{n_k} / n_k| \geqslant x\Big) = \mathbf{P}\Big(\sup_{k\geqslant 1}\max_{l_{k-1}<n_j\leqslant l_k}|S_{n_j} / n_j| \geqslant x\Big) \leqslant$$

$$\leqslant \mathbf{P}\Big(\sup_{k\geqslant 1}|S_{m_{k-1}} / m_{k-1}| \geqslant x/2\Big) + \mathbf{P}\Big(\sup_{k\geqslant 1}\max_{m_{k-1}<n_j\leqslant l_k}|S_{n_j} / n_j| \geqslant x/2\Big). \tag{8}$$

Since $l_k \leqslant t_k \leqslant m_k \leqslant t_{k+1}$, using Lemma, we obtain, for $b_k = 1/m_{k-1}$,

$$\mathbf{P}\Big(\sup_{k\geqslant 1}\max_{m_{k-1}<n_j\leqslant l_k}|S_{n_j} / n_j| \geqslant x/2\Big) \leqslant$$

$$\leqslant \mathbf{P}\Big(\sup_{k\geqslant 1} b_k \max_{m_{k-1}<j\leqslant t_k}|S_j| \geqslant x/2c\Big).$$

Now (8) implies that

$$\mathbf{P}\Big(\sup_{k\geqslant 1}|S_{n_k} / n_k| \geqslant x\Big) \leqslant$$

$$\leqslant \mathbf{P}\Big(\sup_{k\geqslant 1}|S_{m_k} / m_k| \geqslant x/2\Big) + 2\mathbf{P}\Big(\sup_{k\geqslant 1}|S_{t_k} / t_k| \geqslant x/2c\Big). \tag{9}$$

Since $t_{k-1} / t_k \leqslant 1/c < 1$, we have, by Gut's theorem,

$$\mathbf{E}|X_1|\,\Psi(\{t_k\}, |X_1|) < \infty \iff \mathbf{E}\sup_{k\geqslant 1}|S_{t_k} / t_k| < \infty. \tag{10}$$

Furthermore,

$$\mathbf{E}|X_1|\,\Psi(\{t_k\}, |X_1|) < \infty.$$

The last inequality follows from

$$\Psi(\{t_k\}, cz) \geqslant \Psi(\{m_k\}, z) + 1 \geqslant \Psi(\{t_k\}, cz - 1).$$

Finally, integrating both parts of inequality (9) over $x > 0$, we cor ,lete the proof of Theorem for symmetric X_1.
Now let $\{X'_n, n \geqslant 1\}$ be i.i.d. random variables and sequences $\{X_n\}$ and $\{X'_n\}$ be independent. We assume that X'_n has the same distribution as X_n and put $T_n = X'_1 + \ldots + X'_n$, $X^s_n = X_n - X'_n$, $S^s_n = X^s_1 + \ldots + X^s_n$. Denote by $\mathfrak{F}$ the σ-algebra generated by the sequence $\{X'_n, n \geqslant 1\}$, $m = \mathbf{E}X_1$. Theorem holds true if

$$\mathbf{E}\sup_{k\geqslant 1}|S_{n_k}/n_k| \leqslant m + \mathbf{E}\sup_{k\geqslant 1}|S^s_{n_k}/n_k|.$$

But this inequality follows from the inequalities

$$\mathbf{E}\sup_{k\geqslant 1}\left|S^s_{n_k}/n_k\right| \geqslant \mathbf{E}\left[\mathbf{E}\left(\sup_{k\geqslant 1}\left|S^s_{n_k}/n_k\right|\right)\middle|\mathfrak{F}\right] \geqslant$$

$$\geqslant \mathbf{E}\left[\sup_{k\geqslant 1}\left|\mathbf{E}\left(S^s_{n_k}/n_k\right)\middle|\mathfrak{F}\right|\right] \geqslant \mathbf{E}\sup_{k\geqslant 1}\left|S_{n_k}/n_k\right| - m.$$

REFERENCES

Burkholder, D.L. (1962). Succesive conditional expectations of an integrable functions. *Ann. Math. Statist.* **33**, 887–893.

Gut, A. (1979). On the integrability of $\sup|S_n/n|$ for subsequences. *Ann. Probab.* **7**, 1059–1062.

Marcinkiewicz, D.L. and Zygmund, A. (1937). Sur les fonctions independantes. *Fund. Math.* **29**, 60–90.

New Trends in Probab. and Statist., pp. 43 – 56
V. Sazonov and T. Shervashidze (Eds.)
1991 VSP/Mokslas

ASYMPTOTIC EXPANSIONS IN LARGE DEVIATION ZONES FOR THE DISTRIBUTION DENSITY OF SUMS OF INDEPENDENT RANDOM VARIABLES

L. SAULIS
Institute of Mathematics and Informatics, Lithuanian Academy of Sciences, Akademijos 4, 232600 Vilnius, Lithuania

ABSTRACT

The papers (Cramér, 1938; Linnik, 1961; 1962; Petrov, 1954; 1965; Ibragimov, 1967; Borovkov, 1962; Zolotarev, 1962; Prokhorov, 1956; Nagaev, 1979) and others enable to obtain asymptotic expansions in large deviation theorems. The present paper employing certain general lemma for distribution density of arbitrary random variable deals with theorems related to large deviations in the Cramér and the Linnik zones for the distribution density of sums of independent random variables.

1. ASYMPTOTIC EXPANSIONS FOR DISTRIBUTION DENSITY OF RANDOM VARIABLE

Suppose that a random variable (r.v.) ξ with $\mathbf{E}\xi = 0$, $\operatorname{var}\xi = 1$, has distribution density $p_\xi(x)$ such that

$$\sup_x p_\xi(x) < \infty. \tag{A}$$

Denote by ε the set of all points of the real line that are either continuity points or points of discontinuity of the first kind of the density $p_\xi(x)$. Moreover, at a discontinuity point x_0 we will agree that

$$p(x_0) := \frac{1}{2}\Big(p_\xi(x_0 - 0) + p_\xi(x_0 + 0)\Big).$$

Suppose that there exist constants $\gamma \geqslant 0$ and $\Delta > 0$ such that the k–th order cumulants of r.v. ξ satisfy the condition:

$$\big|\Gamma_k(\xi)\big| \leqslant (k!)^{1+\gamma}\Big/\Delta^{k-2}, \quad k = 3, 4, \dots \tag{S$_\gamma$}$$

Denote

$$f_\xi(t) = \mathbf{E}\exp\{it\xi\}, \quad \varphi(x) = \frac{1}{\sqrt{2\pi}}\exp\Big\{-\frac{1}{2}x^2\Big\}, \tag{1.1}$$

$$\Delta_\gamma = c_\gamma \Delta^{1/(1+2\gamma)}, \quad c_\gamma = \frac{1}{6}\left(\frac{\sqrt{2}}{6}\right)^{1/(1+2\gamma)} \tag{1.2}$$

$$\varepsilon(\Delta,\gamma) = \frac{1}{12}\left(1 - \frac{x}{\Delta_\gamma}\right)\Delta_\gamma, \quad 0 \leqslant x < \Delta_\gamma; \tag{1.3}$$

$$f^*(t) = \begin{cases} \sum\limits_{k=0}^{s} \left(\frac{3}{2}\right)^k \frac{x^k}{k!} \left| f_\xi^{(k)}(t) \right| & \text{if } \gamma > 0, \\ \int\limits_{-\infty}^{\infty} \exp\{itu\} p_h(u)\, du & \text{if } \gamma = 0. \end{cases} \tag{1.4}$$

Here

$$s = 2\left[\frac{1}{2}\left(\frac{\Delta^2}{18}\right)^{1/(1+2\gamma)}\right] - 2, \tag{1.5}$$

$$p_h(u) = \exp\{hu\} p_\xi(u) \Big/ \int\limits_{-\infty}^{\infty} \exp\{hu\} p_\xi(u) du, \tag{1.6}$$

where h is the root of the equation

$$x = \sum_{k=2}^{\infty} \frac{1}{(k-1)!} \Gamma_k(\xi) h^{k-1}.$$

Denote

$$c_1(\gamma) = 2\sqrt{\pi} + 6^\gamma 25^3 \sqrt{2\pi}/\Delta, \tag{1.7}$$

$$q(l,\gamma) = \left(\frac{3\sqrt{2e}}{2}\right)^l 8(l+2)^2 6^{\gamma(l-1)} 4^{3(l+1)} \Big((l+1)!\Big)^{\gamma(l-1)} \Gamma\left(\frac{3l-1}{2}\right), \tag{1.8}$$

and

$$r^*(\Delta, x) = \left(1 + 9((m+2)!)^\gamma 16^{m-1} c_\gamma^{m+1-l} \frac{1}{m+1}\left(\frac{x}{\Delta}\right)^l\right) \times$$
$$\times \left(1 + 46\Delta_\gamma \exp\left\{-\frac{1}{2}\left(1 - \frac{x}{\Delta_\gamma}\right)\sqrt{\Delta_\gamma}\right\} \Big/ \left(1 - \frac{x}{\Delta_\gamma}\right)\right), \tag{1.9}$$

where $m = (1/\gamma) + l + 1$, $\gamma > 0$, $l \geqslant 1$, and $r^*(\Delta, x) \equiv 0$ as $\gamma = 0$.
Let Θ (with or without an index) denote some variable, not always one and the same, not exceeding a unit in absolute value, $[m]$ is the integer part of the number m.

LEMMA 1. *If r.v. ξ satisfies conditions* (S_γ) *and* (A), *then for every* l, $l \geqslant 1$, *in the interval* $0 \leqslant x < \Delta_\gamma$ *the relation*

$$\frac{p_\xi(x)}{\varphi(x)} = \exp\left\{L_m(x)\right\}\left(1+\sum_{\nu=0}^{l-1} M_\nu(x) + \Theta_1 q(l,\gamma)\left(\frac{x+1}{\Delta}\right)^l + \right.$$

$$+\Theta_2 c_1(\gamma)\Delta_\gamma^{3/2}\exp\left\{-\frac{1}{72}\left(1-\frac{x}{\Delta_\gamma}\right)\sqrt{\Delta_\gamma}\right\}+$$

$$\left. +\Theta_3 \int\limits_{|t|\geqslant\varepsilon(\Delta,\gamma)} |f^*(t)|dt\right)\left(1+\Theta r^*(\Delta,x)\right) \tag{1.10}$$

holds. The quantities Δ_γ, $c_1(\gamma)$, $q(l,\gamma)$, $\varphi(x)$, $f^*(t)$ and $\varepsilon(\Delta,x)$ are defined by relations (1.2), (1.7), (1.8), (1.1), (1.4) and (1.3), respectively;

$$L_m(x) = \sum_{3\leqslant k<m} \lambda_k x^k, \quad m = \begin{cases} 1/\gamma + l + 1, & \gamma > 0, \\ \infty, & \gamma = 0, \end{cases} \tag{1.11}$$

and

$$|\lambda_k| \leqslant \frac{2}{k}\left(\frac{16}{\Delta}\right)^{k-2}\left((k+1)!\right)^\gamma, \quad k = 3,4,\ldots \tag{1.12}$$

For calculating the coefficients λ_k, one has to employ the relation

$$\lambda_k = -\frac{1}{k} b_{k-1} \tag{1.13}$$

with b_k solving of the recurrent equations:

$$b_1^{(1)} = 1, \quad b_j^{(1)} = 0, \quad j = 2,3,\ldots, \tag{1.14}$$

$$b_j^{(1)} = \sum_{r=1}^{j} \frac{1}{r!}\Gamma_{r+1}(\xi) \sum_{\substack{j_1+\cdots+j_r=j \\ j_i\geqslant 1}} \prod_{i=1}^{r} b_{j_i}.$$

The following formula is valid for the polynomials $M_\nu(x)$

$$M_\nu(x) = \sum_{k=0}^{\nu} K_k(x) q_{\nu-k}(x), \tag{1.15}$$

$$K_\nu(x) = \sum \prod_{m=1}^{\nu} \frac{1}{k_m!}\left(-\lambda_{m+2}x^{m+2}\right)^{k_m}, \quad K_0(x) \equiv 1,$$

$$q_\nu(x) = \sum H_{\nu+2l}(x) \prod_{m=1}^{\nu} \frac{1}{k_m!}\left(\Gamma_{m+2}(\xi)/(m+2)!\right)^{k_m}, \quad q_0(x) \equiv 1$$

where $H_l(x)$ are the Chebyshev–Hermite polynomials and the sum is taken over all solutions of the equation $k_1 + 2k_2 + \cdots + \nu k_\nu = \nu$.

A complete proof of Lemma 1 can be found in (Saulis and Statulevičius, 1989, pp. 137–153).

2. ASYMPTOTIC EXPANSIONS IN CRAMER ZONE AND IN THE LINNIK POWER ZONES FOR DISTRIBUTION DENSITY OF SUMS OF INDEPENDENT RANDOM VARIABLES

Let ξ_j, $j = 1, 2, \ldots$, be independent r.v. with $\mathbf{E}\xi_j = 0$ and $\sigma_j^2 = \operatorname{var}\xi_j < \infty$, $j = 1, 2, \ldots$
Denote

$$S_n = \sum_{j=1}^{n} \xi_j, \quad B_n^2 = \sum_{j=1}^{n} \sigma_j^2, \quad Z_n = \frac{S_n}{B_n}, \tag{2.1}$$

$L_{k,n} = \sum_{j=1}^{n} \mathbf{E}|\xi_j|^k / B_n^k$ is the k–th Lyapounov fraction and

$$F_{Z_n}(x) = \mathbf{P}\{Z_n < x\}, \quad p_{Z_n}(x) = \frac{d}{dx} F_{Z_n}(x).$$

Suppose that there exist $\gamma \geqslant 0$ and $K > 0$ such that

$$|\mathbf{E}\xi_j^k| \leqslant (k!)^{1+\gamma} K^{k-2} \sigma_j^2, \quad k = 3, 4, \ldots \tag{B$_\gamma$}$$

STATEMENT 1. *If r.v. ξ_j, $j = 1, 2, \ldots, n$, satisfy condition* (B$_\gamma$) *then*

$$|\Gamma_k(Z_n)| \leqslant (k!)^{1+\gamma} / \Delta_n^{k-2}, \quad k = 3, 4, \ldots, \tag{2.2}$$

where

$$\Delta_n = B_n / K_n, \quad K_n = 2\Big(K \vee \max_{1 \leqslant j \leqslant n} \sigma_j\Big), \tag{2.3}$$

$$a \vee b = \max\{a, b\}, \quad a \wedge b = \min\{a, b\}.$$

The proof of Statement 1 can be found in (Saulis and Statulevičius, 1989, p. 46). Suppose that r.v. ξ_j have probability densities $p_{\xi_j}(x)$ such that

$$\sup_x p_{\xi_j}(x) \leqslant C_j < \infty \tag{D}$$

(if the density of ξ_j does not exist, then $C_j = \infty$).

THEOREM 1. *If r.v. ξ_j satisfys conditions* (B$_\gamma$) *with $\gamma = 0$ and* (D), *then, for every l, $l \geqslant 1$, in the interval*

$$0 \leqslant x < \Delta_n^*, \quad \Delta_n^* = \frac{1}{24} \Delta_n,$$

the relation

$$\frac{p_{Z_n}(x)}{\varphi(x)} = \exp\{L_n(x)\} \left(1 + \sum_{\nu=1}^{l-1} M_{\nu,n}(x) + \Theta_1 q(l, 0) \left(\frac{x+1}{\Delta_n^*}\right)^l + \right.$$

$$+\Theta_2 825\pi^2 \exp\left\{-\frac{1}{72}\left(1-\frac{x}{\Delta_n^*}\right)^2 \Delta_n^{*2}\right\} \Big/ \left(1-\frac{x}{\Delta_n^*}\right)+$$

$$+\Theta_3 684 e^4 \pi\sqrt{2\pi} K_n \prod_{i=1}^{4} C_i^{1/4} \exp\left\{-\frac{c}{K_n^2}\sum_{j=1}^{n}\frac{1}{C_j^2}\right\}\Bigg) \tag{2.4}$$

holds. Here

$$L_n(x) = \sum_{k=3}^{\infty} \lambda_{k,n} x^k, \tag{2.5}$$

where the coefficients $\lambda_{k,n}$ *and the polynomials* $M_{\nu,n}(x)$ *are defined according to formulae* (1.13) *and* (1.15) *with the cumulants of r.v.* ξ *replaced by the cumulants of* Z_n; $c = 729/c_0$, $0 < c_0 < 10^{-3}$.

THEOREM 2. *If r.v.* ξ_j *satisfy conditions* (B_γ) *with* $\gamma > 0$, (D) *and*

$$\varliminf_{n\to\infty} \frac{1}{(1 \vee L_{1,n})\Delta_n(\gamma)} \sum_{k=1}^{n} \frac{1}{\left(\sigma_k^2 + N_n^2\right) C_k^2} \geqslant c_1 > 0, \tag{2.6}$$

$N_n = 4B_n L_{3,n}$, *then, for every integer* l, $l \geqslant 1$, *in the interval*

$$0 \leqslant x < \frac{1}{2}\Delta_n(\gamma), \quad \Delta_n(\gamma) = c_\gamma \Delta_n^{1/(1+2\gamma)}$$

the relation

$$\frac{p_{Z_n}(x)}{\varphi(x)} = \exp\left\{L_{m,n}(x)\right\}\left(1 + \sum_{\nu=0}^{l-1} M_{\nu,n}(x) + \Theta_1 q(l,\gamma)\left(\frac{x+1}{\Delta_n}\right)^l +\right.$$

$$+\Theta_2 c_1(\gamma)\Delta_n^{3/2}(\gamma)\exp\left\{-\frac{1}{72}\left(1-\frac{x}{\Delta_n(\gamma)}\right)\sqrt{\Delta_n(\gamma)}\right\}+$$

$$+\Theta_3\left(\frac{5\pi e}{16}x^2 + \frac{\pi^2}{\varepsilon_n}\right)\exp\left\{-\frac{1}{16\pi^2}\varepsilon_n^2\right\} + \Theta_4 42\pi\sqrt{2\pi}e^2 N_n \times$$

$$\times \max_{1\leqslant r_i\leqslant n} \prod_{i=1}^{4} C_{r_i}^{1/4}\left(1 + \frac{\pi\sigma_{r_i}}{2\sqrt{2}N_n}\right)^{1/4} \exp\left\{-c_2\sum_{k=1}^{n}\frac{1}{\left(\sigma_k^2+N_n^2\right)C_k^2}\right\}\Bigg), \tag{2.7}$$

is valid, where

$$\varepsilon_n = \frac{1}{4}\left(1 - \frac{x}{\Delta_n(\gamma)}\right)\Delta_n(\gamma), \quad c_2 \geqslant 2^{-9}, \tag{2.8}$$

and, for $L_{m,n}(x)$, $M_{\nu,n}(x)$, $c_1(\gamma)$ *and* $q(l,\gamma)$, *formulae* (1.11), (1.15), (1.7) *and* (1.8) *hold, where the cumulants of r.v.* ξ *must be replaced by the cumulants of the r.v.* Z_n.

Proof of Theorem 2. According to Lemma 1, we must evaluate

$$\int_{\varepsilon_n}^{\infty} |f_n^*(t)| dt, \quad |f_n^*(t)| = \sum_{k=0}^{s} \frac{1}{k!} \left(\frac{3x}{2}\right)^k \left|f_{Z_n}^{(k)}(t)\right|,$$

$f_{Z_n}^{(0)}(t) = f_{Z_n}(t)$. At the very beginning let us estimate $f_{Z_n}^{(l)}(t)$. We have

$$f_{Z_n}(t) = \prod_{j=1}^{n} f_{\xi_j}(t/B_n). \tag{2.9}$$

Denote

$$f_{Z_n,k_1}(t) = \prod_{\substack{k=1\\k\neq k_1}}^{n} f_{\xi_k}(t/B_n),$$
$$\cdots\cdots\cdots\cdots\cdots\cdots \tag{2.10}$$
$$f_{Z_n,k_1,\dots,k_l}(t) = \prod_{\substack{k=1\\k\neq k_1\neq\cdots\neq k_l}}^{n} f_{\xi_k}(t/B_n).$$

Mathematical induction method (see (Pipiras and Statulevičius, 1968)) implies

$$f_{Z_n}^{(l)}(t) = \sum_{j=1}^{l} \sum_{\substack{m_{1j}+\cdots+m_{jj}=l\\ m_{\nu j}>0,\ \nu=1,\dots,j}} A_{m_{1j},\dots,m_{jj}}(l) \sum_{k_1=1}^{n} f_{\xi_{k_1}}^{(m_{1j})}(t/B_n)\times$$
$$\times \sum_{\substack{k_2=1\\k_2\neq k_1}}^{n} f_{\xi_{k_2}}^{(m_{2j})}(t/B_n)\cdots \sum_{\substack{k_j=1\\k_j\neq k_1\neq\cdots\neq k_{l-1}}}^{n} f_{\xi_{k_j}}^{(m_{jj})}(t/B_n) f_{Z_n,\dots,k_j}(t), \tag{2.11}$$

where $A_{m_{1j},\dots,m_{jj}}(l)$ depend on l only. Due to $\mathbf{E}\xi_k = 0$, and

$$f_{\xi_k}^{(1)}(t/B_n) = \int_{-\infty}^{\infty} \frac{ix}{B_n} \exp\left\{\frac{itx}{B_n}\right\} dF_{xi_k}(x) = -\frac{t}{B_n^2} \int_{-\infty}^{\infty} \exp\left\{\Theta \frac{itx}{B_n}\right\} dF_{\xi_k}(x),$$

we obtain

$$\sum_{k_1=1}^{n} \left|f_{\xi_{k_1}}^{(1)}(t/B_n)\right| \leqslant |t|. \tag{2.12}$$

Similarly, it follows that

$$\left|f_{\xi_k}^{(\nu)}(t/B_n)\right| \leqslant \mathbf{E}|\xi_k|^{\nu}/B_n^{\nu}, \quad \nu = 1, 2, \dots, s \tag{2.13}$$

and finally, we conclude

$$\sum_{k=1}^{n} \left|f_{\xi_k}^{(\nu)}(t/B_n)\right| \leqslant L_{\nu,n}, \quad \nu = 1, \dots, s. \tag{2.14}$$

Relations (2.11)–(2.14) imply

$$\left|f_{Z_n}^{(1)}(t)\right| \leqslant \left(|t| \wedge L_{1,n}\right) \max_{1\leqslant k_1 \leqslant n} \left|f_{Z_n,k_1}(t)\right|, \tag{2.15}$$

$$\left|f_{Z_n}^{(2)}(t)\right| \leqslant \left(1 + (t^2 \wedge L_{1,n}^2)\right) \max_{1\leqslant k_1 \neq k_2 \leqslant n} \left|f_{Z_n,k_1,k_2}(t)\right| \tag{2.16}$$

and

$$\left|f_{Z_n}^{(l)}(t)\right| \leqslant c(l)\left(1 + (|t|^l \wedge L_{1,n}^l) + L_{l,n}\right) \times$$
$$\times \max_{1\leqslant k_1 \neq k_2 \neq \dots \neq k_l \leqslant n} \left|f_{Z_n,k_1,k_2,\dots,k_l}(t)\right|, \quad l = 3, 4, \dots, \tag{2.17}$$

$c(l)$ is a quantity dependent only on l.
Denote

$$\mathcal{J}_k(t) = \frac{1}{2}\left(1 - \left|f_{\xi_k}(2\pi t)\right|^2\right) = \int_{-\infty}^{\infty} \sin^2(\pi t x)\tilde{p}_{\xi_k}(x)\, dx,$$

where $\tilde{p}_\xi$ is the density of the symmetrized distribution function

$$\widetilde{F}_\xi(x) = \int_{-\infty}^{\infty} F_\xi(x+y)\, dF_\xi(y), \quad \tilde{f}_\xi(t) = \left|f_\xi(t)\right|^2,$$

and

$$\mathrm{I}_n(t) = \sum_{k=1}^{n} \mathcal{J}_k(t). \tag{2.18}$$

Then $\left|f_{\xi_k}(2\pi t)\right| \leqslant \exp\left\{-\mathcal{J}_k(t)\right\}$, and

$$\left|f_{S_n}(2\pi t)\right| \leqslant \exp\left\{-\mathrm{I}_n(t)\right\}. \tag{2.19}$$

Consequently,

$$\max_{1\leqslant k_1 \neq k_2 \neq \dots \neq k_l \leqslant n} \left|f_{Z_n,k_1,\dots,k_l}(2\pi B_n t)\right| \leqslant$$
$$\leqslant \exp\left\{-\min_{1\leqslant k_1 \neq k_2 \neq \dots \neq k_l \leqslant n} \sum_{\substack{k=1 \\ k\neq k_1 \neq \dots \neq k_l}} \mathcal{J}_k(t)\right\} =$$
$$= \exp\left\{-\mathrm{I}_n(t) + \max_{1\leqslant k_1 \neq \dots \neq k_l \leqslant n} \left(\mathcal{J}_{k_1}(t) + \dots + \mathcal{J}_{k_l}(t)\right)\right\} \leqslant$$
$$\leqslant e^{l/2} \exp\left\{-\mathrm{I}_n(t)\right\}. \tag{2.20}$$

In its turn,

$$\mathrm{I}_n(t) \geqslant 4t^2 B_n^2 l_n\Big(\frac{1}{2|t|}\Big), \tag{2.21}$$

where

$$l_n(N_n) = \frac{1}{B_n^2}\sum_{k=1}^{n}\int\limits_{|x|\leqslant N_n} x^2 d\widetilde{F}_{\xi_k}(x) \geqslant 2\Big(1 - \frac{2L_{3,n}B_n}{N_n}\Big).$$

So, in the domain $|t| \leqslant \pi B_n/N_n$, the inequality

$$\max_{1\leqslant k_1\neq\ldots\neq k_l\leqslant n}\big|f_{Z_n,k_1,k_2,\ldots,k_l}(t)\big| \leqslant$$

$$\leqslant e^{l/2}\exp\Big\{-\frac{t^2}{\pi^2}l_n\big(\pi B_n/|t|\big)\Big\} \leqslant e^{l/2}\exp\Big\{-\frac{t^2}{\pi^2}l_n(N_n)\Big\}, \tag{2.22}$$

$l = 1, 2, \ldots, s$, holds. If we suppose $N_n = 4B_nL_{3,n}$, then, in the domain $|t| \leqslant$ $\leqslant (1/4)\pi L_{3,n}^{-1}$, we have

$$\big|f_{Z_n}^{(l)}(t)\big| \leqslant c^*(l)\Big(1 + \big(|t|^l \wedge L_{1,n}^l\big) + L_{l,n}\Big)\exp\Big\{-\frac{t^2}{\pi^2}\Big\}, \tag{2.23}$$

where $c^*(l) = c(l)\exp\{l/2\}$.
Further, under condition (B_γ), according to the inequality

$$(l!)^{1+\gamma}/\Delta_n^{l-2} \leqslant (l-2)!/\big(3,8\Delta_n(\gamma)\big)^{l-2}, \quad l = 3, 4, \ldots, s+2,$$

$$\Delta_n(\gamma) = c_\gamma \Delta_n^{1/(1+2\gamma)},$$

we obtain

$$L_{l,n} = B_n^{-l}\sum_{j=1}^{n}\mathbf{E}|\xi_j|^l \leqslant (l!)^{1+\gamma}/(B_n/K)^{l-2} \leqslant$$

$$\leqslant (l!)^{1+\gamma}/\Delta_n^{l-2} \leqslant (l-2)!/\big(3.8\,\Delta_n(\gamma)\big)^{l-2}, \tag{2.24}$$

$$l = 3, 4, \ldots, s+2, \quad \Delta_n = B_n/K_n, \quad K_n = 2\Big(K \vee \max_{1\leqslant j\leqslant n}\sigma_j\Big).$$

Assume $T_n = (\pi/4)L_{3,n}^{-1}$. Then

$$\int\limits_{\varepsilon_n}^{\infty}\big|f_n^*(t)\big|dt = \mathcal{J}_3 + \mathcal{J}_4, \tag{2.25}$$

where

$$\mathcal{J}_3 = \sum_{l=0}^{s}\frac{1}{l!}\Big(\frac{3x}{2}\Big)^l\int\limits_{\varepsilon_n}^{T_n}\big|f_{Z_n}^{(l)}(t)\big|dt, \tag{2.26}$$

$$\mathcal{J}_4 = \sum_{l=0}^{s} \frac{1}{l!}\left(\frac{3x}{2}\right)^l \int_{T_n}^{\infty} |f_{Z_n}^{(l)}(t)|dt. \tag{2.27}$$

Similarly, with respect to (2.26) and (2.23) we obtain

$$\mathcal{J}_3 = \mathcal{J}_3^{(0)} + \mathcal{J}_3^{(1)} + \mathcal{J}_3^{(2)} + \mathcal{J}_3^{(3)}, \tag{2.28}$$

where

$$\mathcal{J}_3^{(0)} = \int_{\varepsilon_n}^{T_n} |f_{Z_n}(t)|dt \leqslant \int_{\varepsilon_n}^{T_n} \exp\{-t^2/\pi^2\}dt \leqslant$$

$$\leqslant \frac{\pi^2}{2\varepsilon_n} \exp\{-\varepsilon_n^2/\pi^2\}, \quad \varepsilon_n = \frac{1}{4}\left(1 - \frac{x}{\Delta_n(\gamma)}\right)\Delta_n(\gamma); \tag{2.29}$$

$$\mathcal{J}_3^{(1)} = \frac{3x}{2} \int_{\varepsilon_n}^{T_n} |f'_{Z_n}(t)|dt \leqslant \frac{3\sqrt{e}x}{2} \int_{\varepsilon_n}^{T_n} (t \wedge L_{1,n}) \exp\{-t^2/\pi^2\}dt \leqslant$$

$$\leqslant \frac{3\sqrt{e}\pi^2 x}{2} \exp\left\{-\frac{1}{\pi^2}\varepsilon_n^2\right\}; \tag{2.30}$$

$$\mathcal{J}_3^{(2)} = \frac{1}{4}\left(\frac{3x}{2}\right)^2 \int_{\varepsilon_n}^{T_n} |f''_{Z_n}(t)|dt \leqslant \frac{9ex^2}{16} \int_{\varepsilon_n}^{T_n} \left(1 + (t^2 \wedge L_{1,n}^2)\right) \exp\left\{-\frac{t^2}{\pi^2}\right\}dt \leqslant$$

$$\leqslant \frac{9e\pi^2 x^2}{32}\left(\varepsilon_n + \frac{\pi^2}{2\varepsilon_n}\right) \exp\left\{-\frac{1}{\pi^2}\varepsilon_n^2\right\}; \tag{2.31}$$

$$\mathcal{J}_3^{(3)} = \sum_{l=3}^{s} \frac{1}{l!}\left(\frac{3x}{2}\right)^l \int_{\varepsilon_n}^{T_n} |f_{Z_n}^{(l)}(t)|dt \leqslant$$

$$\leqslant \sum_{l=3}^{s} \frac{1}{l!}\left(\frac{3x}{2}\right)^l c_1(l) \int_{\varepsilon_n}^{T_n} \left(1 + (|t|^l \wedge L_{1,n}^l) + L_{l,n}\right) \exp\left\{-\frac{1}{\pi^2}t^2\right\}dt \leqslant$$

$$\leqslant \frac{\pi^2}{2\varepsilon_n} \sum_{l=3}^{s} \frac{1}{l!}\left(\frac{3\sqrt{e}x}{4}\right)^l \left\{1 + \left(L_{1,n}^l \wedge \left(\varepsilon_n^l + \left(\frac{\pi}{\sqrt{2}}\right)^l\right)\right) + (l-2)!/(3,8\Delta_n(\gamma))^{l-2}\right\} \times$$

$$\times \exp\left\{-\frac{1}{\pi^2}\varepsilon_n^2\right\} \leqslant \frac{\pi^2}{2\varepsilon_n} \exp\left\{-\frac{1}{16\pi^2}\varepsilon_n^2\right\} + \frac{\varepsilon_n}{48\pi^2} \exp\left\{-\frac{1}{\pi^2}\varepsilon_n^2\right\} \tag{2.32}$$

if $0 \leqslant x \leqslant 5\varepsilon_n/(8\pi^2\sqrt{e})$.

It remains to estimate the $\mathcal{J}_4$ which is defined by equality (2.27). For this aim, using expression (2.11) for $f_{Z_n}^{(l)}(t)$, at first let us estimate the integral

$$\widetilde{\mathcal{J}}_4 = \int\limits_{|t|\geqslant T_n} \left|f_{Z_n,k_1,k_2,\dots,k_j}(t)\right|dt, \tag{2.33}$$

for $j \leqslant l,\ l = 0, 1, \dots, s$. We have

$$\widetilde{\mathcal{J}}_4 \leqslant 2\pi B_n \int\limits_{|t|\geqslant 1/(2N_n)} \prod_{r=1}^{4}{}'' \left|f_{\xi_{r_i}}(2\pi t)\right| \prod_{i=1}^{n-j-4}{}' \left|f_{\xi_{l_i}}(2\pi t)\right|dt \leqslant$$

$$\leqslant 2\pi B_n \int\limits_{|t|\geqslant 1/(2N_n)} \exp\Big\{-\big[\mathrm{I}_n(t) - (\mathcal{J}_{k_1}(t) + \dots + \mathcal{J}_{k_j}(t)) +$$

$$+\mathcal{J}_{r_1}(t) + \dots + \mathcal{J}_{r_4}(t)\big]\Big\} \prod_{i=1}^{4}{}'' \left|f_{\xi_i}(2\pi t)\right|dt \leqslant$$

$$\leqslant 2\pi B_n \exp\left\{\frac{j+4}{2}\right\} \int\limits_{|t|\geqslant 1/(2N_n)} \exp\Big\{-\mathrm{I}_n(t)\Big\} \prod_{i=1}^{4}{}'' \left|f_{\xi_{r_i}}(2\pi t)\right|dt, \tag{2.34}$$

$j \leqslant l,\ l = 0, 1, 2, \dots, s$. Here Π'' is the product over all $r_i,\ i = 1, 2, 3, 4$, which do not coincide with $k_1, k_2, \dots, k_j$ and Π' is the product over all indices l_i which do not coincide with $k_1, k_2, \dots, k_j,\ r_1, r_2, r_3, r_4$.
Using Lemma 2 (see (Statulevičius, 1965)) and the fact that $N_n = 4B_nL_{3,n}$, we obtain

$$\mathrm{I}_n(t) \geqslant 2\left(t - t_{k_0}^{(n)}\right)^2 B_n^2 \tag{2.35}$$

for $t \in \left[t_k^{(n)},\ t_{k+1}^{(n)}\right]$, $(1/6N_n) \leqslant t_{k+1}^{(n)} - t_k^{(n)} \leqslant (1/4N_n)$, where $t_{k_0}^{(n)}$, for any fixed n, equals either to $t_k^{(n)}$ or to $t_{k+1}^{(n)}$depending on the value of k.
Let $\mathfrak{M}_k = \{\Delta_{ki},\ C_{ki},\ i = 1, 2, \dots\}$ be a collection of nonoverlaping intervals Δ_{ki} of length $|\Delta_{ki}|$ and positive constants $C_{ki} \leqslant \infty,\ k = 1, 2, \dots$
Denote

$$\alpha(\mathfrak{M}_k, N) = \sum_{i=1}^{\infty} \frac{Q_{ki}^3}{\left(|\Delta_{ki}| + 2N\right)^2 C_{ki}^2}, \tag{2.36}$$

where $N > 0$ and

$$Q_{ki} = \int\limits_{\Delta_{ki}} (C_{ki} \wedge \tilde{p}_{\xi_k}(x))dx.$$

Under requirement $\mathfrak{M}_k = \{\Delta_{k1},\ C_{k1}\}$, $\Delta_{k1} =]-2\sigma_k, 2\sigma_k[$, $C_{k1} = C_k$, we have

$$\alpha(\mathfrak{M}_k, N_n) = \frac{Q_{k1}^3}{(4\sigma_k + 2N_n)^2 C_k^2} \leqslant \frac{1}{256} \frac{1}{(\sigma_k^2 + N_n^2)C_k^2}, \tag{2.37}$$

because

$$Q_{k1} = \int_{-2\sigma_k}^{2\sigma_k} \tilde{p}_{\xi_k}(x)dx \geqslant \frac{1}{2}.$$

Applying Lemma 1 of (Statulevičius, 1965) for the quantity $\mathrm{I}_n(t)$ defined in (2.18), we obtain the estimate

$$\mathrm{I}_n(t) \geqslant \frac{t^2}{3} \sum_{k=1}^{n} \alpha\left(\mathfrak{M}_k, \frac{1}{2|t|}\right), \tag{2.38}$$

for all $-\infty < t < \infty$. The minimal value of the right-hand side of (2.38) on the interval $|t| \geqslant 1/(2N_n)$ is

$$M_n = \frac{1}{3} \sum_{k=1}^{n} \alpha(\mathfrak{M}_k, N_n) = \frac{1}{3.256} \sum_{k=1}^{n} \frac{1}{(\sigma_k^2 + N_n^2)C_k^2}. \tag{2.39}$$

Decompose $\mathrm{I}_n(t) = (3/4)\mathrm{I}_n(t) + (1/4)\mathrm{I}_n(t)$, and applying (2.35) and (2.39), we obtain

$$\tilde{\mathcal{J}}_4 \leqslant 2\pi B_n \exp\left\{\frac{j+4}{2}\right\} \exp\left\{-c\sum_{j=1}^{n} \frac{1}{(\sigma_j^2 + N_n^2)C_j^2}\right\} \times$$

$$\times \sum_k \int_{t_k^{(n)}}^{t_{k+1}^{(n)}} \exp\left\{-\frac{1}{2}(t - t_{k0}^{(n)})^2 B_n^2\right\} \prod_{i=1}^{4}{}'' |f_{\xi_{r_i}}(2\pi t)| dt \leqslant$$

$$\leqslant \pi\sqrt{2\pi} \exp\left\{\frac{j+4}{2}\right\} \exp\left\{-c\sum_{j=1}^{n} \frac{1}{(\sigma_j^2 + N_n^2)C_j^2}\right\} U_n, \tag{2.40}$$

where

$$U_n = \sum_k \sup_{t_k^{(n)} \leqslant t \leqslant t_{k+1}^{(n)}} \prod_{i=1}^{4} |f_{\xi_{r_i}}(2\pi t)| \leqslant$$

$$\leqslant \prod_{i=1}^{4}{}'' \left(\sum_k \sup_{t_k^{(n)} \leqslant t \leqslant t_{k+1}^{(n)}} |f_{\xi_{r_i}}(2\pi t)|^4\right)^{1/4}, \tag{2.41}$$

$0 < c \leqslant 2^{-10}$. To estimate the quantity U_n we need the following statement (see (Statulevičius, 1965)).
Let a non-negative function $g(t)$ defined on $[a, \infty[$ satisfy the Lipschitz condition

$$|g(t+s) - g(t)| \leqslant K|s|. \tag{2.42}$$

Moreover, assume that

$$V = \int_a^\infty g(t)dt < \infty.$$

Then, for any $\varepsilon > 0$ and any subdivision $a = t_0 < t_1 < t_2 < \ldots$ of the interval $[a, \infty[$ with $\max\big(t_{k+1} - t_k\big) \leqslant \varepsilon$, we have

$$\sum_{k=0}^{\infty}\Big(\max_{t_k<t<t_{k+1}} g^2(t)\Delta t_k\Big) \leqslant V\big(2K\varepsilon + 4\sup_{a\leqslant t<\infty} g(t)\big), \tag{2.43}$$

where $\Delta t_k = t_{k+1} - t_k$.
If we suppose $g_{r_i}(t) = |f_{r_i}(2\pi t)|^2$ then it is immediate that inequality (2.43) is valid with $V_{r_i} = \tilde{p}_{\xi_{r_i}}(0) \leqslant C_{r_i}$ and $K_{r_i} \leqslant 2\sqrt{2}\pi\sigma_{r_i}$. Taking into account that $(1/6N_n) \leqslant t_{k+1}^{(n)} - t_k^{(n)} \leqslant (1/4N_n)$, we obtain

$$U_n \leqslant 6N_n {\prod_{i=1}^{4}}'' C_{r_i}^{1/4}\Big(4 + \frac{\sqrt{2}\pi\sigma_{r_i}}{N_n}\Big)^{1/4}. \tag{2.44}$$

Together with relation (2.40) this implies

$$\widetilde{\mathcal{J}}_4 \leqslant 24\pi\sqrt{2\pi}\exp\Big\{\frac{j+4}{2}\Big\}N_n {\prod_{i=1}^{4}}'' C_{r_i}^{1/4}\Big(1 + \frac{\pi\sigma_{r_i}}{2\sqrt{2}N_n}\Big)^{1/4} \times$$

$$\times \exp\Big\{-c\sum_{k=1}^{n}\frac{1}{(\sigma_k^2 + N_n^2)C_k^2}\Big\}. \tag{2.45}$$

Now, employing (2.27) and (2.45), we obtain

$$\mathcal{J}_4 \leqslant 24\pi\sqrt{2\pi}e^2 N_n \max_{1\leqslant r_i\leqslant s+4}\prod_{i=1}^{4} C_{r_i}^{1/4}\Big(1 + \frac{\pi\sigma_{r_i}}{2\sqrt{2}N_n}\Big)^{1/4} \times$$

$$\times \sum_{k=0}^{s}\frac{1}{l!}\Big(\frac{3\sqrt{e}x}{4}\Big)^l\Big(1 + L_{1,n}^l + L_{l,n}\Big)\exp\Big\{-c\sum_{k=1}^{n}\frac{1}{(\sigma_k^2 + N_n^2)C_k^2}\Big\}. \tag{2.46}$$

Note that $N_n = 4B_nL_{3,n} \leqslant 96\cdot 3^\gamma K \leqslant 96\cdot 3^\gamma K_n$, where $K_n = 2\big(K \vee \max_{1\leqslant j\leqslant n}\sigma_j\big)$. This implies

$$\sum_{l=0}^{s}\frac{1}{l!}\Big(\frac{3\sqrt{e}(1\vee L_{1,n})x}{4}\Big)\exp\Big\{-c\sum_{k=1}^{n}\frac{1}{(\sigma_k^2 + N_n^2)C_k^2}\Big\} \leqslant$$

$$\leqslant \exp\Big\{\frac{1\vee L_{1,n}}{85}\Delta_n(\gamma) - c\sum_{k=1}^{n}\frac{1}{(\sigma_k^2 + N_n^2)C_k^2}\Big\} \leqslant$$

$$\leqslant \exp\Big\{-c_2\sum_{k=1}^{n}\frac{1}{(\sigma_k^2 + N_n^2)C_k^2}\Big\}, \quad c_2 = \frac{1}{2}c, \tag{2.47}$$

if

$$\lim_{n\to\infty} \frac{1}{(1 \vee L_{1,n})\Delta_n(\gamma)} \sum_{k=1}^{n} \frac{1}{(\sigma_k^2 + N_n^2)C_k^2} \geqslant c_1,$$

$c_1 = 170/c$, and

$$\sum_{l=3}^{s} \frac{1}{l!}\Big(\frac{3\sqrt{e}x}{4}\Big)^l L_{l,n} \leqslant \Big(\frac{3\sqrt{e}c^*}{2}\Big)^2, \quad \sum_{l=3}^{s} \frac{1}{l(l-1)}\Big(\frac{3\sqrt{e}x}{14,2\Delta_n(\gamma)}\Big)^{l-2} \leqslant \frac{1}{6},$$

for $x \leqslant 5\varepsilon_n/(8\pi^2\sqrt{e})$, where the quantity ε_n is defined by (2.8). Relations (2.27), (2.46) and (2.47) enable to conclude that

$$\mathcal{J}_4 \leqslant 42\pi\sqrt{2\pi}e^2 N_n \max_{1\leqslant r_i\leqslant s+4} \prod_{i=1}^{4} C_{r_i}^{1/4}\Big(1 + \frac{\pi\sigma_{r_i}}{2\sqrt{2}N_n}\Big)^{1/4} \times$$

$$\times \exp\Big\{ - c_2 \sum_{k=1}^{n} \frac{1}{(\sigma_k^2 + N_n^2)C_k^2}\Big\}, \quad c_2 = \frac{1}{2}c, \tag{2.48}$$

where $N_n = 4B_nL_{3,n} \leqslant 96 \cdot 3^\gamma K$. Relations (2.25), (2.26), (2.32) and (2.38) imply

$$\int_{\varepsilon_n}^{\infty} |f_n^*(t)|dt = \sum_{l=0}^{s} \frac{1}{l!}\Big(\frac{3x}{2}\Big)^l \int_{\varepsilon_n}^{\infty} |f_{Z_n}^{(l)}(t)|dt \leqslant$$

$$\leqslant \frac{5e\pi\varepsilon_n x^2}{16} \exp\Big\{-\frac{\varepsilon_n^2}{\pi^2}\Big\} + \frac{\pi^2}{\varepsilon_n}\exp\Big\{-\frac{1}{16\pi^2}\varepsilon_n^2\Big\} +$$

$$+ 42\pi\sqrt{2\pi}e^2 N_n \max_{1\leqslant r_i\leqslant n} \prod_{i=1}^{4} C_{r_i}^{1/4}\Big(1 + \frac{\pi\sigma_{r_i}}{2\sqrt{2}N_n}\Big)^{1/4} \times$$

$$\times \exp\Big\{ - c_2 \sum_{k=1}^{n} \frac{1}{(\sigma_k^2 + N_n^2)C_k^2}\Big\}. \tag{2.49}$$

Applying Statement 1, estimation (2.49) and Lemma 1, we obtain the statement of Theorem 2.2.

The proof of Theorem 1 can be found in (Saulis, 1986). It is worth noting that this proof does not demand estimation of derivatives of the characteristic function of r.v. Z_n.

REFERENCES

Borovkov, A.A. (1962). New limit theorems in boundary problems for sums of independent variables. *Sibirian Math. J.* **3**, 645–694 (in Russian).

Cramér, H. (1938). Sur un nonveau theoreme limite de la theorie des probabilities. *Actual. Sc. et Ind.* 736.

Ibragimov, J.A. (1967). On the Chebyshev–Cramér asymptotic expansions. *Theory Probab. Appl.* **12**, 455–469.

Linnik, Yu.V. (1961/1962). Limit theorems for the sums of independent variables taking into account large deviations. *Theory Probab. Appl.* **6/7**, 131–149, 345–359/ 115–127.

Nagaev, S.V. (1979). Large deviations of sums of independent random variables. *Ann. Probab.* **7**, 745–789.

Petrov, V.V. (1954). Generalization of the limit theorem of Cramér. *Adv. Math. Sci.* **9**, 23–27 (in Russian).

Petrov, V.V. (1965). On the probabilities of large deviations for sums of independent random variables. *Theory Probab. Appl.* **10**, 287–298.

Pipiras, V. and Statulevičius, V. (1968). Asymptotic expansions for sums of independent random variables. *Lithuanian Math. J.* **8**, 137–151 (in Russian).

Prokhorov, Yu.V. (1956). Convergence of random processes and limit theorems of probability theory. *Theory Probab. Appl.* **1**, 157–214.

Saulis, L. (1980). General lemma for the distribution density function. *Lithuanian Math. J.* **20**, 165–185 (in Russian).

Saulis, L. (1986). On large deviations of sums of independent random variables. In: *Probab. Theory and Mathematical Statist.* Proceedings of the Fourth Vilnius Conference, **2**, Ed. by Yu.V. Prokhorov et al. VNU Sci. Press, Utrecht, pp. 541–559.

Saulis, L. and Statulevičius, V. (1989). *Limit Theorems for Large Deviations.* Mokslas, Vilnius, 1989 (in Russian).

Statulevičius, V. (1965). Limit theorems for the density functions and asymptotic expansions for the distributions of sums of independent random variables. *Theory Probab. Appl.* **10**, 645–659.

Statulevičius, V. (1966). On large deviations. *Z. Wahrsch. verw. Geb.* **6**, 133–144.

Zolotarev, V.M. (1962). On one approach to the limit theorems considering large deviations. In: *Proceedings of the Fourth Vilnius All-Union Conference,* Vilnius, 1962. pp. 43–48 (in Russian).

New Trends in Probab. and Statist., pp. 57 - 69
V. Sazonov and T. Shervashidze (Eds.)
1991 VSP/Mokslas

STRONG LAW OF LARGE NUMBERS FOR OPERATOR NORMALIZED SUMS OF INDEPENDENT RANDOM VECTORS

S. SOLNTSEV
Kiev Polytechnical Institute, Pr. Pobedi 37, 252056 Kiev, USSR

ABSTRACT

In this paper the almost sure convergence to zero of operator normalized sums of independent random vectors taking values in finite dimensional space is studied. In the case when the summands are identically distributed with zero mean and finite second moments the equivalence of operator and scalar normalizing is established.

INTRODUCTION

Let $(X_k,\ k \geqslant 1)$ be a sequence of independent random vectors taking values in the Euclidean space $\mathbf{R}^m$ equipped with the usual inner product $\langle\cdot\,,\ \cdot\rangle$ and induced norm $\|\cdot\|$; $S_n = X_1 + X_2 + \ldots + X_n$, $n \geqslant 1$; $L(\mathbf{R}^m, \mathbf{R}^d)$ be the collection of linear operators from $\mathbf{R}^m$ to $\mathbf{R}^d (d \geqslant 1)$; $(A_n,\ n \geqslant 1) \subset L(\mathbf{R}^m, \mathbf{R}^d)$. If $A \in L(\mathbf{R}^m, \mathbf{R}^d)$ then A^* denotes the operator conjugate A; I is the identity operator; $S_1(0) = \{x \in \mathbf{R}^m\colon\ \|x\| = 1\}$; $\mathfrak{N}_\infty$ is the class of all nondecreasing and unbounded sequences of natural numbers; $I\{\cdot\}$ is the indicator function. We denote by $\mathcal{L}(X)$ the subspace of minimal dimension on which the distribution of a random vector X is supported (i.e., $X \in \mathcal{L}(X)$ a.s.); $\mathcal{L}\{X_k,\ k \geqslant 1)\}$ is the subspace of minimal dimension which contains all $\mathcal{L}(X_k)$, $k \geqslant 1$ (i.e., $X_k \in \mathcal{L}\{(X_k,\ k \geqslant 1)\}$ a.s. for all $k \geqslant 1$). A random vector X is symmetrical in $\mathbf{R}^m$ if X and $(-X)$ have the same distribution. $R(X)$ denotes the covariance operator of a random vector X.

In the study of limit theorems there are two possible types of normalizations: scalar and operator. Limit theorems with scalar normalizations are investigated quite well and form the basis of the classical probability theory. At the same time the operator normalizations have become the object of study first in weak limit theorems (Kandelaki and Sazonov, 1964; Hahn, 1979; Hahn and Klass, 1981; 1985; Sémovskii, 1984; Griffin, 1986) and much later in strong limit theorems (Jadrenko, 1982; Buldygin et al., 1984; Weiner, 1986; Buldygin and Solntsev, 1987a; 1987b; 1989). In particular, the strong law of large numbers (SLLN) in the Prokhorov–Loève form for sums of independent random vectors normed by operators have been established in (Buldygin and Solntsev, 1987a; 1987b; 1989). Note that the above mentioned form of SLLN dates back to (Prokhorov, 1950; Loève, 1960).

SLLN in the Prokhorov–Loève form for operator normalizations (cf. (Buldygin and Solntsev, 1987a; 1987b; 1989)).
Let $(X_k,\ k \geqslant 1)$ be a sequence of independent symmetric random vectors taking values in $\mathbf{R}^m$. Then

$$\|A_n S_n\| \longrightarrow 0 \text{ a.s.}, \quad n \to \infty,$$

iff

I) for each $k \geqslant 1$,

$$\|A_n X_k\| \longrightarrow 0 \text{ a.s.}, \quad n \to \infty;$$

II) for all $(n_j,\ j \geqslant 1) \in \mathfrak{N}_\infty$,

$$\left\|A_{n_{j+1}}(S_{n_{j+1}} - S_{n_j})\right\| \to 0 \text{ a.s.}, \quad n \to \infty.$$

Because of independence of $A_{n_{j+1}}(S_{n_{j+1}} - S_{n_j}),\ j \geqslant 1$, it is clear that II) is equivalent to

II*) for any $\varepsilon > 0$ and for all $(n_j,\ j \geqslant 1) \in \mathfrak{N}_\infty$,

$$\sum_j \mathbf{P}\left(\left\|A_{n_{j+1}}(S_{n_{j+1}} - S_{n_j})\right\| > \varepsilon\right) < \infty.$$

It is quite difficult to express $\mathbf{P}(\|A_{n_{j+1}}(S_{n_{j+1}} - S_{n_j})\| > \varepsilon)$ via characteristics of summands $A_{n_{j+1}} X_k$. Therefore the verification of II*) is rather complicated. In the Gaussian case we have a fortunate exception.

LEMMA 1 (Buldygin and Solntsev, 1989). *Let $(Y_k,\ k \geqslant 1)$ be a sequence of independent Gaussian random vectors in $\mathbf{R}^m$; $\mathbf{E}Y_k \equiv 0,\ k \geqslant 1$. Then*

$$\left\|A_n \sum_{k=1}^{n} Y_k\right\| \longrightarrow 0 \text{ a.s.}, \quad n \to \infty,$$

iff

C$_1$) *for each $k \geqslant 1$,*

$$\mathbf{E}\left\|A_n Y_k\right\|^2 \longrightarrow 0, \quad n \to \infty;$$

C$_2$) *for arbitrary sequence $(n_j,\ j \geqslant 1) \in \mathfrak{N}_\infty$,*

$$\sum_j \exp\left(-\varepsilon \Big/ \sum_{k=n_j+1}^{n_{j+1}} \mathbf{E}\left\|A_{n_{j+1}} X_k\right\|^2\right) < \infty$$

for any $\varepsilon > 0$.

Our purpose is to extend the above assertion to the non–Gaussian case and to study the conditions under which such integral type criterion for the SLLN remains valid.

1. INTEGRAL TYPE CRITERION FOR THE SLLN

To simplify the notation we assume that

$$\mathcal{L}\{(X_k,\ k \geqslant 1)\} \equiv \mathbf{R}^m. \tag{1}$$

This is not a loss in generality since otherwise we may consider the random vectors $X_k,\ k \geqslant 1$, as vectors taking values in $\mathbf{R}^{\ell}$, where $\ell = \dim \mathcal{L}\{(X_k,\ k \geqslant 1)\}$.
Note that the following assertion is a multidimensional analogue of a one-dimensional result in (Martikainen, 1984).

THEOREM 1. *Let $(X_k,\ k \geqslant 1)$ be a sequence of independent symmetric random vectors in $\mathbf{R}^m$; $\delta > 0$, $\Delta \in (0,1]$ denote some constants. Suppose that* (1) *and the following condition hold:*

$\mathbf{C}_3$) *for all $(n_j,\ j \geqslant 1) \in \mathfrak{N}_\infty$,*

$$\sum_{j\geqslant 1} \sum_{k=n_j+1}^{n_{j+1}} \mathbf{E}\left\|A_{n_{j+1}}X_k\right\|^{2+\Delta} I\left\{\left\|A_{n_{j+1}}X_k\right\| \leqslant \delta\right\} < \infty.$$

Then

$$\|A_n S_n\| \longrightarrow 0 \text{ a.s.}, \quad n \to \infty, \tag{2}$$

if and only if for all sequences $(n_j,\ j \geqslant 1) \in \mathfrak{N}_\infty$ the following conditions hold:

$\mathbf{C}_4$) $\|A_n\| \to 0,\ n \to \infty$;

$\mathbf{C}_5$) $\sum_j \sum_{k=n_j+1}^{n_{j+1}} \mathbf{P}\left(\|A_{n_{j+1}}X_k\| > \delta\right) < \infty$;

$\mathbf{C}_6$) *for any $\varepsilon > 0$*

$$\sum_j \exp\left(-\varepsilon \Big/ \sum_{k=n_j+1}^{n_{j+1}} \mathbf{E}\left\|A_{n_{j+1}}X_k\right\|^2 I\left\{\left\|A_{n_{j+1}}X_k\right\| \leqslant \delta\right\}\right) < \infty.$$

2. PROOF OF THEOREM 1

Rewrite the SLLN in the Prokhorov–Loève form under the assumption (1).

LEMMA 2. *Let $(X_k,\ k \geqslant 1)$ be a sequence of independent symmetric random vectors in $\mathbf{R}^m$ satisfying* (1). *Then* (2) *is true iff conditions* II) *and* $\mathbf{C}_4$) *hold.*

Proof. It is enough to prove the necessity of $\mathbf{C}_4$).
Let (2) holds true. Then I) implies for all $x \in \mathcal{L}\{(X_k,\ k \geqslant 1)\}$,

$$A_n x \to 0, \quad n \to \infty. \tag{3}$$

In view of (1) the convergence (3) occurs for all $x \in \mathbf{R}^m$ which leads to $\mathbf{C}_4$).

LEMMA 3. *Let $(X_k,\ k \geqslant 1)$ be the same as in Lemma 2. Set*

$$\widetilde{X}_k = X_k I\left\{\left\|A_{n_{j+1}}X_k\right\| \leqslant \delta\right\}, \quad n_j < k \leqslant n_{j+1}, \quad j \geqslant 1;$$

$$\widetilde{S}_n = \sum_{k=1}^{n} \widetilde{X}_k, \quad n \geqslant 1,$$

where $(n_j,\ j \geqslant 1) \in \mathfrak{N}_\infty$. *Then* (2) *holds true iff for all sequences* $(n_j,$ $j \geqslant 1) \in \mathfrak{N}_\infty$ *conditions* $\mathbf{C}_4$), $\mathbf{C}_5$) *and the following condition hold:*

$$\mathbf{C}_7) \quad \left\| A_{n_{j+1}}(\widetilde{S}_{n_{j+1}} - \widetilde{S}_{n_j}) \right\| \to 0 \text{ a.s.}, \quad j \to \infty.$$

Proof. Necessity. Suppose (2) is true. Then, by Lemma 2, conditions $\mathbf{C}_4$) and II*) hold. Using Lévy's inequality we get

$$\mathbf{P}\Big(\max_{n_j < k \leqslant n_{j+1}} \|A_{n_{j+1}} X_k\| > \varepsilon \Big) \leqslant 2\mathbf{P}\Big(\|A_{n_{j+1}}(S_{n_{j+1}} - S_{n_j})\| > \varepsilon \Big).$$

Hence, applying the Borel–Cantelli lemma, we conclude that

$$\max_{n_j < k \leqslant n_{j+1}} \left\| A_{n_{j+1}} X_k \right\| \longrightarrow 0 \text{ a.s.}, \quad j \to \infty.$$

Now $\mathbf{C}_5$) holds.
Furthermore,

$$\mathbf{P}\Big(A_{n_{j+1}}(S_{n_{j+1}} - S_{n_j}) \neq A_{n_{j+1}}(\widetilde{S}_{n_{j+1}} - \widetilde{S}_{n_j}) \Big) \leqslant$$

$$\leqslant \sum_{k=n_j+1}^{n_{j+1}} \mathbf{P}\Big(A_{n_{j+1}} X_k I\Big\{ \|A_{n_{j+1}} X_k\| > \delta \Big\} \neq 0 \Big) =$$

$$= \sum_{k=n_j+1}^{n_{j+1}} \mathbf{P}\Big(\|A_{n_{j+1}} X_k\| > \delta \Big) < \infty. \tag{4}$$

Having in mind $\mathbf{C}_5$) and using the Borel–Cantelli lemma, we get $\mathbf{C}_7$).

Sufficiency. Let conditions $\mathbf{C}_4$), $\mathbf{C}_5$ and $\mathbf{C}_7$) hold. In view of inequality (4), condition $\mathbf{C}_5$) and the Borel–Cantelli lemma imply II). Lemma 3 now follows from Lemma 2.

LEMMA 4. *Let* $(A_n,\ n \geqslant 1)$ *and* $(X_k,\ k \geqslant 1)$ *satisfy the conditions of Theorem 1. Define* $(\widetilde{X}_k,\ k \geqslant 1)$ *as in Lemma 3. Then, for an arbitrary sequence* $(n_j,\ j \geqslant 1) \in \mathfrak{N}_\infty$, *the following conditions are equivalent:*

$\mathbf{C}_8$) $\quad \|A_{n_{j+1}}(\widetilde{S}_{n_{j+1}} - \widetilde{S}_{n_j})\| \to 0$ a.s., $\quad j \to \infty$,

$\mathbf{C}_9$) $\quad \|A_{n_{j+1}}(T_{n_{j+1}} - T_{n_j})\| \to 0$ a.s., $\quad j \to \infty$,

where

$$T_n = \sum_{k=1}^{n} Y_k,$$

$(Y_k,\ k \geqslant 1)$ *is a sequence of independent Gaussian random vectors in* $\mathbf{R}^m$, $\mathbf{E}Y_k \equiv \equiv 0, R(Y_k) = R(\widetilde{X}_k),\ k \geqslant 1.$

Proof. To prove the equivalence of $\mathbf{C}_8$) and $\mathbf{C}_9$), it is sufficient to show that, for any $e \in S_1(0)$,

$$\langle e, A_{n_{j+1}}(\widetilde{S}_{n_{j+1}} - \widetilde{S}_{n_j})\rangle \longrightarrow 0 \text{ a.s.}, \quad j \to \infty,$$

iff

$$\langle e, A_{n_{j+1}}(T_{n_{j+1}} - T_{n_j})\rangle \longrightarrow 0 \text{ a.s.}, \quad j \to \infty.$$

Set

$$\eta_k = \langle e, A_{n_{j+1}}\widetilde{X}_k\rangle, \quad \xi_k = \langle e, A_{n_{j+1}}Y_k\rangle, \quad n_j < k \leqslant n_{j+1}, \quad j \geqslant 1.$$

Note, that $\mathbf{E}\eta_k = \mathbf{E}\xi_k$, $\mathbf{E}\eta_k^2 = \mathbf{E}\xi_k^2$. Using nonuniform estimates in the central limit theorem (Bikyalis, 1966), for every $\varepsilon > 0$, we get

$$\left|\mathbf{P}\left(\left|\sum_{k=n_j+1}^{n_{j+1}} \eta_k\right| > \varepsilon\right) - \mathbf{P}\left(\left|\sum_{k=n_j+1}^{n_{j+1}} \xi_k\right| > \varepsilon\right)\right| \leqslant C \sum_{k=n_j+1}^{n_{j+1}} \mathbf{E}|\eta_k|^{2+\Delta} \leqslant$$

$$\leqslant C \sum_{k=n_j+1}^{n_{j+1}} \mathbf{E}\left\|A_{n_{j+1}}X_k\right\|^{2+\Delta} I\left\{\|A_{n_{j+1}}X_k\| \leqslant \delta\right\}.$$

Here $C > 0$ is some absolute constant. Condition $\mathbf{C}_3$) and the Borel–Cantelli lemma imply Lemma 4.
Now Theorem 1 obviously follows from Lemmas 1–4.

3. EQUIVALENCE OF OPERATOR AND SCALAR NORMALIZATIONS

Consider several corollaries of Theorem 1 in the case of independent identically distributed (i.i.d.) summands. Note that in this case condition (1) may be rewritten as

$$\mathcal{L}(X_1) \equiv \mathbf{R}^m. \tag{5}$$

The following assertion shows that the operator sequence $(A_n,\ n \geqslant 1)$ we can replace by the scalar sequence $(\|A_n\|,\ n \geqslant 1)$ if $(X_k,\ k \geqslant 1)$ is a sequence of i.i.d. random vectors with $\mathbf{E}\|X_1\|^2 < \infty$.

THEOREM 2. *Let $(X_k,\ k \geqslant 1)$ be a sequence of i.i.d. random vectors in $\mathbf{R}^m$ (here we do not assume X_k to be symmetrical), $\mathbf{E}X_1 = 0$, $\mathbf{E}\|X_1\|^2 < \infty$. Then*

$$\|A_nS_n\| \longrightarrow 0 \text{ a.s.}, \quad n \to \infty,$$

iff

$$\|A_n\|\|S_n\| \longrightarrow 0 \text{ a.s.}, \quad n \to \infty.$$

4. PROOF OF THEOREM 2

In order to prove Theorem 2 we will need some preliminary results.

LEMMA 5 (Buldygin and Solntsev, 1989). *Let X be a random vector in $\mathbf{R}^m$. Then*

$$\|A_n X\| \xrightarrow{P} 0, \quad n \to \infty,$$

implies

$$\|A_n X\| \longrightarrow 0, \text{ a.s.}, \quad n \to \infty.$$

LEMMA 6. *Let $(X_k,\ k \geqslant 1)$ be the same as in Theorem 2. Then*

$$\|A_n S_n\| \xrightarrow{P} 0, \quad n \to \infty,$$

implies

$$\sqrt{n}\|A_n\| \longrightarrow 0, \quad n \to \infty.$$

Proof. Let $(X'_k,\ k \geqslant 1)$ be an independent copy of $(X_k,\ k \geqslant 1)$. Set

$$S'_n = \sum_{k=1}^{n} (X_k - X'_k), \quad n \geqslant 1.$$

Observe that $\mathcal{L}(X_1 - X'_1) \equiv \mathbf{R}^m$ and

$$\|A_n S'_n\| \xrightarrow{P} 0, \quad n \to \infty.$$

To simplify the notation we just write X_k and S_n, respectively, for $X_k - X'_k$ and S'_n. We can choose a sequence of vectors $(e_n,\ n \geqslant 1) \subset S_1(0)$ such that

$$\|A_n^* e_n\| = \|A_n^*\|, \quad n \geqslant 1.$$

Set

$$a_n = \|A_n\|, \quad e'_n = A_n^* e_n / a_n, \quad n \geqslant 1.$$

Due to $\|A_n^*\| = \|A_n\|$, it is obvious that $(e'_n,\ n \geqslant 1)$ is a sequence of unit vectors and

$$\langle e_n,\ A_n S_n \rangle = a_n \langle e'_n,\ S_n \rangle, \quad n \geqslant 1.$$

Moreover, $a_n \langle e'_n, S_n \rangle \xrightarrow{\mathbf{P}} 0,\ n \to \infty$.
By the criterion for the convergence in probability (Petrov, 1975),

$$\sum_{k=1}^{n} \int_{|t| \leqslant 1} t^2 d\mathbf{P}\big(a_n \langle e'_n, X_k \rangle < t\big) \longrightarrow 0, \quad n \to \infty.$$

Hence,

$$n \int_{|t| \leqslant 1} t^2 d\mathbf{P}\big(\langle e'_n, X_1 \rangle < t/a_n\big) =$$

$$= na_n^2 \mathbf{E}(\langle e'_n, X_1\rangle)^2 I\Big\{|\langle e'_n, X_1\rangle| \leqslant 1/a_n\Big\} \longrightarrow 0, \quad n \to \infty.$$

Now it is sufficient to show that

$$\liminf_{n\to\infty} \mathbf{E}(\langle e'_n, X_1\rangle)^2 I\Big\{|\langle e'_n, X_1\rangle| \leqslant 1/a_n\Big\} > 0. \tag{6}$$

Indeed,

$$\mathbf{E}(\langle e'_n, X_1\rangle)^2 I\Big\{|\langle e'_n, X_1\rangle| \leqslant 1/a_n\Big\} \geqslant$$

$$\geqslant \langle e'_n,\ R(X_1)e'_n\rangle - \mathbf{E}\|X_1\|^2 I\Big\{\|X_1\| > 1/a_n\Big\} \geqslant$$

$$\geqslant \inf_{e\in S_1(0)} \langle e, R(X_1)e\rangle - \mathbf{E}\|X_1\|^2 I\Big\{\|X_1\| > 1/a_n\Big\}.$$

By Lévy's inequality,

$$\|A_n X_1\| \xrightarrow{P} 0, \quad n \to \infty.$$

Using Lemma 5 and repeating the proof of Lemma 2, we obtain $\|A_n\| \equiv a_n \to 0$, $n \to \infty$. Hence,

$$\mathbf{E}\|X_1\|^2 I\Big\{\|X_1\| > 1/a_n\Big\} \longrightarrow 0,\ n \to \infty.$$

Since $\langle h, R(X_1)h\rangle$ is a strictly positive continuous function on $S_1(0)$, it is obvious that (6) holds and this concludes the proof of Lemma 6.

A random vector in $\mathbf{R}^m$ is called *spherically symmetrical* if its distribution is invariant with respect to any transformation from the rotation group of the space $\mathbf{R}^m$.

LEMMA 7 (Buldygin and Solntsev, 1989). *Let $(Z_k,\ k \geqslant 1)$ be a sequence of independent spherically symmetrical random vectors in $\mathbf{R}^m$. Then*

$$\left\| A_n \sum_{k=1}^{n} Z_k \right\| \longrightarrow 0 \text{ a.s.}, \quad n \to \infty,$$

iff

$$\|A_n\| \left\| \sum_{k=1}^{n} Z_k \right\| \longrightarrow 0 \text{ a.s.}, \quad n \to \infty.$$

LEMMA 8 (Solntsev, 1985). *Let $(\tau_k,\ k \geqslant 1)$ be a sequence of i.i.d. random variables, $\mathbf{E}\tau_1 = 0$, $0 < \mathbf{E}\tau_1^2 < \infty$; $(a_n,\ n \geqslant 1) \subset \mathbf{R}^1$; $c > 1$ denote some constant. Then*

$$a_n(\tau_1 + \tau_2 + \ldots + \tau_n) \longrightarrow 0 \text{ a.s.}, \quad n \to \infty,$$

iff, for any $\varepsilon > 0$,

$$\sum_k \exp\Big\{ -\varepsilon \Big/ c^{2k} \max_{c^k \leqslant j < c^{k+1}} a_j^2 \Big\} < \infty.$$

Moreover, in the necessity part the series converges for all $c > 1$.

Observe that by Lemma 8 the convergence of $(\|A_n\| S_n,\ n \geqslant 1)$ to zero a.s. depends only on the covariance operator $R(X_1)$, i.e.,

$$\|A_n\|\|S_n\| \longrightarrow 0 \text{ a.s.}, \quad n \to \infty,$$

iff

$$\|A_n\|\left\|\sum_{k=1}^{n} W_k\right\| \longrightarrow 0 \text{ a.s.}, \quad n \to \infty,$$

where $(W_k,\ k \geqslant 1)$ is a sequence of i.i.d. random vectors in $\mathbf{R}^m$, $\mathbf{E}W_1 = 0$, $R(W_1) = R(X_1)$.

Proof of Theorem 2. In order to prove Theorem 2 it is sufficient to show that

$$\|A_n S_n\| \longrightarrow 0 \text{ a.s.} \Longrightarrow \|A_n\|\|S_n\| \longrightarrow 0 \text{ a.s.}, \quad n \to \infty. \tag{7}$$

We first prove (7) in the case when the random vectors $(X_k,\ k \geqslant 1)$ are all symmetrical and $R(X_1) = I$.
Set

$$\widehat{X}_k = X_k I\left\{\|X_k\| \leqslant \sqrt{k}\right\}, \quad k \geqslant 1; \quad \widehat{S}_n = \sum_{k=1}^{n} \widehat{X}_k,\ n \geqslant 1.$$

We shall show that (2) implies

$$\|A_n \widehat{S}_n\| \longrightarrow 0 \text{ a.s.}, \quad n \to \infty. \tag{8}$$

Indeed, in view of $\mathbf{E}\|X_1\|^2 < \infty$, for any sequence $(n_j,\ j \geqslant 1) \in \mathfrak{N}_\infty$,

$$\sum_j \mathbf{P}\Big(A_{n_{j+1}}(S_{n_{j+1}} - S_{n_j}) \neq A_{n_{j+1}}(\widehat{S}_{n_{j+1}} - \widehat{S}_{n_j})\Big) \leqslant$$

$$\leqslant \sum_j \sum_{k=n_j+1}^{n_{j+1}} \mathbf{P}(A_{n_{j+1}} X_k \neq A_{n_{j+1}} \widehat{X}_k) \leqslant$$

$$\leqslant \sum_j \sum_{k=n_j+1}^{n_{j+1}} \mathbf{P}(X_k \neq \widehat{X}_k) = \sum_k \mathbf{P}(\|X_k\| > \sqrt{k}) < \infty.$$

Hence, applying the Borel–Cantelli lemma, we conclude that

$$\left\|A_{n_{j+1}}(S_{n_{j+1}} - S_{n_j})\right\| \longrightarrow 0 \text{ a.s.}, \quad j \to \infty,$$

iff

$$\left\|A_{n_{j+1}}(\widehat{S}_{n_{j+1}} - \widehat{S}_{n_j})\right\| \longrightarrow 0 \text{ a.s.}, \quad j \to \infty.$$

Then, by Lemma 2, (2) implies (8).
Now, by Lemma 6,

$$\sqrt{n}\|A_n\| \longrightarrow 0, \quad n \to \infty,$$

and thus, for all sufficiently large n,

$$\|A_n\| \leqslant \frac{\delta}{\sqrt{n}}.$$

Without loss of generality we assume that the above inequality holds true for all $n \geqslant 1$. Then, for all $k \geqslant 1$ $(n_j < k \leqslant n_{j+1})$,

$$\left\{\|A_{n_{j+1}}\widehat{X}_k\| > \delta\right\} \subset \left\{\|\widehat{X}_k\| > \sqrt{n_{j+1}}\right\} =$$

$$= \left\{\|X_k\| > \sqrt{n_{j+1}},\ \|X_k\| \leqslant \sqrt{k}\right\} = \varnothing.$$

Hence, for $n_j < k \leqslant n_{j+1}$,

$$I\left\{\|A_{n_{j+1}}\widehat{X}_k\| \leqslant \delta\right\} I\left\{\|X_k\| \leqslant \sqrt{k}\right\} = I\left\{\|X_k\| \leqslant \sqrt{k}\right\}. \tag{9}$$

Consequently, for an arbitrary sequence $(n_j,\ j \geqslant 1) \in \mathfrak{N}_\infty$, we get

$$\sum_j \sum_{k=n_j+1}^{n_{j+1}} \mathbf{E}\|A_{n_{j+1}}\widehat{X}_k\|^3 I\{\|A_{n_{j+1}}\widehat{X}_k\| \leqslant \delta\} \leqslant$$

$$\leqslant \sum_j \|A_{n_{j+1}}\|^3 \sum_{k=n_j+1}^{n_{j+1}} \mathbf{E}\|\widehat{X}_k\|^3 I\{\|A_{n_{j+1}}\widehat{X}_k\| \leqslant \delta\} =$$

$$= \sum_j \|A_{n_{j+1}}\|^3 \sum_{k=n_j+1}^{n_{j+1}} \mathbf{E}\|X_k\|^3 I\left\{\|X_k\| \leqslant \sqrt{k}\right\} \leqslant$$

$$\leqslant \delta \sum_j n_{j+1}^{-3/2} \sum_{k=n_j+1}^{n_{j+1}} \int\limits_{|t| \leqslant \sqrt{k}} |t|^3 d\mathbf{P}(\|X_1\| < t) \leqslant$$

$$\leqslant \delta \sum_n n^{-3/2} \int\limits_{|t| \leqslant \sqrt{n}} |t|^3 d\mathbf{P}(\|X_1\| < t) =$$

$$= \delta \sum_n \int\limits_{\sqrt{n-1} < |t| \leqslant \sqrt{n}} |t|^3 d\mathbf{P}(\|X_1\| < t) \sum_{k=n}^{\infty} k^{-3/2} \leqslant$$

$$\leqslant \delta \sum_n \int\limits_{\sqrt{n-1} < |t| \leqslant \sqrt{n}} t^2 d\mathbf{P}(\|X_1\| < t) \sqrt{n} \sum_{k=n}^{\infty} k^{-3/2} < \infty,$$

where we have used $\sup_{n \geqslant 1} \sqrt{n} \sum_{k=n}^{\infty} k^{-3/2} < \infty$ and $\mathbf{E}\|X_1\|^2 < \infty$ in obtaining the last inequality. Hence, sequences $(A_n,\ n \geqslant 1)$ and $(\widehat{X}_k,\ k \geqslant 1)$ satisfy condition $\mathbf{C}_3$). By Theorem 1, $\mathbf{C}_6$) holds true. Having in mind (9), note that, for $j \to +\infty$,

$$\sum_{k=n_j+1}^{n_{j+1}} \mathbf{E}\|A_{n_{j+1}}\widehat{X}_k\|^2 I\{\|A_{n_{j+1}}\widehat{X}_k\| \leqslant \delta\} =$$

$$= \sum_{k=n_j+1}^{n_{j+1}} \mathbf{E}\|A_{n_{j+1}}X_k\|^2 I\{\|X_k\| \leqslant \sqrt{k}\} \sim \sum_{k=n_j+1}^{n_{j+1}} \mathbf{E}\|A_{n_{j+1}}X_k\|^2.$$

Using Lemma 1, we get

$$\left\|A_n \sum_{k=1}^{n} \Gamma_k\right\| \longrightarrow 0 \text{ a.s.}, \quad n \to \infty,$$

where $(\Gamma_n,\ n \geqslant 1)$ is a sequence of independent centered Gaussian random vectors in $\mathbf{R}^m$ with $R(\Gamma_n) = I$. By Lemma 7,

$$\|A_n\|\left\|\sum_{k=1}^{n} \Gamma_n\right\| \longrightarrow 0 \text{ a.s.}, \quad n \to \infty.$$

Finally, by Lemma 8,

$$\|A_n\|\|S_n\| \longrightarrow 0 \text{ a.s.}, \quad n \to \infty,$$

thereby completing the proof of (7) in our case.

In order to prove (7) in the general case we observe that (5) implies that there exist the inverse operator $R^{-1}(X_1)$.

Let $(X'_k,\ k \geqslant 1)$ be an independent copy of $(X_k,\ k \geqslant 1)$. Assume that

$$\|A_n S_n\| \longrightarrow 0 \text{ a.s.}, \quad n \to \infty.$$

Then

$$\left\|A_n R^{1/2}(X_1) \sum_{k=1}^{n} R^{-1/2}(X_1)\, \frac{X_k - X'_k}{\sqrt{2}}\right\| \longrightarrow 0 \text{ a.s.}, \quad n \to \infty.$$

Since

$$\left(R^{-1/2}(X_1)\, \frac{X_k - X'_k}{\sqrt{2}},\ k \geqslant 1\right)$$

is a sequence of symmetric i.i.d. random vectors in $\mathbf{R}^m$ with identity covariance operator, by the result proved above, we get

$$\|A_n R^{1/2}(X_1)\|\left\|\sum_{k=1}^{n} R^{-1/2}(X_1)(X_k - X'_k)\right\| \longrightarrow 0 \text{ a.s.}, \quad n \to \infty.$$

Hence,

$$\|A_n\|\left\|\sum_{k=1}^{n}(X_k - X'_k)\right\| \longrightarrow 0 \text{ a.s.}, \quad n \to \infty.$$

Then there exists a sequence of centering vectors $(b_n,\ n \geqslant 1)$ such that

$$\|A_n\|\|S_n - b_n\| \longrightarrow 0 \text{ a.s.}, \quad n \to \infty.$$

Combining Markov's inequality and Lemma 6, we get

$$\|A_n\|\|S_n\| \xrightarrow{\mathbf{P}} 0, \quad n \to \infty.$$

Consequently, $\|A_n\|\|b_n\| \to 0, \quad n \to \infty.$
This concludes the proof of (7) and thus the proof of Theorem 2.
The fact that condition $\mathbf{E}\|X_1\|^2 < \infty$ cannot be weakened is clarified by the following example, which shows that this condition can not be replaced even by

$$\mathbf{E}\frac{\|X_1\|^2}{\log^+ \log^+ \|X_1\|} < \infty.$$

(Here $\log^+ x = \max\{1, \log x\}$.)

Example. Let $(\gamma_k,\ k \geqslant 1)$ be a sequence of independent $\mathcal{N}(0,1)$–distributed random variables, $(\eta_k,\ k \geqslant 1)$ be a sequence of i.i.d. random variables with density $f(x)$, where $f(x) = C/|x|^3 \log|x|$ if $|x| > 2$ and $f(x) = 0$ otherwise. Here C denotes a suitable norming constant.
Let $X_k^T = (\gamma_k, \eta_k)^T$ and $A_n = \operatorname{diag}\,(1/\sqrt{n \log n},\ 1/n)$ be a diagonal matrix. In view of the classical law of the iterated logarithm we conclude that

$$\frac{\gamma_1 + \gamma_2 + \ldots + \gamma_n}{\sqrt{n \log n}} \longrightarrow 0 \text{ a.s.}, \quad n \to \infty.$$

At the same time by the SLLN we have

$$\frac{\eta_1 + \ldots + \eta_n}{n} \longrightarrow 0 \text{ a.s.}, \quad n \to \infty.$$

Hence, $\|A_n S_n\| \to 0$ a.s., $n \to \infty.$
Now to prove that

$$\|A_n\|\|S_n\| \nrightarrow 0 \text{ a.s.}, \quad n \to \infty, \tag{10}$$

holds, suppose the contrary. Thus

$$\frac{\eta_1 + \ldots + \eta_n}{\sqrt{n \log n}} \longrightarrow 0 \text{ a.s.}, \quad n \to \infty.$$

Hence, using necessary conditions for the SLLN (Petrov, 1975), we have

$$\sum_n \mathbf{P}\left(|\eta_1| > \sqrt{n \log n}\right) < \infty. \tag{11}$$

But, at the same time,

$$\sum_n \int\limits_{|x| > \sqrt{n \log n}} \frac{C}{|x|^3 \log|x|}\, dx = \infty,$$

and the series in (11) diverges. This contradiction establishes (10).

Observe that

$$\mathbf{E}\|X_1\|^2 = \infty, \quad \mathbf{E}\frac{\|X_1\|^2}{\log^+ \log^+ \|X_1\|} < \infty.$$

5. INTEGRAL TYPE CRITERION FOR THE SLLN

Together Theorem 2 and Lemma 8 lead to the following

THEOREM 3. *Let $(X_k, k \geqslant 1)$ be a sequence of i.i.d. random vectors in $\mathbf{R}^m$, $\mathbf{E}X_1 = 0$, $\mathbf{E}\|X_1\|^2 < \infty$. Let $c > 1$ denote some constant. Then*

$$\|A_n S_n\| \longrightarrow 0 \text{ a.s.}, \quad n \to \infty,$$

iff, for any $\varepsilon > 0$,

$$\sum_k \exp\Big(- \varepsilon / c^{2k} \max_{c^k \leqslant j < c^{k+1}} \|A_j\|^2 \Big) < \infty.$$

Moreover, in the necessity part the series converges for all $c > 1$.

COROLLARY. *Let the sequence $(X_k,\ k \geqslant 1)$ be the same as in Theorem 3 and*

$$\|A_n\| \Big/ \sqrt{n \log\log n} \longrightarrow 0, \quad n \to \infty.$$

Then

$$\|A_n S_n\| \longrightarrow 0 \text{ a.s.}, \quad n \to \infty. \tag{12}$$

In addition, if $\|A_n\| / \sqrt{n \log\log n}$ tends to zero monotonically then (12) holds true.

REFERENCES

Bikyalis, A. (1966). On estimate of the rest term in the central limit theorem. *Lith. Math. J.* **7**, 303–308 (in Russian).

Buldygin, V.V. and Solntsev, S.A. (1987a). The contraction principle for c_0–summing operators and SLLN for weighted sums. In: *Probability Theory and Mathematical Statistics. Proc. Fourth Vilnius Conf.* **1**, VNU Press, Utrecht, pp. 269–290.

Buldygin, V.V. and Solntsev, S.A. (1987b). SLLN with operator normalizations and convergence to zero of Gaussian sequences. *Theory Probab. Appl.* **32**, 243–256.

Buldygin, V.V. and Solntsev, S.A. (1989). *The Functional Methods in Problems of Summing of Random Variables.* Naukova Dumka, Kiev (in Russian).

Buldygin, V.V., Solntsev, S.A. and Jadrenko, O.M. (1984). Strong laws of large numbers with operator normalization. *Preprint* No. 84.34. Inst. Math. Academy Sci. Ukrainian SSR, Kiev (in Russian).

Griffin, P.S. (1986). Matrix normalized sums of independent identically distributed random vectors. *Ann. Probab.* **1**, 224–246.

Hahn, M. (1979). The generalized domain of attraction of a Gaussian law on Hilbert space. *Lect. Notes Math.* **709**, 125–144.

Hahn, M. and Klass, M. (1981). The multidimensional central limit theorem for arrays normed by affine transformations. *Ann. Probab.* **9**, 611–623.

Hahn, M. and Klass, M. (1985). Affine normability of partial sums of i.i.d. random vectors: a characterization. *Z. Wahrsch. verw. Geb.* **69**, 479–505.

Jadrenko, O.M. (1982). Strong laws of large numbers with matrix normalizations and condition for convergence of some class of Gaussian sequences to zero. In: *Some Questions on the Theory of Stochastic Processes.* Inst. Math. Academy. Sci. Ukrainian SSR, Kiev, pp. 117–133 (in Russian).

Kandelaki, N.P. and Sazonov, V.V. (1964). On a central limit theorem for random elements with values in Hilbert space. *Theory Probab. Appl.* **9**, 38–46.

Loève, M. (1960). *Probability Theory.* 2nd ed., Van Nostrand, Princeton etc.

Martikainen, A.I. (1984). Criteria for strong convergence of a normalized sums of independent random variables and their applications. *Theory Probab. Appl.* **29**, 519–533.

Petrov, V.V. (1975). *Sums of Independent Random Variables.* Springer, Berlin etc.

Prohorov, Yu.V. (1950). On the strong law of large numbers. *Izv. Acad. Sci. USSR* **14**, 523–536 (in Russian).

Semovski, S.V. (1984). The Bernstein–Feller central limit theorem in $\mathbf{R}^p$. *Theory Probab. Appl.* **29**, 586–590.

Solntsev, S.A. (1985). On oscillation of sums of independent random variables. *Ukrainian Math. J.* **37**, 637-642.

Weiner, D.C. (1986). A law of the iterated logarithm for distributions in the generalized domain of attraction of a nondegenerate Gaussian law. *Probab. Th. Rel. Fields.* **72**, 337-357.

Part Two

LIMIT THEOREMS IN GENERAL SPACES

New Trends in Probab. and Statist., pp. 73 – 77
V. Sazonov and T. Shervashidze (Eds.)
1991 VSP/Mokslas

ON THE DISTRIBUTION OF THE SUP–NORM FOR A STABLE MOTION AND THE RATE OF CONVERGENCE

M. BLOZNELIS
Vilnius University, Naugarduko str. 24, 232006 Vilnius, Lithuania

ABSTRACT

The distribution of the sup–norm for a real stable motion with a parameter $\alpha \in (1,2)$ is investigated. We show that the sup–norm has a bounded density. It allows us to obtain new results on the rate of convergence.

1. INTRODUCTION

Let $Y = \{Y(t),\ t \in [0,1]\}$ be a random process (r.p.) with independent increments. We are interested when the distribution of the sup–norm of Y has a bounded density. We will present here (Proposition 1) a simple approach (based on the ideas of (Davydov and Lifshits, 1984) and (Lifshits, 1986)) which allows us to obtain results like the following. Let $(E, \|\cdot\|)$ be an uniformly convex separable Banach space and the modulus of convexity

$$\alpha(\varepsilon) = 1 - \sup \left\{\|(x+y)/2\| \colon \|x\|, \|y\| \leqslant 1,\ \|x-y\| \geqslant \varepsilon\right\}$$

satisfies the condition $\alpha(\varepsilon) \geqslant \beta\varepsilon^p$ for some β, $p > 0$ and all $\varepsilon \in (0,2)$. Let $Y = W$ be an E valued Wiener process (a Gaussian r.p. with independent increments and such that $\mathbf{E}W(t) = 0$, $\operatorname{cov} W(t) = t \operatorname{cov} W(1)$, $t \in [0,1]$). Then the random variable (r.v.) $\|W(t)\|_\infty = \sup\{\|W(t)\|,\ t \in [0,1]\}$ has a bounded density. A complete proof of this fact can be found in (Bloznelis, 1990). Here (Proposition 1) we will demonstrate our approach in a more simple situation. As an application we obtain a new result on the rate of convergence (Proposition 2).

2. RESULTS

Let Y be a real stable motion with a parameter $\alpha \in (1,2)$ (i.e., a stable r.p. with independent and stationary increments with $\mathbf{E}Y(t) = 0$, $t \in [0,1]$).

PROPOSITION 1. *Let* $\mathbf{P}(Y \in \mathbf{D}[0,1]) = 1$. *Then the* r.v. $\|Y(t)\|_\infty = \sup\{|Y(t)|,\ t \in [0,1]\}$ *has a bounded density.*

Let $X_1, X_2, \dots$ be i.i.d. $\mathbf{R}$–valued random variables with the (same) distribution, F and G denotes the distribution of the strictly stable r.v. $Y(1)$ with parameter

$\alpha \in (1,2)$. Let $\tau, \tau_1, \tau_2, \ldots$ be i.i.d. random variables uniformly distributed on $[0,1]$ and the sequences of r.v. $\{\tau_i\}$ and $\{X_i\}$ are independent. We define the r.p.

$$\left\{\overline{S}_n(t) = n^{-1/\alpha} \sum_{i=1}^{n} \mathbf{1}\{\tau_i \leqslant t\} X_i\right\} \in \mathbf{D}[0,1],$$

where $\mathbf{1}\{\cdot\}$ is the indicator function. Denote $\Delta_n = \sup\{\Delta_n(a),\ a \geqslant 0\}$,

$$\Delta_n(a) = \Big|\mathbf{P}\big(\|\overline{S}_n(t)\|_\infty < a\big) - \mathbf{P}\big(\|Y(t)\|_\infty < a\big)\Big|.$$

PROPOSITION 2. *Assume* $\mathbf{E}X_1 = \mathbf{E}Y(1) = 0$ *and*

$$\int x^2 (F - G)(dx) = 0,$$

then

$$\Delta_n \leqslant \bar{c}(Y) \max\left\{n^{-1/(2\alpha+1)},\ (\nu_{1+\alpha})^{1/(\alpha+1)} n^{-1/(\alpha(\alpha+1))}\right\},$$

where

$$\nu_{1+\alpha} = \int |x|^{1+\alpha} |F - G|(dx)$$

is the pseudomoment of order $1+\alpha$ *and* $0 < \bar{c}(Y) < \infty$ *is a constant.*

The conditions $EX = EY(1) = 0$, $\int x^2(F-G)(dx) = 0$ (and $\nu_{1+\alpha} < \infty$) are usual in Berry–Esseen type theorems in case of stable limit distributions (see, e.g., (Paulauskas, 1975)).
The weak convergence: $\overline{S}_n \Rightarrow Y$ in $\mathbf{D}[0,1]$, when r.v. X_1 is from domain of normal attraction of symmetric stable r.v. $Y(1)$, has been established recently in (Dehling et al., 1989).

Proof of Proposition 1. Let $(\Omega, \mathcal{F}, \mathbf{P})$ be the common probability space where all our random elements are defined. For $t_i = (1+i)^{-4}$ we define $T_i = [t_{i+1}, t_i]$ and the measurable subsets

$$A_i = \Big\{\omega \in \Omega \colon \|Y(t)\|_\infty = \sup\big\{|Y(t)|,\ t \in T_i\big\}\Big\}, \quad i \geqslant 0.$$

For given $a, \delta > 0$, we have

$$\mathbf{P}\big(\|Y(t)\|_\infty \in [a, a+\delta]\big) \leqslant \sum_{i=0}^{\infty} \mathbf{P}\Big(\big\{\|Y(t)\|_\infty \in [a, a+\delta]\big\} \cap A_i\Big),$$

since $\mathbf{P}(\bigcup_{i \geqslant 0} A_i) = 1$. Let us define the distributions

$$\mathbf{P}_i(B) = \mathbf{P}\Big(\big\{\|Y(t)\|_\infty \in B\big\} \cap A_i\Big), \quad i \geqslant 0,$$

where $B \subset \mathbf{R}$ is a measurable subset. We will show that for every $i \geqslant 0$ the distribution $\mathbf{P}_i$ has a bounded density $p_i(u)$, $u > 0$, i.e., $p_i(u) < c(i)$ with an absolute constant $c(i) > 0$. Moreover, the constants $c(i)$ satisfy condition $\sum_{i \geqslant 0} c(i) < \infty$. Therefore,

$$\mathbf{P}\Big(\|Y(t)\|_\infty \in [a, a+\delta]\Big) \leqslant \sum \mathbf{P}_i([a, a+\delta]) \leqslant \delta \sum c(i).$$

Now, for fixed i and $a > 0$, we estimate $\mathbf{P}_i([a, a+\varepsilon]) \leqslant c(i)\varepsilon$ with $0 < \varepsilon < a$:

$$\mathbf{P}_i([a, a+\varepsilon]) \leqslant$$
$$\leqslant \mathbf{P}\Big(\Big\{\sup\{|Y(t)|,\ t \in T_i\} \in [a, a+\varepsilon]\Big\} \bigcap \Big\{|Y_j| \leqslant 2a,\ 1 \leqslant j \leqslant k\Big\}\Big). \quad (1)$$

Here $Y_j = Y(t_1 + (t_0 - t_1)j/k) - Y(t_1 + (t_0 - t_1)(j-1)/k)$ are independent stable r.v. For $i \geqslant 1$, the right-hand side in (1) does not exceed

$$\mathbf{P}\Big(\sup\ \{|Y(t)|,\ t \in T_i\} \in [a, a+\varepsilon]\Big)\mathbf{P}\Big(|Y_j| < 2a,\ 1 \leqslant j \leqslant k\Big). \quad (2)$$

The first factor here equals to

$$\int \mathbf{P}\Big(\sup\{|Y(t_{i+1}) + y(t)|,\ t \in T_i\} \in [a, a+\varepsilon]\Big)\, d\mathbf{P}_{T_i}(y(t)). \quad (3)$$

By $\mathbf{P}_{T_i}$ we denote the distribution of the r.p. $\{Y(t) - Y(t_{i+1}),\ t \in T_i\}$. It is easy to verify that the probability under the integral does not exceed $\varepsilon c_1(Y) t_{i+1}^{-1/\alpha}$. Moreover, if, in addition, $\sup\{|y(t)|,\ t \in T_i\} < a/2$, then this probability does not exceed

$$c_2(Y)\varepsilon t_{i+1}^{-1/\alpha}\left(a t_{i+1}^{-1/\alpha}\right)^{-1-\alpha} = c_2(Y)\varepsilon a^{-1-\alpha} t_{i+1}.$$

Therefore, the integral (3) is bounded by

$$c_1(Y)\varepsilon t_{i+1}^{-1/\alpha}\mathbf{P}\Big(\sup\ \Big\{|Y(t) - Y(t_{i+1})|,\ t \in T_i\Big\} \geqslant a/2\Big) +$$
$$+ c_2(Y)\varepsilon a^{-1-\alpha} t_{i+1} \leqslant c_3(Y)\big(a^{-\alpha} + a^{-1-\alpha}\big)\big(i^{-4} + i^{-5} i^{4/\alpha}\big). \quad (4)$$

Now we return to inequality (1) and observe that

$$\mathbf{P}\big(|Y_j| < 2a,\ 1 \leqslant j \leqslant k\big) < c(Y, k) a^k,$$

for $a < 1$. Put $k = 3 > 1 + \alpha$. Formulas (1)–(4) imply

$$\mathbf{P}_i\big([a, a+\varepsilon]\big) \leqslant c_4(Y)\varepsilon\big(i^{-4} + i^{-5+4/\alpha}\big), \quad i \geqslant 1.$$

For $i = 0$, the integral (3) equals to $\mathbf{P}_0([a, a+\varepsilon])$ and we use the first estimate $\varepsilon t_{i+1}^{-1/\alpha}$ of the integral. This completes the proof of Proposition 1.

Proof of Proposition 2. Let us denote by $S_n = S_n(t)$, $0 \leqslant t \leqslant 1$, the continuous process such that

$$S_n(k/n) = n^{-1/\alpha} \sum_{i \leqslant k} X_i, \quad S_n(0) = 0,$$

and S_n is linear on the intervals $[(k-1)/n,\ k/n]$, $1 \leqslant k \leqslant n$. Let $Y_1, Y_2, \ldots \in \mathbf{R}$ be i.i.d. strictly stable r.v. with the distribution G. Denote $Z_n = Z_n(t)$, $0 \leqslant t \leqslant 1$, the random polygonal line with vertexes at the points

$$(0;0), \quad \Big(k/n; n^{-1/\alpha}\sum_{i\leqslant k} Y_i\Big), \quad 1\leqslant k\leqslant n.$$

We observe that

$$\mathbf{P}\Big(\|\overline{S}_n(t)\|_\infty < a\Big) = \mathbf{P}\Big(\|S_n(t)\|_\infty < a\Big) \quad \forall a>0.$$

Further, $\Delta_n(a) \leqslant I_n(a) + I(a)$, where

$$I_n(a) = \Big|\mathbf{P}\Big(\|S_n(t)\|_\infty < a\Big) - \mathbf{P}\Big(\|Z_n(t)\|_\infty < a\Big)\Big|,$$

$$I(a) = \mathbf{P}\Big(\max_{1\leqslant k\leqslant n} |Y(k/n)| < a\Big) - \mathbf{P}\Big(\|Y(t)\|_\infty < a\Big).$$

To estimate $I(a)$, let us denote

$$A = \Big\{\|Y(t)\|_\infty > a+\varepsilon\Big\}, \quad \varepsilon>0.$$

We have

$$I(a) \leqslant \mathbf{P}\Big(\Big\{\max_{1\leqslant k\leqslant n}|Y(k/n)| < a\Big\}\cap A\Big) + \mathbf{P}\Big(\|Y(t)\|_\infty \leqslant a+\varepsilon\Big) -$$

$$-\mathbf{P}\Big(\|Y(t)\|_\infty < a\Big) \leqslant \mathbf{P}\Big(W(Y)>\varepsilon\Big) + \mathbf{P}\Big(\|Y(t)\|_\infty \in [a, a+\varepsilon]\Big). \tag{5}$$

Here

$$W(Y) = \max_{1\leqslant k\leqslant n} R_k, \quad R_k = \sup_{t_{k-1}\leqslant t\leqslant t_k} \min\Big\{|Y(t_k)-Y(t)|,\ |Y(t)-Y(t_{k-1})|\Big\}$$

with $t_k = k/n$, $1\leqslant k\leqslant n$ (see, e.g., (Paulauskas and Juknevičiene, 1988)). It follows from the independence and the stationarity of increments of Y that, for $s<t<u$,

$$\mathbf{P}\Big(|Y(t)-Y(s)|>\varepsilon,\ |Y(u)-Y(t)|>\varepsilon\Big) \leqslant$$

$$\leqslant \mathbf{P}\Big(|Y(1)| > \varepsilon\,(t-s)^{-1/\alpha}\Big)\mathbf{P}\Big(|Y(1)| > \varepsilon(u-t)^{-1/\alpha}\Big) \leqslant c_5(Y)\varepsilon^{-2\alpha}(t-s)(u-t).$$

Now, the well-known Theorem 12.1 from (Billingsley, 1968) implies

$$\mathbf{P}(R_k>\varepsilon) \leqslant c_6(Y)\varepsilon^{-2\alpha}n^{-2}.$$

Therefore,

$$\mathbf{P}\big(W(Y)>\varepsilon\big) \leqslant \sum \mathbf{P}(R_k>\varepsilon) \leqslant c_6(Y)\varepsilon^{-2\alpha}n^{-1}. \tag{6}$$

We choose $\varepsilon = n^{-1/(2\alpha+1)}$. It follows from (5), (6) and Proposition 1 that $I(a) \leqslant c_7(Y)n^{-1/(2\alpha+1)}$. To estimate $I_n(a)$ we could use the so called "strong invariance principle" together with Proposition 1. However, here we have chosen another way. We apply the methods developed in (Paulauskas and Račkauskas,

1989). We use the Sakhanenko's Lemma (see (Sakhanenko, 1988) or (Bentkus and Liubinskas, 1987)) instead of the smoothing Lemma 5.1.1 and decomposition formula (5.1.2) in (Paulauskas and Račkauskas, 1989). Furthermore, using the induction as in the proof of Theorem 5.1.7 (ibidem), we obtain

$$I_n(a) < \bar{c}(Y) \max \left\{ n^{-1/(2\alpha+1)},\ \nu_{1+\alpha}^{1/(1+\alpha)} n^{-1/\alpha(\alpha+1)} \right\}.$$

A similar estimate (in the case of invariance principle for Banach space valued r.v. with the Wiener limit) was obtained recently by V. Bentkus. Proposition 2 is proved.

REFERENCES

Bentkus, V. and Liubinskas, K. (1987). Rate of convergence in the invariance principle in Banach space. *Lithuanian Math. J.*. **27**, pp. 205—213.

Billingsley, P. (1968). *Convergence of Probability Measures.* Wiley, New York.

Bloznelis, M. (1990). On the distribution of multidimensional Wiener process. *Lithuanian Math. J.* **29** (in Russian).

Davydov, I.A. and Lifshits, M.A. (1984). Method of stratification in some probabilistics problems. In: *Itogi nauki i tekhniki. Ser. Theory Probab. Math. Statist.* **22**, pp. 61—157 (in Russian).

Dehling, H., Denker, M. and Woyczynski, W.A. (1989). Resampling U-statistics using p-stable laws. *Preprint Nr.* 86-50. Western Reserve University, Department of Mathematics and Statistics, Cleveland, Ohio.

Lifshits, M.A. (1986). On the distribution of the maximum of Gaussian processes. *Theory Probab. Appl.* **31**, 125—131.

Paulauskas V. (1975). On the rate of convergence in the multidimensional limit theorem with a stable limiting law. *Lithunian Math. J.* **15**, 168-184.

Paulauskas, V. and Juknevičiene, D. (1988). On the rate of convergence in the central limit theorem in D[0,1]. *Lithuanian Math. J.* **28**, 229-239 (in Russian).

Paulauskas, V. and Račkauskas, A. (1989). *Approximation Theory in the Central Limit Theorem. Exact Results in Banach Spaces.* Kluwer Academic Publisher, Dordrecht, Boston, London.

Sakhanenko, A.I. (1988). Simple method of obtaining estimates in the invariance principle. *Lect. Notes Math.* **1299**, 430—443.

New Trends in Probab. and Statist., pp. 78 – 92
V. Sazonov and T. Shervashidze (Eds.)
1991 VSP/Mokslas

SEMI–INVARIANT CONDITIONS OF WEAK CONVERGENCE OF RANDOM PROCESSES IN THE SPACE OF CONTINUOUS FUNCTIONS

V. BULDYGIN
Kiev Polytechnical Institute, Pr. Pobedy 37, 252056 Kiev, USSR

ABSTRACT

Conditions of weak convergence of random processes in the space of continuous functions in terms of semi–invariants of increments of these processes are given. As an example, a central limit theorem for stochastic integrals with respect to Poisson random measure is considered.

INTRODUCTION

Let $\{\mu_\alpha\}$ be a family of probabilistic Borel measures on a metric space X. Recall that the family $\{\mu_\alpha\}$ is tight (in the space X) if for every $\varepsilon > 0$ there exists a compact set $K_\varepsilon \subset X$ such that

$$\sup_\alpha \mu_\alpha(X \backslash K_\varepsilon) < \varepsilon.$$

The famous Prokhorov theorem (Prokhorov, 1956) establishes the relation between the weak convergence of a sequence of probabilistic measures and the tightness of this sequence in the corresponding metric space. For a number of functional metric spaces, general criteria of tightness of a family of probabilistic measures are known. In particular, if $C(S)$ is a space of continuous real functions defined on a metric compact (S,d) with the metric of uniform convergence and $\{(\xi_\alpha(s),\ s \in S)\}$ is a family of almost surely (a.s.) continuous random processes, then the family of distributions of these processes is tight in $C(S)$ iff the following conditions are fulfilled:

T1) there exists $s_0 \in S$ such that

$$\lim_{c\to\infty} \sup_\alpha \mathbf{P}\Big\{\big|\xi_\alpha(s_0)\big| > c\Big\} = 0;$$

T2) for every $\varepsilon > 0$,

$$\lim_{h\to 0} \sup_\alpha \mathbf{P}\Big\{\sup_{\{s,t\in S:\ d(s,t)<h\}} \big|\xi_\alpha(s) - \xi_\alpha(t)\big| > \varepsilon\Big\} = 0.$$

Verification of these conditions is an essential part of many investigations on weak convergence of random processes. One needs to use different characteristics which reflect specific features of the considered processes. In (Prokhorov, 1956) the tightness conditions based on the moments of increments of random processes were given. These conditions can be made much more precise under additional restrictions on the processes considered. In this paper, the conditions of tightness of distributions in the space of continuous functions based on local behavior of semi-invariants of increments of random processes are found. As an example, a central limit theorem in the space of continuous functions for stochastic processes represented by stochastic integrals with respect to a Poisson measure is considered. The results of this paper are closely related to (Buldygin and Yarovaya, 1983; 1990).

1. EXPONENTIAL BOUNDS AND ENTROPY TIGHTNESS CONDITION

Let (S, ρ) be a compact metric or pseudometric space.
If $\varepsilon > 0$, $t \in S$, then the set

$$Q_\varepsilon(t) = \{s \in S\colon\ \rho(t,s) < \varepsilon\}$$

is the open ρ–ball of radius ε with the center t. Denote by $N_\rho(S,\varepsilon)$ the minimal number of ρ–balls of radius not greater than ε, with centers in S, covering the set S.
The quantity

$$H_\rho(S,\varepsilon) = \log N_\rho(S,\varepsilon)$$

is called the ε–entropy of the set S with respect to the pseudometric ρ.

LEMMA 1. *Let (S,d) be a compact metric space and $\{(\xi_\alpha(s),\ s \in S)\}$ be a family of random processes on (S,d) having a.s. continuous sample paths. Let the following conditions be fulfilled:*

1) *There exist positive constants β, ε_0, C and for every α there exists pseudometric ρ_α on S such that, for any $\varepsilon \in (0,\varepsilon_0)$ and $s,t \in S$, the following inequality holds*

$$\mathbf{P}\Big\{\big|\xi_\alpha(s) - \xi_\alpha(t)\big| > \varepsilon\Big\} \leqslant C\exp\left(-\frac{\varepsilon^2}{\beta\rho_\alpha^2(t,s)}\right); \tag{1}$$

2) *The pseudometric $\sup_\alpha \rho_\alpha(t,s)$ is continuous on (S,d);*

3) $$\lim_{u\to 0}\sup_\alpha \int_0^u H_{\rho_\alpha}^{1/2}(S,\varepsilon)\,d\varepsilon = 0.$$

Then, for every $\varepsilon > 0$,

$$\lim_{u\to 0} \sup_{\alpha} \mathbf{P}\Big\{ \sup_{\{s,t\in S:\ d(s,t)<h\}} |\xi_\alpha(t) - \xi_\alpha(s)| > \varepsilon \Big\} = 0.$$

Looking at the tightness conditions for the family of distributions of $\{\xi_\alpha\}$ in $C(S)$ formulated in the Introduction, one can see that conditions 1)–3) imply **T2)** and together with **T1)** they give sufficient conditions of tightness. Note that the exponential inequality (1) holds not for every $\varepsilon > 0$ but only for $\varepsilon \in (0, \varepsilon_0)$. This is rather natural because the probability appearing in the condition **T2)** does not decrease as $\varepsilon \downarrow 0$. This fact permits to extend the class of processes satisfying condition 1). This will be done further. The method of the proof of Lemma 1 goes back to (Dudley, 1967).

Proof of Lemma 1. Since $\xi_\alpha \in C(S,d)$ a.s., so S can be replaced by its countable subset dense with respect to the metric d.
Let $S_m^{(\alpha)}$ be a minimal net in S with respect to the pseudometric ρ_α. Then

$$\operatorname{card} S_m^{(\alpha)} = N_{\rho_\alpha}(S, 2^{-m}),$$

and

$$\operatorname{card}\Big\{(s,t) \in S \times S:\ s,t \in S_{m-1}^{(\alpha)} \cup S_m^{(\alpha)}\Big\} \leqslant 4N_{\rho_\alpha}^2(S, 2^{-m}).$$

Since the pseudometric ρ_α is continuous with respect to d and (S,d) is compact, so $N_{\rho_\alpha}(S,\varepsilon) < \infty$, for every $\varepsilon > 0$. If $b_m^{(\alpha)} > 0$, then

$$\mathbf{P}\Big\{ \sup_{\substack{t,s\in S_{m-1}^{(\alpha)} \cup S_m^{(\alpha)} \\ \sup_\alpha \rho_\alpha(t,s) < 2^{-m+2}}} \Big|\xi_\alpha(t) - \xi_\alpha(s)\Big| > b_m^{(\alpha)} \Big\} \leqslant$$

$$\leqslant \mathbf{P}\Big\{ \sup_{\substack{t,s\in S_{m-1}^{(\alpha)} \cup S_m^{(\alpha)} \\ \rho_\alpha(t,s) < 2^{-m+2}}} \Big|\xi_\alpha(t) - \xi_\alpha(s)\Big| > b_m^{(\alpha)} \Big\} \leqslant$$

$$\leqslant 4N_{\rho_\alpha}^2(S, 2^{-m}) \sup_{\substack{t,s\in S_{m-1}^{(\alpha)} \cup S_m^{(\alpha)} \\ \rho_\alpha(t,s) < 2^{-m+2}}} \mathbf{P}\Big\{ |\xi_\alpha(t) - \xi_\alpha(s)| > b_m^{(\alpha)} \Big\}.$$

If $b_m^{(\alpha)} \in (0, \varepsilon_0)$ then, using the exponential bound (1), we obtain

$$\mathbf{P}\Big\{ \sup_{\substack{t,s\in S_{m-1}^{(\alpha)} \cup S_m^{(\alpha)} \\ \sup_\alpha \rho_\alpha(t,s) < 2^{-m+2}}} \Big|\xi_\alpha(t) - \xi_\alpha(s)\Big| > b_m^{(\alpha)} \Big\} \leqslant$$

$$\leqslant 4CN_{\rho_\alpha}^2(S, 2^{-m}) \exp\Big(- 2^{2m-4} \big(b_m^{(\alpha)}\big)^2 / \beta \Big) = D_m^{(\alpha)},$$

where

$$D_m^{(\alpha)} = 4C \exp\left(2H_{\rho_\alpha}(S, 2^{-m}) - 2^{2m-4}\left(b_m^{(\alpha)}\right)^2/\beta\right).$$

Suppose that $\left(b_m^{(\alpha)},\ m \geqslant 1\right)$ are sequences of positive numbers such that, for every α and all m, $b_m^{(\alpha)} \in (0, \varepsilon_0)$ and

$$\lim_{M\to\infty} \sup_\alpha \sum_{m=M}^{\infty} b_m^{(\alpha)} = 0, \qquad \lim_{M\to\infty} \sup_\alpha \sum_{m=M}^{\infty} D_m^{(\alpha)} = 0. \tag{2}$$

Then, for any $\varepsilon > 0$, $\Delta > 0$, there exists a natural number $m_0 = m_0(\varepsilon, \Delta)$ such that, for all α,

$$\sum_{m=m_0}^{\infty} b_m^{(\alpha)} < \frac{\varepsilon}{3}, \qquad \sum_{m=m_0}^{\infty} D_m^{(\alpha)} < \Delta.$$

Fix arbitrary $\varepsilon > 0$, $\Delta > 0$. Put

$$\Omega_m^{(\alpha)} = \left\{\omega: \sup_{\substack{t,s \in S_{m-1}^{(\alpha)} \cup S_m^{(\alpha)} \\ \sup_\alpha \rho_\alpha(t,s) < 2^{-m+2}}} \left|\xi_\alpha(t,\omega) - \xi_\alpha(s,\omega)\right| > b_m^{(\alpha)}\right\}.$$

For $\tau \in S$ denote by $\tau_m^{(\alpha)}$ the point from $S_m^{(\alpha)}$ nearest to τ in the sense of the pseudometric ρ_α. In the case when there are several such points let $\tau_m^{(\alpha)}$ be any of them. If $t, s \in S$, $\sup_\alpha \rho_\alpha(t,s) < 2^{-m_0}$ and $m > m_0$, then

$$\rho_\alpha\left(s_{m_0}^{(\alpha)},\ t_{m_0}^{(\alpha)}\right) \leqslant 2^{-m_0+2},$$

$$\rho_\alpha\left(s_m^{(\alpha)},\ s_{m+1}^{(\alpha)}\right) \leqslant 2^{-m+2}, \quad \rho_\alpha\left(t_m^{(\alpha)},\ t_{m+1}^{(\alpha)}\right) \leqslant 2^{-m+2}.$$

For any point $\tau \in S$,

$$\rho_\alpha\left(\tau, \tau_m^{(\alpha)}\right) \to 0, \qquad m \to \infty,$$

and by inequality (1), for every $\delta > 0$,

$$\mathbf{P}\left\{\left|\xi_\alpha(\tau) - \xi_\alpha(\tau_m^{(\alpha)})\right| > \delta\right\} \leqslant \mathbf{P}\left\{\left|\xi_\alpha(\tau) - \xi_\alpha(\tau_m^{(\alpha)})\right| > \min\left(\delta, \varepsilon_0/2\right)\right\} \leqslant$$

$$\leqslant \exp\left(-\frac{\left(\min\left(\delta, \varepsilon_0/2\right)\right)^2}{\beta\rho_\alpha^2\left(\tau, \tau_m^{(\alpha)}\right)}\right).$$

Hence, for any $\delta > 0$,

$$\lim_{m\to\infty} \mathbf{P}\left\{\left|\xi_\alpha(\tau) - \xi_\alpha\left(\tau_m^{(\alpha)}\right)\right| > \delta\right\} = 0,$$

i.e., the convergence in probability takes place. This implies that, for any $t, s \in S$,

$$\xi_\alpha(t_m^{(\alpha)}) \xrightarrow{\mathbf{P}} \xi_\alpha(t), \quad \xi_\alpha(s_m^{(\alpha)}) \xrightarrow{\mathbf{P}} \xi_\alpha(s), \quad m \to \infty.$$

Therefore, the series

$$\sum_{m=m_0}^{\infty} \left(\xi_\alpha(t_m^{(\alpha)}) - \xi_\alpha(t_{m+1}^{(\alpha)})\right), \quad \sum_{m=m_0}^{\infty} \left(\xi_\alpha(s_m^{(\alpha)}) - \xi_\alpha(s_{m+1}^{(\alpha)})\right)$$

converge in probability and a.s., for any t, s from a countable subset of S,

$$\left|\xi_\alpha(t) - \xi_\alpha(s)\right| \leqslant \left|\xi_\alpha(s) - \xi_\alpha(s_{m_0}^{(\alpha)})\right| +$$

$$+\left|\xi_\alpha(s_{m_0}^{(\alpha)}) - \xi_\alpha(t_{m_0}^{(\alpha)})\right| + \left|\xi_\alpha(t_{m_0}^{(\alpha)}) - \xi_\alpha(t)\right| \leqslant$$

$$\leqslant \left|\xi_\alpha(s_{m_0}^{(\alpha)}) - \xi_\alpha(t_{m_0}^{(\alpha)})\right| + \sum_{m=m_0}^{\infty} \left|\xi_\alpha(s_m^{(\alpha)}) - \xi_\alpha(s_{m+1}^{(\alpha)})\right| +$$

$$+ \sum_{m=m_0}^{\infty} \left|\xi_\alpha(t_m^{(\alpha)}) - \xi_\alpha(t_{m+1}^{(\alpha)})\right|.$$

Hence, there exists Ω' such that $\mathbf{P}(\Omega') = 1$ and such that, for all $t, s \in S$, satisfying the inequality $\sup_\alpha \rho_\alpha(t,s) < 2^{-m_0}$ as soon as

$$\omega \notin \left(\Omega' \cap \bigcup_{m=m_0}^{\infty} \Omega_m^{(\alpha)}\right),$$

we have

$$\left|\xi_\alpha(t,\omega) - \xi_\alpha(s,\omega)\right| \leqslant b_{m_0}^{(\alpha)} + 2\sum_{m=m_0}^{\infty} b_m^{(\alpha)} < \varepsilon.$$

Since, for any α,

$$\mathbf{P}\Big(\bigcup_{m=m_0}^{\infty} \Omega_m^{(\alpha)}\Big) \leqslant \sum_{m=m_0}^{\infty} \mathbf{P}(\Omega_m^{(\alpha)}) \leqslant \sum_{m=m_0}^{\infty} D_m^{(\alpha)} < \Delta,$$

so

$$\sup_\alpha \mathbf{P}\left\{ \sup_{\substack{s,t \in S \\ \sup_\alpha \rho_\alpha(t,s) < 2^{-m_0}}} \left|\xi_\alpha(t) - \xi_\alpha(s)\right| > \varepsilon \right\} < \Delta,$$

and by the continuity of $\sup_\alpha \rho_\alpha(t,s)$ with respect to d, we obtain

$$\lim_{u \to 0} \sup_\alpha \mathbf{P}\left\{ \sup_{\substack{s,t \in S \\ d(s,t) < u}} \left|\xi_\alpha(t) - \xi_\alpha(s)\right| > \varepsilon \right\} = 0.$$

To complete the proof of Lemma 1, it remains to show existence of sequences $\left(b_m^{(\alpha)},\ m \geqslant 1\right)$ satisfying condition 2). Put

$$b_m^{(\alpha)} = \min\left\{\frac{\varepsilon_0}{2},\ \sqrt{\beta}\max\left\{2^{4-m}H_{\rho_\alpha}^{1/2}(S,2^{-m}),\ m^{-2}\right\}\right\}.$$

Then, by condition 3), one has, for sufficiently large M,

$$\sum_{m=M}^{\infty} b_m^{(\alpha)} \leqslant \sqrt{\beta}\left[2^{-4}\sum_{m=M}^{\infty} H_{\rho_\alpha}^{1/2}\left(S,2^{-m}\right)2^{-m} + \sum_{m=M}^{\infty}\frac{1}{m^2}\right],$$

and

$$\sum_{m=M}^{\infty} D_m^{(\alpha)} \leqslant 4C\exp\left(-2^{2m-8}\left(b_m^{(\alpha)}\right)^2/\beta\right) \leqslant$$

$$\leqslant 4C\sum_{m=M}^{\infty}\exp\left(-2^{2m-8}/m^2\right).$$

These inequalities and condition 3) of Lemma 1 imply (2). Lemma 1 is proved. As a simple corollary of Lemma 1, let us formulate a condition of weak convergence in a form, convenient for future exposition.

COROLLARY 1. *Let*

$$\xi_\alpha = \left(\xi_\alpha(s),\ s \in S\right) \in C(S) \text{ a.s.}, \quad \alpha \in (0,+\infty].$$

If finite-dimensional distributions of the processes ξ_α converge as $\alpha \to +\infty$ to the corresponding distributions of a process $\xi_\infty = \left(\xi_\infty(s),\ s \in S\right)$, and the conditions of Lemma 1 are fulfilled, then $\xi_\alpha \Rightarrow \xi_\infty$ in $C(S)$, i.e., the distributions of the random fields ξ_α converge weakly as $\alpha \to +\infty$ in $C(S)$ to the distribution of the field ξ_∞.

Remark. If $\left(\xi_0(s),\ s \in S\right)$ is separable random process which satisfies the conditions of Lemma 1 (for $\alpha = 0$), then this process has a.s. continuous sample paths on (S,d).
As an example illustrating the above results consider Gaussian processes. Let $\left(\xi_\alpha(s),\ s \in S\right)$ be a separable Gaussian process with zero mean. Take the structural deviation

$$\sigma_\alpha(t,s) = \sqrt{\mathbf{E}\left|\xi_\alpha(t) - \xi_\alpha(s)\right|^2}, \quad t,s \in S,$$

then condition 1 of Lemma 1 is satisfied with $C = 2,\ \beta = 2,\ \varepsilon_0 = +\infty$. So, Lemma 1 and the remark following it imply the statement below which agrees with the well-known statement in (Dudley, 1967).

COROLLARY 2. *Let $\{(\xi_\alpha(s),\ s \in S)\}$ be a family of separable Gaussian processes with zero means. If*

1) *the pseudo-metric $\sup_\alpha \sigma_\alpha(t,s)$ is continuous on (s,d);*

2) *for every* α,

$$\int_{0+}^{\infty} H_{\sigma_\alpha}^{1/2}(S,\varepsilon)\,d\varepsilon < \infty,$$

3) $$\lim_{u\to 0}\sup_{\alpha}\int_{0}^{u} H_{\sigma_\alpha}^{1/2}(S,\varepsilon)\,d\varepsilon = 0,$$

then $\xi_\alpha \in C(S)$ a.s. *for all* α *and, moreover, the corresponding family of distributions is tight in* $C(S)$.

2. SEMI–INVARIANT CHARACTERISTICS OF RANDOM VARIABLES

If ξ is a random variable, then

$$\chi_{\xi(u)} = \mathbf{E}\exp(iu\xi), \quad u \in \mathbf{R},$$

is its characteristic function,

$$\varphi_{\xi(u)} = \log \chi_{\xi(u)}$$

is the principal value of natural logarithm defined at least in a neighbourhood of the point $u = 0$. The quantity

$$\varkappa_k(\xi) = \frac{1}{i^k}\frac{d^k}{du^k}\varphi_{\xi(u)}\Big|_{u=0}$$

(provided the derivative exists) is called the semi–invariant (cumulant) of order k of the random variable ξ.
Put

$$b(\xi) = \sup_{k\geqslant 2}\left(\frac{|\varkappa_k(\xi)|}{(k-2)!}\right)^{1/2(k-1)}.$$

Put $b(\xi) = \infty$ if $\varkappa_k(\xi)$ does not exist for some $k \geqslant 2$.

Example. Let ξ be a $\mathcal{N}(0,\sigma^2)$–random variable. Then $\varkappa_2(\xi) = \sigma^2$, $\varkappa_k(\xi) = 0$, $k > 2$, hence $b^2(\xi) = \sigma^2$.
The following exponential bound is related to the quantity $b(\xi)$ introduced above.

LEMMA 2. *Let* ξ *be a random variable with* $\mathbf{E}\xi = 0$ *and* $b(\xi) < \infty$. *Then, for any* $\varepsilon \in (0,2)$,

$$\mathbf{P}\Big\{|\xi| > \varepsilon\Big\} \leqslant 2\exp\left(-\frac{\varepsilon^2}{8b^2(\xi)}\right).$$

To prove Lemma 2, some auxiliary statements are necessary.
If ξ is a random variable such that, for some $U \in (0,\infty]$,

$$\mathbf{E}\exp(u\xi) < \infty, \quad |u| < U, \tag{3}$$

then put

$$\Psi_\xi(u) = \sup_{|x|\leqslant u} \frac{2\log \mathbf{E}\exp(x\xi)}{x^2}, \quad u \in (0, U).$$

LEMMA 3. *Let $\mathbf{E}\xi = 0$, and let condition (1) be satisfied. If there exist numbers $a \in (1/U, \infty)$, $\gamma \in (0, \infty)$ such that*

$$\Psi_\xi\left(\frac{1}{a}\right) \leqslant \gamma a, \tag{4}$$

then, for every $\varepsilon \in (0, \gamma)$,

$$\mathbf{P}\Big\{|\xi| > \varepsilon\Big\} \leqslant 2\exp\left(-\frac{\varepsilon^2}{2\gamma a}\right). \tag{5}$$

Proof. The definition of the function $\Psi_\xi(u)$ and condition (2) imply that, for $|x| \leqslant 1/a$,

$$\mathbf{E}\exp(x\xi) \leqslant \exp\left(\frac{\gamma a x^2}{2}\right).$$

Then, making use of the Chebyshev–Markov inequality, one has, for $|x| \leqslant 1/a$, $\varepsilon > 0$,

$$\mathbf{P}\{\xi > \varepsilon\} \leqslant \frac{\mathbf{E}\exp(x\xi)}{\exp(x\xi)} \leqslant \exp\left(-\varepsilon x + \frac{\gamma a}{2}x^2\right).$$

The function $g(x) = -\varepsilon x + \gamma a x^2/2$ has its minimum at the point $x_0 = \varepsilon/\gamma a$. If $\varepsilon \in (0, \gamma)$, then $x_0 \in \big(0, 1/a\big)$. Hence,

$$\mathbf{P}\{\xi > \varepsilon\} \leqslant \exp\left(-\frac{\varepsilon^2}{2\gamma a}\right).$$

An analogous reasoning shows that

$$\mathbf{P}\{\xi < -\varepsilon\} = \mathbf{P}\{-\xi > \varepsilon\} \leqslant \exp\left(-\frac{\varepsilon^2}{2\gamma a}\right).$$

These relations imply inequality (5). Lemma 3 is proved.
Denote

$$m_k(\xi) = \mathbf{E}\xi^k, \quad k \geqslant 1.$$

Note that moments $m_k(\xi)$, $k = 1, \ldots, n$, exist iff semi-invariants $\varkappa_k(\xi)$ exist. We'll need the following relation between moments and semi-invariants.

LEMMA 4. *Let $m_k(\xi)$, $\varkappa_k(\xi)$, $k = 1, \ldots, l$, exist. Then, for any $n = 1, \ldots, l-1$, the following relation takes place*

$$m_{n+1}(\xi) = \sum_{k=0}^{n} C_n^k \varkappa_{k+1}(\xi) m_{n-k}(\xi), \tag{6}$$

where

$$C_n^k = \frac{n!}{k!(n-k)!}, \quad m_0 = 1.$$

In particular, if semi–invariants of any order of the random variable ξ exist, then the relation (6) *holds for any $k \geqslant 1$.*

For the proof of (6) see e.g., (Riordan, 1958).

LEMMA 5. *Let the random variable ξ be such that, for some $A \in (0, \infty)$ and all $k \geqslant 1$,*

$$\left|\varkappa_k(\xi)\right| \leqslant A^k (k-1)! \tag{7}$$

Then, for any $k \geqslant 1$,

$$\left|m_k(\xi)\right| \leqslant A^k (k+1)! \tag{8}$$

Proof. Use the induction on $k \geqslant 1$. Let $k = 1$, then

$$\left|m_1(\xi)\right| = \left|\varkappa_1(\xi)\right|$$

and inequality (8) is a corollary of inequality (7). Suppose that the statement of Lemma 5 is true for all $k = 1, \ldots, n$ $(n \geqslant 1)$. We shall show that it is also true for $k = n+1$. Making use of relation (6) and inequality (7), one has

$$\left|m_{n+1}(\xi)\right| \leqslant \sum_{k=0}^{n} C_n^k \left|\varkappa_{k+1}(\xi)\right| \left|m_{n-k}(\xi)\right| \leqslant$$

$$\leqslant \sum_{k=0}^{n} \frac{n!}{k!(n-k)!} \left(A^{k+1} k!\right) \left(A^{n-k} (n-k+1)!\right) =$$

$$= A^{n+1} n! \sum_{j=1}^{n+1} j = A^{n+1} n! \, \frac{(n+1)(n+2)}{2} < A^{n+1} (n+2)!$$

Therefore, (6) holds for $k = n+1$. Lemma 5 is proved.

Proof of Lemma 2. Consider the formal power series in the complex plane $\mathbf{C}$:

$$\Lambda(z) = \sum_{k=1}^{\infty} \frac{\varkappa_k(\xi)}{k!} z^k, \quad z \in \mathbf{C}. \tag{9}$$

Since $b(\xi) < \infty$, and for all $k \geqslant 2$,

$$\left|\varkappa_k(\xi)\right| \leqslant b^{2(k-1)}(\xi)(k-2)!, \tag{10}$$

then, for sufficiently large k,

$$\left|\varkappa_k(\xi)\right| \leqslant b^{2k}(\xi)k!$$

and

$$L_1 = \limsup_{k\to\infty} \left(\frac{\left|\varkappa_k(\xi)\right|}{k!}\right)^{1/k} \leqslant b^2(\xi) < \infty. \tag{11}$$

Consequently, the series (9) is convergent in the circle $\{|z| < R_1\}$, where

$$R_1 = \frac{1}{L_1} \geqslant \frac{1}{b^2(\xi)} > 0. \tag{12}$$

Therefore, the both functions $\Lambda(z)$ (given by series (9)) and $\Gamma(z) = \exp\left(\Lambda(z)\right)$ are analytic in the circle $\{|z| < R_1\}$.
Further, the relation (11) implies, by Lemma 4,

$$L_2 = \limsup_{k\to\infty} \left(\frac{\left|m_k(\xi)\right|}{k!}\right)^{1/k} < \infty.$$

If is well known (see, e.g., (Feller, 1971)) that this condition implies analyticity of the function $h(z) = \mathbf{E}\exp(iz\xi)$ in some neighbourhood of the real axis including the strip $\{|\mathrm{Im}\, z| < 1/L_2\}$. Hence, in some neighbourhood of zero the principal branch of logarithm $\log h(z)$ is an analytic function and coincides with the function $\Lambda(z)$ by the definition of the series (9). That is in some neighbourhood of the origin, we have

$$h(z) = \Gamma(z). \tag{13}$$

Then, by the theorem on the uniqueness of analytic extension, one concludes that the function $h(z)$ is analytic in $\{|z| < 1/L_1\}$ and there the equality

$$h(z) = \exp\left(\Lambda(z)\right)$$

holds. Hence, for all z such that $|z| < 1/b^2(\xi)$,

$$\Lambda(z) = \log \mathbf{E}\exp(iz\xi).$$

In particular, for all $x \in \mathbf{R}$ such that $|x| < 1/b^2(\xi)$,

$$\log \mathbf{E}\exp(x\xi) = \Lambda(-ix) = \sum_{k=1}^{\infty} \frac{\varkappa_k(\xi)}{k!}(-ix)^k.$$

Since $\varkappa_1 = 0$ then

$$\log \mathbf{E}\exp(x\xi) = \sum_{k=2}^{\infty} \frac{(-i)^k \varkappa_k(\xi)}{k!}\, x^k$$

for $|x| < 1/b^2(\xi)$.

This relation together with (10) imply the following inequality for $0 < u < 1/b^2(\xi)$:

$$\Psi_\xi(u) \leqslant 2 \sup_{|x|\leqslant u} \sum_{k=2}^{\infty} \frac{|\varkappa_k(\xi)|}{k!}|x|^{k-2} \leqslant \frac{2}{u}\sum_{k=2}^{\infty}\frac{|\varkappa_k(\xi)|}{k!}u^{k-1} \leqslant$$

$$\leqslant \frac{2}{u}\sum_{k=2}^{\infty}\frac{1}{k!}\frac{|\varkappa_k(\xi)|}{b^{2(k-1)}(\xi)} \leqslant \frac{2}{u}\sum_{k=2}^{\infty}\frac{1}{k(k-1)} = \frac{2}{u}.$$

Put $u = 1/2b^2(\xi)$, then

$$\Psi_\xi\left(\frac{1}{2b^2(\xi)}\right) \leqslant 4b^2(\xi).$$

So condition (4) of Lemma 3 holds with $\gamma = 2$ and $a = 2b^2(\xi)$. According to this lemma,

$$\mathbf{P}\Big\{|\xi| > \varepsilon\Big\} \leqslant 2\exp\left(-\frac{\varepsilon^2}{8b^2(\xi)}\right)$$

for all $\varepsilon \in (0,2)$. Lemma 2 is proved.

Remark. The inequality in Lemma 2 can be refined as follows:

$$\mathbf{P}\Big\{|\xi| > \varepsilon\Big\} \leqslant 2\exp\left(-\frac{\varepsilon^2}{cb^2(\xi)}\right),$$

where $c \in (4,\infty)$.

3. SEMI–INVARIANT CONDITIONS OF TIGHTNESS AND WEAK CONVERGENCE IN $C(S)$

If $\xi(s)$, $s \in S$, is a random process, denote by $\varkappa_k(t,s)$ the semi–invariant of k–th order of the increment $\xi(t) - \xi(s)$, i.e.,

$$\varkappa_k(t,s) = \varkappa_k\big(\xi(t) - \xi(s)\big).$$

Put also

$$b(t,s) = b\Big(\xi(t) - \xi(s)\Big) = \left(\frac{|\varkappa_k(t,s)|}{(k-2)!}\right)^{1/2(k-1)}.$$

Suppose hereafter that for any $t, s \in S$ there exists $U_{t,s} > 0$ such that

$$\mathbf{E}\exp\Big(u\big(\xi(t) - \xi(s)\big)\Big) < \infty, \quad |u| < U_{t,s}.$$

As it was shown in Section 2, this condition is equivalent to the fact that $b(t,s) < \infty$ for any $t, s \in S$. Hence, by Lemma 2 for each $\varepsilon \in (0,2)$, we have

$$\mathbf{P}\Big\{|\xi(t) - \xi(s)| > \varepsilon\Big\} \leqslant 2\exp\left(-\frac{\varepsilon^2}{8b^2(t,s)}\right).$$

This inequality together with Lemma 1 leads to the following statement.

THEOREM 1. *Let* $\{(\xi_\alpha(s),\ s \in S)\}$ *be a family of separable random processes such that, for every* α *and* $t, s \in S$,

$$b_\alpha(t,s) = b\big(\xi_\alpha(t) - \xi_\alpha(s)\big) < \infty.$$

If

1) *for any* α *there exists a pseudometric* $\rho_\alpha(t,s)$ *such that*

$$b_\alpha(t,s) \leqslant \rho_\alpha(t,s) \quad (t, s \in S),$$

2) *the pseudo-metric* $\sup_\alpha \rho_\alpha(t,s)$ *is continuous on* (S,d);

3) *for every* α,

$$\int_{0+}^{\infty} H_{\rho_\alpha}^{1/2}(S,\varepsilon)\, d\varepsilon < \infty,$$

4)
$$\lim_{u\to 0} \sup_\alpha \int_0^u H_{\rho_\alpha}^{1/2}(S,\varepsilon)\, d\varepsilon = 0,$$

then $\xi_\alpha \in C(S)$ *a.s. for any* α *and, moreover, the family of the distributions of these processes is tight in* $C(S)$.

Remark 1. In the case when ξ_α are Gaussian processes with zero means, then (see the example in Section 2)

$$b_\alpha(t,s) = \sigma_\alpha(t,s) = \sqrt{\mathbf{E}\big|\xi(t) - \xi(s)\big|^2}$$

and the statement of Theorem 1 coincides with that of Corollary 2 (see Section 1).

Remark 2. If S is a compact set in $\mathbf{R}^n$ and $|\cdot|_n$ stands for the Euclidean norm in $\mathbf{R}^n$, then conditions of Theorem 1 can be replaced by the following one: there exists a non-negative continuous function $\varphi(x)$ which increases monotonously on some interval $(0, x_0)$ and satisfies the conditions:

1) there exists $h > 0$ such that

$$\sup_\alpha b_\alpha(t,s) \leqslant \varphi\big(|t-s|_n\big),$$

for all $|t-s|_n < h$;

2)
$$\int_0^{x_0} \frac{\varphi(u)}{u|\log u|^{1/2}}\, du < \infty.$$

COROLLARY. *Let* $\alpha \in (0,\infty)$, *and let, in addition to the conditions of Theorem 1, finite-dimensional distributions of the processes* ξ_α *converge as* $\alpha \to \infty$ *to the corresponding distributions of a separable random process* $\xi_\infty = (\xi_\infty(s),\ s \in S)$. *Then* $\xi_\infty \in C(S)$ a.s. and $\xi_\alpha \Rightarrow \xi_\infty$, $\alpha \to \infty$ *in* $C(S)$.

4. A CENTRAL LIMIT THEOREM IN $C(S)$ FOR THE SCHOTTKY EFFECT PROCESSES

Let $\pi_\nu(\Delta)$, $\Delta \in K$, be a Poisson homogeneous random measure with intensity ν defined on the ring K of bounded Borel subsets of the Euclidean space $\mathbf{R}^n$, $n \geqslant 1$. Consider a function $(f(s),\ s \in \mathbf{R}^n)$, which is continuous and bounded on $\mathbf{R}^n$ and, for all $q \geqslant 1$,

$$\int_{\mathbf{R}^n} |f(s)|^q ds < \infty.$$

The homogeneous random field

$$\xi_\nu(s) = \int_{\mathbf{R}^n} f(s-t)\pi_\nu(dt), \quad s \in \mathbf{R}^n,$$

is often called the Schottky effect process.
Hereafter we shall consider separable versions of such processes.
The logarithm of characteristic functional χ_{ξ_ν} of the process ξ_ν can be written in the following way (see, e.g., (Prokhorov and Rozanov, 1969)):

$$\log \chi_{\xi_\nu}(\mu(\cdot)) = \nu\left[\exp\left(i\int_{\mathbf{R}^n} f(s)\mu(ds)\right) - 1\right],$$

where μ is a Borel measure on $\mathbf{R}^n$ of bounded variation. This implies that the semi-invariant of k-th order of the increment $\xi_\nu(t) - \xi_\nu(s)$ can be written as

$$\varkappa_k\left(\xi_\nu(t) - \xi_\nu(s)\right) = \nu\int_{\mathbf{R}^n} \left(f(t-s+u) - f(u)\right)^k du.$$

Hence, if $\nu > 0$, then

$$b_{\xi_\nu}(t,s) = \sup_{k\geqslant 2}\left(\frac{|\varkappa_k(\xi_\nu(t) - \xi_\nu(s))|}{(k-2)!}\right)^{1/2(k-1)} \leqslant \rho(t,s),$$

where

$$\rho(t,s) = \sup_{k\geqslant 2}\left(\frac{\int_{\mathbf{R}^n} |f(t-s+u) - f(u)|^k du}{(k-2)!}\right)^{1/2(k-1)}$$

One can see that $\rho(t,s)$, $t,s \in \mathbf{R}^n$, is a pseudometric on $\mathbf{R}^n$. Therefore (see Theorem 1), if the following conditions are fulfilled:

C1) ρ is continuous on $\mathbf{R}^n$;

C2) $\int_0^1 H_\rho^{1/2}(T,\varepsilon)d\varepsilon < \infty$, where T denotes the unit cube in $\mathbf{R}^n$, then $\xi_\nu \in C(\mathbf{R}^n)$ a.s. for any $\nu > 0$.

Consider the processes

$$\zeta_\nu = \left(\zeta_\nu(t) = \frac{\xi_\nu(t) - E\xi_\nu(t)}{\sqrt{\nu}}, \quad t \in \mathbf{R}^n\right), \quad \nu \geqslant 1.$$

It is well known that finite-dimensional distributions of processes ξ_ν converge as $\nu \to \infty$ to the corresponding distributions of the Gaussian process $\gamma = (\gamma(t), t \in \mathbf{R}^n)$ with zero mean and covariance function

$$B_\gamma(t,s) = \int_{\mathbf{R}^n} f(t-u)f(s-u)\,du.$$

(The process γ can be regarded as a separable one.)
Theorem 1 permits us to make this fact more precise.

THEOREM 2. *Let S be a compact set in $\mathbf{R}^n$. If conditions* C1) *and* C2) *are fulfilled, then $\gamma \in C(S)$* a.s. *and $\zeta_\nu \Rightarrow \gamma$, $\nu \to \infty$, in $C(S)$.*

Proof. Since

$$\mathbf{E}\xi_\nu(t) \equiv \nu \int_{\mathbf{R}^n} f(s)\,ds,$$

then, for all $t, s \in \mathbf{R}^n$,

$$\zeta_\nu(t) - \zeta_\nu(s) = \frac{1}{\sqrt{\nu}}\big(\xi_\nu(t) - \xi_\nu(s)\big).$$

Therefore,

$$\varkappa_k\big(\zeta_\nu(t) - \zeta_\nu(s)\big) = \int_{\mathbf{R}^n} \big(f(t-s+u) - f(u)\big)^k du$$

and

$$b_{\zeta_\nu}(t,s) = \sup_{k\geqslant 2}\left(\frac{\big|\varkappa_k\big(\zeta_\nu(t) - \zeta_\nu(s)\big)\big|}{(k-2)!}\right)^{1/2(k-1)}$$

As

$$\sigma_\gamma(t,s) \leqslant \rho(t,s) \quad \big(t,s \in \mathbf{R}^n\big),$$

then by conditions C1), C2) and Dudley's theorem (Dudley, 1967) (see also the remark to Theorem 1), $\gamma \in C(S)$ a.s. Moreover, conditions C1), C2) imply the conditions of Theorem 1. Theorem 2 is proved.

Remark. Conditions C1), C2) are satisfied, for example, if there exist constants $h > 0,\ C > 0$ such that

$$\sup_{0<|t-s|_n<h} \sup_{u\in\mathbf{R}^n} \left|f(t-s+u)-f(u)\right| < \frac{C}{|\ln h|^{1+\varepsilon}} .$$

REFERENCES

Buldygin, V.V. and Yarovaya, N.V. (1983). A functional limit theorem for Schottky effect processes. In: *Problems of Theory of Probability Distributions.* Institute of Mathematics of the Academy Sci. Ukrainian SSR, Kiev, pp. 25–41 (in Russian).

Buldygin, V.V. and Yarovaya, N.V. (1990). Semi-invariant conditions of continuity of random processes. *Dopovidi AN Ukrainian SSR,* Ser. A, **2**, 3–6 (in Russian).

Dudley, R.M. (1967). The sizes of compact subsets of Hilbert space and continuity of Gaussian processes. *J. Funct. Anal.* **1**, 290–330.

Feller, W. (1971). *An Introduction to Probability Theory and its Applications.* Wiley, New York.

Prokhorov, Yu.V. (1956). Convergence of random processes and limit theorems in probability theory. *Theory Probab. Appl.* **1**, 157–214.

Prokhorov, Yu.V. and Rozanov, Yu.A. (1969). *Probability Theory: Basic Concepts, Limit Theorems, Random Processes.* Springer, Berlin etc.

New Trends in Probab. and Statist., pp. 93 - 105
V. Sazonov and T. Shervashidze (Eds.)
1991 VSP/Mokslas

ALMOST SURE PERMUTATIONAL CONVERGENCE OF VECTOR RANDOM SERIES AND KOLMOGOROV'S PROBLEM

S.A. CHOBANYAN and G.J. GIORGOBIANI
Muskhelishvili Institute of Computational Mathematics of the Georgian Academy of Sciences, Akuri str. 8, 380093 Tbilisi, Georgia, USSR

ABSTRACT

We investigate the problem of existence of a rearrangement of a random vector series such that the rearranged series converges a.s. We present also a result concerning a problem due to Kolmogorov on existence of a rearrangement of any orthonormal basis in L_2 such that the Fourier series of any function with respect to the rearranged basis is a.s. convergent.

INTRODUCTION

The main problem we deal with is as follows. Let $\sum \xi_k$ be a series of random variables with values in a normed space. Suppose that this series converges in probability to a random variable S. One has to find conditions under which there exists a rearrangement of the series which is a.s. convergent to S.
The following example show that some conditions are actually needed, i.e., that the convergence of series in probability does not imply existence of a.s. convergent rearrangement of the series.

Example. We consider a sequence of packets M_n^+ and M_n^- of square integrable random variables defined on [0,1] with the Lebesgue measure. The packet M_n^+ consists of all random variables

$$f_{nk}^+(x) = I_{[(k-1)/2^n, k/2^n]}(x), \quad k = 1, ..., 2^n, \quad x \in [0,1],$$

while the packet M_n^- consists of the variables

$$f_{nk}^-(x) = -f_{nk}^+(x), \quad k = 1, ..., 2^n, \quad x \in [0,1],$$

where I_B denotes the indicator function of a measurable set B. The series we consider consist of all the functions of all the packets M_n^+ and M_n^-, $n = 1, 2, ...$ In other words, we consider the series consisting of the indicator functions, taken with plus and minus signs, of sets obtained by dividing [0,1] into 2^n equal parts. It is easy to see that the terms of this series can be arranged in such a way that it is mean square convergent (e.g., to zero). Indeed, we arrange the terms as follows

$$f_{11}^+ + f_{11}^- + f_{12}^+ + f_{12}^- + ... + f_{n1}^+ + f_{n1}^- + ... + f_{n2^n}^+ + f_{n2^n}^- + ...$$

Clearly this series converges to zero in mean square (and consequently in probability). However, as it can easily be seen, none of its rearrangements can converge to zero a.s.

In (Garsia, 1964) it was shown (Garsia's theorem) that if real random variables ξ_i, $i \in \mathbf{N}$, are mutually orthogonal, $\mathbf{E}|\xi_i|^2 < \infty$ and $\sum_{i=1}^{\infty} \mathbf{E}|\xi_i|^2 < \infty$, then for some rearrangement σ the series $\sum_{i=1}^{\infty} \xi_{\sigma(i)}$ converges a.s. (the initial series $\sum_{i=1}^{\infty} \xi_i$ obviously converges in probability, since the condition $\sum_{i=1}^{\infty} \mathbf{E}|\xi_i|^2 < \infty$, for an orthogonal series $\sum_{i=1}^{\infty} \xi_i$, is a necessary and sufficient condition for its mean square convergence).

Later in (Nikishin, 1967) the following generalization of Garsia's statement was obtained. Nikishin's theorem: If a series $\sum_{i=1}^{\infty} \xi_i$ of real random variables converges in probability to a random variable S and if $\sum_{i=1}^{\infty} |\xi_i|^2 < \infty$ a.s., then there exists a permutation σ such that $\sum_{i=1}^{\infty} \xi_{\sigma(i)}$ converges a.s. to S.

One of the main aims of the present paper is to find an analogue of Nikishin's condition ($\sum_{i=1}^{\infty} |\xi_i|^2 < \infty$ a.s.) for series of random variables taking values in a normed space X. We will show that such an analogue is the a.s. convergence of the series $\sum \xi_i \varepsilon_i$, where (ε_i) is a sequence of independent Bernoulli random variables independent of the sequence (ξ_i). The latter condition in the case of reals is obviously equivalent to Nikishin's condition, while in the case of type 2 Banach space it is ensured by the following straightforward analogue of Nikishin's condition: $\sum \|\xi_i\|^2 < \infty$ a.s. We would like to emphasize that our condition in many cases can be effectively verified. What we need is nothing more than just conditions for a.s. convergence of the series $\sum a_k \varepsilon_k$ for nonrandom a_k, and for a wide class of Banach spaces such conditions do exist.

In Section 1 we state an inequality we have established in (Chobanyan and Giorgobiani, 1989) which plays a crucial role in the proof of the analogue of the Nikishin theorem. We give also a simple proof of an inequality due to Garsia which is the main tool in the proof of the aforementioned Garsia theorem. Our proof is valid for the case of Hilbert space.

Section 2 deals with the generalization of the Nikishin theorem. In Section 3 we analyze the following problem. Let (φ_k) be a sequence of mutually orthogonal real random variables defined on a probability space $(\Omega, \mathcal{A}, \mathbf{P})$, and $\mathbf{E}\varphi_k^2 = 1, k \in \mathbf{N}$ (in other words, (φ_k) is an orthonormal sequence in $L_2(\Omega, \mathcal{A}, \mathbf{P})$). The question is: for which subsets $M \subset l_2$ there exists a common rearrangement $(\varphi_{\sigma(k)})$ of (φ_k) such that the series $\sum_{k=1}^{\infty} \alpha_{\sigma(k)} \varphi_{\sigma(k)}$ is a.s. convergent for any sequence $(\alpha_k) \in M$? In the early twenties Kolmogorov conjectured that such a rearrangement does exist for any orthonormal system (φ_k) and for $M = l_2$. This is one of the oldest problems in the theory of orthogonal series remaining open so far. The Garsia theorem actually asserts that for singletons the desired rearrangement always exists.

We show using Nikishin's theorem that the desired common rearrangement exists in a sense for "almost all" $(\alpha_k) \in l_2$; more precisely, if (φ_k) is an arbitrary system, and if μ is a probability measure on l_2, then there exists a permutation π: $\mathbf{N} \to \mathbf{N}$ and a subset $V \subset l_2$, $\mu(V) = 1$, such that the series $\sum_{k=1}^{\infty} \alpha_{\pi(k)} \varphi_{\pi(k)}$ converges a.s. for each sequence $(\alpha_k) \in V$. One of the important corollaries of the result is connected with the use of the Wiener measure as μ. It says that if (φ_k) is an arbitrary orthogonal basis in $L_2[0,1]$, then there exists a permutation σ: $\mathbf{N} \to$ $\to \mathbf{N}$ such that, for any function f on [0,1] with continuous derivative, its Fourier expansion

$$f = \sum_{k=1}^{\infty} \alpha_{\sigma(k)} \varphi_{\sigma(k)}$$

converges to f_1 a.s. with respect to the Lebesgue measure (here $\alpha_{\sigma(k)} = = \int_0^1 f(t)\varphi_{\sigma(k)}(t)dt$).

1. INEQUALITIES RELATED TO REARRANGEMENTS OF SUMMANDS

Besides Theorem 1.1 which is used below in Section 2, we present here some recent closely related results which may have an independent interest.
The proof of Theorem 2.1 is based on the following inequality (Chobanyan and Giorgobiani, 1989).

THEOREM 1.1. *Let X be a normed space (real or complex), and let $a_1, \ldots, a_n \in \in X$ be such that $\sum_{i=1}^n a_i = 0$. Then for any increasing convex function $f\colon \mathbf{R}_+ \to \to \mathbf{R}_+$*

$$\mathbf{E} f\left(\frac{1}{2}\left\|\sum_{i=1}^n a_i \varepsilon_i\right\|\right) \leqslant \frac{1}{n!}\sum_{\sigma} \max_{1 \leqslant k \leqslant n} f\left(\left\|a_{\sigma(1)} + \ldots + a_{\sigma(k)}\right\|\right) \leqslant$$

$$\leqslant 2\mathbf{E} f\left(\left\|\sum_{i=1}^n a_i \varepsilon_i\right\|\right), \tag{1.1}$$

where (ε_i) is a sequence of independent Bernoulli random variables, and the sum in the middle expression is taken over all permutations σ: $[1, \ldots, n] \to [1, \ldots, n]$.

In the sequel we will be interested in the right-hand side of (1.1). In the particular case of $f(x) = x$ it yields

$$\frac{1}{n!}\sum_{\sigma} \max_{1 \leqslant k \leqslant n} \|a_{\sigma(1)} + \ldots + a_{\sigma(k)}\| \leqslant 2\mathbf{E}\left\|\sum_{i=1}^n a_i \varepsilon_i\right\|.$$

In the case when $\sum_{i=1}^n a_i$ does not vanish the last inequality together with the arguments of the triangular inequality immediately imply

$$\frac{1}{n!}\sum_{\sigma} \max_{1 \leqslant k \leqslant n} \left\|a_{\sigma(1)} + \ldots + a_{\sigma(k)}\right\| \leqslant 3\left(\mathbf{E}\left\|\sum_{i=1}^n a_i \varepsilon_i\right\| + \left\|\sum_{i=1}^n a_i\right\|\right). \tag{1.2}$$

The right-hand side of (1.1) goes back to the book (Garsia, 1970), where it has been proved for the case $X = \mathbf{R}$ and $f(t) = t^p$, $1 \leqslant p < \infty$. As far as the left-hand side of (1.1) is concerned, apparently it remained unnoticed for a long time in spite of its simplicity. It has been noticed independently by Kadec and Kwapien in the process of discussion with them of our paper (Chobanyan and Giorgobiani, 1989). Does inequality (1.1) hold for any increasing function f (not necessarily convex), may be with some other constants? If this is the case, then one could take as f

the indicator function of the set (t, ∞) thus getting two-sided estimates for the tales of the distributions of the random variables $\| \sum_{i=1}^{n} a_i \varepsilon_i \|$ and

$$S_\sigma = \max_{1 \leqslant k \leqslant n} \| a_{\sigma(1)} + \ldots + a_{\sigma(k)} \|$$

(the latter being considered as a random variable on the finite probability space of all permutations σ: $[1, \ldots, n] \to [1, \ldots, n]$ with the uniform distribution on it). However it is not difficult to prove that such estimates can not be obtained as a corollary of Theorem 1.1. Estimates of this kind have been found in (Chobanyan, 1990). Namely the following statement was proved.

THEOREM 1.2. *Let $a_1, \ldots, a_n$ be a collection of elements of a normed space X such that $\sum_{i=1}^{n} a_i = 0$. Then, for any $t > 0$,*

$$\mathbf{P}\left\{\omega: \left\| \sum_{i=1}^{n} a_i \varepsilon_i(\omega) \right\| > 2t \right\} \leqslant \frac{1}{n!} \operatorname{card} \{\sigma: S_\sigma > t\} \leqslant$$

$$\leqslant 11 \mathbf{P}\left\{\omega: \left\| \sum_{i=1}^{n} a_i \varepsilon_i(\omega) \right\| > t/6 \right\}.$$

The right-hand side of this inequality in the case $X = \mathbf{R}$ has been found earlier in (Kashin and Saakian, 1984) with some absolute constants C_1 and C_2 instead of 11 and 6. Theorem 1.2 immediately implies the following

COROLLARY. *Let f: $\mathbf{R}_+ \to \mathbf{R}_+$ be an increasing (not necessarily convex) left continuous function, $f(0) = 0$. Then, under the condition of Theorem 1.2,*

$$\mathbf{E} f\left(\frac{1}{2} \left\| \sum_{i=1}^{n} a_i \varepsilon_i \right\| \right) \leqslant \frac{1}{n!} \sum_{\sigma} f(S_\sigma) \leqslant 11 \mathbf{E} f\left(6 \left\| \sum_{i=1}^{n} a_i \varepsilon_i \right\| \right).$$

Now we suggest a simple proof of an inequality representing the basic tool in the proof of the Garsia theorem.

THEOREM 1.3 (Garsia's inequality). *Let $\xi_1, \ldots, \xi_n$ be pairwise orthogonal real variables, $\mathbf{E}\xi_i^2 = 1$. Then for each collection of real numbers $\alpha_1, \ldots, \alpha_n$ there exists a permutation π: $[1, \ldots, n] \to [1, \ldots, n]$ such that*

$$\int_\Omega \max_{1 \leqslant k \leqslant n} \left| \alpha_{\pi(1)} \xi_{\pi(1)} + \ldots + \alpha_{\pi(k)} \xi_{\pi(k)} \right|^2 d\mathbf{P} \leqslant 6 \sum_{i=1}^{n} \alpha_i^2.$$

Proof. Let σ: $[1, \ldots, n] \to [1, \ldots, n]$ be a permutation, $\theta = (\theta_1, \ldots, \theta_n)$ be a collection of signs ($\theta_i = \pm 1$). Denote

$$|\alpha_\sigma \xi_\sigma| = \max_{1 \leqslant k \leqslant n} \left| \alpha_{\sigma(1)} \xi_{\sigma(1)} + \ldots + \alpha_{\sigma(k)} \xi_{\sigma(k)} \right|,$$

$$|\alpha_\sigma \xi_\sigma| = \max_{1 \leqslant k \leqslant n} \left|\alpha_{\sigma(1)}\xi_{\sigma(1)}\theta_1 + \ldots + \alpha_{\sigma(k)}\xi_{\sigma(k)}\theta_k\right|, \quad s = -\sum_{i=1}^{n} \alpha_i \xi_i,$$

$$|s, \alpha_\sigma \xi_\sigma \theta| = \max \left(|s|, |s + \alpha_{\sigma(1)}\xi_{\sigma(1)}\theta_1|, \ldots, \left|s + \sum_{i=1}^{n} \alpha_{\sigma(i)}\xi_{\sigma(i)}\theta_i\right| \right).$$

Then

$$\int_\Omega |s, \alpha_\sigma \xi_\sigma \theta|^2 d\mathbf{P} = \int_\Omega |s, \alpha_\sigma, \xi_\sigma \theta| + |s, \alpha_\sigma, \xi_\sigma| - |s, \alpha_\sigma \xi_\sigma|^2 d\mathbf{P} \geqslant$$

$$\geqslant \int_\Omega \left[2 \max \left(|s, (\alpha_\sigma \xi_\sigma)^+|, \ |(\alpha_\sigma, \xi_\sigma)^-|\right) - |s, \alpha_\sigma \xi_\sigma|\right]^2 d\mathbf{P},$$

where $(\alpha_\sigma, \xi_\sigma)^+$ is the collection of those $\alpha_{\sigma(i)}\xi_{\sigma(i)}$ for which θ_i are positive and the ordering is preserved; and $(\alpha_\sigma \xi_\sigma)^-$ is the collection of those $\alpha_{\sigma(i)}\xi_{\sigma(i)}$ for which θ_i are negative. It is easy to see that

$$\max \left(|s, (\alpha_\sigma \xi_\sigma)^+|, |(\alpha_\sigma \xi_\sigma)^-|\right) = |s, \alpha_{\sigma^*} \xi_{\sigma^*}|,$$

where σ^*: $[1, \ldots, n] \to [1, \ldots, n]$ is a permutation defined by given σ and θ as follows. First there are those indices $\sigma(i)$ for which θ_i are positive, and then in the reverse order follow those $\sigma_{(i)}$ for which θ_i are negative. Now (1.3) yields

$$\int_\Omega |s, \alpha_\sigma \xi_\sigma \theta|^2 d\mathbf{P} \geqslant \int_\Omega \left(2|s, \alpha_{\sigma^*} \xi_{\sigma^*}| - |s, \alpha_\sigma \xi_\sigma|\right)^2 d\mathbf{P} \geqslant$$

$$\geqslant 2 \int_\Omega |s, \alpha_{\sigma^*} \xi_{\sigma^*}|^2 d\mathbf{P} - \int_\Omega |s, \alpha_\sigma, \xi_\sigma|^2 d\mathbf{P}.$$

A permutation σ and a collection θ of signs were arbitrary so far. Let now $\sigma = \sigma_0$ be a permutation for which $\int_\Omega |s, \alpha_\sigma \xi_\sigma|)^2 d\mathbf{P}$ attains its minimum. Then

$$\int_\Omega |s, \alpha_{\sigma_0} \xi_{\sigma_0} \theta^2| d\mathbf{P} \geqslant \int_\Omega |s, \alpha_{\sigma_0} \xi_{\sigma_0}|^2 d\mathbf{P} = \int_\Omega |\alpha_\pi \xi_\pi|^2 d\mathbf{P},$$

where the permutation π is defined by σ_0 as follows: $\pi(k) = \sigma_0(n - k + 1)$. Hence (we agree that formally $\sum_{k=1}^{0} = 0$),

$$\int_\Omega \max_{1 \leqslant k \leqslant n} \left|\alpha_{\pi(1)}\xi_{\pi(1)} + \ldots + \alpha_{\pi(k)}\xi_{\pi(k)}\right|^2 d\mathbf{P} \leqslant$$

$$\leqslant \int_\Omega \max_{1 \leqslant k \leqslant n} \left| -\sum_{i=1}^{n} \alpha_i \xi_i + \sum_{i=1}^{k} \alpha_{\sigma_0(i)}\xi_{\sigma_0(i)}\theta_i \right|^2 d\mathbf{P} \leqslant$$

$$\leqslant 2\int_\Omega \left|\sum_{i=1}^n \alpha_i\xi_i\right|^2 d\mathbf{P} + 2\int_\Omega \max_{1\leqslant k\leqslant n}\left|\sum_{i=1}^k \alpha_{\sigma_0(i)}\xi_{\sigma_0(i)}\theta_i\right|^2 d\mathbf{P} \leqslant$$

$$\leqslant 2\sum_{i=1}^n \alpha_i^2 + 2\int_\Omega \max_{1\leqslant k\leqslant n}\left|\sum_{i=1}^k \alpha_{\sigma_0(i)}\xi_{\sigma_0(i)}\theta_i\right|^2 d\mathbf{P}.$$

Taking the average of both sides of this inequality with respect to all signs and using the Levy inequality, we get finally

$$\int_\Omega \max_{1\leqslant k\leqslant n} |\alpha_{\pi(1)}\xi_{\pi(1)} + \ldots + \alpha_{\pi(1)}\xi_{\pi(k)}|^2 d\mathbf{P} \leqslant$$

$$\leqslant 2\sum_{i=1}^n \alpha_i^2 + 2\int_\Omega \mathbf{E}\max_{1\leqslant k\leqslant n}\left|\sum_{i=1}^k \alpha_{\sigma_0(i)}\xi_{\sigma_0(i)}\varepsilon_i\right|^2 d\mathbf{P} \leqslant$$

$$\leqslant 2\sum_{i=1}^n \alpha_i^2 + 4\int_\Omega \mathbf{E}\left|\sum_{i=1}^n \alpha_{\sigma_0(i)}\xi_{\sigma_0(i)}\varepsilon_i\right|^2 d\mathbf{P} = 6\sum_{i=1}^n \alpha_i^2.$$

Theorem 1.3 is proved.

Remark 1. In the original Garsia's proof the coefficient on the right-hand side was equal to 16.

Remark 2. Theorem 1.3 remains valid if α_i are elements of a real Hilbert space instead of reals. To this end one should understand by $|\cdot|$ the norm that the same proof goes through.

2. GENERALIZATION OF NIKISHIN'S THEOREM

THEOREM 2.1. *Let $\sum_{k=1}^\infty \xi_k$ be a series of separable-valued random variables taking values in a real normed space X. If the series converges in probability to an X-valued random variable S and if, for a sequence (ε_k) of independent Bernoulli random variables independent of (ξ_k), the series $\sum_{k=1}^\infty \xi_k\varepsilon_k$ converges a.s. in X, then there exists a permutation σ: $\mathbf{N}\to\mathbf{N}$ such that the rearranged series $\sum_{k=1}^\infty \xi_{\sigma(k)}$ converges to S a.s.*

We need the following

LEMMA 2.1. *Let (η_n) be a sequence of non-negative random variables and suppose that, for a sequence (η_n) of integers,*

$$f = \sum_{k=1}^\infty \frac{1}{n_{k+1}-n_k}\sum_{i=n_k+1}^{n_{k+1}} \eta_i < \infty \quad a.s. \tag{2.1}$$

Then there exists a sequence (m_k) of integers such that

$$n_{k+1} > m_k \geqslant n_k + 1 \quad \text{and} \quad \sum_{k=1}^\infty \eta_{m_k} < \infty \quad a.s.$$

Proof. Suppose at first that $\mathbf{E}f < \infty$. Then

$$\mathbf{E}\sum_{k=1}^{\infty}\frac{1}{n_{k+1}-n_k}\sum_{i=n_k+1}^{n_{k+1}}\eta_i=\sum_{k=1}^{\infty}\frac{1}{n_{k+1}-n_k}\sum_{i=n_k+1}^{n_{k+1}}\mathbf{E}\eta_i<\infty. \tag{2.2}$$

It is clear that for each k there exists m_k such that

$$n_{k+1} > m_k \geqslant n_k + 1 \quad \text{and} \quad \mathbf{E}\eta_{m_k} \leqslant \frac{1}{n_{k+1}-n_k}\sum_{i=n_k+1}^{n_{k+1}}\mathbf{E}\eta_i.$$

This together with (2.2) implies $\sum_{k=1}^{\infty}\eta_{m_k} < \infty$ a.s. So the lemma is proved under the condition $\mathbf{E}f < \infty$. If this is not the case one should consider the random variables $\eta_i' = \eta_i/f$ instead of η_i (if f is equal to zero, then the ratio is mean to be zero). Lemma 2.1 is proved.

Proof of Theorem 2.1. Without loss of generality we may assume that

$$(\Omega, \mathcal{A}, \mathbf{P}) = (\Omega_1 \times \Omega_2, \mathcal{A}_1 \times \mathcal{A}_2, \mathbf{P}_1 \times \mathbf{P}_2)$$

and $\xi_k = \xi_k(\Omega_1, \Omega_2)$ depends only on ω_1 and $\varepsilon_k = \varepsilon_k(\omega_1, \omega_2)$ depends only on ω_2. The convergence of $\sum_{k=1}^{\infty}\xi_k$ in probability implies the existence of a subsequence of partial sums $S_{p_k} = \sum_{i=1}^{p_k}\xi_i$ which is convergent to S a.s. Further, the convergence of $\sum \xi_k\varepsilon_k$ a.s. implies that

$$\mathbf{E}_2\left\|\sum_{k=m}^{n}\xi_k(\omega_1)\varepsilon_k\right\| \to 0 \quad \text{as} \quad m, n \to \infty$$

for almost all $\omega_1 \in \Omega_1$ ($\mathbf{E}_2$ is the expectation with respect to $\mathbf{P}_2$).
According to the Egorov theorem, one can choose a sequence of measurable sets $H_j \subset \Omega_1$ such that

$$H_j \subset H_{j+1}, \quad \lim_{j\to\infty}\mathbf{P}_1(H_j) = 1,$$

and S_{p_m} and $\mathbf{E}_2\left\|\sum_{k=m}^{n}\xi_k(\omega)\varepsilon_k\right\|$ are uniformly convergent on each H_j, $j \in \mathbf{N}$ (as $m, n \to \infty$). Now we choose numbers k_j in such a way that on each H_j the following inequalities hold

$$\left\|\sum_{k=k_j}^{l}\sum_{i=p_k+1}^{p_{k+1}}\xi_i\right\| \leqslant \frac{1}{2^j} \tag{2.3}$$

for every $l > k_j$, and

$$\mathbf{E}_2\left\|\sum_{i=p_{k_j}+1}^{q}\xi_i(\omega)\varepsilon_i\right\| \leqslant \frac{1}{2^j} \tag{2.4}$$

for every $q > p_{k_j}$. Let us prove that the series

$$\sum_{j=1}^{\infty}\left\|\sum_{i=p_{k_j}+1}^{p_{k_{j+1}}}\xi_i(\omega)\right\|, \quad \sum_{j=1}^{\infty}\mathbf{E}_2\left\|\sum_{i=p_{k_j}+1}^{p_{k_{j+1}}}\xi_i(\omega)\varepsilon_i\right\| \tag{2.5}$$

are a.s. convergent on Ω_1. To this end it is enough to show that they converge on each H_j. Suppose $\omega_0 \in H_j$. Then $\omega_0 \in H_s$ for all $s \geqslant j$. According to (2.3),

$$\left\| \sum_{k=k_s}^{k_{s+1}-1} \sum_{i=p_k+1}^{p_{k+1}} \xi_i(\omega_0) \right\| \leqslant \frac{1}{2^s}$$

if $s \geqslant j$. Therefore,

$$\sum_{s=j}^{\infty} \left\| \sum_{i=p_{k_s}}^{p_{k_{s+1}}} \xi_i(\omega_0) \right\| \leqslant 1.$$

An analogous reasoning gives

$$\sum_{s=j}^{\infty} \mathbf{E}_2 \left\| \sum_{i=p_{k_s}+1}^{p_{k_{s+1}}} \xi_i(\omega_0)\varepsilon_i \right\| \leqslant 1.$$

This implies the a.s. convergence of the series (2.5) on Ω_1. We set for simplicity $p_{k_j} = n_j$. Now we summarize the facts obtained up to now as follows: a.s. on Ω_1

$$\left\| S_{n_j} - S \right\| \to 0, \quad \sum_{j=1}^{\infty} \left\| \sum_{i=n_j+1}^{n_{j+1}} \xi_i \right\| < \infty,$$

$$\sum_{j=1}^{\infty} \mathbf{E}_2 \left\| \sum_{i=n_j+1}^{n_{j+1}} \xi_i \varepsilon_i + \right\| < \infty. \tag{2.6}$$

Consider the random variable

$$\eta_j(\sigma, \omega) = \max_{n_j < i \leqslant n_{j+1}} \left\| \sum_{k=n_j+1}^{i} \xi_{\sigma(k)}(\omega) \right\|,$$

where σ: $[n_j+1, ..., n_{j+1}] \to [n_j+1, ..., n_{j+1}]$ is a permutation. Taking into account (1.2), we get

$$\frac{1}{(n_{j+1} - n_j!)} \sum_{\sigma} \eta_j(\sigma, \omega) \leqslant 3\left(\mathbf{E}_2 \left\| \sum_{i=n_j+1}^{n_{j+1}} \xi_i(\omega)\varepsilon_i \right\| + \left\| \sum_{i=n_j+1}^{n_{j+1}} \xi_i(\omega) \right\| \right).$$

Now according to (2.6),

$$\sum_{j=1}^{\infty} \frac{1}{(n_{j+1} - n_j)!} \sum_{\sigma} \eta_j(\sigma, \omega) < \infty$$

for almost all $\omega \in \Omega_1$. We conclude by virtue of Lemma 2.1 that there exists a permutation

$$\sigma_j\colon \ [n_j + 1, ..., n_{j+1}] \to [n_j + 1, ..., n_{j+1}]$$

such that

$$\sum_{j=1}^{\infty} \eta_j(\sigma_j, \omega) < \infty \quad \text{a.s. on } \Omega_1. \tag{2.7}$$

Let σ: $\mathbf{N} \to \mathbf{N}$, $\sigma = (\sigma_j)$, be the permutation defined as follows: if $k \in [n_j + 1, ..., n_{j+1}]$, then $\sigma(k) = \sigma_j(k)$. Let us prove that the series $\sum_{i=1}^{\infty} \xi_{\sigma(i)}$ converges to S a.s. It is clear that

$$\sum_{i=1}^{n_j} \xi_{\sigma(i)} = \sum_{i=1}^{n_j} \xi_i, \quad j \in \mathbf{N},$$

hence, $\| \sum_{i=1}^{n_j} \xi_{\sigma(i)} - S\| \to 0$ a.s. as $j \to \infty$. Further, according to (2.7),

$$\lim_{j \to \infty} \max_{n_j < k \leqslant n_{j+1}} \left\| \sum_{i=n_j+1}^{k} \xi_{\sigma(i)} \right\| = 0 \quad \text{a.s.}$$

This implies that the series $\sum_{j=1}^{\infty} \xi_{\sigma(i)}$ converges to S a.s.
Theorem 2.1 is proved. The question of possibility of effective verification of the condition of Theorem 2.1 arises. For a wide class of Banach spaces the problem of convergence of the series $\sum a_k \varepsilon_k$ with nonrandom $a_k \in X$, $k \in \mathbf{N}$, is fully investigated. For instance if X is a Banach space of type p, $1 \leqslant p \leqslant 2$, then $\sum a_k \varepsilon_k$ converges if $\sum \|a_k\|^p < \infty$. Further, if X is a separable Banach sublattice of some $L_0(T, \Sigma, \nu)$ with a σ-finite measure ν, and if X in addition is of some cotype q, $2 \leqslant q < \infty$, then a necessary and sufficient condition for $\sum a_k \varepsilon_k$, $a_k \in X$, to converge in X is

$$\left(\sum_{k=1}^{\infty} |a_k(t)|^2 \right)^{1/2} \in X$$

(see (Maurey, 1974)). Now we can deduce from Theorem 2.1 the following

COROLLARY 2.1. *Theorem 2.1 remains valid in each of the following cases.*

(a) *X is a Banach space of type p, $1 \leqslant p \leqslant 2$, and the condition of convergence of $\sum \xi_k \varepsilon_k$ is replaced by*

$$\sum_{k=1}^{\infty} \|\xi_k\|^p < \infty \quad \text{a.s.}$$

(b) *X is a separable Banach sublattice of some $L_0(T, \Sigma, \nu)$ with a σ-finite measure ν, and the condition of convergence of $\sum \xi_k \varepsilon_k$ is replaced by*

$$\left(\sum_{k=1}^{\infty} \xi_k^2 \right)^{1/2} \in X \quad \text{a.s.}$$

Recall that the spaces l_p, L_p are of the type $\min(2, p)$, and that among Banch sublattices having some cotype are all the spaces l_p, L_p, $1 \leqslant p < \infty$, Orlicz spaces L_Φ with convex Φ satisfying the Δ_2-condition.

We note in conclusion that the technique used in the proof of Theorem 2.1 also applicable in the following situation. Let $L_\Phi(\Lambda, X)$ be an Orlicz space of functions defined on a measurable space $(\Lambda, \mathcal{B}, \rho)$ with a finite measure ρ, and taking values in a Banach space X.

THEOREM 2.2. *Suppose that the series $\sum_{k=1}^{\infty} \xi_k$, $\xi_k \in L_\Phi(\Lambda, X)$, converges in $L_\Phi(\Lambda, X)$ to some $S \in L_\Phi(\Lambda, X)$, where Φ is a convex Orlicz function satisfying the Δ_2-condition. If for the sequence (r_k) of Rademacher functions the series $\sum_{k=1}^{\infty} \xi_k r_k(\omega)$ converges for almost all $\omega \in [0,1]$ in the metric of $L_\Phi(\Lambda, X)$, there exists a permutation σ: $\mathbf{N} \to \mathbf{N}$ such that the series $\sum_{k=1}^{\infty} \xi_{\sigma(i)}$ is convergent to S both a.s. on Λ and in $L_\Phi(\Lambda, X)$.*

We are able also to prove an analogous result for concave Φ with the following additional requirement: there exists a constant $c > 1$ such that

$$\Phi(2t) \geqslant c\Phi(t), \quad t \in \Lambda.$$

Theorem 2.2 for the case $X = \mathbf{R}$ has been obtained for first time in (Megrabian, 1985).

3. KOLMOGOROV'S PROBLEM

Let $(\varphi_k)_{k=1}^{\infty}$ be an arbitrary orthonormal system of real random variables. Does there exist a permutation π: $\mathbf{N} \to \mathbf{N}$ such that for any sequence (α_k) of real numbers with $\sum \alpha_k^2 < \infty$ the series $\sum_{k=1}^{\infty} \alpha_k \varphi_{\pi(k)}$ converges a.s.? This question posed by Kolmogorov in the twenties still remains open and carries the name of Kolmogorov's problem.
It is clear for any permutation π the series $\sum \alpha_k \varphi_{\pi(k)}$ (under the condition $\sum \alpha_k^2 < < \infty$) is mean square convergent, and the problem consists in finding a permutation π which is common for all $(\alpha_k) \in l_2$ and for which the series $\sum \alpha_k \varphi_{\pi(k)}$ converges a.s. for each sequence $(\alpha_k) \in l_2$. The celebrated Carleson theorem asserts actually that the trigonometric system does not need any rearrangements natural arrangement ensures the a.s. convergence.
The Garsia theorem formulated in Introduction supports the positive answer to the Kolmogorov question: it says that for any single sequence $(\alpha_k) \in l_2$ a desired rearrangement always exists. Attempts of solving the Kolmogorov problem have led Garsia to the following conjecture (Garsia, 1970).

Conjecture of Garsia. There exists a universal constant $C > 0$ such that for any finite orthonormal system $\varphi_1, ..., \varphi_n \in L_2(\Omega, \mathcal{A}, \mathbf{P})$ there exists a permutation σ: $[1, ..., n] \to [1, ..., n]$ such that

$$\sup_{\sum \alpha_i^2 = 1} \int_\Omega \max_{1 \leqslant k \leqslant n} \left|\alpha_1 \varphi_{\sigma(1)} + ... + \alpha_k \varphi_{\sigma(k)}\right|^2 d\mathbf{P} \leqslant C. \tag{3.1}$$

The validity of the Conjecture of Garsia implies the positive solution of the Kolmogorov problem. That is why the problem itself is often called the Garsia problem. However, as it has been found out by Bourgain (Bourgain, 1989) the problem has been posed by Kolmogorov much earlier. Incidentally Bourgain has succeeded

in getting (3.1) with the additional factor $\log\log n$ in the right-hand side. Garsia himself has shown the validity of his conjecture for the case when the system $\varphi_1, ..., \varphi_n$ is rearrangement invariant, i.e., when the distribution of the vector $(\varphi_1, ..., \varphi_n)$ does not change under rearrangements of its components. This enabled him to answer the Kolmogorov problem in affirmative in the case of rearrangement invariant (φ_k).
We will show here that for any orthonormal system there exists its rearrangement which suits in a sense almost all $(\alpha_k) \in l_2$ (however this fact is far from being the "almost solution" of the Kolmogorov problem). Namely the following statement holds.

THEOREM 3.1 (Giorgobiani, 1990). *Let (φ_k) be an arbitrary orthonormal system and let μ be an arbitrary probability measure on the Borel σ-algebra of l_2. Then there exists a measurable vector subspace $V \subset l_2$ with $\mu(V) = 1$ and a permutation π: $\mathbf{N} \to \mathbf{N}$ such that the series $\sum \alpha_{\pi(k)}\varphi_{\pi(k)}$ converges a.s.for any sequence $(\alpha_k) \in V$.*

Proof. Denote $\xi_k(\alpha, \omega) = \alpha_k\varphi_k(\omega)$. We can consider ξ_k as a random variable defined on the probability space $l_2 \times \Omega$ with the probability $\mu \times \mathbf{P}$. It is easy to check that $\sum \xi_k^2 < \infty$ a.s. (with respect to $\mu \times \mathbf{P}$). Besides, as it is easy to see, the series $\sum \xi_k$ converges in measure $\mu \times \mathbf{P}$ (to its sum $S = \sum \xi_k$ which obviously does not depend here on the ordering of the summands). Therefore, the conditions of the Nikishin theorem are satisfied, and consequently there exists a permutation π: $\mathbf{N} \to \mathbf{N}$ such that the series $\sum \xi_{\pi(k)}$ converges $\mu \times \mathbf{P}$-a.s. to S. This implies the existence of a Borel set $H_0 \subset l_2$ such that $\mu(H_0) = 1$ and for any $(\alpha_k) \in H_0$ the series $\sum \alpha_{\pi(k)}\varphi_{\pi(k)}$ converges $\mathbf{P}$-a.s. This series clearly is a.s. convergent also for any (α_k) belonging to $\widetilde{H}_0$, the linear span of H_0. The theorem would be proved, if $\widetilde{H}_0$ were measurable. However the linear span of a measurable set in general is not measurable. This obstacle can be avoided as follows. Since the measure μ is Radon, there exists a σ-compact set $H_1 \subset H_0$ such that $\mu(H_1) = \mu(H_0) = 1$. Now $\widetilde{H}_1$ is measurable (since the linear span of any σ-compact set is measurable). Consequently, we get that $\widetilde{H}_1$ is a measurable vector subspace of l_2 with $\mu(\widetilde{H}_1) = 1$, and for any sequence $(\alpha_k) \in \widetilde{H}_1$ the series $\sum \alpha_{\pi(k)}\varphi_{\pi(k)}$ converges a.s. Theorem 3.1 is proved.
Let now (φ_k) be an orthonormal basis in $L_2 = L_2(\Omega, \mathcal{A}, \mathbf{P})$. Since any measure on L_2 generates a measure on l_2 (the space of the Fourier coefficients of functions from L_2 with respect to (φ_k)), and vice versa, we can reformulate Theorem 3.1 as follows.

THEOREM 3.2. *Let (φ_k) be an orthonormal basis in L_2 and let μ be an arbitrary probability measure on the Borel σ-algebra of L_2. Then there exist a measurable vector subspace $V \subset L_2$ with $\mu(V) = 1$ and a permutation π: $\mathbf{N} \to \mathbf{N}$ such that the Fourier series*

$$\sum_{k=1}^{\infty} \left(f, \varphi_{\pi(k)}\right) \varphi_{\pi(k)}$$

of any function $f \in V$ with respect to the rearranged system $(\varphi_{\pi(k)})$ converges a.s. to f.

Theorem 3.2 enables us to point out subsets $V \subset L_2$ for which the desired common rearrangement of an orthonormal system exists. For instance, taking a measure concentrated on countable subsets of L_2 as μ, one gets the existence of a common rearrangement of an orthonormal system for an arbitrary countable set $V \subset l_2$. Consider now a Gaussian measure as μ. The class of covariance operators of Gaussian measures in L_2 coincides with that of symmetric positive nuclear operators. This means that an operator $R: L_2 \to L_2$ is the covariance operator of a Gaussian measure if and only if it has the form $R = T^*T$, where $T: L_2 \to L_2$ is a Hilbert-Schmidt operator. It is well-known that if μ is a centered Gaussian measure with the covariance operator $R = T^*T$, and if $H_0 \subset L_2$ is a vector subspace (not necessarily closed) with $\mu(H_0) = 1$, then H_0 contains the whole subspace T^*L_2. It is evident that every Hilbert-Schmidt operator can be taken as T^* to get the Gaussian covariance operator R. Consequently we have the following

COROLLARY 3.1. *Let (φ_k) be any orthonormal basis in L_2 and let G be any Hilbert-Schmidt operator in L_2. Then there exists a permutation $(\varphi_{\pi(k)})$ of the basis such that for any function $f \in GL_2$ the Fourier series of f with respect to the system $(\varphi_{\pi(k)})$ converges to f a.s.*

Finally take the Wiener measure on $L_2[0,1]$ as μ (more precisely, the natural image of the Wiener measure on $C[0,1]$). Its covariance operator $R: L_2 \to L_2$ is defined by

$$(Rf)(t) = \int_0^1 \min(s,t)f(s)ds, \quad f \in L_2.$$

As T^* corresponding R, one can take

$$(T^*f)(t) = \int_0^t f(s)ds.$$

Now Corollary 3.1 implies

COROLLARY 3.2. *For any orthonormal basis (φ_k) in $L_2[0,1]$ there exists its rearrangement $(\varphi_{\pi(k)})$ such that the Fourier series with respect to $(\varphi_{\pi(k)})$ of any absolutely continuous function f on [0,1] with $f' \in L_2$ (in particular, the Fourier series of any differentiable function f with continuous derivative f') converges to f a.s.*

REFERENCES

Bourgain, J. ryear1989 On Kolmogorov's rearrangment problem for orthogonal systems and Garsia's conjecture. *Lect. Notes Math.* **1376**, 209–250.

Chobanyan, S.A. (1990). On some inequalities related to permutations of summands in a normed space. *Preprint*. Inst. Comp. Math. Georgian Academy Sci., Tbilisi, pp. 1–21.

Chobanyan, S.A. and Giorgobiani, G.J. (1989). A problem on rearrangement of summands in normed spaces and Rademacher sums. *Lect. Notes Math.* **1391**, 33–46.

Garsia, A. (1964). Existence of almost everywhere convergent rearrangements for Fourier series of L_2 functions. *Ann. Math.* **79**, 623–629.

Garsia, A. (1970). *Topics in Almost Everywhere Convergence.* Markham Publishers, Chicago.

Giorgobiani, G.J. (1990). Almost everywhere convergent rearrangements of expansions with respect to an orthogonal system. *Preprint,* Inst. Comp. Math., Georgian Academy Sci., Tbilisi, pp. 1–6.

Kashin, B.S. and Saakian, A.A. (1984). *Orthogonal Series.* Nauka, Moscow.

Maurey, B. (1974). Type et cotype dans les espaces munis de structures locales inconditionnelles *Sem. Maurey-Schwartz 1973–1974.* Exposes 24, 25, Centre Math., Ecole Polytech., Paris.

Megrabian, R.M. (1985). On the set of sums of function series in the spaces L_Φ. *Theory Probab. Appl.* **30**, 542–556.

Nikishin, E.M. (1967). On convergent rearrangements of functional series. *Math. Notes* **1**, 129–136.

New Trends in Probab. and Statist., pp. 106 - 110
V. Sazonov and T. Shervashidze (Eds.)
1991 VSP/Mokslas

LIMIT THEOREMS FOR STABLE LAWS IN BANACH SPACES

A.N. CHUPRUNOV
Kazan State University, Chebotarev Institute of Mathematics and Mechanics, University str., 17, 420008 Kazan, Tataria, USSR

ABSTRACT

The paper deals with the weak convergence of normed sums of independent Banach space valued random elements to the p-stable law and Banach spaces in which this convergence holds. It is shown that such property of Banach spaces is a modification of the central limit theorem property.

Let B be a Banach space with dual B', $(\Omega, \mathfrak{A}, \mathbf{P})$ be a probability space, $0 < p \leqslant 2$, Y, Y_n be independent identically distributed random elements in B with mean 0, such that $\mathbf{E}|x'(Y)|^p < \infty$ for all $x' \in B'$, where $\mathbf{E}$ denotes the expectation (that is the Bochner integral with respect to $\mathbf{P}$), γ_n^p be standard p-stable independent random variables (i.e., random variables with the characteristic function $\hat{\gamma}_n^p(t) = \exp(-|t|^p)$, $t \in \mathbf{R}$). We will assume that $\{Y_n\}$ and $\{\gamma_n^p\}$ are independent families. We denote $S_n^p(Y) = n^{-1/p} \sum_{k=1}^n \gamma_k^p Y_k$, $n \in \mathbf{N}$. By Kolmogorov's strong law of large numbers and Lebesgue's theorem the following relation holds for all $x' \in B'$

$$\widehat{S}_n^p(Y)(x') = \mathbf{E} \exp\left(ix'(S_n^p(Y)\right) = \mathbf{E} \prod_{k=1}^n \mathbf{E}^{\gamma_k^p} \exp\left(ix'(n^{-1/p}\gamma_k^p Y_k\right) =$$

$$= \mathbf{E} \prod_{k=1}^n \exp(-n^{-1}|x'(Y_k)|^p) = \mathbf{E} \exp(-n^{-1} \sum_{k=1}^n |x'(Y_k)|^p) \to$$

$$\to \exp(-\mathbf{E}|x'(Y)|^p). \tag{1}$$

So, if $S_n^p(Y)$ weakly converges in B, then its limit is a p-stable random element γ_Y^p in B with the characteristic functional $\hat{\gamma}_Y^p(x') = \exp(-\mathbf{E}|x'(Y)|^p)$, $x' \in B'$. Thus we are led to the following definition.

Definition 1. (1) A random element Y in B is said to be p-prestable if γ_y^p is the characteristic functional of a tight probability and $\mathbf{E}\|Y\|^p < \infty$.
(2) A random element Y is said to satisfy the central limit theorem property for p-stable law (CLTp) if it is p-prestable and the sequence $S_n^p(Y)$ is weakly convergent.
(3) A Banach space B is said to satisfy the central limit theorem property for

p-stable law (CLTp) if every p-prestable random element in B has the CLTp. Note that by (Mouchtari, 1989), if $0 < p < 2$, the tightness of γ_Y^p implies $\mathbf{E}\|Y\|^p < \infty$. Therefore if $0 < p < 2$, then the condition $\mathbf{E}\|Y\|^p < \infty$ is not needed.
We denote by CLTp(B) the family of random elements with the CLTp.

THEOREM 1. *Let $0 < q < p$. A random element $Y \in$ CLTp(B) if and only if it is a limit in the quasinorm*

$$\|X\|_{p,q} = \sup_{n \in N} (\mathbf{E}\|S_n^p(X)\|^q)^{1/q}.$$

of simple random elements X.

Proof. The sufficiency is proved in the same way as Theorem 3.1 in (Pisier, 1975).
Necessity. Let $S_n^p(Y) \to \gamma_Y^p$ weakly in B, $0 < q < p$, $\varepsilon > 0$. Then, by (Acosta and Gine, 1979), there exists $n_0 \in \mathbf{N}$ such that $|\mathbf{E}\|S_n^p(Y)\|^q - \mathbf{E}\|\gamma_Y^p\|^q| < \varepsilon/3$ for all $n > n_0$. Let us choose a finite-codimension projection Q in B such that $\sup_{1 \leqslant n \leqslant n_0} \mathbf{E}\|S_n^p(QY)\|^q < \varepsilon/3$ and $\mathbf{E}\|Q\gamma_Y^p\|^q < \varepsilon/3$.
Then

$$\sup_{1 \leqslant n < \infty} \mathbf{E}\|S_n^p(QY)\|^q < \sup_{1 \leqslant n \leqslant n_0} \|S_n^p(QY)\|^q +$$

$$+ \sup_{n > n_0} |\mathbf{E}\|S_n^p(QY)\|^q - \mathbf{E}\|\gamma_{QY}^p\|^q| + \mathbf{E}\|\gamma_{QY}^p\|^q < \varepsilon.$$

Therefore, the random element Y is a limit in the quasinorm $\|\cdot\|_{p,q}$ of finite-dimensional random elements. It follows from the finite-dimensional theorem that any finite-dimensional random element $X \in$ CLTp(B) is a limit of simple random elements in the quasinorm $\|\cdot\|_{p,q}$. This completes the proof of the theorem.
Thus, CLTp(B), by Theorem 1, is a complete metrizable linear space with the topology defined by the quasinorm $\|\cdot\|_{p,q}$.
In (Hoffmann–Jorgensen and Pisier, 1976; Pisier, 1975) random elements Y in a Banach space B with the central limit theorem property (CLT) are studied. A random element Y with mean 0 in B is a pregaussian random element if $\mathbf{E}\|Y\|^2 <$ $< \infty$ and γ_Y^2 is the characteristic functional of a tight probability. A random element Y in B satisfies the CLT if it is a pregaussian random element and the sequence $S'_n(Y) = n^{-1/2} \sum_{k=1}^n Y_k$ weakly converges in B. We denote CLT(B) the family of random elements Y in B with the CLT.

THEOREM 2. *A random element $Y \in$ CLT(B) if and only if $Y \in$ CLT2(B) and* $\mathbf{E}Y = 0$.

Proof. Let $Y \in$ CLT(B). Since $\int_0^\infty \sqrt{\mathbf{P}\{|\gamma_1^2| > t\}}dt < \infty$, by (Ledoux and Talagrand, 1986), we have $\gamma_1^2 Y \in$ CLT(B). Therefore, $Y \in$ CLT2(B).
Let the random element $\gamma_1^2 Y \in$ CLT2(B) and $\mathbf{E}Y = 0$. Using the inequality in (Buldygin, 1980, p.167), Lemma 3.4 in (Mouchtari, 1989) and Theorem 3.1 in (Pisier, 1975), we have

$$\inf_{Q \in \mathfrak{F}(B)} \sup_{n \in \mathbf{N}} \mathbf{E}\|QS'_n(Y)\| = \inf_{Q \in \mathfrak{F}(B)} \sup_{n \in \mathbf{N}} n^{-1/2} \mathbf{E}\| \sum_{k=1}^n QY_k\| <$$

$$< 8 \inf_{Q \in \mathfrak{F}(B)} \sup_{n \in \mathbf{N}} n^{-1/2} \mathbf{E} \| \sum_{k=1}^{n} \gamma_k^p Q Y_k \| = 0,$$

where $\mathfrak{F}(B)$ is the family of finite-codimensional projections in B. Then, by Theorem 3.1 in (Pisier, 1975), we have $Y \in CLT(B)$.
Let the numbers a_k^n, $1 \leqslant k \leqslant k_n$, $n \in \mathbf{N}$, be such that

$$\max_{1 \leqslant k \leqslant k_n} |a_k^n| \to 0, \quad n \to \infty. \tag{2}$$

We set

$$S_n^a(Y) = \sum_{k=1}^{k_n} a_k^n \gamma_k^p Y_k, \qquad \widehat{S}_n^a(Y)(x') = \mathbf{E} \exp\big(ix'(S_n^a(Y)\big), \qquad x' \in B', \ n \in \mathbf{N}.$$

Let ξ be a random variable, $C > 0$. We define

$$(\xi)_C = \xi I_{\{|\xi| > C\}}, \qquad (\xi)^C = \xi I_{\{|\xi| \leqslant C\}}.$$

LEMMA. *The sequence $\widehat{S}_n^a(Y)(x')$, $n \in \mathbf{N}$, converges for every $x' \in B'$ if and only if*

$$\sum_{k=1}^{k_n} |a_k^n|^p \to C, \quad n \to \infty, \tag{3}$$

for some $C > 0$. Moreover, from (3) it follows that

$$\lim_{n \to \infty} \widehat{S}_n^a(Y)(x') = \gamma_{c^{1/p} Y}^p(x'), \quad x' \in B'.$$

Proof. In the same way as in (1), we have

$$\widehat{S}_n^a(Y)(x') = \mathbf{E} \exp\left(- \sum_{k=1}^{k_n} |a_k^n x'(Y_k)|^p \right), \quad x' \in B'.$$

Let (3) be true. We define $b_k^n = |a_k^n|$, $\xi_k = |x'(Y_k)|$. Then, by (2), (3), for all $\varepsilon > 0$, we obtain

$$\sum_{k=1}^{k_n} \mathbf{P}\{b_k^n \xi_k > \varepsilon\} < \sum_{k=1}^{k_n} \mathbf{E}\Big(\frac{b_k^n}{\varepsilon} \xi_k\Big)_1 = \frac{1}{\varepsilon} \sum_{k=1}^{k_n} b_k^n \mathbf{E}(\xi_k)_{\varepsilon / b_k^n} \to 0, \quad n \to \infty,$$

and

$$\sum_{k=1}^{k_n} \mathbf{E}(b_k^n \xi_k)^\varepsilon = \sum_{k=1}^{k_n} b_k^n \mathbf{E}(\xi_k)^{\varepsilon / b_k^n} \to C\mathbf{E}\, \xi_1, \quad n \to \infty.$$

By Theorem 28.2 in (Gnedenko and Kolmogorov, 1954), we have $\sum_{k=1}^{k_n} b_k^n \xi_k \to$ $\to C\mathbf{E}\, \xi_1$ in probability as $n \to \infty$. Therefore, by Lebesgue's theorem, we have $\lim_{n \to \infty} \widehat{S}_n^a(Y)(x') = \exp\big(-C\mathbf{E}|x'(Y)|^p\big)$, $x' \in B'$.

Let $x' \in B'$ be such that $x'(Y_k)$ are nondegenerate random variables and the sequence $x'(S_n^a(Y))$ is weakly convergent. Then, since the random variables $x'(S_n^a(Y))$ have the same distribution as $\left(\sum_{k=1}^{k_n} |a_k^n x'(Y_k)|^p\right)^{1/p} \gamma_1^p$, for any $\varepsilon > 0$, there exists $C > 0$ such that $\mathbf{P}\{\sum_{k=1}^{k_n} |a_k^n x'(Y_k)|^p > C\} < \varepsilon$, $n \in \mathbf{N}$. Therefore, $\sup_{n\in\mathbf{N}} \sum_{k=1}^{k_n} \mathbf{E}(|a_k^n x'(Y_k)|^p)^1 < \infty$. Hence,

$$\sup_{n\in\mathbf{N}} \sum_{k=1}^{k_n} |a_k^n|^p = C_1^p < \infty.$$

Then, by Theorem 28.2 in (Gnedenko and Kolmogorov, 1954),we have

$$\mathbf{E}\exp\left(ix'(S_n^a(Y)\right) = \exp\left(-\sum_{k=1}^{k_n} a_k^n \mathbf{E}|x'(Y_k)|^p\right)\exp(o(1)), \qquad n \to \infty.$$

Consequently, (3) is true.

Families of Banach spaces connected with p-stable distributions on them have been studied in (Mouchtari, 1989; Pisier, 1975). A Banach space B is of stable type p, $0 < p \leqslant 2$, if convergence of the series $\sum_{k=1}^{\infty} \|x_k\|^p$, $x_k \in B$, implies almost sure convergence of the series $\sum_{k=1}^{\infty} \gamma_k^p x_k$ in B.

We denote in the same way as in (Mouchtari, 1989) by $\gamma_0(B)$ (correspondingly, $\gamma_p(B)$) the weakest topology in B' in which all the characteristic functionals of tight probabilities in B (correspondingly, the seminorms $|x'|_{y,p} = (\mathbf{E}|x'(Y)|^p)^{1/p}$, $x' \in B'$, where Y is a p-prestable random element in B) are continuous. A Banach space B is a S-space (correspondingly, of cotype p) if the continuity in topology $\gamma_0(B)$ (correspondingly, $\gamma_p(B)$) of a characteristic functional implies the tightness of the corresponding cylindrical probability.

THEOREM 3. *Let the sequence $S_n^a(Y)$ satisfy the conditions of Lemma and one of the following conditions:*

(1) *the Banach space B is of stable type p and $\|Y\|^p < \infty$;*

(2) *the Banach space B is of cotype p, $p > 1$, and the random element Y is p-prestable;*

(3) *the Banach space B is an S-space and the random element Y is p-prestable. Then $S_n^a(Y) \to \gamma_{C^{1/p}Y}^p$ as $n \to \infty$, weakly in B.*

Proof. Let us prove the tightness of the distributions of the sequence $S_n^a(Y)$. Let the Banach space B be of stable type p, $q < p$. Then, by Theorem 2.9 in (Mouchtari, 1989), we have

$$\inf_{Q\in\mathfrak{F}(B)} \sup_{n\in\mathbf{N}} (\mathbf{E}\|S_n^a(QY)\|^q)^{1/q} < C_q(B) C_1 \inf_{Q\in\mathfrak{F}(B)} (\mathbf{E}\|QY\|^p)^{1/p} = 0,$$

where the constant $C_q(B)$ depends only on B, and C_1 is from Lemma. Therefore, the family of distributions of the sequence $S_n^a(Y)$ is tight in B.

Let the Banach space B have cotype p, $p > 1$, or be a S-space. Because

$$1-\widehat{S}^a_n(Y)(x')=\mathbf{E}\Big(1-\exp\Big(-\sum_{k=1}^{k_n}|a^n_k x'(Y_k)|^p\Big)\Big)<$$

$$<\mathbf{E}\Big(\sum_{k=1}^{k_n}|a^n_k x'(Y_k)|^p\Big)<C_1\mathbf{E}|x'(Y)|^p,$$

$n\in\mathbf{N}$, $x'\in B'$, the family of characteristic functionals $\{\widehat{S}^a_n(Y)\}$ is uniformly continuous in topology γ_p. By Theorem 3.7 in (Mouchtari, 1989), the family of distributions of the sequence $S^a_n(Y)$ is tight in B.
By Theorem 2.2.2 in (Buldygin, 1980), the tightness of the distributions of the sequence $S^a_n(Y)$ and Lemma imply the weak convergence $S^a_n(Y)\to\gamma^p_{C^{1/p}Y}$ as $n\to\infty$.
If $a^n_k=n^{-1/p}$, $1\leqslant k\leqslant n$, Theorem 3 implies

COROLLARY. (1) *A Banach space B of stable type p has the* CLTp.
(2) *A Banach space B of cotype p, $p>1$, has the* CLTp.
(3) *An S-space B has the* CLTp.

Taking into consideration Theorem 2, from Corollary 1 we have the following known result.

PROPOSITION. (Hoffmann-Jorgensen and Pisier, 1976; Chobanjan and Tarieladze, 1977).

(1) *A Banach space B of type* 2 *has the* CLTp.
(2) *A Banach space B of cotype* 2 *has the* CLTp.

REFERENCES

Acosta, A. de, Gine E. (1979). Convergence of moments and related functionals in general central limit theorem in Banach spaces. *Z. Wahrsch. verw. Geb.* **48**, 213–231.

Buldygin, V.V. (1980). *Convergence of Random Elements in Banach Spaces.* Naukova dumka, Kiev (in Russian).

Chobanjan, S.A. and Tarieladze, V.I. (1977). Graussian characterizations of some Banach spaces. *J. Multivar. Anal.* **7**, 183–203.

Gnedenko, B.V. and Kolmogorov, A.N. (1954). *Limit Distributions for Sums of Independent Random Variables.* Addison Wesley, Cambridge.

Ledoux, M. and Talagrand, M. (1986). Conditions d'integrabilite pour les multiplicateur dans les TLC Banachique. *Ann. Probab.* **11**, 916–921.

Hoffmann-Jorgensen, J. and Pisier, G. (1976). The law of large numbers and the central limit theorem in Banach spaces. *Ann. Probab.* **4**, 587–599.

Mouchtari, D.H. (1989). Probabilities and Topologies in Banach Spaces. Kazan University Press, Kazan (in Russian).

Pisier, G. (1975). Le théorème de la limit centrale et la loi du logarithme iteré dans les espaces de Banach. In: *Seminaire Maurey–Schwartz,* (1975–1976). Paris. Exp. 3–4.

New Trends in Probab. and Statist., pp. 111 - 122
V. Sazonov and T. Shervashidze (Eds.)
1991 VSP/Mokslas

AN ESTIMATE OF THE RATE OF CONVERGENCE IN THE CLT FOR DEPENDENT HILBERT SPACE VALUED RANDOM VARIABLES

P. GUDYNAS
Institute of Mathematics and Informatics, Lithuanian Academy Sci., Akademijos str. 4, 232600 Vilnius, Lithuania

ABSTRACT

We consider distributions of sums of dependent random variables with values in the Hilbert space. The aim of our investigation is to estimate the accuracy of the normal approximation to these distributions in terms of conditional moments of the summands.

1. INTRODUCTION AND RESULTS

In this paper we extend to the Hilbert space some results on the rate of convergence in the Central limit theorem for real semimartingales obtained in (Liptser and Shiryaev, 1982). Let H be a real separable Hilbert space, $(\Omega, \mathcal{F}, \mathbf{P})$ be a probability space, $X_1, \ldots, X_r$ be H-valued random variables (r.v.'s), $Y_1, \ldots, Y_r$ be independent Gaussian r.v.'s. Let $\mathcal{F}_0, \mathcal{F}_1, \ldots, \mathcal{F}_r$ be an increasing sequence of sub-σ-fields of events such that X_i is $\mathcal{F}_i$-measurable for every $i = 1, \ldots, r$. Denote by $L(V, Z)$ the Lévy distance between the distributions of real r.v.'s V and Z, by $\operatorname{Cov} X$ the covariance operator of a r.v. X, by $\operatorname{Cov}(X|\ \cdot)$ the conditional covariance operator of X. Finally, set

$$\overline{X}_i := X_i - \mathbf{E}(X_i|\ \mathcal{F}_{i-1}), \quad i = 1, \ldots, r. \tag{1}$$

Our main result is

THEOREM 1. *If, for every* $i = 1, \ldots, r$,

$$\mathbf{E}|X_i|^3 < \infty, \quad \mathbf{E}Y_i = \mathbf{E}X_i, \quad \operatorname{Cov} Y_i = \operatorname{Cov} \overline{X}_i, \tag{2}$$

then there exists an absolute constant K such that

$$\sup\Big\{L\Big(|X_1 + \ldots + X_r + a|,\ |Y_1 + \ldots + Y_r + a|\Big):\ a \in H\Big\} \leqslant$$

$$\leqslant K\Big\{\Big(\mathbf{E}\Big|\sum_{i=1}^{r}\Big(\mathbf{E}(X_i|\ \mathcal{F}_{i-1}) - \mathbf{E}X_i\Big)\Big|\Big)^{1/2} + \Big(\sum_{i=1}^{r}\mathbf{E}|\overline{X}_i|^3\Big)^{1/4} +$$

$$+\left(\mathbf{E}\sup\left\{\sum_{j=1}^{\infty}\left|\sum_{i=1}^{r}\left(\mathbf{E}((\overline{X}_i,e_j)^2|\mathcal{F}_{i-1})-\mathbf{E}(\overline{X}_i,e_j)^2\right)\right|:\ \{e_j\}\right\}\right)^{1/3}\Bigg\}, \tag{3}$$

where the supremum on the right-hand side is taken over the set of all orthonormal bases $\{e_j\}$ *of* H.

Let us note that under additional assumptions a result of such type has been obtained by the author in (Gudynas, 1984). We formulate it below for comparison.

THEOREM 2. *Let*

$$\mathbf{E}(X_i|\ \mathcal{F}_{i-1})=0,\quad \mathbf{E}|X_i|^3<\infty,\quad i=1,\dots,r, \tag{4}$$

$$\sum_{i=1}^{r}\operatorname{Cov}(X_i|\ \mathcal{F}_{i-1})=\sum_{i=1}^{r}\operatorname{Cov}X_i. \tag{5}$$

Let $Y_1,\dots,Y_r$ *be independent Gaussian r.v.'s such that*

$$\mathbf{E}Y_i=0,\quad \operatorname{Cov}Y_i=\operatorname{Cov}X_i,\quad i=1,\dots,r. \tag{6}$$

Then there exists an absolute constant K_1 *such that*

$$\sup\left\{L\left(|X_1+\dots+X_r+a|,\ |Y_1+\dots+Y_r+a|\right):\ a\in H\right\}\leqslant$$

$$\leqslant K\left(\sum_{i=1}^{r}\mathbf{E}|X_i|^3\right)^{1/4}. \tag{7}$$

Unlike similar results in one-dimensional case (Liptser and Shiryaev, 1982), in this theorem we have additional restriction (5). It turned out that even in the two-dimensional Euclidean space it is not easy to give up this restriction (Gudynas, 1988). It is still more difficult to give up the restriction (5) in the case of H-valued random variables. The problem is complicated because the Hilbert space is not completely ordered. Our goal here is to overcome the before mentioned difficulty. From Theorem 1 it can be derived a similar statement for continuous time martingales.

THEOREM 3. *Let* $M=(M_s,\mathcal{F}_s)$ *be a locally square integrable* H*-valued martingale with predictable quadratic characteristic* $\langle M\rangle$ *and let* Y *be Gaussian r.v. with* $\mathbf{E}Y=0$, $\operatorname{Cov}Y=\mathbf{E}\langle M\rangle_t$. *Define* $\Delta M_s:=M_s-M_{s-}$, *where* $M_0:=M_0$, $M_{s-}:=\lim_{p\uparrow s}M_p$ *for* $s>0$. *Then*

$$\sup\left\{L\left(|M_t+a|,\ |Y+a|\right):\ a\in H\right\}\leqslant$$

$$\leqslant K_2\left(\left(\mathbf{E}\sup\left\{\sum_{j=1}^{\infty}\left|(\langle M\rangle_t e_j,e_j)-\mathbf{E}(\langle M\rangle_t e_j,e_j)\right|:\ \{e_j\}\right\}\right)^{1/3}+\right.$$

$$+\left(\mathbf{E}\sum_{0<s\leqslant t}|\Delta M_s|^3\right)^{1/4}\Bigg),$$

where K_2 is an absolute constant and the supremum on the right-hand side is taken over the set of all orthonormal bases $\{e_j\}$ of H.

2. PRELIMINARY RESULTS

Let us consider real symmetric matrices. We write $A > 0$ ($B \geqslant 0$) if a real symmetric matrix A is positive definite (B is non-negative definite). It is easy to verify the following statement.

LEMMA 1. *If $A > 0$ and $B \geqslant 0$ are symmetric $n \times n$-matrices there exists a symmetric matrix $Y \geqslant 0$ such that*

$$YAY = B. \tag{8}$$

Proof. Denote $Y := A^{-1/2}(A^{1/2}B\,A^{1/2})^{1/2}A^{-1/2}$. It is clear that $Y \geqslant 0$. We have

$$\begin{aligned} YAY &= A^{-1/2}(A^{1/2}B\,A^{1/2})^{1/2}A^{-1/2}A\,A^{-1/2}(A^{1/2}B\,A^{1/2})^{1/2}A^{-1/2} = \\ &= A^{-1/2}(A^{1/2}B\,A^{1/2})^{1/2}I(A^{1/2}B\,A^{1/2})^{1/2})A^{-1/2} = \\ &= A^{-1/2}(A^{1/2}B\,A^{1/2})A^{-1/2} = B, \end{aligned}$$

where I is the unit matrix. Hence (8) holds.
The next lemma is very important in the sequel.

LEMMA 2. *Suppose $A \geqslant 0$ and $B \geqslant 0$ are (symmetric) $n \times n$-matrices. Then there exists a $n \times n$-matrix $0 \leqslant \Lambda \leqslant I$ such that the matrices*

$$C_1 := B - \Lambda A\Lambda, \quad C_2 := (I - \Lambda)A(I - \Lambda)$$

are non-negative definite and $C_1C_2 = C_2C_1 = 0$.

Proof. Without loss of generality, let us suppose that $A > 0$. Denote by Y a solution of equation (8). It is well known that Y can be expressed in the form $S\,V\,S'$, where S is an orthogonal matrix and $V = \operatorname{diag}\{\lambda_1, \lambda_2, \ldots, \lambda_n\}$ is the diagonal matrix of the eigenvalues of Y. Denote

$$W := \operatorname{diag}\big\{\min\{1,\lambda_1\}, \min\{1,\lambda_2\}, \ldots, \min\{1,\lambda_n\}\big\}, \quad \Lambda := S\,W\,S'.$$

It is clear that $0 \leqslant \Lambda \leqslant I$ and, therefore, $C_2 \geqslant 0$. Observe next that $\Lambda = S\,Q_1 V\,S' = S\,Q_1 S' S\,V\,S' = Q\,Y$, where $0 \leqslant Q_1, Q_2 \leqslant I$. Thus, $B = Y\,\Lambda\,Y \geqslant Q_2 Y\,A\,Y\,Q_2 = \Lambda\,A\,\Lambda$. Hence, $C_1 \geqslant 0$.
Let us consider now the matrix C_1C_2. We have

$$C_1C_2 = (B - \Lambda A\Lambda)(I - \Lambda)A(I - \Lambda). \tag{9}$$

It is clear that

$$(B - \Lambda A \Lambda)(I - \Lambda) = 0 \tag{10}$$

implies $C_1 C_2 = 0$. Therefore, in order to complete the proof of the theorem it suffices to prove (10). As $C_1 \geqslant 0$ and $I - \Lambda \geqslant 0$, equality (10) follows from

$$\operatorname{tr}\big((B - \Lambda A \Lambda)(I - \Lambda)\big) = 0. \tag{11}$$

Let us prove (11). Suppose $\{e_i\}$ is the orthonormal system of the eigenelements of Y and Λ. Then

$$\operatorname{tr}\Big((B - \Lambda A\Lambda)(I - \Lambda)\Big) = \sum_{i=1}^{n} \Big((B - \Lambda A\Lambda)(I - \Lambda)e_i, e_i\Big) =$$

$$= \sum_{i=1}^{n} \Big((I - \Lambda)e_i, (B - \Lambda A\Lambda)e_i\Big). \tag{12}$$

If $\lambda_i \geqslant 1$ for some i then $(I - \Lambda)e_i = 0$ and, hence,

$$\Big((I - \Lambda)e_i, (B - \Lambda A\Lambda)e_i\Big) = 0.$$

If $\lambda_j < 1$ for some j then

$$\big(e_j, (B - \Lambda A\Lambda)e_j\big) = (e_j, Be_j) - (e_j, \Lambda A \Lambda e_j) =$$

$$= (e_j, Y A Y e_j) - (e_j, \Lambda A \Lambda e_j) = (Y e_j, A Y e_j) - (\Lambda e_j, A \Lambda e_j) =$$

$$= \lambda_j^2 (e_j, Ae_j) - \lambda_j^2 (e_j, Ae_j) = 0$$

and, hence,

$$\Big((I - \Lambda)e_j, (B - \Lambda A \Lambda)e_j\Big) = (1 - \lambda_j)\Big(e_j, (B - \Lambda A \Lambda)e_j\Big) = 0.$$

Now, from (12) it follows (11). This ends the proof of the theorem.
If $\mathcal{F}$ is the σ–field generated by a finite partition $\{D_i\}$ of Ω, we call $\mathcal{F}$ a finite σ–field.

COROLLARY 1. *Let $\mathcal{G}$ be a finite sub–σ–field of the σ–field $\mathcal{F}$, X be a simple (with the finite set of values) $\mathcal{F}$–measurable H–valued* r.v. *and let* $\mathbf{E}(X|\,\mathcal{G}) = 0$. *Let $G = G(\omega)$ be a random $\mathcal{G}$–measurable operator taking its values in the space of non–negative symmetric bounded linear operators of finite rank on H. Then there exists a random $\mathcal{G}$–measurable symmetric operator Θ of finite rank such that*

(a) $0 \leqslant \Theta \leqslant I$;

(b) $\operatorname{Cov}(\Theta X|\,\mathcal{G}) \leqslant G$;

(c) $\operatorname{Cov}\Big((I - \Theta)X|\,\mathcal{G}\Big)\Big(G - \operatorname{Cov}(\Theta X|\,\mathcal{G})\Big) = 0$.

Remark. It is clear that $\mathbf{E}(\Theta X|\,\mathcal{G}) = 0$.

Proof of the corollary. $\mathcal{G}$ is a finite σ–field. Therefore, G and $\operatorname{Cov}(X|\,\mathcal{G})$ are simple r.v.'s. There exists a finite partition $\{D_i,\ i = 1,\dots,m\}$ of Ω such that on each D_i the r.v.'s G and $\operatorname{Cov}(X|\,\mathcal{G})$ take constant values.
By Lemma 2, for nonrandom values $A_i = \operatorname{Cov}\big(X|\,\mathcal{G}\big)(w_i)$, $B_i = G(w_i)$, $w_i \in D_i$, there exists a symmetric operator Q_i of finite rank such that $0 \leqslant Q_i \leqslant I$,

$$\operatorname{Cov}\big(Q_i X|\,\mathcal{G}\big)(w_i) = Q_i A_i Q_i \leqslant B_i = G(w_i), \tag{13}$$

and

$$\Big(\operatorname{Cov}\big((I - Q_i)X|\,\mathcal{G}\big)\big(G - \operatorname{Cov}(Q_i X|\,\mathcal{G})\big)\Big)(w_i) =$$

$$= \Big((I - Q_i)A_i(I - Q_i)\Big)\big(B_i - Q_i A\, Q_i\big) = 0. \tag{14}$$

To complete the proof, put $\Theta(w) := Q_i$ for all $w \in D_i$, $i = 1,\dots,m$. It is not hard to see that the operator Θ defined in this way satisfies conditions **(a)**, **(b)** and **(c)**. The corollary is proved.
Below we shall need some results obtained in (Gudynas, 1984). Let us formulate them here in a slightly modified form.
Let s be a positive number, m be an integer such that $m < s \leqslant m + 1$ and $\alpha := s - m$. Define the distance η between the distributions of r.v.'s X and Y (see (Gudynas, 1983))

$$\eta(X,Y;s) := \sup\Big\{\big|\mathbf{E}\big(f(X) - f(Y)\big)\big| :\ f \in \mathcal{C}(s)\Big\}, \tag{15}$$

where $\mathcal{C}(s)$ is the set of all real functions f defined on H and satisfying the following condition: There exist derivatives $f^{(j)}(x)$, $j = 1,\dots,m$, (in the Frechet sense) at all points $x \in H$ and

$$\sup\Big\{\big|f^{(j)}(x)\big|:\ x \in H\Big\} \leqslant 1, \quad j = 0,\dots,m, \quad \big(f^{(0)} := f\big), \tag{16}$$

$$\sup\Big\{\big|f^{(m)}(x) - f^{(m)}(y)\big| / \big|x - y\big|^{\alpha}:\ x, y \in H\Big\} \leqslant 2^{1-\alpha}. \tag{17}$$

Let us also define the distance

$$\bar{\eta}(F,G;s) = \sup\left\{\left|\int\limits_H f(x)(F - G)(dx)\right| :\ f \in \mathcal{C}(s)\right\}, \tag{18}$$

between every pair F, G of probability distributions on H.

LEMMA 3 (Gudynas, 1984). *Let $X_1,\dots,X_r$, $Y_1,\dots,Y_r$ be H–valued r.v.'s. Let there exist a system of σ–fields $\mathcal{F}_0 \subset \mathcal{F}_1 \subset \dots \subset \mathcal{F}_r$ such that X_i is $\mathcal{F}_i$–measurable, for every $i = 1,\dots,r$, and, for every $i = 0,1,\dots,i-2$, the random variables X_{i+1} and Y_{i+1} are $\mathcal{F}_i$–conditionally independent of the r.v.'s $Y_{i+2} + {}+Y_{i+3}\dots + Y_r$. Then, for each $s > 0$,*

$$\eta(X_1 + \ldots + X_r,\ Y_1 + \ldots + Y_r;\ s) \leqslant$$

$$\leqslant \sum_{i=1}^{r} \mathbf{E}\Big(\bar{\eta}(F_{X_i}(\cdot \mid \mathcal{F}_{i-1}),\ F_{Y_i}(\cdot \mid \mathcal{F}_{i-1});\ s)\Big), \tag{19}$$

where $F_{X_i}(\cdot \mid \mathcal{F}_{i-1})$ *and* $F_{Y_i}(\cdot \mid \mathcal{F}_{i-1})$ *are the conditional distributions of* X_i *and* Y_i.

Let X, Y be H-valued r.v.'s, with distributions F_X, F_Y, respectively. Suppose $\mathbf{E}X = \mathbf{E}Y$, $\operatorname{Cov} X = \operatorname{Cov} Y$. Then (see (Gudynas, 1984))

$$\bar{\eta}(F_X, F_Y; 3) = \eta(X, Y; 3) \leqslant (\mathbf{E}|X|^3 + \mathbf{E}|Y|^3)/3!. \tag{20}$$

For any r.v.'s X, Y and any real number $s > 0$, we have (see (Gudynas, 1984))

$$\sup\Big\{L^{1+s}(|X + a|,\ |Y + a|)\colon\ a \in H\Big\} \leqslant K_3 \eta(X, Y; s), \tag{21}$$

where K_3 depends only on s.
If X and Y are real r.v.'s, then (see (Liptser and Shiryaev, 1982))

$$L(X, Y) \leqslant K(s)(\mathbf{E}|X - Y|^s)^{1/(s+1)}, \tag{22}$$

where $K(s)$ depends only on s.

LEMMA 4 (Gudynas, 1988). *Let* X *be a* H*-valued r.v.,* $\mathbf{E}X = 0$, Θ *be a symmetric linear operator,* $0 \leqslant \Theta \leqslant I$. *Put* $Z := \Theta X$. *Then*

$$\operatorname{tr}\operatorname{Cov}(X - Z) \leqslant \operatorname{tr}\operatorname{Cov} X - \operatorname{tr}\operatorname{Cov} Z.$$

Proof. Put $A := \operatorname{Cov} X$. We have

$$\operatorname{Cov}(X - Z) = \operatorname{Cov}\big((I - \Theta)X\big) = (I - \Theta)A(I - \Theta) =$$

$$= A - \Theta A - A\Theta + \Theta A \Theta.$$

As $\operatorname{tr} BC = \operatorname{tr} CB$, so

$$\operatorname{tr}\operatorname{Cov}(X - Z) = \operatorname{tr} A - \operatorname{tr}(\Theta A \Theta) - 2\operatorname{tr}(\Theta A) + 2\operatorname{tr}(\Theta^2 A) =$$

$$= \operatorname{tr} A - \operatorname{tr}(\Theta A \Theta) - 2\operatorname{tr}\big((\Theta - \Theta^2)A\big). \tag{23}$$

Since $\Theta - \Theta^2 \geqslant 0$, there exists $(\Theta - \Theta^2)^{1/2}$. Hence,

$$\operatorname{tr}\big((\Theta - \Theta^2)A\big) = \operatorname{tr}\big((\Theta - \Theta^2)^{1/2} A (\Theta - \Theta^2)^{1/2}\big) \geqslant 0.$$

In view of the last inequality, the lemma follows from (23).

3. PROOF OF THEOREM 1

Let us begin with less complicated case of simple random variables and finite σ-fields. This will help us to avoid problems connected with measurability. Later we shall extend the proof to the general case.

1. *Discrete case*

Let in Theorem 1, $\mathcal{F}_1, \dots, \mathcal{F}_r$ be finite σ-fields. Then, obviously, $X_1, \dots, X_r$ and $\overline{X}_1, \dots, \overline{X}_r$ are simple r.v.'s.

Let us define new random variables $Z_1, \dots, Z_r$. Put $G_1 := \operatorname{Cov}(\overline{X}_1 + \dots + \overline{X}_r)$. It is clear that the operator G_1 is of finite rank. By Corollary 1, there exists a $\mathcal{F}_0$-measurable random operator Θ_1 taking values in the set of symmetric operators of finite rank such that $0 \leqslant \Theta_1 \leqslant I$ and the random variable $Z_1 := \Theta_1 \overline{X}_1$ satisfies the conditions

$$\operatorname{Cov}(Z_1 |\, \mathcal{F}_0) \leqslant G_1, \quad \operatorname{Cov}\Big(\overline{X}_1 - Z_1 |\, \mathcal{F}_0\Big)\Big(G_1 - \operatorname{Cov}(Z_1 |\, \mathcal{F}_0)\Big) = 0.$$

Suppose that we have defined r.v.'s $Z_1, \dots, Z_{i-1}$, $1 < i \leqslant r$. Put

$$G_i := G_1 - \Big(\operatorname{Cov}(Z_1 |\, \mathcal{F}_0) + \dots + \operatorname{Cov}(Z_{i-1} |\, \mathcal{F}_{i-2})\Big).$$

Obviously all values of G_i are symmetric operators of finite rank.

Therefore, by Corollary 1, there exists a $\mathcal{F}_{i-1}$-measurable random operator Θ_i taking values in the set of symmetric operators of finite rank such that $0 \leqslant \Theta_i \leqslant I$ and the random variable $Z_i := \Theta_i X_i$ satisfies the conditions

$$\operatorname{Cov}(Z_i |\, \mathcal{F}_{i-1}) \leqslant G_1 - \Big(\operatorname{Cov}(Z_1 |\, \mathcal{F}_0) + \dots + \operatorname{Cov}(Z_{i-1} |\, \mathcal{F}_{i-2})\Big) = G_i,$$

$$\operatorname{Cov}(\overline{X}_i - Z_i |\, \mathcal{F}_{i-1})\Big(G_i - \operatorname{Cov}(Z_i |\, \mathcal{F}_{i-1})\Big) = 0.$$

Thus, we have defined random variables $Z_1, \dots, Z_r$ with the following properties: For every $i = 1, \dots, r$, Z_i is $\mathcal{F}_i$-measurable and

$$\mathbf{E}(Z_i |\, \mathcal{F}_{i-1}) = 0, \quad \mathbf{E}|Z_i|^3 \leqslant \mathbf{E}|\overline{X}_i|^3, \tag{24}$$

$$\operatorname{Cov}(\overline{X}_i - Z_i |\, \mathcal{F}_{i-1})\Big(\sum_{i=1}^{r} \operatorname{Cov} \overline{X}_j - \sum_{i=1}^{i} \operatorname{Cov}(Z_j |\, \mathcal{F}_{j-1})\Big) = 0, \tag{25}$$

$$\sum_{i=1}^{r} \operatorname{Cov}(Z_i |\, \mathcal{F}_{i-1}) \leqslant \sum_{i=1}^{r} \operatorname{Cov} \overline{X}_i. \tag{26}$$

Let W be independent Gaussian r.v. and let $\mathbf{E}\, W = 0$, $\operatorname{Cov} W = I$. Define

$$Z_{r+1} := G_{r+1}^{1/2} W = \Big(G_1 - \sum_{i=1}^{r} \operatorname{Cov}(Z_i |\, \mathcal{F}_{i-1})\Big)^{1/2} W. \tag{27}$$

Obviously,

$$\mathbf{E}\big(Z_{r+1}|\ \mathcal{F}_r\big) = 0,$$

$$\sum_{i=1}^{r+1} \mathrm{Cov}\left(Z_i|\ \mathcal{F}_{i-1}\right) = \mathrm{Cov} \sum_{i=1}^{r+1} Z_i = \mathrm{Cov}\left(\overline{X}_1 + \ldots + \overline{X}_r\right). \tag{28}$$

Let us now return directly to the proof. Put $a_1 := a + \mathbf{E}X_1 + \ldots + \mathbf{E}X_r$, $\overline{Y}_i := Y_i - \mathbf{E}X_i = Y_i - \mathbf{E}Y_i$. By the triangle inequality,

$$L\Big(|X_1 + \ldots + X_r + a|,\ |Y_1 + \ldots + Y_r + a|\Big) =$$

$$= L\Big(\Big|\sum_{i=1}^{r} (X_i - \mathbf{E}X_i) + a_1\Big|,\ \Big|\sum_{i=1}^{r} (Y_i - \mathbf{E}Y_i) + a_1\Big|\Big) \leqslant J_1 + J_2 + J_3 + J_4, \tag{29}$$

where

$$J_1 = L\Big(\Big|\sum_{i=1}^{r} (X_i - \mathbf{E}X_i) + a_1\Big|,\ \Big|\sum_{i=1}^{r} \overline{X}_i + a_1\Big|\Big),$$

$$J_2 = L\Big(\big|\overline{X}_1 + \ldots + \overline{X}_r + a_1\big|,\ \big|Z_1 + \ldots + Z_r + a_1\big|\Big),$$

$$J_3 = L\Big(\big|Z_1 + \ldots + Z_r + a_1\big|,\ \big|Z_1 + \ldots + Z_r + Z_{r+1} + a_1\big|\Big),$$

$$J_4 = L\Big(\big|Z_1 + \ldots + Z_{r+1} + a_1\big|,\ \big|Y_1 + \ldots + Y_r + a_1\big|\Big).$$

By (22) and the inequality

$$\big||a + c| - |b + c|\big| \leqslant |a - b|, \tag{30}$$

which holds for all $a, b, c \in H$, it follows

$$J_1 \leqslant K(1)\left(\mathbf{E}\Big|\Big|\sum_{i=1}^{r}(X_i - \mathbf{E}X_i) + a_1\Big| - \Big|\sum_{i=1}^{r} \overline{X}_i + a_1\Big|\Big|\right)^{1/2} \leqslant$$

$$\leqslant K(1)\left(\mathbf{E}\Big|\sum_{i=1}^{r} (X_i - \mathbf{E}X_i) - \sum_{i=1}^{r} \overline{X}_i\Big|\right)^{1/2} =$$

$$= K(1)\left(\mathbf{E}\Big|\sum_{i=1}^{r} (\mathbf{E}(X_i|\ \mathcal{F}_{i-1}) - \mathbf{E}X_i)\Big|\right)^{1/2}. \tag{31}$$

Similarly,

$$J_2 \leqslant K(2)\Big(\mathbf{E}\big||\overline{X}_1 + \ldots + \overline{X}_r + a_1| - |Z_1 + \ldots + Z_r + a_1|\big|^2\Big)^{1/3} \leqslant$$

$$\leqslant K(2)\Big(\mathbf{E}\big|\overline{X}_1 + \ldots + \overline{X}_r - (Z_1 + \ldots + Z_r)\big|^2\Big)^{1/3} =$$

$$= K(2)\left(\mathbf{E}\left|\sum_{i=1}^{r}(\overline{X}_i - Z_i)\right|^2\right)^{1/3}. \tag{32}$$

Since $\mathbf{E}(\overline{X}_i - Z_i |\ \mathcal{F}_{i-1}) = 0,\ i = 1, \ldots, r$, so

$$\mathbf{E}\left|\sum_{i=1}^{r}(\overline{X}_i - Z_i)\right|^2 = \sum_{i=1}^{r}\mathbf{E}|\overline{X}_i - Z_i|^2 = \mathbf{E}\sum_{i=1}^{r}\mathbf{E}\Big(|\overline{X}_i - Z_i|^2 |\ \mathcal{F}_{i-1}\Big). \tag{33}$$

From Lemma 4,

$$\mathbf{E}\sum_{i=1}^{r}\mathbf{E}\Big(|\overline{X}_i - Z_i|^2 |\ \mathcal{F}_{i-1}\Big) = \mathbf{E}\sum_{i=1}^{r}\operatorname{tr}\operatorname{Cov}(\overline{X}_i - Z_i |\ \mathcal{F}_{i-1}) \leqslant$$

$$\leqslant \mathbf{E}\sum_{i=1}^{r}\Big(\operatorname{tr}\operatorname{Cov}(\overline{X}_i |\ \mathcal{F}_{i-1}) - \operatorname{tr}\operatorname{Cov}(Z_i |\ \mathcal{F}_{i-1})\Big) =$$

$$= \mathbf{E}\operatorname{tr}\left(\sum_{i=1}^{r}\operatorname{Cov}\overline{X}_i - \sum_{i=1}^{r}\operatorname{Cov}(Z_i |\ \mathcal{F}_{i-1})\right). \tag{34}$$

Let $N_i = N_i(\omega)$ be the zero–subspace of the operator

$$D_i := \sum_{j=1}^{r}\operatorname{Cov}\overline{X}_j - \sum_{j=1}^{i}\operatorname{Cov}(Z_j |\ \mathcal{F}_{j-1}). \tag{35}$$

By (26), $N_1 \subset N_2 \subset \ldots \subset N_{r-1} \subset N_r$. Let Q_i be the support of D_i. Note that $N_i \times Q_i = H$. Then, by (35),

$$Q_r \subset Q_{r-1} \subset \ldots \subset Q_1.$$

Let $\{e_j\}'$ be an orthonormal basis of Q_r. It follows from (25) that, for every $e_j \in \{e_j\}'$, $i = 1, \ldots, r$,

$$\mathbf{E}\Big((Z_i, e_j)^2 |\ \mathcal{F}_{i-1}\Big) = \mathbf{E}\Big((\overline{X}_i, e_j)^2 |\ \mathcal{F}_{i-1}\Big).$$

Therefore,

$$\operatorname{tr} D_r = \operatorname{tr}\left(\sum_{i=1}^{r}\operatorname{Cov}\overline{X}_i - \sum_{i=1}^{r}\operatorname{Cov}(Z_i |\ \mathcal{F}_{i-1})\right) =$$

$$= \sum_{\{e_j\}'}\sum_{i=1}^{r}\Big(\mathbf{E}(\overline{X}_i |\ e_j)^2 - \mathbf{E}((Z_i, e_j)^2 |\ \mathcal{F}_{i-1})\Big) =$$

$$= \sum_{\{e_j\}'}\sum_{i=1}^{r}\Big(\mathbf{E}(\overline{X}_i, e_j)^2 - \mathbf{E}((\overline{X}_i, e_j)^2 |\ \mathcal{F}_{i-1})\Big) \leqslant$$

$$\leqslant \sum_{\{e_j\}'} \left| \sum_{i=1}^{r} \left(\mathbf{E}((\overline{X}_i, e_j)^2 | \mathcal{F}_{i-1}) - \mathbf{E}(\overline{X}_i, e_j)^2 \right) \right|.$$

Hence,

$$\operatorname{tr} D_r \leqslant \sup \left\{ \sum_{j=1}^{\infty} \left| \sum_{i=1}^{r} \left(\mathbf{E}((\overline{X}_i, e_j)^2 | \mathcal{F}_{i-1}) - \mathbf{E}(\overline{X}_i, e_j)^2 \right) \right| : \{e_j\} \right\}, \tag{36}$$

where the supremum is taken over the set of all orthonormal bases $\{e_j\}$ of H. From (32), (33), (34) and (36), it follows that

$$J_2 \leqslant K(2) \left(\mathbf{E} \sup \left\{ \sum_{j=1}^{\infty} \left| \sum_{i=1}^{r} \left(\mathbf{E}(\overline{X}_i, e_j)^2 | \mathcal{F}_{i-1}) - \mathbf{E}(\overline{X}_i, e_j)^2 \right) \right| : \{e_j\} \right\} \right)^{1/3}. \tag{37}$$

Next, from (22), (27), (30) and (36), we conclude that

$$J_3 \leqslant K(2) \left(\mathbf{E} |Z_{r+1}|^3 \right)^{1/3} = K(2) \left(\operatorname{tr} \operatorname{Cov} Z_{r+1} \right)^{1/3} =$$

$$= K(2) \left(\mathbf{E} \operatorname{tr} \left(\sum_{i=1}^{r} \operatorname{Cov} \overline{X}_i - \sum_{i=1}^{r} \operatorname{Cov} (Z_i | \mathcal{F}_{i-1}) \right) \right)^{1/3} \leqslant$$

$$\leqslant K(2) \left(\mathbf{E} \sup \left\{ \sum_{j=1}^{\infty} \left| \sum_{i=1}^{r} \left(\mathbf{E}((\overline{X}_i, e_j)^2 | \mathcal{F}_{i-1}) - \mathbf{E}(\overline{X}_i, e_j)^2 \right) \right| : \{e_j\} \right\} \right)^{1/3}. \tag{38}$$

Now, it remains to evaluate J_4 in (29).
Let $W_1, \ldots, W_{r+1}$ be independent Gaussian r.v.'s and let $\mathbf{E}W_i = 0$, $\operatorname{Cov} W_i = I$, $i = 1, \ldots, r+1$. Put

$$V_i := \left(\operatorname{Cov} (Z_i | \mathcal{F}_{i-1}) \right)^{1/2} W_i, \quad i = 1, \ldots, r+1. \tag{39}$$

It can be easily verified that the r.v.'s $Y_1 + \ldots + Y_r$ and $V_1 + \ldots + V_{r+1}$ have the same Gaussian distribution with zero expectation and the covariance operator G_1. Therefore, from (21) and (39), we obtain

$$J_4 = L\left(|Z_1 + \ldots + Z_{r+1} + a_1|, \ |Y_1 + \ldots + Y_r + a_1| \right) \leqslant$$

$$\leqslant K_2^{1/4} \eta^{1/4} (Z_1 + \ldots + Z_{r+1}, \ V_1 + \ldots + V_{r+1}; 3). \tag{40}$$

Observe that

$$\mathbf{E} \left(\sum_{j=i+2}^{r+1} V_j \middle| \mathcal{F}_i \right) = 0, \tag{41}$$

$$\operatorname{Cov}\Big(\sum_{j=i+2}^{r+1} V_j \Big| \mathcal{F}_i\Big) = \sum_{j=i+2}^{r+1} \operatorname{Cov}\big(Z_j | \mathcal{F}_{j-1}\big). \tag{42}$$

By (28),

$$\operatorname{Cov}\Big(\sum_{j=i+2}^{r+1} V_j \Big| \mathcal{F}_r\Big) = \sum_{j=1}^{r} \operatorname{Cov}\overline{X}_j - \sum_{j=1}^{i+1} \operatorname{Cov}\big(Z_j | \mathcal{F}_{j-1}\big). \tag{43}$$

Hence, the $\mathcal{F}_i$–conditional distribution of the r.v. $V_{i+2}+V_{i+3}\ldots+V_{r+1}$ is Gaussian and this r.v. depends only on the σ–field $\mathcal{F}_i$.
Therefore, for every $i = 0, 1, \ldots, r-1$, the random variable $V_{i+2} + \ldots + V_{r+1}$ is $\mathcal{F}_i$–conditionally independent of the random variables Z_{i+1} and V_{i+1}. Lemma 3 shows that

$$\eta(Z_1 + \ldots + Z_{r+1},\ V_1 + \ldots + V_{r+1}; 3) \leqslant$$

$$\leqslant \sum_{i=1}^{r+1} \mathbf{E}\overline{\eta}\Big(F_{Z_i}(\,\cdot \mid \mathcal{F}_{i-1}),\ F_{V_i}(\,\cdot \mid \mathcal{F}_{i-1}); 3\Big), \tag{44}$$

where $F_{Z_i}(\cdot \mid \cdot)$ and $F_{V_i}(\cdot \mid \cdot)$ are the conditional distributions of Z_i and V_i. By (20), for $i = 1, \ldots, r$,

$$\mathbf{E}\overline{\eta}\Big(F_{Z_i}(\,\cdot \mid \mathcal{F}_{i-1}),\ F_{V_i}(\,\cdot \mid \mathcal{F}_{i-1}); 3\Big) \leqslant$$

$$\leqslant \mathbf{E}\Big(\mathbf{E}\big(|Z_i|^3 \mid \mathcal{F}_{i-1}\big) + \mathbf{E}\big(|V_i|^3 \mid \mathcal{F}_{i-1}\big)\Big)/3! = \big(\mathbf{E}|Z_i|^3 + \mathbf{E}|V_i|^3\big)/3! \tag{45}$$

As the $\mathcal{F}_{i-1}$–conditional distribution of V_i is Gaussian with the covariance operator $\operatorname{Cov}\big(Z_i | \mathcal{F}_{i-1}\big)$, so there exists an absolute constant K_4 such that

$$\mathbf{E}|V_i|^3 = \mathbf{E}\mathbf{E}\big(|V_i|^3 \mid \mathcal{F}_{i-1}\big) \leqslant K_4 \mathbf{E}\mathbf{E}^{3/2}\big(|Z_i|^2 \mid \mathcal{F}_{i-1}\big) \leqslant$$

$$\leqslant K_4 \mathbf{E}\mathbf{E}^{3/2}\big(|\overline{X}_i|^2 \mid \mathcal{F}_{i-1}\big) \leqslant K_4 \mathbf{E}|\overline{X}_i|^3.$$

Therefore, by (24) and (45),

$$\mathbf{E}\overline{\eta}\Big(F_{Z_i}(\,\cdot \mid \mathcal{F}_{i-1}),\ F_{V_i}(\,\cdot \mid \mathcal{F}_{i-1}); 3\Big) \leqslant K_5 \mathbf{E}|\overline{X}_i|^3, \tag{46}$$

where K_5 is an absolute constant. Obviously,

$$\mathbf{E}\overline{\eta}\Big(F_{Z_{r+1}}(\,\cdot \mid \mathcal{F}_r),\ F_{V_{r+1}}(\,\cdot \mid \mathcal{F}_r); 3\Big) = 0. \tag{47}$$

From (40), (44), (46) and (47), it follows

$$J_4 \leqslant K_6 \Big(\sum_{i=1}^{r} \mathbf{E}|\overline{X}_i|^3\Big)^{1/4}. \tag{48}$$

Now, from (29), (31), (37), (38) and (48), it follows the theorem in the discrete case.

2. General case

There are two possibilities. First, let all the σ–fields be of countable type. Then there exist sequences $\{\mathcal{F}_i^{(n)}\}$ of finite σ–fields such that $\mathcal{F}_i$ is generated by $\{\mathcal{F}_i^{(n)}, n = 1,2,\dots\}$, $i = 0,\dots,r$. Moreover, it is possible to choose such sequences $\{\mathcal{F}_i^{(n)}\}$ that, for every $n = 1,2,\dots$, $\mathcal{F}_0^{(n)} \subset \mathcal{F}_1^{(n)} \subset \dots \subset \mathcal{F}_{r-1}^{(n)} \subset \mathcal{F}_r^{(n)}$.
Put $X_i^{(n)} := \mathbf{E}(X_i |\, \mathcal{F}_i^{(n)})$. Then $\lim_{n\to\infty} X_i^{(n)} = X_i$ in $L^3(\Omega)$. Analogously converge the conditional expectations

$$\mathbf{E}\Big(X_i^{(n)} |\, \mathcal{F}_{i-1}^{(n)}\Big) = \mathbf{E}\Big(X_i |\, \mathcal{F}_{i-1}^{(n)}\Big),$$

$$\mathbf{E}\Big((X_i^{(n)}, e_j) |\, \mathcal{F}_{i-1}^{(n)}\Big) = \mathbf{E}\Big((\overline{X}_i, e_j) |\, \mathcal{F}_{i-1}^{(n)}\Big)$$

and expectation $\mathbf{E}|\overline{X}_i^{(n)}|^3$. Therefore, the statement of theorem in this case follows from earlier proved statement as $n \to \infty$.
Consider now the case of arbitrary σ–fields. Put

$$\mathcal{C}_i := \sigma\Big\{X_i,\ \mathbf{E}(X_i |\, \mathcal{F}_{i-1}),\ \mathrm{Cov}\,(\overline{X}_i |\, \mathcal{F}_{i-1})\Big\}, \quad i = 0,\dots,r-1,$$

where $X_0 := 0$. Define the σ–fields

$$\widehat{\mathcal{F}}_i := \sigma\{\mathcal{C}_0, \mathcal{C}_1, \dots, \mathcal{C}_i\}, \quad i = 0, 1, \dots, r.$$

Obviously, they are of countable type. Moreover, $\widehat{\mathcal{F}}_0 \subset \widehat{\mathcal{F}}_1 \subset \dots \subset \widehat{\mathcal{F}}_{r-1} \subset \widehat{\mathcal{F}}_r$, and for every $i = 1,\dots,r$,

$$\mathbf{E}(X_i |\, \widehat{\mathcal{F}}_i) = X_i, \quad \mathbf{E}(X_i |\, \mathcal{F}_{i-1}) = \mathbf{E}(X_i |\, \mathcal{F}_{i-1}),$$

$$\mathbf{E}\Big((\overline{X}_i, e_j) |\, \widehat{\mathcal{F}}_{i-1}\Big) = \mathbf{E}\Big((\overline{X}_i, e_j) |\, \mathcal{F}_{i-1}\Big), \quad j = 1, 2\dots$$

Thus the theorem in the general case follows immediately from the foregoing part of the proof.

REFERENCES

Liptser, R. and Shiryaev, A. (1982). On the rate of convergence in the central limit theorem for semimartingales. *Theory Probab. Appl.* **27**, 1–13.

Gudynas, P. (1983). On approximation of distributions of sums of dependent Banach space valued random variables. *Lithuanian Math. J.* **23**, 265–275.

Gudynas, P. (1984). On the approximation by means of distributions of sums of conditionaly independent random variables. *Lithuanian Math. J.* **24**, 320–325.

Gudynas, P. (1988). An estimation of the rate of convergence in CLT for dependent two–dimensional random vectors. *Lithuanian Math. J.* **28**, 47–54.

New Trends in Probab. and Statist., pp. 123 - 136
V. Sazonov and T. Shervashidze (Eds.)
1991 VSP/Mokslas

AN IMPROVED ESTIMATE OF THE ACCURACY OF GAUSSIAN APPROXIMATION IN HILBERT SPACE

V.V. SAZONOV and V.V. ULYANOV
Steklov Mathematical Institute, 117966 Moscow, USSR,
Moscow State University, 119899 Moscow, USSR

ABSTRACT

We construct a new estimate of the rate of convergence in the Central Limit Theorem in Hilbert space for the class of all balls. This estimate is considerably better than our previous result (Sazonov et al., 1989a) for balls with large distances between the origin and the boundary of the ball.

1. INTRODUCTION AND THE MAIN RESULT

This paper continues our investigations carried jointly with B.A. Zalesskii and published in (Sazonov et al., 1989a).
Let $X_1, X_2, \ldots$ be independent identically distributed random variables with values in a separable Hilbert space H. Assume that $\mathbf{E}|X_1|^2 < \infty$ and denote by V the covariance operator of X_1: $(Vx, y) = \mathbf{E}(X_1 - \mathbf{E}X_1, x)(X_1 - \mathbf{E}X_1, y)$. Define $S_n = n^{-1/2}\sigma^{-1}\sum_{i=1}^n (X_i - \mathbf{E}X_i)$, where $\sigma^2 = \mathbf{E}|X_1 - \mathbf{E}X_1|^2$ and let Y be $(0, \sigma^{-2}V)$ Gaussian H-valued random variable. For $a \in H$, $r \geqslant 0$ put $B_r(a) = \{x \in H: |x - a| < r\}$. In (Sazonov et al., 1989a) it was proved that

$$\Delta_n(a,r) = \left|\mathbf{P}(S_n \in B_r(a)) - \mathbf{P}(Y \in B_r(a))\right| \leqslant$$

$$\leqslant c\left(\prod_{i=1}^{6}\sigma_i^{-1}\right)\sigma^3\beta\left(1 + |a|^3\right)n^{-1/2}, \tag{1}$$

where c is an absolute constant, $\sigma_1^2 \geqslant \sigma_2^2 \geqslant \ldots$ are the eigenvalues of V and $\beta = \mathbf{E}|X_1 - \mathbf{E}X_1|^3$. In (Sazonov et al., 1989a) it was also pointed out that (1) is a precise estimate in the sense that it is impossible to construct an estimate which would be better than (1) for all balls $B_r(a)$. This does not mean of course that it is impossible to improve (1) for some classes of balls. In the present paper we construct an estimate which is considerably better than (1) when the distance $d = \big|r - |a|\big|$ between the origin and the boundary of $B_r(a)$ is large. Namely, we prove the following

THEOREM. *For any $a \in H$, $r \geqslant 0$,*

$$\Delta_n(a,r) \leqslant c_1\Bigg(\sigma^{-3}\beta(1+d)^{-3}n^{-1/2}+$$

$$+\exp\{-c_2d^\alpha\}\left(\prod_{i=1}^{6}\sigma_i^{-1}\right)\sigma^3\beta\Big(1+\min\left(|a|^3,r^3\right)\Big)n^{-1/2}\Bigg), \tag{2}$$

where α, c_1, c_2 are absolute constants.

From our proof it follows that $\alpha \geqslant 1/5$.

COROLLARY. *For any $a \in H$, $r \geqslant 0$,*

$$\Delta_n(a,r) \leqslant c\left(\prod_{i=1}^{6}\sigma_i^{-1}\right)\left(1+\min\left(|a|^3,r^3\right)\right)\sigma^3\beta(1+d)^{-3}n^{-1/2},$$

where c is an absolute constant.

Remarks. 1. The above Corollary was announced in (Sazonov et al., 1989b). 2. Without loss of generality, we can assume $\sigma = 1$. It is enough to prove Theorem in the case

$$c \leqslant d \leqslant 12\ln\Big(n^{1/2}\Lambda/\left(\beta\left(1+\min\left(|a|^3,r^3\right)\right)\right)\Big), \tag{3}$$

where c is an absolute constant, $c \geqslant 45$, and $\Lambda = \prod_{i=1}^{6}\sigma_i$.
Indeed, if $d \leqslant c$ then (2) follows from the Theorem in (Sazonov et al., 1989a). Assume now that the second inequality in (3) is violated. By Lemmas 1 and 2 in (Sazonov et al., 1988), we have (cf. (63) in (Sazonov et al., 1988))

$$\Delta_n(a,r) \leqslant c\beta(1+d)^{-3}n^{-1/2} + \left|\mathbf{P}\left(S_n\left(\sqrt{n}\right) \in B_r(a)\right) - \mathbf{P}\big(Y \in B_r(a)\big)\right|, \tag{4}$$

where $S_n(u) = n^{-1/2}\sum_1^n X_i\chi_i(u)$ and $\chi_i(u)$ is the indicator of $\{|X_i| < u\}$, $u > 0$. Obviously,

$$\left|\mathbf{P}\left(S_n\left(\sqrt{n}\right) \in B_r(a)\right) - \mathbf{P}\big(Y \in B_r(a)\big)\right| \leqslant$$

$$\leqslant \mathbf{P}\left(\left|S_n\left(\sqrt{n}\right)\right| \geqslant d\right) + \mathbf{P}\left(|Y| \geqslant d\right). \tag{5}$$

Furthermore,

$$\mathbf{P}\left(|Y| \geqslant d\right) \leqslant 2\exp\left(-d^2/2\right) \tag{6}$$

(see, e.g., Lemma 3 in (Sazonov et al., 1988)) and, by Lemma 2.1 in (Bentkus, 1984),

$$\mathbf{E}\exp\left(\left|S_n\left(\sqrt{n}\right)\right|/11\right) \leqslant c < \infty. \tag{7}$$

Inequalities (4)–(7) imply (2) when the second inequality in (3) is violated.
In what follows we will assume that condition (3) is satisfied.
The proof of the Theorem is based on a series of lemmas which will be proved in the next section.

2. MAIN NOTATION AND AUXILIARY LEMMAS

Recall the notation from (Sazonov et al., 1989a) which will be used below.
Let $\mathbf{P}$ be the probability measure on H corresponding to X_1. Fix n and define

$$P_1(E) = \mathbf{P}(E \cap B_{n^{1/2}}) + \mathbf{P}(B^c_{n^{1/2}})\,\delta_0(E),$$

where δ_x, $x \in H$, is the probability measure concentrated at x, and E^c denotes the complement of a set E, $B_r = B_r(0)$. Put

$$s = 3\sigma_5^{-2}\beta, \quad R = \left(\sigma_6^{-8}\beta^3 n\right)^{1/5} \tag{8}$$

and

$$P_2(E) = \mathbf{P}(E \cap B_R)/\mathbf{P}(B_R),$$
$$P_2'(E) = \mathbf{P}(E \cap B_s)/\mathbf{P}(B_s),$$
$$P_{21}(E) = P_2(E \cap B_{n^{1/2}}) + P_2(B^c_{n^{1/2}})\,\delta_0(E).$$

If

$$\Lambda^{-1}\beta n^{-1/2} \leqslant 3^{-5/2}, \tag{9}$$

then

$$P_2 = (P_2' + P_3')/2, \quad P_{21} = (P_2' + P_3'')/2;$$

and, if (9) is satisfied and $R \leqslant n^{1/2}$, then

$$P_1 = (P_2 + P_3)/2,$$

where P_3, P_3', P_3'' are some probability measures (see (Sazonov et al., 1989a)). Note that (9) follows from (3) which is assumed to be satisfied. Below we will always assume if something else is not explicitly stated that s and R are defined by (8). Furthermore, we will say that we have

the case **A**, if $R \leqslant n^{1/2}$,

and

the case **B**, if $R > n^{1/2}$.

Note that if V_2' (respectively, V_2) is the covariance operator corresponding to P_2' (respectively, $\mathbf{P}_2$) and $(\sigma_{21}')^2 \geqslant (\sigma_{22}')^2 \geqslant \dots$ (respectively, $\sigma_{21}^2 \geqslant \sigma_{22}^2 \geqslant \dots$) are its eigenvalues, then (see Sazonov et al., 1989a))

$$(\sigma_{2i}')^2 \leqslant 2\sigma_i^2, \quad i = 1, 2, \dots; \qquad (\sigma_{2i}')^2 \geqslant \frac{2}{3}\sigma_i^2, \quad i = 1, \dots, 5,$$

and

$$\sigma_{2i}^2 \leqslant 2\sigma_i^2, \quad i = 1, 2, \dots; \qquad \sigma_{2i}^2 \geqslant \frac{2}{3}\sigma_i^2, \quad i = 1, \dots, 6.$$

Let P_4 be the probability measure corresponding to the $H-$ valued random variable $Z' = \xi(Z - \mathbf{E}Z) + Y' + \mathbf{E}Z$, where ξ is a bounded real random variable such that

$\mathbf{E}\xi = 0$, $\mathbf{E}\xi^2 = 1/2$, $\mathbf{E}\xi^3 = 1$ (see (Sazonov, 1981, p.84)), Z is distributed as P_2, Y' is a $(0, V_2/2)$ Gaussian random variable and ξ, Z, Y' are independent. Note that Z' and Z are $(\mathbf{E}Z, V_2)$ Gaussian random variables and, for any $h_i \in \in H$, $i = 1,2,3$,

$$\mathbf{E}\prod_{i=1}^{3}(Z', h_i) = \mathbf{E}\prod_{i=1}^{3}(Z, h_i)$$

(see (Sazonov, 1981, p.85)), i.e., Z and Z' have the same moments of the first three orders. If $R \leqslant n^{1/2}$ we define $P_5 = (P_4 + P_3)/2$. Denote by P_6 the probability measure corresponding to $\xi(Z - \mathbf{E}Z) + \mathbf{E}Z$.
Finally, define

$$P_1'(E) = \mathbf{P}(E \cap B_R) + \mathbf{P}(B_R^c)\,\delta_0(E)$$

and let $P_7 = pP_2 + q\delta_0$, where $p = 2\mathbf{P}(B_R) - 1$, $q = 2\mathbf{P}(B_R^c)$. Since by (9), $p \geqslant 0$ and $p + q = 1$, P_7 is a probability measure. Note that $P_1' = (P_2 + P_7)/2$ and denote

$$P_8 = (P_4 + P_7)/2.$$

In what follows $X_1^{(k)}, X_2^{(k)}, \ldots$ is a sequence of independent random variables with the same probability measure P_k, $k = 1, \ldots, 8$. The measures P_k and random variables $X_i^{(k)}$, $k = 3, 5$, are considered only in the case **A**. Finally let $\overline{P}_k(\cdot) = = P_k(n^{1/2})$.

LEMMA 1. *Let X_i, $\chi_i(u)$, $u \geqslant 0$, $i = 1, 2, \ldots$, and S_n, $S_n(u)$ be the same as in Theorem and Remark 2. Then, for any real measurable function f on H, any u_1, u_2, $0 \leqslant u_2 \leqslant u_1 < \infty$, and any H-valued random variable Z,*

$$\Delta_1 = \left|\mathbf{E}\Big(f(S_n(u_1) + Z) - f(S_n(u_2) + Z)\Big)\right| \leqslant$$

$$\leqslant n\mathbf{E}\left|f(S_n(u_1) + Z) - f(S_n(u_2) + Z)\right|(1 - \chi_1(u_2))$$

provided the expectations exist.

Proof. For a set A let $\mathbf{1}_A$ be the indicator of A. We have

$$\Delta_1 = \Big|\mathbf{E}(f(S_n(u_1) + Z) - f(S_n(u_2) + Z))\times$$

$$\times\left(\mathbf{1}_{\cap\{|X_i| < u_2\}} + \mathbf{1}_{\cup\{|X_i| \geqslant u_2\}}\right)\Big| \leqslant$$

$$\leqslant \mathbf{E}\left|f(S_n(u_1) + Z) - f(S_n(u_2) + Z)\right|\sum_{i=1}^{n}(1 - \chi_i(u_2)) \leqslant$$

$$\leqslant n\mathbf{E}|f(S_n(u_1) + Z) - f(S_n(u_2) + Z)|\,(1 - \chi_1(u_2)).$$

LEMMA 2. *For any integer $k \geqslant 0$,*

$$\Delta_2 = \left|\mathbf{P}(|S_n' - a| < r) - \mathbf{P}(|S_n'' - a| < r)\right| \leqslant$$

$$\leqslant c(k)(1+d)^{-k}\Lambda\big(1+\min\big(|a|^3,r^3\big)\big)\beta n^{-1/2},$$

where

$$S_n' = S_n\left(n^{1/2}\right), \quad S_n'' = n^{-1/2}\sum_1^n X_i^{(5)} \quad \text{in the case } \mathbf{A},$$

$$S_n' = S_n(R), \quad S_n'' = n^{-1/2}\sum_1^n X_i^{(8)} \quad \text{in the case } \mathbf{B}.$$

Proof. Consider, first, the case **B**. We have

$$\Delta_2 = \left|\left(((\overline{P}_2+\overline{P}_7)/2)^n - ((\overline{P}_4+\overline{P}_7)/2)^n\right)(B_r(a))\right| \leqslant I_1 + I_2, \tag{10}$$

where

$$I_i = 2^{-n}\textstyle\sum_1 \binom{n}{q}\left|\left(\overline{P}_2^q - \overline{P}_4^q\right) * \overline{P}_7^{n-q}(B_r(a))\right|, \quad i=1,2,$$

$\sum_1$ is the summation over all integer q such that

$$|q-(n+1)/2| < 2^{-1}n^{-7/10} \tag{11}$$

and $\sum_2$ is the summation over the remaining q, $0 \leqslant q \leqslant n$. By Lemma 6 in (Sazonov et al., 1989a) and (3),

$$I_2 \leqslant cn^{-1/2}\exp(-d). \tag{12}$$

From (16), (18), it follows that to prove Lemma 2 in the case **B** it is enough to prove that, for q satisfying (11),

$$\left|\left(\overline{P}_2^q - \overline{P}_4^q\right) * \overline{P}_7^{n-q}(B_r(a))\right| \leqslant$$

$$\leqslant c(k)(1+d)^{-k}\left(1+\min\left(|a|^3,r^3\right)\right)\beta n^{-1/2}. \tag{13}$$

Let P_{71} be the distribution of $X_1^{(7)}\mathbf{1}_{\{|X_1^{(7)}|<n^{1/2}\}}$. By Lemma 1,

$$\left|\left(\overline{P}_2^q - \overline{P}_4^q\right) * \left(\overline{P}_7^{n-q} - \overline{P}_{71}^{n-q}\right)(B_r(a))\right| \leqslant$$

$$\leqslant n\mathbf{P}\left(|X_1^{(7)}| > n^{1/2}\right) \leqslant \mathbf{E}|X_1^{(2)}|^4/n. \tag{14}$$

From (3) and the definition of R (see (8)), it follows that

$$\mathbf{E}|X_1^{(2)}|^4/n \leqslant 2\beta R/n \leqslant 2\beta\sigma_6^{-1}n^{-1/2}\exp(-d/20). \tag{15}$$

Note that the same reasoning as in Lemma 2.1 in (Bentkus,1984), shows that $\overline{P}_{71}^n$ has an exponentional moment, namely

$$\int \exp(|x|/15)\,\overline{P}_{71}^n(dx) \leqslant c < \infty. \tag{16}$$

Assume now that, for all q satisfying (11), we can prove that

$$\left|\left(\overline{P}_2^q - \overline{P}_4^q\right)\left(B_r(a)\right)\right| \leqslant$$
$$\leqslant c(k)(1+d)^{-k}\Lambda\left(1+\min\left(|a|^3, r^3\right)\right)\beta n^{-1/2}. \tag{17}$$

Then from (22), (23) it follows that

$$\left|\left(\overline{P}_2^q - \overline{P}_4^q\right) * \overline{P}_{71}^{n-q}\left(B_r(a)\right)\right| \leqslant$$
$$\leqslant \left|\left(\int\limits_{|y|<d/2} + \int\limits_{|y|\geqslant d/2}\right)\left(\overline{P}_2^q - \overline{P}_4^q\right)\left(B_r(a+y)\right)\overline{P}_{71}^{n-q}(dy)\right| \leqslant$$
$$\leqslant c(k)(1+d)^{-k}\Lambda\left(1+\min\left(|a|^3, r^3\right)\right)\beta n^{-1/2}. \tag{18}$$

Thus, by (13)–(15), (18), it remains only to prove (17), where, by (11), we can take q to be equal to n.

The proof of (17) is based on ideas from (Sazonov et al., 1988) and (Sazonov et al., 1989a).

Let ξ be a real random variable with density function

$$p_q(x) = c(l)\, l^{-1}\left[\sin(x/l)/(x/l)\right]^l,$$

where $q = \max(35, K+3)$, $l \geqslant q+3$, is an even number and

$$c(l) = \left(\int (\sin x/x)^l\, dx\right)^{-1}.$$

As in (Sazonov et al., 1989a), put

$$T = c\left(s^{-2}n\right)^{5/9}\left(\ln\left(s^{-2}n\right)\right)^{-1/2},$$

where s is defined in (8). Denote $\lambda = T^{1/36}$. Then, by (3),

$$\mathbf{P}(|\xi| \geqslant \lambda) \leqslant T^{-1}\mathbf{E}|\xi|^{36} \leqslant c(k)\beta\sigma_5^{-2}n^{-1/2}\exp(-cd) \tag{19}$$

and

$$\lambda T^{-1} \leqslant c_1\beta\sigma_5^{-2}n^{-1/2}\exp(-c_2 d). \tag{20}$$

One can show (cf. (44), (45) in (Sazonov et al., 1989a)) that the density function of $|n^{-1/2}\sum_{i=1}^n X_i^{(4)} - a|^2$ is bounded by $c\sigma_1^{-1}\sigma_2^{-1}$. Thus, from Lemmas 7, 9 in (Sazonov et al., 1989a), (25), (26), it follows that to prove (17) it is enough to prove that

$$|\Delta_T(a,r)| \leqslant c(k)(1+d)^{-k}\Lambda\left(1+\min\left(|a|^3, r^3\right)\right)\beta n^{-1/2}, \tag{21}$$

where

$$\Delta_T(a,r) = \mathbf{P}\left(\left|n^{-1/2}\sum_{i=1}^{n} X_i^{(2)} - a\right|^2 + \xi T^{-1} < r^2\right) -$$

$$-\mathbf{P}\left(\left|n^{-1/2}\sum_{1}^{n} X_i^{(4)} - a\right|^2 + \xi T^{-1} < r^2\right).$$

Let $\Delta_T'(a,r)$ be $\Delta_T(a,r)$ with $X_i^{(2)}$, $X_i^{(4)}$ replaced by

$$X_i^{(21)} = X_i^{(2)}\mathbf{1}_{\{|X_i^{(2)}|<n^{1/2}\}}, \qquad X_i^{(41)} = X_i^{(4)}\mathbf{1}_{\{|X_i^{(4)}|<n^{1/2}\}}$$

, respectively, $i = 1,2,\ldots,n$. Using (3) and Lemma 1, in the same way as (14), (15), we obtain

$$|\Delta_T(a,r) - \Delta_T'(a,r)| \leqslant c\beta\sigma_6^{-1}n^{-1/2}\exp(-d/20).$$

Hence, instead of (21) it suffices to prove that

$$|\Delta_T'(a,r)| \leqslant \frac{c(k)\Lambda}{1+d^k}\left(1+\min|a|^3, r^3)\right)\frac{\beta}{\sqrt{n}}. \tag{22}$$

Note that if P_{21}, P_{41} are the probability measures corresponding to $X_1^{(21)}$, $X_1^{(41)}$, respectively, then, since the moments of the first three orders of P_2 and P_4 coincide, we have

$$\int |y|^{2m}(x,y)^l\left(\overline{P}_{21} - \overline{P}_{41}\right)(dy) =$$

$$= n^{-m-l/2}\int\limits_{|y|>n^{1/2}} |y|^{2m}(x,y)^l(P_4 - P_2)(dy), \tag{23}$$

where m, l are non-negative integers such that $2m + l \leqslant 3$.
Now to estimate (22) we rewrite $\Delta_T'(a,r)$, using the inversion formula, and then integrate by parts k times (cf. Lemma 12 in (Sazonov et al., 1988)). After that it remains essentially to repeat the proof of Lemma 7 in (Sazonov et.al., 1989a). Using (15), (23), the Chebyshev inequality and the relations (cf. (22)), we have

$$\int |x|^k\left(\overline{P}_{21}^n + \overline{P}_{41}^n\right)(dx) \leqslant c(k) < \infty.$$

This finishes the proof of Lemma 2 in the case **B**.
In the case **A** the proof is simpler, since Lemma 9 from (Sazonov et al., 1988) can be applied directly to

$$\left(\overline{P}_2^q - \overline{P}_4^q\right) * \overline{P}_3^{n-q}(B_r(a)),$$

and, by Lemma 2 from (Sazonov et al., 1989a), no additional truncations of random variables with probability measures P_2, P_3, P_4 are needed. Thus, we do not have relations like (23), but we use the coincidence of moments of the first three orders of P_2 and P_4.

Lemma 2 is proved.

Before we formulate the next lemma, we remind the necessary notation from (Sazonov et al., 1989a).

Let Y_i be $(0,\sigma_i^2)$ Gaussian real random variables, $i=1,\ldots,6$, $1\geqslant\sigma_1^2\geqslant\ldots\geqslant\sigma_6^2$. Assume that Y_i, $i=1,\ldots,m$, are independent and denote by $g_m=g_m(x,u)$ (respectively, $\bar{g}_m=\bar{g}_m(x,u)$) the density function of $\sum_{i=1}^m(Y_i+x_i)^2$ (respectively, $\sum_{i=1}^m(\sqrt{2}Y_i+x_i)^2$), $x=(x_1,\ldots,x_m)\in\mathbf{R}^m$. Let η be an independent of Y_i, $i=1,\ldots,6$, real random variable with continuously differentiable density function $f_\eta(u)$ such that $f_\eta(u)=0$ if $|u|\geqslant 1$. For $T\geqslant 1$ denote $\widetilde{g}_m(u)=\widetilde{g}_m(x,u)$ (respectively $\widetilde{\bar{g}}_m(u)=\widetilde{\bar{g}}_m(x,u)$) the density function of

$$\sum_1^m(Y_i+x_i)^2+\eta T^{-1}$$

(respectively, $\sum_1^m(\sqrt{2}Y_i+x_i)^2+\eta T^{-1}$), $m=1,\ldots,6$.

LEMMA 3. *The density function $\widetilde{g}_4(u)$ is continuously differentiable, $\widetilde{g}_6(u)$ is twice continuously differentiable and, for $m=4,6$,*

$$\widetilde{g}_m(u)\leqslant c\sigma_1^{-1}\sigma_2^{-1}\exp(-d_m^2/(3\bar{\sigma}_m^2)),\tag{24}$$

$$\left|\frac{\partial\widetilde{g}_m(u)}{\partial u}\right|\leqslant c\prod_{i=1}^4\sigma_i^{-1}\exp(-d_m^2/(3\bar{\sigma}_m^2)),\tag{25}$$

$$\left|\frac{\partial^2\widetilde{g}_6(u)}{\partial u}\right|\leqslant c\prod_{i=1}^6\sigma_i^{-1}\exp(-d_6^2/(3\bar{\sigma}_6^2)),\tag{26}$$

$$\left|\frac{\partial^2\widetilde{g}_m(u)}{\partial x_j\partial u}\right|\leqslant c\sigma_m^{-1}\prod_{i=1}^4\sigma_i^{-1}\exp(-d_m^2/(3\bar{\sigma}_m^2)),\tag{27}$$

$$\left|\frac{\partial^{\delta_1+\delta_2+\delta_3}}{\partial x_i^{\delta_1}\partial x_j^{\delta_2}\partial x_k^{\delta_3}}\widetilde{g}_m(u)\right|\leqslant c\sigma_i^{-\delta_1}\sigma_j^{-\delta_2}\sigma_k^{-\delta_3}\widetilde{\bar{g}}_m,\tag{28}$$

where $i,j,k=1,\ldots,6$, $\delta_1+\delta_2+\delta_3\leqslant 3$,

$$\bar{\sigma}_m^2=\sum_{i=1}^m\sigma_i^2;\qquad d_m=\sqrt{u}-\sqrt{x_1^2+\ldots+x_m^2}.$$

Remark. Estimates (24)–(27) can be improved, namely $3\bar{\sigma}_m^2$ in them can be replaced by $(2+\varepsilon)\bar{\sigma}_m^2$ (cf. (Ulyanov, 1990a)). However, in this case c has to be replaced by $c(\varepsilon)$ and $c(\varepsilon)\to\infty$ as $\varepsilon\to 0$. For our purposes estimates (24)–(27) as stated are sufficient.

Proof. Since

$$\widetilde{g}_m(u)=\int_{-T^{-1}}^{T^{-1}}g_m(u-v)Tf_\eta(Tv)\,dv$$

with $T \geqslant 1$, by Lemma 11 in (Sazonov et al., 1989a), it suffices to construct estimates of $g_m(x,u)$ similar to (24)–(28). To this end it is enough to repeat the reasoning from the proof of Lemma 9 in (Sazonov et al., 1989a) and:
1. Observe that, for any $a \geqslant 0,\ b \geqslant 0,\ c \geqslant d \geqslant 0$,

$$\left(\sqrt{c-d}-a\right)^2 + \left(\sqrt{d}-b\right)^2 \geqslant \left(\sqrt{c}-\sqrt{a^2+b^2}\right)^2.$$

This inequality is easily proved by squaring the both sides of it.
2. To use instead of the estimate

$$p_{ij}(u) \leqslant c\ \min\left\{\left(\sigma_i^{-1}+\sigma_j^{-1}\right)u^{-1/2}, \sigma_i^{-1}\sigma_j^{-1}\right\}$$

(see (65) in (Sazonov et al., 1989a)) the estimate

$$p_{ij}(u) \leqslant c\ \min\left\{\left(\sigma_i^{-1}+\sigma_j^{-1}\right)u^{-1/2}, \sigma_i^{-1}\sigma_j^{-1}\right\} \times$$
$$\times \exp\left(-\frac{3}{5}\left(\sqrt{u}-\sqrt{x_i^2+x_j^2}\right)^2 / \left(\sigma_i^2+\sigma_j^2\right)\right). \tag{30}$$

The last estimate follows from (29), the representation

$$p_i(u) = (2\pi u)^{-1/2}\,\sigma_i^{-1} d_i(u),$$

where

$$d_i(u) = d\left(\sigma_i^2, u\right) = \frac{1}{2}\left(\exp\left\{-\left(u^{1/2}-a_i\right)^2 / \left(2\sigma_i^2\right)\right\} + \right.$$
$$\left. + \exp\left\{-\left(u^{-1/2}+a_i\right)^2 / \left(2\sigma_i^2\right)\right\}\right),$$

and the inequality

$$d_i(u) \leqslant \exp\left\{-\left(u^{-1/2}-|a_i|\right)^2 / \left(2\lambda^2\right)\right\} \times$$
$$\times d\left(\lambda^2\sigma_i^2 / \left(\lambda^2-\sigma_i^2\right), u\right), \tag{31}$$

which is valid for any $\lambda^2 > \sigma_i^2$.

LEMMA 4. *Let Z be an H-valued $(0,W)$ Gaussian random variable with $\operatorname{tr} W \leqslant 2$. Let $\tau_1^2 \geqslant \tau_2^2 \geqslant \ldots$ denote the eigenvalues and $e_1, e_2, \ldots$ the corresponding eigenvectors of W. Let G be $(0,V)$ Gaussian probability measure on H and put $\overline{G}(\cdot) = G(n^{1/2}\cdot)$. Finally, define $R(dx) = \left(\overline{P}_5 - \overline{G}\right)(dx)$ in the case* **A** *and $R(dx) = \left(\bar{p}_8 - \overline{G}\right)(dx)$ in the case* **B**. *Then, for any $r \geqslant 0$, $b \in H$,*

$$\left|\int \mathbf{P}(|Z+x-b|<r)R(dx)\right| \leqslant$$

$$\leqslant c\left(\prod_{i=2}^{6}\tau_i^{-1}+\sigma_6^{-1}\prod_{i=1}^{5}\tau_i^{-1}\right)\times$$

$$\times\beta n^{-3/2}\left(1+\min\left(|b|^3,r^3\right)\right)\exp\left(-c\big|r-|b|\big|\right),$$

where $|r-|b||$ satisfies (3) with a replaced by b.

Proof. Let $M\colon H\to H$ be a projection operator and put $\rho=\big|r-|b|\big|$. Note, that in the case **A**,

$$\int|x|^l|R|(dx)\leqslant c\beta n^{-3/2}\quad\text{for}\quad l\geqslant 3,\tag{32}$$

$$\int\limits_{|Mx|\geqslant c\rho}|x|^m|R|(dx)\leqslant c\beta n^{-3/2}\exp\left(-c\rho^2\right)\quad\text{for}\quad m\leqslant 3,\tag{33}$$

and in the case **B**, for $l\geqslant 3,\ m\leqslant 3$,

$$\int\limits_{|Mx|\leqslant c\rho}|Mx|^l|R|(dx)\leqslant c\beta\rho^{l-3}n^{-3/2},\tag{34}$$

$$\int\limits_{|Mx|\geqslant c\rho}|x|^m|R|(dx)\leqslant c\beta\sigma_6^{-1}n^{-3/2}\exp\left(-c\rho\right).\tag{35}$$

Estimates (32), (34) follow easily from the definition of measure R. Inequalities (33), (35) follow from the Chebyshev inequality, (6), and (15).
Note that, for any $a\geqslant b\geqslant 0,\ c\geqslant 0$, we have

$$\left|\sqrt{a-b}-c\right|\geqslant\left|\sqrt{a}-\sqrt{c^2+b}\right|.\tag{36}$$

The proof of Lemma 4 is basically the same as the proof of Lemma 12 in (Sazonov et al., 1989a). We will indicate only the necessary changes. The set E_1 now should be defined as (cf. (83) in (Sazonov et al., 1989a))

$$E_1=\{x\in H\colon\ |x|<\rho/8\}.$$

As an example, let us show how to estimate $I_{31}+I_{32}$. We have

$$I_{31}=\int\limits_{E_1}Q_3R(dx)=\int\limits_{E_1}Q_3\Big(\mathbf{1}_{\{|\bar Z_3|<\rho/8\}}+$$

$$+\mathbf{1}_{\{|\bar Z_3|\geqslant\rho/8\}}\Big)R(dx)=I_{311}+I_{312},\tag{37}$$

where Q_3 is the same as in (Sazonov et al., 1989a). To estimate I_{311}, we use (25), (26), (32), (34), (36) and obtain

$$|I_{311}|\leqslant c\left(\prod_{i=1}^{6}\tau_i^{-1}\right)\beta\left(1+|b|^3\right)\exp\left(-c\rho^2\right)n^{-3/2}.\tag{38}$$

To estimate I_{312}, we apply (75), (76) from (Sazonov et al., 1989a), (32), (34), (36) and (6), which results in

$$|I_{312}| \leqslant c\left(\prod_{i=1}^{6}\tau_i^{-1}\right)\beta\left(1+|b|^3\right)\exp\left(-c\rho^2\right)n^{-3/2}. \tag{39}$$

To estimate I_{32} (we preserve the notation of (Sazonov et al., 1989a) and use the new definition of E_1),

$$I_{32} = \int_{E_2}\left(Q - \sum_{j=0}^{2}(j!)^{-1}Q_j\right)R(dx)$$

we have, by (75) from (Sazonov et al., 1989a) and (33), (35),

$$|I_{32}| \leqslant c\prod_{i=1}^{5}\tau_i^{-1}\left(\tau_6^{-1}+\sigma_6^{-1}\right)\beta\left(1+|b|^3\right)\exp\left(-c\rho\right)n^{-3/2}. \tag{40}$$

Finally, from (37)–(40) it follows that

$$|I_{31}+I_{32}| \leqslant c\prod_{i=1}^{5}\tau_i^{-1}\left(\tau_6^{-1}+\sigma_6^{-1}\right)\beta\left(1+|b|^3\right)\exp\left(-c\rho\right)n^{-3/2}. \tag{41}$$

Estimates similar to (41) hold true also for I_2, I_{12}, I_{112}, I_{111} (see p.34 in (Sazonov et al., 1989a)). To estimate I_{113} one has to use also the following inequalities

$$\left|\int_0^v \frac{\partial}{\partial u}\widetilde{q}_4(v-t)\,p(t)\,dt\right| \leqslant$$

$$\leqslant c\left(\prod_{i=1}^{4}\tau_i^{-1}\right)\exp\left(-c\left(\sqrt{v}-|\bar{b}_1+\bar{b}_2|\right)^2\right) \tag{42}$$

and

$$\left|\int_0^v \frac{\partial}{\partial u}\widetilde{q}_4(v-t)\,\frac{\partial}{\partial x_i}p(x_5,x_6,t)\,dt\right| \leqslant$$

$$\leqslant c\tau_6^{-1}\left(\prod_{i=1}^{4}\tau_i^{-1}\right)\exp\left(-c\left(\sqrt{v}-|\bar{b}_1+\bar{b}_2-\bar{x}_2|\right)^2\right), \tag{43}$$

where $\widetilde{q}_4(v)$ (respectively, $p(t), p(x_5,x_6,t)$) is the density function of $\xi_1(0)+\eta T^{-1}$ (respectively, $|Z_{22}-\bar{b}_2|^2$ and $|Z_{22}-\bar{b}_2+\bar{x}_2|^2$).
The estimate (42) follows from (75) in (Sazonov et al., 1989a) and (25), (29) and (31). To prove (43), one has to use also (89) from (Sazonov et al., 1989a).
Together (42), (43) lead to an estimate of I_{113} similar to (41).

To estimate I_0 we use now a somewhat different definition of $F_2(\lambda)$ (cf. p.36 in (Sazonov et al.,1989a)). Put now

$$F_2(\lambda) = \begin{cases} \int\limits_{-T^{-1}}^{\eta_2(\lambda)} \tilde{q}_4(u)\,du & \text{if } r \leqslant |b|, \\ \int\limits_{\eta_2(\lambda)}^{\infty} \tilde{q}_4(u)\,du & \text{if } r > |b| \end{cases}$$

and take $E_1 = \{x \in H\colon\ |x| < \rho/8\}$ instead of B in the definition of I_{02}, I'_{02}. The rest of the proof of Lemma 4 is essentially the same as that of Lemma 12 in (Sazonov et al., 1989a), the difference be the same as in the proof of (41).

LEMMA 5. *In the notation of Lemmas* 3, 4,

$$\Delta_3 = \left|\mathbf{P}\left(|S''_n - a| < r\right) - \mathbf{P}\left(|Y - a| < r\right)\right| \leqslant$$

$$\leqslant c\left(\prod_{i=1}^{6}\sigma_i^{-1}\right)\beta\left(1+|a|^3\right)n^{-1/2}\exp\left(-c\big|r-|a|\big|\right). \tag{44}$$

Proof. Without loss of generality we may assume that n is even. We have

$$\Delta_3 = \left|\left(\overline{P}_k^n - \overline{G}^n\right)\left(B_r(a)\right)\right| \leqslant \Delta_{31} + \Delta_{32},$$

where

$$\Delta_{31} = \left|\overline{P}_k^{n/2} * \left(\overline{P}_k^{n/2} - \overline{G}^{n/2}\right)\left(B_r(a)\right)\right|,$$

$$\Delta_{32} = \left|\overline{G}^{n/2} * \left(\overline{P}_k^{n/2} - \overline{G}^{n/2}\right)\left(B_r(a)\right)\right|$$

and $k = 5$ in the case **A**, $k = 8$ in the case **B**.
Consider the case **B** which is more complicated.
Applying Lemma 1, we see that instead of Δ_{32} it suffices to estimate

$$\Delta'_{32} = \left|\overline{G}^{n/2} * \left(\overline{P}_{81}^{n/2} - \overline{G}^{n/2}\right)\left(B_r(a)\right)\right|,$$

where P_{81} is the probability measure corresponding to $X_1^{(8)}\,\mathbf{1}_{\{|X_1^{(8)}|<\sqrt{n}\}}$. This reduction to the random variables truncated at $n^{1/2}$ makes it possible to use the following property of P_{81}: there exists $\alpha > 0$ such that (cf. (16) above)

$$\int \exp\left(\alpha|x|\right)\overline{P}_{81}^{n/2}(dx) < \infty. \tag{45}$$

Using (45), from Lemma 4 we deduce

$$\Delta'_{32} \leqslant c\left(\prod_{i=1}^{6}\sigma_i^{-1}\right)\beta\left(1+|a|^3\right)n^{-1/2}\exp\left(-c\big|r-|a|\big|\right).$$

To estimate Δ_{31}, we again reduce the problem to the estimation of

$$\Delta'_{31} = \left|\overline{P}_{81}^{n/2} * \left(\overline{P}_{81}^{n/2} - \overline{G}^{n/2}\right)(B_r(a))\right|.$$

Furthermore, we can write (cf. p.46 in (Sazonov et al., 1989a))

$$\overline{P}_8^m = \left[\left(\overline{P}_4 + \overline{P}_7\right)/2\right]^m = \sum_{i=1}^{2} \sum\nolimits_i 2^{-m} \binom{m}{l} \overline{P}_4^l * \overline{P}_7^{m-l},$$

where $m = n/2$, $\sum_1$ is the summation over all l such that $|2l - m| < n^{7/10}$ and $\sum_2$ is the summation over the remaining l from $\{0, 1, \ldots, m\}$. By (3), it suffices to obtain an estimate like (44) for

$$I_l = \left|\overline{P}_4^l * \overline{P}_7^{m-l} * \left(\overline{P}_{81}^m - \overline{G}^m\right)(B_r(a))\right| =$$

$$= \left|G_l * \overline{P}_6^l * \overline{P}_7^{m-l} * \left(\overline{P}_{81}^m - \overline{G}^m\right)(B_r(a))\right|.$$

Here we use Lemma 1 again and reduce the problem to the case of truncated measures P_{61}, P_{71} which lead to a quantity like Δ'_{32}. This proves Lemma 5 in the case **B**.

In the case **A** it is not necessary to make a second truncation since (cf. (16) above), for some $\alpha > 0$ and all n', $0 \leqslant n' \leqslant n$,

$$\mathbf{E}\exp\left(\alpha\left|n^{-1/2}\sum_{i=1}^{n'} X_i^{(5)}\right|\right) \leqslant c(\alpha) < \infty$$

and also if $|n_3 - n_6| \leqslant n^{7/10}$, then

$$\mathbf{E}\exp\left(\alpha\left|n^{-1/2}\sum_{k=3,6}\sum_{i=1}^{n_k} X_i^{(k)}\right|\right) \leqslant c(\alpha) < \infty.$$

These inequalities can be proved in the same way as Lemma 2.1 in (Bentkus, 1984). In the proof, (12) and (14) from (Sazonov et al., 1989a) with $p = 1$ are used.

3. PROOF OF THE THEOREM

In the usual way (see, e.g., (Sazonov et al., 1988)) from Lemmas 1, 2, and 5, we deduce that, for any non-negative integer k,

$$\Delta_n(a, r) \leqslant c\beta(1 + d)^{-3} n^{-1/2} +$$

$$+ c(k)(1 + d)^{-k} \Lambda\left(1 + \min\left(|a|^3, r^3\right)\right)\beta n^{-1/2}. \tag{46}$$

From the proof of Lemma 2 it follows that $c(k)(1 + d)^{-k}$ appears here in the same way as in Theorem 1 in (Sazonov et al., 1988). Now the same reasoning which allows to replace $c(k)(1 + d)^{-k}$ by $c\exp\left(-d^{1/5}\right)$ in Theorem 1 in (Sazonov et al.,1988) (see (Ulyanov, 1990b)) leads from (46) to the theorem.

REFERENCES

Bentkus, V.Yu. (1984). Asymptotic expansions for distributions of sums of independent random elements of Hilbert space. *Lithuanian Math.J.* **24**, 305–319.

Sazonov, V.V. (1981). Normal Approximation. Some Recent Advances. *Lect.Notes Math.* **879.**

Sazonov, V.V., Ulyanov, V.V. and Zalesskii, B.A. (1988). Normal approximation in Hilbert space.I,II. *Theory Probab.Appl.* **33**, 207–227, 473–487.

Sazonov, V.V., Ulyanov, V.V. and Zalesskii, B.A. (1989a). A precise estimate of the rate of conference in the central limit theorem in Hilbert space. *Preprint of Steklov Mathematical Institute.* N 6, 1–42.

Sazonov, V.V., Ulyanov, V.V. and Zalesskii, B.A. (1989b). Some new results on the central limit theorem in Hilbert and Banach space. In: *Abstracts of Communications. Fifth Vilnius Conference on Probab. Theory and Math. Statist.* **2.** IMC, Vilnius, pp. 139–140.

Ulyanov, V.V. ((1990a)). Non–uniform estimates of the density of the squared norm of a Gaussian vector in Hilbert space. *Preprint of Universität Bielefeld,* SFB, **343**, 1–11.

Ulyanov, V.V. (1990b). Letter to the Editors. *Theory Probab.Appl.* **35**, 187.

New Trends in Probab. and Statist., pp. 137 – 164
V. Sazonov and T. Shervashidze (Eds.)
1991 VSP/Mokslas

ON LARGE DEVIATIONS FOR L–ESTIMATES

R. ZITIKIS
Institute of Mathematics and Informatics, Lithuanian Academy of Sciences, Akademijos str. 4, 232600 Vilnius, Lithuania

ABSTRACT

In this paper we prove a theorem on Cramér type large deviations for linear combinations of order statistics. Both the moment conditions and the rate of convergence are of right order.

1. INTRODUCTION AND MAIN RESULTS

Let $Y, Y_1, Y_2, \ldots$ be a sequence of independent and identically distributed real–valued random variables. Suppose that $\mathbf{E}Y = 0$ and $\mathbf{E}Y^2 = 1$. Furthermore, let $(\mathbf{H}, \|\cdot\|)$ be a real separable Hilbert space and let $\mathcal{Y}, \mathcal{Y}_1, \mathcal{Y}_2, \ldots$ be a sequence of independent and identically distributed $\mathbf{H}$–valued random variables. Suppose that $\mathbf{E}\langle \mathcal{H}, \mathcal{Y}\rangle = 0$ for all (nonrandom) $\mathcal{H} \in \mathbf{H}$. Throughout the paper we do not put any condition on the dependence between Y and $\mathcal{Y}$ (consequently, on the dependence between Y_i and $\mathcal{Y}_i$, $i = 1, 2, \ldots$, as well); we only require that the pairs $\{Y, \mathcal{Y}\}, \{Y_1, \mathcal{Y}_1\}, \ldots$ be independent random variables. The last random variable we want to introduce here is D. We set no a priori conditions on it except that there is a real constant $d \geqslant 0$ such that $\mathbf{P}(|D| \leqslant d) = 1$.
The main result of the present paper is the following theorem on large deviations for

$$L_n := \frac{1}{\sqrt{n}} \sum_{i=1}^{n} Y_i + \frac{D}{\sqrt{n}} \left\| \frac{1}{\sqrt{n}} \sum_{i=1}^{n} \mathcal{Y}_i \right\|^2, \ n \geqslant 1.$$

THEOREM 1. *Let $a \in (0, 1/2]$ be a real number. Suppose there exist $b > 0$ such that*

$$\mathbf{E} \exp\left\{b|Y|^a\right\} + \mathbf{E} \exp\left\{b\|\mathcal{Y}\|^a\right\} < \infty. \tag{1}$$

Then there exist constants $A > 0$ and $c \geqslant 0$ such that

$$\left|\mathbf{P}(L_n > r)/\mathbf{P}(G > r) - 1\right| \leqslant c(1 + r)^3/\sqrt{n} \tag{2}$$

for all $0 \leqslant r \leqslant An^{a/(4-2a)}$, where G denotes a Gaussian random variable with mean 0 and variance 1.

It is known (see the proof of Theorem 3, for references) that L–estimates with Lipschitzian scores generating functions can be represented as L_n. We will reformulate Theorem 1 for L–estimates below (see Theorem 3). Now we shall formulate a theorem concerning Berry–Esséen bound for L_n, which is a preliminary for establishing Theorem 1 and might be of independent interest itself (c.f. (Friedrich, 1989; Helmers, 1982; van Zwet, 1984); see, also (Götze, 1989)).

THEOREM 2. *Let*

$$\mathbf{E}|Y|^3 + \mathbf{E}\|\mathcal{Y}\|^3 < \infty. \tag{3}$$

Then there exists a constant $c \geqslant 0$ *such that*

$$\sup_{r\in\mathbf{R}} \left|\mathbf{P}(L_n > r) - \mathbf{P}(G > r)\right| \leqslant c/\sqrt{n}, \tag{4}$$

where G *is the same as in Theorem* 1.

Using this, we derive the following result.

COROLLARY. *Let* $A > 0$ *be an arbitrary, but fixed, real number, and let the condition* (3) *be satisfied. Then there exists a constant* $c \geqslant 0$ *such that*

$$\left|\mathbf{P}(L_n > r)/\mathbf{P}(G > r) - 1\right| \leqslant c/\sqrt{n} \tag{5}$$

for all $0 \leqslant r \leqslant A$.

Consider now the case of L–estimates. Let $X, X_1, X_2, \ldots$ be independent and identically distributed random variables with distribution function F. The ordered values of $X_1, X_2, \ldots, X_n$ will be denoted by $X_{1:n} \leqslant X_{2:n} \leqslant \ldots \leqslant X_{n:n}$. Furthermore, let $c_{1n}, c_{2n}, \ldots, c_{nn}$ be real constants and let

$$\widehat{L}_n := \sum_{i=1}^{n} c_{in} X_{i:n}$$

which is called L–estimate. Assume that $c_{1n}, c_{2n}, \ldots, c_{nn}$ are generated by a Lipschitzian function, say J. More exactly, let

$$c_{in} = \int\limits_{\left((i-1)/n,\, i/n\right)} J(u)du,$$

where the function J: $[0,1] \to \mathbf{R}$ is such that for some $K \geqslant 0$,

$$\left|J(u) - J(v)\right| \leqslant K|u - v| \quad \forall u, v \in [0,1]. \tag{6}$$

Denote

$$\mu := \int\limits_{\mathbf{R}} J\big(F(u)\big)u\, dF(u),$$

$$\sigma^2 := \int\limits_{\mathbf{R}}\int\limits_{\mathbf{R}} J\big(F(u)\big)J\big(F(v)\big)\big\{F(u \wedge v) - F(u)F(v)\big\}\, dudv,$$

and assume that $\sigma^2 > 0$.

THEOREM 3. *Let $a \in (0, 1/2]$ and suppose that there exists $b > 0$ such that*

$$\mathbf{E}\exp\left\{b|X|^a\right\} < \infty. \tag{7}$$

Then there exist constants $A > 0$ and $c \geqslant 0$ such that

$$\Delta_{n,r} := \mathbf{P}\left(\frac{\sqrt{n}}{\sigma}\{\widehat{L}_n - \mu\} > r\right)/\mathbf{P}(G > r) = 1 + \theta c\frac{(1+r)^3}{\sqrt{n}} \tag{8}$$

for all $0 \leqslant r \leqslant An^{a/(4-2a)}$. Here G denotes a Gaussian random variable with mean 0 and variance 1, and $\theta \in [-1, 1]$ is a real number.

Now we give some historical references on large deviations for L–estimates. To simplify the discussion let us consider the case $a = 1/2$ only. Obviously, in the case $J(u) \equiv 1$, Theorem 3 coincide with the well–known one for the empirical mean (see, e.g., (Ibragimov and Linnik, 1971; Petrov, 1975; Saulis and Statulevičius, 1989)). Aside from the empirical mean (the case $c_{in} = 1/n$) the first paper concerning large deviations of Cramér type for L–estimates is (Vandemaele and Veraverbeke, 1982), where the following result is proved.

THEOREM (Vandemaele and Veraverbeke, 1982). *If*

$$\mathbf{P}(|X| \leqslant c) = 1$$

for some real $c \geqslant 0$, then

$$\Delta_{n,r} = 1 + o(1) \tag{9}$$

in the range $0 \leqslant r \leqslant o(n^{1/6})$ as $n \to \infty$.

For a wider class of statistics a similar result is proved in (Puri and Seoh, 1988). There, and in (Vandemaele and Veraverbeke, 1982) as well, the results on large deviations for L–estimates are proved using the so–called "method of moments". In (Aleškevičienė, 1989), using the "method of cumulants" (for this method the reader is referred to (Saulis and Statulevičius, 1989)), the relation (9) is proved under a less restrictive condition. Namely, it is proved that the relation (9) holds, if there exists $b > 0$ such that $\mathbf{E}\exp\{b|X|\} < \infty$. A theorem from (Bentkus and Zitikis, 1990) allows the X to be such that $\mathbf{E}\exp\{b\sqrt{|X|}\} < \infty$ for some $b > 0$. It is well known (see, e.g., (Ibragimov and Linnik, 1971; Bentkus and Račkauskas, 1990)) that the latter condition cannot be improved in general case (at least in the case $J(u) \equiv 1$). The technique of (Bentkus and Zitikis, 1990) is based on results concerning large deviations in one–dimensional and infinite–dimensional spaces. (It has been recently announced that in (Aleškevičienė, 1990) the result of (Bentkus and Zitikis, 1990) is proved without using methods of infinite–dimensional spaces, namely, by using the "method of moments".)
In the present paper we discuss the rate of convergence in the theorem of large deviations for L–estimates. We use the ideas of (Bentkus, 1984; Götze, 1979; 1984; Paulauskas and Račkauskas, 1989; Yurinskii, 1982), where the central limit theorem in Banach spaces is discussed, and the methods of (Bentkus, 1986; Bentkus and Račkauskas, 1990; Račkauskas, 1988; Zalesskii, 1989) on large deviations in Banach spaces. By the way, using the technique of (Bentkus and Zitikis, 1990),

the present author managed to obtain the rate of convergence in (9) of order $(1+r)^3 \log n/\sqrt{n}$.
Concerning the so-called "Chernoff type" large deviations, we refer the reader to (Groeneboom and Shorack, 1981; Shorack and Wellner, 1986).
Throughout the paper non-negative real constants, not always the same from line to line, will be denoted by c. If necessary to distinguish between them, we will use also the notations $c_1, c_2, \dots$. The inner product in $\mathbf{H}$ is denoted by $\langle \cdot, \cdot \rangle$. If V is a random variable, then, for simplicity of notation, we shall denote $(V_1+\ldots+ +V_k)/\sqrt{n}$ and $(V_{k+1}+\ldots+V_n)/\sqrt{n}$ by $V_n(k)$ and $V_n^*(n-k)$, respectively ($V_1, V_2, \ldots$ are independent copies of V). Symbols $\mathbf{P}_V$ and $\mathbf{E}_V$ denote the conditional probability and conditional expectation with respect to V. The ∂_t is the partial derivative with respect to t, the indicator will be denoted by $\mathbf{1}$.

2. PROOF OF THEOREM 2

First note that there are no difficulties to deduce Corollary from Theorem 2. Let us note only that there is $c > 0$ (for example, $c := \mathbf{P}(G > A)$) such that $\mathbf{P}(G > r) \geqslant c$ for all $0 \leqslant r \leqslant A$.
Now let us prove Theorem 2. Straightforward calculations lead us to

$$\big|\mathbf{P}(L_n > r) - \mathbf{P}(G > r)\big| \leqslant \max\big(P(1), P(2)\big),$$

where

$$P(1) := \left|\mathbf{P}\left(Y_n(n) + \frac{d}{\sqrt{n}}\|\mathcal{Y}_n(n)\|^2 > r\right) - \mathbf{P}(G > r)\right|,$$

$$P(2) := \left|\mathbf{P}\left(Y_n(n) - \frac{d}{\sqrt{n}}\|\mathcal{Y}_n(n)\|^2 > r\right) - \mathbf{P}(G > r)\right|.$$

Thus our aim is to show that both $P(1)$ and $P(2)$ do not exceed $c/\sqrt{n}$ for some $c \geqslant 0$, but we restrict ourselves to the case of the first of them only (similar arguments prove the same for $P(2)$ as well).
Let us introduce the following notation:

$$\delta_1 := \frac{d}{\sqrt{n}}\left(\|\mathcal{Y}_n(k)\|^2 - \frac{1}{n}\sum_{i=1}^{k}\|\mathcal{Y}_i\|^2\right),$$

$$\delta_2 := \frac{d}{\sqrt{n}}\left(\frac{1}{n}\sum_{i=1}^{k}\|\mathcal{Y}_i\|^2 + 2\big\langle \mathcal{Y}_n(k), \mathcal{Y}_n^*(n-k)\big\rangle + \|\mathcal{Y}_n^*(n-k)\|^2\right),$$

where $k := n - [\sqrt{n}]$ (for real x, $[x]$ denotes the greatest integer less or equal to x). Thus $P(1)$ can be rewritten as $\big|\mathbf{P}\big(Y_n(n) + \delta_1 + \delta_2 > r\big) - \mathbf{P}\big(G > r\big)\big|$ and, therefore, it is not difficult to see that, for any real $c_1 > 0$, $P(1) \leqslant S(1)+ +\max\big(S(2), S(3)\big)$, where

$$S(1) := \mathbf{P}\left(|\delta_2| > \frac{c_1}{\sqrt{n}}\right),$$

$$S(2) := \left|\mathbf{P}\left(Y_n(n) + \delta_1 > r - \frac{c_1}{\sqrt{n}}\right) - \mathbf{P}(G > r)\right|,$$

$$S(3) := \left|\mathbf{P}\left(Y_n(n) + \delta_1 > r + \frac{c_1}{\sqrt{n}}\right) - \mathbf{P}(G > r)\right|.$$

PROPOSITION 1. *If c_1 is large enough (say, $c_1 \geqslant 3d(1+\mathbf{E}\|\mathcal{Y}\|^2)$), then, for some real $c \geqslant 0$, $S(1) \leqslant c/\sqrt{n}$.*

Proof. The case $d = 0$ is obvious. Therefore, from now on we suppose $d > 0$. Plainly, $S(1)$ is bounded above by $S^*(1) + S^{**}(2) + S^{***}(2)$,

$$S^*(1) := \mathbf{P}\left(\frac{d}{n}\sum_{i=1}^{k}\|\mathcal{Y}_i\|^2 > \frac{c_1}{3}\right),$$

$$S^{**}(1) := \mathbf{P}\left(2d\left|\langle\mathcal{Y}_n(k), \mathcal{Y}_n^*(n-k)\rangle\right| > \frac{c_1}{3}\right),$$

$$S^{***}(1) := \mathbf{P}\left(d\left\|\mathcal{Y}_n^*(n-k)\right\|^2 > \frac{c_1}{3}\right)$$

and the further work is to show that for some fixed $c \geqslant 0$, these quantities are not greater than $c/\sqrt{n}$.
For c_1 as in the proposition,

$$S^*(1) \leqslant \mathbf{P}\left(\frac{1}{k}\sum_{i=1}^{k}\{\|\mathcal{Y}_i\|^2 - \mathbf{E}\|\mathcal{Y}\|^2\} > 1\right).$$

Thus we can infer from $\mathbf{E}\|\mathcal{Y}\|^3 < \infty$ and a theorem concerning the rate of convergence in the law of large numbers (see, e.g., (Petrov, 1975)) that $c/\sqrt{n}$ bounds $S^*(1)$ for some $c \geqslant 0$.
By the Chebyshev inequality and

$$\mathbf{E}\langle\mathcal{Y}_n(k), \mathcal{Y}_n^*(n-k)\rangle^2 = \frac{1}{n^2}\sum_{i=1}^{k}\sum_{j=k+1}^{n}\mathbf{E}\langle\mathcal{Y}_i, \mathcal{Y}_j\rangle^2 \leqslant \frac{c}{\sqrt{n}},$$

we have the desired result for $S^{**}(1)$. Arguments similar to those we used above show that $S^{***}(1)$ is not greater than $c/\sqrt{n}$.

PROPOSITION 2. *There is $c \geqslant 0$ such that, for all $r \in \mathbf{R}$, $c/\sqrt{n}$ bounds $S(2)$ and $S(3)$.*

Proof. Clearly, it is enough to prove that

$$\sup_{r\in\mathbf{R}}\left|\mathbf{P}(Y_n(n) + \delta_1 > r) - \mathbf{P}(G > r)\right| \leqslant \frac{c}{\sqrt{n}}$$

which, by the Esséen theorem, holds if for some fixed $c_2 > 0$,

$$\int_{|t|\leqslant c_2\sqrt{n}} \frac{1}{|t|}\left|\mathbf{E}\exp\left\{it(Y_n(n) + \delta_1)\right\} - \mathbf{E}\exp\left\{itY_n(n)\right\}\right| dt \leqslant \frac{c}{\sqrt{n}}, \qquad (10)$$

$$\int\limits_{|t|\leqslant c_2\sqrt{n}} \frac{1}{|t|}\left|\mathbf{E}\exp\{itY_n(n)\} - \mathbf{E}\exp\{itG\}\right| dt \leqslant \frac{c}{\sqrt{n}}. \tag{11}$$

Here one can choose $c_2 > 0$ arbitrary close to zero, in which case (11) is well known (see, e.g., the proof of the Berry–Esséen theorem in (Petrov, 1975)). Therefore, it remains to prove (10). Put $g(t) := \mathbf{E}\exp\{itY/\sqrt{n}\}$ and note that, by the inequality $|\exp\{ix\} - 1 - ix| \leqslant x^2/2$, the bound (10) follows from

$$\int\limits_{|t|\leqslant c_2\sqrt{n}} |g(t)|^{n-k}\left|\mathbf{E}\delta_1\exp\{itY_n(k)\}\right| dt \leqslant \frac{c}{\sqrt{n}}, \tag{12}$$

$$\int\limits_{|t|\leqslant c_2\sqrt{n}} |g(t)|^{n-k}|t|\, dt\, \mathbf{E}\delta_1^2 \leqslant \frac{c}{\sqrt{n}}. \tag{13}$$

In order to prove these inequalities, let us recall (see, e.g., (Petrov, 1975)) that for a sufficiently small $c_2 > 0$ there exists $c > 0$ such that $|g(t)| \leqslant 1 - ct^2/n$ and, consequently,

$$|g(t)| \leqslant \exp\{-ct^2/n\} \tag{14}$$

in the range $|t| \leqslant c_2\sqrt{n}$.

We start from (12). Clearly,

$$\left|\mathbf{E}\delta_1\exp\{itY_n(k)\}\right| \leqslant \frac{dk^2}{n^{3/2}}|g(t)|^{k-2}\left|\mathbf{E}\langle\mathcal{Y}_1,\mathcal{Y}_2\rangle\exp\{it(Y_1+Y_2)/\sqrt{n}\}\right| \leqslant$$

$$\leqslant d\sqrt{n}|g(t)|^{k-2}\Big|\mathbf{E}\langle\mathcal{Y}_1,\mathcal{Y}_2\rangle\left(\exp\{itY_1/\sqrt{n}\} - 1\right)\times$$

$$\times\left(\exp\{itY_2/\sqrt{n}\} - 1\right)\Big| \leqslant \frac{c}{\sqrt{n}}|g(t)|^{k-2}t^2.$$

This estimation, together with (14), completes the proof of (12).

Note that

$$\int\limits_{|t|\leqslant c_2\sqrt{n}} |g(t)|^{n-k}|t|dt \leqslant c\sqrt{n},$$

hence, (13) follows from

$$\mathbf{E}\delta_1^2 = \frac{d^2}{n^3}\mathbf{E}\Big\{\sum\langle\mathcal{Y}_i,\mathcal{Y}_j\rangle\Big\}^2 = \frac{2d^2}{n^3}\sum\mathbf{E}\langle\mathcal{Y}_i,\mathcal{Y}_j\rangle^2 \leqslant \frac{c}{n},$$

where the sum is taken over all $i, j = 1, 2, \ldots, k,\ i \neq j$. This completes the proof of Theorem 2.

3. PROOF OF THEOREM 1

First, let us make some remarks. Without loss of generality, we can take n large enough, i.e., $n \geqslant c_3$, where $c_3 \geqslant 1$ is a sufficiently large fixed constant. We will also assume that r is greater than, for example, 10. (For $r \leqslant 10$, Corollary obviously

holds.) And the last remark here is: every random variable to be introduced below will be independent from the others.
Let $\mathcal{G}$ be a Gaussian **H**-valued random variable with $\mathbf{E}\langle\mathcal{H},\mathcal{G}\rangle = 0$ and $\mathbf{E}\langle\mathcal{H},\mathcal{G}\rangle^2 = \mathbf{E}\langle\mathcal{H},\mathcal{Y}\rangle^2$ for all (nonrandom) $\mathcal{H} \in \mathbf{H}$. Then straightforward computation gives

$$\left|\mathbf{P}(L_n > r) - \mathbf{P}(G > r)\right| \leqslant \max(S(4), S(5)) + S(6), \tag{15}$$

where

$$S(4) := \left|\mathbf{P}\left(Y_n(n) + \frac{d}{\sqrt{n}}\|\mathcal{Y}_n(n)\|^2 > r\right) - \mathbf{P}\left(G + \frac{d}{\sqrt{n}}\|\mathcal{G}\|^2 > r - \frac{1}{\sqrt{n}}\right)\right|,$$

$$S(5) := \left|\mathbf{P}\left(Y_n(n) - \frac{d}{\sqrt{n}}\|\mathcal{Y}_n(n)\|^2 > r\right) - \mathbf{P}\left(G - \frac{d}{\sqrt{n}}\|\mathcal{G}\|^2 > r - \frac{1}{\sqrt{n}}\right)\right|,$$

$$S(6) := \mathbf{E}1\left(r - \frac{1}{\sqrt{n}} - \frac{d}{\sqrt{n}}\|\mathcal{G}\|^2 \leqslant G \leqslant r + \frac{1}{\sqrt{n}} + \frac{d}{\sqrt{n}}\|\mathcal{G}\|^2\right).$$

Using the estimate

$$\mathbf{P}(s - t \leqslant G \leqslant s + t) \leqslant c(1+s)t\exp\{cst\}\mathbf{P}(G > s)$$

and noting that $\mathbf{E}\exp\{c_4\|\mathcal{G}\|^2\}$ is finite for a sufficiently small $c_4 > 0$ (see, e.g., (Kuo, 1975)), we obtain

$$S(6) \leqslant (1+r)\mathbf{P}(G > r)/\sqrt{n}.$$

Thus, the main task is to show that $S(4)$ and $S(5)$ do not exceed

$$c(1+r)^3\mathbf{P}(G > r)/\sqrt{n}.$$

It is sufficient to prove this for $S(4)$ only, because $S(5)$ can be bounded quite similarly. For this we need a function $f_{n,r}\colon \mathbf{R} \to [0,1]$ such that

$$f_{n,r}(t) = \begin{cases} 0, & t \leqslant r - 1/\sqrt{n}, \\ 1, & t \geqslant r, \end{cases} \tag{16}$$

$$\left|\partial_t^i f_{n,r}(t)\right| \leqslant cn^{i/2}, \quad t \in \mathbf{R}, \tag{17}$$

for $i = 1, 2, 3$. So, using (16), in the inequality (15) we can replace $S(4)$ by $\max(S^*(4), S^{**}(4))$, where

$$S^*(4) := \left|\mathbf{E}f_{n,r}\left(Y_n(n) + \frac{d}{\sqrt{n}}\|\mathcal{Y}_n(n)\|^2\right) - \mathbf{E}f_{n,r}\left(G + \frac{d}{\sqrt{n}}\|\mathcal{G}\|^2\right)\right|,$$

$$S^{**}(4) := \left|\mathbf{E}f_{n,r-1/\sqrt{n}}\left(Y_n(n) + \frac{d}{\sqrt{n}}\|\mathcal{Y}_n(n)\|^2\right) - \mathbf{E}f_{n,r-1/\sqrt{n}}\left(G + \frac{d}{\sqrt{n}}\|\mathcal{G}\|^2\right)\right|.$$

Since the investigation of $S^{**}(4)$ does not really differs from that of $S^*(4)$, the remaining part of the proof of Theorem 1 consist of estimation of the latter one.

In the following, $G_1, G_2, \ldots$ and $\mathcal{G}_1, \mathcal{G}_2, \ldots$ denote independent copies of G and G, respectively. Denote also, for $k = 0, 1, \ldots, n$,

$$W_n(k) := Y_n(k) + G_n^*(n-k) + \frac{d}{\sqrt{n}} \big\| \mathcal{Y}_n(k) + \mathcal{G}_n^*(n-k) \big\|^2.$$

Then, evidently,

$$S^*(4) = \Big| \mathbf{E} f\big(W_n(n)\big) - \mathbf{E} f\big(W_n(0)\big) \Big| \quad (f := f_{n,r}).$$

Furthermore, let $\alpha_1, \alpha_2, \alpha_3$ be independent random variables distributed uniformly in (0,1), and set

$$T_n^\alpha := \frac{1}{\sqrt{n}} \sum_{i=1}^{3} \alpha_i T_{n+i},$$

where $T \in \{Y, G, \mathcal{Y}, \mathcal{G}\}$. Then denote, for $k = 0, 1, \ldots, n$,

$$W_n^s(k, \alpha) := Y_n(k) + G_n^*(n-k) + S_n^\alpha + \frac{d}{\sqrt{n}} \big\| \mathcal{Y}_n(k) + \mathcal{G}_n^*(n-k) + \mathcal{S}_n^\alpha \big\|^2,$$

where $s := \{S, \mathcal{S}\} \in \{y := \{Y, \mathcal{Y}\},\ g := \{G, \mathcal{G}\}\}$.
Now we return to the estimation of $S^*(4)$. Namely, we have

$$\begin{aligned} S^*(4) \leqslant \Theta^y(n) + \Lambda^y(n, [n/2]) + \Theta^y([n/2]) + \\ + \Theta^g([n/2]) + \Lambda^g([n/2], 0) + \Theta^g(0), \end{aligned} \tag{18}$$

where

$$\Theta^s(k) := \Big| \mathbf{E} f\Big(W_n(k)\Big) - \mathbf{E} f\Big(W_n^s(k,\alpha)\Big) \prod_{i=1}^{3} \Big\{ \sqrt{n}\, \partial_{\alpha_i} W_n^s(k,\alpha) \Big\}^2 \Big|,$$

$$\begin{aligned} \Lambda^s(u,v) := \Big| \mathbf{E} f\Big(W_n^s(u,\alpha)\Big) \prod_{i=1}^{3} \Big\{ \sqrt{n}\, \partial_{\alpha_i} W_n^s(u,\alpha) \Big\}^2 - \\ - \mathbf{E} f\Big(W_n^s(v,\alpha)\Big) \prod_{i=1}^{3} \Big\{ \sqrt{n}\, \partial_{\alpha_i} W_n^s(v,\alpha) \Big\}^2 \Big|. \end{aligned}$$

Thus, to estimate $S^*(4)$, we have to bound the six quantities in the right-hand side of (18). But we will consider $\Theta^y(k)$, $k = [n/2], \ldots, n$, and $\Lambda^y(n, [n/2])$ only, because the other summands in (18) can be estimated in a similar (even in a simpler) way. Let us formulate two statements concerning the bounds of the mentioned quantities.

STATEMENT 1. *For all* $k = [n/2], \ldots, n$

$$\Theta^y(k) \leqslant c(1+r)\mathbf{P}(G > r)/\sqrt{n}. \tag{19}$$

STATEMENT 2. *For every* $n \geqslant 1$

$$\Lambda^y([n/2], n) \leqslant c(1+r)^3 \mathbf{P}(G > r)/\sqrt{n}. \tag{20}$$

Obviously, the proof of these statements ends the proof of Theorem 1.

Proof of Statement 1. Note that

$$\sqrt{n}\partial_{\alpha_i} W_n^y(k,\alpha) = Y_{n+i} + \frac{2d}{\sqrt{n}}\Big\langle \mathcal{Y}_{n+i}, \mathcal{Y}_n(k) + \mathcal{G}_n^*(n-k) + \mathcal{Y}_n^\alpha \Big\rangle \tag{21}$$

approximates Y_{n+i}. Therefore, we can write $\Theta^y(k) \leqslant \Theta_1^y(k) + \Theta_2^y(k)$, where

$$\Theta_1^y(k) := \left|\mathbf{E}f(W_n(k))\left\{\prod_{i=1}^3 Y_{n+1}^2 - \prod_{i=1}^3\left\{\sqrt{n}\partial_{\alpha_i}W_n^y(k,\alpha)\right\}^2\right\}\right|,$$

$$\Theta_2^y(k) := \left|\mathbf{E}f(W_n(k)) - f(W_n^y(k,\alpha))\prod_{i=1}^3\left\{\sqrt{n}\partial_{\alpha_i}W_n^y(k,\alpha)\right\}^2\right|,$$

and the proof of (19) follows from the two propositions below.

PROPOSITION 3. *For all* $k = [n/2], \dots, n$,

$$\Theta_1^y(k) \leqslant c(1+r)\mathbf{P}(G > r)/\sqrt{n}. \tag{22}$$

Proof. Since

$$\prod_{q=1}^3 \big(b_{q,1} + b_{q,2} + b_{q,3}\big) - b_{1,1}b_{2,1}b_{3,1} = \sum b_{1,i}b_{2,j}b_{3,l},$$

where the sum is taken over all $i,j,l \in \{1,2,3\}$, $i+j+l \geqslant 4$, then in order to prove (22) it is sufficient to show this bound for the quantities

$$\mathbf{E}\mathbf{1}\left(W_n(k) > r - \frac{1}{\sqrt{n}}\right)\left|B_{n+1,i}\right|\left|B_{n+2,j}\right|\left|B_{n+3,l}\right|. \tag{23}$$

Here, for $q \in \{1,2,3\}$,

$$B_{n+q,1} := Y_{n+q}^2, \quad B_{n+q,2} := \frac{4d}{\sqrt{n}}Y_{n+q}\Big\langle \mathcal{Y}_{n+q}, \mathcal{Y}_n(k) + \mathcal{G}_n^*(n-k) + \mathcal{Y}_n^\alpha\Big\rangle,$$

$$B_{n+q,3} := \frac{4d^2}{n}\Big\langle \mathcal{Y}_{n+q}, \mathcal{Y}_n(k) + \mathcal{G}_n^*(n-k) + \mathcal{Y}_n^\alpha\Big\rangle^2.$$

We can write, for arbitrary $c_5 > 0$,

$$1 \leqslant I_1 + I_2 + I_3,$$

where

$$I_1 = \mathbf{1}\big(\|\mathcal{Y}_n(k)\| > c_5 r\big), I_2 = \mathbf{1}\big(\|\mathcal{G}_n^*(n-k)\| > c_5 r\big),$$
$$I_3 = \mathbf{1}\big(\|\mathcal{Y}_n(k)\| \leqslant c_5 r\big)\mathbf{1}\big(\|\mathcal{G}_n^*(n-k)\| \leqslant c_5 r\big).$$

Therefore, the sum of the quantities

$$B(m) := \mathbf{E}I_m\mathbf{1}\left(W_n(k) > r - \frac{1}{\sqrt{n}}\right)\left|B_{n+1,i}\right|\left|B_{n+2,j}\right|\left|B_{n+3,l}\right|,$$

for $m = 1, 2, 3$, bounds (23).

Before proceeding further, one remark should be made. Namely, for any real $p \geqslant 0$, there exists $c \geqslant 0$ such that

$$\mathbf{E}|Y_{n+1}|^p, \mathbf{E}\|\mathcal{Y}_{n+1}\|^p, \mathbf{E}\|\mathcal{G}_n^*(n-.k)\|^p, \mathbf{E}\|\mathcal{Y}_n(k)\|^p \leqslant c. \tag{24}$$

It is evident that the first three quantities in the left–hand side of (24) are finite, due to (1) and well-known properties of Gaussian random variables. Using, for example, a result from (Zalesskii, 1982), one can easily prove that the fourth one is finite also.

Now let us return to $B(m)$. Using the Hölder inequality, we have

$$B(1) \leqslant \frac{c}{\sqrt{n}}\{\mathbf{E}I_1^2\}^{1/2} = \frac{c}{\sqrt{n}}\Big\{\mathbf{P}(\|\mathcal{Y}_n(k)\| > c_5 r)\Big\}^{1/2}. \tag{25}$$

At this point we need two lemmas.

LEMMA 1 (Bentkus and Račkauskas, 1990). *There exist A_1 and $c \geqslant 0$ such that*

$$\mathbf{P}(\|\mathcal{Y}_n(k)\| > x) \leqslant c\mathbf{P}(\|\mathcal{G}\| > x)$$

for all $0 \leqslant x \leqslant A_1 n^{a/(4-2a)}$.

LEMMA 2 (see, e.g., (Pisier, 1985)). *There exist $c_6 > 0$ and $c \geqslant 0$ such that*

$$\mathbf{P}(\|\mathcal{G}\| > x) \leqslant c\exp\{-c_6 x^2\}$$

for all $x \geqslant 0$.

Using the above lemmas, we choose c_5 and A such that $c_5 \geqslant \sqrt{2/c_6}$ and $A \leqslant A_1/c_5$ and, since $c_5 r \leqslant A_1 n^{a/(4-2a)}$, we have

$$\Big\{\mathbf{P}(\|\mathcal{Y}_n(k)\| > c_5 r)\Big\}^{1/2} \leqslant c\exp\{-r^2\} \leqslant c\mathbf{P}(G > r),$$

which, together with (25), proves that $B(1) \leqslant c\mathbf{P}(G > r)/\sqrt{n}$. The same bound is true for $B(2)$, since

$$\Big\{\mathbf{P}(\|\mathcal{G}_n^*(n-k)\| > c_5 r)\Big\}^{1/2} \leqslant \Big\{\mathbf{P}(\|\mathcal{G}\| > c_5 r)\Big\}^{1/2} \leqslant c\mathbf{P}(G > r)$$

due to Lemma 2. In the case of $B(3)$, we have

$$B(3) \leqslant \mathbf{E}\Big\{I_3 \mathbf{1}\Big(Y_n(k) + G_n^*(n-k) > r - \frac{1 + 4dc_5^2 r^2}{\sqrt{n}}\Big)|B_{n+1,i}|\,|B_{n+2,j}|\,|B_{n+3,l}|\Big\} \leqslant$$

$$\leqslant c\frac{1+r}{\sqrt{n}}\mathbf{P}\Big(Y_n(k) + G_n^*(n-k) > r - \frac{1 + 4dc_5^2 r^2}{\sqrt{n}}\Big). \tag{26}$$

Thus to prove the estimate

$$B(3) \leqslant c(1+r)\mathbf{P}(G > r)/\sqrt{n}$$

it remains to show that, for some $c \geqslant 0$, $c\mathbf{P}(G > r)$ bounds the probability in the right-hand side of (26). But this follows from Lemma 3 below and the estimate $\mathbf{P}(G > s-t) \leqslant c\exp\{cst\}\mathbf{P}(G > s)$.

LEMMA 3. *There exist $A_2 > 0$ and $c \geqslant 0$ such that*

$$\mathbf{P}\big(Y_n(k) + G_n^*(n-k) > x\big) \leqslant c\mathbf{P}(G > x)$$

for all real $x \leqslant A_2 n^{a/(4-2a)}$.

Of course, to be correct we have to assume that $A \leqslant A_2$. This ends the proof of Proposition 3.

PROPOSITION 4. *For all $k = [n/2], \ldots, n$,*

$$\Theta_2^y(k) \leqslant c(1+r)\mathbf{P}(G > r)/\sqrt{n}.$$

Proof. Denote

$$D_{n+i} := |Y_{n+i}| + \frac{2d}{\sqrt{n}}\|\mathcal{Y}_{n+i}\|\,\|\mathcal{Y}_n(k)\| + \frac{2d}{\sqrt{n}}\|\mathcal{Y}_{n+i}\|\,\|\mathcal{G}_n^*(n-k)\| + \frac{9d}{n}\|\mathcal{Y}_{n+i}\|^2.$$

Then

$$\big|W_n(k) - W_n^y(k,\alpha)\big| \leqslant \frac{1}{\sqrt{n}}\sum_{i=1}^{3} D_{n+i}$$

and, therefore,

$$\Big|f\big(W_n(k)\big) - f\big(W_n^y(k,\alpha)\big)\Big| \leqslant \sum_{i=1}^{3} \mathbf{1}\Big(\big|W_n^y(k,\alpha) - r\big| \leqslant \frac{1+3D_{n+i}}{\sqrt{n}}\Big).$$

Using the last inequality we obtain

$$\Theta_2^y(k) \leqslant c\mathbf{E}\mathbf{1}\Big(\big|W_n^y(k,\alpha) - r\big| \leqslant \frac{1+3D_{n+1}}{\sqrt{n}}\Big)\prod_{i=1}^{3}\{\sqrt{n}\partial_{\alpha_i}W_n^y(k,\alpha)\}^2.$$

However, if τ denotes a random variable distributed uniformly in the interval $\{x\colon |x-r| \leqslant (1+3D_{n+1})/\sqrt{n}\}$, then

$$\mathbf{1}\Big(\big|W_n^y(k,\alpha) - r\big| \leqslant \frac{1+3D_{n+1}}{\sqrt{n}}\Big)\partial_{\alpha_2}W_n^y(k,\alpha) = \frac{2+6D_{n+1}}{\sqrt{n}}\partial_{\alpha_2}\mathbf{P}\big(\tau \leqslant W_n^y(k,\alpha)\big).$$

Therefore,

$$\Theta_2^y(k) \leqslant c\mathbf{E}\Big\{\partial_{\alpha_2}\{\mathbf{P}_\tau\big(\tau \leqslant W_n^y(k,\alpha)\big)\}(1+3D_{n+1})D_n^y(k,\alpha)\Big\},$$

where

$$D_n^y(k,\alpha) := \{\sqrt{n}\partial_{\alpha_1}W_n^y(k,\alpha)\}^2\{\sqrt{n}\partial_{\alpha_2}W_n^y(k,\alpha)\}\{\sqrt{n}\partial_{\alpha_3}W_n^y(k,\alpha)\}^2.$$

Thus, integrating by parts and recalling the definition of D_{n+1}, we arrive at the following bound:

$$\Theta_2^y(k) \leqslant c\{S(7) + \ldots + S(10)\},$$

where

$$S(7) := \mathbf{E}\Big\{\mathbf{P}_\tau(\tau \leqslant W_n^y(k,\alpha))(1+3|Y_{n+1}|)D_n^y(k,\alpha)\Big\}\Big|_{\alpha_2=1},$$

$$S(8) := \frac{1}{\sqrt{n}}\Big|\mathbf{E}\Big\{\mathbf{P}_\tau(\tau \leqslant W_n^y(k,\alpha))\Big(2\|\mathcal{Y}_{n+1}\|\|\mathcal{Y}_n(k)\|+$$

$$+2\|\mathcal{Y}_{n+1}\|\|\mathcal{G}_n^*(n-k)\| + \frac{9}{\sqrt{n}}\|\mathcal{Y}_{n+i}\|^2\Big)D_n^y(k,\alpha)\Big\}\Big|_{\alpha_2=1}\Big|,$$

$$S(9) := \Big|\mathbf{E}\Big\{\mathbf{P}_\tau(\tau \leqslant W_n^y(k,\alpha))(1+3D_{n+1})D_n^y(k,\alpha)\Big\}\Big|_{\alpha_2=0}\Big|,$$

$$S(10) := \Big|\mathbf{E}\Big\{\mathbf{P}_\tau(\tau \leqslant W_n^y(k,\alpha))(1+3D_{n+1})\partial_{\alpha_2}D_n^y(k,\alpha)\Big\}\Big|.$$

We shall prove that

$$S(7), \ldots, S(10) \leqslant c(1+r)\mathbf{P}(G > r)/\sqrt{n}.$$

We begin with a trick we shall use often in the sequel. Namely, if h is a measurable function and $\lambda_1, \lambda_2, \ldots$ is a sequence of independent and identically distributed random variables, then, for any $m \geqslant 1$,

$$\mathbf{E}h\Big(\sum_{i=1}^m \lambda_i\Big)\lambda_m = \frac{1}{m}\mathbf{E}h\Big(\sum_{i=1}^m \lambda_i\Big)\Big(\sum_{i=1}^m \lambda_i\Big) \tag{27}$$

and, consequently,

$$\mathbf{E}h\Big(\sum_{i=1}^{2m} \lambda_i\Big)\lambda_m\lambda_{2m} = \frac{1}{m^2}\mathbf{E}h\Big(\sum_{i=1}^{2m} \lambda_i\Big)\prod_{j=0}^{1}\sum_{i=jm+1}^{(j+1)m}\lambda_i, \tag{28}$$

and so on.

Let us examine $S(7)$. By the above trick we have that $S(7) \leqslant |\mathbf{E}H|\sqrt{n}/(k+1)$ and, therefore,

$$S(7) \leqslant c\mathbf{E}|H|/\sqrt{n}, \tag{29}$$

where

$$H := \mathbf{P}_\tau(\tau \leqslant W_n^y(k,\alpha))\prod_{i\in\{1,3\}}\Big\{\sqrt{n}\partial_{\alpha_i}W_n^y(k,\alpha)\Big\}^2\Big|_{\alpha_2=1}\times$$

$$\times\Big(1+3|Y_{n+1}|\Big)\Big(Y_n(k) + \frac{1}{\sqrt{n}}Y_{n+2} + \frac{2d}{\sqrt{n}}\Big\langle\mathcal{Y}_n(k) + \frac{1}{\sqrt{n}}\mathcal{Y}_{n+2}, \mathcal{Y}_n(k)+$$

$$+\frac{1}{\sqrt{n}}\mathcal{Y}_{n+2} + \mathcal{G}_n^*(n-k) + \frac{1}{\sqrt{n}}\alpha_1 Y_{n+1} + \frac{1}{\sqrt{n}}\alpha_3 Y_{n+3}\Big\rangle\Big).$$

For every $c_7 > 0$, we have $1 \leqslant \sum_{i=4}^{14} I_i$, where

$$I_4 = \mathbf{1}\Big(Y_n(k) + G_n^*(n-k) > c_7 r\Big), I_5 = \mathbf{1}\Big(|Y_n(k)| > c_7 r\Big),$$

$$I_6 = \mathbf{1}\Big(\|\mathcal{Y}_n(k)\| > c_7 r\Big), I_7 = \mathbf{1}\Big(\|\mathcal{G}_n^*(n-k)\| > c_7 r\Big),$$

$$I_{7+1} = \mathbf{1}\Big(|Y_{n+i}| > c_7 r\sqrt{n}\Big), I_{10+i} = \mathbf{1}\Big(\|\mathcal{Y}_{n+i}\| > c_7 r\sqrt{n}\Big),\ i = 1,2,3,$$

$$I_{14} = \mathbf{1}\Big(Y_n(k) + G_n^*(n-k) \leqslant c_7 r\Big)\mathbf{1}\Big(|Y_n(k)| \leqslant c_7 r\Big)\times$$

$$\times \mathbf{1}\Big(\|\mathcal{Y}_n(k)\| \leqslant c_7 r\Big)\mathbf{1}\Big(\|\mathcal{G}_n^*(n-k)\| \leqslant c_7 r\Big)\times$$

$$\times \prod_{i=1}^{3}\Big\{\mathbf{1}\Big(|Y_{n+i}| \leqslant c_7 r\sqrt{n}\Big)\mathbf{1}\Big(\|\mathcal{Y}_{n+i}\| \leqslant c_7 r\sqrt{n}\Big)\Big\}.$$

Thus, returning to (29), we see that $S(7)$ will be estimated if we show that, for some $c_7 > 0$ and all $i = 4, \dots, 14$,

$$H(i) := \mathbf{E}I_i|H| \leqslant c(1+r)\mathbf{P}(G > r). \tag{30}$$

Combining the Hölder inequality, (24) and Lemma 3, we can write, for $c_7 \geqslant 2$ and $A \leqslant A_2/c_7$,

$$H(4) \leqslant c\{\mathbf{E}I_4^2\}^{1/2} = c\big\{\mathbf{P}(Y_n(k) + G_n^*(n-k) > c_7 r)\big\}^{1/2} \leqslant$$

$$\leqslant c\big\{\mathbf{P}(G > c_7 r)\big\}^{1/2} \leqslant c\mathbf{P}(G > r).$$

To estimate $H(5)$ we need the following lemma.

LEMMA 4 (see, e.g., (Petrov, 1975; Saulis and Statulevičius, 1989)). *There exist A_3 and $c \geqslant 0$ such that*

$$\mathbf{P}\big(|Y_n(k)| > x\big) \leqslant c\mathbf{P}\big(|G| > x\big)$$

for all $x \leqslant A_3 n^{a/(4-2a)}$.

If we choose c_7 and A such that $c_7 \geqslant 2$ and $A \leqslant A_3/c_7$, then, by the Hölder inequality, (24) and Lemma 4,

$$H(5) \leqslant c\big\{\mathbf{P}(|Y_n(k)| > c_7 r)\big\}^{1/2} \leqslant c\big\{\mathbf{P}(|G| > c_7 r)\big\}^{1/2} \leqslant c\mathbf{P}(G > r).$$

Arguments similar to those above show that

$$H(6), H(7) \leqslant c\mathbf{P}(G > r)$$

(see the estimation of $B(1)$ and $B(2)$ for more details). The quantity $H(8)$ can be estimated as follows. Choose $c_7 \geqslant \{2A^{2-a}/b\}^{1/a}$. Then $b(c_7 r\sqrt{n})^a/2 \geqslant r^2$, and so

$$H(8) \leqslant c\{\mathbf{P}(|Y| > c_7 r\sqrt{n})\}^{1/2} \leqslant$$

$$\leqslant c\{\mathbf{E}\exp\{b|Y|^a\}\}^{1/2} \exp\{-b(c_7 r\sqrt{n})^a/2\} \leqslant c\exp\{-r^2\} \leqslant c\mathbf{P}(G > r).$$

Analogously, one can prove that $H(9), \ldots, H(13) \leqslant c\mathbf{P}(G > r)$. The estimation of $H(14)$ is more complicated. Firstly, we obtain, for some $p \geqslant 0$,

$$H(14) \leqslant c(1+r)\mathbf{E}\prod_{i=1}^{3}\Big\{(1+|Y_{n+i}|)(1+\|\mathcal{Y}_{n+i}\|)\Big\}^p I_{14}\times$$

$$\times\mathbf{1}\left(Y_n(k) + G_n^*(n-k) > r - \frac{1+cr^2}{\sqrt{n}} - c\sum_{j=1}^{3}\left\{\frac{1}{\sqrt{n}}|Y_{n+j}| + \frac{1}{n^{3/2}}\|\mathcal{Y}_{n+j}\|^2\right\}\right).$$

Thus, for $A \leqslant A_2$, Lemma 3 yields

$$H(14) \leqslant c(1+r)\mathbf{E}\prod_{i=1}^{3}\Big\{(1+|Y_{n+i}|)^p(1+\|\mathcal{Y}_{n+i}\|)^p\times$$

$$\times\mathbf{1}\Big(|Y_{n+i}| \leqslant c_7 r\sqrt{n}\Big)\mathbf{1}\Big(\|\mathcal{Y}_{n+i}\| \leqslant c_7 r\sqrt{n}\Big)\Big\}\times$$

$$\times\mathbf{P}_G\left(G > r - \frac{1+cr^2}{\sqrt{n}} - c\sum_{j=1}^{3}\left\{\frac{1}{\sqrt{n}}|Y_{n+j}| + \frac{1}{n^{3/2}}\|\mathcal{Y}_{n+j}\|^2\right\}\right),$$

which, together with

$$\mathbf{P}(G > s - t) \leqslant c\exp\{cst\}\mathbf{P}(G > s),$$

gives the bound

$$H(14) \leqslant c(1+r)\mathbf{P}(G > r)\{h_1 h_2\}^{3/2},$$

where

$$h_1 := \mathbf{E}\mathbf{1}\Big(|Y| \leqslant c_7 r\sqrt{n}\Big)(1+|Y|)^{2p}\exp\Big\{c_8 r|Y|/\sqrt{n}\Big\},$$

$$h_2 := \mathbf{E}\mathbf{1}\Big(\|\mathcal{Y}\| \leqslant c_7 r\sqrt{n}\Big)(1+\|\mathcal{Y}\|)^{2p}\exp\Big\{c_8 r\|\mathcal{Y}\|^2/n^{3/2}\Big\}.$$

Consider now h_1. If we choose A such that $A \leqslant \{b/(2c_8 c_7^{1-a})\}^{1/(2-a)}$, then $c_8 r|Y|/\sqrt{n} \leqslant b|Y|^a/2$. But, for any $q \geqslant 0$, $(1+|Y|)^q \leqslant c\exp\{b|Y|^a/2\}$ and $\mathbf{E}\exp\{b|Y|^a\} < \infty$. Therefore, $h_1 \leqslant c$. To prove $h_2 \leqslant c$ one can use $\|\mathcal{Y}\| \leqslant cn$ and the analysis above. This completes the estimation of $S(7)$.

To estimate $S(8)$, let us write $1 \leqslant \sum_{i=15}^{23} I_i$, where

$$I_{15} = \mathbf{1}(\|\mathcal{Y}_n(k)\| > c_9 r), I_{16} = \mathbf{1}(\|\mathcal{G}_n^*(n-k)\| > c_9 r),$$

$$I_{16+i} = \mathbf{1}\Big(|Y_{n+i}| > c_9 r\sqrt{n}\Big), \quad I_{19+i} = \mathbf{1}\Big(\|\mathcal{Y}_{n+i}\| > c_9 r\sqrt{n}\Big), \ i = 1,2,3,$$

$$I_{23} = \mathbf{1}\Big(\|\mathcal{Y}_n(k)\| \leqslant c_9 r\Big)\mathbf{1}\Big(\|\mathcal{G}_n^*(n-k)\| \leqslant c_9 r\Big)\times$$

$$\times \prod_{i=1}^{3}\Big\{\mathbf{1}\Big(|Y_{n+i}| \leqslant c_9 r\sqrt{n}\Big)\mathbf{1}\Big(\|\mathcal{Y}_{n+i}\| \leqslant c_9 r\sqrt{n}\Big)\Big\}.$$

Thus we shall prove the required bounds for

$$K(i) := \mathbf{E}I_i\Big\{\mathbf{P}_\tau\big(\tau \leqslant W_n^y(k,\alpha)\big)\big(\|\mathcal{Y}_{n+1}\|\,\|\mathcal{Y}_n(k)\|+$$

$$+\|\mathcal{Y}_{n+1}\|\,\|\mathcal{G}_n^*(n-k)\| + \|\mathcal{Y}_{n+1}\|\big)\big|D_n^y(k,\alpha)\big|\Big\}\Big|_{\alpha_2=1}, \ i = 15,\ldots,23.$$

Proceeding as above, we obtain that

$$K(15),\ldots,K(22) \leqslant c\mathbf{P}(G > r),$$

quite easily. To estimate $K(23)$, note that, for some $p \geqslant 0$,

$$K(23) \leqslant c(1+r)\mathbf{E}\prod_{i=1}^{3}\Big\{\mathbf{1}\Big(|Y_{n+i}| \leqslant c_9 r\sqrt{n}\Big)\times$$

$$\times\mathbf{1}\Big(\|\mathcal{Y}_{n+i}\| \leqslant c_9 r\sqrt{n}\Big)\big(1+|Y_{n+i}|\big)^p\big(1+\|\mathcal{Y}_{n+i}\|\big)^p\Big\}\times$$

$$\times\mathbf{1}\left(Y_n(k) + G_n^*(n-k) > r - \frac{1+cr^2}{\sqrt{n}} - c\sum_{j=1}^{3}\left\{\frac{1}{\sqrt{n}}|Y_{n+j}| + \frac{1}{n^{3/2}}\|\mathcal{Y}_{n+j}\|^2\right\}\right).$$

Now, analogously as in the case of $H(14)$, we obtain

$$K(23) \leqslant c(1+r)\mathbf{P}(G > r),$$

completing the estimation of $S(8)$.
Let us examine $S(9)$. Returning to (21), we see that

$$\mathbf{E}\Big\{\sqrt{n}\partial_{\alpha_2}W_n^y(k,\alpha)\Big\}\Big|_{\alpha_2=0} = 0$$

and, consequently, $S(9) = 0$.
To estimate $S(10)$, one can use the above analysis. Let us note only that

$$\sqrt{n}\partial_{\alpha_i\alpha_j}W_n^y(k,\alpha) = \frac{2d}{n}\langle\mathcal{Y}_{n+i},\mathcal{Y}_{n+j}\rangle \tag{31}$$

for all $i,j \in \{1,2,3,\}$. This ends the proof of Proposition 4 and Statement 1 as well.

Proof of Statement 2. We have

$$\Lambda^y([n/2],n) \leqslant \sum_{k=[n/2]+1}^{n} \left|\Lambda_1^y(k) + \Lambda_2^y(k)\right|,$$

where

$$\Lambda_1^y(k) := \mathbf{E}\Big\{f\big(W_n^y(k,\alpha)\big) - f\big(W_n^y(k-1,\alpha)\big)\Big\} \prod_{i=1}^{3} \Big\{\sqrt{n}\partial_{\alpha_i} W_n^y(k-1,\alpha)\Big\}^2,$$

$$\Lambda_2^y(k) := \mathbf{E} f\big(W_n^y(k,\alpha)\big)\left\{ \prod_{i=1}^{3} \Big\{\sqrt{n}\partial_{\alpha_i} W_n^y(k,\alpha)\Big\}^2 - \right.$$

$$\left. - \prod_{i=1}^{3} \Big\{\sqrt{n}\partial_{\alpha_i} W_n^y(k-1,\alpha)\Big\}^2 \right\}.$$

First we prove that

$$\Lambda_1^y(k) \leqslant c(1+r)^3 \mathbf{P}(G > r)/n^{3/2}. \tag{32}$$

If we set

$$V_n := Y_n(k-1) + G_n^*(n-k) + Y_n^\alpha + \frac{d}{\sqrt{n}} \|\mathcal{Y}_n(k-1) + \mathcal{G}_n^*(n-k) + \mathcal{Y}_n^\alpha\|^2,$$

$$R_n^s := \frac{1}{\sqrt{n}} S_k + \frac{2d}{n} \langle \mathcal{S}_k, \mathcal{Y}_n(k-1) + \mathcal{G}_n^*(n-k) + \mathcal{Y}_n^\alpha \rangle + \frac{d}{n^{3/2}} \|\mathcal{S}_k\|^2,$$

then

$$W_n^y(k,\alpha) = V_n + R_n^y, \quad W_n^y(k-1,\alpha) = V_n + R_n^g. \tag{33}$$

So, we can write

$$\Lambda_1^y(k) = \mathbf{E}\Big\{f(V_n + R_n^y) - f(V_n + R_n^g)\Big\} \prod_{i=1}^{3} \Big\{\sqrt{n}\partial_{\alpha_i} V_n + \sqrt{n}\partial_{\alpha_i} R_n^g\Big\}^2. \tag{34}$$

Note that

$$\sqrt{n}\partial_{\alpha_i} V_n = Y_{n+i} + \frac{2d}{\sqrt{n}} \Big\langle \mathcal{Y}_{n+i}, \mathcal{Y}_n(k-1) + \mathcal{G}_n^*(n-k) + \mathcal{Y}_n^\alpha \Big\rangle \tag{35}$$

and

$$\sqrt{n}\partial_{\alpha_i} R_n^s = \frac{2d}{n} \langle \mathcal{S}_k, \mathcal{Y}_{n+i} \rangle. \tag{36}$$

Therefore, from now on we can direct our attention to

$$\mathbf{E}\Big\{f(V_n + R_n^y) - f(V_n + R_n^g)\Big\} \prod_{i=1}^{3} \big\{\sqrt{n}\partial_{\alpha_i} V_n\big\}^2 \tag{37}$$

and

$$\mathbf{E}\Big\{f(V_n + R_n^y) - f(V_n + R_n^g)\Big\} \prod_{i=1}^{2} \{\sqrt{n}\partial_{\alpha_i} V_n\}^2 \{\sqrt{n}\partial_{\alpha_3} V_n\}\{\sqrt{n}\partial_{\alpha_3} R_n^g\}, \quad (38)$$

only. (The remaining components of (34) are not greater than c/n^2 for some $c \geqslant 0$.) Moreover, we can decompose (37) and (38) using

$$f(V_n + R_n^y) - f(V_n + R_n^g) = f'(V_n)\{R_n^y - R_n^g\} + \frac{1}{2} f''(V_n)\{(R_n^y)^2 - (R_n^g)^2\} +$$

$$+ \frac{1}{2}\mathbf{E}_\nu (1-\nu)^2 \Big\{ f'''(V_n + \nu R_n^y)(R_n^y)^3 - f'''(V_n + \nu R_n^g)(R_n^g)^3 \Big\},$$

where ν denotes a random variable distributed uniformly in (0,1). As a result we have six quantities $F(1)\dots F(6)$, the estimation of which leads to (32). These quantities are exactly defined and estimated in the following six propositions.

PROPOSITION 5. *For any* $n \geqslant 1$

$$F(1) := \left| \mathbf{E} f'(V_n)\{R_n^y - R_n^g\} \prod_{i=1}^{3} \{\sqrt{n}\partial_{\alpha_i} V_n\}^2 \right| \leqslant c(1+r)\mathbf{P}(G > r)/n^{3/2}.$$

Proof. Since $\mathbf{E}Y_k = \mathbf{E}G_k = 0$ and $\mathbf{E}\langle \mathcal{H}, \mathcal{Y}_k \rangle = \mathbf{E}\langle \mathcal{H}, \mathcal{G}_k \rangle = 0$ for all $\mathcal{H} \in \mathbf{H}$, we have

$$F(1) = \frac{c}{n^{3/2}} \left| \mathbf{E} f'(V_n) \prod_{i=1}^{3} \{\sqrt{n}\partial_{\alpha_i} V_n\}^2 \right|.$$

Replace now $f'(V_n)\partial_{\alpha_1} V_n$ by $\partial_{\alpha_1} f(V_n)$ and use integration by parts. We get

$$nF(1) \leqslant c\{S(11) + S(12) + S(13)\},$$

where

$$S(11) := \left| \mathbf{E}\Big(f(V_n)\{\sqrt{n}\partial_{\alpha_1} V_n\} \prod_{i=2}^{3} \{\sqrt{n}\partial_{\alpha_i} V_n\}^2 \Big)\Big|_{\alpha_1 = 1} \right|,$$

$$S(12) := \left| \mathbf{E}\Big(f(V_n)\{\sqrt{n}\partial_{\alpha_1} V_n\} \prod_{i=2}^{3} \{\sqrt{n}\partial_{\alpha_i} V_n\}^2 \Big)\Big|_{\alpha_1 = 0} \right|,$$

$$S(13) := \left| \mathbf{E} f(V_n)\partial_{\alpha_1} \Big(\{\sqrt{n}\partial_{\alpha_1} V_n\} \prod_{i=2}^{3} \{\sqrt{n}\partial_{\alpha_i} V_n\}^2 \Big) \right|.$$

From (27) we deduce that in $S(11)$ the quantity $\sqrt{n}(\partial_{\alpha_1} V_n)\big|_{\alpha_1 = 1}$ can be replaced by

$$\frac{1}{k}\left(\sum_{i=1}^{k-1} Y_i + Y_{n+1} \right) + \frac{2d}{k\sqrt{n}} \left\langle \sum_{i=1}^{k-1} \mathcal{Y}_i + \mathcal{Y}_{n+1}, \mathcal{Y}_n(k-1) + \mathcal{G}_n^*(n-k) + \mathcal{Y}_n^\alpha \Big|_{\alpha_1 = 1} \right\rangle.$$

The obvious inequality

$$f(x) \leqslant \mathbf{1}(x > r - 1/\sqrt{n}) \tag{39}$$

and arguments used above yield the bound $c(1+r)\mathbf{P}(G > r)/\sqrt{n}$ for $S(11)$. Next, $S(12) = 0$ because the conditional expectation of $(\partial_{\alpha_1} V_n)|_{\alpha_1=0}$ with respect to Y_{n+1} and $\mathcal{Y}_{n+1}$ vanishes.
Finally, noting that

$$\sqrt{n}\partial^2_{\alpha_i\alpha_j} V_n = \frac{2d}{n}\langle \mathcal{Y}_{n+i}, \mathcal{Y}_{n+j}\rangle, \tag{40}$$

recalling (39) and following the ideas used in the proof of Propositions 3 and 4, we see that $S(13)$ is not greater than $c\mathbf{P}(G > r)/n$ for some $c \geqslant 0$.

PROPOSITION 6. *For any* $n \geqslant 1$

$$F(2) := \left|\mathbf{E}f'(V_n)\{R^y_n - R^g_n\}\prod_{i=1}^{2}\{\sqrt{n}\partial_{\alpha_i}V_n\}^2\{\sqrt{n}\partial_{\alpha_3}V_n\}\{\sqrt{n}\partial_{\alpha_3}R^g_n\}\right| \leqslant$$

$$\leqslant c(1+r)\mathbf{P}(G > r)/n^{3/2}.$$

Proof. We start by noting that

$$F(2) = \frac{2d}{n}\left|\mathbf{E}\left\{\partial_{\alpha_3}f(V_n)\right\}\left\{\sqrt{n}R^y_n - \sqrt{n}R^g_n\right\}\prod_{i=1}^{2}\left\{\sqrt{n}\partial_{\alpha_i}V_n\right\}^2\left\langle \mathcal{G}_k, \mathcal{Y}_{n+3}\right\rangle\right|.$$

Thus, integrating by parts, we obtain

$$nF(2) \leqslant 2d\{S(14) + S(15) + S(16)\}$$

$$S(14) := \left|\mathbf{E}\left(f(V_n)\left\{\sqrt{n}R^y_n - \sqrt{n}R^g_n\right\}\prod_{i=1}^{2}\left\{\sqrt{n}\partial_{\alpha_i}V_n\right\}^2\right)\Big|_{\alpha_3=1}\langle \mathcal{G}_k, \mathcal{Y}_{n+3}\rangle\right|,$$

$$S(15) := \left|\mathbf{E}\left(f(V_n)\left\{\sqrt{n}R^y_n - \sqrt{n}R^g_n\right\}\prod_{i=1}^{2}\left\{\sqrt{n}\partial_{\alpha_i}V_n\right\}^2\right)\Big|_{\alpha_3=0}\langle \mathcal{G}_k, \mathcal{Y}_{n+3}\rangle\right|,$$

$$S(16) := \left|\mathbf{E}f(V_n)\partial_{\alpha_3}\left(\left\{\sqrt{n}R^y_n - \sqrt{n}R^g_n\right\}\prod_{i=1}^{2}\left\{\sqrt{n}\partial_{\alpha_i}V_n\right\}^2\right)\langle \mathcal{G}_k, \mathcal{Y}_{n+3}\rangle\right|.$$

First of all, note that (27) implies

$$S(14) = \frac{1}{k}\left|\mathbf{E}\left(f(V_n)\left\{\sqrt{n}R^y_n - \sqrt{n}R^g_n\right\}\prod_{i=1}^{2}\left\{\sqrt{n}\partial_{\alpha_i}V_n\right\}^2\right)\Big|_{\alpha_3=1}\times\right.$$

$$\left.\times\left\langle \mathcal{G}_k, \mathcal{Y}_1 + \ldots + \mathcal{Y}_{k-1} + \mathcal{Y}_{n+3}\right\rangle\right|,$$

and the technique above leads us to the estimate $S(14) \leqslant c(1+r)\mathbf{P}(G > r)/\sqrt{n}$.

The quantity $S(15)$ vanishes, because of $\mathbf{E}\langle\mathcal{H},\mathcal{Y}_{n+3}\rangle = 0$. Furthermore, $S(16) \leqslant \leqslant c\mathbf{P}(G > r)/n$, which follows from (39), (40) and the equality

$$\partial_{\alpha_i}\{\sqrt{n}R_n^y - \sqrt{n}R_n^g\} = \frac{2d}{n}\langle\mathcal{Y}_k - \mathcal{G}_k, \mathcal{Y}_{n+1}\rangle.$$

PROPOSITION 7. *For any* $n \geqslant 1$

$$F(3) := \left|\mathbf{E}f''(V_n)\Big\{(R_n^y)^2 - (R_n^g)^2\Big\}\prod_{i=1}^{3}\Big\{\sqrt{n}\partial_{\alpha_i}V_n\Big\}^2\right| \leqslant c(1+r)^3\mathbf{P}(G > r)/n^{3/2}.$$

Proof. Recall that $\mathbf{E}Y_k^2 = \mathbf{E}G_k^2$ and $\mathbf{E}\langle\mathcal{H},\mathcal{Y}_k\rangle^2 = \mathbf{E}\langle\mathcal{H},\mathcal{G}_k\rangle^2$ for all $\mathcal{H} \in \mathbf{H}$. Therefore, $n^{3/2}\{(R_n^y)^2 - (R_n^g)^2\}$ equals to

$$\begin{aligned}R_n := \frac{d^2}{n^{3/2}}\Big(\|\mathcal{Y}_k\|^4 - \|\mathcal{G}_k\|^4\Big) + \frac{2d}{\sqrt{n}}\Big(Y_k\|\mathcal{Y}_k\|^2 - G_k\|\mathcal{G}_k\|^2\Big) + \\ +4d\Big(Y_k\big\langle\mathcal{Y}_k, \mathcal{Y}_n(k-1) + \mathcal{G}_n^*(n-k) + \mathcal{Y}_n^\alpha\big\rangle - \\ -G_k\big\langle\mathcal{G}_k, \mathcal{Y}_n(k-1) + \mathcal{G}_n^*(n-k) + \mathcal{Y}_n^\alpha\big\rangle\Big) + \\ +\frac{4d^2}{n}\Big(\|\mathcal{Y}_k\|^2\big\langle\mathcal{Y}_k, \mathcal{Y}_n(k-1) + \mathcal{G}_n^*(n-k) + \mathcal{Y}_n^\alpha\big\rangle - \\ -\|\mathcal{G}_k\|^2\big\langle\mathcal{G}_k, \mathcal{Y}_n(k-1) + \mathcal{G}_n^*(n-k) + \mathcal{Y}_n^\alpha\big\rangle\Big).\end{aligned}$$

So rewrite

$$F(3) = \frac{1}{n^{3/2}}\left|\mathbf{E}f''(V_n)R_n\prod_{i=1}^{3}\{\sqrt{n}\partial_{\alpha_i}V_n\}^2\right|$$

and, since

$$f''(V_n)\partial_{\alpha_1}(V_n)\partial_{\alpha_2}(V_n) = \partial^2_{\alpha_1\alpha_2}\big(f(V_n)\big) - f'(V_n)\partial_{\alpha_1\alpha_2}V_n, \tag{41}$$

we have

$$\sqrt{n}F(3) \leqslant S(17) + S(18),$$

where

$$S(17) := \left|\mathbf{E}\Big\{\partial^2_{\alpha_1\alpha_2}f(V_n)\Big\}R_n\prod_{i=1}^{2}\{\sqrt{n}\partial_{\alpha_i}V_n\}\{\sqrt{n}\partial_{\alpha_3}V_n\}^2\right|,$$

$$S(18) := \left|\mathbf{E}f'(V_n)\Big\{\partial^2_{\alpha_1\alpha_2}V_n\Big\}R_n\prod_{i=1}^{2}\{\sqrt{n}\partial_{\alpha_i}V_n\}\{\sqrt{n}\partial_{\alpha_3}V_n\}^2\right|.$$

Using

$$\big|f'(x)\big| \leqslant c\mathbf{1}\big(x > r - 1\sqrt{n}\big)\sqrt{n} \tag{42}$$

we obtain that $S(18) \leqslant c(1+r)\mathbf{P}(G > r)/n$ (recall (40) and use arguments above). Let us estimate $S(17)$. Integrating by parts we can change $\partial^2_{\alpha_1\alpha_2} f(V_n)$ by $f(V_n)$. In such a way nine quantities appear. Denote their absolute values by $E(1), \dots, E(9)$; the exact definitions of them can be found below. For convenience, define

$$\Phi := f(V_n), \quad \Psi := R_n \prod_{i=1}^{2} \{\sqrt{n}\partial_{\alpha_i} V_n\} \{\sqrt{n}\partial_{\alpha_3} V_n\}^2$$

and consider the quantity

$$E(1) = \left| \mathbf{E}(\Phi\Psi)|_{\alpha_1=1,\alpha_2=1} \right|$$

firstly. Using (28) and applying the ideas used above, we have that

$$E(1) \leqslant c(1+r)^3 \mathbf{P}(G > r)/n.$$

The quantities

$$E(2) = \left| \mathbf{E}(\Phi\Psi)|_{\alpha_1=1,\alpha_2=0} \right|, \quad E(3) = \left| \mathbf{E}(\Phi\Psi)|_{\alpha_1=0,\alpha_2=1} \right|,$$

$$E(4) = \left| \mathbf{E}(\Phi\Psi)|_{\alpha_1=0,\alpha_2=0} \right|$$

vanish. Furthermore, the quantities

$$E(5) = \left| \mathbf{E}(\Phi\partial_{\alpha_1}\Psi)|_{\alpha_2=1} \right|, \quad E(6) = \left| \mathbf{E}(\Phi\partial_{\alpha_2}\Psi)|_{\alpha_1=1} \right|$$

coincide; choose the first one and decompose it as follows:

$$E(5) \leqslant E^*(5) + E^{**}(5),$$

where

$$E^*(5) := \left| \mathbf{E}\Big(f(V_n)\partial_{\alpha_1}(R_n) \prod_{i=1}^{2} \{\sqrt{n}\partial_{\alpha_i} V_n\} \{\sqrt{n}\partial_{\alpha_3} V_n\}^2 \Big)\Big|_{\alpha_2=1} \right|,$$

$$E^{**}(5) := \left| \mathbf{E}\Big(f(V_n) R_n \partial_{\alpha_1} \prod_{i=1}^{2} \{\sqrt{n}\partial_{\alpha_i} V_n\} \{\sqrt{n}\partial_{\alpha_3} V_n\}^2 \Big)\Big|_{\alpha_2=1} \right|.$$

Recalling (39) and (40), and using the above technique, we arrive at

$$E^{**}(5) \leqslant c(1+r)\mathbf{P}(G > r)/n.$$

To prove the same bound for $E^*(5)$, note that

$$\partial_{\alpha_1} R_n = \frac{4d}{\sqrt{n}} \Big(Y_k \langle \mathcal{Y}_k, \mathcal{Y}_{n+1} \rangle - G_k \langle \mathcal{G}_k, \mathcal{Y}_{n+1} \rangle \Big) +$$

$$+\frac{4d^2}{n^{3/2}}\Big(\|\mathcal{Y}_k\|^2\langle\mathcal{Y}_k,\mathcal{Y}_{n+1}\rangle-\|\mathcal{G}_k\|^2\langle\mathcal{G}_k,\mathcal{Y}_{n+1}\rangle\Big)$$

and use (27) and (39). Arguments similar to those used in the proof of Propositions 3 and 4 prove the required bound.
The quantities

$$E(7)=\left|\mathbf{E}(\Phi\partial_{\alpha_1}\Psi)\big|_{\alpha_2=0}\right|,\quad E(8)=\left|\mathbf{E}(\Phi\partial_{\alpha_2}\Psi)\big|_{\alpha_1=0}\right|$$

vanish.
To prove that $E(9)=\mathbf{E}\Phi\partial^2_{\alpha_1\alpha_2}\Psi$ does not exceed $c(1+r)\mathbf{P}(G>r)/n$, note that $\partial^2_{\alpha_1\alpha_2}R_n$ vanishes, use (39) and (40), and then follow the arguments used in the proof of Propositions 3 and 4. Putting all the parts together, we have the proof of Proposition 7 completed.

PROPOSITION 8. *For any $n\geqslant 1,0$.*

$$F(4):=\left|\mathbf{E}f''(V_n)\Big\{(R_n^y)^2-(R_n^g)^2\Big\}\prod_{i=1}^{2}\{\sqrt{n}\partial_{\alpha_i}V_n\}^2\{\sqrt{n}\partial_{\alpha_3}V_n\}\{\sqrt{n}\partial_{\alpha_3}R_n^g\}\leqslant\right.$$

$$\leqslant c(1+r)^2\mathbf{P}(G>r)/n^2.$$

Proof. From (36) and (41) we have

$$nF(4)\leqslant 2d\{S(19)+S(20)\},$$

where

$$S(19):=\left|\mathbf{E}\partial^2_{\alpha_1\alpha_2}(f(V_n))\Big\{(\sqrt{n}R_n^y)^2-(\sqrt{n}R_n^g)^2\Big\}\prod_{i=1}^{3}\{\sqrt{n}\partial_{\alpha_i}V_n\}\langle\mathcal{G}_k,\mathcal{Y}_{n+3}\rangle\right|,$$

$$S(20):=\left|\mathbf{E}f'(V_n)\Big\{(\sqrt{n}R_n^y)^2-(\sqrt{n}R_n^g)^2\Big\}\partial^2_{\alpha_1\alpha_2}(V_n)\prod_{i=1}^{3}\{\sqrt{n}\partial_{\alpha_i}V_n\}\langle\mathcal{G}_k,\mathcal{Y}_{n+3}\rangle\right|.$$

Using (42) and the technique above, we can prove that $S(20)\leqslant c\mathbf{P}(G>r)/n$. The quantity $S(19)$ does not exceed $c(1+r)^2\mathbf{P}(G>r)/n$. We omit the details, referring the reader to the estimation of $S(17)$.

PROPOSITION 9. *For any $n\geqslant 1$*

$$\left|\mathbf{E}(1-\nu)^2\Big\{f'''(V_n+\nu R_n^y)(R_n^y)^3-f'''(V_n+\nu R_n^g)(R_n^g)^3\Big\}\prod_{i=1}^{3}\{\sqrt{n}\partial_{\alpha_i}V_n\}^2\right|\leqslant$$

$$\leqslant c(1+r)^3\mathbf{P}(G>r)/n^{3/2}.$$

Proof. Set $V_n^* := V_n + \nu R_n^y$ and consider

$$F(5) := \left|\mathbf{E}(1-\nu)^2 f'''(V_n^*)(R_n^y)^3 \prod_{i=1}^{3}\{\sqrt{n}\partial_{\alpha_i} V_n\}^2\right|$$

only; the other quantities can be estimated similarly. We have that $F(5) \leqslant S(21)+$ $+S(22)$, where

$$S(21) := \left|\mathbf{E}(1-\nu)^2 f'''(V_n^*)(R_n^y)^3 \prod_{i=1}^{3}\{\sqrt{n}\partial_{\alpha_i} V_n^*\}^2\right|,$$

$$S(22) := \left|\mathbf{E}(1-\nu)^2 f'''(V_n^*)(R_n^y)^3 \left\{\prod_{i=1}^{3}\{\sqrt{n}\partial_{\alpha_i} V_n\}^2 - \prod_{i=1}^{3}\{\sqrt{n}\partial_{\alpha_i} V_n^*\}^2\right\}\right|.$$

We start by considering $S(21)$. Clearly, the following formula

$$f'''(x)\prod_{i=1}^{3}\partial_{\alpha_i} x = \partial^3_{\alpha_1\alpha_2\alpha_3} f(x) - f''(x)\partial_{\alpha_1}(x)\partial^2_{\alpha_2\alpha_3}(x) -$$

$$-f''(x)\partial_{\alpha_2}(x)\partial^2_{\alpha_1\alpha_3}(x) - f''(x)\partial_{\alpha_3}(x)\partial^2_{\alpha_1\alpha_2}(x) - f'(x)\partial^3_{\alpha_1\alpha_2\alpha_3}(x) \qquad (43)$$

leads us to $S(21) \leqslant S^*(21) + S^{**}(21) + S^{***}(21)$, where

$$S^*(21) := \left|\mathbf{E}(1-\nu)^2 \partial^3_{\alpha_1\alpha_2\alpha_3}(f(V_n^*))\{\sqrt{n}R_n^y\}^3 \prod_{i=1}^{3}\{\sqrt{n}\partial_{\alpha_i} V_n^*\}\right|,$$

$$S^{**}(21) := \frac{3}{\sqrt{n}}\left|\mathbf{E}(1-\nu)^2 f''(V_n^*)\partial^2_{\alpha_2\alpha_3}(V_n^*)\times\right.$$

$$\left.\times\{\sqrt{n}R_n^y\}^3\{\sqrt{n}\partial_{\alpha_1} V_n^*\}^2 \prod_{i=2}^{3}\{\sqrt{n}\partial_{\alpha_i} V_n^*\}\right|,$$

$$S^{***}(21) := \left|\mathbf{E}(1-\nu)^2 f'(V_n^*)\partial^3_{\alpha_1\alpha_2\alpha_3}(V_n^*)\{\sqrt{n}R_n^y\}^3 \prod_{i=1}^{3}\{\sqrt{n}\partial_{\alpha_i} V_n^*\}\right|.$$

Computations similar (but more tedious) to those used for $S(17)$ show that

$$S^*(21) \leqslant c(1+r)^3 \mathbf{P}(G > r)/n^{3/2}.$$

The quantity $S^{***}(21)$ vanishes, since $\partial^3_{\alpha_1\alpha_2\alpha_3} V_n^* = 0$ (recall (36) and (40)). To bound $S^{**}(21)$, replace $f''(\ldots)\partial_{\alpha_1}(\ldots)$ by $\partial_{\alpha_1} f'(\ldots)$ and note that

$$\partial^2_{\alpha_i\alpha_j} V_n^* = \frac{2d}{n^{3/2}}\langle \mathcal{Y}_{n+i}, \mathcal{Y}_{n+j}\rangle.$$

So,

$$S^{**}(21) = \frac{c}{n^{3/2}}\left|\mathbf{E}(1-\nu)^2 \partial_{\alpha_1} f'(V_n^*)\langle \mathcal{Y}_{n+2}, \mathcal{Y}_{n+3}\rangle\{\sqrt{n}R_n^y\}^3 \prod_{i=1}^{3}\{\sqrt{n}\partial_{\alpha_i} V_n^*\}\right|$$

and the technique above allow us to obtain the bound

$$S^{**}(21) \leqslant c(1+r)\mathbf{P}(G>r)/n^{3/2},$$

completing the estimation of $S(21)$.
To estimate $S(22)$, let us firstly observe that the main task is to prove the bound

$$S(22) \leqslant cQ/n^{3/2} + c\mathbf{P}(G>r)/n^2,$$

where

$$Q := \Big|\mathbf{E}(1-\nu)^2\nu f'''(V_n^*)\{\sqrt{n}R_n^y\}^3 \prod_{i=1}^{2}\{\sqrt{n}\partial_{\alpha_i}V_n\}^2\{\sqrt{n}\partial_{\alpha_3}V_n\}\{\sqrt{n}\partial_{\alpha_3}R_n^y\}\Big|.$$

This can be achieved quite easily using the bound

$$|f'''(x)| \leqslant c\mathbf{1}\Big(x > r - 1/\sqrt{n}\Big)n^{3/2} \tag{44}$$

and the argument above. On the other hand, $Q \leqslant S(23) + S(24)$, where

$$S(23) := \Big|\mathbf{E}(1-\nu)^2 f'''(V_n^*)\{\sqrt{n}R_n^y\}^3 \prod_{i=1}^{2}\{\sqrt{n}\partial_{\alpha_i}V_n\}^2\{\sqrt{n}\partial_{\alpha_3}V_n^*\}\{\sqrt{n}\nu\partial_{\alpha_3}R_n^y\}\Big|,$$

$$S(24) := \Big|\mathbf{E}(1-\nu)^2 f'''(V_n^*)\{\sqrt{n}R_n^y\}^3 \prod_{i=1}^{2}\{\sqrt{n}\partial_{\alpha_i}V_n\}^2\{\sqrt{n}\nu\partial_{\alpha_3}R_n^y\}^2\Big|.$$

So, we shall direct our attention to these last two quantities.
If we recall (36) and (44), arguments used in previous propositions show that

$$S(24) \leqslant c\mathbf{P}(G>r)/\sqrt{n}.$$

To estimate $S(23)$, let us firstly replace $f'''(\ldots)\partial_{\alpha_3}(\ldots)$ by $\partial_{\alpha_3}f'''(\ldots)$. Then

$$S(23) = \sqrt{n}\Big|\mathbf{E}(1-\nu)^2\nu\partial_{\alpha_3}(f''(V_n^*))\{\sqrt{n}R_n^y\}^3 \prod_{i=1}^{2}\{\sqrt{n}\partial_{\alpha_i}V_n\}^2\{\sqrt{n}\partial_{\alpha_3}R_n^y\}\Big|.$$

From this, integrating by parts, we have that $S(23) \leqslant S^*(23) + S^{**}(23) + S^{***}(23)$, where

$$S^*(23) := \sqrt{n}\Big|\mathbf{E}(1-\nu)^2\nu\Big(f''(V_n^*)\{\sqrt{n}R_n^y\}^3 \prod_{i=1}^{2}\{\sqrt{n}\partial_{\alpha_i}V_n\}^2\{\sqrt{n}\partial_{\alpha_3}R_n^y\}\Big)\Big|_{\alpha_3=1}\Big|,$$

$$S^{**}(23) := \sqrt{n}\Big|\mathbf{E}(1-\nu)^2\nu\Big(f''(V_n^*)\{\sqrt{n}R_n^y\}^3 \prod_{i=1}^{2}\{\sqrt{n}\partial_{\alpha_i}V_n\}^2\{\sqrt{n}\partial_{\alpha_3}R_n^y\}\Big)\Big|_{\alpha_3=0}\Big|,$$

$$S^{***}(23) := \sqrt{n}\left|\mathbf{E}(1-\nu)^2\nu f''(V_n^*)\partial_{\alpha_3}\Big(\{\sqrt{n}R_n^y\}^3\prod_{i=1}^{2}\{\sqrt{n}\partial_{\alpha_i}V_n\}^2\{\sqrt{n}\partial_{\alpha_3}R_n^y\}\Big)\right|.$$

Recalling (36), using (27) and

$$\left|f''(x)\right| \leqslant c\mathbf{1}(x > r - 1/\sqrt{n})n \tag{45}$$

we obtain that $S^*(23)$ does not exceed $c(1+r)\mathbf{P}(G > r)$.
Furthermore, $S^{**}(23) = 0$ because the conditional expectation of $\langle \mathcal{Y}_k, \mathcal{Y}_{n+3}\rangle$ with respect to $\mathcal{Y}_{n+3}$ vanishes. If we recall (36) and (40), then it is not difficult to prove that $S^{***}(23)$ does not exceed $c\mathbf{P}(G > r)/\sqrt{n}$, which completes the proof of the bound $Q \leqslant c(1+r)\mathbf{P}(G > r)$ and the proposition as well.

PROPOSITION 10. *For any* $n \geqslant 1$

$$\Big|\mathbf{E}(1-\nu)^2\Big\{f'''(V_n+\nu R_n^y)\big(R_n^y\big)^3 - f'''(V_n+\nu R_n^g)\big(R_n^g\big)^3\Big\}\times$$

$$\times\prod_{i=1}^{2}\{\sqrt{n}\partial_{\alpha_i}V_n\}^2\{\sqrt{n}\partial_{\alpha_3}V_n\}\{\sqrt{n}\partial_{\alpha_3}R_n^g\}\Big| \leqslant c(1+r)\mathbf{P}(G > r)/n^{3/2}.$$

Proof. As in the previous proposition, set $V_n^* := V_n + \nu R_n^y$. Consider the quantity

$$F(6) := \frac{1}{n^{5/2}}\Big|\mathbf{E}(1-\nu)^2 f'''(V_n^*)\{\sqrt{n}R_n^y\}^3\prod_{i=1}^{2}\{\sqrt{n}\partial_{\alpha_i}V_n\}^2\times$$

$$\times\{\sqrt{n}\partial_{\alpha_3}V_n\}\langle \mathcal{G}_k, \mathcal{Y}_{n+3}\rangle\Big|,$$

which we decompose as follows: $n^{5/2}\ F(6) \leqslant S(25) + S(26)$, where

$$S(25) := \Big|\mathbf{E}(1-\nu)^2 f'''(V_n^*)\{\sqrt{n}R_n^y\}^3\prod_{i=1}^{2}\{\sqrt{n}\partial_{\alpha_i}V_n\}^2\{\sqrt{n}\partial_{\alpha_3}V_n^*\}\langle \mathcal{G}_k, \mathcal{Y}_{n+3}\rangle\Big|,$$

$$S(26) := \Big|\mathbf{E}(1-\nu)^2 f'''(V_n^*)\{\sqrt{n}R_n^y\}^3\prod_{i=1}^{2}\{\sqrt{n}\partial_{\alpha_i}V_n\}^2\{\sqrt{n}\nu\partial_{\alpha_3}R_n^y\}\langle \mathcal{G}_k, \mathcal{Y}_{n+3}\rangle\Big|.$$

The bound $S(26) \leqslant c\mathbf{P}(G > r)/\sqrt{n}$ can be proved quite easily if we recall (36) and use (44). Furthermore, if we replace $f'''(\ldots)\partial_{\alpha_3}(\ldots)$ by $\partial_{\alpha_3}f''(\ldots)$, then integrating by parts we obtain $S(25) \leqslant S^*(25) + S^{**}(25) + S^{***}(25)$, where

$$S^*(25) := \sqrt{n}\Big|\mathbf{E}(1-\nu)^2\Big(f''(V_n^*)\{\sqrt{n}R_n^y\}^3\prod_{i=1}^{2}\{\sqrt{n}\partial_{\alpha_i}V_n\}^2\Big)\Big|_{\alpha_3=1}\langle \mathcal{G}_k, \mathcal{Y}_{n+3}\rangle\Big|,$$

$$S^{**}(25) := \sqrt{n}\Big|\mathbf{E}(1-\nu)^2\Big(f''(V_n^*)\{\sqrt{n}R_n^y\}^3\prod_{i=1}^{2}\{\sqrt{n}\partial_{\alpha_i}V_n\}^2\Big)\Big|_{\alpha_3=0}\langle \mathcal{G}_k, \mathcal{Y}_{n+3}\rangle\Big|,$$

$$S^{***}(25) := \sqrt{n}\left|\mathbf{E}(1-\nu)^2 f''(V_n^*)\partial_{\alpha_3}\left(\left\{\sqrt{n}R_n^y\right\}^3 \prod_{i=1}^{2}\left\{\sqrt{n}\partial_{\alpha_i}V_n\right\}^2\right)\langle \mathcal{G}_k, \mathcal{Y}_{n+3}\rangle\right|.$$

Using (27) and (45) we prove that $S^*(25) \leqslant c(1+r)\mathbf{P}(G > r)n$. The quantity $S^{**}(25)$ vanishes due to $\mathbf{E}\langle \mathcal{H}, \mathcal{Y}_{n+3}\rangle = 0$ for all $\mathcal{H} \in \mathbf{H}$. If we recall (36) and (40), then using (45) we obtain that $S^{***}(25)$ does not exceed $c\mathbf{P}(G > r)n$. This completes the proof.
Now we shall estimate $\Lambda_2^y(k)$.

PROPOSITION 11. *For all* $k = [n/2], \ldots, n$,

$$\left|\Lambda_2^y(k)\right| \leqslant c(1+r)\mathbf{P}(G > r)/n^{3/2}.$$

Proof. We know, from (33), that

$$W_n^y(k-1,\alpha) = W_n^y(k,\alpha) + R_n^g - R_n^y.$$

Thus $\Lambda_2^y(k)$ equals to

$$\mathbf{E}f(W_n^y(k,\alpha))\left\{\prod_{i=1}^{3}\left\{\sqrt{n}\partial_{\alpha_i}W_n^y(k,\alpha)\right\}^2 - \right.$$

$$\left. - \prod_{i=1}^{3}\left\{\sqrt{n}\partial_{\alpha_i}W_n^y(k,\alpha) - \frac{2d}{n}\langle \mathcal{G}_k - \mathcal{Y}_k, \mathcal{Y}_{n+1}\rangle\right\}^2\right\}$$

which, by the technique used in Propositions 3 and 4, leads us to

$$\left|\Lambda_2^y(k)\right| \leqslant c|M|/n + c\mathbf{P}(G > r)/n^2,$$

where

$$M := \mathbf{E}f(W_n^y(k,\alpha))\prod_{i=1}^{2}\left\{\sqrt{n}\partial_{\alpha_i}W_n^y(k,\alpha)\right\}^2\langle \mathcal{G}_k - \mathcal{Y}_k, \mathcal{Y}_{n+3}\rangle.$$

Since the expectation $\mathbf{E}\langle \mathcal{H}, \mathcal{G}_k\rangle$ vanishes for all $\mathcal{H} \in \mathbf{H}$, M equals to

$$\mathbf{E}f(W_n^y(k,\alpha))\prod_{i=1}^{2}\left\{\sqrt{n}\partial_{\alpha_i}W_n^y(k,\alpha)\right\}^2\langle \mathcal{Y}_k, \mathcal{Y}_{n+3}\rangle.$$

Therefore, using (27) we arrive at

$$|M| \leqslant c(1+r)\mathbf{P}(G > r)/\sqrt{n},$$

which completes the proof of the proposition and of Statement 2 as well.

4. PROOF OF THEOREM 3

It is well known (see, e.g., (Govindarajulu and Mason, 1983; Helmers, 1982)) that

$$\widehat{L}_n - \mu = \int_{\mathbf{R}} \Big\{\psi(F_n(x)) - \psi(F(x))\Big\}dx, \tag{46}$$

where F_n denotes the empirical distribution function, i.e.,

$$F_n(x) = \frac{1}{n}\sum_{i=1}^{n} \mathbf{1}(X_i \leqslant x)$$

for all $x \in \mathbf{R}$, and

$$\psi(x) := \int_{(x,1)} J(u)du$$

for all $x \in (0,1)$. If we use (6) in (46), we obtain

$$\frac{\sqrt{n}\{\widehat{L}_n - \mu\}}{\sigma} = \frac{1}{\sqrt{n}}\sum_{i=1}^{n} Y_i + \frac{\varepsilon k}{\sigma}\frac{1}{\sqrt{n}}\left\|\frac{1}{\sqrt{n}}\sum_{i=1}^{n}\mathcal{Y}_i\right\|^2$$

for some $\varepsilon \in [-1,1]$, where $\mathcal{Y}_i\big(\in \mathbf{L}_2(\mathbf{R})\big)$ is defined by

$$\mathcal{Y}_i(x) := \mathbf{1}(X_i \leqslant x) - F(x)$$

for all $x \in \mathbf{R}$, and Y_i equals to

$$\frac{1}{\sigma}\int_{\mathbf{R}} J\big(F(x)\big)\mathcal{Y}_i(x)dx.$$

This result, together with Theorem 1, completes the proof.

Acknowledgments
I wish to express my thanks to my teachers and scientific supervisors V. Bentkus and V. Paulauskas for discussion and remarks. I also thank D. Surgailis and E. Žalys for their comments and suggestions on the manuscript.

REFERENCES

Aleškevičienė, A. (1989). On large deviations for linear combinations of order statistics. *Lithuanian Math. J.*, **29**, 212–222 (in Russian).

Aleškevičienė, A. (1990). *Private Communication.*

Bentkus, V. (1984). Asymptotic expansions for the distributions of sums of independent random elements of a Hilbert space. *Lithuanian Math. J.*, **24**, 305–319.

Bentkus, V. (1986). Large deviations in Banach spaces. *Theory Probab. Appl.*, **31**, 627–633.

Bentkus, V. and Račkauskas, A. (1990). On probabilities of large deviations in Banach spaces. *Probab. Theory and Rel. Fields,* **86**, 131–154.

Bentkus, V. and Zitikis, R. (1990). Probabilities of large deviations for L–statistics. *Lithuanian Math. J.,* **30**, 479–488.

Callaert, H., Vandemaele, M. and Veraverbeke, N. (1982). A Crámer type large deviation theorem for trimmed linear combinations of order statistics. *Commun. Statist.-Theor. Meth.,* **11**, 2689–2698.

Friedrich, K.O. (1989). A Berry–Esséen bound for functions of independent random variables. *Ann. Statist.,* **17**, 170–183.

Ghosh, M. (1972). On the representation of linear functions of order statistics. *Sankhya* A, **34**, 349–356.

Götze, F. (1979). Asymptotic expansions for bivariate von Mises functionals. *Z. Wahr. Verw. Gebiete,* **50**, 333–355.

Götze, F. (1984). Expansions for von Mises functionals. *Z. Wahr. Verw. Gebiete,* **65**, 599–625.

Götze, F. (1989). Edgeworth expansions in functional limit theorems. *Ann. Probab.* **17**, 1602–1634.

Govindarajulu, Z. and Mason, D.M. (1983). A strong representation for linear combinations of order statistics with application to fixed–width confidence intervals for location and scale parameters. *Scand. J. Statist.,* **10**, 97–115.

Groeneboom, P. and Shorack, G.R. (1981). Large deviations of fit statistics and linear combinations of order statistics. *Ann. Probab.,* **9**, 971–987.

Helmers, R. (1982). *Edgeworth Expansions for Linear Combinations of Order Statistics.* Mathematical Centre Tracts 105, Amsterdam.

Ibragimov, I.A. and Linnik, Yu.V. (1971). *Independent and Stationary Sequences of Random Variables.* Springer, Berlin etc.

Kuo, H.-H. (1975). Gaussian Measures in Banach Spaces. *Lect. Notes in Math.* Springer, Berlin etc.

Paulauskas, V. and Račkauskas, A. (1989). *Approximation Theory in the Central Limit Theorem. Exact Results in Banach Spaces.* Kluwer, Dordrecht.

Petrov, V.V. (1975). *Sums of Independent Random Variables.* Springer, Berlin etc.

Pisier, G. (1985). Probabilistic Methods in the Geometry of Banach Spaces. *Lect. Notes in Math.* **1206**, pp. 167–241.

Puri, M.L. and Seoh, M. (1988). On the rate of convergence in normal approximation and large deviation probabilities for a class of statistics. *Theory Probab. Appl.,* **33**, 682–697.

Račkauskas, A. (1988). Probabilities of large deviations in Linnik zones in a Hilbert space. *Lithuanian Math. J.,* **28**, 239–247.

Saulis, L. and Statulevičius, V. (1989). *Limit Theorems for Large Deviations.* Mokslas, Vilnius (in Russian).

Seoh, M., Ralescu, S.S. and Puri, M.L. (1985). Cramér type large deviations for generalized rank statistics. *Ann. Probab.,* **13**, 115–125.

Serfling, R.J. (1980). *Approximation Theorems of Mathematical Statistics.* Wiley, New York.

Shorack, G.R. and Wellner, J.A. (1986). *Empirical Processes with Applications to Statistics.* Wiley, New York.

Vandemaele, M. and Veraverbeke, N. (1982). Cramér type large deviations for linear combinations of order statistics. *Ann. Probab.,* **10**, 433–434.

Yurinskii, V.V. (1976). Exponential inequalities for sums of random vectors. *J. Multivar. Anal.,* **6**, 473–499.

Yurinskii, V.V. (1982). On the accuracy of normal approximation of the probability of hitting a ball. *Theory Probab. Appl.*, **27**, 280–289 (in Russian).

Yurinskii, V.V. (1990). Osipov's Theorem Revisited. In: *Proc. Fifth Vilnius Conf. on Probab. Theory and Math. Statist.*, VSP/Mokslas, pp. 585–589.

Zalesskii, B.A. (1982). Estimate of the accuracy of normal approximation in Hilbert space. *Theory Probab. Appl.*, **27**, 290–298.

Zalesskii, B.A. (1989a). Probabilities of large deviations in Hilbert space. *Theory Probab. Appl.*, **34**, 650–655 (in Russian).

Zalesskii, B.A. (1989b). Probabilities of Large Deviations for Balls with Arbitrary Centres in Hilbert Space. Preprint 41(391), Institute of Math. Belorussian Academy Sci., Minsk.

van Zwet W.R. (1984). A Berry–Esséen bound for symmetric statistics. *Z. Wahr. verw. Gebiete*, **66**, 425–440.

Part Three

LIMIT THEOREMS FOR STOCHASTIC PROCESSES

New Trends in Probab. and Statist., pp. 167 - 182
V. Sazonov and T. Shervashidze (Eds.)
1991 VSP/Mokslas

RATES OF CONVERGENCE IN THE INVARIANCE PRINCIPLE FOR EMPIRICAL PROCESSES

V.I. KOLCHINSKII
Kiev University, Vladimirskaya str. 64,
252017 Kiev, Ukraine, USSR

ABSTRACT

New estimates of convergence rates in the invariance principle for empirical processes on classes of sets and functions are presented. The conditions are formulated in terms of the empirical entropy and β-approximability of the classes of sets. Some recent results of Borisov and Massart are obtained as a corollary. Applications to the rates of convergence for empirical characteristic functions are also considered.

1. INTRODUCTION

Let $(X, \mathfrak{A}, \mathbf{P})$ be a probability space, $\{\xi_n\}_{n\geqslant 1}$ be a sequence of independent random elements, taking values in X with the distribution $\mathbf{P}$ and defined on a "rich" enough probability space $(\Omega, \overline{\Sigma}, \Pr)$.
The empirical distribution, based on sample $(\xi_1, \ldots, \xi_n)$ is the discrete random measure

$$\mathbf{P}_n^* = n^{-1} \sum_1^n \delta_{\xi_k},$$

where

$$\delta_x(A) = \begin{cases} 1, & x \in A, \\ 0, & x \notin A, \end{cases} \quad A \in \mathfrak{A}, \quad x \in X,$$

and the empirical process is the signed random measure on $(X, \mathfrak{A})$

$$Z_n = n^{1/2}(\mathbf{P}_n^* - \mathbf{P}).$$

For a signed measure ν on $(X, \mathfrak{A})$ and an $\mathfrak{A}$-measurable function $f: X \to \mathbf{R}^1$ let

$$\nu(f) = \int_X f \, d\nu$$

(assuming that the integral on the right-hand side of the equality exists).

Usually the empirical process Z_n is considered as a random function, defined on some class of measurable sets $\mathfrak{C} \subset \mathfrak{A}$,

$$Z_n(C) = n^{1/2}(\mathbf{P}_n^*(C) - \mathbf{P}(C)), \quad C \in \mathfrak{C},$$

or on some class of measurable functions $\mathcal{F} \subset L_2(X, d\mathbf{P})$,

$$Z_n(f) = n^{1/2}(\mathbf{P}_n^*(f) - \mathbf{P}(f)), \quad f \in \mathcal{F}.$$

If the space X coincides with the interval $I = [0,1]$, $\mathfrak{A}$ is the σ-algebra of Borel subsets of I, $\mathbf{P}$ is the Lebesgue measure, then the class of halfintervals $\{[0,x)\colon x \in I\}$ is often choosen as a class $\mathfrak{C}$. In this case $\mathbf{P}_N^*([0,x))$ coincides with the empirical distribution function $F_n^*(x)$ and the empirical process is defined by the formula:

$$\zeta_n(x) = n^{1/2}\big(F_n^*(x) - x\big) = Z_n\big([0,x)\big), \quad x \in I.$$

It was proved in (Komlos, Major and Tusnady, 1975) that a copy $\overset{\circ}{W}_n$ of Brownian bridge can be defined, for each n, on a common probability space $(\Omega, \overline{\Sigma}, \Pr)$ with a sequence $\{\xi_n\}_{n\geqslant 1}$ in a such way, that

$$\Pr\Big(\{\sup_{x\in I} |\zeta_n(x) - \overset{\circ}{W}_n(x)| \geqslant n^{-1/2}(t + C\log n)\}\Big) \leqslant \Lambda e^{-\theta t}, \quad t > 0,$$

$\Lambda > 0$, $C > 0$ and $\theta > 0$ being constants.

In (Borisov, 1983, 1985, 1987; Massart, 1986, 1989) methods of obtaining the rates of convergence estimates of the Komlos–Major–Tusnady type for general empirical processes on classes of measurable sets and functions are developed. This paper continues the investigations in the same direction. In the paper some new rates of convergence estimates are presented, generalizing and strengthening the recent results of Borisov and Massart. The paper deals mainly with the methods of checking the conditions of the main theorem and with its corollaries. The proof of this theorem will be published in details in a separate paper.

2. THE BASIC NOTIONS AND DEFINITIONS

Let $\mathcal{B}_I$ denote the σ-algebra of Lebesgue measurable subsets of the closed interval $I = [0,1]$, λ be the Lebesgue measure on I. Let $\Omega = X^\infty \times I$, $\Sigma = \mathfrak{A}^\infty \times \mathcal{B}_I$, $\Pr = \mathbf{P}^\infty \times \lambda$ and $\overline{\Sigma}$ be the completion of Σ with respect to $\Pr$. In what follows the role of the basic probability space will be played by $(\Omega, \overline{\Sigma}, \Pr)$.

For $B \subset \Omega$ we set

$$\Pr^*(B) = \inf\{\Pr(B')\colon B' \supset B,\ B' \in \overline{\Sigma}\},$$

and for $\zeta\colon \Omega \to \mathbf{R}^1$ we set

$$\mathbf{E}^*\zeta = \inf\{\mathbf{E}\zeta'\colon \zeta' \geqslant \zeta,\ \zeta' \text{ is } \overline{\Sigma}\text{-measurable}\}.$$

($\mathbf{E}$ denotes the mathematical expectation with respect to $\Pr$).

For $\omega \in \Omega$, $\omega = (x_1, x_2, \dots, u)$, $x_n \in X$, $u \in I$, we define $\xi_n(\omega) = x_n$, $n \geqslant 1$. We set

$$\rho_{\mathbf{P}}(f,g) = \left(\mathbf{P}(|f-g|^2) - (\mathbf{P}(f-g))^2\right)^{1/2}, \quad f,g \in L_2(X, d\mathbf{P}).$$

Let $\mathcal{F} \subset L_2(X, d\mathbf{P})$. Denote $C(\mathcal{F}; \rho_{\mathbf{P}})$ the space of uniformly bounded and $\rho_{\mathbf{P}}$-uniformly continuous functions $Y\colon \mathcal{F} \to \mathbf{R}^1$. We define

$$\|Y\|_{\mathcal{F}} = \sup\left\{|Y(f)|\colon f \in \mathcal{F}\right\}, \quad Y : \mathcal{F} \to \mathbf{R}^1.$$

If $\mathfrak{C} \subset \mathfrak{A}$, then the space $C(\mathfrak{C}; \rho_{\mathbf{P}})$ and the norm $\|\cdot\|_{\mathfrak{C}}$ are defined in a similar way. In this case

$$\rho_{\mathbf{P}}(A,B) = \rho_{\mathbf{P}}(\chi_A, \chi_B), \quad A, B \in \mathfrak{A}.$$

A centered Gaussian random function $\overset{\circ}{W}_{\mathbf{P}}(f)$, $f \in L_2(X, d\mathbf{P})$, with covariance

$$\mathbf{E}\overset{\circ}{W}_{\mathbf{P}}(f)\overset{\circ}{W}_{\mathbf{P}}(g) = \mathbf{P}(fg) - \mathbf{P}(f)\mathbf{P}(g), \quad f,g \in L_2(X, d\mathbf{P})$$

is called a $\mathbf{P}$-Brownian bridge. The values of a $\mathbf{P}$-Brownian bridge on measurable sets are defined by

$$\overset{\circ}{W}_{\mathbf{P}}(A) = \overset{\circ}{W}_{\mathbf{P}}(\chi_A), \quad A \in \mathfrak{A}.$$

The class $\mathcal{F} \subset L_2(X, d\mathbf{P})$ (respectively, the class $\mathfrak{C} \subset \mathfrak{A}$) is called $\mathbf{P}$-pregaussian if there exist a copy of $\mathbf{P}$-Brownian bridge $\overset{\circ}{W}_{\mathbf{P}}$ such that $\overset{\circ}{W}_{\mathbf{P}} \in C(\mathcal{F}; \rho_{\mathbf{P}})$ a.s. (respectively, $\overset{\circ}{W}_{\mathbf{P}} \in C(\mathfrak{C}; \rho_{\mathbf{P}})$ a.s.). In (Dudley, 1978; Kolchinskii, 1981; Gine and Zinn, 1984) the notion of empirical measurability of a class $\mathfrak{C} \in \mathfrak{A}$ with respect to $\mathbf{P}$ is used. The collection of empirically measurable with respect to $\mathbf{P}$ classes $\mathfrak{C} \subset \mathfrak{A}$ will be denoted EM($\mathbf{P}$) (Gine and Zinn denote such collection NLDM($\mathbf{P}$)). Let $\mathfrak{C} \subset \mathfrak{A}$. For a finite set $F \subset X$ we define

$$\Delta^{\mathfrak{C}}(F) = \operatorname{card}\left(\{C \cap F : C \in \mathfrak{C}\}\right),$$

Where *card* denote the number of elements of the set.
Let us define the empirical entropy of a class $\mathfrak{C}$ by sample of the size n by

$$h_{\mathfrak{C}}(n) = \log \Delta^{\mathfrak{C}}\left(\{\xi_1, \dots \xi_n\}\right).$$

In (Vapnik and Chervonenkis, 1971) the Glivenko–Cantelli classes were characterized in terms of $h_{\mathfrak{C}}(n)$, in (Kolchinskii, 1981, 1985, 1986; Le Cam, 1983; Gine and Zinn, 1984, 1985; Talagrand, 1985) the Donsker classes (introduced in (Dudley, 1978)) and the classes $\mathfrak{C}$, on which the uniform strong law of large numbers of Marcinkiewicz type for empirical processes is fulfiled, were characterized. It was proved, that if $\mathfrak{C} \in$ EM($\mathbf{P}$), then

(i) for $\alpha \in (1/2, 1]$ the condition

$$\Pr{}^*\left(\{h_{\mathfrak{C}}(n) \geqslant \varepsilon n^{\alpha}\}\right) \to 0, \quad n \to \infty, \quad \varepsilon > 0, \tag{1}$$

is necessary and sufficient for

$$\|\mathbf{P}_n^* - \mathbf{P}\|_{\mathfrak{C}} = o(n^{\alpha-1}), \quad n \to \infty \text{ a.s.}; \tag{2}$$

(ii) the conditions $\mathfrak{C}$ is **P**-pregaussian and (1), for $\alpha = 1/2$, are satisfied if and only if $\mathfrak{C}$ is a **P**-Donsker class.

For $E \in \mathcal{B}_I$, $h \in I$ let us denote by E_h the shift of the set E:

$$E_h = \{u \dotplus h\colon\ u \in E\}$$

(summation $\dotplus$ is modulo 1).

Let J_n be the class of subsets of I, which are the union of no more then N closed intervals.

For $E \in \mathcal{B}_I$ we set

$$w(E;h) = \lambda(E \bigtriangleup E_h),$$

$$d(E;N) = \inf_{E' \in J_N} \lambda(E \bigtriangleup E').$$

For a class of measurable sets $\mathcal{E} \subset \mathcal{B}_I$ let us denote

$$w(\mathcal{E};h) = \sup_{E \in \mathcal{E}} w(E;h),$$

$$d(\mathcal{E};N) = \sup_{E \in \mathcal{E}} d(E;N).$$

The quantity $d(\mathcal{E};N)$ which characterizes the accuracy of approximation of sets of the class $\mathcal{E}$ by sets from J_N, plays the role similar to Kolmlgorov's N-dimensional diameter in linear spaces.

PROPOSITION 1. *If $\beta \in (0,1)$, then the following conditions are equivalent:*

$$w(\mathcal{E};h) = O(h^\beta), \quad h \to 0, \tag{3}$$

and

$$d(\mathcal{E};N) = O(N^{-\beta/(1-\beta)}), \quad N \to \infty.$$

Condition (2), for $\beta = 1$, is fulfilled if and only if $d(\mathcal{E};N) = 0$ for sufficiently large N. It is easy to obtain the proof from the following statements in (Zhdanov and Sevastyanov, 1983). If $E \in \mathcal{B}_I$, $h \in I$, then a set $E' \in J_N$ exists, where

$$N = \left[\sup_{0 \leqslant t \leqslant h} w(E;t)/h\right],$$

such that $\lambda(E \bigtriangleup E') \leqslant 4 \sup\{w(E;t)\colon 0 \leqslant t \leqslant h\}$.

On the other hand, for any $E \in \mathcal{B}_I$ the following estimate is true

$$w(E;h) \leqslant 2\big(d(E;N) + Nh\big), \quad N \geqslant 1,\ h \in I.$$

Definition 1. A class $\mathfrak{C} \subset \mathfrak{A}$ is called β-approximable ($\beta \in (0,1]$) with respect to $\mathbf{P}$ (we shall write $\mathfrak{C} \in \mathrm{AP}(\mathbf{P};\beta)$), if a random element ξ: $(I, \mathcal{B}_I) \to (X, \mathfrak{A})$ with the distribution $\mathbf{P}$ exists such that the class

$$\mathcal{E} = \{\xi^{-1}C\colon\ C \in \mathfrak{C}\} = \big\{\{u \in I\colon\ \xi(u) \in C\}\colon\ C \in \mathfrak{C}\big\}$$

satisfies condition (3).

3. THE KOMLOS–MAJOR–TUSNADY CLASSES

Let $\{\delta_n\}_{n\geqslant 1}$ be a monotone decreasing sequence, $\delta_n > 0$, $n \geqslant 1$, $\delta_n \to 0$, $n \to \infty$, and for some $\varepsilon > 0$

$$\delta_n = O(n^{-\varepsilon}), \quad n \to \infty.$$

Definition 2. A uniformly bounded class $\mathcal{F}$ of $\mathfrak{A}$-measurable functions is called a Komlos–Major–Tusnady class with respect to $\mathbf{P}$ and with the rate of convergence $\{\delta_n\}_{n\geqslant 1}$ (we shall write $\mathcal{F} \in \mathrm{KMT}(\mathbf{P};\ \{\delta_n\}_{n\geqslant 1})$), if $\mathcal{F}$ is a $\mathbf{P}$-pregaussian class and for each n one can define a copy $\overset{\circ}{W}_n$ of $\mathbf{P}$-Brownian bridge on the probability space $(\Omega, \overline{\Sigma}, \mathrm{Pr})$ in such way, that

$$\mathrm{Pr}^*\big(\{\|Z_n - \overset{\circ}{W}_n\|_{\mathcal{F}} \geqslant \delta_n(t + C\log n)\}\big) \leqslant \Lambda e^{-\theta t}, \quad t > 0, \tag{4}$$

where $C > 0$, $\Lambda > 0$, $\theta > 0$ are constants depending only on $\mathcal{F}$.
The Komlos–Major–Tusnady classes of sets are defined similarly. In this case $\mathfrak{C} \in \mathrm{KMT}(\mathbf{P}; \{\delta_n\}_{n\geqslant 1})$ if and only if the class of indicator functions

$$\mathcal{F} = \{\chi_C :\ C \in \mathfrak{C}\} \in \mathrm{KMT}(\mathbf{P}; \{\delta_n\}_{n\geqslant 1}).$$

The following proposition is proved in a similar way as Theorem 1.9 in (Dudley, 1981).

PROPOSITION 2. *Let $\mathcal{F}$ be a uniformly bounded class of $\mathfrak{A}$-measurable functions, $\mathfrak{C} \subset \mathfrak{A}$. Suppose that*

$$\{x\colon\ f(x) > c\} \in \mathfrak{C}, \quad f \in \mathcal{F}, \quad c \in \mathbf{R}^1. \tag{5}$$

Under these conditions, if $\mathfrak{C} \in \mathrm{KMT}(\mathbf{P};\ \{\delta_n\}_{n\geqslant 1})$, for some sequence $\{\delta_n\}_{n\geqslant 1}$, then $\mathcal{F} \in \mathrm{KMT}(\mathbf{P};\ \{\delta_n\}_{n\geqslant 1})$.

Proof. Without loss of generality one can assume that $f\colon X \to [0,1]$, $f \in \mathcal{F}$. Let

$$f_N = N^{-1}\sum_{j=1}^{N-1} \chi_{C_{N,j}(f)},$$

$$C_{N,j}(f) = \{x\colon\ f(x) > j/N\} \in \mathfrak{C}, \quad f \in \mathcal{F}, \quad j = 0,1,\ldots,N-1, \quad N \geqslant 1.$$

Obviously,

$$\sup_{x\in X} |f_N(x) - f(x)| \leqslant N^{-1}, \quad f \in \mathcal{F}, \quad N \geqslant 1. \tag{6}$$

We set

$$\tilde{\mathcal{F}} = \mathcal{F} \cup \bigcup_{n\geqslant 1} \{f_N\colon\ f \in \mathcal{F}\}.$$

Obviously, the class $\tilde{\mathcal{F}}$, as well as $\mathcal{F}$, satisfies condition (5).

Since $\mathfrak{C} \in \mathrm{KMT}(\mathbf{P};\ \{\delta_n\}_{n\geqslant 1})$, $\delta_n \to 0$, $n \to \infty$, $\mathfrak{C}$ is a Donsker class with respect to $\mathbf{P}$. By Theorem 1.9 in (Dudley, 1981), the class $\tilde{\mathcal{F}}$ is also $\mathbf{P}$-Donsker. Particularly, it means, that $\tilde{\mathcal{F}}$ is a $\mathbf{P}$-pregaussian class.

By definition 2, a sequence $\overset{\circ}{W}_n$ of copies of $\mathbf{P}$-Brownian bridges exists such that estimate (4) takes place for the class $\{\chi_C\colon\ C \in \mathfrak{C}\}$.

By the linearity of the functions $Z_n(f)$, $\overset{\circ}{W}_n(f)$, we have, for each $f \in \mathcal{F}$,

$$\big|Z_n(f_N) - \overset{\circ}{W}_n(f_N)\big| \leqslant N^{-1} \sum_{j=1}^{N-1} \big|Z_n(\chi_{C_{N,j}(f)}) - \overset{\circ}{W}_n(\chi_{C_{N,j}(f)})\big| \leqslant$$

$$\leqslant \sup_{C\in\mathfrak{C}} \big|Z_n(\chi_C) - \overset{\circ}{W}_n(\chi_C)\big|. \tag{7}$$

By (6), we have

$$\mathbf{P}(f_N) \to \mathbf{P}(f), \quad \mathbf{P}_n^*(f_N) \to \mathbf{P}_n^*(f),$$

$$\rho_{\mathbf{P}}(f_N, f) \to 0, \quad N \to \infty, \quad f \in \mathcal{F}.$$

Since $\tilde{\mathcal{F}}$ is a $\mathbf{P}$-pregaussian class, it follows from the last relation that a.s.

$$\overset{\circ}{W}_n(f_N) \to \overset{\circ}{W}_n(f), \quad N \to \infty, \quad f \in \mathcal{F}.$$

Passing in (7) to the limit as $N \to \infty$, we obtain for each $f \in \mathcal{F}$

$$|Z_n(f) - \overset{\circ}{W}_n(f)| \leqslant \sup_{C\in\mathfrak{C}} |Z_n(C) - \overset{\circ}{W}_n(C)|.$$

Therefore, a.s.

$$\|Z_n - \overset{\circ}{W}_n\|_{\mathcal{F}} \leqslant \|Z_n - \overset{\circ}{W}_n\|_{\mathfrak{C}}. \tag{8}$$

Since $\mathfrak{C} \in \mathrm{KMT}(\mathbf{P};\ \{\delta_n\}_{n\geqslant 1})$, by (8) we have $\mathcal{F} \in \mathrm{KMT}(\mathbf{P};\ \{\delta_n\}_{n\geqslant 1})$.

The proposition is proved.

Our main result is the following theorem.

THEOREM 1. *Let* $\mathfrak{C} \in \mathfrak{A}$, $\mathfrak{C} \in \mathrm{EM}(\mathbf{P})$ *and* $\mathfrak{C} \in \mathrm{AP}(\mathbf{P};\beta)$ *for some* $\beta \in (0,1]$.

(i) *If, for some constants* $A, B > 0$ *and* $\alpha \in (0, 1/2)$,

$$\Pr^*(\{h_{\mathfrak{C}}(n) \geqslant tn^{\alpha}\}) \leqslant Ae^{-Bt}, \quad t > 0, \quad n \geqslant 1, \tag{9}$$

then $\mathfrak{C} \in \mathrm{KMT}(\mathbf{P};\ \{\delta_n\}_{n\geqslant 1})$, *where*

$$\delta_n = n^{-\gamma}, \quad n \geqslant 1, \quad \gamma = (1/2 - \alpha)\min\big(\beta/(1-\alpha), 1\big).$$

(ii) *If, for some constants $A, B > 0$ and $\rho > 0$,*

$$\Pr^*(\{h_{\mathfrak{C}}(n) \geqslant t \log^{\rho} n\}) \leqslant Ae^{-Bt}, \quad t > 0, \quad n \geqslant 1, \tag{10}$$

then $\mathfrak{C} \in \mathrm{KMT}(\mathbf{P};\ \{\delta_n\}_{n\geqslant 1})$, *where*

$$\delta_n = n^{-\beta/2} \log^{\rho/2} n, \quad n \geqslant 1.$$

The proof of this theorem, based on the methods of estimation of the modulus continuity of empirical processes, uses the empirical entropy, developed in (Kolchinskii, 1981, 1985, 1986), and the methods of estimation of the rate of convergence in invariance principle, developed in (Borisov, 1983, 1985, 1987; Massart, 1987). It will be published in a separate paper. In what follows, we shall consider different methods of checking the conditions of Theorem 1 as well as its particular cases.

4. THE ESTIMATE OF THE EMPIRICAL ENTROPY USING THE ENTROPY WITH "INCLUSION"

Let $\mathfrak{C} \in \mathfrak{A}$. Denote $N_I(\mathfrak{C}; \varepsilon)$ the smallest number $N \geqslant 1$, for which there exist sets A_j^-, A_j^+, $A_j^- \subset A_j^+$, $j = 1, \ldots, N$, such that

$$\mathbf{P}(A_j^+ \backslash A_j^-) < \varepsilon, \quad 1 \leqslant j \leqslant N,$$

and for each $A \in \mathfrak{C}$ there is j, $1 \leqslant j \leqslant N$, for which $A_j^- \subset A \subset A_j^+$.
We set

$$H_I(\mathfrak{C}; \varepsilon) = \log N_I(\mathfrak{C}; \varepsilon), \quad \varepsilon > 0.$$

The quantity $H_I(\mathfrak{C}; \varepsilon)$ is called ε-entropy with "inclusion" of the class $\mathfrak{C}$ (Dudley, 1978).

THEOREM 2. (i) If, for some $\alpha \in (0, 1)$,

$$H_I(\mathfrak{C}; \varepsilon) = O(\varepsilon^{-\alpha/(1-\alpha)}), \quad \varepsilon \to 0, \tag{11}$$

then the class $\mathfrak{C}$ satisfies condition (9).

(ii) If, for some $\rho > 0$,

$$H_I(\mathfrak{C}; \varepsilon) = O(\log^{\rho} 1/\varepsilon), \quad \varepsilon \to 0, \tag{12}$$

then the class $\mathfrak{C}$ satisfies condition (10).

The proof is based on the following lemma.

LEMMA 1. *If, for some $u > 0$,*

$$H_I(\mathfrak{C}; 0.25u) \leqslant 0.02nu, \tag{13}$$

then

$$\Pr^*(\{h_{\mathfrak{C}}(n) \geqslant nu\}) \leqslant e^{-0.03nu}. \tag{14}$$

Proof. Let $N = N_I(\mathfrak{C}; 0.25u)$. Then there exist sets C_j^-, $C_j^+ \in \mathfrak{A}$, $1 \leqslant j \leqslant N$, such that

$$C_j^- \subset C_j^+, \quad \mathbf{P}(C_j^+ \backslash C_j^-) \leqslant 0.25u, \quad j = 1, \dots, N,$$

and $\mathfrak{C} \subset \bigcup_{j=1}^N \mathfrak{C}_j$, where

$$\mathfrak{C}_j = \{C \in \mathfrak{C}\colon\ C_j^- \subset C \subset C_j^+\}.$$

Obviously,

$$\Delta^{\mathfrak{C}}(\{\xi_1, \dots \xi_n\}) \leqslant \sum_{j=1}^N \Delta^{\mathfrak{C}_j}(\{\xi_1, \dots \xi_n\}).$$

Consequently,

$$h_{\mathfrak{C}}(n) \leqslant \log\left(N N^{-1} \sum_{j=1}^N \Delta^{\mathfrak{C}_j}(\{\xi_1, \dots \xi_n\})\right) \leqslant$$

$$\leqslant \log N + \max_{1 \leqslant j \leqslant N} h_{\mathfrak{C}_j}(n).$$

Since

$$\log N = H_I(\mathfrak{C}; 0.25u) \leqslant 0.25nu,$$

we have

$$\Pr{}^*(\{h_{\mathfrak{C}}(n) \geqslant nu\}) \leqslant N \max_{1 \leqslant j \leqslant N} \Pr{}^*(\{h_{\mathfrak{C}_j}(n) \geqslant 0.5nu\}). \tag{15}$$

It is easy to check that

$$\Delta^{\mathfrak{C}_j}(\{\xi_1, \dots \xi_n\}) \leqslant 2^\kappa,$$

where $\kappa = \sum_{k=1}^n \chi_{C_j^+ \backslash C_j^-}(\xi_k)$ which implies

$$h_{\mathfrak{C}_j}(n) \leqslant \kappa. \tag{16}$$

Since, for each $j = 1, \dots, N$,

$$n\mathbf{P}(C_j^+ \backslash C_j^-) \leqslant 0.25nu,$$

using (15) and (16), we get

$$\Pr{}^*(\{h_{\mathfrak{C}}(n) \geqslant nu\}) \leqslant$$

$$\leqslant N \max_{1 \leqslant j \leqslant N} \Pr\left(\left\{\sum_{k=1}^n \left[\chi_{C_j^+ \backslash C_j^-}(\xi_k) - \mathbf{P}(C_j^+ \backslash C_j^-)\right] \geqslant 0.25nu\right\}\right). \tag{17}$$

Note that $\mathbf{E}[\chi_{C_j^+ \backslash C_j^-}(\xi)]^2 = \mathbf{P}(C_j^+ \backslash C_j^-) \leqslant 0.25u$. Using the Bernstein inequality, we obtain, for each $j = 1, \dots, N$,

$$\Pr\left(\left\{\sum_{k=1}^n \left[\chi_{C_j^+ \backslash C_j^-}(\xi_k) - \mathbf{P}(C_j^+ \backslash C_j^-)\right] \geqslant 0.25nu\right\}\right) \leqslant \exp\{-0.05nu\}. \tag{18}$$

Estimates (17) and (18) give (in view of (13))

$$\Pr^*(\{h_{\mathfrak{C}}(n) \geqslant nu\}) \leqslant \exp\{-0.05nu + H_I(\mathfrak{C}; 0,25u)\} \leqslant \exp\{-0.03nu\}.$$

Lemma 1 is proved.

Proof of Theorem 2. (i) Condition (11) means that, for some constant $C > 0$,

$$H_I(\mathfrak{C}; \varepsilon) \leqslant C\varepsilon^{-\alpha/(1-\alpha)}, \quad \varepsilon > 0.$$

Let $D = \max(50C, 4)$, $t > D$, $u = n^{\alpha-1}t$. Then

$$H_I(\mathfrak{C}; 0.25u) \leqslant H_I(\mathfrak{C}; n^{\alpha-1}) \leqslant Cn^{\alpha} \leqslant 0.02nu.$$

Using Lemma 1, we have

$$\Pr^*(\{h_{\mathfrak{C}}(n) \geqslant tn^{\alpha}\}) \leqslant \exp\{-0.03tn^{\alpha}\} \leqslant \exp\{-0.03t\}, \quad t \geqslant D, \quad n \geqslant 1.$$

Therefore, condition (9) holds with $A = \exp\{0.03D\}$, $B = 0.03$.
(ii) Condition (12) means that, for some $C > 0$,

$$H_I(\mathfrak{C}; \varepsilon) \leqslant C \log^{\rho}(1/\varepsilon), \quad \varepsilon > 0.$$

Setting $D = \max(50C, 4)$, $t > D$, $u = t(\log^{\rho} n)/n$, we see that (13) holds. By Lemma 1, for $n \geqslant 3$,

$$\Pr^*(\{h_{\mathfrak{C}}(n) \geqslant t \log^{\rho} n\}) \leqslant e^{-0.03t \log^{\rho} n} \leqslant e^{-0.03t}, \quad t > D.$$

Consequently, condition (10) holds with $A = \exp\{0.03D\}$, $B = 0.03$.
Theorem 2 is proved.

5. β-APPROXIMABILITY OF CLASSES OF BOREL SETS IN METRIC SPACES

Let (x, ρ) be a complete separable metric space, $\mathfrak{A}$ the Borel σ-algebra in X, $\mathbf{P}$ a Borel probability measure on $\mathfrak{A}$. The diameter of a set $A \subset X$ is defined as

$$\mathrm{Diam}(A) = \sup\{\rho(x, y) : x, y \in A\}.$$

Let $\partial_\delta A$ denote the δ-neighbourhood of the boundary of a set A:

$$\partial_\delta A = \{x \in X\colon B(x,\delta) \cap A \neq \varnothing, \quad B(x,\delta) \backslash A \neq \varnothing\}, \quad \delta > 0,$$

where $B(x, \delta)$ is a ball with radius δ and center at $x \in X$.
A binary expansion of the space X with respect to the distribution $\mathbf{P}$ is a family

$$\{\{\Delta_{i_1,\dots,i_n}\}\colon (i_1, \dots, i_n) \in \{0,1\}^n, \ n \geqslant 1\}$$

of subsets of X, satisfying the following conditions:

(i) $\Delta_{i_1,\dots,i_n} \in \mathfrak{A}$, $\quad n \geqslant 1$, $\quad (i_1, \dots, i_n) \in \{0,1\}^n$;

(ii) $\Delta_0 \cup \Delta_1 = X$, $\Delta_0 \cap \Delta_1 = \varnothing$,
$\Delta_{i_1,\dots,i_n,0} \cup \Delta_{i_1,\dots,i_n,1} = \Delta_{i_1,\dots,i_n}$, $\Delta_{i_1,\dots,i_n,0} \cap \Delta_{i_1,\dots,i_n,1} = \varnothing$,
$(i_1,\dots,i_n) \in \{0,1\}^n$, $n \geqslant 1$;

(iii) $\mathbf{P}(\Delta_0) = \mathbf{P}(\Delta_1)$, $\mathbf{P}(\Delta_{i_1,\dots,i_n,0}) = \mathbf{P}(\Delta_{i_1,\dots,i_n,1})$,
$(i_1,\dots,i_n) \in \{0,1\}^n$, $n \geqslant 1$.

The space X we shall call β-expansible with respect to the distribution $\mathbf{P}(\beta \in (0,1])$ if there is a binary expansion of X with respect to $\mathbf{P}$ such that

$$\operatorname{Diam}(\Delta_{i_1,\dots,i_n}) \leqslant D2^{-\beta n}, \quad (i_1,\dots,i_n) \in \{0,1\}^n, \quad n \geqslant 1, \tag{19}$$

where $D > 0$ is a constant.

THEOREM 3. *If, for some $\beta \in (0,1)$, the space X is β-expansible with respect to $\mathbf{P}$ and a class $\mathfrak{C} \subset \mathfrak{A}$ satisfies the condition*

$$\sup\{\mathbf{P}(\partial_\delta C)\colon C \in \mathfrak{C}\} = O(\delta), \quad \delta \to 0, \tag{20}$$

then $\mathfrak{C} \in \mathrm{AP}(\mathbf{P};\beta)$.

Proof. For each $u \in I$ we have

$$u = \sum_{j=1}^{\infty} u_j 2^{-j},$$

where $u_j = 0$ or 1, $j \geqslant 1$.
We set

$$I_{i_1,\dots,i_n} = \{u \in I\colon u_j = i_j,\ j = 1,\dots,n\}, \quad (i_1,\dots,i_n) \in (0,1)^n, \quad n \geqslant 1.$$

Then $\{I_{i_1,\dots,i_n}\colon (i_1,\dots,i_n) \in (0,1)^n,\ n \geqslant 1\}$ is a binary expansion of the space I with respect to the Lebesgue measure λ, $I_{i_1,\dots,i_n}$ being an interval of the length 2^{-n}. Let $\{\Delta_{i_1,\dots,i_n}\colon (i_1,\dots,i_n) \in \{0,1\}^n,\ n \geqslant 1\}$ be a binary expansion of X with respect to $\mathbf{P}$, for which (19) holds. Then a Borel isomorphism $\xi : I \mapsto X$ exists such that $\xi(I_{i_1,\dots,i_n}) = \Delta_{i_1,\dots,i_n}$.
Let $h \in (2^{-r-1}, 2^{-r}]$, $r \geqslant 1$.
We set

$$\nu_r(u) = \min\{j \geqslant 0\colon u_{r-j} = 0\}, \quad u \in I, \quad E_{r,j} = \{u \in I\colon \nu_r(u) = r - j\}.$$

Then

$$E_{r,j} = \{u \in I\colon u_r = 1, \quad u_{r-1} = 1, \ \dots, \quad u_{j+1} = 1, \quad u_j = 0\}.$$

If $u \in E_{r,j}$, then $(u \dot{+} h)_k = u_k$, $k = 1,\dots,j-1$. Therefore, $u \dot{+} h \in I_{u_1,\dots,u_{j-1}}$ and so $\xi(u \dot{+} h) \in \Delta_{u_1,\dots,u_{j-1}}$. Consequently, for $u \in E_{r,j}$, by (19),

$$\rho(\xi(u \dot{+} h), \xi(u)) \leqslant \operatorname{Diam}(\Delta_{u_1,\dots,u_{j-1}}) \leqslant \delta_j = D2^{-(j-1)\beta}.$$

Let us denote

$$\mathcal{E} = \{\xi^{-1}C\colon\ C \in \mathfrak{C}\}.$$

Let $E \in \mathcal{E}$, i.e., $E = \xi^{-1}C$ for some $C \in \mathfrak{C}$,. If $u \in E_{r,j}$, $\xi(u) \in C$ and $\xi(u\dot{+}h) \notin C$, then, by (20), we have $\xi(u) \in \partial_{\delta_j} C$. Consequently,

$$\lambda(E\backslash E_h) = \lambda\big(\{u \in I\colon\ \xi(u) \in C, \quad \xi(u\dot{+}h) \notin C\}\big) \leqslant$$

$$\leqslant \sum_{j=0}^{r} \lambda\big(\{u \in I\colon\ \xi(u) \in \partial_{\delta_j} C\} \bigcap E_{r,j}\big). \tag{21}$$

We define for $j \leqslant r$ the random variable $\tau_j : I \to I$, setting for $u \in I$

$$\big(\tau_j(u)\big)_k = u_k, \quad \text{if } k < j \text{ and } k > r,$$

$$\big(\tau_j(u)\big)_k = 1, \quad \text{if } j < k \leqslant r,$$

$$\big(\tau_j(u)\big)_j = 0.$$

It is easy to see, that, for each $j \leqslant r$, the random variable τ_j and the event $E_{r,j}$ are independent.
We set

$$\xi_j'(u) = \xi\big((\tau_j(u)\big), \quad u \in I.$$

Then $\xi_j'\colon\ I \to X$ is a random element, ξ_j' and $E_{r,j}$ are independent, $j \leqslant r$.
Since $\tau_j(u) \in I_{u_1,\ldots,u_{j-1}}$, we have $\xi_j'(u) \in \Delta_{u_1,\ldots,u_{j-1}}$ and, in view of (19), we obtain

$$\rho\big(\xi_j(u), \xi_j'(u)\big) \leqslant \operatorname{Diam}(\Delta_{u_1,\ldots,u_{j-1}}) \leqslant \delta_j. \tag{22}$$

For $u \in E_{r,j}$, $\tau_j(u) = u$ and $\xi_j'(u) = \xi(u)$. Therefore, (21) and (22), in view of the independence of ξ_j' and $E_{r,j}$, imply

$$\lambda(E\backslash E_h) \leqslant \sum_{j=0}^{r} \lambda\big(\{u\colon\ \xi_j' \in \partial_{\delta_j} C\} \bigcap E_{r,j}\big) =$$

$$= \sum_{j=0}^{r} \lambda(E_{r,j})\lambda\big(\{u\colon\ \xi_j' \in \partial_{\delta_j} C\}\big) \leqslant \sum_{j=0}^{r} \lambda(E_{r,j})\lambda\big(\{u\colon\ \xi \in \partial_{2\delta_j} C\}\big). \tag{23}$$

Note that

$$\lambda(E_{r,j}) = \lambda\big(\{u \in I:\ u_r = 1, \ldots, u_{j+1} = 1, \quad u_j = 0\}\big) \leqslant 2^{-r+j-1}, \quad j \leqslant r. \tag{24}$$

Under condition (20) there is a constant $Q > 0$, for which

$$\mathbf{P}(\partial_\delta C) \leqslant Q\delta, \quad \delta > 0, \quad C \in \mathfrak{C}.$$

Therefore,

$$\lambda\big(\{u\colon\ \xi \in \partial_{2\delta_j} C\}\big) = \mathbf{P}(\partial_{2\delta_j} C) \leqslant 2Q\delta_j = 2^{1+\beta} Q D 2^{-j\beta}. \tag{25}$$

It follows from (23), (24) and (25), in view of the condition $h > 2^{-r-1}$, that

$$\lambda(E\backslash E_h) \leqslant 2^{1+\beta}QD2^{-r-1}\sum_{j=0}^{r}2^{j(1-\beta)} \leqslant 2^{1+\beta}QD(2^{1-\beta}-1)^{-1}2^{-(r+1)\beta} \leqslant 0.5Rh^{\beta},$$

where $R > 0$ is a constant.
Thus, we have

$$\sup_{E\in\mathcal{E}} \lambda(E\backslash E_h) \leqslant 0.5Rh^{\beta}, \quad h > 0.$$

Similarly,

$$\sup_{E\in\mathcal{E}} \lambda(E_h\backslash E) \leqslant 0.5Rh^{\beta}, \quad h > 0,$$

which implies

$$w(\mathcal{E}; h) = O(h^{\beta}), \quad h \to 0,$$

i.e., $\mathfrak{C} \in \mathrm{AP}(\mathbf{P};\beta)$.
Theorem 3 is proved.

6. EXAMPLES AND APPLICATIONS

Let (X,ρ) be a complete separable metric space, $\mathfrak{A}$ the Borel σ-algebra in X, $\mathbf{P}$ the Borel probability measure on $\mathfrak{A}$.
Theorems 1, 2 and 3 imply the following generalization of one of the main results in (Massart, 1989).

THEOREM 4. *Let, for some $\beta \in (0,1)$, the space X be β-expansible with respect to $\mathbf{P}$, a class $\mathfrak{C} \subset \mathfrak{A}$ satisfy condition* (20) *and $\mathfrak{C} \in \mathrm{EM}(\mathbf{P})$.*

(i) *If, for some $\rho \in (0,1)$,*

$$H_I(\mathfrak{C};\varepsilon) = O(\varepsilon^{-\rho}), \quad \varepsilon \to 0, \tag{26}$$

then $\mathfrak{C} \in \mathrm{KMT}(\mathbf{P};\{\delta_n\}_{n\geqslant 1})$, where

$$\delta_n = n^{-\gamma}, \quad n \geqslant 1, \quad \gamma = \frac{1-\rho}{2}\min\left(\beta, \frac{1}{1+\rho}\right).$$

(ii) *If, for some $\rho > 0$, condition* (12) *holds, then $\mathfrak{C} \in \mathrm{KMT}(\mathbf{P};\{\delta_n\}_{n\geqslant 1})$, where*

$$\delta_n = n^{-\beta/2}\log^{\rho/2} n, \quad n \geqslant 1.$$

If $X = [0,1]^d$, $d \geqslant 2$, $\mathbf{P}$ is the Lebesgue measure on X (the case, considered in (Massart, 1989)) then it is easy to check that X is a β-expansible space with respect to $\mathbf{P}$ for $\beta = 1/d$. In this case Massart's estimates follows from Theorem 4. For the class of convex subsets of $[0,1]^2$ condition (26) holds with $\rho = 1/2$ (Dudley, 1974). Therefore, this class belongs to $\mathrm{KMT}(\mathbf{P};\{\delta_n\}_{n\geqslant 1})$ with $\delta_n = n^{-1/8}$, $n \geqslant 1$. For $d > 2$ the class of convex subsets of $[0,1]^d$ is not a Donsker class (Dudley, 1978).

Condition (21) holds with $\rho = (d-1)/q$ for the class of subsets of $[0,1]^d$, $d \geqslant 2$, having q-differentiable boundary, where $q > d-1$ (Dudley, 1974). Thus, this class belongs to $\mathrm{KMT}(\mathbf{P}; \{\delta_n\}_{n\geqslant 1})$ with

$$\delta_n = n^{-(q-d+1)/2qd}, \quad n \geqslant 1.$$

For $q \leqslant d-1$ this class is not a Donsker class (Dudley, 1978).
Let as before $(X, \mathfrak{A}, \mathbf{P})$ be an arbitrary probability space. The following theorem generalizes one of results in (Massart, 1989).

THEOREM 5. *Let* $\mathfrak{C} \subset \mathfrak{A}$, $\mathfrak{C} \in \mathrm{EM}(\mathbf{P})$ *and* $\mathfrak{C} \in \mathrm{AP}(\mathbf{P}; \beta)$ *for some* $\beta \in (0,1]$. *If* $\mathfrak{C}$ *is a Vapnik–Chervonenkis class, then* $\mathfrak{C} \in \mathrm{KMT}(\mathbf{P}; \{\delta_n\}_{n\geqslant 1})$, *where*

$$\delta_n = n^{-\beta/2} \log^{1/2} n, \quad n \geqslant 1.$$

For a definition of Vapnik–Chervonenkis classes see (Dudley, 1978). The proof of Theorem 5 follows from Theorem 1 (ii) and the fact that, for a Vapnik–Chervonenkis class $\mathfrak{C}$,

$$h_{\mathfrak{C}}(n) \leqslant V(\mathfrak{C}) \log n, \quad n \geqslant 2,$$

where $V(\mathfrak{C})$ is a constant depending only on $\mathfrak{C}$ (Dudley, 1978), i.e., condition (10) holds for $\rho = 1$.
Let us consider in more details another example.
Let $X = \mathbf{R}^1$, $\mathfrak{A}$ be the σ-algebra of Borel subsets of $\mathbf{R}^1$, $\mathbf{P}$ be a Borel probability measure on $\mathfrak{A}$ satisfying, for some $\sigma \in (0,1]$ and $D > 0$, the condition

$$\mathbf{P}(\Delta) \leqslant D\lambda(\Delta)^{\sigma}, \quad \Delta \in \mathfrak{A} \tag{27}$$

(λ is the Lebesgue measure in $\mathbf{R}^1$).
Let $\varphi(t)$ be the characteristic function of $\mathbf{P}$,

$$\varphi(t) = \int_{\mathbf{R}^1} e^{itx} \mathbf{P}(dx), \quad t \in [-1,1],$$

and let $\varphi_n^*(t)$ be the empirical characteristic function, based on sample $(\xi_1, \ldots \xi_n)$,

$$\varphi_n^*(t) = \int_{\mathbf{R}^1} e^{itx} \mathbf{P}_n^*(dx), \quad t \in [-1,1],$$

We denote

$$\zeta_n^*(t) = n^{1/2}[\varphi_n^*(t) - \varphi(t)], \quad \zeta(t) = \int_{\mathbf{R}^1} e^{itx} \mathring{W}_{\mathbf{P}}(dx), \quad t \in [-1,1].$$

The estimate from (Csörgö, 1981) is improved in the following theorem.

THEOREM 6. *If, for some $\gamma > 0$,*

$$\mathbf{P}(\{x: |x| > A\}) = O(A^{-\gamma}), \quad A \to \infty, \tag{28}$$

then a continuous a.s. on $[-1,+1]$ copy ζ_n of the process ζ can be defined, for each n, on the probability space $(\Omega, \overline{\Sigma}, \mathrm{Pr})$ such that

$$\mathrm{Pr}^*\Big(\{\sup_{t\in[-1,1]} |\zeta_n^*(t) - \zeta_n(t)| \geqslant n^{-\gamma/2(\gamma+1)} \log^{1/2} n(t + C\log n)\}\Big) \leqslant$$

$$\leqslant \Lambda e^{-\theta t}, \quad t > 0, \tag{29}$$

where $C > 0$, $\Lambda > 0$, $\theta > 0$ are constants.

Proof. Let us consider the class of functions

$$\mathcal{F} = \{\cos(tx): t \in [-1,1]\} \bigcup \{\sin(tx): t \in [-1,1]\}.$$

Let $\mathfrak{C}$ denote the class of subsets of $\mathbf{R}^1$, which include the empty set, $\mathbf{R}^1$ and all subsets of the form

$$\bigcup_{k=-\infty}^{\infty} (\Delta + kT),$$

where $T \geqslant 2\pi$, $\Delta \subset \mathbf{R}^1$ is an open interval of length less or equal to T.
Then, for each $f \in \mathcal{F}$, $c \in \mathbf{R}^1$,

$$\{x: f(x) > c\} \in \mathfrak{C}.$$

It is easy to see that $\mathfrak{C} \in \mathrm{EM}(\mathbf{P})$ (Dudley, 1978).
Now we shall prove that $\mathfrak{C} \in \mathrm{AP}(\mathbf{P}; \beta)$, where $\beta = \gamma/(\gamma+1)$.
We define the map $F\colon \mathbf{R}^1 \to I$, by setting

$$F(x) = \mathbf{P}((-\infty, x)), \quad x \in \mathbf{R}^1.$$

By condition (27), map F is continuous. Let us define the random element $\xi\colon (I, \mathcal{B}_I) \to (\overline{\mathbf{R}}^1, \mathfrak{A})$, by setting

$$\xi(u) = F^{-1}(u), \quad u \in I,$$

where $\overline{\mathbf{R}}^1 = \mathbf{R}^1 \cup \{-\infty, +\infty\}$. Then ξ has the distribution $\mathbf{P}$. We set

$$\mathcal{E} = \big\{\{u \in I: \xi(u) \in C\}: C \in \mathfrak{C}\big\} = \big\{F(C): C \in \mathfrak{C}\big\}.$$

Let $Q_N = [-\pi N, \pi N]$. Obviously, for each $C \in \mathfrak{C}$ and $N \geqslant 1$ the set $C \cap Q_N$ consists of not more then N intervals. So, the set $F(C \cap Q_N)$ also consists of not more then N intervals.
Note that

$$F(C) \,\Delta\, F(C \cap Q_N) = F(C \backslash Q_N) \subset F(\mathbf{R}^1 \backslash Q_N).$$

Consequently, for each $E \in \mathcal{E}$,

$$d(E; N) \leqslant \lambda\big(F(\mathbf{R}^1 \backslash Q_N)\big) = \lambda\big(\{u\colon\ |\xi(u)| \geqslant \pi N\}\big) = \mathbf{P}(\mathbf{R}^1 \backslash Q_N),$$

i.e., $d(\mathcal{E}; N) \leqslant \mathbf{P}(\mathbf{R}^1 \backslash Q_N)$. By condition (28) we obtain

$$d(\mathcal{E}; N) = O(N^{-\gamma}), \quad N \to \infty,$$

which, in view of Proposition 1, implies $\mathfrak{C} \in \mathrm{AP}(\mathbf{P}; \beta)$, $\beta = \gamma/(\gamma+1)$. Approximating sets of the class $\mathfrak{C}$ by unions of disjoint intervals (of equal length and with equal distances between the intervals), the ends of which belong to the grating $\{k\varepsilon\colon\ k = 0, \pm 1, \pm 2, \ldots\}$ with the appropriate step $\varepsilon > 0$, one can prove that, under conditions (27), (28),

$$H_I(\mathfrak{C}; \delta) = O\big(\log(1/\delta)\big), \quad \delta \to 0.$$

By Theorem 2, this means that the class $\mathfrak{C}$ satisfies condition (10) with $\rho = 1$. Consequently, by Theorem 1 (ii), $\mathfrak{C} \in \mathrm{KMT}(\mathbf{P}; \{\delta_n\}_{n\geqslant 1})$, where

$$\delta_n = n^{-\gamma/2(\gamma+1)} \log^{1/2} n, \quad n \geqslant 2.$$

Thus, Proposition 2 implies that $\mathcal{F} \in \mathrm{KMT}(\mathbf{P}; \{\delta_n\}_{n\geqslant 1})$, which in tern implies estimate (29) with some constants $C > 0$, $\Lambda > 0$, $\theta > 0$.
Theorem 6 is proved.

REFERENCES

Borisov, I.S. (1983). Rate of convergence in invariance principle in linear spaces. Application to empirical measures. *Lect. Notes Math.* **1021**, 45–58.

Borisov, I.S. (1985). New approach to the problem of approximation of distributions of summs of independent random variables. *Trans. Math. Inst. Siberian Branch USSR Academy Sci.* **5**, 3–27 (in Russian).

Borisov, I.S. (1987). Rate of convergence in the invariance principle for empirical measures. In: *Proc. 1-st World Congress Bernoulli Soc.* VNU Sci. Press, Amsterdam, pp. 833–836.

Csörgö, S. (1981). Limit behaviour of the empirical characteristic function. *Ann. Probab.* **9**, 130–144.

Dudley, R.M. (1974). Metric entropy of some classes of sets with differentiable boundaries. *J. Approx. Theory.* **10**, 227–236.

Dudley, R.M. (1978). Central limit theorems for empirical measures. *Ann. Probab.* **6**, 899–929.

Dudley, R.M. (1981). Donsker classes of functions. In: *Statistics and Related Topics (Proc. Symp. Ottawa, 1980).* North-Holland, New York, pp. 341–352.

Gine, E. and Zinn, J. (1984). Some limit theorems for empirical processes. *Ann. Probab.* **12**, 929–998.

Gine, E. and Zinn, J (1986). Lectures on the central limit theorem for empirical processes. *Lect. Notes Math.* **1221**, 50–113.

Kolchinskii, V.I. (1981). On central limit theorem for empirical measures. *Theory Probab. Math. Statist.* **24**, 63–75.

Kolchinskii, V.I. (1985; 1986). Functional limit theorems and empirical entropy. *Theory Probab.Math. Statist.* **33**, 35–45; **34**, 81-93.

Komlos, J., Major, P. and Tusnady G. (1975; 1976). An approximation of partial sums of independent RV's and the sample DF. *Z. Wahrsch. verw. Geb.* **32**, 111–131; **34**, 33–58.

Le Cam, L. (1983). A remark on empirical processes. In: *Festschrift for E.L. Lehman in Honour of his Sixty-Fifth Birthday,* Wadsworth, Belmont (Calif.), pp. 305–327.

Massart, P. (1986). Rates of convergence in the central limit theorem for empirical processes. *Ann. Inst. Henri Poincare.* **22**, 381–423.

Massart, P. (1989). Strong approximation for multivariate empirical and related processes, Via KMT constructions. *Ann. Probab.* **17**, 266–291.

Talagrand, M. (1985). Classes de Donsker et ensembles pulverises. *C.r. Acad. Sci. Paris.* **300,** Ser I, 161–163.

Vapnik, V.N. and Chervonenkis, A.Ya. (1971). On the uniform convergence of frequencies of events to their probabilities. *Theory Probab. Appl.* **16**, 264–279.

Zhdanov, A.N. and Sevastyanov, E.A. (1983). Approximatial and diffirential properties of measurable sets. *Math. USSR Sbornik.* **121**, 403–422.

New Trends in Probab. and Statist., pp. 183 – 189
V. Sazonov and T. Shervashidze (Eds.)
1991 VSP/Mokslas

ERGODIC THEOREM FOR SEMIMARTINGALE-HELIX

R.Sh. LIPTSER
Institute of Control Sciences of the USSR Academy of Sciences,
Profsoyuznaya str. 65, 117342 Moscow, USSR.

ABSTRACT

If $(\xi_t)_{t\geqslant 0}$ is a strictly stationary ergodic random process with $\mathbf{E}|\xi_0| < \infty$, then for the random process $(X_t)_{t\geqslant 0}$ with

$$X_t = \int_0^t \xi_s \, ds$$

the Birkhoff–Khintchine theorem holds:

$$\lim_{t\to\infty} \frac{1}{t} X_t = \mathbf{E}\xi_0 \; (\mathbf{P}\text{-a.s.}) \quad \text{and} \quad \lim_{t\to\infty} \mathbf{E}\left|\frac{1}{t}X_t - \mathbf{E}\xi_0\right| = 0.$$

This result is extended to strictly stationary processes $(X_t)_{t\geqslant 0}$ which do not have the above mentioned integral representation. For example, when $(X_t)_{t\geqslant 0}$ is a semimartingale with strictly stationary increments, the Birkhoff–Khintchine theorem remains true if

$$\mathbf{E} \sup_{0\leqslant t\leqslant 1} |X_t| < \infty.$$

In particular, if $(X_t)_{t\geqslant 0}$ is a martingale with strictly stationary increments and $X_0 = 0$ then

$$\mathbf{E}[X,X]_1^{1/2} < \infty \Longrightarrow \lim_{t\to\infty} \frac{1}{t} X_t = 0 \; (\mathbf{P}\text{-a.s.}) \quad \text{and} \quad \lim_{t\to\infty} \mathbf{E}\left|\frac{X_t}{t}\right| = 0.$$

1. Let $\xi = \xi(\omega)$ be an integrable random variable $(\mathbf{E}|\xi| < \infty)$ defined on some probability space $(\Omega, \mathcal{F}, \mathbf{P})$ and let $\theta = (\theta_t)_{t\in\mathbf{R}}$ be a measure preserving group of transformations from Ω to Ω with the group operation: $\theta_t\theta_s = \theta_{t+s}(\theta_0\omega = \omega)$; moreover, let the mapping $(t,\omega) \to \theta_t\omega$ be $\mathcal{B}(\mathbf{R}) \otimes \mathcal{F}/\mathcal{F}$ measurable. Under this assumption, the following measurable random processes are defined: $(\xi(\theta_t\omega))_{t\in\mathbf{R}}$ and $(X_t(\omega))_{t\in\mathbf{R}}$ with

$$X_t(\omega) = \int_0^t \xi(\theta_s\omega)\, ds. \tag{1}$$

By virtue of the classical Birkhoff–Khintchine theorem (Birkhoff, 1931; Khintchine, 1932), we have

$$\lim_{t\to\infty} \frac{1}{t} X_t(\omega) = \mathbf{E}\big(\xi(\omega)|J\big) \quad \mathbf{P}\text{-a.s.} \tag{2}$$

and

$$\lim_{t\to\infty} \mathbf{E}\left|\frac{1}{t} X_t(\omega) - \mathbf{E}\big(\xi(\omega)|J\big)\right| = 0, \tag{3}$$

where J is the σ-algebra of invariant sets corresponding to the group θ:

$$J = \left\{A \in \mathcal{F}^P\colon\ \mathbf{P}(I_A(\omega)) = \big(I_A(\theta_t\omega)\big) = 1,\ \forall t \in \mathbf{R}\right\}$$

and $\mathcal{F}^P$ is the completion of $\mathcal{F}$ with respect to $\mathbf{P}$. The random process $(X_t(\omega))_{t\in\mathbf{R}}$ defined in (1) is a helix, i.e., (see (De Sam Lazaro and Meyer, 1971; Liptser and Shiryayev, 1989)) $X_0(\omega) = 0$, $X_t(\omega)$ is right continuous in t,

$$X_{t+h}(\omega) - X_{s+h}(\omega) = X_t(\theta_h\omega) - X_s(\theta_h\omega). \tag{4}$$

The question is: do the ergodic relations (2) and (3) hold true for the helix $\big(X_t(\omega)\big)$ which does not admit the integral representation (1)?
The partial answer is given by Theorem 4.11.8 in (Liptser and Shiryayev, 1989) for a helix whose trajectories are right continuous, have limits on the left and satisfy

$$\mathbf{E} \sup_{0\leqslant t\leqslant 1} |X_t(\omega)| < \infty. \tag{5}$$

Namely, an ergodic relation of the type (3) holds:

$$\lim_{t\to\infty} \mathbf{E}\left|\frac{1}{t} X_t(\omega) - \mathbf{E}\big\{X_1(\omega)|J\big\}\right| = 0. \tag{6}$$

The aim of this paper is to show that under assumption (5) the ergodic relation of type (2) holds for the helix too.

THEOREM 1. *Let $(X_t(\omega))_{t\in\mathbf{R}}$ be a helix whose trajectories are right continuous, have limits on the left and satisfy* (5).
Then

$$\lim_{t\to\infty} \frac{1}{t} X_t(\omega) = \mathbf{E}\big\{X_1(\omega)|J\big\} \quad \mathbf{P}\text{-a.s.}$$

Proof. Denote by $[\cdot]$ the greatest integer function. By virtue of the helix property (4) we have

$$X_{[t]}(\omega) = \sum_{j=1}^{[t]} \Big[X_j(\omega) - X_{j-1}(\omega)\Big] = \sum_{j=1}^{[t]} X_1(\theta_{j-1}\omega).$$

Consequently, by the Birkhoff–Khintchine theorem for discrete time (see, for example, (Shiryayev, 1984))

$$\lim_{t\to\infty} \frac{1}{t} X_{[t]}(\omega) = \mathbf{E}\Big\{X_1(\omega)|J\Big\} \quad \mathbf{P}\text{-a.s.}$$

From this, taking into account that $\lim_{t\to\infty} t/[t] = 1$, we obtain

$$\lim_{t\to\infty} \frac{1}{t} X_{[t]}(\omega) = \mathbf{E}\Big\{ X_1(\omega)|J \Big\} \quad \mathbf{P}\text{-a.s.}$$

The desired result follows now from the relation $\lim_{t\to\infty}(1/t)|X_t(\omega) - X_{[t]}(\omega)| = 0 \quad \mathbf{P}$-a.s.
To prove this, we use the estimate

$$|X_t(\omega) - X_{[t]}(\omega)| \geqslant \sup_{0\leqslant s\leqslant 1} |X_{[t]+s}(\omega) - X_{[t]}(\omega)|$$

and show that

$$\lim_{t\to\infty} \frac{1}{t} \sup_{0\leqslant s\leqslant 1} |X_{[t]+s}(\omega) - X_{[t]}(\omega)| = 0 \quad \mathbf{P}\text{-a.s.}. \tag{7}$$

By the criterion of the convergence with probability one (see (Shiryayev, 1984)), equality (7) follows from

$$\lim_{t\to\infty} \mathbf{P}\Big\{ \sup_{j\geqslant t} \frac{1}{j} \sup_{0\leqslant s\leqslant 1} |X_{j+s}(\omega) - X_j(\omega)| > \varepsilon \Big\} = 0, \quad \varepsilon > 0. \tag{8}$$

Now we will show that (5) implies (8). Using (4), we derive

$$\mathbf{P}\Big\{ \sup_{j\geqslant t} \frac{1}{j} \sup_{0\leqslant s\leqslant 1} |X_{j+s}(\omega) - X_j(\omega)| > \varepsilon \Big\} = \mathbf{P}\Big\{ \sup_{j\geqslant t} \frac{1}{j} \sup_{0\leqslant s\leqslant 1} |X_s(\theta_j\omega)| > \varepsilon \Big\} =$$

$$= \mathbf{P}\Big\{ \bigcup_{j\geqslant t} \frac{1}{j} \sup_{0\leqslant s\leqslant 1} |X_s(\theta_j\omega)| > \varepsilon \Big\} \leqslant \sum_{j\geqslant t} \mathbf{P}\Big\{ \frac{1}{j} \sup_{0\leqslant s\leqslant 1} |X_s(\theta_j\omega)| > \varepsilon \Big\} =$$

$$= \sum_{j\geqslant t} \mathbf{P}\Big\{ \frac{1}{j} \sup_{0\leqslant s\leqslant 1} |X_s(\omega)| > \varepsilon \Big\} \sum_{j\geqslant t} \int_{j-1}^{j} \mathbf{P}\Big\{ \sup_{0\leqslant s\leqslant 1} |X_s(\omega)| > \varepsilon j \Big\} dx \leqslant$$

$$\leqslant \int_{t-1}^{\infty} \mathbf{P}\Big\{ \sup_{0\leqslant s\leqslant 1} |X_s(\omega)| > \varepsilon x \Big\} dx.$$

Evidently, the last term in these inequalities tends to zero as $t \to \infty$ if

$$\int_0^\infty \mathbf{P}\Big\{ \sup_{0\leqslant s\leqslant 1} |X_s(\omega)| > \varepsilon x \Big\} dx < \infty. \tag{9}$$

Therefore, (8) follows from (9). But (5) implies (9) because

$$\int_0^\infty \mathbf{P}\Big\{ \sup_{0\leqslant s\leqslant 1} |X_s(\omega)| > \varepsilon x \Big\} dx = \frac{1}{\varepsilon} \mathbf{E} \sup_{0\leqslant s\leqslant 1} |X_s(\omega)|.$$

2. Now we consider the semimartingale-helix case. For this we define first a stochastic basis. Let $\mathcal{F}_0$ be a sub-σ-algebra of $\mathcal{F}$ such that $\theta_t^{-1}(\mathcal{F}_0) \subset \mathcal{F}_0$ for $t < 0$. Define σ-algebras $\mathcal{F}_t$, $t \in \mathbf{R}$, setting

$$\mathcal{F}_t = \theta_t^{-1}(\mathcal{F}_0).$$

The family $\mathsf{F} = (\mathcal{F}_t)_{t\in\mathbf{R}}$ is nondecreasing. For a completion of F we use the notation $\mathsf{F}^p = (\mathcal{F}_t^p)_{t\in\mathbf{R}}$ where $\mathcal{F}_t^p = \mathcal{F}_t \vee \mathcal{N}$, $\mathcal{N}$ is a system from $\mathcal{F}^p$ of $\mathbf{P}$-measure zero. Set $\mathcal{F}_\infty^p = \sigma(\bigcup_{t\in\mathbf{R}} \mathcal{F}_t^p)$ and further on assume that $\mathcal{F}_\infty^p = \mathcal{F}^p$. The family F^p is right continuous (see (Liptser and Shiryayev, 1989, Theorem 4.11.1)). Let $X = (X_t)_{t\geqslant 0}$ be a semimartingale-helix with $X_0 = 0$ with respect to F^p (defined for negative t by $X_t(\omega) = X_{-t}(\theta_t\omega)$). By the canonical decomposition for semimartingales, we have

$$X_t = A_t + M_t \tag{10}$$

with a bounded local variation process $A = (A_t)_{t\geqslant 0}$ and a local martingale $M = (M_t)_{t\geqslant 0}$. By Theorem 4.11.5 (ibidem, see also (De Sam Lazaro and Meyer, 1971)), the processes A and M can be chosen as helices.
Evidently,

$$\sup_{0\leqslant s\leqslant 1} |A_s| \leqslant \int_0^1 |dA_s| = \mathrm{Var}(A)_1 \tag{11}$$

and by the Davis inequality (see (Liptser and Shiryayev, 1989, Theorem 1.9.6)),

$$\mathbf{E} \sup_{0\leqslant s\leqslant 1} |M_s| \leqslant C\mathbf{E}[M,M]_1^{1/2}. \tag{12}$$

Thus, Theorem 1 implies

THEOREM 2. *Let X be a semimartingale-helix admitting decomposition* (10) *with helices A and M.*
If

$$\mathbf{E}\,\mathrm{Var}\,(A)_1 < \infty \quad \text{and} \quad \mathbf{E}\,[M,M]_1^{1/2} < \infty, \tag{13}$$

then $\mathbf{P}$-a.s.

$$\lim_{t\to\infty} \frac{1}{t} X_t = \mathbf{E}\,\{A_1|J\}.$$

Proof. The assumptions of the theorem imply (5). In this case by Theorem 1, $\lim_{t\to\infty}(1/t)X_t = \mathbf{E}\{X_1|J\}$ $\mathbf{P}$-a.s. Taking into account decomposition (10), it is sufficient to show that $\mathbf{E}\{M_1|J\} = 0$. To this end, we use the facts that $J \subseteq \mathcal{F}_0^p$ (see (Liptser and Shiryayev, 1989, Problem 4.11.1) and (Rozanov, 1967)), $\mathbf{E}\{M_1|\mathcal{F}_0^p\} = 0$ $\mathbf{P}$-a.s. and $\mathbf{E}\{M_1|J\} = \mathbf{E}\{\mathbf{E}(M_1|\mathcal{F}_0^p)|J\}$.

Remark. If a semimartingale X is a process with strictly stationary ergodic increments, which admits decomposition (10), where A and M have the same property (possessing strictly stationary ergodic increments) and satisfy (13), then

$$\lim_{t\to\infty} \frac{1}{t} X_t = \mathbf{E}A_1, \quad \mathbf{P}\text{-a.s.} \tag{14}$$

This result follows easily from Theorem 2.

3. Consider two examples of semimartingales which are processes with strictly stationary increments.

Example 1. Let $(\xi_t)_{t\geqslant 0}$ be a strictly stationary ergodic (measurable) process and $Z = (Z_t)_{t\geqslant 0}$ be a cadlag process with independent and homogeneous increments. The processes (ξ_t) and (Z_t) are supposed to be independent. By Corollary 11.4.18 in (Jacod and Shiryayev, 1987), the triplet of predictable characteristics of Z is deterministic and has the form

$$B_t = bt, \qquad C_t = ct, \qquad \nu(dt, dx) = dt\, K(dx)$$

with constants b and c ($c \geqslant 0$), while the measure $K(dx)$ on $\mathbf{R}_0 = \mathbf{R}\backslash\{0\}$ is such that $\int_{\mathbf{R}_0}(x^2 \wedge 1)K(dx) < \infty$. Assume that

$$\int\limits_{\mathbf{R}_0}(x^2 \wedge |x|)K(dx) < \infty, \quad b = -\int\limits_{|x|>1} xK(dx).$$

In this case $\mathbf{E}(Z_t - Z_s) = 0$, $t, s \geqslant 0$. Under an appropriate choice of some filtration $\mathrm{G} = (G_t)_{t\geqslant 0}$, the process Z is a martingale with respect to G and ξ_t is G_0-measurable random variable for any $t \geqslant 0$. In addition, assume that

$$\mathbf{E}\left\{\int\limits_0^1 \xi_s^2 ds\right\}^{1/2} < \infty, \qquad \mathbf{E}\left\{\int\limits_{R_0} \frac{\xi_0^2 x^2}{1+|\xi_0 x|} K(dx)\right\} < \infty.$$

These facts allow us to define a stochastic integral (see (Liptser and Shiryayev, 1989, Chapter 2))

$$X_t = \int\limits_0^t \xi_s\, dZ_s, \quad t \geqslant 0,$$

which is a martingale with strictly stationary ergodic increments, and to show that

$$\mathbf{E}\,[X, X]_1^{1/2} < \infty.$$

By the definition of the quadratic variation (see ibidem), we have (ΔZ_s is the jump of Z at the point s)

$$[X, X]_1 = c\int\limits_0^1 \xi_s^2\, ds + \sum_{0<s\leqslant 1} \xi_s^2(\Delta Z_s)^2$$

and so, taking into account the inequality $(a+b)^{1/2} \leqslant a^{1/2} + b^{1/2}$, $\quad a, b \geqslant 0$,

$$[X, X]_1^{1/2} \leqslant c^{1/2}\left\{\int\limits_0^1 \xi_s^2\, ds\right\}^{1/2} + \left\{\sum_{0<s\leqslant 1} \xi_s^2(\Delta Z_s)^2\right\}^{1/2}.$$

Now we use the inequality

$$\Big(\sum_i a_i^2\Big)^{1/2} \leqslant 1/4 + 2\sum_i \frac{a_i^2}{1+|a_i|}$$

(Lemma 2.2.1, ibidem) and obtain

$$[X,X]_1^{1/2} \leqslant c^{1/2}\Big\{\int_0^1 \xi_s^2\,ds\Big\}^{1/2} + \frac{1}{4} + 2\sum_{0<s\leqslant 1}\frac{\xi_s^2(\Delta Z_s)^2}{1+|\xi_s\Delta Z_s|} =$$

$$= \frac{1}{4} + c^{1/2}\Big\{\int_0^1 \xi_s^2 ds\Big\}^{1/2} + 2\int_0^1\int_{\mathbf{R}_0} \frac{\xi_s^2 x^2}{1+|\xi_s x|}\mu(ds,dx),$$

where $\mu(ds,dx)$ is the measure of jumps of Z.
This inequality implies

$$\mathbf{E}[X,X]_1^{1/2} \leqslant \frac{1}{4} + c^{1/2}\mathbf{E}\Big\{\int_0^1 \xi_s^2\,ds\Big\}^{1/2} + 2\mathbf{E}\int_0^1\int_{\mathbf{R}_0}\frac{\xi_s^2x^2}{1+|\xi_s x|}K(dx)ds < \infty,$$

since

$$2\mathbf{E}\int_0^1\int_{\mathbf{R}_0}\frac{\xi_s^2x^2}{1+|\xi_0 x|}K(dx)\,ds = \mathbf{E}\int_{\mathbf{R}_0}\frac{\xi_0^2x^2}{1+|\xi_o x|}K(dx).$$

By Remark to Theorem 2,

$$\lim_{t\to\infty}\frac{1}{t}\int_0^t \xi_s\,dZ_s = 0 \quad \mathbf{P}\text{-a.s.}$$

Example 2. Let $(\xi_t)_{t\geqslant 0}$ and $Z=(Z_t)_{t\geqslant 0}$ be independent processes. The strictly stationary process $(\xi_t)_{t\geqslant 0}$ is ergodic and $\mathbf{E}|\xi_0|<\infty$. The cadlag increasing process $(Z_t)_{t\geqslant 0}$ has independent and homogeneous increments, $Z_0=0$, and $\mathbf{E}Z_t\equiv t$. So,

$$\mathbf{E}\sup_{0\leqslant t\leqslant 1}\Big|\int_0^t \xi_s\,dZ_s\Big| \leqslant \mathbf{E}\int_0^1|\xi_s|dZ_s = \mathbf{E}\int_0^1|\xi_s|ds = \mathbf{E}|\xi_0| < \infty$$

and, by the remark to Theorem 2,

$$\lim_{t\to\infty}\frac{1}{t}\int_0^1 \xi_s\,dZ_s = \mathbf{E}\int_0^1\xi_s\,dZ_s = \mathbf{E}\int_0^t \xi_s ds = \mathbf{E}\xi_0 \quad \mathbf{P}\text{-a.s.}$$

REFERENCES

Birkhoff, G.D. (1931). Proof of the ergodic theorem. *Proc. Nat. Acad. Sci. USA* **17**, 656–660.

De Sam Lazaro and Meyer, P.-A. (1971). Méthodes de martingales et théorie des flots. *Z. Wahrsch. verw. Geb.* **18**, 116–140.

Jacod, J. and Shiryayev, A.N. (1987). *Limit Theorems for Stochastic Processes.* Springer, Berlin etc.

Khintchine, A.Ja. (1932). Zur Birkhoffs Lösung des Ergodenproblems. *Math. Ann.* **107**, 485–488.

Liptser, R.Sh. and Shiryayev, A.N. (1989). *Theory of Martingales.* Kluwer, Dordrecht etc.

Rozanov, Yu.A. (1967). *Stationary Random Processes.* Holden-Day, San Francisco.

Shiryayev, A.N. (1984). *Probability.* Springer, Berlin etc.

New Trends in Probab. and Statist., pp. 190 - 197
V. Sazonov and T. Shervashidze (Eds.)
1991 VSP/Mokslas

ERGODIC THEOREMS FOR DISCRETE TIME RANDOM PROCESSES

S.V. NAGAEV
Institute of Mathematics, Universitetskii pr. 4., 690090 Novosibirsk, USSR

ABSTRACT

Ergodic theorems for discrete time random processes which are stationary or asymptotically stationary in a certain generalized sense are proved. In particular, homogeneous Markov chains are treated from this point of view. Some connections with the theory of random transformations and the queueing theory are discussed.

Let $X_0, X_1, \dots, X_n, \dots$ be a sequence of random variables (r.v.) taking values in a measurable space (Y, S). In other words, X_k is a measurable mapping of the basic probability space Ω into (Y, S). Further, let $\xi_0, \xi_1, \dots, \xi_n, \dots$ be a sequence of r.v. on Ω taking values 0 and 1. Denote by A_k the event $\xi_k = 1$. Let $\overline{A}_k = \Omega - A_k$. Define

$$Q_j(B_1, B_2, \dots, B_j) = \sum_{k=0}^{\infty} \mathbf{P}\left(A_0 \prod_{i=1}^{k} \overline{A}_i \prod_{l=1}^{j} \{X_{l+k} \in B_l\}\right).$$

Here and in the sequel $\prod_m^{m-1} \overline{A}_j = \Omega$.
The process X_n is said to be conditionally stationary with respect to the sequence A_k and a subalgebra $S' \in S$ if the equality

$$\mathbf{P}\left(A_m \prod_{j=m+1}^{j=m+k} \overline{A}_j \prod_{l=1}^{j} \{X_{l+m+k} \in B_l\}\right) =$$

$$= \mathbf{P}\left(A_0 \prod_{j=1}^{k} \overline{A}_j \prod_{l=1}^{j} \{X_{l+k} \in B_l\}\right) \tag{1}$$

holds for any $m \geqslant 1$, $k \geqslant 1$, $j \geqslant 0$ and $B_l \in S'$.

PROPOSITION. *If the process X_n is conditionally stationary, then*

$$\left|\mathbf{P}\left(\prod_{l=n+1}^{n+j} \{X_l \in B_l\}\right) - Q_j(B_1, B_2, \dots, B_j)\right| < \mathbf{P}\left(\prod_{i=0}^{n} \overline{A}_i\right), \quad B_l \in S'. \tag{2}$$

Proof. Applying condition (1), we have

$$\mathbf{P}\left(\prod_{l=1}^{j}\{X_{l+n}\in B_l\}\right)=\sum_{k=0}^{n}\mathbf{P}\left(A_{n-k}\prod_{i=n-k+1}^{n}\overline{A}_i\prod_{l=1}^{j}\{X_{l+n}\in B_l\}\right)+$$

$$+\mathbf{P}\left(\prod_{i=0}^{n}\overline{A}_i\prod_{l=1}^{j}\{X_{l+n}\in B_l\}\right)=\sum_{k=0}^{n}\mathbf{P}\left(A_0\prod_{i=1}^{k}\overline{A}_i\prod_{l=1}^{j}\{X_{l+k}\in B_l\}\right)+$$

$$+\mathbf{P}\left(\prod_{i=0}^{n}\overline{A}_i\prod_{l=1}^{j}\{X_{n+l}\in B_l\}\right). \tag{3}$$

In particular,

$$\sum_{k=0}^{n}\mathbf{P}\left(A_0\prod_{i=1}^{k}\overline{A}_i\right)+\mathbf{P}\left(\prod_{i=0}^{n}\overline{A}_i\right)=1.$$

Hence,

$$\sum_{k=n+1}^{\infty}\mathbf{P}\left(A_0\prod_{i=1}^{k}\overline{A}_i\prod_{l=1}^{j}\{X_{l+k}\in B_l\}\right)\leqslant$$

$$\leqslant 1-\sum_{k=0}^{n}\mathbf{P}\left(A_0\prod_{i=1}^{k}\overline{A}_{i=1}\right)=\mathbf{P}\left(\prod_{i=0}^{n}\overline{A}_i\right). \tag{4}$$

From (3) and (4) follows (2).

THEOREM 1. *If the process X_n is conditionally stationary and*

$$\mathbf{P}\left(\prod_{i=0}^{\infty}\overline{A}_i\right)=0, \tag{5}$$

then

$$\lim_{n\to\infty}\sup_{B_1,B_2,\ldots,B_j\in S'}\left|\mathbf{P}\left(\prod_{l=1}^{j}\{X_{l+n}\in B_l\}\right)-Q_j(B_1,B_2,\ldots,B_j)\right|=0, \tag{6}$$

and the distributions Q_j satisfy the conditions

$$Q_{j+1}(B_1,B_2,\ldots,B_j,Y)=Q_j(B_1,B_2,\ldots,B_j), \tag{7}$$

$$Q_{j+1}(Y,B_1,B_2,\ldots,B_j)=Q_j(B_1,B_2,\ldots,B_j). \tag{8}$$

In other words, the distributions Q_j are consistent, and the corresponding infinite dimensional distribution is invariant with respect to a shift.

Proof. Equality (6) follows from (2). By (2) Q_l is a distribution on Y^l and, therefore, the condition (7) holds. It remains to prove (8). By (2),

$$\mathbf{P}\left(\prod_{l=1}^{j}\{X_{n+l}\in B_l\}\right)=\mathbf{P}\left(\{X_{n-1}\in Y\}\prod_{l=n}^{n+j}\{X_l\in B_l\}\right)\longrightarrow$$

$$\longrightarrow Q_{j+1}(Y, B_1, B_2, \ldots, B_j), \quad n \to \infty.$$

On the other hand,

$$\mathbf{P}\left(\prod_{l=n}^{n+j} \{X_l \in B_l\}\right) \longrightarrow Q_j(B_1, B_2, \ldots, B_j), \quad n \to \infty.$$

So,

$$Q_{j+1}(Y, B_1, B_2, \ldots, B_j) = Q_j(B_1, B_2, \ldots, B_j).$$

This completes the proof.

THEOREM 2. *Let, for any $j \geqslant 1$, $k \geqslant 0$,*

$$\lim_{n\to\infty} \mathbf{P}\left(A_n \prod_{j=n+1}^{n+k} \overline{A}_j \prod_{l=1}^{j} \{X_{l+n+k} \in B_l\}\right) = G_j^k(B_1, B_2, \ldots, B_j), \quad B_l \in S, \tag{9}$$

and, for any k,

$$\mathbf{P}\left(\prod_{j=k}^{k+n} \overline{A}_j\right) < \lambda_n, \tag{10}$$

where $\lambda_n \to 0$ as $n \to \infty$.
Then

$$\lim_{n\to\infty} \mathbf{P}\left(\prod_{l=1}^{j} \{X_{l+n} \in B_l\}\right) = \sum_{k=0}^{\infty} G_j^k(B_1, \ldots, B_j). \tag{11}$$

Proof. Applying the equality

$$\bigcup_{k=N}^{n} A_{n-k} \prod_{n-k+1}^{n-N} \overline{A}_i = \Omega - \prod_{i=0}^{n-N} \overline{A}_i,$$

we have

$$\sum_{k=N}^{n} \mathbf{P}\left(A_{n-k} \prod_{i=n-k+1}^{n} \overline{A}_i \prod_{l=1}^{j} \{X_{l+n} \in B_l\}\right) \leqslant$$

$$\leqslant \sum_{k=N}^{n} \mathbf{P}\left(A_{n-k} \prod_{i=n-k+1}^{n} \overline{A}_i\right) \leqslant \mathbf{P}\left(\prod_{i=n-N+1}^{n} \overline{A}_i\right) \leqslant \lambda_{N-1}. \tag{12}$$

It follows from (3), (9), (10) and (12) that

$$\sum_{k=0}^{\infty} G_j^k(B_1, \ldots, B_j) < \infty$$

and

$$\lim_{n\to\infty} \left| \mathbf{P}\left(\prod_{l=1}^{j} \{X_{l+n} \in B_l\}\right) - \sum_{k=0}^{n} G_j^k(B_1, \ldots, B_j) \right| = 0.$$

From the two last relations we obtain the assertion of the theorem.

Remark. If the process ξ_k is stationary (in the strong sense), then condition (10) follows from condition (5). As Y one may consider a group or semi-group not necessary topological.

Ergodic properties of the sequence X_n are studied in (Borovkov, 1978); see also (Borovkov, 1980) in the case when Y is a semi-group of transformations of a vector space into itself. (The semi-group operation is defined by the superposition of transformations.) More exactly,

$$X_n = \prod_{k=1}^{n} Z_k, \tag{13}$$

where $Z_k = f(\cdot, \tau_k)$, f is a map of a vector space into itself (for a fixed τ_k), τ_k is a stationary process.

The conditions providing the ergodic theorem are formulated in (Borovkov, 1978) in terms of the so called renovating events. The notion of a renovating event in an explicit form was defined for the first time in (Leontjeva, 1976) for queueing systems controlled by a sequence of independent random variables, although the phenomenon of renovation was known before in the queueing theory. This definition was extended in (Ahmarov and Leontjeva, 1976) to systems with stationary controlling sequences. Borovkov in (Borovkov, 1978) extended the notion of renovating event to the Loynes construction (Loynes, 1962).

Some aspects of the theory of processes with innovation were discussed in (Kalashnikov and Rachev, 1988). It should be noted that the concept of renovating events is used for the present time only in queueing theory.

It is not hard to see that if a sequence of renovating events A_n is stationary (Borovkov, 1980), then our condition (1) holds. Hence in this case the ergodicity of the recurrent sequence (13) follows from our Theorem 1 though in a slightly weaker sense than in (Borovkov, 1978) but with an estimate of the rate of convergence. Note that bound (2) is close by appearance to bound (11) in (Borovkov, 1978, p. 322) which was obtained for a single-server queue without any connection with the concept of renovation.

Simultaneously with a stationary sequence of renovating events Borovkov considered a nonstationary one. However the condition imposed in this case was in fact that the corresponding sequence of distributions were fundamental in the topology of uniform convergence. Therefore it can not be considered as efficient. (By the way this concerns also our conditions (9) and (10).)

Let the process X_n be a homogenious Markov chain with a transition function $p(x, E)$, $E \in S$. Put

$$P_n(B) = \mathbf{P}(X_n \in B), \quad P_E^{(n)}(B) = P(X_n \in B,\ X_j \in E,\ j = 0, 1, \ldots, n-1),$$

$$P_E^{(n)}(x, B) = \mathbf{P}\left(X_n \in B,\ X_j \in E,\ j = 1, 2, \ldots, n-1 \mid X_0 = x\right), \quad \overline{B} = Y - B.$$

THEOREM 3. *Suppose that there exists a set $B_0 \in S$ such that, for any $B \in B_0 S$,*

$$p_n(B) = \mu(B), \qquad n = 0, 1, \ldots, \tag{14}$$

where μ is some measure on B_0S. Then

$$\sup_{B\in S}\left|p^{(n)}(B)-\sum_{k=1}^{n}\int p_{\overline{B}_0}^{(k)}(y,B)\,\mu(dy)\right|<\mathbf{P}\left(\prod_{j=0}^{n-1}\{X_j\in\overline{B}_0\}\right). \tag{15}$$

Proof. Let $A_n=\{X_n\in B_0\}$. Then

$$\mathbf{P}\left(A_n\prod_{k=1}^{m-1}\overline{A}_{n+k}\{X_{n+m}\in B\}\right)=\int_{B_0}p_{\overline{B}_0}^{(m)}(y,B)\,\mu(dy).$$

It remains to apply Theorem 1.
Condition (14) holds, for instance, if

$$p_0(B)=\mu(B),\qquad p(x,B)=\mu(B),\qquad x\in Y,\quad B\in B_0S.$$

Indeed, in this case,

$$\mathbf{P}\left(\prod_{k=1}^{n-1}\overline{A}_k\{X_n\in B_0\}\middle|X_0=x\right)=$$

$$=\mu(B_0)\,p_{\overline{B}_0}^{(n-1)}(x,\overline{B}_0)=\mu(B_0)\,(1-\mu(B_0))^{n-1}.$$

So, the probability that the first hit of B_0 occurs at the moment n equals the probability that the first success in the Bernoulli trials with probability of success $\mu(B_0)$ occurs at n-th trial. Hence, $p(B_0)=\mu(B_0)$. Exactly in the same way the equality $p_n(B_0)=\mu(B_0)$ is proved for any $B\in B_0S$.
Note that bound (15) conforms to the estimates of error in renewal type theorems (see, e.g., (Rogozin, 1973)).

THEOREM 4. *Assume that there exists a set B_0 such that*

$$\lim_{n\to\infty}\sup_{B\in B_0S}|p_n(B)-\mu(B)|=0, \tag{16}$$

where μ is some measure on B_0S with $\mu(B_0)>0$ and one of the following two conditions holds:

$$\mathbf{P}\left(\prod_{j=n}^{n+k}\{X_j\in\overline{B}_0\}\right)<a_k \tag{17}$$

uniformly with respect to n, where $a_k\to 0$ as $k\to 0$, or

$$p_n(B)<c\mu(B),\quad B\in SB_0,\quad \lim_{n\to\infty}p_{\overline{B}_0}^{(n)}(\overline{B}_0)=0, \tag{18}$$

where c is a constant not depending on n, $\overline{B}_0=Y-B_0$.
Then

$$\lim_{n\to\infty}\sup_{B\in S}|p_n(B)-\lambda(B)|=0, \tag{19}$$

where

$$\lambda(B) = \sum_{k=1}^{\infty} \int_{B_0} p_{\overline{B}_0}^{(k)}(x, B)\mu(dx). \tag{20}$$

Proof. It is not difficult to see that

$$p_n(B) = \sum_{k=1}^{n} \lambda_{nk}(B) + p_{\overline{B}_0}^{(n)}(B), \tag{21}$$

where $\lambda_{nk}(B) = \int_{B_0} p_{\overline{B}_0}^{(k)}(x, B)p_{n-k}(dx)$. By condition (16), for any fixed k,

$$\lim_{n\to\infty} \sup_{B\in S} \left| \lambda_{nk}(B) - \int_{B_0} p_{\overline{B}_0}^{(k)}(x, B)\mu(dx) \right| = 0. \tag{22}$$

Assume that condition (17) holds. Then, for any $\varepsilon > 0$ and $N > N(\varepsilon)$ (cf.(12)),

$$\sum_{k=N}^{n} \lambda_{nk}(Y) < \mathbf{P}\left(\prod_{n-N+1}^{n-1} \{X_j \in \overline{B}_0\} \right) < a_{N-2} < \varepsilon. \tag{23}$$

From (20)–(22) equality (19) follows. Assume now that condition (18) holds. By (21) and (16), for any fixed N,

$$\sum_{k=1}^{N} \int_{B_0} p_{\overline{B}_0}^{(k)}(x, Y)\mu(dx) \leqslant 1.$$

Hence,

$$\sum_{k=1}^{\infty} \int_{B_0} p_{\overline{B}_0}^{(k)}(x, Y)\mu(dx) \leqslant 1. \tag{24}$$

It follows from (24) and condition (18) that, for $N > N(\varepsilon)$,

$$\sum_{k=N}^{n} \lambda_{nk}(Y) < c \sum_{k=N}^{n} \int_{B_0} p_{\overline{B}_0}^{(k)}(x, Y)\mu(dx) < \varepsilon. \tag{25}$$

From (21), (22), and (25) we deduce (19).

Let Y be a locally compact separable space, S be the Borel σ-algebra, C be the space of bounded continuous real functions on Y.

A transition function $p(x, E)$ is said to be Fellerian if the operator P defined by the equality

$$Pf(x) = \int_Y f(y)p(x, dy)$$

maps C into C (see, e.g., (Skorohod, 1986)).

A transition function is said to be Fellerian on $E \in S$ if, for any $f \in C$,

$$\int_E f(y)p(x,dy) \in C.$$

THEOREM 5. *Let the transition function $p(\cdot,\cdot)$ be Fellerian and there exists a set $B_0 \in S$ such that, for any $f \in C$,*

$$\lim_{n\to\infty} \int_{B_0} f(y)p_n(dy) = \int_{B_0} f(y)\mu(dy), \tag{26}$$

where μ is some measure on B_0S. Furthermore, let the transition function $p(\cdot,\cdot)$ be Fellerian on B_0 and

$$\mathbf{P}\left\{\bigcap_{j=1}^{\infty}\{X_j \in \overline{B}_0\}\right\} = 0.$$

Then, for any $f \in C$,

$$\lim_{n\to\infty} \int_Y f(y)p_n(dy) = \int_Y f(y)\mu(dy),$$

where λ is defined by equality (20).

Proof. First of all note that the transition function $p(\cdot,\cdot)$ is Fellerian on $\overline{B}_0$. Therefore, for $f \in C$,

$$\int_Y f(y)p^{(k)}_{\overline{B}_0}(x,dy) \in C \tag{27}$$

for any $k \geqslant 1$. It follows from (26) and (27) that

$$\lim_{n\to\infty} \int_Y f(y)\lambda_{nk}(dy) = \int_{B_0}\left(\int_Y f(y)p^{(k)}_{\overline{B}_0}(x,dy)\right)\mu(dx) \tag{28}$$

(for the definition of λ_{nk} see (21)). Combining (21) and (28) we obtain the assertion of the theorem.

Remark 1. Expression (20) for the limit distribution λ is close by appearance and essentially to the representation of invariant measure obtained in (Harris, 1956).

Remark 2. Theorems 4 and 5 are associated with the known Rossberg's result (Rossberg, 1979) that the convergence of distributions of normalized and centered sums of independent r.v.'s on a half-line yields their convergence on the whole line.

Remark 3. If the state space Y of a Markov chain is topological and S is the Borel σ-algebra, then the transition function $p(\cdot,\cdot)$ admits the representation

$$p(x, B) - \quad f\colon f(x) \in B), \quad B \in S, \tag{29}$$

where m is a probability measure on the space B of Borel maps of Y into itself (see (Kifer, 1986, p.7–10)). For $Y = \mathbf{R}$ a representation of such type was obtained in (Borovkov, 1988).

Thus, in this case, the representation $X_n = \prod_{j=1}^{n} Z_j x$ holds for a Markov chain X_n, where Z_j are i.i.d. and take values in B , x is an initial state of the chain X_N. (The multiplication $\prod_{j=1}^{n} f_j$, $f_j \in B$ is defined as a superposition of the transformations f_j.)

This representation permits to reduce the study of ergodic properties of the chain X_n to the corresponding problem for the adjoint chain $\prod_{j=1}^{n} Z_j$.

Thus, the traditional ergodic theorem for a homogenious Markov chain is a version of a limit theorem for products of independent r.v.'s taking values in a group or semi-group. However, the traditional mixing conditions are too weak for this purpose.

In conclusion note that equality (29) may be treated as a realization of the Choquet principle.

REFERENCES

Ahmarov, I. and Leontjeva, N.P. (1976). Conditions for the converqence to limit processes and the strong law of large number in the queuing systems. *Theory Probab. Appl.* **21**, 559–570.

Borovkov, A.A. (1978). Ergodic and stability theorems for a class of stochastic equations and their applications. *Theory Probab. Appl.* **23**, 241–262.

Borovkov, A.A. (1984). *Asymptotic Methods in the Queuing Theory.* Wiley, New York.

Borovkov, A.A. (1988). On ergodicity and stability of the sequence $w_{n+1}=f(w_n,\xi_n)$. Applications to communication networks. *Theory Probab. Appl.* **33**, 641–658.

Harris, T.E. (1956). The existence of stationary measures for certain Markov processes. In: *Proceedings of the Third Berkeley Symposium on Mathematical Statistics and Probability,* **2**, Univ. Calif. Press, pp. 113–124.

Kalashnikov, V.V. and Rachev, S.T. (1988). *Mathematical Methods of Constructing Stochastic Models of Service.* Nauka, Moscow (in Russian).

Kifer, Y. (1986). *Ergodic Theory of Random Transformations.* Birkhäuser, Boston, etc.

Leontjeva, N.P. (1976). Theorems on convergence of processes of queues and of expectation time with continuous time. *Ukrainian Math. J.* **28**, 612–621.

Loynes, R. (1962). The stability of queue with non-independent inter-arrival and service times. *Proc. Cambridge Phil. Soc.* **58**, 424–439.

Rogozin, B.A. (1973). An estimate of the remainder term in limit theorems of renewal theory. *Theory Probab. Appl.* **18**, 703–717.

Rossberg, H.-J. (1979). Limit theorems for identically distributed summands assuming the convergence of the distributions function on a half-axis. *Theory Probab. Appl.* **24**, 692–709.

Skorokhod, A.V. (1986). Topologically recurrent Markov chains. Ergodic properties. *Theory Probab. Appl.* **31**, 641–650.

New Trends in Probab. and Statist., pp. 198 - 218
V. Sazonov and T. Shervashidze (Eds.)
1991 VSP/Mokslas

ON FUNCTIONAL PRINCIPLE OF LARGE DEVIATIONS

A. PUHALSKII
Institute of Control Sciences, Profsoyuznaya str. 65, Moscow GSP–312, USSR

ABSTRACT

We show that basic facts of the theory of weak convergence of probability measures have counterparts in the theory of large deviations. The main result is an analog of the Prokhorov theorem: a sequence of probability measures on a metric space contains a subsequence obeying the large deviation principle if a tightness condition holds (called strong tightness). On that basis we develop a method to establish the large deviation principle for stochastic processes analogous to the method of finite dimensional distributions in the theory of weak convergence. It consists, first, in proving that finite dimensional distributions satisfy the large deviation principle and, second, in checking the strong tightness.

1. Let S be a metric space. A function I: $S \longrightarrow [0,\infty]$ is called a rate function (see (Freidlin and Wentzell, 1984; Varadhan, 1984; Strook, 1984)) if it is lower semicontinuous and, moreover, for any $a \geqslant 0$, the set $\{x \in S\colon\ I(x) \leqslant a\}$ is compact. A sequence (P_n) of probability measures on the Borel σ–field $\mathcal{B}(S)$ is said to obey the large deviation principle (l.d.p.) if the following conditions hold:

DV1. For any open set $G \subset S$,

$$\underline{\lim}_{n} \frac{1}{n} \log P_n(G) \geqslant -\inf_{G} I(x),$$

DV2. For any closed set $F \subset S$,

$$\overline{\lim}_{n} \frac{1}{n} \log P_n(F) \leqslant -\inf_{F} I(x).$$

This definition resembles that of weak convergence of probability measures (Billingsley, 1968). We are going to show that this likeliness has far reaching consequences: main results of the theory of weak convergence admit counterparts in the theory of large deviations. Our major goal is an analog of the Prokhorov theorem on the equivalence of weak relative compactness and tightness of probability measures (Billingsley, 1968).

We say that a sequence (P_n) satisfies the partial large deviation principle if, for any subsequence (n') of (n), there exists a subsequence (n'') of (n') such that the sequence $(P_{n''})$ satisfies the large deviation principle with a rate function I''.

We say that a sequence (P_n) is strongly tight if, for any $\varepsilon > 0$, there exists a compact $K \subset S$ such that

$$P_n(K) > 1 - \varepsilon^n$$

for all n (another term is "exponential tightness" (Deuschel and Strook, 1989)). Then we have the following

THEOREM (P). 1. *If a sequence (P_n) is strongly tight, then there exists a subsequence $(P_{n'})$ obeying the large deviation principle.*
2. *In a complete separable metric space the partial large deviation principle and strong tightness are equivalent.*

This theorem is proved in Section 3. In Section 2 we describe some properties of the rate function. In particular we prove an analog of the Daniel–Kolmogorov theorem on the continuation of a consistent family of rate functions (cf. (Deuschel and Strook, 1989)). In Section 4 the conditions are derived for strong tightness in functional spaces **C** and **D** similar to those in the theory of weak convergence (see (Billingsley, 1968; Liptser and Shiryaev, 1989)). In Section 4 we also show that the validity of the l.d.p. can be proved by first proving the l.d.p. for finite dimensional distributions and then verifying the strong tightness. Section 5 contains examples of applications of the approach.

2. Let $h\colon S \to S'$ be a map from S to a metric space S'. Given a rate function I on S define the function $I \circ h^{-1}$ on S' by

$$I \circ h^{-1}(x') = \inf_{x:\, h(x)=x'} I(x). \tag{2.1}$$

It is known (Freidlin and Wentzell, 1984) that if h is continuous, then $I \circ h^{-1}$ is a rate function on S'. We somewhat generalize this result (cf. (Varadhan, 1984)). Denote by

$$D_h = \{x \in S'\colon\ h(x_n) \nrightarrow h(x) \text{ for some } x_n \to x\}$$

the set of discontinuity of h.

LEMMA 2.1. *If $I(x) = \infty$, for every $x \in D_h$, then $I \circ h^{-1}$ is a rate function on S'.*

Proof. We have to prove that the set $\{x'\colon\ I \circ h^{-1}(x') \leqslant a\}$ is compact for any $a \geqslant 0$. Note that

$$\{x'\colon\ I \circ h^{-1}(x') \leqslant a\} = \bigcap_{n=1}^{\infty} h\Big(\Big\{x\colon\ I(x) \leqslant a + \frac{1}{n}\Big\}\Big).$$

The restriction of h to the set $A_n = \{x\colon\ I(x) \leqslant a + 1/n\}$ is a continuous map from A_n endowed with the relative topology into S'. Since A_n is a compact space in the relative topology, $h(A_n)$ is compact in S'. Hence $\bigcap_{n=1}^{\infty} A_n$ is compact. Lemma 2.1 is proved.

Remark 1. It is easy to see that, under the conditions of the lemma, inf in (2.1) is attained.

Remark 2. Under the conditions of lemma for any $A \subset S'$,

$$\inf_A I \circ h^{-1}(x') = \inf_{h^{-1}(A)} I(x).$$

Let h: $S \to S'$ be a measurable map with respect to the Borel σ–fields $\mathcal{B}(S), \mathcal{B}(S')$. For a measure P on $\mathcal{B}(S)$ we denote $P \circ h^{-1}$ the image of P under h:

$$P \circ h^{-1}(A) = P\big(h^{-1}(A)\big), \quad A \in \mathcal{B}(S').$$

The following assertion is an analog of Theorem 5.1 in (Billingsley, 1968). Its version for a continuous h can be found in (Freidlin and Wentzell, 1984).

THEOREM 2.2. *Let h satisfy the conditions of Lemma 2.1. If a sequence (P_n) of probability measures on S obeys the large deviation principle with a rate function I, then the sequence $(P_n \circ h^{-1})$ of probability measures on S' obeys the large deviation principle with the rate function $I \circ h^{-1}$.*

Proof is similar to the proof of Theorem 5.1 in (Billingsley, 1968). Let us verify for example, **DV2**. Denote by $\overline{A}$ the closure of a set $A \subset S$. Since $\overline{h^{-1}F} \subset (h^{-1}F) \cup D_h$ it is easily seen that, for a closed $F \subset S'$,

$$\overline{\lim_n} \frac{1}{n} \log P_n \circ h^{-1}(F) \leqslant - \inf_{\overline{h^{-1}F}} I(x) = \inf_F I \circ h^{-1}(x')$$

(see Remark 2 to Lemma 2.1).
The lower bound is proved analogously using the inclusion $h^{-1}G \subset (h^{-1}G)^\circ \cup D_h$ (A° is the interior of A).

LEMMA 2.3. *Let I be a rate function on S. If (A_n) is a decreasing sequence of closed sets and $A = \bigcap_n A_n$ then*

$$\sup_n \inf_{A_n} I(x) = \inf_A I(x) \quad (\inf_{\varnothing} = \infty).$$

Proof. We can assume A_n to be nonempty. It is obvious that

$$\inf_{A_n} I(x) \leqslant \inf_A I(x).$$

Denote $c = \sup_n \inf_{A_n} I(x)$ and suppose that $c < \inf_A I(x)$. In this case $c < \infty$. Because A_n are closed and I is a rate function, we can choose $x_n \in A_n$ such that $I(x_n) = \inf_{A_n} I(x)$. Since $I(x_n) \leqslant c$, there exists a subsequence $(x_{n'})$ converging to a point $\hat{x} \in \bigcap_n A_n = A$. Hence, A is nonempty. By the lower semicontinuity of I,

$$\underline{\lim_n} I(x_n) \geqslant I(\hat{x}) \geqslant \inf_A I(x),$$

which proves the lemma.
We now state and prove an analog of the Daniel–Kolmogorov theorem on the continuation of a consistent family of probability measures (see also (Deuschel and Strook, 1989)).

Let $(S_i)_{i\in\mathbf{N}}$ be a sequence of metric spaces and let mappings $\pi_i^j\colon S_j \to S_i$, $j > i$, be continuous and consistent, i.e.,

$$\pi_i^{i+1} \circ \pi_{i+1}^{i+2} = \pi_i^{i+2}.$$

Let $\widetilde{S} \subset \prod_i S_i$ be the projective limit of $(S_i)_{i\in\mathbf{N}}$, i.e., $x = (x_i,\ i \in \mathbf{N}) \in \widetilde{S}$ if $x_i \in S_i$ and $\pi_i^{i+1} x_{i+1} = x_i$. The topology on $\widetilde{S}$ is the Tikhonov topology which makes $\widetilde{S}$ a metric space. We define projections $\pi_i\colon \widetilde{S} \to S_i$ by $\pi_i x = x_i$ for $x = (x_i) \in \widetilde{S}$. They are continuous by the definition of the topology on $\widetilde{S}$.
Let I_i be rate functions on S_i. We call the family $(I_i)_{i\in\mathbf{N}}$ consistent, if

$$I_i = I_{i+1} \circ \left(\pi_i^{i+1}\right)^{-1}, \quad i \in \mathbf{N}.$$

THEOREM 2.4. *If the family $(I_i)_{i\in\mathbf{N}}$ is consistent then there exists a unique rate function $\widetilde{I}$ on $\widetilde{S}$ satisfying $I_i = \widetilde{I} \circ \pi_i^{-1}$.*

Proof. For $x \in \widetilde{S}$ define

$$\widetilde{I}(x) = \sup_{i\in\mathbf{N}} I_i(\pi_i x). \tag{2.2}$$

Then $\widetilde{I}$ is a rate function on $\widetilde{S}$. Indeed, for $a \geqslant 0$, the set

$$A = \{x \in \widetilde{S}\colon\ \widetilde{I}(x) \leqslant a\} = \bigcap_i \pi_i^{-1}\{x_i \in S_i\colon\ I_i(x_i) \leqslant a\}$$

is closed as an intersection of closed sets. Next, if $\{x(n)\}$ is a sequence of elements of A then, since all the sets $\{x_i \in S_i\colon\ I_i(x_i) \leqslant a\}$ are compact, by the diagonal procedure we can choose a subsequence (n') such that $\pi_i x(n') \to \hat{x}_i \in S_i$, $\forall i \in \mathbf{N}$. Since $\hat{x} = (\hat{x}_i,\ i \in \mathbf{N})$ obviously belongs to $\widetilde{S}$, by the definition of the Tikhonov topology $x(n') \to \hat{x}$, where $\hat{x} \in A$ because A is closed.
Let us prove now that $\widetilde{I}$ from (2.2) meets the requirements of the theorem. Choose $j \in \mathbf{N}$ and $x_j \in S_j$, then, by (2.2),

$$\widetilde{I} \circ \pi_j^{-1}(x_j) = \inf_{\pi_j^{-1} x_j} \widetilde{I}(x) \geqslant I_j(x_j).$$

Conversely, by the consistency condition and Remark 1 after Lemma 2.1, we can construct a sequence $(x_i)_{i\geqslant j}$, $x_i \in S_i$, such that

$$\pi_i^{i+1} x_{i+1} = x_i, \quad I_{i+1}(x_{i+1}) = I_i(x_i).$$

Then the element $x = (x_l,\ l \in \mathbf{N})$, where, for $l < j$, we put $x_l = \pi_l^j x_j$, belongs to $\widetilde{S}$ and

$$\widetilde{I}(x) = \sup_i I_i(\pi_i x) = I_j(x_j).$$

Since $\pi_j x = x_j$, the function $\widetilde{I}$ satisfies the requirements of the theorem.

To prove the uniqueness, note that if I' is a rate function with $I_i = I' \circ \pi_i^{-1}$, $i \in \mathbf{N}$, then, by Lemma 2.3, for any $x \in \widetilde{S}$, we have

$$I'(x) = \inf_{\bigcap_{i\in\mathbf{N}} \left(\pi_i^{-1}(\pi_i - x)\right)} I'(x) = \sup_{i\in\mathbf{N}} \inf_{\pi_i^{-1}(\pi_i x)} I'(x) = \sup_{i\in\mathbf{N}} I_i(\pi_i x).$$

Remark 2.5. From **DV1**, **DV2** and Lemma 2.3 it follows immediately (see also (Freidlin and Wentzell, 1984)), that if a sequence (P_n) on $\mathcal{B}(S)$ satisfies the l.d.p. with a rate function I then, for all $x \in S$,

$$I(x) = -\overline{\lim_{\delta\to 0}}\ \overline{\lim_{n}} \frac{1}{n} \log P_n\big(y\colon\ d(x,y) \leqslant \delta\big) =$$

$$= -\underline{\lim_{\delta\to 0}}\ \underline{\lim_{n}} \frac{1}{n} \log P_n\big(y\colon\ d(x,y) < \delta\big), \tag{2.3}$$

where $d(\cdot,\cdot)$ is the metric on S. In particular, I is determined uniquely.

3. In this section we prove Theorem (P). We present here a proof along the lines of the proof of the Prokhorov theorem in (Billingsley, 1968). Another proof, based on a counterpart of the Riesz representation theorem, will be published elsewhere.

LEMMA 3.1. *Let $\mathcal{U}$ be a base of the topology on S. If for lower semicontinuous non-negative function I on S we have, for any $U \in \mathcal{U}$,*

$$\underline{\lim_{n}} \frac{1}{n} \log P_n(U) \geqslant -\inf_{U} I(x), \tag{3.1}$$

$$\overline{\lim_{n}} \frac{1}{n} \log P_n(\overline{U}) \leqslant -\inf_{\overline{U}} I(x), \tag{3.2}$$

and, for any $C > 0$, there exists a compact $K \subset S$ such that, for all n,

$$\frac{1}{n} \log P_n(S \backslash K) < -C, \tag{3.3}$$

i.e., the family (P_n) is strongly tight, then I is a rate function and (P_n) satisfies the l.d.p. with the rate function I.

Proof. If $G \subset S$ is open then $G = \bigcup_\alpha U_\alpha$, $U_\alpha \in \mathcal{U}$, and, by (3.1), we have

$$\underline{\lim_{n}} \frac{1}{n} \log P_n(G) \geqslant \underline{\lim_{n}} \frac{1}{n} \sup_\alpha \log P_n(U_\alpha) \geqslant$$

$$\geqslant \sup_\alpha \underline{\lim_{n}} \frac{1}{n} \log P_n(U_\alpha) \geqslant -\inf_{G} I(x). \tag{3.4}$$

This proves **DV1**.
Taking in (3.4) $G = S\backslash K$, where K is a compact satisfying (3.3), we get

$$\inf_{S\backslash K} I(x) > C.$$

This means that, for $a \leqslant C$, the set $A = \{x\colon I(x) \leqslant a\} \subset K$. Since I is lower semicontinuous, A is a closed subset of a compact and hence is a compact itself. Since C is arbitrary, I is a rate function.
We now verify condition **DV2**. First, let F be compact. For $\delta > 0$ denote by F^δ the open δ–neighbourhood of F. Choose $U_i \in \mathcal{U}$, $1 \leqslant i \leqslant m$, satisfying

$$F \subset \bigcup_{i=1}^{m} U_i \subset F^\delta.$$

We have

$$\overline{\lim_n} \frac{1}{n} \log P_n(F) \leqslant \overline{\lim_n} \frac{1}{n} \log\Big(\sum_{i=1}^{m} P_n(U_i)\Big) \leqslant$$

$$\leqslant \overline{\lim_n} \frac{1}{n} \log\Big[m \max_{1\leqslant i\leqslant m} P_n(U_i)\Big] \leqslant \max_{1\leqslant i\leqslant m}\Big(-\inf_{\bar U_i} I(x)\Big) \leqslant -\inf_{\bar F^\delta} I(x).$$

Letting $\delta \to 0$ and using Lemma 2.3, we get **DV2** for F compact. For an arbitrary closed F, we have choosing a compact K as in (3.3),

$$\overline{\lim_n} \frac{1}{n} \log P_n(F) \leqslant \overline{\lim_n} \frac{1}{n} \log\Big[2\big(P_n(K \cap F) \vee P_n(S\backslash K)\big)\Big] \leqslant$$

$$\leqslant \Big[\overline{\lim_n} \frac{1}{n} \log P_n(K \cap F)\Big] \vee (-C) \leqslant \Big[-\inf_F I(x)\Big] \vee (-C).$$

Since C is arbitraiy, **DV2** is proved. Lemma 3.1 is proved.

LEMMA 3.2. *Let (P_n) be a strongly tight sequence of probability measures on $\mathbf{R}^k$. Then (P_n) contains a subsequence obeying the l.d.p.*

Proof. Denote by $\mathbf{Q}^k$ the set of elements of $\mathbf{R}^k$ with rational coordinates. For $x = (x^1, \ldots, x^k)$, $y = (y^1, \ldots, y^k) \in \mathbf{R}^k$, we say that $x < y$ if $x^i < y^i$, $1 \leqslant i \leqslant k$. For $x < y$ denote

$$]x, y[= \{z \in \mathbf{R}^k\colon\ x < z < y\}.$$

Using the diagonal procedure we can choose a subsequence (n') such that, for all $r_1, r_2 \in \mathbf{Q}^k$, $r_1 < r_2$, the limits (possibly infinite) exist:

$$\Phi(r_1, r_2) = -\lim_{n'} \frac{1}{n'} \log P_{n'}\big(]r_1, r_2[\big). \tag{3.5}$$

Define, for $x \in \mathbf{R}^k$,

$$I(x) = \sup_{\substack{]r_1, r_2[\ni x \\ r_1, r_2 \in \mathbf{Q}^k}} \Phi(r_1, r_2). \tag{3.6}$$

The function I is lower semicontinuous. Indeed, let $x_n \to x$ $(n \to \infty)$. Given $\varepsilon > 0$ there exist $r_1, r_2 \in \mathbf{Q}^k$, $]r_1, r_2[\ni x$, such that $I(x) \leqslant \Phi(r_1, r_2) + \varepsilon$. Then $x_n \in]r_1, r_2[$, for large n, and consequently

$$I(x_n) = \sup_{\substack{]t_1,t_2[\ni x_n \\ t_1,t_2\in \mathbf{Q}^k}} \Phi(t_1,t_2) \geqslant \Phi(r_1,r_2) \geqslant I(x) - \varepsilon.$$

Thus the lower semicontinuity of I is proved.
By Lemma 3.1 remains to prove **DV1** and **DV2** for $G =]r_1, r_2[$ and $F = [r_1, r_2]$, $r_1, r_2 \in \mathbf{Q}^k$, $r_1 < r_2$. By the definition of I (see (3.6)),

$$\inf_{]r_1,r_2[} I(x) \geqslant \Phi(r_1, r_2),$$

and **DV1**, for $G =]r_1, r_2[$, follows by (3.5).
To prove **DV2** for $F = [r_1, r_2]$, $r_1 < r_2$, fix arbitrary $\varepsilon > 0$ and rational $\delta > 0$. Then, by (3.5),

$$\overline{\lim_{n'}} \frac{1}{n'} \log P_{n'}\big([r_1, r_2]\big) \leqslant -\Phi(r_1 - \delta\mathbf{e}, r_2 + \delta\mathbf{e}), \tag{3.7}$$

where $\mathbf{e} = (1, \ldots, 1) \in \mathbf{R}^k$. By (3.6), for every $x \in [r_1 - \delta\mathbf{e}, r_2 + \delta\mathbf{e}]$, we can choose $\delta_1(x,\varepsilon), \delta_2(x,\varepsilon) \in \mathbf{R}^k$ such that $0 < \delta_i(x,\varepsilon) < \delta\mathbf{e}$, $i = 1, 2$, $x - \delta_1(x,\varepsilon) \in \mathbf{Q}^k$, $x + \delta_2(x,\varepsilon) \in \mathbf{Q}^k$ and

$$I(x) \leqslant \Phi\Big(x - \delta_1(x,\varepsilon), x + \delta_2(x,\varepsilon)\Big) + \varepsilon. \tag{3.8}$$

Since

$$\big[r_1 - \delta\mathbf{e}, r_2 + \delta\mathbf{e}\big] \subset \bigcup \Big]x - \delta_1(x,\varepsilon), x + \delta_2(x,\varepsilon)\Big[,$$

where the union is taken over all $x \in [r_1 - \delta\mathbf{e}, r_2 + \delta\mathbf{e}]$, there exists $x_1, \ldots, x_m \in$ $\in [r_1 - \delta\mathbf{e}, r_2 + \delta\mathbf{e}]$ such that

$$\big[r_1 - \delta\mathbf{e}, r_2 + \delta\mathbf{e}\big] \subset \bigcup_{i=1}^{m} \Big]x_i - \delta_1(x_i,\varepsilon), x_i + \delta_2(x_i,\varepsilon)\Big[$$

and, hence,

$$P_{n'}\Big(\big]r_1 - \delta\mathbf{e}, r_2 + \delta\mathbf{e}\big[\Big) \leqslant m \max_{1\leqslant i\leqslant m} P_n\Big(\big]x_i - \delta_1(x_i,\varepsilon), x_i + \delta_2(x_i,\varepsilon)\big[\Big).$$

This implies, by (3.5), that

$$\Phi(r_1 - \delta\mathbf{e}, r_2 + \delta\mathbf{e}) \geqslant -\underline{\lim_{n'}} \frac{1}{n'} \log\Big(m \max_{1\leqslant i\leqslant m} P_{n'}(]x_i - \delta_1(x_i,\varepsilon), x_i + \delta_2(x_i,\varepsilon)[)\Big) =$$

$$= -\max_{1\leqslant i\leqslant m} \lim_{n'} \log P_{n'}(]x_i - \delta_1(x_i,\varepsilon), x_i + \delta_2(x_i,\varepsilon)[) =$$

$$= \min_{1\leqslant i\leqslant m} \Phi\big(x_i - \delta_1(x_i,\varepsilon), x_i + \delta_2(x_i,\varepsilon)\big).$$

Applying (3.8) to $x = x_i$, $1 \leqslant i \leqslant m$, we obtain

$$\Phi(r_1-\delta\mathbf{e},r_2+\delta\mathbf{e})\geqslant \inf_{[r_1-\delta\mathbf{e},r_2+\delta\mathbf{e}]} I(x)-\varepsilon.$$

Now using (3.7) and Lemma 2.3, we get the required relation **DV2**.
Lemma 3.2 is proved.

LEMMA 3.3. *The assertion of Lemma 3.2 holds for* $\mathbf{R}^\infty$.

Proof. Define projections

$$\pi_k\colon \mathbf{R}^\infty\to\mathbf{R}^k,\qquad \pi_k^{k+1}\colon \mathbf{R}^{k+1}\to\mathbf{R}^k,$$

if $x=(x_1,x_2,\dots)\in\mathbf{R}^\infty$ then $\pi_k x=(x_1,x_2,\dots,x_k)$, if $x^{k+1}=(x_1,\dots,x_{k+1})\in\mathbf{R}^{k+1}$ then $\pi_k^{k+1}x^{k+1}=(x_1,\dots,x_k)\in\mathbf{R}^k$. It is clear that, for all $k=1,2\dots$, the sequence of measures $(P_n\circ\pi_k^{-1})$ on $\mathcal{B}(\mathbf{R}^k)$ is strongly tight. Then using Lemma 3.2, we can choose a subsequence (n') such that, for $k=1,2,\dots$, the sequence $(P_{n'}\circ\pi_k^{-1})$ satisfies the l.d.p. with a rate function I_k. Let us show that the family (I_k) of rate functions is consistent: $I_{k+1}\circ(\pi_k^{k+1})^{-1}=I_k$. Denote by $d_k(\cdot,\cdot)$ the metric in $\mathbf{R}^k$ and let

$$A(u,x^k)=\{y^k\in\mathbf{R}^k\colon d_k(x^k,y^k)<u\},\quad u>0,\quad x^k\in\mathbf{R}^k.$$

By Remark 2.5, **DV1** and Lemma 2.3, we have

$$I_k(x^k)=-\underline{\lim}_{\delta\to0}\ \underline{\lim}_{n'}\ \frac{1}{n'}\log P_{n'}\circ\pi_k^{-1}(A(\delta,x^k))=$$

$$=-\underline{\lim}_{\delta\to0}\ \underline{\lim}_{n'}\ \frac{1}{n'}\log P_{n'}\circ\pi_{k+1}^{-1}\left((\pi_k^{k+1})^{-1}A(\delta,x^k)\right)\leqslant$$

$$\leqslant-\underline{\lim}_{\delta\to0}\Big(-\inf_{(\pi_k^{k+1})^{-1}A(\delta,x^k)} I_{k+1}(y^{k+1})\Big)\leqslant$$

$$\leqslant\overline{\lim}_{\delta\to0}\ \inf_{\overline{(\pi_k^{k+1})^{-1}A(\delta/2,x^k)}} I_{k+1}(y^{k+1})=\inf_{(\pi_k^{k+1})^{-1}x^k} I_{k+1}(y^{k+1}).$$

The opposite inequality is proved analogously.
By Theorem 2.4, there exists a rate function I on $\mathbf{R}^\infty$ such that $I\circ\pi_k^{-1}=I_k$. This means that **DV1** and **DV2** hold for sets $\pi_k^{-1}G_k$ and $\pi_k^{-1}\overline{G}_k$, respectively, where G_k is an open set in $\mathbf{R}^k$. But the sets $\pi_k^{-1}G_k$ form a base of the topology on $\mathbf{R}^\infty$ and $\pi_k^{-1}\overline{G}_k=\overline{\pi_k^{-1}G_k}$. Applying Lemma 3.1, we finish the proof.
We now prove Theorem (P) in full generality.
Let us start with part 1. Assume first that S is a σ-compact metric space. Then it is separable and, by the Uryson lemma, is homeomorphic to a subset of $\mathbf{R}^\infty$. The strong tightness and the l.d.p. are easily seen to be preserved under homeomorphisms. So we may assume that S is a subset of $\mathbf{R}^\infty$ with the relative topology. Since the Borel σ-field $\mathcal{B}(S)$ of S consists of sets $S\cap A$, where A is a Borel set in $\mathbf{R}^\infty$, we can define a family of probability measures (Q_n) on $\mathcal{B}(\mathbf{R}^\infty)$ by

$$Q_n(A)=P_n(A\cap S),\quad A\in\mathcal{B}(\mathbf{R}^\infty).\tag{3.9}$$

If K is a compact in S then it is a compact in $\mathbf{R}^\infty$, since the natural injection $S \to \mathbf{R}^\infty$ is continuous. So, the sequence (Q_n) is strongly tight. By Lemma 3.3, a subsequence $(Q_{n'})$ satisfies the l.d.p. with a rate function J. It is easily seen that

$$J(x) = \infty, \quad x \in \mathbf{R}^\infty \backslash S. \tag{3.10}$$

Indeed, if K is a compact in S with $P_n(K) > 1 - \varepsilon^n$, then by the l.d.p. for $(Q_{n'})$ we have

$$-\inf_{\mathbf{R}^\infty \backslash K} J(x) \leqslant \underline{\lim}_{n'} \frac{1}{n'} \log Q_{n'}(\mathbf{R}^\infty \backslash K) = \underline{\lim}_{n'} \frac{1}{n'} \log P_{n'}(S \backslash K) \leqslant \log \varepsilon.$$

So,

$$\inf_{\mathbf{R}^\infty \backslash S} J(x) \geqslant \inf_{\mathbf{R}^\infty \backslash K} J(x) \geqslant -\log \varepsilon.$$

Since ε is arbitrary, (3.10) is proved.

Define a function I on S by $I(x) = J(x)$, $x \in S$. Let us show that I is a rate function on S. By (3.10), the set $A = \{x\colon I(x) \leqslant a\}$, $a \geqslant 0$, is compact in $\mathbf{R}^\infty$. As $A \subset S$, A is compact in S. Further, the sequence $(P_{n'})$ is easily seen to obey the l.d.p. with the rate function I. Let us verify, for instance, **DV1**. For $G \subset S$, where G is open, we have $G = \widetilde{G} \cap S$, where $\widetilde{G}$ is open in $\mathbf{R}^\infty$. By (3.9), (3.10) and the l.d.p. for $(Q_{n'})$, we have

$$\underline{\lim}_{n'} \frac{1}{n'} \log P_{n'}(G) = \underline{\lim}_{n'} \frac{1}{n'} \log Q_{n'}(\widetilde{G}) \geqslant -\inf_{\widetilde{G}} J(x) = -\inf_{G} I(x).$$

DV2 is proved analogously. The case of σ-compact S is completed.

If S is not σ-compact then we can choose a σ-compact $M \subset S$ so that $P_n(M) = 1$ for all n (see Theorem 6.1 in (Billingsley, 1968)). The set M is a σ-compact metric space in the relative topology. If $\widetilde{B}$ is a Borel set in M then $\widetilde{B} = B \cap M$, $B \in \mathcal{B}(S)$, and we can define $\widetilde{P}_n(\widetilde{B}) = P_n(B)$. It is easily seen that the sequence $(\widetilde{P}_n)$ is strongly tight and, therefore, a subsequence $(\widetilde{P}_{n'})$ obeys the l.d.p. But $P_n = \widetilde{P}_n \circ \pi^{-1}$, where $\pi\colon M \to S$ is the natural injection. Applying Theorem 2.2, we derive that $(P_{n'})$ satisfies the l.d.p. and part 1 of Theorem (P) is proved.

Now let S be complete and separable and let a sequence (P_n) of probabilities on $\mathcal{B}(S)$ obey the partial l.d.p. As in the proof of Theorem 6.2 in (Billingsley, 1968), it can be shown that if, for all $\varepsilon, \delta > 0$, there exist open δ-spheres $A_1, \ldots, A_m$ such that $P_n(\bigcup_{i \leqslant m} A_i) > 1 - \varepsilon^n$ for all n, then the family (P_n) is strongly tight. Suppose that the family (P_n) is not strongly tight. Then, for some $\varepsilon > 0$, $\delta > 0$ we have, for any sequence $A_1, A_2, \ldots$ of open δ-spheres and any m, that there exists n' such that

$$P_{n'}\Big(\bigcup_{i \leqslant m} A_i\Big) \leqslant 1 - \varepsilon^{n'}. \tag{3.11}$$

Since S is separable, we can choose $A_1, A_2, \ldots$ such that $\bigcup_i A_i = S$. Denote

$$B_m = \bigcup_{i \leqslant m} A_i.$$

For every m choose n' to be the smallest index with the property (3.11). Then $n' \to \infty$ as $m \to \infty$. As the sequence (P_n) obeys the partial l.d.p., for a subsequence (n'') of (n') and a rate function I'', we have, for any m, by **DV2**,

$$- \inf_{S\setminus B_m} I''(x) \geqslant \overline{\lim_{n''}} \frac{1}{n''} \log\Big(1 - P_{n''}(B_m)\Big) \geqslant$$

$$\geqslant \overline{\lim_{n''}} \frac{1}{n''} \log\Big(1 - P_{n''}(B_{m''})\Big) \geqslant \log\varepsilon,$$

where (m'') is a subsequence of (m) corresponding to the subsequence (n''). Thus $\inf_{S\setminus B_m} I''(x) \leqslant -\log\varepsilon$ for all m. But, since $\bigcup_m B_m = S$ and I'' is a rate function, we have, by Lemma 2.3,

$$\lim_{m\to\infty} \inf_{S\setminus B_m} I''(x) = \infty.$$

This contradiction concludes the proof.

COROLLARY 3.4. *If a sequence (P_n) is strongly tight and a function I on S satisfies* (2.3), *for all $x \in S$, then I is a rate function on S and the sequence (P_n) obeys the l.d.p. with the rate function I.*

This corollary is a straightforward consequence of Theorem (P) and Remark 2.5.

4. In this section we consider the l.d.p. in spaces $\mathbf{D}$ and $\mathbf{C}$. Here $\mathbf{D} = \mathbf{D}[0,\infty[$ is the Skorokhod space of right continuous, having left–hand limits functions $x = \big(x(t),\ t \geqslant 0\big)$ defined on $\mathbf{R}^+$ and taking values in $\mathbf{R}$. It is endowed with the Skorokhod–Lindvall topology (Liptser and Shiryaev, 1989) which converts $\mathbf{D}$ into a complete separable metric space. The Borel σ–field $\mathcal{D}$ is generated by cylinders $\{x \in \mathbf{D}\colon\ x(t_i) \in B_i\}$, $B_i \in \mathcal{B}(\mathbf{R})$; $\mathbf{C} = \mathbf{C}[0,\infty[$ is a subspace of $\mathbf{D}$ consisting of continuous functions. With the relative topology which is the topology of uniform convergence on compact intervals it is a complete separable metric space. Its Borel σ–field $\mathcal{C}$ is generated by cylinders as well. All the measures considered are defined either on $\mathcal{D}$ or on $\mathcal{C}$.

First, we give criteria for strong tightness similar to those which are known for tightness (Billingsley, 1968).

Consider the case of the space $\mathbf{C}$. For $x \in \mathbf{C}$ and $T > 0$ define the modulus of continuity

$$w_T(x,\delta) = \sup_{\substack{|t-s|<\delta \\ 0\leqslant t,s\leqslant T}} \big|x(t) - x(s)\big|.$$

THEOREM 4.1. *A sequence (P_n) of probability measures on $\mathcal{C}$ is strongly tight if and only if the following conditions hold:*

(i)

$$\lim_{A\to\infty} \overline{\lim_{n}} \frac{1}{n} \log P_n\big(x\colon\ |x(0)| \geqslant A\big) = -\infty,$$

(ii) *for any $\eta > 0$ and any $T > 0$,*

$$\lim_{\delta\to 0}\overline{\lim_{n}}\frac{1}{n}\log P_n\big(x:\ w_T(x,\delta)\geqslant\eta\big)=-\infty.$$

Proof is analogous to the proof of Theorem 8.2 in (Billingsley, 1968). Consider, e.g., the sufficiency. For any $C>0,\ T>0,\ k=1,2,\ldots$, choose $A,\ \delta_k$ such that, for all n,

$$P_n\Big\{x:\ |x(0)|<A\Big\}\geqslant 1-\frac{1}{2}\,e^{-nC},$$

$$P_n\Big\{x:\ w_T\big(x,\delta_k^T\big)<\frac{1}{k}\Big\}\geqslant 1-2^{-(T+k+1)}e^{-nC}.$$

Then the set

$$K=\big\{x:\ |x(0)|<A\big\}\bigcap\Big(\bigcap_{T=1}^{\infty}\bigcap_{k=1}^{\infty}\Big\{x:\ w_T\big(x,\delta_k^T\big)<\frac{1}{k}\Big\}\Big)$$

is totally bounded and, therefore, relatively compact and

$$P_n(K)\geqslant 1-e^{-nC},$$

which finishes the proof.

Now consider the space $\mathbf{D}$. For $x=\big(x(t),\ t\geqslant 0\big)\in\mathbf{D}$ and $T>0,\ \delta>0$, define

$$w_x[s,t[\ =\sup_{s\leqslant u,v<t}|x(u)-x(v)|,\quad s<t,$$

$$w'_T(x,\delta)=\inf_{t_j}\max_{0<j\leqslant k}w_x[t_{j-1},t_j[\,,$$

where $(t_j)_{0\leqslant j\leqslant k}$ are points of $[0,T]$ such that $0=t_0<t_1<\ldots<t_k=T$, $t_j-t_{j-1}>\delta,\ j=1,\ldots,k$;

$$w''_T(x,\delta)=\sup_{\substack{t_1,t_2\in[0,T]\\ t_2-t_1\leqslant\delta}}\ \sup_{t\in[t_1,t_2]}\Big(|x(t_1)-x(t)|\wedge|x(t_2)-x(t)|\Big).$$

THEOREM 4.2. *A sequence (P_n) of probability measures on $\mathcal{D}$ is strongly tight if and only if:*

(i) *for any $T>0$,*

$$\lim_{A\to\infty}\overline{\lim_{n}}\frac{1}{n}\log P_n\big(x:\ \sup_{t\leqslant T}|x(t)|\geqslant A\big)=-\infty,$$

and one of the following conditions holds

(ii) *for any $\eta>0,\ T>0$,*

$$\lim_{\delta\to 0}\overline{\lim_{n}}\frac{1}{n}\log P_n\big(x:\ w'_T(x,\delta)\geqslant\eta\big)=-\infty,$$

or

(ii)′ *for any* $\eta > 0,\ T > 0,$

$$\lim_{\delta\to 0}\overline{\lim_{n}}\ \frac{1}{n}\log P_n\big(x:\ w''_T(x,\delta)\geqslant\eta\big) = -\infty,$$

$$\lim_{\delta\to 0}\overline{\lim_{n}}\ \frac{1}{n}\log P_n\big(x:\ w_x[0,\delta[\ \geqslant\eta\big) = -\infty.$$

Proof is just a modification of the corresponding proof for tightness criteria in **D**. It is known that the tightness criteria of the kind given by Theorems 4.1 and 4.2, are rather difficult to check. More convenient is the Aldous sufficient condition (see, for example, Theorem 6.3.1 in (Liptser and Shiryaev, 1989)). Now we give an analog of the Aldous condition for strong tightness. First we prove a simple lemma.

LEMMA 4.3. *Let* $(\xi_i,\ i\geqslant 1)$ *be a sequence of positive random variables on a probability space* $(\Omega,\mathcal{F},\mathbf{P})$. *Define the point process*

$$A_t = \max\Big(k:\ \sum_{i=1}^{k}\xi_i\leqslant t\Big).$$

Then, for any $\delta > 0,\ t > 0,\ N = 1,2\ldots,$ *such that* $t/N\delta < 1$, *the following inequality holds*

$$\mathbf{P}(A_t\geqslant N)\leqslant\max_{n\leqslant N}\mathbf{P}\Big(\xi_n\leqslant\delta,\ A_t\geqslant n\Big)\Big/\Big(1-\frac{t}{N\delta}\Big). \tag{4.1}$$

Proof. Let $m\leqslant N$. Applying the Chebyshev inequality, we have

$$\mathbf{P}(A_t\geqslant N)\leqslant\mathbf{P}(A_t\geqslant N,\ \xi_m\leqslant\delta)+\mathbf{P}(A_t\geqslant N,\ \xi_m\geqslant\delta)\leqslant$$

$$\leqslant\max_{n\leqslant N}\mathbf{P}(A_t\geqslant N,\ \xi_n\leqslant\delta)+\frac{1}{\delta}\mathbf{E}\xi_m\mathbf{1}(A_t\geqslant N), \tag{4.2}$$

where $\mathbf{1}(B)$ equals 1 on B and 0 outside B.
Now choose m, satisfying

$$\mathbf{E}\xi_m\mathbf{1}(A_t\geqslant N)=\min_{n\leqslant N}\mathbf{E}\xi_n\mathbf{1}(A_t\geqslant N).$$

Then

$$\mathbf{E}\xi_m\mathbf{1}(A_t\geqslant N)\leqslant\frac{1}{N}\sum_{n=1}^{N}\mathbf{E}\xi_n\mathbf{1}(A_t\geqslant N)=$$

$$=\frac{1}{N}\mathbf{E}\sum_{n=1}^{N}\xi_n\mathbf{1}(A_t\geqslant N)\leqslant\frac{1}{N}\mathbf{E}t\mathbf{1}(A_t\geqslant N)=\frac{t}{N}\mathbf{P}(A_t\geqslant N). \tag{4.3}$$

Substituting (4.3) into (4.2), we obtain (4.1).

Remark. If we continued (4.3) to get $\mathbf{E}\xi_m \mathbf{1}(A_t \geqslant N) \leqslant t/N$, then instead of (4.1) we would have

$$\mathbf{P}(A_t \geqslant N) \leqslant \max_{n \leqslant N} \mathbf{P}(A_t \geqslant n,\ \xi_n \leqslant \delta) + \frac{t}{N\delta}\,. \tag{4.4}$$

For $n = 1, 2, \ldots$ let $\mathfrak{F}^n = (\mathcal{F}^n_u,\ u \geqslant 0)$ be a filtration on the measurable space $(\mathbf{D}, \mathcal{D})$ such that $x(t)$ is $\mathcal{F}^n_u$-measurable, for all $t \geqslant 0$, and $x = \big(x(t),\ t \geqslant 0\big) \in \mathbf{D}$. Denote by $\mathcal{T}_T(\mathfrak{F}^n)$ the set of $\mathfrak{F}^n$-stopping times not exceeding T.

THEOREM 4.4. *Let a sequence (P_n) of probability measures on $\mathcal{D}$ satisfy the following conditions:*

(i) *for any $T > 0$,*

$$\lim_{A \to \infty} \overline{\lim_{n}} \frac{1}{n} \log P_n(x\colon\ \sup_{t \leqslant T} |x(t)| \geqslant A) = -\infty,$$

(ii) *for any $T > 0$ and $\eta > 0$,*

$$\lim_{\delta \to 0} \overline{\lim_{n}} \frac{1}{n} \sup_{\tau \in \mathcal{T}_T(\mathfrak{F}^n)} \log P_n\big(x\colon\ \sup_{0 < t \leqslant \delta} |x(\tau + t) - x(\tau)| \geqslant \eta\big) = -\infty.$$

Then the sequence (P_n) is strongly tight.

Proof. According to Theorem 4.2, it is sufficient to verify condition (ii) of that theorem. For $x = \big(x(t),\ t \geqslant 0\big) \in \mathbf{D}$, define

$$\sigma_0 = 0, \quad \sigma_k = \inf\big(t > \sigma_{k-1}\colon\ |x(t) - x(\sigma_{k-1})| \geqslant \eta/2\big),$$

$$\xi_k = \sigma_k - \sigma_{k-1}, \quad A_T = \max(k\colon\ \sigma_k \leqslant T).$$

Obviously, $\sigma_k \wedge T \in \mathcal{T}_T(\mathfrak{F}^n)$. By the definition of w', we have, for any $N = 1, 2, \ldots$,

$$P_n\big(w'_T(x, \delta) \geqslant \eta\big) \leqslant P_n\Big(\min_{1 \leqslant k \leqslant A_T} \xi_k \leqslant \delta\Big) \leqslant$$

$$\leqslant P_n(A_T \geqslant N) + P_n\Big(\min_{1 \leqslant k \leqslant A_T} \xi_k \leqslant \delta,\ A_T < N\Big) \leqslant$$

$$\leqslant P_n(A_T \geqslant N) + \sum_{i=1}^{N} \sum_{k=1}^{i} P_n(\xi_k \leqslant \delta,\ A_T = i) \leqslant$$

$$\leqslant P_n(A_T \geqslant N) + N^2 \max_{k \leqslant N} P_n(\xi_k \leqslant \delta,\ A_T \geqslant k), \tag{4.5}$$

since $w_x[\sigma_{k-1}, \sigma_k[\leqslant \eta/2$.
Note that

$$P_n(\xi_k \leqslant \delta,\ A_T \geqslant k) = P_n\Big(\sup_{t \leqslant \delta} \big|x(\sigma_{k-1} + t) - x(\sigma_{k-1})\big| \geqslant \eta/2,\ \sigma_k \leqslant T\Big) \leqslant$$

$$\leqslant P_n\Big(\sup_{t\leqslant\delta}\big|x(\sigma_{k-1}\wedge T)-x(\sigma_{k-1}\wedge T+t)\big|\geqslant\eta/2\Big)\leqslant$$

$$\leqslant\sup_{\tau\in\mathcal{T}_T(\mathfrak{F}^n)} P_n\Big(\sup_{t\leqslant\delta}\big|x(\tau+t)-x(\tau)\big|\geqslant\eta/2\Big). \tag{4.6}$$

Since, by Lemma 4.3, for $\varepsilon_N = 2T/N$,

$$P_n(A_T\geqslant N)\leqslant 2\max_{k\leqslant N}P_n(\xi_k\leqslant\varepsilon_N,\ A_T\geqslant k),$$

applying to the latter expression the same reasoning as in (4.6), from (4.5), we obtain

$$\frac{1}{n}\log P_n(w'_T(x,\delta)\geqslant\eta)\leqslant$$

$$\Big[\frac{2}{n}\log N+\sup_{\tau\in\mathcal{T}_T(\mathfrak{F}^\backslash)}\frac{1}{n}\log P_n\Big(\sup_{t\leqslant\delta}\big|x(\tau+t)-x(\tau)\big|\geqslant\eta/2\Big)\Big]\bigvee$$

$$\bigvee\Big[\sup_{\tau\in\mathcal{T}_T(\mathfrak{F}^\backslash)}\frac{1}{n}\log P_n\Big(\sup_{t\leqslant\varepsilon_N}\big|x(\tau+t)-x(\tau)\big|\geqslant\eta/2\Big)+\frac{1}{n}\log 2\Big]+\frac{1}{n}\log 2.$$

Letting $n\to\infty$ and then $N\to\infty$, we get, using the conditions of Theorem 4.4,

$$\overline{\lim_n}\,\frac{1}{n}\log P_n(w'_T(x,\delta)\geqslant\eta)\leqslant$$

$$\leqslant\overline{\lim_n}\sup_{\tau\in\mathcal{T}_T(\mathfrak{F}^\backslash)}\frac{1}{n}\log P_n\Big(\sup_{t\leqslant\delta}\big|x(t+\tau)-x(\tau)\big|\geqslant\eta/2\Big).$$

Letting $\delta\to 0$, we obtain the desired result.

Remark. If instead of (4.1) we used (4.4), we could obtain one more proof of the Aldous condition for tightness.

Now we present the method of finite dimensional distributions for the l.d.p. in $\mathbf{D}$ (and $\mathbf{C}$). For $x=(x(t),\ t\geqslant 0)\in\mathbf{D},\ \mathbf{t_1},\dots,\mathbf{t_k}\geqslant\mathbf{0}$, denote

$$\pi_{t_1\dots t_k}x=\big(x(t_1),\dots,x(t_k)\big),\quad \Delta x(t)=x(t)-x(t-).$$

Say that a rate function I on $\mathbf{D}$ is quasi left-continuous at t if $I(x)=\infty$ for any $x\in\mathbf{D}$ with $|\Delta x(t)|>0$.

THEOREM 4.5. *Let $\mathbf{T_0}$ be a dense cou table set in $\mathbf{R}^+$. Assume that a sequence (P_n) of probability measures on $\mathbf{D}$ is strongly tight and, for all $k=1,2,\dots$ and $t_1,\dots,t_k\in\mathbf{T_0}$, the finite dimensional projections $(P_n\circ\pi^{-1}_{t_1\dots t_k})$ satisfy the l.d.p. (in $\mathbf{R^k}$) with rate functions $I_{t_1\dots t_k}(x_1,\dots,x_k)$. Furthermore, assume that the following condition holds:*

(QL) *if $\widetilde{I}$ is such that a subsequence $(P_{n'})$ satisfies the l.d.p. with $\widetilde{I}$, as the rate function, then $\widetilde{I}$ is quasi left-continuous at t for all $t\in\mathbf{T_0}$.*

Then the sequence (P_n) satisfies the l.d.p. with the rate function

$$I(x) = \sup_{t_1,\ldots,t_k \in \mathbf{T_0}} I_{t_1\ldots t_k}\big(x(t_1),\ldots,x(t_k)\big). \tag{4.7}$$

Proof. By strong tightness for any subsequence (n') there exists its subsequence (n'') such that $(P_{n''})$ obeys the l.d.p. with a rate function J. It is enough to prove that J coincides with I given by (4.7). Indeed, then I is a rate function and all the subsequences $(P_{n''})$ obey the l.d.p. with the same rate function I. To prove the equality $J = I$ note, that because the projection $x \to x(t)$ is continuous at x if x is continuous at t, by Theorem 2.2 and quasi left-continuity of J at $t \in \mathbf{T_0}$, the sequence $(P_{n''} \circ \pi^{-1}_{t_1\ldots t_k})$ satisfies the l.d.p. with the rate function $J \circ \pi^{-1}_{t_1\ldots t_k}$ for all $t_1,\ldots,t_k \in \mathbf{T_0}$. By Remark 2.5, $J \circ \pi^{-1}_{t_1\ldots t_k} = I_{t_1\ldots t_k}$ and so, for $x \in \mathbf{D}$,

$$J(x) \geqslant \sup_{t_1,\ldots,t_k \in \mathbf{T_0}} I_{t_1\ldots t_k}\big(x(t_1),\ldots,x(t_k)\big). \tag{4.8}$$

Define

$$A_{t_1\ldots t_k} = \big\{y \in \mathbf{D}\colon |\Delta \mathbf{y}(\mathbf{t_i})| > 0 \text{ for some } i,\ 1 \leqslant \mathbf{i} \leqslant \mathbf{k}\big\}, \quad \mathbf{A} = \bigcup_{\mathbf{t_1,\ldots,t_k \in T_0}} \mathbf{A_{t_1\ldots t_k}},$$

$$B_{t_1\ldots t_k} = \overline{\pi^{-1}_{t_1\ldots t_k}\big(x(t_1),\ldots,x(t_k)\big)}, \quad B = \bigcap_{t_1,\ldots,t_k \in \mathbf{T_0}} B_{t_1\ldots t_k}.$$

Since $\mathbf{T_0}$ is dense in $\mathbf{R}^+$, we have

$$\{x\} = \bigcap_{t_1,\ldots,t_k \in \mathbf{T_0}} \pi^{-1}_{t_1\ldots t_k}\big(x(t_1),\ldots,x(t_k)\big)$$

and, hence,

$$B \subset A \cup \{x\}. \tag{4.9}$$

Furthermore, by quasi left-continuity of J, for any $y \in A_{t_1,\ldots,t_k}$, $J(y) = \infty$, from (4.9), we have

$$J(x) \leqslant \inf_B J(y). \tag{4.10}$$

But, by Lemma 2.3,

$$\inf_B J(y) = \sup_{t_1,\ldots,t_k \in \mathbf{T_0}} \inf_{B_{t_1\ldots t_k}} J(y)$$

and thus we get, from (4.10) and the equality $J \circ \pi^{-1}_{t_1\ldots t_k} = I_{t_1\ldots t_k}$,

$$J(x) \leqslant \sup_{t_1,\ldots,t_k \in \mathbf{T_0}} I_{t_1\ldots t_k}\big(x(t_1),\ldots,x(t_k)\big)$$

which together with (4.8) proves that $I = J$.
Theorem 4.5 is proved.

Remark 1. Let (τ_n) be an increasing sequence of k–tuples from

$$\mathbf{T_0}:\ \tau_n = (\mathbf{t}_1^n, \dots, \mathbf{t}_{k_n}^n), \quad 0 \leqslant \mathbf{t}_1^n < \dots < \mathbf{t}_{k_n}^n, \quad \mathbf{t}_i^n \in \mathbf{T_0},$$

and, for $p < l$, all the elements of τ_p are contained in τ_l. If $\bigcup_n \tau_n = \mathbf{T_0}$ then I in (4.7) is equal to

$$I(x) = \lim_{n\to\infty} I_{\tau_n}(\pi_{\tau_n} x).$$

Remark 2. If $P_n(\mathbf{C}) = 1$ for all n, then condition (**QL**) holds.
A convenient condition for verifying (**QL**) gives Theorem 4.4.

THEOREM 4.6. *If condition* (ii) *of Theorem* 4.4 *holds then condition* (**QL**) *holds for any* $\mathbf{T_0} = \mathbf{R}^+$.

Proof. Take $x \in \mathbf{D}$ with $|\Delta x(t)| \geqslant \eta > 0$. Denote by $\rho(\cdot,\cdot)$ the Lindvall–Skorokhod metric in $\mathbf{D}$ (see (Liptser and Shiryaev, 1989, Chapter 6)). It is easily seen that, for all sufficiently small $\delta > 0$, we have

$$\{y:\ \rho(x,y) \leqslant \delta\} \subset \Big\{y:\ \sup_{|t-s|\leqslant\delta} |\Delta y(s)| \geqslant \eta/2\Big\} \subset$$
$$\subset \Big\{y:\ \sup_{|t-s|\leqslant\delta} |y(t) - y(s)| \geqslant \eta/4\Big\}. \tag{4.11}$$

By (4.11) and Remark 2.5,

$$-I(x) = \overline{\lim_{\delta\to0}}\ \overline{\lim_{n}}\ \frac{1}{n}\log P_n\big(y:\ \rho(x,y) \leqslant \delta\big) \leqslant$$
$$\leqslant \overline{\lim_{\delta\to0}}\ \overline{\lim_{n}}\ \frac{1}{n}\log P_n\Big(y:\ \sup_{|t-s|\leqslant\delta} \big|y(t) - y(s)\big| \geqslant \eta/4\Big) \leqslant$$
$$\leqslant \overline{\lim_{\delta\to0}}\ \overline{\lim_{n}}\ \frac{1}{n}\log\Big[2\sup_{u\leqslant t} P_n\Big(y:\ \sup_{0<s\leqslant\delta} \big|y(u+s) - y(u)\big| \geqslant \eta/4\Big)\Big].$$

But the latter term equals $-\infty$, by the conditions of the theorem. Thus $I(x) = \infty$. Since,

$$\big\{x:\ |\Delta x(t)| > 0\big\} = \bigcup_K \Big\{x:\ |\Delta x(t)| \geqslant \frac{1}{K}\Big\},$$

Theorem 4.6 is proved.

5. EXAMPLES

According to Theorem 4.5, the proof of the l.d.p. in **D** may be carried out in two steps. First, prove the l.d.p. for finite dimensional distributions and then verify strong tightness. The second task can be solved with the help of Theorem 4.4. To prove the l.d.p. for finite dimensional distributions, we can use the method of exponential moments developed by Gartner (see (Freidlin and Wentzell, 1984, Chapter 5)). The following examples demonstrate the possibilities of our approach.

Example 1 (the invariance principle). Let $(\xi_{in},\ 1 \leqslant i \leqslant n)$, $n \geqslant 1$, be a triangular array consisting, for each n, of independent random variables. Assume that, for all $\lambda \in \mathbf{R}$,

$$\mathbf{E}\exp(\lambda\xi_{in}) < \infty, \quad 1 \leqslant i \leqslant n, \quad n \geqslant 1.$$

Let (a_n) be a sequence of integers such that

$$\frac{a_n}{n} \to \infty \quad (n \to \infty). \tag{5.1}$$

Let the following conditions hold:

$$\mathbf{E}\xi_{in} = 0, \quad \frac{1}{n}\sum_{i=1}^{\lfloor a_n t\rfloor} \mathbf{E}\xi_{in}^2 \to t \quad (n \to \infty), \quad t \geqslant 0. \tag{5.2}$$

Define

$$X_n(t) = \frac{1}{n}\sum_{i=1}^{\lfloor a_n t\rfloor} \xi_{in}, \quad t \geqslant 0, \quad X_n(0) = 0.$$

Then, if a Lindeberg-type condition holds,

$$\frac{1}{n}\sum_{i=1}^{\lfloor a_n t\rfloor} \mathbf{E}e^{\lambda\xi_{in}}\mathbf{1}(|\xi_{in}| > \varepsilon) \to 0 \quad (n \to \infty), \quad \varepsilon > 0, \quad \lambda \in \mathbf{R}, \tag{5.3}$$

the sequence (X_n) with $X_n = \big(X_n(t)\big)_{t\geqslant 0}$ satisfies the l.d.p. in $\mathbf{D}$ with the rate function

$$I(x) = \frac{1}{2}\int_0^\infty \big(\dot{x}(s)\big)^2 ds,$$

if $x = (x(t),\ t \geqslant 0)$ is absolutely continuous (with respect to the Lebesgue measure) and $x(0) = 0$, and $I(x) = \infty$ otherwise.
To prove this statement, first note that under conditions (5.2) and (5.3) it can be proved just in the same way as in the proof of the central limit theorem under the Lindeberg condition that, for $t_1 < t_2$,

$$\lim_n \frac{1}{n}\sum_{i=\lfloor a_n t_1\rfloor+1}^{\lfloor a_n t_2\rfloor} \log \mathbf{E}e^{\lambda\xi_{in}} = \frac{\lambda^2}{2}(t_2 - t_1). \tag{5.4}$$

Then, for $0 = t_0 < t_1 < \ldots < t_k$, we have

$$\lim_n \frac{1}{n}\log \mathbf{E}\Big[\exp\Big(n\sum_{i=1}^{k}\lambda_i(X_n(t_i) - X_n(t_{i-1}))\Big)\Big] = \frac{1}{2}\sum_{i=1}^{k}\lambda_i^2(t_i - t_{i-1}).$$

Taking the Legendre transform, we have, by the Gartner theorem, that the sequence $\big(X_n(t_1), \ldots, X_n(t_k)\big)$ of random vectors satisfies the l.d.p. with the rate function

$$I_{t_1\dots t_k}(x_1,\dots,x_k)=\frac{1}{2}\sum_{i=1}^{k}\frac{(x_i-x_{i-1})^2}{t_i-t_{i-1}},\qquad x_0=0.$$

Now we prove that, for $x=\big(x(t),\ t\geqslant 0\big)\in\mathbf{D}$,

$$I(x)=\sup_{t_1,\dots,t_k} I_{t_1\dots t_k}\big(x(t_1),\dots,x(t_k)\big). \tag{5.5}$$

Let $x=(x(t),\ t\geqslant 0)$ be absolutely continuous, $x(0)=0$.
Using the Schwartz inequality, it is easy to see that

$$I(x)\geqslant \sup_{t_1,\dots,t_k} I_{t_1\dots t_k}\big(x(t_1),\dots,x(t_k)\big).$$

To prove the opposite inequality, define, for $T>0$,

$$t_i^k=iT/k,\quad 1\leqslant i\leqslant k,\quad t_0^k=0,$$

$$x^k(t)=x(t_{i-1}^k)+\frac{x(t_i^k)-x(t_{i-1}^k)}{t_i^k-t_{i-1}^k}(t-t_{i-1}^k),\quad t\in[t_{i-1}^k,\ t_i^k],$$

$$x^k(t)=x(T),\quad t\geqslant T.$$

Then the function $x^k=(x^k(t),\ t\geqslant 0)$ is absolutely continuous and

$$I_{t_1^k\dots t_k^k}\big(x(t_1^k),\dots,x(t_k^k)\big)=\frac{1}{2}\int_0^T\big(\dot{x}^k(s)\big)^2ds.$$

Since, for almost all $t\leqslant T$ with respect to the Lebesgue measure $\lim\limits_{k\to\infty}\dot{x}^k(t)=$
$=\dot{x}(t)$, by the Fatou lemma,

$$\varliminf_{k\to\infty}\int_0^T\big(\dot{x}^k(s)\big)^2ds\geqslant\int_0^T\big(\dot{x}(s)\big)^2ds,$$

and thus, for any $T>0,\ t_1,\dots,t_k\leqslant T$,

$$\sup_{t_1\dots t_k} I_{t_1\dots t_k}\big(x(t_1),\dots,x(t_k)\big)\geqslant\frac{1}{2}\int_0^T\big(\dot{x}(s)\big)^2ds$$

which proves (5.5).
For not absolutely continuous x or if $x(0)\neq 0$, the same reasoning shows that sup in (5.5) is ∞.
Now we check conditions of Theorem 4.4. Introduce the filtration $\mathfrak{F}^n=(\mathcal{F}_\sqcup^\backslash)_{t\geqslant 0}$ with $\mathcal{F}_\sqcup^\backslash=\sigma\{\mathcal{X}_\backslash(\int),\ \int\leqslant\sqcup\}$. We shall show that, for any $T>0,\ \eta>0$,

$$\lim_{\delta\to 0}\varlimsup_{n}\frac{1}{n}\log\sup_{\tau\in\mathcal{T}_T(\mathfrak{F}^\backslash)}\mathbf{P}\Big(\sup_{0<\mathbf{t}\leqslant\delta}\big(\mathbf{X}_\mathbf{n}(\tau+\mathbf{t})-\mathbf{X}_\mathbf{n}(\tau)\big)\geqslant\eta\Big)=-\infty. \tag{5.6}$$

An analogous equality, for $X_n(\tau)-X_n(\tau+t)$ and condition (i), are proved similarly. To prove (5.6), note that, by the Doob inequality, for $\tau \in \mathcal{T}_T(\mathfrak{F}^{\setminus})$ and $\lambda > 0$,

$$\mathbf{P}\Big(\sup_{0<t\leqslant\delta}(\mathbf{X_n}(\tau+\mathbf{t})-\mathbf{X_n}(\tau))\geqslant\eta\Big)\leqslant\frac{\mathbf{E}\exp\Big(\mathbf{n}\lambda\big(\mathbf{X_n}(\mathbf{t}+\tau)-\mathbf{X_n}(\tau)\big)\Big)}{\exp(\mathbf{n}\lambda\eta)}$$

and hence, by (5.4),

$$\overline{\lim_n}\,\frac{1}{n}\log\sup_{\tau\in\mathcal{T}_T(\mathfrak{F}^{\setminus})}\mathbf{P}\Big(\sup_{0<t\leqslant\delta}(\mathbf{X_n}(\tau+\mathbf{t})-\mathbf{X_n}(\tau))\geqslant\eta\Big)\leqslant\lambda^2\delta/2-\lambda\eta.$$

Since λ is arbitrary, (5.6) is proved.
This invariance principle seems to be new. Earlier it was known (Borovkov, 1967) for partial sums of random variables $\xi_{in}=(n/a_n)^{1/2}\zeta_i$, where $\{\zeta_i\}$ are i.i.d. with $\mathrm{E}\exp(\lambda\zeta_1)<\infty$, $|\lambda|<\lambda_0$, $\lambda_0>0$.
Another example is an analog of the convergence to the Brownian bridge for empirical distribution functions (Billingsley, 1968).

Example 2. Let $(\xi_i)_{i\geqslant1}$ be a sequence of independent random variables uniformly distributed on the interval $[0,1]$. Define

$$X_n(t)=\frac{1}{\sqrt{na_n}}\sum_{i=1}^{a_n}\big(\mathbf{1}(\xi_i\leqslant t)-t\wedge1\big),\quad X_n=\big(X_n(t)\big)_{t\geqslant0},$$

where (a_n) is as in Example 1.
Again, by the Gartner theorem, we see that

$$I_{t_1\dots t_k}(x_1,\dots,x_k)=\frac{1}{2}\sum_{i=1}^{k+1}\frac{(x_i-x_{i-1})^2}{t_i-t_{i-1}},$$

where $0=t_0<t_1<\dots<t_k<t_{k+1}=1$, $x_0=0$, $x_{k+1}=1$. By the same arguments as above, it can be shown that

$$\sup_{t_1,\dots,t_k}I_{t_1\dots t_k}\big(x(t_1),\dots,x(t_k)\big)=\frac{1}{2}\int_0^1\big(\dot{x}(s)\big)^2ds \tag{5.7}$$

if $x=\big(x(t),\ t\geqslant0\big)$ is absolutely continuous, $x(0)=0$, $x(t)=0$, $t\geqslant1$, and $I(x)=\infty$ otherwise.
To prove that (5.7) gives the rate function for the sequence (X_n), we need strong tightness. Again we check only condition (ii) of Theorem 4.4.
Introduce the filtration $\mathfrak{F}^n=(\mathcal{F}^{\setminus}_{\sqcup})_{t\geqslant0}$, $\mathcal{F}^{\setminus}_{\sqcup}=\sigma\big(\mathcal{X}_{\setminus}(\int),\ \int\leqslant\sqcup\big)$, and define

$$M_n(t)=\frac{1}{\sqrt{na_n}}\sum_{i=1}^{a_n}\Big[\mathbf{1}(\xi_i\leqslant t)-\int_0^{t\wedge\xi_i}\frac{dx}{1-x}\Big].$$

Then $M_n=\big(M_n(t)\big)_{t\geqslant0}$ is an $\mathfrak{F}^n$–square integrable martingale and it is easy to see that

$$X_n(t) = -\int_0^t \frac{X_n(s)}{1-s}\,ds + M_n(t), \quad 0 \leqslant t \leqslant 1.$$

Hence, for $s \leqslant t \leqslant 1-\varepsilon$, $\varepsilon > 0$,

$$\left|X_n(t) - X_n(s)\right| \leqslant \frac{1}{\varepsilon}\, e^{1/\varepsilon}|t-s| \sup_{u \leqslant t}\left|M_n(u)\right| + \left|M_n(t) - M_n(s)\right|.$$

Applying the Doob inequality, we have, for $T \leqslant 1-\varepsilon$, $\tau \in \mathcal{T}_T(\mathfrak{F}^n)$, $\lambda \in \mathbf{R}$,

$$\frac{1}{n}\log \mathbf{P}\Big(\sup_{\mathbf{t}\leqslant\delta}|\mathbf{X_n}(\mathbf{t}+\tau) - \mathbf{X_n}(\tau)| \geqslant \eta\Big) \leqslant$$

$$\leqslant \left[\frac{1}{n}\log\frac{\mathbf{E}\exp\left(\mathbf{n}\lambda|\mathbf{M_n}(\tau+\delta) - \mathbf{M_n}(\tau)|\right)}{\exp\left(n\lambda\eta/2\right)}\right] \bigvee$$

$$\bigvee \left[\frac{1}{n}\log\frac{\mathbf{E}\exp\left(\mathbf{n}\lambda|\mathbf{M_n}(\mathbf{1}-\varepsilon)|\delta\right)}{\exp\left(n\lambda\eta/2\right)} - \log\left(\frac{1}{\varepsilon}e^{1/\varepsilon}\right)\right] + \frac{1}{n}\log 2.$$

Now it is not difficult to see that

$$\lim_{n\to\infty}\sup_{\tau\in\mathcal{T}_T(\mathfrak{F}^n)}\frac{1}{n}\log\mathbf{E}\exp\left(\mathbf{n}\lambda|\mathbf{M_n}(\tau+\delta) - \mathbf{M_n}(\tau)|\right) \leqslant \lambda^2\delta/2,$$

which goes to zero as $\delta \to 0$. Analogously, one can prove that

$$\overline{\lim_{n\to\infty}}\,\frac{1}{n}\log\mathbf{E}\exp\left(\mathbf{n}\lambda|\mathbf{M_n}(\mathbf{1}-\varepsilon)|\delta\right) \to \mathbf{0} \quad (\delta \to \mathbf{0}).$$

Thus we have, for $T \leqslant 1-\varepsilon$,

$$\lim_{\delta\to 0}\overline{\lim_{n}}\,\frac{1}{n}\sup_{\tau\in\mathcal{T}_T(\mathfrak{F}^n)}\log\mathbf{P}\Big(\sup_{\mathbf{t}\leqslant\delta}|\mathbf{X_n}(\tau+\mathbf{t}) - \mathbf{X_n}(\tau)| \geqslant \eta\Big) = -\infty. \tag{5.8}$$

Denote by $\mathcal{T}^T(\mathfrak{F}^n)$ the set of $\mathfrak{F}^n$-stopping times τ with $T \leqslant \tau \leqslant 1$. It is easy to see that if we prove that, for all $T < 1$,

$$\lim_{\delta\to 0}\overline{\lim_{n}}\,\frac{1}{n}\sup_{\tau\in\mathcal{T}^T(\mathfrak{F}^n)}\log\mathbf{P}\Big(\sup_{\mathbf{t}\leqslant\delta}|\mathbf{X_n}(\mathbf{t}+\tau) - \mathbf{X_n}(\tau)| \geqslant \eta\Big) = -\infty, \tag{5.9}$$

then condition (ii) of Theorem 4.4 will be proved. But since $X_n(t)$ and $\left(-X_n(1-t)\right)$ are identically distributed, for the process $\widetilde{X}_n = \left(\widetilde{X}_n(t)\right)$ with $\widetilde{X}_n(t) = -X_n(1-t)$, from (5.8) follows (5.9).

Remark. Using Theorem 2.2, it is easy to show that if (ξ_i) are arbitrarily distributed with a distribution function $F(u)$ then the sequence (X_n), $X_n = \left(X_n(t)\right)_{t\geqslant 0}$, where

$$X_n(t) = \frac{1}{\sqrt{na_n}}\sum_{i=1}^{a_n}\Big(\mathbf{1}(\xi_i \leqslant t) - F(t)\Big),$$

obeys the l.d.p. with the rate function

$$I(x) = \frac{1}{2}\int\limits_0^\infty \left(\frac{dx}{dF}(u)\right)^2 dF(u),$$

if $x = \big(x(t),\ t \geqslant 0\big)$ is a function of locally bounded variation, absolutely continuous with respect to F and $x(t) \to 0$ $(t \to \infty)$, and $I(x) = \infty$ otherwise.

Acknowledgments
The author is indebted to the referee for pointing at an error in Example 1.

REFERENCES

Billingsley, P. (1968). *Convergence of Probability Measures.* Wiley, New York.

Borovkov, A.A. (1967). Boundary-value problems for random walks and large deviations in function spaces. *Theory Probab. Appl.* **12**, 575–595.

Deuschel, J.-D. and Strook, D.W. (1989). *Large Deviations.* Academic Press, Boston etc.

Freidlin, M.I. and Wentzell, A.D. (1984). *Random Perturbations of Dynamical Systems.* Springer, New York etc.

Liptser, R.Sh. and Shiryaev, A.N. (1989). *Theory of Martingales.* Reidel, Dordrecht.

Strook, D.W. (1984). *An Introduction to the Theory of Large Deviations.* Springer, Berlin etc.

Varadhan, S.R.S. (1984). *Large Deviations and Applications.* SIAM, Philadelphia.

Part Four

PROBABILITIES IN GENERAL SPACES

New Trends in Probab. and Statist., pp. 221 - 232
V. Sazonov and T. Shervashidze (Eds.)
1991 VSP/Mokslas

SOME PROPERTIES OF GAUSSIAN MEASURES AND POTENTIALS IN MARTINGALE SPACES WITH MIXED NORM

A.D. BENDIKOV
Rostov Civil-Engineering Institute, Sotsialisticheskaya str. 162,
344022 Rostov-on-Don, USSR

ABSTRACT

Some new estimates of semigroups of Gaussian measures and associated Bessel potentials on the compact group T^∞ are obtained complementing the author's earlier works.

INTRODUCTION

We consider Gaussian product-measures on the compact Abelian group T^∞, where $T = \mathbf{R}^1/2\pi\mathbf{Z}$ and estimate their norms in so-called martingale spaces with mixed norm (Bendikov and Pavlov, 1987). With the help of these estimates, we prove an inequality of the Sobolev-Nikolsky type for the Bessel potentials. This inequality is applied to the investigation of weak solutions of the elliptic equation on the group T^∞ (Bendikov, 1987).

1. MARTINGALE SPACES WITH MIXED NORM

Let $\mathcal{B}$ be the σ-algebra of Borel subsets of T^∞ and $\mathcal{B}_n \subset \mathcal{B}$ be the σ-algebra of sets of the form $A \otimes T \otimes T \otimes \ldots$, where A is a Borel subset of T^n, $n = 1, 2, \ldots$ It is clear that $\mathcal{B}_n \uparrow$ and $\mathcal{B} = \bigvee_{n=1}^{\infty} \mathcal{B}_n$. Denote the Haar measure on T^∞ by

$$dx = \bigotimes_{n=1}^{\infty} dx_n,$$

where dx_n is the normed Lebesgue measure on $T_n = T$, $n = 1, 2, \ldots$

Definition 1.1. Let $\vec{p} = (p_1, p_2, \ldots)$ be an infinite dimensional vector such that $1 \leqslant p_k \leqslant \infty$, $k = 1, 2, \ldots$ (we shall write $1 \leqslant \vec{p} \leqslant \infty$). Denote $M_{\vec{p}}$ the set of martingales $f = (f_n, \mathcal{B}_n, dx)$ on $(T^\infty, \mathcal{B})$ such that

$$\|f\|_{\vec{p}} := \sup \left\{ \|f_n\|_{\vec{p}},\ n = 1, 2, \ldots \right\} < \infty,$$

where

$$\left\|f_n\right\|_{\vec{p}} := \left\| \cdots \left\| \left\| f_n(x_1, \ldots, x_{n-1} x_n) \right\|_{p_n, x_n} \right\|_{p_{n-1}, x_{n-1}} \cdots \right\|_{p_1, x_1},$$

$$\left\|f_n(x_1\dots x_n)\right\|_{p_n,x_n} = \begin{cases} \left(\int_T \left|f_n\left(x_1\dots x_n\right)\right|^{p_n} dx_n\right)^{1/p_n}, & p_n<\infty,\\ \operatorname{ess\,sup}_{x_n\in T} \left|f_n(x_1\dots x_n)\right|, & p_n=\infty. \end{cases}$$

Let $L_{\vec{p}}$ be the subset of $M_{\vec{p}}$ consisting of uniformly integrable martingales. It is easy to prove that $M_{\vec{p}}$ and $L_{\vec{p}}$ are Banach spaces and $L_{\vec{p}}$ is a subspace of $M_{\vec{p}}$. For any $n=1,2,\dots$, (T^n, B_n) is a standard Borel space. So, according to (Parthasarathy, 1967), the space $M_{\vec{p}}$ is naturally identified with the space of signed Borel measures on (T^∞, B) which are (B_n)–locally, absolutely continuous with respect to the Haar measure and such that the associated martingales belong to $M_{\vec{p}}$.

THEOREM 1.1. *The spaces $L_{\vec{p}}$ and $M_{\vec{p}}$ coincide iff*

$$\liminf_{k\to\infty} p_k > 1.$$

Consider the following properties:

(A) If a sequence $(f_k)\subset L_{\vec{p}}$ converges in probability to some f and $|f_k|\leqslant g\in$ $\in L_{\vec{p}}$, $k=1,2,\dots$, then $f\in L_{\vec{p}}$ and $\|f-f_k\|_{\vec{p}}\to 0$.

(B) The space $L_{\vec{p}}$ is separable.

(C) The set D of cylindric infinitely differentiable functions is dense in $L_{\vec{p}}$.

(D) For any function $f\in L_{\vec{p}}$ and any $\varepsilon>0$ there exists some neighbourhood V of the neutral element such that

$$\|f(x+y)-f(x)\|_{\vec{p},x}<\varepsilon, \quad \forall y\in V.$$

THEOREM 1.2. *The properties* (A), (B) *and* (D) *are equivalent. These properties hold iff* $\sup_k p_k<\infty$.
In this case, for any $f\in L_{\vec{p}}$ and δ–sequence (φ_n), we have $\|f*\varphi_n-f\|_{\vec{p}}\to 0$ as $n\to\infty$.

THEOREM 1.3 (Minkowski's inequality). *Let $\varphi\in L_1(T^\infty\times X)$, where X is a measurable space with a σ–finite measure μ. Then*

$$\left\|\int_X \varphi(\omega,x)\,d\mu(x)\right\|_{\vec{p},\omega} \leqslant \int_X \left\|\varphi(\omega,x)\right\|_{\vec{p},\omega} d\mu(x).$$

THEOREM 1.4 (Young's inequality). *Let $1\leqslant \vec{p},\vec{q},\vec{r}\leqslant\infty$ and $1/\vec{p}+1/\vec{q}=$ $=1+1/\vec{r}$. If $\mu_1\in M_{\vec{p}}$, $\mu_2\in M_{\vec{q}}$, then $\mu_1*\mu_2\in M_{\vec{r}}$ and*

$$\|\mu_1*\mu_2\|_{\vec{r}}\leqslant \|\mu_1\|_{\vec{p}}\,\|\mu_2\|_{\vec{q}}.$$

Other properties of the spaces $M_{\vec{p}}$ and $L_{\vec{p}}$ can be found in (Bendikov and Pavlov, 1987).

2. GAUSSIAN SEMIGROUPS

Consider Gaussian measure $dn_t(x) = n_t(x)\,dx$ on $T = [-\pi, \pi]$, where

$$n_t(x) = \sqrt{\frac{\pi}{t}} \sum_{k=-\infty}^{\infty} \exp\left(- \frac{(x-2\pi k)^2}{4t} \right).$$

For a given sequence (a_k), $a_k > 0$, consider the product–measure μ_t on T^∞ defined by

$$\mu_t = \bigotimes_{k=1}^{\infty} n_{a_k t}.$$

For any $t > 0$, the measure μ_t is (B_n)–locally, absolutely continuous with respect to the Haar measure, hence $\mu_t \in M_1$. Under additional conditions on (a_k) and the vector $\vec{p}$, the stronger result is true.

THEOREM 2.1. *Let* $1 \leqslant \vec{p} \leqslant \infty$, $1/\vec{p} + 1/\vec{q} = 1$. *Denote* $\tilde{a}_k = \min(a_k, \log^{-1} k)$, $k = 2, 3, \ldots$, *and suppose that*

$$\prod_{k=2}^{\infty} \tilde{a}_k^{1/q_k} > 0. \tag{2.1}$$

Then the following properties are satisfied:
1) $\mu_t \in M_{\vec{p}}$, *for any* $t > 0$,
2) $\alpha := \sum_{k=1}^{\infty} 1/q_k < \infty$ *and, for any* $\varepsilon > 0$,

$$c_1 t^{-(\alpha/2-\varepsilon)} < \|\mu_t\|_{\vec{p}} < c_2 t^{-\alpha/2}, \quad t \downarrow 0.$$

Remark 2.1. Let condition (2.1) be satisfied, then

$$1 > \prod_{k=2}^{\infty} (\log k)^{-1/q_k} \geqslant \prod_{k=2}^{\infty} \tilde{a}_k^{1/q_k} > 0.$$

These inequalities imply the convergence of the infinite product $\prod (\log k)^{1/q_k}$. This is equivalent to

$$\sum_{k=2}^{\infty} \log\log k / q_k < \infty. \tag{2.2}$$

In particular, we have that condition (2.1) implies the convergence of the series $\alpha = \sum_{k=1}^{\infty} 1/q_k$. Note one more particular case. Let the sequence (a_k) satisfy the following condition

$$\liminf_{k\to\infty} a_k \log k = c > 0, \tag{2.3}$$

then condition (2.2) implies condition (2.1).

Proof of the theorem relies on the estimates of the function $t \longrightarrow \|n_t\|_p$ given in the following lemma.

LEMMA 2.1. *The following inequalities are true for any* $1 \leqslant p, q \leqslant \infty$, $1/p + +1/q = 1$,

$$\|n_t\|_p \leqslant \left(\sqrt{\frac{\pi}{t}}\right)^{1/q} \left(\frac{1}{p}\right)^{1/2p} \left[1 + \exp\left(-\frac{\pi^2}{8t}\right)\right], \quad 0 < t < \frac{1}{2}, \tag{2.4}$$

$$\|n_t\|_p \sim \left(\sqrt{\frac{\pi}{t}}\right)^{1/q} \left(\frac{1}{p}\right)^{1/2p}, \quad t \downarrow 0, \tag{2.5}$$

$$\|n_t\|_p - 1 \sim \frac{p}{2q} \exp(-2t), \quad t \uparrow \infty. \tag{2.6}$$

Proof. We have

$$n_t(x) = \sqrt{\frac{\pi}{t}} \sum_{k=-\infty}^{\infty} \exp\left(-\frac{(x - 2\pi k)^2}{4t}\right),$$

hence,

$$\|n_t\|_p \leqslant \sqrt{\frac{\pi}{t}} \sum_{k=-\infty}^{\infty} \left\|\exp\left(-\frac{(x - 2\pi k)^2}{4t}\right)\right\|_{p,x}.$$

Let us estimate the summands on the right–hand side of the above inequality:

$$\left\|\exp\left(-\frac{x^2}{4t}\right)\right\|_p^p = \frac{1}{2\pi} \int_{-\pi}^{\pi} \exp\left(-\frac{x^2 p}{4t}\right) dx \leqslant$$

$$\leqslant \frac{1}{2\pi} \int_{-\infty}^{\infty} \exp\left(-\frac{x^2 p}{4t}\right) dx = \sqrt{\frac{t}{\pi}} \left(\frac{1}{p}\right)^{1/2},$$

$$\left\|\exp\left(-\frac{(x - 2\pi k)^2}{4t}\right)\right\|_p^p = \frac{1}{2\pi} \int_{-\pi(2k+1)}^{-\pi(2k-1)} \exp\left(-\frac{x^2 p}{4t}\right) dx \leqslant$$

$$\leqslant \begin{cases} \exp\left(-\dfrac{\pi^2(2k-1)^2 p}{4t}\right), & k \geqslant 1, \\ \exp\left(-\dfrac{\pi^2(2k+1)^2 p}{4t}\right), & k \leqslant -1. \end{cases}$$

It follows that, for $0 < t < 1/2$,

$$\|n_t\|_p \leqslant \sqrt{\frac{\pi}{t}} \left(\left(\sqrt{\frac{t}{\pi}}\right)^{1/p} \left(\frac{1}{p}\right)^{1/2p} + 2\sum_{k=0}^{\infty} \exp\left(-\frac{\pi^2(2k+1)^2}{4t}\right)\right) \leqslant$$

$$\leqslant \sqrt{\frac{\pi}{t}} \left(\left(\sqrt{\frac{t}{\pi}}\right)^{1/p} \left(\frac{1}{p}\right)^{1/2p} + 2\sum_{k=1}^{\infty} \exp\left(-\frac{\pi^2 k}{4t}\right)\right) =$$

$$= \sqrt{\frac{\pi}{t}}\left(\left(\sqrt{\frac{t}{\pi}}\right)^{1/p}\left(\frac{1}{p}\right)^{1/2p} + 2\exp\left(-\frac{\pi^2}{4t}\right)\left(1-\exp\left(-\frac{\pi^2}{4t}\right)\right)^{-1}\right) \leqslant$$

$$\leqslant \left(\sqrt{\frac{\pi}{t}}\right)^{1/q}\left(\frac{1}{p}\right)^{1/2p}\left(1+3p^{1/2p}\left(\sqrt{\frac{\pi}{t}}\right)^{1/p}\exp\left(-\frac{\pi^2}{4t}\right)\right) \leqslant$$

$$\leqslant \left(\sqrt{\frac{\pi}{t}}\right)^{1/q}\left(\frac{1}{p}\right)^{1/2p}\left(1+\exp\left(-\frac{\pi^2}{8t}\right)\right),$$

thus (2.4) is proved.
Applying the inequality $(a+b)^p \geqslant a^p + b^p$, $a, b \geqslant 0$, $p \geqslant 1$, we get

$$\|n_t\|_p \geqslant \left(\sqrt{\frac{\pi}{t}}\right)\left(\sum_{k=-\infty}^{\infty}\left\|\exp\left(-\frac{(x-2\pi k)^2}{4t}\right)\right\|_p^p\right)^{1/p} =$$

$$= \left(\sqrt{\frac{\pi}{t}}\right)\left(\sum_{k=-\infty}^{\infty}\frac{1}{2\pi}\int_{-\pi}^{\pi}\exp\left(-\frac{(x-2\pi k)^2 p}{4t}\right)dx\right)^{1/p} =$$

$$= \left(\sqrt{\frac{\pi}{t}}\right)\left(\frac{1}{2\pi}\int_{-\infty}^{\infty}\exp\left(-\frac{x^2 p}{4t}\right)dx\right)^{1/p} = \left(\sqrt{\frac{\pi}{t}}\right)^{1/q}\left(\frac{1}{p}\right)^{1/2p}.$$

So, with the help of (2.4), we have

$$\left(\sqrt{\frac{\pi}{t}}\right)^{1/q}\left(\frac{1}{p}\right)^{1/2p} \leqslant \|n_t\|_p \leqslant \left(\sqrt{\frac{\pi}{t}}\right)^{1/q}\left(\frac{1}{p}\right)^{1/2p}\left(1+\exp\left(-\frac{\pi^2}{8t}\right)\right),$$

this proves (2.5).
For $t > 0$ and $x \in [-\pi, \pi]$, denote

$$\varphi(t,x) = \sum_{k=1}^{\infty}\exp(-tk^2)\cos kx.$$

We have

$$n_t(x) = \sum_{k=-\infty}^{\infty}\exp(-tk^2)\cos kx = 1 + 2\varphi(t,x), \tag{2.7}$$

$$|\varphi(t,x)| \leqslant \sum_{k=1}^{\infty}\exp(-tk^2) \leqslant \int_0^{\infty}\exp(-tx^2)\,dx = \frac{1}{2}\sqrt{\frac{\pi}{t}}. \tag{2.8}$$

It follows from (2.8) that, for $t > \pi$,

$$|\varphi(t,x)| \leqslant \frac{1}{2}\sqrt{\frac{\pi}{t}} < \frac{1}{2},$$

so the following series

$$\left(n_t(x)\right)^p = \sum_{k=0}^{\infty} \binom{p}{k} 2^k \varphi(t,x)^k$$

converges uniformly for $x \in [-\pi, \pi]$.
After integration we have

$$\left\|n_t\right\|_p^p = 1 + \sum_{k=2}^{\infty} \binom{p}{k} 2^k \frac{1}{2\pi} \int_{-\pi}^{\pi} \varphi(t,x)^k \, dx.$$

Furthermore,

$$e^t \varphi(t,x) = \cos x + \sum_{k=2}^{\infty} \exp\left(-tk(k-1)\right) \cos kx = \cos x + A(t,x),$$

where

$$\left|A(t,x)\right| \leqslant \sum_{k=1}^{\infty} \exp\left(-tk^2\right) \leqslant \frac{1}{2}\sqrt{\frac{\pi}{t}},$$

hence, uniformly in $x \in [-\pi, \pi]$,

$$\lim_{t\to\infty} e^t \varphi(t,x) = \cos x.$$

It follows that

$$\lim_{t\to\infty} e^{2t}\left(\left\|n_t\right\|_p^p - 1\right) = \frac{p(p-1)}{2}$$

and, consequently,

$$\left\|n_t\right\|_p = \left[1 + \frac{p(p-1)}{2} e^{-2t}\left(1 + o(1)\right)\right]^{1/p}.$$

Finally, as $t \uparrow \infty$ we have

$$\left\|n_t\right\|_p - 1 \sim \frac{1}{p} \frac{p(p-1)}{2} e^{-2t} = \frac{p}{2q} e^{-2t}.$$

Lemma 2.1 is proved.

Proof of Theorem 2.1. According to the Young inequality, we have that function $t \to \|n_t\|_p$ decreases, so that

$$\left\|n_{a_k t}\right\|_p \leqslant \left\|n_{\bar{a}_k t}\right\|_p.$$

Using (2.4) together with definition of the measure μ_t and the mixed norm, for $0 < t < 1/2$, we have

$$\left\|\mu_t\right\|_{\bar{p}} = \prod_{k=1}^{\infty} \left\|n_{a_k t}\right\|_{p_k} \leqslant \prod_{k=1}^{\infty} \left\|n_{\bar{a}_k t}\right\|_{p_k} \leqslant \Pi_1 \Pi_2 \Pi_3,$$

where

$$\Pi_1 = \prod_{k=1}^{\infty} \left(\sqrt{\frac{\pi}{\tilde{a}_k t}}\right)^{1/q_k}, \ \Pi_2 = \prod_{k=1}^{\infty} \left(\frac{1}{p_k}\right)^{1/2p_k}, \ \Pi_3 = \prod_{k=1}^{\infty} \left(1 + \exp\left(-\frac{\pi^2}{8\tilde{a}_k t}\right)\right).$$

Condition (2.1) implies

$$\Pi_1 = \left(\prod_{k=1}^{\infty} \tilde{a}_k^{1/q_k}\right)^{-1/2} \left(\frac{\pi}{t}\right)^{1/2 \sum_{k=1}^{\infty} 1/q_k} = c_1 t^{-\alpha/2},$$

$$\Pi_2 = \left(\prod_{k=1}^{\infty} \left(1 - \frac{1}{q_k}\right)^{1/p_k}\right)^{1/2} < e^{\alpha/2} = c_2,$$

$$\Pi_3 \leqslant \exp\left(\sum_{k=1}^{\infty} \exp\left(-\frac{\pi^2}{8\tilde{a}_k t}\right)\right) \leqslant \exp\left(\sum_{k=1}^{\infty} \exp\left(-\frac{\pi^2}{8t}\log k\right)\right) =$$

$$= \exp\left(\sum_{k=1}^{\infty} \frac{1}{k^{\pi^2/8t}}\right) \leqslant \exp\left(\sum_{k=1}^{\infty} \frac{1}{k^{3/2}}\right) = c_3.$$

Therefore, for $0 < t < 1/2$, we have

$$\|\mu_t\|_{\bar{p}} \leqslant c_4 t^{-\alpha/2}.$$

On the other hand, for any fixed $\varepsilon > 0$, if we choose a number $n \geqslant 1$ such that $\sum_{k=n}^{\infty} 1/q_k < \varepsilon$, then, according to (2.5), we have

$$\|\mu_t\|_{\bar{p}} = \prod_{k=1}^{\infty} \|n_{a_k t}\|_{p_k} \geqslant \prod_{k=1}^{n} \|n_{a_k t}\|_{p_k} \sim c_5 t^{-\frac{1}{2}\sum_{k=1}^{n} 1/q_k} \geqslant c_5 t^{-(\alpha/2-\varepsilon)}.$$

Theorem 2.1 is proved.

Remark 2.2. Suppose that condition (2.2) and the condition

$$0 < \prod_{k=1}^{\infty} a_k^{1/q_k} < \infty \tag{2.9}$$

are satisfied, then, for $0 < t < 1/2$,

$$c_1 t^{-\alpha/2} \leqslant \|\mu_t\|_{\bar{p}} \leqslant c_2 t^{-\alpha/2}. \tag{2.10}$$

Indeed, conditions (2.2), (2.9) imply that (2.1) is satisfied. So, according to Theorem 2.1, the second inequality in (2.10) is true; the first one is also true because of

$$\|\mu_t\|_{\bar{p}} = \prod_{k=1}^{\infty} \|n_{a_k t}\|_{p_k} \geqslant \prod_{k=1}^{\infty} \left(\sqrt{\frac{\pi}{a_k t}}\right)^{1/q_k} \left(\frac{1}{p_k}\right)^{1/2p_k} =$$

$$= \left(\prod_{k=1}^{\infty} a_k^{1/q_k}\right)^{-1/2} \pi^{\alpha/2} \left(\prod_{k=1}^{\infty} \frac{1}{p_k}\right)^{1/2} t^{-\alpha/2} = c_1 t^{-\alpha/2}.$$

3. BESSEL POTENTIALS

For any bounded Borel function φ, define the Bessel potential $J_\beta[\varphi]$, $\beta > 0$, as $J_\beta[\varphi] = J_\beta * \varphi$, where the probability measure J_β is given by

$$J_\beta(dx) = \frac{1}{\Gamma(\beta/2)} \int_0^\infty e^{-t} t^{\beta/2-1} \mu_t(dx)\, dt.$$

THEOREM 3.1. *Let $1 \leqslant \vec{p}, \vec{q} \leqslant \infty$ and $1/\vec{p} + 1/\vec{q} = 1$. Suppose that in addition to* (2.1), *the following condition*

$$\alpha := \sum_{k=1}^{\infty} 1/q_k < \beta \tag{3.1}$$

is satisfied. Then $J_\beta \in M_{\vec{p}}$.

Proof. Write the semigroup property for (μ_t): $\mu_{t+\Delta} = \mu_t * \mu_\Delta$. According to the Young inequality,

$$\|\mu_{t+\Delta}\|_{\vec{p}} \leqslant \|\mu_t\|_{\vec{p}} \|\mu_\Delta\|_1 = \|\mu_t\|_{\vec{p}},$$

i.e., $t \to \|\mu_t\|_{\vec{p}}$ is a decreasing function. Using the Minkowski's inequality for (B_n)-locally absolute continuous measures and Theorem 2.1, we have

$$\|J_\beta\|_{\vec{p}} \leqslant \Gamma^{-1}(\beta/2) \int_0^\infty e^{-t} t^{\beta/2-1} \|\mu_t\|_{\vec{p}}\, dt \leqslant$$

$$\leqslant c_1 \int_0^1 e^{-t} t^{(\beta-\alpha)/2-1}\, dt + c_2 \int_1^\infty e^{-t} t^{\beta/2-1}\, dt < \infty.$$

Theorem 3.1 is proved.

Remark 3.1. According to (3.1) the components p_k of the vector $\vec{p}$ tend to 1 as $k \to \infty$. We would like to note that $J_\beta \notin M_{\vec{r}}$ if the vector $\vec{r} = (r_1, r_2, \ldots)$ is such that $\inf_{k \geqslant 1} r_k = r > 1$. In fact, if $J_\beta \in M_{\vec{r}}$, then the measure J_β is absolutely continuous with respect to the Haar measure (Theorem 1.1) and the corresponding density $J_\beta(x)$ belongs to $L_r \subset L_\Delta$ for some Δ, $1 < \Delta \leqslant \min(r, 2)$. Hence according to the Hausdorf–Young theorem $\widehat{J}_\beta \in L_s(\mathbf{Z}^{(\infty)})$, $1/s + 1/\Delta = 1$, where $\widehat{J}_\beta$ is the Fourier transform of the measure J_β. It is easy to prove that $\widehat{J}_\beta(\theta) = (1 + \psi(\theta))^{-\beta/2}$, where $\psi(\theta) = \sum_{k=1}^{\infty} a_k \theta_k^2$ and $\theta = (\theta_1, \theta_2, \ldots) \in \mathbf{Z}^{(\infty)}$. In the consequence, for some number $\nu > 0$ the function $(1 + \psi)^{-\nu/2}$ is summable on $\mathbf{Z}^{(\infty)}$, which is impossible. In fact, choose a number $n \geqslant \nu$ and let $\mathbf{Z}^{(n)} = \{\theta \in$

$\in \mathbf{Z}^{(\infty)}$: $\theta_k = 0,\ k > n\}$ be the finite dimensional sub-lattice of $\mathbf{Z}^{(\infty)}$, then we have

$$\int\limits_{\mathbf{Z}^{(\infty)}} \left(1+\psi(\theta)\right)^{-\nu/2} d\theta \geqslant \int\limits_{\mathbf{Z}^{(n)}} \left(1+\psi(\theta)\right)^{-\nu/2} d\theta =$$

$$= \sum_{\theta_1 \dots \theta_n = -\infty}^{\infty} \left[1 + a_1\theta_1^2 + \dots + a_n\theta_n^2\right]^{-\nu/2} \geqslant c_1 \sum_{\theta_1 \dots \theta_n = -\infty}^{\infty} \left[1 + \theta_1^2 + \dots + \theta_n^2\right]^{-n/2} =$$

$$= c_1 \sum_{\theta_1 \dots \theta_{n-1} = -\infty}^{\infty} \sum_{\theta_n = -\infty}^{\infty} \left[1 + \theta_1^2 + \dots + \theta_n^2\right]^{-n/2} \geqslant$$

$$\geqslant c_1 \sum_{\theta_1 \dots \theta_{n-1} = -\infty}^{\infty} \int\limits_1^{\infty} \left[1 + \theta_1^2 + \dots + \theta_{n-1}^2 + \theta^2\right]^{-n/2} d\theta \geqslant$$

$$\geqslant c_2 \sum_{\theta_1 \dots \theta_{n-1} = -\infty}^{\infty} \left[1 + \theta_1^2 + \dots + \theta_{n-1}^2\right]^{-(n-1)/2} \geqslant \dots \geqslant c_n \sum_{\theta_1 = -\infty}^{\infty} \left[1 + \theta_1^2\right]^{-1/2} = \infty.$$

Remark 3.2. Absolute continuity with respect to the Haar measure of all measures (J_β) is equivalent to the absolute continuity with respect to the Haar measure of all measures (μ_t) and this last property by the Fukushima's theorem is equivalent to the absolute continuity with respect to the Haar measure of the measure $J_2 = r^1$, where (r^λ) is the resolvent of the semigroup (μ_t). According to (Berg, 1976; Bendikov, 1987) the measures μ_t for all $t > 0$ are absolutely continuous with respect to the Haar measure if $\sum_{k=1}^{\infty} \exp(-a_k t) < \infty$ for all $t > 0$. In particular, if the last condition and (3.1) are satisfied, then $J_\alpha \in L_{\vec{p}}$. Contrary to this situation, for $a_1 = a_2 = \dots = 1$, all measures (J_β) are singular with respect to the Haar measure and so, in this case, $J_\alpha \notin L_{\vec{p}}$.
The following theorem is the infinite dimensional generalization of the Sobolev–Nikolsky inequality (see the classical case in (Besov et al., 1978)).

THEOREM 3.2. *Let* $1 \leqslant \vec{p}, \vec{q}, \vec{\Delta} \leqslant \infty$, $1/\vec{p} = 1/\vec{q} - 1/\vec{\Delta}$, $\tilde{a}_k = \min(a_k, \log^{-1} k)$. *Suppose that*

$$\prod_{k=2}^{\infty} \tilde{a}_k^{1/\Delta_k} > 0 \quad \text{and} \quad \sum_{k=1}^{\infty} 1/\Delta_k < \beta,$$

then

$$\left\|J_\beta[\varphi]\right\|_{\vec{p}} \leqslant c \left\|\varphi\right\|_{\vec{q}}.$$

Proof. According to Theorem 3.2, the measure J_β belongs to $M_{\vec{\Delta}'}$, $1/\vec{\Delta} + 1/\vec{\Delta}' = 1$. Applying Young inequality to the function $J_\beta[\varphi] = J_\beta * \varphi$, $\varphi \in L_{\vec{q}}$, we obtain the desired inequality.
Note that in the theorem $\vec{p} > \vec{q}$ and so, the Bessel potential $J_\beta[f]$ improves integrability properties of functions from $L_{\vec{q}}$.
For a vector $1 \leqslant \vec{p} \leqslant \infty$ and a number $\alpha > 0$, denote

$$L_{\vec{p},\alpha} = \begin{cases} L_{\vec{p}} & \text{if } \alpha = 0, \\ \{J_\alpha[\varphi]: \varphi \in L_{\vec{p}}\} & \text{if } \alpha > 0, \end{cases}$$

$$L^0_{\vec{p},\alpha} = \left\{ u \in L_{\vec{p},\alpha}: \exists \varphi_n \in D: \varphi_n \xrightarrow{L_{\vec{p},\alpha}} u \right\}.$$

THEOREM 3.3. *Let the vector $1 \leqslant \vec{p} \leqslant \infty$, the number $\alpha \geqslant 0$ and the sequence (a_k) satisfy the following conditions*

$$\prod_{k=1}^{\infty} \tilde{a}_k^{1/p_k} > 0, \quad \sum_{k=1}^{\infty} 1/p_k \leqslant \alpha.$$

Then:

1) *If $\sum_{k=1}^{\infty} \exp(-a_k t) = \infty$ $(\forall t > 0)$, then each function $u \in L_{\vec{p},\alpha}$ coincides dx–almost surely with some continuous function $\tilde{u}$;*
2) *If $\sum_{k=1}^{\infty} \exp(-a_k t) = \infty$ $(\exists t > 0)$, then each function $u \in L^0_{\vec{p},\alpha}$ coincides dx–almost surely with some continuous function $\tilde{u}$.*

Proof. 1) Let $u = J_\alpha[\varphi]$. Choose a number $\tilde{\alpha}$ such that $\sum_{k=1}^{\infty} 1/p_k < \tilde{\alpha} < \alpha$ and write

$$u = J_\Delta\big[J_{\tilde{\alpha}}[\varphi]\big], \quad \Delta = \alpha - \tilde{\alpha}.$$

According to Theorem 3.2 the function $J_{\tilde{\alpha}}[\varphi] \in L_\infty$. Let $\tilde{\varphi}$ be a Borel function such that $\tilde{\varphi} = J_{\tilde{\alpha}}[\varphi]$ dx–a.s. Denote $\tilde{u} = J_\Delta[\tilde{\varphi}]$, then $u = \tilde{u}$ dx–a.s. We prove that the function $\tilde{u}$ is continuous. In fact, the condition

$$\sum_{k=1}^{\infty} \exp\,(-a_k t) < \infty \quad (\forall t > 0)$$

implies absolute continuity with respect to the Haar measure of all measures (μ_t) and so all measures (J_α) are absolutely continuous with respect to the Haar measure. In particular, $J_\Delta(dx) = J_\Delta(x)\,dx$, $J_\Delta(x) \in L_1$. Therefore the function $\tilde{u} = J_\Delta * \tilde{\varphi}$ is continuous.
2) Let $(\varphi_n) \subset D$: $\varphi_n \to \varphi$ in $L_{\vec{p}}$ as $n \to \infty$. Consider a sequence (u_n) of continuous functions $u_n = J_\alpha[\varphi_n]$. According to Theorem 3.2,

$$\|u_n - u_m\|_\infty \leqslant c\|\varphi_n - \varphi_m\|_{\vec{p}} \longrightarrow 0$$

when $m, n \to \infty$.
Hence the sequence (u_n) is fundamental in the space of continuous functions $C(T^\infty)$; denote the limit of this sequence by $\tilde{u}$. It is easy to prove that $u = \tilde{u}$ dx–a.s. Theorem 3.3 is proved.

4. WEAK SOLUTIONS

We begin with some facts which can be found in (Bendikov, 1987).
Consider an infinite dimensional elliptic differential operator $\mathcal{L} = \sum_{k=1}^{\infty} a_k \partial_k^2$, which is well defined on the set D of the cylindric infinitely differentiable functions.

Definition 4.1. Let V be an open subset of T^∞ and $p > 1$. A function $u \in (L_p)_{\text{loc}}(V)$ is called a weak solution of the equation $\mathcal{L}u = 0$ on the set V if, for all functions $\varphi \in D$ such that supp $\varphi \subset V$, the equality

$$\int\limits_{T^\infty} u(x)\mathcal{L}\varphi(x)\,dx = 0$$

holds.

THEOREM 4.1. *Let u be a weak solution of the equation $\mathcal{L}u = 0$ on the set V. Then:*

1) *the function u is infinitely differentiable in L_p-sense;*
2) *$\sum_{k=1}^\infty a_k \partial_k^2 u \equiv 0$ and the series converges in the norm of $L_p(K)$ for any compact $K \subset V$.*

THEOREM 4.2. *The following statements are equivalent:*

(A) *any weak solution u concides dx–a.s. with some continuous function $\tilde{u}$;*

(B) *for any $t > 0$ the measure $\mu_t = \bigotimes_{k=1}^\infty n_{a_k t}$ is absolutely continuous with respect to the Haar measure, has continuous density $\mu_t(x)$ and, for all $c > 0$,*

$$\lim_{t\downarrow 0} \exp\Big(-\frac{c}{t}\Big)\mu_t(0) = 0.$$

THEOREM 4.3. *Consider the following conditions*

(C) $\lim\limits_{n\to\infty} a_n/n = \infty$;

(C*) $\limsup\limits_{n\to\infty} a_n/n = \infty$.

Then the implications (C) $\Rightarrow$ (B) $=$ (C*) *are true.*

The following theorem complements Theorems 4.2 and 4.3 (for example, $a_n = n^\alpha$, $\alpha \leqslant 1$).

THEOREM 4.4. *Suppose that the vector $1 \leqslant \vec{p} \leqslant \infty$ and the sequence (a_k) satisfy the following conditions*

$$\prod_{k=2}^\infty \tilde{a}_k^{1/p_k} > 0, \quad \tilde{a}_k = \min\big(a_k, \log^{-1} k\big), \quad \sum_{k=1}^\infty 1/p_k < 2,$$

and let u be a weak solution of the equation $\mathcal{L}u = 0$ on the open set V. In each of the two case

(a) $\forall t > 0 \quad \sum\limits_{k=1}^\infty \exp(-a_k t) < \infty, \quad \{u, \partial_1 u, \partial_2 u, \ldots\} \subset L_{\vec{p}}(V)$,

(b) $\exists t > 0 \quad \sum\limits_{k=1}^\infty \exp(-a_k t) < \infty, \quad \{u, \partial_1 u, \partial_2 u, \ldots\} \subset L_{\vec{p}}^0(V)$

there exists some continuous function $\tilde{u}$ such that $u = \tilde{u}$ dx–a.s. on the set V.

Proof. Choose a function $\beta \in D$ such that supp $\beta \subset V$ and set $u_\beta = \beta u$. It is easy to prove that $\mathcal{L}u_\beta - u_\beta = g$, where

$$g = (\mathcal{L}\beta - \beta)u + 2\sum_{k=1}^{\infty} a_k \partial_k \beta \partial_k u.$$

So we have $u_\beta = -J_2[g]$. Now it remains to use Theorem 3.3.

REFERENCES

Bendikov, A.D. and Pavlov, I.V. (1987). Spaces $L_{\bar{p}}$ with mixed norm on infinite Cartesian product of probabilistic spaces. *Anal. Math.* **13**, 231–250.

Bendikov, A.D. (1987). Markov processes and partial differential equations on a group: the space–homogeneous case. *Russian Math. Surveys* **42**, 49–94.

Parhasarathy, K.R. (1967). *Probability Measures on Metric Spaces.* Academic Press, New York.

Berg, Ch. (1976). Potential theory on the infinite dimensional torus. *Invent. Math.* **32**, 49–100.

Besov, O.V., Ilyin, V.P. and Nikolskii, S.M. (1978). *Integral Representations of Functions and Embedding Theorems.* Wiley, New York.

New Trends in Probab. and Statist., pp. 233 - 240
V. Sazonov and T. Shervashidze (Eds.)
1991 VSP/Mokslas

UNEXPECTED LIMIT BEHAVIOUR OF THE GAUSSIAN TAIL PROBABILITIES

M.A. LIFSHITS
Leningrad Institute of Finance and Economy,
Department of Economical Informatics,
Griboedov channel 30/32, 191023 Leningrad, USSR

ABSTRACT

Let $\sigma_i > 0$, ξ_i be independent standard normal random variables. We develop a general approach for the investigation of tail probabilities

$$P(r) = \mathbf{P}\left\{ \sum_{i=1}^{\infty} |\sigma_i \xi_i| \geqslant r \right\}, \quad r \to \infty.$$

In particular, for $\sigma_i = \sigma_1 B^{1-i}$ the behaviour of the tails turns out to be unusual:

$$P(r) \sim \chi\big(\log(r/\sigma^2)\big)(r/\sigma^2)^{\beta}\big(1 - \Phi(r/\sigma)\big), \quad r \to \infty,$$

where $\beta = \log 2/\log B$, $\sigma^2 = \sum_{i=1}^{\infty} \sigma_i^2$, χ is an explicitly calculated *periodical* function, Φ is the standard normal distribution function.

INTRODUCTION

We investigate the tail probabilities

$$\mathbf{P}\left\{x \colon \|x\| \geqslant r\right\}$$

for a Gaussian measure $\mathbf{P}$ in a Banach space $(E, \|\cdot\|)$. This problem has been intensively analyzed by Landau, Shepp, Sudakov, Dudley, Tsyrelson, Fernique, Borell, Marcus, Pickands, Piterbarg, Ehrhard, Dmitrovsky, Hwang and many others. This paper develops the research initiated in (Dobric, Marcus and Weber, 1988; Linde, 1989), which deal with the case $E = \mathbf{l}^p$ with $\mathbf{P}$-independent coordinates. The Hilbert space case $p = 2$ was earlier completely investigated in (Zolotarev, 1961; Hoeffding, 1964). Our subject is $E = \mathbf{l}^1$ and the aim is to show that for $\mathbf{l}^1$ the rate of tails decrease may vary in the broad range. In particular, the asymptotics may include an unexpected "strange" periodical factor (this phenomenon is impossible for $p > 2$).

NOTATION AND KNOWN RESULTS

Let $\mathcal{N}$ be the standard normal distribution. Let ξ_i, $i = 1, 2, \ldots$, be a sequence of independent $\mathcal{N}$-distributed random variables. Let $p \geqslant 1$ and σ_i, $i = 1, 2, \ldots$, be a sequence of numbers, $\sigma_1 \geqslant \sigma_2 \geqslant \ldots \geqslant 0$. We call it a spectral sequence or a spectrum, and always assume that $\sigma_1 > 0$ and

$$\sum_{i=1}^{\infty} |\sigma_i|^p < \infty.$$

We investigate the tail probabilities

$$P(r) = \mathbf{P}\{\mathfrak{z} \geqslant r\}$$

for the $\mathbf{l}^p$-norm

$$\mathfrak{z} = \left(\sum_{i=1}^{\infty} |\sigma_i \xi_i|^p \right)^{1/p}.$$

Let the functions f and g be defined on $\mathbf{R}^+$. We write $f \sim g$ or, respectively, $f < \sim g$ if

$$\lim_{r \to \infty} f(r)/g(r) = 1 \qquad \text{or} \qquad \limsup_{r \to \infty} f(r)/g(r) \leqslant 1.$$

Let

$$\Phi(r) = (2\pi)^{-1/2} \int_{-\infty}^{r} \exp\{-u^2/2\}\, du.$$

The known results, in short, are as follows. For $p = 2$ the tail behaviour depends on the number of largest terms of the spectral sequence. In (Zolotarev, 1961) was proved that if

$$\sigma_1 = \sigma_2 = \ldots = \sigma_k > \sigma_{k+1}, \tag{1}$$

then

$$P(r) \sim c r^{k-1} \bigl(1 - \Phi(r/\sigma_1)\bigr), \quad r \to \infty, \tag{2}$$

and the constant c depends on the entire spectrum.

For $p > 2$ the situation is more simple because the intersection of the unit $\mathbf{l}^p$-sphere with the ellipsoid generated by the Gaussian measure, inscribed in this sphere, consists of a finite number of points. Namely, if $p > 2$ and (1) holds, then in (Dobric, Marcus and Weber, 1988) it was proved that

$$P(r) \sim 2k\bigl(1 - \Phi(r/\sigma_1)\bigr). \tag{3}$$

For $1 \leqslant p < 2$ the situation is more complicated because the intersection mentioned above is not countable. In earlier works only the finite-dimensional case has been investigated. If $\sigma_k > 0$ and $\sigma_{k+1} = \ldots = 0$, then (cf. (Linde, 1989))

$$P(r) \sim 2^k (2 - p)^{(1-k)/2} \bigl(1 - \Phi(r/\sigma)\bigr) \tag{4}$$

with

$$\sigma = \left(\sum_{i=1}^{\infty} \sigma_i^m\right)^{1/m}, \quad m = 2p/(2-p).$$

In particular, for $p = 1$

$$\sigma = \left(\sum_{i=1}^{\infty} \sigma_i^2\right)^{1/2}. \tag{5}$$

According to (5) the main parameter σ depends now on the entire spectrum, so the change of a single σ_i may have an essential influence on the tail probabilities. From now on we restrict our consideration to the case of $p = 1$ and provide some interesting examples. They will show that unified tail behaviour as in (2), (3) or (4) does not take place for $p = 1$. The intermediate case $1 < p < 2$, which is basically identical to $p = 1$, may be a subject for a separate article.

BINARY-STATIONARY PROCESSES

Let $T = [0,1]$. For $t \in T$, we denote by t_i the i-th figure in the binary decomposition of t.
Let us introduce the family of functions θ_i by the formulae $\theta_0 = 1$ and

$$\theta_i(t) = 2t_i - 1, \quad i > 0.$$

It is clear that possible values of θ_i are $+1$ and -1 and the family $\{\theta_i, i \geqslant 0\}$ is an orthogonal system in $\mathbf{L}^2(T)$. We define now the Gaussian process

$$\xi(t) = \sum_{i=1}^{\infty} \sigma_i \theta_i(t) \xi_i. \tag{6}$$

We call it a *binary-stationary* process because its finite-dimensional distributions are stable with respect to the inversion of any set of positions in the binary code of the time parameter t.
The obvious relation

$$\mathfrak{z} = \sum_{i=1}^{\infty} \sigma_i |\xi_i| = \sum_{i=1}^{\infty} \sigma_i \mathrm{sign}(\xi_i)\xi_i = \sup_{t\in T} \xi(t) \tag{7}$$

shows that for the case $p = 1$ our problem is exactly the investigation of the tail probabilities for the supremum of ξ. It is worthy of note that the class of binary-stationary processes is as large as the class of usual stationary processes. In fact, the representation (6) is very close to the general form of decomposition for a periodical stationary process $\tilde{\xi}$:

$$\tilde{\xi}(t) = \sum_{i=0}^{\infty} \left(\sigma_{i,1} \cos\left(\frac{i}{2\pi}t\right)\xi_{i,1} + \sigma_{i,2} \sin\left(\frac{i}{2\pi}t\right)\xi_{i,2}\right).$$

This analogy explains our spectral terminology for σ_i. We can now keep in mind the thoroughly investigated stationary case and expect that the properties of

the tail probabilities essentially depend on the rate of decrease of the spectral sequence $\{\sigma\}$.
The variance of $\xi(t)$ does not depend on t and is equal to $\sum_{i=1}^{\infty} \sigma_i^2$; this value coincides with σ^2 proposed in (5). So, the first advantage we obtain from representation (7) is the possibility to apply the Ylvisaker theorem (Ylvisaker, 1965) on the distribution density of the supremum of a Gaussian process with constant variance. This theorem states that the distribution of $\mathfrak{z} = \sup \xi$ has the density

$$\Pi(r) = \frac{g(r/\sigma)}{(2\pi)^{1/2}\sigma} \exp\{-r^2/2\sigma^2\} \tag{8}$$

and g is a nondecreasing function on $\mathbf{R}^+$.

LAPLACE TRANSFORM

We define the Laplace transform of $\mathfrak{z}$ as

$$\Psi(\lambda) = \mathbf{E} \exp\{\lambda \mathfrak{z}\}. \tag{9}$$

Later on we shall obtain sharp estimates for $\Psi(\lambda)$ for large λ and use them to analyze the tail probabilities. Our approach is not identical to the usual Tauberian approach since the latter uses small negative λ whereas we use large positive λ.

LEMMA 1. *Let the functions g and Ψ be defined in* (8) *and* (9). *Then*

$$\Psi(\lambda) = \bar{g}(\sigma\lambda) \exp\{\sigma^2\lambda^2/2\} \tag{10}$$

where

$$\bar{g}(\mu) = (2\pi)^{-1/2} \int g(\mu + u) \exp\{-u^2/2\}\, du. \tag{11}$$

Proof follows immediately from the equalities

$$\Psi(\lambda) = \int \exp\{\lambda r\}\Pi(r)\, dr = (2\pi)^{-1/2}\sigma^{-1} \int g(r/\sigma) \exp\{\lambda r - r^2/2\sigma^2\}\, dr =$$

$$= \exp\{\sigma^2\lambda^2/2\}(2\pi)^{-1/2}\sigma^{-1} \int g(r/\sigma) \exp\big\{-(r - \sigma^2\lambda)^2/2\sigma^2\big\}\, dr =$$

$$= \exp\{\sigma^2\lambda^2/2\}(2\pi)^{-1/2} \int g(u + \sigma\lambda) \exp\{-u^2/2\}\, du.$$

Now the outline of the further investigation is clear:

$$\Psi \to \bar{g} \to g \to \Pi \to P(t).$$

So we start with the analysis of Ψ.
By definition,

$$\Psi(\lambda) = \prod_{i=1}^{\infty} \mathbf{E} \exp\big\{\lambda\sigma_i|\xi_i|\big\}.$$

We may write each separate factor in the form

$$2(2\pi)^{-1/2}\int_0^\infty \exp\{\sigma_i x\lambda - x^2/2\}\,dx = 2\Phi(\sigma_i\lambda)\exp\{\sigma_i^2\lambda^2/2\}.$$

Hence, for σ defined by (5) we have

$$\Psi(\lambda) = \left[\prod_{i=1}^{\infty} 2\Phi(\sigma_i\lambda)\right]\exp\{\sigma^2\lambda^2/2\}. \qquad (12)$$

We will investigate (12) for two specific situations.

EXPONENTIALLY DECREASING SPECTRUM

Let $B > 1,\ \sigma_1 > 0,\ \sigma_i = \sigma_1 B^{1-i}\quad (i = 1, 2, \ldots)$.
We put $\beta = \log 2/\log B$, use the old notation (5) for σ and introduce the periodical function $\chi : \mathbf{R}^1 \to \mathbf{R}^1$ with period $\log B$ by the formula

$$\chi(l) = \exp(-\beta l)\prod_{i=1}^{\infty}\Big[2\Phi\big(\sigma_1 B^{1-i}\exp(l)\big)\Phi\big(\sigma_1 B^{i}\exp(l)\big)\Big], \quad 0 \leqslant l \leqslant \log B.$$

THEOREM 1. *If $r \to \infty$, then*

$$\mathbf{P}\{\mathfrak{z} \geqslant r\} \sim \chi(\log r - 2\log\sigma)(r/\sigma^2)^{\beta}\big(1 - \Phi(r/\sigma)\big). \qquad (13)$$

Proof. Let us represent any $\lambda \geqslant 1$ in the form

$$\lambda = \lambda_0 B^n, \quad \lambda_0 \in [1, B), \quad n \geqslant 0,$$

and rewrite (12) as

$$\Psi(\lambda) = \left[\prod_{i=1}^{\infty} 2\Phi(\sigma_1 B^{1-i+n}\lambda_0)\right]\exp\{\sigma^2\lambda^2/2\} =$$

$$= \left[\prod_{j=1}^{n}\Phi(\sigma_1 B^{j}\lambda_0)\right]\left[\prod_{i=1}^{\infty}2\Phi(\sigma_1 B^{1-i}\lambda_0)\right]2^n\exp\{\sigma^2\lambda^2/2\} =$$

$$= \left[\prod_{j=1}^{\infty}\Phi(\sigma_1 B^{j}\lambda_0)\right]\left[\prod_{i=1}^{\infty}2\Phi(\sigma_1 B^{1-i}\lambda_0)\right]2^n\exp\{\sigma^2\lambda^2/2\}\Big/\prod_{j=n+1}^{\infty}\Phi(\sigma_1 B^{j}\lambda_0).$$

Clearly, the denominator of the latter expression is uniformly close to unity as $\lambda \to \infty$.
We also note that

$$n = (\log\lambda - \log\lambda_0)/\log B,$$

hence,

$$2^n = (\lambda/\lambda_0)^\beta.$$

Our expression for Ψ takes now a simple form

$$\Psi(\lambda) \sim \left\{ \prod_{i=1}^{\infty} \left[\Phi(\sigma_1 B^i \lambda_0) 2\Phi(\sigma_1 B^{1-i}\lambda_0) \right] \lambda_0^{-\beta} \right\} \lambda^\beta \exp\{\sigma^2\lambda^2/2\} =$$

$$= \chi(\log\lambda)\lambda^\beta \exp\{\sigma^2\lambda^2/2\}. \qquad (14)$$

Comparing (14) and (10), we obtain

$$\bar{g}(\sigma\lambda) \sim \chi(\log\lambda)\lambda^\beta, \quad \lambda \to \infty. \qquad (15)$$

The most technical (but also simple) step of the proof is the transition from $\bar{g}$ to g.

LEMMA 2. *Let the functions g and $\bar{g}$ be connected by relation (11), g be non-decreasing and $\bar{g}$ satisfy the "weak condition of slow variation": for some function $\varepsilon(\mu) \to 0 \quad (\mu \to \infty)$,*

$$\delta(\mu) = \sup_{u>0} \left\{ \bar{g}(\mu+u)\exp\{-\varepsilon(\mu)u\}/\bar{g}(\mu) \right\} \sim 1. \qquad (16)$$

Then $g \sim \bar{g}$.

Proof. It follows from (11) that, for an arbitrary $A > 0$,

$$\bar{g}(\mu + A) \geqslant \Phi(A)g(\mu).$$

We also obtain from (16) that

$$\bar{g}(\mu + A) \leqslant \bar{g}(\mu)\exp\{\varepsilon(\mu)A\}\delta(\mu) \sim \bar{g}(\mu).$$

So

$$g(\mu) <\sim \bar{g}(\mu)/\Phi(A).$$

We can choose A arbitrary large, consequently, $g <\sim \bar{g}$.
To obtain the opposite inequality, we again use (11):

$$\bar{g}(\mu - A) \leqslant g(\mu) + (2\pi)^{-1/2} \int_A^\infty g(\mu - A + u)\exp\{-u^2/2\}\,du \leqslant$$

$$\leqslant g(\mu) + \left[\sup_{\omega \geqslant \mu} \frac{g}{\bar{g}}(\omega)\right] \int_0^\infty \bar{g}(\mu+v)\exp\{-(A+v)^2/2\}\,dv \leqslant$$

$$\leqslant g(\mu) + \left[\sup_{\omega \geqslant \mu} \frac{g}{\bar{g}}(\omega)\right] \delta(\mu)\bar{g}(\mu) \int_0^\infty \exp\{\varepsilon(\mu)v - (A+v)^2/2\}\,dv.$$

We may rewrite this inequality in the form

$$g(\mu) \geqslant \bar{g}(\mu - A) - \left(A - \varepsilon(\mu)\right)^{-1} \exp\{-A^2/2\}\bar{g}(\mu). \tag{17}$$

It follows from (16) that

$$\bar{g}(\mu - A) \sim \bar{g}(\mu).$$

Hence (17) implies $\bar{g} <\sim \bar{g}$.
The application of this lemma to g and $\bar{g}$ defined by (8) and (11) gives, in combination with (15),

$$g(\sigma\lambda) \sim \bar{g}(\sigma\lambda) \sim \chi(\log\lambda)\lambda^{\beta}, \quad \lambda \to \infty,$$

$$g(r/\sigma) \sim \chi(\log r - 2\log\sigma)\left(\frac{r}{\sigma^2}\right)^{\beta}, \cdot\ r \to \infty. \tag{18}$$

The desired equivalence (13) is clearly implied by (18) and the density representation (8).

Remark. For $\beta > 1$ the obtained expression for the tail probabilities (13) should be compared with the behaviour of the supremum tail probabilities of a usual stationary process $\tilde{\xi}$ with "moderate nondifferentiability": $\mathbf{E}|\tilde{\xi}(t) - \tilde{\xi}(s)|^2 \sim$ $\sim |t-s|^{2/\beta},\ t - 1 \to 0.$
It is well known that for $\sigma^2 = \mathbf{E}\tilde{\xi}^2(t)$

$$\mathbf{P}\Big\{\sup_{0\leqslant t\leqslant 1} |\tilde{\xi}(t)| \geqslant r\Big\} \sim cr^{\beta}\left(1 - \Phi(r/\sigma)\right), \quad r \to \infty. \tag{19}$$

The right-hand side of (19) is the same as in (13), except for the periodical factor χ.

POWER-TYPE DECREASING SPECTRUM

Let $B > 1,\ \sigma_1 > 0,\ \sigma_i = \sigma_1 i^{-B};\ \beta = B^{-1};$

$$I = \beta\sigma_1^{\beta} \int\limits_0^{\infty} z^{-1-\beta}\Big[\log\left(2\Phi(z)\right)\Big]\, dz.$$

THEOREM 2. *If $r \to \infty$ then*

$$\mathbf{P}\{\mathfrak{z} \geqslant r\} \sim \exp\Big\{I\sigma^{-2\beta}r^{\beta}\Big\}\left(1 - \Phi(r/\sigma)\right)/\sqrt{2}. \tag{20}$$

Proof. As in the proof of the previous theorem, we begin with the investigation of (12). In our case the neighbour points $\sigma_i\lambda$ are close enough and we can interpret the logarithm of (12) as a Riemann sum and give the integral approximation for it.

LEMMA 3. *The following equality is valid*

$$\lim_{\lambda\to\infty}\Big\{\sum_{i=1}^{\infty}\Big[\log\left(2\Phi(\sigma_i\lambda)\right)\Big] - \beta(\lambda\sigma_1)^{\beta}\int\limits_0^{\sigma_1\lambda} z^{-1-\beta}\Big[\log\left(2\Phi(z)\right)\Big]\, dz\Big\} = \frac{\log 2}{2}. \tag{21}$$

We omit the proof of this lemma which is purely arithmetical.
It is also easy to check that

$$\int_{\sigma_1\lambda}^{\infty} z^{-1-\beta}\Big[\log\big(2\Phi(z)\big)\Big]\,dz = \beta^{-1}(\sigma_1\lambda)^{-\beta}\log 2 + o\big(\exp(-\lambda)\big).$$

After the exponentiation of (21) we have

$$\prod_{i=1}^{\infty} 2\Phi(\sigma_i\lambda) \sim \exp\Big\{I\lambda^{\beta} - \log 2/2\Big\}.$$

Fortunately, $\beta < 1$ and condition (16) is satisfied. So we can apply Lemma 2 to obtain

$$g\Big(\frac{r}{\sigma}\Big) \sim \bar{g}\Big(\frac{r}{\sigma}\Big) \sim \exp\Big\{I(r/\sigma^2)^{\beta} - \log 2/2\Big\}, \quad r \to \infty,$$

$$\Pi(r) \sim \frac{1}{(2\pi)^{1/2}\sigma}\exp\Big\{I(r/\sigma^2)^{\beta} - r^2/z\sigma^2 - \log 2/2\Big\},$$

$$P(r) \sim \exp\Big\{I(r/\sigma^2)^{\beta}\Big\}\big(1 - \Phi(r/\sigma)\big)/\sqrt{2}.$$

Acknowledgement
I would like to thank Werner Linde who suggested the subject of this paper for discussion.

REFERENCES

Dobric, V., Marcus, M.B. and Weber, M. (1988). The distribution of large values of the supremum of a Gaussian process. *Asterisque.* No.157–158, 95–127.

Hoeffding, W. (1964). On a theorem of V.M. Zolotarev. *Theory Probab. Appl.* **9**, 89–92.

Linde, W. (1989). Gaussian measure of large balls in l^p. *Preprint,* Madrid.

Ylvisaker, N.D. (1965). The expected number of zeros of a stationary Gaussian process. *Ann. Math. Stat.* **36**, 1043–1046.

Zolotarev, V.M. (1961). Concerning a certain probability problem. *Theory Probab. Appl.* **6**, 201–204.

New Trends in Probab. and Statist., pp. 241 - 245
V. Sazonov and T. Shervashidze (Eds.)
1991 VSP/Mokslas

TOPOLOGIES ON THE SPACE OF σ-ADDITIVE CYLINDRICAL PROBABILITIES

D.H. MUSHTARI
Kazan State University, Chebotarev Institute of Mathematics and Mechanics, University str. 17, 420008 Kazan, Tataria, USSR

ABSTRACT

A generalization of the notion of weak convergence is introduced for probabilities in linear topological spaces with a compactology. For the case of a compactology consisting of all balls in the dual of a Banach space, an analog of the Prokhorov tightness criterion of weak compactness is obtained.

Let X, Y be real linear spaces in duality, $\mathfrak{C}(y_1, \ldots, y_n)$ be the smallest σ-algebra on X such that $y_1, \ldots, y_n$ are measurable, $\mathfrak{C}_Y(X)$ be the union of all σ-algebras $\mathfrak{C}(y_1, \ldots, y_n)$, where $\{y_1, \ldots, y_n\}$ runs over all finite subsets of Y. A cylindrical probability μ is a finitely additive set function on $\mathfrak{C}_Y(X)$ provided the restriction of μ on every σ-algebra $\mathfrak{C}(y_1, \ldots, y_n)$ is a probability measure. We use the notation

$$\mu^* A = \inf \{\mu(C)\colon\ C \supset A,\ C \in \mathfrak{C}_Y(X)\}.$$

It is well known that μ is regular (i.e., for every $C \in \mathfrak{C}_Y(X)$ and for any $\varepsilon > 0$ there exist a closed set F and an open set $\mathcal{G}$ from $\mathfrak{C}_Y(X)$ such that $F \subset C \subset \mathcal{G}$ and $\mu(\mathcal{G}) - \mu(F) < \varepsilon$). Sets from $\mathfrak{C}_Y(X)$ will be called cylinders.

Definition 1. By a compactology $\mathfrak{K}$ on X we mean a collection of $\sigma(X, Y)$-compact starlike (i.e., $tK \subset K$ for all $t \in [-1, 1]$) sets K in X such that

(K1) every finite dimensional compact starlike set in X belongs to $\mathfrak{K}$,

(K2) every $\sigma(X, Y)$-compact starlike subset of some element of $\mathfrak{K}$ belongs to $\mathfrak{K}$ too,

(K3) if $K_1, K_2 \in \mathfrak{K}$, then $K_1 \cup K_2 \in \mathfrak{K}$, $K_1 + K_2 \in \mathfrak{K}$,

(K4) if $K \in \mathfrak{K}$, $t \in \mathbf{R}$, then $tK \in \mathfrak{K}$.

Examples of compactologies: i) $\mathfrak{N}$ contains all compact starlike subsets of a Banach space X, $Y = X'$; ii) Y is a Banach space, $X = Y'$, $\mathfrak{C}$ contains all starlike $\sigma(Y', Y)$-compact subsets in Y'.

Definition 2. A family $\{\mu_\alpha\}$ of cylindrical probabilities is said to be uniformly $\mathfrak{K}$-tight if, for every $\varepsilon > 0$, there exists $K \in \mathfrak{K}$ such that

$$\mu_\alpha^*(K) > 1 - \varepsilon \quad \text{for every } \alpha. \tag{1}$$

If the relation (1) holds for one cylindrical probability μ, then μ is called $\mathfrak{K}$-tight. A $\mathfrak{K}$-tight cylindrical probability μ may be extended to a regular Borel probability measure $\bar{\mu}$ (Prokhorov, 1960).
Notation: $\mathfrak{P}_{\mathfrak{K}}(X)$ is the set of all $\mathfrak{K}$-tight cylindrical probabilities on $\mathfrak{C}_Y(X)$.
In the case of the compactology $\mathfrak{N}$ the Prokhorov theorem (Prokhorov, 1956) implies that a family $\{\mu_\alpha\}$ of $\mathfrak{N}$-tight probabilities is weakly compact iff it is uniformly $\mathfrak{N}$-tight. Our aim is to prove an analog of this theorem for the compactology $\mathfrak{E}$. The first step is to define an analog of the weak topology on the space $\mathfrak{P}_{\mathfrak{E}}(Y')$, associated with the compactology $\mathfrak{E}$.

Definition 3. $C_Y(X)$ is the space of all $\sigma(X, Y)$-continuous functions f on X which can be represented in the form

$$f(x) = \varphi\big(y_1(x), \dots, y_n(x)\big), \qquad x \in X, \quad y_i \in Y,$$

where φ is a continuous bounded function on $\mathbf{R}^n$. $C_{\mathfrak{K}}(X)$ is the closure of the set $C_Y(X)$ in the space of all bounded functions on X with the topology of uniform convergence on all sets K from $\mathfrak{K}$.

LEMMA. *$C_{\mathfrak{K}}(X)$ is the set of all functions f on X such that the restriction of f on every $K \in \mathfrak{K}$ is $\sigma(X, Y)$-continuous.*

Proof. Obviously, $f \in C_{\mathfrak{K}}(X)$ is continuous on $K \in \mathfrak{K}$. The converse statement follows from the Stone–Weierstrass theorem (Dunford and Schwartz, 1958). In fact, $C_Y(X)$ is an algebra separating the points of X. So, the closure of the restriction of $C_Y(X)$ on $K \in \mathfrak{K}$ is equal to the space of all $\sigma(X, Y)$-continuous functions on K.

Definition 4. The topology on the space $\mathfrak{P}_{\mathfrak{K}}(X)$ defined by the family of semimetrics

$$d_f(\mu, \nu) = \left| \int_X f(x)\, d\bar{\mu}(x) - \int_X f(x)\, d\bar{\nu}(x) \right|, \tag{2}$$

$\mu, \nu \in \mathfrak{P}_{\mathfrak{K}}(X)$, where f runs over the set $C_{\mathfrak{K}}(X)$, will be called $\mathfrak{K}$-weak topology.

Remark 1. The semimetrics d_f on $\mathfrak{P}_{\mathfrak{K}}(X)$ are well defined since the restriction of f on every compact K in $\mathfrak{K}$ is continuous, therefore it is Lebesgue measurable with respect to the measure $\bar{\mu}$.

Definition 5. The cylindrical topology on the space $\mathfrak{P}_Y(X)$ of all cylindrical probabilities is defined by the set of all semimetrics d_f, $f \in C_Y(X)$.
Notation: $\xrightarrow{\text{cyl}}$ is the convergence in the cylindrical topology, $\xrightarrow{\mathfrak{K}}$ is the convergence in the $\mathfrak{K}$-weak topology.

PROPOSITION 1. *Let a set $\{\mu_\alpha\}$ of $\mathfrak{K}$-tight cylindrical probabilities be uniformly $\mathfrak{K}$-tight and $\mu_\alpha \xrightarrow{\text{cyl}} \mu$. Then μ is $\mathfrak{K}$-tight too and $\mu_\alpha \xrightarrow{\mathfrak{K}} \mu$.*

Proof. Let K and $\varepsilon > 0$ satisfy (1) for $\{\mu_\alpha\}$. We will prove that $\mu^*(K) \geqslant 1-\varepsilon$. Let $C \supset K$. Then $K \subset F \subset C$, where $C \in \mathfrak{C}(y_1, \dots, y_n)$,

$$F = \{x\colon\ (y_1, \dots, y_n)(x) \in (y_1, \dots, y_n)(K)\}.$$

F is a closed set from $C_Y(X)$, so, by the Alexandrov theorem (Alexandrov, 1941)

$$\mu(C) \geqslant \mu(F) \geqslant \overline{\lim_{\alpha}}\, \mu_\alpha(F) \geqslant 1-\varepsilon.$$

Further, let $f \in C_{\mathfrak{K}}(X)$, $|f| < M$, $\varepsilon > 0$, $g \in C_Y(X)$ be such that $|f-g| < \varepsilon$ on K, $|g| < 2M$. Then $\mu_\alpha \xrightarrow{\text{cyl}} \mu$ implies

$$\int\limits_X g(x)\, d\mu_\alpha(x) \longrightarrow \int\limits_X g(x)\, d\mu(x).$$

We have

$$\int\limits_X |f-g|\, d\bar{\mu} \leqslant \int\limits_X |f-g|\, d\bar{\mu} + 3M\bar{\mu}(K^c) \leqslant \varepsilon + 3M\varepsilon.$$

The same inequality holds for all μ_α, so

$$\int\limits_X f(x)\, d\bar{\mu}_\alpha(x) \longrightarrow \int\limits_X f(x)\, d\bar{\mu}(x).$$

Remark 2. Let X be a Banach space, $Y = X'$, then the $\mathfrak{N}$-weak topology is equivalent to the weak topology $\mathfrak{P}_{\mathfrak{N}}(X)$. In fact, $f \in C_{\mathfrak{N}}(X)$ is continuous on every convergent sequence from X, so it is continuous.
We see, that the $\mathfrak{K}$-weak topology may be used as an analog of the weak topology for a compactology $\mathfrak{K}$.

THEOREM 1. *Let Y be a Banach space, $X = Y'$. Then a sequence (μ_n) of $\mathfrak{C}$-tight cylindrical probabilities on $\mathfrak{C}_Y(X)$ is $\mathfrak{C}$-weak convergent to μ iff the following conditions are fulfilled:*

a) $\mu_n \xrightarrow{\text{cyl}} \mu$,

b) *$\{\mu_n\}$ is uniformly $\mathfrak{C}$-tight.*

Proof. *'If'* follows from Proposition 1.
'Only if'. 1) Let us suppose the contrary: there exists $\varepsilon > 0$ such that, for every $n_1 \in \mathbf{N}$, a nonuniformly $\mathfrak{C}$-tight subsequence $(\mu_s^{(1)}) \subset (\mu_n)$ can be chosen such that, for all $s \in \mathbf{N}$,

$$(\mu_s^{(1)})^*(K_1) < 1-\varepsilon, \quad \text{where} \quad K_1 = \{x \in X\colon\ \|x\| \leqslant n_1\}. \tag{3}$$

2) Choose n_1 such that $\mu^*(K_1/2) > 1 - \varepsilon/8$. According to (3), there exist closed and open sets $F_1 \subset \mathcal{G}_1$ in $\mathfrak{C}_Y(X)$ disjoint with K_1 such that

$$\mu_1^{(1)}(F_1) > \varepsilon, \qquad \mu(\overline{\mathcal{G}}_1) < \varepsilon/8.$$

By the Urysohn theorem, there exists a function $f_1 \in C_Y(X)$ such that $f_1 = 0$ on $\mathcal{G}_1^c$, $f_1 = 1$ on F_1, and $0 \leqslant f_1 \leqslant 1$.
3) We describe only the next step of the construction by induction. Choose $K_2 \in \mathfrak{C}$ such that

$$\mu^*(K_2/2) > 1 - \varepsilon/16$$

and a subsequence $(\mu_s^{(2)}) \subset (\mu_s^{(1)})$ such that:

i) the family $(\mu_s^{(2)})$ is not uniformly $\mathfrak{C}$-tight;

ii) $\mu_s^{(2)}(\mathcal{G}_1) < \varepsilon/8$ for all $s \in \mathbf{N}$;

iii) $(\mu_s^{(2)})^*(K_2) < 1 - \varepsilon, \quad s \in \mathbf{N}$.

We have $(\mu_s^{(2)})^*(K_2 \cup \mathcal{G}_1) < 1 - \varepsilon + \varepsilon/8$ for all s. Therefore, there exist open and closed cylinders $\mathcal{G}_2 \supset F_2$ disjoint with $K_2 \cup \mathcal{G}_1$ such that $\mu_2^{(2)}(F_2) > \varepsilon - \varepsilon/8$. Further we construct a function f_2 as in 2) above and so on.
4) Thus, we have

i) the subsequences $(\mu_s^{(n+1)}) \subset (\mu_s^{(n)})$ convergent to μ in the $\mathfrak{C}$-weak topology,

ii) the cylinders $\mathcal{G}_n \supset F_n$ such that

$$\mu(\overline{\mathcal{G}}_n) < \varepsilon/2^{n+2}, \qquad \mu_n^{(n)}(F_n) > \varepsilon - \varepsilon/8 - \varepsilon/16 - \ldots,$$

$$\mathcal{G}_n \cap \mathcal{G}_m = \varnothing \text{ for } n \neq m, \qquad \mathcal{G}_n \cap K_m = \varnothing \text{ for } n > m,$$

iii) the functions $f_n \in C_Y(X)$ with values in [0, 1] equal to 0 on $\mathcal{G}_n^c$ and to 1 on F_n.

We denote $f = \sum_n f_n$. So, we have $f \leqslant 1$,

$$\int_X f(x)\, d\bar{\mu}_n^{(n)}(x) \geqslant \varepsilon - \varepsilon/8 - \varepsilon/16 - \ldots \geqslant 3\varepsilon/4,$$

and

$$\int_X f(x)\, d\bar{\mu}(x) \leqslant \varepsilon/8 + \varepsilon/16 + \ldots \leqslant \varepsilon/4.$$

On every compact $K \in \mathfrak{C}$ there is only a finite number of functions f_n not equal to 0, so $f \in C_{\mathfrak{C}}(X)$. Nevertheless,

$$\int_X f(x)\, d\bar{\mu}_n^{(n)}(x) \not\to \int_X f(x)\, d\bar{\mu}(x).$$

Remark 3. The $\mathfrak{C}$-weak convergence implies the convergence of the integrals of bounded functions whose restrictions on every weak compact in Y' are continuous in the weak topology. An example of such function is $f(x) = \min\{1, \sup x(y_n)\}$, (y_n) is convergent in Y.

Problem

Is an analog of the Prokhorov theorem true for the compactology of all weak compacts in a nonreflexive Banach space X?

REFERENCES

Alexandrov, A.D. (1940). Additive set functions on abstract spaces. *Soviet Math. Sbornik* 8, 307–348.

Dunford, N. and Schwartz, J.T. (1958). *Linear Operators. I: General Theory.* Wiley, New-York

Prokhorov, Ju. V. (1956). Convergence of random processes and limit theorems in probability theory. *Theory Probab. Appl.* 1, 177–238.

Prokhorov, Ju. V. (1960). The method of characteristic functionals. In: *Proceedings of the Fourth Berkeley Symposium on Mathematical Statistics and Probability.* 2. Univ. California Press, Los Angeles–Berkeley, pp. 403–419.

New Trends in Probab. and Statist., pp. 246 - 254
V. Sazonov and T. Shervashidze (Eds.)
1991 VSP/Mokslas

TWO-SIDED EXPONENTIAL INEQUALITIES FOR THE DISTRIBUTION OF THE NORM OF BANACH SPACE VALUED RANDOM VARIABLES

E.I. OSTROVSKII
Obninsk Institute of Atomic Power Engineering, 249020 Obninsk, USSR

ABSTRACT

We derive almost unimprovable two-sided estimates of exponential type for the distribution of the norm of Banach space valued random variables. The results are obtained by means of introduction of some spaces of random variables.

INTRODUCTION

Let ξ be a centered random vector in a linear space B_0 and B be a linear Banach subspace of B_0. Let $|x|$ be the norm in B. The methods of this paper permit to deduce:

a) sufficient conditions for the equality

$$\mathbf{P}(\xi \in B) = 1;$$

b) estimations of the following type:

$$u \geqslant u_0 \implies \mathbf{P}(|\xi| > u) \leqslant C_1 u^{C_2} \exp(-C_3 u^{C_4});$$

$$u \geqslant u_0 \implies \mathbf{P}(|\xi| > u) \geqslant C_5 u^{C_6} \exp(-C_3 u^{C_4});$$

here C_i, $C_i \in (0, \infty)$, denote constants depending on u_0;

c) conditions for convergence of the distributions of ξ_n in the Prokhorov topology of B.

NOTATION AND DEFINITIONS

On the set of all centered random variables defined on a fixed probability space define

$$\|\xi\|_\varphi = \sup_{\lambda \neq 0} \varphi^{-1}(\log \mathbf{E} \exp \lambda\, \xi)/|\lambda|.$$

Here $\varphi(\lambda)$ is a convex function increasing when $\lambda > 0$ and such that $\varphi(0) = 0$, $\varphi(\lambda) \sim C\lambda^2$ as $\lambda \to 0$. On the set

$$B_\varphi = \{\xi\colon\ \mathbf{E}\xi = 0,\ \|\xi\| < \infty\}$$

the functional $\xi \to \|\xi\|_\varphi$ is a norm and B_φ is a Banach space (Ostrovskii, 1982); a r.v. ξ belongs to B_φ if and only if $\mathbf{E}\xi = 0$ and, for some constant $C_7 > 0$,

$$\mathbf{P}(|\xi| > x) \leqslant 2\exp\big(-\varphi^*(C_7 x)\big).$$

Here $\varphi^*(x)$ denotes the Young–Fenchel transformation

$$\varphi^*(x) = \sup_{y>0}\big(xy - \varphi(y)\big).$$

An analogous transformation $\varphi_*(x)$ is defined by the equality

$$\varphi_*(x) = \inf_{y>0}\big(xy + \varphi(y)\big).$$

If the function $\lambda \to \varphi(\sqrt{|\lambda|})$ is convex and $\xi_i \in B_\varphi$ are independent random variables then

$$\Big\|\sum_{i=1}^{n}\xi_i\Big\|_\varphi^2 \leqslant \Big\|\sum_{i=1}^{n}\xi_i\Big\|_\varphi^2.$$

Let now $\xi(t)$ be a random field, which is defined on an arbitrary set T, and is such that $\xi(t) \in B_\varphi$ uniformly in t:

$$\sup_{t\in T}\|\,\xi(t)\,\|_\varphi = 1.$$

(The case $\sup\|\xi(t)\| \in (0,\infty)$ can be similarly treated by scaling). The function $\varphi = \varphi(\lambda)$ can be defined, for example, by the equality

$$\varphi(\lambda) = \sup_{t\in T}\log\mathbf{E}\exp\lambda\,\xi(t).$$

The function $\varphi(\lambda)$ is not trivial if and only if $\xi(t)$ satisfies Kramer's condition uniformly in $t \in T$.

Denote by $H(\varepsilon)$ the Kolmogorov ε-entropy of the space T with respect to the distance

$$d(t,s) = \|\xi(t) - \xi(s)\|_\varphi.$$

RESULTS

THEOREM 1. *If the series*

$$\sum_{n=0}^{\infty} 2^{-n}\varphi_*(H(2^{-n}))$$

converges then a d-separable version of $\xi(t)$ belongs to the space $\mathbf{C}(T,d)$ with probability 1.

For example, if $\varphi(\lambda) \leqslant C\lambda^p$, $\lambda \geqslant 1$, $p > 1$, then the condition of Theorem 1 is equivalent to

$$\sum_{n=1}^{\infty} 2^{-n} H^{1/q}(2^{-n}) < \infty, \quad p^{-1} + q^{-1} = 1.$$

In the case of Gaussian $\xi(t)$ we have $p = q = 2$ and then the condition of Theorem 1 coincides with the well-known Dudley condition.

THEOREM 2. *If the condition of Theorem 1 is satisfied then, for any $\varepsilon \in (0,1)$ and all $x > 1$,*

$$\mathbf{P}\Big(\sup_{t\in T} |\xi(t)| > x\Big) \leqslant C_8(\varepsilon) \exp\Big(-\varphi^*\big(x(1-\varepsilon)\big)\Big).$$

THEOREM 3. *If the following series converges*

$$\sum_{n=1}^{\infty} 2^{-n} H(2^{-n}),$$

then, for all $u > 0$,

$$\log \mathbf{P}\Big(\sup_{t\in T} |\xi(t)| > u\Big) \leqslant \inf_{p\in(0,1)} \Big((1-p)\sum_{n=1}^{\infty} p^{n-1} H(p^n) - \varphi^*\big(u(1-p)\big)\Big).$$

Here p may depend on u.

For example, let $\varphi(x) = x^p/p$, $x > 1$, $p > 1$, and let us assume that $H(\varepsilon) \leqslant$ $\leqslant H_0 + \varkappa|\log \varepsilon|$. Then Theorem 3 reduces to

$$u \geqslant C_9 \varkappa^{1/q} \Longrightarrow \mathbf{P}\big(\sup|\xi(t)| > u\big) \leqslant C_{11} C_{10}^{\varkappa} \varkappa^{-\varkappa} u^{q\varkappa} \exp\big(H_0 - u^q/q\big),$$

where C_9, C_{10}, C_{11} are absolute constants. We shall apply this result to estimate the norm of the sum of independent centered random vectors.

THEOREM 4. *Let ξ_i be independent identically distributed centered random vectors in the space $\mathbf{R}^k$, $u > 10$, $p > 1$, $|y|$ be an arbitrary norm in $\mathbf{R}^k$ and assume that*

$$\mathbf{P}(|\xi_i| > u) \leqslant \exp(-u^p).$$

Then

$$\sup_{n\geqslant 1} \mathbf{P}\Big(n^{-1/2}\Big|\sum_{i=1}^{n} \xi_i\Big| > u\Big) \leqslant C_{12} u^r \exp\big(-C_{13} u^{\min(p,2)}\big).$$

Here r is the dimension of the set of angular points of the unit ball dual to the ball $\{y,\ |y| \leqslant 1\}$, C_{11}, C_{12} are constants depending only on k, $|\cdot|$.

The inequality of Theorem 4 is exact as regards the exponential term $\min(p,2)$ and C_{13}.

Our methods can be also applied in the case of summation of weakly dependent vectors. Let, for example, ξ_i be a strongly stationary sequence of random vectors in the space $\mathbf{R}^k$, $|\xi_i| \leqslant 1$, $\mathbf{E}\xi_i = 0$.
Let us assume that the strong mixing coefficient $\alpha(k)$ satisfies the following condition:

$$\sum_{k=1}^{\infty} \alpha(k) < \infty.$$

Then

$$\sup_{n\geqslant 1} \mathbf{P}\left(n^{-1/2}\left|\sum_{i=1}^{n} \xi_i\right| > u\right) \leqslant \exp\left(-C(\alpha, k)u^{-4/3}\right).$$

The following result gives a lower bound for $\mathbf{P}(|\xi| > u)$.

THEOREM 5. *Let us assume that the following conditions are satisfied:*

1) $\varphi(\lambda) = \lambda^p/p$, $\lambda \geqslant 1$, $p > 1$;

2) $\|\xi(t)\|_\varphi \leqslant 1$, $t \in T$;

3) *the space T is locally homogeneous, with respect to the distance $d(t,s) = \|\xi(t) - \xi(s)\|$,*

$$\sup_t H\big(K(t,\delta),\varepsilon\big) \leqslant C_{14} + \varkappa|\log(\delta/\varepsilon)|,$$

$$K(t,\delta) = \{s\colon\ s \in T,\ d(t,s) \leqslant \delta\};$$

4) $H(\varepsilon) \geqslant C_{15} + \varkappa|\log\varepsilon|$;

5) *for all $t, s \in T$ and any $m \in [1,\infty)$*

$$\|\xi(t) + \xi(s)\| \leqslant \|\xi(t)\| + \|\xi(s)\| - C_{16}\|\xi(t) - \xi(s)\|^m;$$

6) $\exists\Delta \geqslant 0$, $\forall u \geqslant u_0$,

$$\mathbf{P}\big(|\,\xi(t)| > u\big) \geqslant C_{17}\, u^{-\Delta} \exp\left(-u^q/q\right).$$

Then, if $\Delta = 0$, there exists $C_{18}(u_0,\ p,\ \varkappa)$ such that, for all $u \geqslant C_{18}$,

$$\mathbf{P}\big(\sup_t |\xi(t)| > u\big) \geqslant C_{19} u^{\varkappa q/m} \exp\left(-u^q/q\right).$$

If $\Delta > 0$, then for $u \geqslant C_{20}(u_0, \Delta, p, \varkappa)$,

$$\mathbf{P}\big(\sup|\xi(t)| > u\big) \geqslant C_{21}(\log^{-\varkappa/m} u)u^{(\varkappa q/m)-\Delta} \exp\left(-u^q/q\right).$$

The formulated result shows that the estimate of Theorem 3 is exact. As a possible application, we mention the following statistical result.

THEOREM 6. *Let $\mathcal{F}_n(x)$, $x \in \mathbf{R}^k$, be the empirical distribution function (d.f.) for independent identically distributed random vectors ξ_i, $i = 1, 2, \ldots, n$, having d.f. $\mathcal{F}(x)$; let us denote*

$$P_k(u) = \sup_{\mathcal{F}} \sup_{n} \mathbf{P}\Big(\sqrt{n} \sup_{x} \big(\mathcal{F}_n(x) - \mathcal{F}(x)\big)' > u\Big).$$

Then there exist constants A_k, B_k depending on k such that, for any $u \geqslant 0$,

$$P_k(u) \leqslant A_k \exp\left(-2u^2 + B_k u\right).$$

The author is grateful to E.I. Gaivoronsky for his help in computer calculation and including in particular imitation modelling.
For other applications we refer to (Matuchin and Ostrovsky, 1990).

PROOFS

We start from Theorem 1. The continuity of $\xi(t)$ in d-metric follows from (Dmitrov sky et al., 1983) if one notes that, for $x > 0$,

$$\mathbf{P}\Big(|\xi(t) - \xi(s)|/d(t,s) > x\Big) \leqslant 2\exp\big(-\varphi^*(x)\big).$$

Now let us estimate the distribution on $\sup \xi(t)$. Let $S(p^n)$ be a minimal p^n-net in T in terms of the distance $d(t,s)$; then $\log|S(p^n)| \leqslant H(p^n)$, where $|S|$ denotes the number of elements in S. If

$$\eta_n = \sup_{t \in S_n} \big(\xi(t) - \xi(\theta_{n-1} t)\big), \quad n \geqslant 1,$$

where $\theta_{n-1} t \in S_n$, $d(t, \theta_{n-1} t) \leqslant p^{n-1}$, then

$$\mathbf{E}\exp \lambda\,\eta_n \leqslant \sum_{t \in S_n} \mathbf{E}\exp \lambda\Big(\xi(t) - \xi\big(\theta_{n-1}(t)\big)\Big) \leqslant$$

$$\leqslant \sum_{t \in S_n} \exp\left(\varphi(\lambda p^{n-1})\right) \leqslant \exp\left(H(p^n) + \varphi(\lambda p^{n-1})\right);$$

If $n = 0$ then

$$\eta_0 = \sup_{t \in S_0} \xi(t), \qquad \mathbf{E}\exp \lambda\eta_0 \leqslant \exp \varphi(\lambda),$$

since it is possible to choose any point t_0^* as S_{-1} and $\xi(t_0^*) = 0$. Because

$$\sup \xi(t) = \lim_{n\to\infty} \sup_{t \in S_n} \xi(t) =$$

$$= \lim_{n\to\infty} \sum_{k=1}^{n} \sup_{t \in S_k} \big(\xi(t) - \xi(\theta_{k-1} t)\big) \leqslant \sum_{n=0}^{\infty} \eta_n,$$

by the Hölder inequality,

$$\log \mathbf{E} \exp \lambda \sup_{t\in T} \xi(t) \leqslant \sum_{n=0}^{\infty} \log \left(\mathbf{E} \exp (\lambda\, \eta_n r_n)\right)^{1/r_n} \leqslant$$

$$\leqslant \sum_{n=0}^{\infty} \left(H(p^n) r_n^{-1} + r_n^{-1} \varphi(\lambda\, r_n p^{n-1})\right).$$

Here $r_n > 1$ and one can assume that $\sum r_n^{-1} \geqslant 1$. The minimization of the right-hand side in $\{r_n\}$ proves Theorem 2. If $\sum 2^{-n} H(2^{-n}) < \infty$ then, for any $p \in (0,1)$,

$$\sum p^n H(p^n) < \infty;$$

and in the above estimate one can choose $r_n^{-1} = p^n(1-p)$ to obtain

$$\log \mathbf{E} \exp \lambda \sup \xi(t) \leqslant \sum_{n=0}^{\infty} H(p^n) p^n (1-p) + \varphi\big(\lambda/(1-p)\big).$$

According to the Chebyschev inequality,

$$\log \mathbf{P}\big(\sup \xi(t) > u\big) \leqslant -\lambda u + \varphi\big(\lambda/(1-p)\big) + (1-p) \sum_{0}^{\infty} p^n H(p^n).$$

Substituting $\lambda = u$ and minimizing in p, we prove Theorem 3 which we use to prove Theorem 4. Namely, in our case $|\xi| = \sup_{t\in T}(\xi, t)$, T is the unit ball of the dual space. Then

$$n^{-1/2} \left| \sum_{i=1}^{n} \xi_i \right| = \sup_{t\in T} (S_n, t); \quad S_n = n^{-1/2} \sum_{m=1}^{n} \xi_m.$$

The condition of Theorem 4 is equivalent to $\|(\xi_i, t)\|_q < C < \infty$, where the norm $\|(\xi_i, t)\|_q$ is defined by B_φ norm with $\varphi(\lambda) = \lambda^q/q$, $\lambda \geqslant 1$, $p^{-1} + q^{-1} = 1$. If $q \leqslant 2$, we have $\|(S_n, t)\|_q < \widetilde{C} < \infty$, and if $q > 2$ (Kozachenko and Ostrovskii, 1985) then $\|(S_n, t)\|_2 \leqslant \widetilde{\widetilde{C}}$. Similarly,

$$\max_n \|(S_n, t) - (S_n, s)\|_{\max(q,2)} \leqslant C|t-s|.$$

We then apply Theorem 4 to the random field (S_n, t).
A more special result is contained in (Ostrovskii, 1989).
Let us prove Theorem 5. Let $T(\varepsilon)$ be ε-separable subset of T (i.e., $\forall t_i$, $t_j \in T(\varepsilon)$, $i \neq j \Rightarrow d(t_i, t_j) > \varepsilon$) having maximal number of elements.
According to (Vitushkin, 1959, p. 26), $L(\varepsilon) = |T(\varepsilon)| \geqslant H(2\varepsilon)$. We have

$$\mathbf{P}\Big(\sup_T \xi(t) > u\Big) \geqslant \mathbf{P}\big(\sup_{t\in T(\varepsilon)} \xi(t) > u\big) =$$

$$= \mathbf{P}\Big(\bigcup_{t_i \in T(\varepsilon)} \{\xi(t_i) > u\} \Big) \geqslant P_1 - P_2,$$

where

$$P_1 = \sum_{i:\ t_i \in T(\varepsilon)} \mathbf{P}\big(\xi(t_i) > u\big),$$

$$P_2 = \sum_{i \neq j} \sum \mathbf{P}\big(\xi(t_i) > u, \quad \xi(t_j) > u\big).$$

Let $\psi(u) = \inf_t \mathbf{P}\big(\xi(t) > u\big)$, then

$$P_1 \geqslant L(\varepsilon)\,\psi(u) \geqslant H(2\varepsilon)\,\psi(u).$$

It is necessary to estimate the sum P_2 from above

$$\mathbf{P}\big(\xi(t_i) > u,\ \xi(t_j) > u\big) \leqslant \mathbf{P}\big(\xi(t_i) + \xi(t_j) > 2u\big) \leqslant$$

$$\leqslant \exp\Big(-\varphi^*\Big(2u/\big(\|\xi(t_i) + \xi(t_j)\|\big)\Big)\Big).$$

By condition 3,

$$\|\xi(t_i) + \xi(t_j)\| \leqslant 2 - C_1 d^m(t_i, t_j).$$

Therefore,

$$\mathbf{P}\big(\xi(t_i) > u,\ \xi(t_j) > u\big) \leqslant$$

$$\leqslant \exp\Big(-\varphi^*\Big(2u/\big(2 - C_1 d^m(t_i, t_j)\big)\Big)\Big).$$

We represent the sum P_2 as

$$P_2 = \sum_k \sum{}^* \mathbf{P}\big(\xi(t_i) > u,\ \xi(t_j) > u\big) \leqslant$$

$$\leqslant \sum_k \sum_{i,j} \exp\Big(-\varphi^*\Big(u/\big(1 - C_2 d^m(t_i, t_j)\big)\Big)\Big) \leqslant Q_0 + Q_1,$$

where Q_0 is the sum over $k \leqslant k_0$, Q_1 is the sum over $k > k_0$ and the summation $\sum^*$ is over indices i, j such that

$$d(t_i, t_j) \in \big[k\varepsilon, (k+1)\varepsilon\big).$$

Here $k_0 + 1$ is the least integer larger then $(2C_2)^{1/q}\varepsilon^{-1}$. The sum Q_1 is estimated as

$$|Q_1| \leqslant |L(\varepsilon)|^2 \exp\big(-\varphi^*(2u)\big).$$

As to the sum Q_0, we note that when $k \leqslant k_0$

$$u/\big(1 - C_2 d^m(t_i, t_j)\big) \geqslant u\big(1 + C_3 d^m(t_i, t_j)\big) \geqslant$$

$$\geqslant u(1 + C_3 \varepsilon^m k^m).$$

Hence,

$$Q_0 \leqslant \sum_{k=1}^{k_0} \sum_{i,j} \exp\Big(- \varphi^*(u(1 + C_3 \varepsilon^m k^m))\Big).$$

Here the summation is over those indices i, j, for which $d(t_i, t_j) \leqslant (k+1)\varepsilon$, $d(t_i, t_j) \geqslant \varepsilon$.

For a fixed i the number of the points t_j, $t_j \in T(\varepsilon)$, for which $d(t_i, t_j) \geqslant \varepsilon$, and $d(t_i, t_j) \leqslant (k+1)\varepsilon$ does not exceed the entropy $H\big(k(t_i, (k+1)\varepsilon), 2\varepsilon\big)$ and, by condition 3), is less than $\exp(C + \rho \log(k+1))$, hence

$$Q_0 \leqslant C \sum_{k=1}^{k_0} \sum_{t_i \in T(\varepsilon)} L(\varepsilon)(k+1)^\rho \exp\Big(- \varphi^*(u(1 + C_3 \varepsilon^m k^m))\Big).$$

Choosing now $\varepsilon = \varepsilon(u)$ such that

$$\varphi^*\big(u(1 + \varepsilon^m C_3 k^m)\big) \geqslant \varphi^*(u) + C_5 k^\gamma, \ \gamma > 0,$$

we see that when C_5 is small enough,

$$Q_0 + Q_1 \leqslant 4^{-1} \exp\Big(H\big(\varepsilon(u)\big) - \varphi^*(u)\Big).$$

Hence,

$$P_2 \leqslant 2^{-1} P_1, \quad P_1 - P_2 \geqslant 2^{-1} P_1.$$

Replacing $\psi(u)$ by the function from the conditions of Theorem 5 and omitting transformations, we obtain the desired bound. The proof of the last statement, i.e., Theorem 6, is based on (Kiefer, 1962). The following important result improves Kiefer's Lemma 1.

LEMMA. *Let, for* $p \in (0, 2^{-1}]$,

$$\mathbf{P}(\mu = k) = \binom{n}{k} p^k (1-p)^{n-k}, \quad k = 0, 1, \ldots, n.$$

Then

$$\mathbf{P}(n^{1/2} |n^{-1}\mu - p| \geqslant x) \leqslant 2 \exp(-x^2 |\log p|).$$

Proof. Let us choose $\varphi(\lambda) = \lambda^2/2$; the norm $\|\xi\|_\varphi$, called subgaussian, has been studied in (Buldygin and Kozatchenko, 1980); it is defined by

$$\|\mu\| = \sup_{\lambda \neq 0} (2\lambda^{-2} \log \mathbf{E} \exp \lambda\mu)^{1/2}.$$

From the properties of the subgaussian norm it follows that it is sufficient to consider the case of $n = 1$. Define $\tau(p) = \|n^{-1/2}(\mu - np)\|$. We have

$$\tau^2(p) = \sup_{\lambda \neq 0} 2\lambda^2 \big[- \lambda p + \log\big(1 + p(\exp \lambda - 1)\big)\big] \leqslant 0,5 |\log^{-1} p|.$$

Therefore, according to the properties of B_φ-norm, we have

$$\mathbf{P}\big(n^{1/2}|n^{-1}\mu - p| > x\big) \leqslant 2\exp\big(-x^2/|\log^{-1} p|\big).$$

REFERENCES

Buldygin, V.V and Kozachenko, Yu.V. (1980). On subgaussian random variables. *Ukrainian Math. J.* **32**, 723–730.

Dmitrovsky, V.A., Ermakov, S.V. and Ostrovskii, E.I. (1983). The central limit theorem for weakly dependent Banach-valued variables. *Theory Probab. Appl.* **28**, 89–104.

Kiefer, J. (1961). On large deviation of the empiric a.f. of vector chance variables and a law of the iterated logarithm. *Pacif. J. Math.* **17**, 649–660.

Kozachenko, Yu.V. and Ostrovskii, E.I. (1985). Banach spaces of subgaussian type random variables. *Theory Probab. Math. Statist.* **32**, 42–53.

Matukhin, P.G. and Ostrovskii, E.I (1990). Nonparametrical estimation of density according to the multifactor tests. *Theory Probab. Appl.* **35**, 72–82.

Ostrovskii, E.I. (1982). Buldygin–Kozachenko norm generalization and CLT in Banach space. *Theory Probab. Appl.* **27**, 618.

Ostrovskii,E.I. (1989). Two-sided exponential estimations for sums of random vectors. *Leningrad Univ. Vestnik*, 108–109.

Vitushkin, A.G. (1959). *The Difficulty of the Task to Estimate Tabulation.* Nauka, Moscow (in Russian).

New Trends in Probab. and Statist., pp. 255 - 263
V Sazonov and T. Shervashidze (Eds.)
1991 VSP/Mokslas

NONUNIFORM ESTIMATES OF THE DENSITY OF THE SQUARED NORM OF A GAUSSIAN VECTOR IN HILBERT SPACE

V.V. ULYANOV
Moscow State University, Len. Gory, 119899 Moscow, USSR

ABSTRACT

A Hoeffding (Hoeffding, 1964) estimate of the density function of the square of the norm of a Gaussian random variable in a Hilbert space with zero mean is extended to the case of Gaussian vector with an arbitrary mean.

1. INTRODUCTION AND RESULTS

Let H be a separable Hilbert space with inner product $(\cdot,\cdot)$ and norm $|x| = (x,x)^{1/2}$, $x \in M$. We denote by Y a Gaussian H-valued random variable with zero mean and covariance operator V, i.e., the characteristic functional of Y has the form

$$\mathbf{E} \exp\{i(x,Y)\} = \exp\{-(Vx,x)/2\}.$$

Let $\sigma_1^2 = \ldots. = \sigma_k^2 \geqslant \sigma_{k+1}^2 \geqslant \ldots$ be the eigenvalues of V. Here and everywhere below k is the multiplicity of σ_1^2. Put

$$\sigma^2 = \mathbf{E}|Y|^2 = \sum_{i=1}^{\infty} \sigma_i^2.$$

We denote by $p(u,a)$, $u \geqslant 0$, $a \in H$, the density function of $|Y-a|^2$. The symbol c denotes constants, the same symbol in various formulas may stand for different constants.

In the case $a = 0$ the following result was obtained in (Hoeffding, 1964).

If $k \geqslant 2$ then

$$p(u,0) \leqslant M(k)\left(2\sigma_1^2\Gamma(k/2)\right)^{-1}\left(u/(2\sigma_1^2)\right)^{(k/2)-1}\exp\left(-u/(2\sigma_1^2)\right), \qquad (1)$$

where $\Gamma(\cdot)$ is the gamma function and

$$M(k) = \prod_{j=k+1}^{\infty}\left(1-\sigma_j^2/\sigma_1^2\right)^{-1/2}.$$

Remark 1. Hoeffding proved also that the estimate (1) is sharp in the sense that if we divide the left-hand side of (1) by the right-hand side then the ratio tends to 1 as $u \to \infty$.
The Hoeffding inequality may be extended to the general case $a \neq 0$.

THEOREM. a) *If* $k \geqslant 2$, *then, for any* $a \in H$,

$$p(u,a) \leqslant M(k)\left(2\sigma_1^2\Gamma(k/2)\right)^{-1}\left(u/\left(2\sigma_1^2\right)\right)^{(k/2)-1} \times$$

$$\times \exp\left(-\left(u^{1/2}-|a|\right)^2 / \left(2\sigma_1^2\right)\right), \tag{2}$$

where $M(k)$ *is the same as in* (1).
b) *If* $k = 1$ *then*

$$p(u,a) \leqslant M(2)\left(2\sigma_1\sigma_2\right)^{-1} \exp\left(-\left(u^{1/2}-|a|\right)^2 / \left(2\sigma_1^2\right)\right). \tag{3}$$

c) *For all* $k \geqslant 1$ *and any* λ: $\lambda^2 > \sigma_1^2$

$$p(u,a) \leqslant \prod_{j=3}^{\infty}\left(1-\sigma_j^2/\lambda^2\right)^{-1/2}\left(2\sigma_1\sigma_2\right)^{-1} \times$$

$$\times \exp\left(-\left(u^{1/2}-|a|\right)^2 / \left(2\lambda^2\right)\right). \tag{4}$$

COROLLARY 1. *For all* $k \geqslant 1$ *and any* $\varepsilon > 0$,

$$p(u,a) \leqslant (1+2/\varepsilon)^{1/2}\left(2\sigma_1\sigma_2\right)^{-1} \exp\left(-\left(u^{1/2}-|a|\right)^2 / \left((2+\varepsilon)\sigma^2\right)\right). \tag{5}$$

Proof. It suffices to take $\lambda^2 = (1+\varepsilon/2)\sigma^2$ in (4) and to note that

$$\prod_{j=3}^{\infty}\left(1-\sigma_j^2/\lambda^2\right)^{-1/2} \leqslant \left(1-\lambda^{-2}\sum_{j=3}^{\infty}\sigma_j^2\right)^{-1/2} \leqslant (1+2/\varepsilon)^{1/2}.$$

Remark 2. Corollary 1 gives a nonuniform version of the well-known estimate (see, e.g., (Sazonov, 1981, p.72))

$$p(u,a) \leqslant c\left(\sigma_1\sigma_2\right)^{-1}.$$

We say that the estimates (2)–(5) are nonuniform because their right-hand sides tend to 0 as $d = \left|u^{1/2}-|a|\right| \to \infty$. Geometrically d is equal to the distance between the boundary of ball $\left\{x \in M\colon |x-a| < u^{1/2}\right\}$ and the origin. Therefore, it is natural to use d as a characteristic of nonuniformity. We note also that in (5) we use the information about the first two eigenvalues of V only. For this reason the exponentional factor in (5) is larger than the corresponding term in (2) or (3).

Remark 3. It is easy to see that if $a = 0$, then (2) implies (1).

Remark 4. Let $p_1(u, a)$ be the density function of $|Y - a|$. Then

$$p_1(u, a) = 2up(u^2, a).$$

Hence from (2)–(5) we can get nonuniform estimates for $p_1(u, a)$ and then get the estimates of probabilities $\mathbf{P}(x < |Y - a| < x + \delta)$, $\delta > 0$, which are used in the proof of the central limit theorem in a Hilbert space (see, for example, (Paulauskas, 1977; Zalesskii, Sazonov and Ulyanov, 1988)). We denote

$$p_1(a) = \sup_{u \geqslant 0} p_1(u, a).$$

Now we recall some earlier results for $p_1(a)$ and $p_1(u, a)$. In (Paulauskas, 1977) the inequality

$$p_1(a) \leqslant c(\sigma_1\sigma_2)^{-1}(\sigma + |a|) \tag{6}$$

was proved. Moreover (see (Paulauskas and Račkauskas, 1989, Proposition 4.2.4)), $|a|$ in (6) cannot be replaced by $|a|^\lambda$ with $\lambda < 1$. Also it turns out (see (Chebotarev, 1979)) that it is impossible to obtain an upper estimate for $p_1(a)$ which depends only on the first. eigenvalue σ_1. However, (6) can be improved in some way (Paulauskas and Račkauskas 1989, Theorem 4.2.3.). In (Osipov and Rotar, 1984) the following estimate based on Lemma 1 from (Sazonov, 1981, p. 71) was proved

$$p_1(u, a) \leqslant cL(V)(\sigma_1\sigma_3)^{-1}\left((\sigma_1\sigma_2\sigma_3)^{1/3} + \min(|a|, u)\right), \tag{7}$$

where $L(V) = \prod_{j=3}^{\infty}\left(1 - \sigma_j^2/(2\sigma_3^2)\right)^{-1/2}$. It is important to point out that in (7) there is $\min(|a|, u)$ instead of $|a|$ in (6). Sazonov proved (see inequality (8) in (Sazonov, 1981, p. 73)) that if $u \geqslant 8|a|^2$, then

$$p(u, a) \leqslant cL(V)(\sigma_1\sigma_2)^{-1}\exp\left(-\left(8\sigma_1^2\right)^{-1}\left((u/2)^{1/2} - 2|a|\right)^2\right), \tag{8}$$

where $L(V)$ is the same as in (7).

Note that (8) is proved only for a special set of values $|a|$ and u. Moreover, the exponent factor is not the best one (cf. (2) and (3) above) while the information of all eigenvalues of V is used.

Later in (Ulyanov, 1984) (see also Corollary 13 and Lemma 6 in (Zalesskii, Sazonov and Ulyanov, 1988)) it was shown that, for any $\varepsilon > 0$,

$$p_1(u, a) \leqslant c\varepsilon^{-1/2}\psi(\sigma + \min(|a|, u))\exp\left(-(u - |a|)^2/\left((2 + \varepsilon)\sigma^2\right)\right), \tag{9}$$

where $\psi = (\psi_1\psi_2\psi_3)^{-2/3}$ and $\psi_j^4 = \sum_{i \geqslant j}\sigma_i^4$, $j \geqslant 1$.

Estimate (9) does not imply (6) because in (9) it has to be $\sigma_3 > 0$ at least. However in some cases the right-hand side in (9) is smaller than the one in (6). In fact if we take $\sigma_i^2 = n^{-1}, i = 1, 2, \ldots, n$, $n \geqslant 3$, and $\sigma_i = 0$ for $i > n$ then $\psi\sigma_1\sigma_2 \to 0$ as $n \to \infty$.

We can resume that since $p_1(u,a)$ depends on many parameters, perhaps it is impossible to construct an estimate which is better than any other. At the same time for nonuniform estimates in the central limit theorem in H it is useful to have nonuniform estimates of $p_1(u,a)$ which imply (6). Our Theorem implies such estimates. Indeed, since for any non-negative numbers b and f,

$$b/(1 + |b - f|) \leqslant 1 + \min(b, f)$$

and, for $g(x) = (1 + x)\exp(-x^2)$, we have for all x

$$g(x) \leqslant g\left(1 + \sqrt{3}\right)$$

for all x, Theorem implies

COROLLARY 2. *For any γ, ε $(\gamma > \varepsilon > 0)$,*

$$p_1(u,a) \leqslant \left(2 + \sqrt{3}\right)\exp\left(-\left(1 + \sqrt{3}\right)^2\right)(1 + 2/\varepsilon)^{1/2}(\sigma_1\sigma_2)^{-1} \times$$

$$\times \left(\sigma\left[(2 + \varepsilon)(2 + \gamma)/(\gamma - \varepsilon)\right]^{1/2} + \min(u, |a|)\right) \times$$

$$\times \exp\left(-(u - |a|)^2/((2 + \gamma)\sigma^2)\right).$$

COROLLARY 3. *If*

$$\left|u/\log^{1/2}(u + \sigma) - |a|/\log^{1/2}(|a| + \sigma)\right| \geqslant \sqrt{3}\sigma, \tag{10}$$

then

$$p_1(u,a) \leqslant (\sigma_1\sigma_2)^{-1}.$$

Proof. Without loss of generality we may assume that

$$u/\log^{1/2}(u + \sigma) \geqslant |a|/\log^{1/2}(|a| + \sigma) + \sqrt{3}\sigma.$$

Then it suffices to use Corollary 2 with $\gamma = 1$ and $\varepsilon = 1/2$.

Remark 5. From Corollary 3 we conclude that if u and $|a|$ are slightly different (see (10)), then $p_1(u,a)$ can be estimated only by $(\sigma_1\sigma_2)^{-1}$ which does not depend on u and $|a|$ (cf. (6) and remark after (6)).

Remark 6. Recently in (Lifshits, 1989) the two-sided estimates for $p_1(u,a)$ were announced illustrating the dependence between the density and the tail probability of norm distribution. However these results are not suitable for our purposes.

2. PROOF OF THEOREM

It is easy to see that for any non-negative numbers a, b, c, d, $a \geqslant b$, the inequality

$$\left((a-b)^{1/2}-c\right)^2+\left(b^{1/2}-d\right)^2 \geqslant \left(a^{1/2}-\left(c^2+d^2\right)^{1/2}\right)^2 \tag{11}$$

holds. Without loss of generality we can assume that V is an injective operator. Let $e_1, e_2 \ldots$ be an orthogonal system of eigenvectors of V with eigenvalues $\sigma_1^2, \sigma_2^2, \ldots$ respectively. Put $a_i = (a, e_i)$ and $Y_i = (Y, e_i)$, $i = 1, 2, \ldots$. Then Y_i, $i = 1, 2, \ldots$, are independent real Gaussian random variables with zero means and variances σ_i^2 respectively. We denote by $g(m,u)$, $m = 1, 2, \ldots$, (respectively, $g_i(u)$, $i = 1, 2, \ldots$) the density function of $\sum_{j=1}^{m}(Y_j - a_j)^2$)(respectively, $(Y_i - a_i)^2$). For all $i = 1, 2, \ldots$ we have

$$g_i(u) = (2\pi u)^{-1/2}\,\sigma_i^{-1} d_i(u), \tag{12}$$

where

$$d_i(u) = d\left(\sigma_i^2, u\right) =$$

$$= \left(\exp\left\{-\left(u^{1/2}-a_i\right)^2/\left(2\sigma_i^2\right)\right\}+\exp\left\{-\left(u^{1/2}+a_i\right)^2/\left(2\sigma_i^2\right)\right\}\right)/2. \tag{13}$$

At first we prove part a) of the Theorem, i.e. we consider the case $k \geqslant 2$. Note that for all $i = 1, 2, \ldots$

$$d_i(u) \leqslant \exp\left\{-\left(u^{1/2}-|a_i|\right)^2/\left(2\sigma_1^2\right)\right\}. \tag{14}$$

By induction, we can show that

$$g(k,u) \leqslant \left(2\sigma_1^2\Gamma(k/2)\right)^{-1}\left(u/\left(2\sigma_1^2\right)\right)^{(k/2)-1} \times$$

$$\times \exp\left\{-\left(u^{1/2}-\left(a_1^2+\ldots+a_k^2\right)^{1/2}\right)^2/\left(2\sigma_1^2\right)\right\}. \tag{15}$$

In fact, for $k = 2$ from (11), (12), (14), we have

$$g(2,u) = \int_0^u g_1(u-v)\,g_2(v)\,dv \leqslant$$

$$\leqslant \left(2\pi\sigma_1^2\right)^{-1}\int_0^u \exp\left\{-\left[\left((u-v)^{1/2}-|a_1|\right)^2+\right.\right.$$

$$+\left(v^{1/2}-|a_2|\right)^2\Big]/\left(2\sigma_1^2\right)\Big\}\left((u-v)v\right)^{-1/2}dv \leqslant$$

$$\leqslant \left(2\pi\sigma_1^2\right)^{-1} \exp\left\{-\left(u^{1/2}-\left(a_1^2+a_2^2\right)^{1/2}\right)^2/\left(2\sigma_1^2\right)\right\}\int_0^1 \left((1-z)z\right)^{-1/2}dz =$$

$$= \left(2\sigma_1^2\right)^{-1} \exp\left\{-\left(u^{1/2}-\left(a_1^2+a_2^2\right)^{1/2}\right)^2/\left(2\sigma_1^2\right)\right\}.$$

Thus (15) is proved for $k=2$. Now suppose $k \geqslant 3$. Using the inductive hypothesis, (11) and (14), we get

$$g(k,u)=\int_0^u g(k-1,u-v)\,g_k(v)\,dv \leqslant$$

$$\leqslant \left(2\sigma_1^2\Gamma((k-1)/2)\right)^{-1}\left(2\pi\sigma_1^2\right)^{-1/2}\left(2\sigma_1^2\right)^{1-(k-1)/2}\times$$

$$\times \exp\left\{-\left(u^{1/2}-\left(a_1^2+\ldots+a_k^2\right)^{1/2}\right)^2/\left(2\sigma_1^2\right)\right\}\int_0^u (u-v)^{-1+(k-1)/2}v^{-1/2}dv =$$

$$= \left(2\sigma_1^2\Gamma((k-1)/2)\right)^{-1}\pi^{-1/2}\left(2\sigma_1^2\right)^{1-(k/2)}u^{(k/2)-1}\times$$

$$\times \exp\left\{-\left(u^{1/2}-\left(a_1^2+\ldots+a_k^2\right)^{1/2}\right)^2/\left(2\sigma_1^2\right)\right\}\int_0^1 (1-x)^{-1+(k-1)/2}x^{-1/2}dx. \tag{16}$$

Since

$$\int_0^1 (1-x)^{-1+(k-1)/2}x^{-1/2}dx=\Gamma((k-1)/2)\Gamma(1/2)/\Gamma(k/2),$$

(16) implies (15) for any $k \geqslant 2$.

Now we estimate $g(m,u)$ for $m>k$. Recall that by k we denoted the multiplicity of σ_1^2. Since (see the definition of $d(\sigma_i^2,u)$ in (13))

$$d_{k+1}(u) \leqslant \exp\left\{-\left(u^{1/2}-|a_{k+1}|\right)^2/\left(2\sigma_1^2\right)\right\}d(\sigma_1^2\sigma_{k+1}^2/\left(\sigma_1^2-\sigma_{k+1}^2\right),u)$$

and

$$\left(2\pi u\right)^{-1/2}\left(\sigma_1^2-\sigma_{k+1}^2\right)^{1/2}\left(\sigma_1\sigma_{k+1}\right)^{-1}d\left(\sigma_1^2\sigma_{k+1}^2/\left(\sigma_1^2-\sigma_{k+1}^2\right),u\right)$$

is the density function of

$$\left(Y_{k+1}-a_{k+1}\right)^2\sigma_1/\left(\sigma_1^2-\sigma_{k+1}^2\right)^{1/2},$$

from (11), (12), (14) we obtain

$$g(k+1,u) = \int_0^u g(k,u-v)\,g_{k+1}(v)\,dv \leqslant$$

$$\leqslant \left(2\sigma_1^2\Gamma(k/2)\right)^{-1}\left(2\sigma_1^2\right)^{1-(k/2)}\left(2\pi\sigma_{k+1}^2\right)^{-1/2} \times$$

$$\times \exp\left\{-\left(u^{1/2}-\left(a_1^2+\ldots+a_{k+1}^2\right)^{1/2}\right)^2/\left(2\sigma_1^2\right)\right\} \times$$

$$\times \int_0^u (u-v)^{(k/2)-1}\,v^{-1/2}d\left(\sigma_1^2\sigma_{k+1}^2/\left(\sigma_1^2-\sigma_{k+1}^2\right),v\right)dv \leqslant$$

$$\leqslant \left(2\sigma_1^2\Gamma(k/2)\right)^{-1}\left(u/\left(2\sigma_1^2\right)\right)^{(k/2)-1}\left(1-\sigma_{k+1}^2/\sigma_1^2\right)^{-}1/2\times$$

$$\times \exp\left\{-\left(u^{1/2}-\left(a_1^2+\ldots+a_{k+1}^2\right)^{1/2}\right)^2/\left(2\sigma_1^2\right)\right\}$$

Reasoning by induction in the same way as in the proof of (15), we can show that, for any $m > k$,

$$g(m,u) \leqslant \left(2\sigma_1^2\Gamma(k/2)\right)^{-1}\left(u/\left(2\sigma_1^2\right)\right)^{(k/2)-1}\prod_{j=k+1}^{m}\left(1-\sigma_j^2/\sigma_1^2\right)^{-1/2} \times$$

$$\times \exp\left\{-\left(u^{1/2}-\left(a_1^2+\ldots+a_m^2\right)^{1/2}\right)^2/\left(2\sigma_1^2\right)\right\}. \qquad (17)$$

Now take an arbitrary $\varepsilon > 0$. Obviously, there exists $M_1 = M_1(\varepsilon)$ such that, for all $m \geqslant M_1$,

$$\sum_{i=m+1}^{\infty} a_i^2 < \varepsilon^2.$$

Therefore, since for any $m \geqslant 1$,

$$\sum_{i=m+1}^{\infty}(Y_i-a_i)^2 \leqslant 2\left(\sum_{i=m+1}^{\infty}Y_i^2+\sum_{i=m+1}^{\infty}a_i^2\right),$$

for $m \geqslant M_1$, we have

$$\mathbf{P}\left(x-\varepsilon < |Y-a|^2 < x+\varepsilon\right) \leqslant$$

$$\leqslant \mathbf{P}\left(x-\varepsilon-2\varepsilon^2-2\sum_{i=m+1}^{\infty}Y_i^2 < \sum_{i=1}^{m}(Y_i-a_i)^2 < x+\varepsilon\right) \leqslant$$

$$\leqslant \mathbf{P}\left(\sum_{i=m+1}^{\infty}Y_i^2 \geqslant \varepsilon^2\right)+\mathbf{P}\left(x-\varepsilon-4\varepsilon^2 < \sum_{i=1}^{m}(Y_i-a_i)^2 < x+\varepsilon\right). \qquad (18)$$

From (Zalesskii, Sazonov, and Ulyanov, 1988, Lemma 3) we get

$$\mathbf{P}\left(\sum_{i=m+1}^{\infty} Y_i^2 \geqslant \varepsilon^2\right) \leqslant 2\exp\left(-\varepsilon^2 \Big/ \left(2\mathbf{E}\sum_{i=m+1}^{\infty} Y_i^2\right)\right) =$$

$$= 2\exp\left(-\varepsilon^2 \Big/ \left(2\sum_{i=m+1}^{\infty} \sigma_i^2\right)\right).$$

There exists $M_2 = M_2(\varepsilon)$ such that, for all $m \geqslant M_2$,

$$\mathbf{P}\left(\sum_{i=m+1}^{\infty} Y_i^2 \geqslant \varepsilon^2\right) \leqslant \varepsilon^2.$$

Hence for any $m \geqslant M_1 + M_2$, inequality (18) implies

$$\mathbf{P}\big(x - \varepsilon < |Y - a|^2 < x + \varepsilon\big) \leqslant \varepsilon^2 + \left(2\varepsilon + 4\varepsilon^2\right) \sup_{y \in T(\varepsilon, x)} g(m, y), \tag{19}$$

where $T(\varepsilon, x) = \left\{y \in \mathbf{R}^1 \colon\ x - \varepsilon - 4\varepsilon^2 < y < x + \varepsilon\right\}$. Dividing the right-hand side of (19) by ε, from (17) we obtain (2) as $\varepsilon \to 0$.
To prove (3) we note that (11) yields

$$g(2, u) = \int_0^u g_1(u - v)\, g_2(v)\, dv \leqslant$$

$$\leqslant (2\pi\sigma_1\sigma_2)^{-1} \exp\left\{-\left(u^{1/2} - \left(a_1^2 + a_2^2\right)^{1/2}\right)^2 / \left(2\sigma_1^2\right)\right\} \int_0^u (u - v)^{-1/2} v^{-1/2} dv =$$

$$= (2\sigma_1\sigma_2)^{-1} \exp\left\{-\left(u^{1/2} - \left(a_1^2 + a_2^2\right)^{1/2}\right)^2 / \left(2\sigma_1^2\right)\right\}. \tag{20}$$

It remains to repeat the same arguments (after (16)) as in the proof of (2). Thus the inequality (3) is proved.
To prove (4) we note that, for any $\lambda^2 > \sigma_1^2$ and for all $i \geqslant 1$,

$$d_i(u) \leqslant \exp\left\{-\left(u^{1/2} - |a_i|\right)^2 / \left(2\lambda^2\right)\right\} d\left(\lambda^2\sigma_i^2 / \left(\lambda^2 - \sigma_i^2\right), u\right).$$

Therefore using (20) and reasoning in the same way as in the proof of (2) (after (16)), we obtain (4). The Theorem is proved.

Acknowledgement
This work was finished while the author was visiting the University of Bielefeld as a guest of the Sonderforschungsbereich 343.
The author is grateful to Prof. Dr. F.Götze for his kind hospitality.

REFERENCES

Chebotarev, V.I. (1979). An estimate of the rate of convergence in the central limit theorem in l_p. *Siberian Math. J.* **20**, 777–789.

Hoeffding, W. (1964). On a theorem of V.M. Zolotarev. *Theory Probab. Appl.* **9**, 89–91.

Lifshits, M.A. (1989). On the norm distribution of Gaussian and other stable vectors. In: *Abst. Commun. 5th Conf. on Probab. Theory and Math. Stat.* **1.** IMC, Vilnius, pp. 314–315.

Osipov, L.V. and Rotar, V.I. (1984). On an infinite dimensional central limit theorem. *Theory Probab. Appl.* **29**, 375–383.

Paulauskas, V. (1977). On the rate of convergence in the central limit theorem in certain Banach spaces. *Theory Probab. Appl.* **21**, 754–769.

Paulauskas, V. and Račkauskas A. (1989). Approximation theory in the central limit theorem. In: *Exact Results in Banach Spaces.* Kluwer Academic Publishers, Dordrecht, pp. 1–156.

Sazonov, V.V. (1981). Normal approximation. Some recent advances. *Lect. Notes Math.* **879**, 1–105.

Ulyanov, V.V. (1984). Estimates of the density of the norm of a Gaussian random variable in H. In: *Proc. of Conf. of Young Scientists of the Faculty of Computing Mathematics and Cybernetics.* Moscow Univ. (in Russian).

Zalesskii, B.A., Sazonov, V.V. and Ulyanov, V.V. (1988). Normal approximation in Hilbert space. *Theory Probab. Appl.* **33**, 207–227; 473–487.

New Trends in Probab. and Statist., pp. 264 - 275
V. Sazonov and T. Shervashidze (Eds.)
1991 VSP/Mokslas

ON SEQUENTIAL CONDITIONS FOR A CYLINDRICAL MEASURE TO BE COUNTABLY ADDITIVE

Yu.N. VLADIMIRSKII
Kostroma Pedagogical Institute, 156601 Kostroma, USSR

ABSTRACT

Criteria for a cylindrical measure to be countably additive are obtained. These criteria are formulated in terms of convergence a.s. or boundness a.s. of sequences of random variables.

INTRODUCTION

Let (E, t) be a locally convex space (l.c.s.), E' the dual of E, λ a cylindrical measure (c.m.) on $\langle E, E'\rangle$ (respectively, on $\langle E', E\rangle$), L: $E' \to S(\Omega, \Sigma, \mathbf{P})$ (respectively, L: $E \to S(\Omega, \Sigma, \mathbf{P})$) a linear random function corresponding to λ. The purpose of this paper is to state criteria to be countably additive or to indicate possibility for λ to be extended to a Radon measure in terms of properties of L.
In Section 2 we investigate the class of l.c.s. E that we call *cylindrically bornological*. If E belong to this class then, for any c.m. λ on $\langle E', E\rangle$, the condition that λ is σ-additive is equivalent to the condition $\sup |L(x_n)| < \infty$ a.s. for any sequence (x_n) in E such that $x_n \to 0$ in (E, t).
The results of Section 2 generalize and strengthen some results of (Muštari, 1970) and (Okazaki and Takahashi, 1987). In Section 3 we consider c.m. λ on $\langle E, E'\rangle$. Let E be a Frechet space or LF-space and, moreover, E is separable (more generally, E may be embedded in a product of a family of Banach WCG-spaces) or E is reflexive. We prove that under these assumptions λ may be extended to a Radon measure iff $L(x'_n) \to 0$ a.s., for any sequence (x'_n) in E', such that $x'_n \to 0$, for $\sigma(E', E)$.

1. DEFINITIONS

Regarding locally convex spaces (l.c.s.) we shall employ the notation and terminology adopted in (Schaefer, 1966) and (Grothendieck, 1973). All vector spaces considered are assumed to be real, all operators are assumed to be linear. Let E, F be a pair of vector spaces in duality $\langle E, F\rangle$.
The symbols $\sigma(E, F)$, $\beta(E, F)$, $\tau(E, F)$ will denote, respectively, the weak topol-

ogy, the strong topology and the Mackey topology on E corresponding to duality $\langle E, F\rangle$.

Following (Grothendieck, 1973), we shall say that a sequence (x_n) in l.c.s. (E,t) tends to 0 in the sense of Mackey, if there exists a bounded set $B \subset E$ such that $\forall \varepsilon > 0\ \exists n \in \mathbf{N}$: $x_k \in \varepsilon B\ \forall k \geqslant n$. An operator T from a topological linear space E into another one, F, is said to be bounded if T transforms every bounded subset of E into a bounded subset of F.

Following (Grothendieck, 1973), we shall understand the expression "the l.c.s. E is the inductive limit of a sequence of l.c.s. E_n" in the wide sense including countable direct sums and quotients. Following (Schaefer, 1966) we shall mean by "*LF-space*" a strict inductive limit of an increasing sequence of Frechet spaces. Regarding cylindrical measures we shall employ the notation and terminology adopted in (Schwartz, 1973) and (Vladimirskii, 1979; 1981). A Borel measure μ on a l.c.s. (E,t) is called a Radon measure if $\mu B = \sup\{\mu K\colon K \subset B, K \text{ is compact}\}$ for every Borel set $B \subset E$. Let E, F be a pair of vector spaces in duality $\langle E, F\rangle$ and λ be a cylindrical measure (c.m.) on the space E with the algebra of Borel cylinders generated by elements of F (briefly: λ be a c.m. on $\langle E, F\rangle$). Usually we shall suppose that $\lambda(E) = 1$. Let $(\Omega, \Sigma, \mathbf{P})$ be a probability space, $S(\Omega, \Sigma, \mathbf{P})$ be the space of all random variables on Ω equipped with the topology of convergence in probability. An operator $L\colon F \to S(\Omega, \Sigma, \mathbf{P})$ is called a linear random function (l.r.f.) corresponding to λ if for any $x \in F$ the random variable $L(x)$ has the distribution $x(\lambda)$. If $A \subset E$, then $\lambda^*(A)$ is defined by

$$\lambda^*(A) = \inf\{\lambda(B)\colon\ A \subset B,\ B \text{ is a borelian cylinder}\}.$$

Let (E,t) be a l.c.s. Then a c.m. λ on $\langle E', E\rangle$ is said to be $\mathcal{E}$-tight if $\forall \varepsilon > 0$ exists a neighbourhood of the origin $U \subset E$ such that $\lambda^*(U^0) > \lambda(E') - \varepsilon$; in this case the corresponding operator $L\colon E \to S(\Omega, \Sigma, \mu)$ is called an $\mathcal{E}$-operator.

Let E be a vector space, $A \subset E$. Following (Vladimirskii, 1979) we denote by E_∞ the space of all finite sequences (x_n) in E, $A_\infty = \{(x_n) \in E_\infty\colon x_n \in A\ \forall n\}$.

To every operator $T\colon E \to F$ naturally corresponds the operator $T_\infty\colon E_\infty \to F_\infty$.

Let (E,t) be a l.c.s. We denote by $(E,t)_\infty$ the space E_∞ equipped with the locally convex topology t_∞ for which the sets U_∞ form a fundamental system of neighbourhoods of 0, where U runs through a fundamental system of neighbourhoods of the origin in E.

We denote by $\widetilde{S}_\infty$ the space $(S(\Omega, \Sigma, \mathbf{P}))_\infty$ equipped with the linear topology for which the sets

$$V_\varepsilon = \left\{(X_n) \in S_\infty\colon\ \mathbf{P}(\max_n |X_n| < \varepsilon) > 1 - \varepsilon\right\}$$

(where $\varepsilon > 0$) form a fundamental system of neighbourhoods of 0.

2. C–BORNOLOGICAL SPACES

Let (E,t) be a l.c.s., λ a c.m. on $\langle E', E\rangle$, $L\colon E \to S(\Omega, \Sigma, \mathbf{P})$ the corresponding l.r.f.

Let us consider the following conditions:

(2.1) λ is $\mathcal{E}$-tight;

(2.2) λ is σ-additive;

(2.3) $L(x_n) \to 0$ a.s. for any sequence (x_n) in E such that $x_n \to 0$ in (E,t);

(2.4) $\sup|L(x_n)| < \infty$ a.s. for any sequence (x_n) in E such that $x_n \to 0$ in (E,t);

(2.5) $\sup\{|L(x)| : x \in B\} < \infty$ a.s. for any bounded subset $B \subset E$;

(2.6) $\sup|L(x_n)| < \infty$ a.s. for any sequence (x_n) in E such that x_n tends to 0 in the sense of Mackey.

PROPOSITION 1. *Let λ be any c.m. on $\langle E', E\rangle$. The following implications hold:* (2.1)$\Rightarrow$(2.2)$\Rightarrow$ (2.3)$\Rightarrow$(2.4)$\Leftrightarrow$(2.5)$\Leftrightarrow$ (2.6).

Proof. (2.1)$\Rightarrow$(2.2) follows from the Prokhorov theorem (see (Schwartz, 1973, p. 81)). The implications (2.2)$\Rightarrow$(2.3) $\Rightarrow$(2.4) and (2.5)$\Rightarrow$(2.4)$\Rightarrow$(2.6) are obvious. (2.6)$\Rightarrow$(2.5) follows from Lemma 1 in (Vladimirskii, 1981).

Definition 1. A l.c.s. (E,t) is said to be cylindrically-bornological (briefly: C-bornological) if (2.5)$\Rightarrow$(2.1) for any c.m. λ on $\langle E', E\rangle$.
We shall show that the class of C-bornological spaces is sufficiently large.

LEMMA 1. *Let λ be a c.m. on $\langle E', E\rangle$, L: $E \to S(\Omega)$ the corresponding l.r.f. Then:*

a) *λ satisfies* (2.1) *iff L_∞: $(E,t)_\infty \to \widetilde{S}_\infty$ is continuous;*

b) *λ satisfies* (2.3) *iff L_∞ is sequentially continuous;*

c) *λ satisfies* (2.5) *iff L_∞ is bounded.*

We omit the proof of this Lemma. Note that assertion **a)** is stated in Theorem 1 (Vladimirskii, 1979).

LEMMA 2. *Let $0 < \varepsilon < 1$ and let G be a vector subspace of $S(\Omega, \Sigma, \mathbf{P})$ such that $\mathbf{P}(X \neq 0) \leqslant \varepsilon$ $\forall X \in G$. Then there exists $\Omega_\varepsilon \in \Sigma$ such that $\mathbf{P}(\Omega_\varepsilon) \geqslant 1 - \varepsilon$ and $X = 0$ a.s. on Ω_ε $\forall X \in G$.*

Proof. Let $X \in S(\Omega, \Sigma, \mathbf{P})$, $\alpha \in (0,1)$. Let us denote $J_\alpha(X, d\mathbf{P}) = \min(C \geqslant 0\colon \mathbf{P}(|X| > C) \leqslant \alpha)$. It follows from the condition of the lemma that $J_\varepsilon(X, d\mathbf{P}) = 0$ $\forall X \in G$. Let us take any $\delta > \varepsilon$ and choose $\eta \in (0,1)$, $\beta \in (0,1)$ such that $\varepsilon + \eta \leqslant \delta\beta$. Let $p \in (0,1)$, $f_1(t)$, $f_2(t), \ldots$ be a sequence of independent random variables on a probability space $(T, \mathcal{A}, \nu)$ such that every $f_n(t)$ has the distribution γ_p with the characteristic function $\varphi(\tau) = \exp(-|\tau|^p)$. Let

$$X_i \in G, \quad a_i \geqslant 0 \quad (1 \leqslant i \leqslant n), \quad \sum_{i=1}^n a_i = 1.$$

Applying the Fubini inequality (Vladimirskii, 1981), we get

$$J_\delta\left\{\left(\sum_{i=1}^n a_i |X_i(\omega)|^p\right)^{1/p},\ d\mathbf{P}\right\} =$$

$$= J_\beta^{-1}(\gamma_p) J_\delta\left\{ J_\beta\left(\sum_{i=1}^n a_i{}^{1/p} f_i(t) X_i(\omega),\ d\nu(t)\right),\ d\mathbf{P}(\omega)\right\} \leqslant$$

$$\leqslant J_\beta^{-1}(\gamma_p) J_\eta \left\{ J_\varepsilon \left(\sum_{i=1}^{n} a_i^{1/p} f_i(t) X_i(\omega),\ d\mathbf{P}(\omega) \right),\ d\nu(t) \right\} = 0.$$

By Nikishin's theorem (see (Maurey, 1973, Theorem 1)), for any $\tilde{\delta} > \delta$, there exists $\Omega_{\tilde{\delta}} \in \Sigma$ with $\mathbf{P}(\Omega_{\tilde{\delta}}) \geqslant 1 - \tilde{\delta}$ such that

$$\int\limits_{\Omega_{\tilde{\delta}}} |X|^p d\mathbf{P} = 0 \quad \forall X \in G.$$

Since $\tilde{\delta}$ may be taken arbitrary close to ε, there exists $\Omega_\varepsilon \in \Sigma$ with $\mathbf{P}(\Omega_\varepsilon) \geqslant 1 - \varepsilon$ such that $X = 0$ a.s. on Ω_ε $\forall X \in G$.

THEOREM 1. *The class $\mathcal{CB}$ of all C–bornological spaces includes all metrizable l.c.s., LF–spaces, bornological DF–spaces, bornological strong duals of LF–spaces. The class $\mathcal{CB}$ is stable with respect to the following operations: countable inductive limit (in the wide sense), countable product, limit of the inverse spectrum*

$$E_1 \xleftarrow{\varphi_1} E_2 \xleftarrow{\varphi_2} E_3 \xleftarrow{\varphi_3} \ldots, \tag{1}$$

where all $E_n \in \mathcal{CB}$, φ_n are continuous operators covering all bounded subsets (that is for any bounded $A_n \subset E_n$ there exists bounded $A_{n+1} \subset E_{n+1}$ such that $\varphi_n(A_{n+1}) \supset A_n$).

Proof. If a l.c.s. E is metrizable, then E_∞ is also metrizable and (2.5)$\Rightarrow$(2.1) by Lemma 1. The stability of the class $\mathcal{CB}$ under countable inductive limits follows from Lemma 1 of this paper and Lemma 1 in (Vladimirskii, 1979).
Any bornological DF–space is C–bornological since it is an inductive limit of a sequence of normed spaces. Now let E be the limit of inverse spectrum (1), where every φ_n is continuous and φ_n covers all bounded subsets of E_n. Let us denote by π_n: $E \to E_n$ the natural projections. Suppose that an operator L: $E \to S(\Omega, \Sigma, \mathbf{P})$ satisfies (2.5). Let us take any $\varepsilon > 0$. Since L is bounded, there exists $n \in \mathbf{N}$ such that $\mathbf{P}(L(x) \neq 0) \leqslant \varepsilon \quad \forall x \in \operatorname{Ker} \pi_n$. It follows from Lemma 2 that there exists $\Omega_\varepsilon \in \Sigma$ with $\mathbf{P}(\Omega_\varepsilon) \geqslant 1 - \varepsilon$ such that there is the following commutative diagram

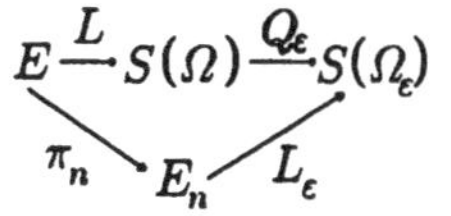

where Q_ε is the natural projection. Since π_n evidently covers bounded subsets of E, also, L_ε satisfies (2.5). Hence, L_ε is an $\mathcal{E}$–operator.
Thus L is an $\mathcal{E}$–operator because ε may be taken arbitrary small. So, $E \in \mathcal{CB}$. If E is the bornological strong dual of a LF–space then the proof is similar. The stability of the class $\mathcal{CB}$ under the countable products follows immediately from the preceding.

COROLLARY 1. *If a l.c.s. E is a reflexive DF–space or the strong dual of a reflexive LF–space, then E is C–bornological.*

Recall that a continuous operator T from a l.c.s. E into a l.c.s. F is called weakly compact (respectively, compact) if T transforms any bounded subset of E (respectively, some neighbourhood of 0 in E) into relatively weakly compact (respectively, relatively compact) subset of F. We omit the proof of the following proposition.

PROPOSITION 2. *Let E be a l.c.s. Suppose that E is a DF–space or that E is a quasi–barrel space and E'_β is C–bornological. Let λ be a c.m. on E, L: $E' \to S(\Omega)$, the corresponding l.r.f., $\tilde{\lambda}$ the corresponding c.m. on $\langle E'', E'\rangle$. The following conditions are equivalent:*

1) *$\tilde{\lambda}$ is $\mathcal{E}$–tight (with respect to the topology $\beta(E', E)$ on E');*
2) *$\tilde{\lambda}$ is σ–additive;*
3) *for any l.c.s. F and any weakly compact operator T: $E \to F$, the c.m. $T(\lambda)$ on F is σ–additive;*
4) *for any Frechet space F and any continuous operator T: $E \to F'_\beta$ the restriction of c.m. $T(\lambda)$ to $\langle F', F\rangle$ is σ–additive;*
5) *for any compact operator T: $E \to c_0$, the c.m. $\widetilde{T(\lambda)}$ on $\langle 1_\infty,\ 1_1\rangle$, corresponding to $T(\lambda)$, is σ–additive;*
6) $\sup|Lx'_n| < \infty$ a.s., *for any sequence (x'_n) in E' such that x'_n tends to 0 uniformly on some neighbourhood of the origin in E.*

COROLLARY 2. *For any infinite dimensional Banach space E there exists a compact non–γ–summing operator from E into c_0.*

Proof follows from Theorems 3 and 5 of (Linde and Pietsch, 1974) and Proposition 2 of this paper.

Remarks and Examples.

1) The implication (2.3)⇒(2.1) is known if E is a metrizable l.c.s. (see (Muštari, 1970, Theorem 7)) or E is a LF–space (see (Okazaki and Takahashi, 1987, Theorem 8)). It is proved in (Vershik and Sudakov, 1969) that if E is a separable Banach space, then (2.2) and (2.5) are equivalent. On the equivalency of (2.2), (2.3), (2.4) and (2.5) in case when **E** is normed see also (Vakhania et al., 1987, Theorem 2.1.2).
2) We shall show below that, in general, (2.4) is essentially weaker than (2.3), and (2.3) is essentially weaker than (2.2) (see Example 3 in Section 3). Let E be the space $S(0,1)$ of Lebesgue measurable functions on $(0,1)$ equipped with the finest locally convex topology. In (Vladimirskii, 1979) an example of the c.m. λ on $\langle E', E\rangle$ is given such that λ is σ–additive but not $\mathcal{E}$–tight. Thus E is bornological and reflexive but is not C–bornological.
3) If a l.c.s. E is C–bornological then $(E, \tau(E, E'))$ is bornological. Indeed, in this case every linear form on E, bounded on the bounded sets, is continuous.

4) In (Grothendieck, 1973, p. 167) an example of a semi-reflexive nonbornological DF-space is given. It is easily seen that this space is C-bornological. Now let Z be any infinite-dimensional reflexive Banach space and let E be the space Z equipped with the topology $c(Z, Z')$ of uniform convergence on compacts of Z'. It is easily seen that E is a C-bornological but is not a Mackey space. These examples give a negative answer to the question in (Vladimirskii, 1979): "Need every l.c.s. possessing the property (A) be a Mackey space?" (a l.c.s. E posses the property (A) if for any σ-additive c.m. λ on $\langle E', E\rangle$ its l.r.f. L: $E \to S(\Omega, \Sigma, P)$ must be continuous).

3. CONDITIONS OF EXTENDABILITY OF A CYLINDRICAL MEASURE TO A RADON MEASURE

THEOREM 2. *Let E be a Frechet space, λ a c.m. on $\langle E, E'\rangle$ and L: $E' \to S(\Omega, \Sigma, \mathbf{P})$ the corresponding l.r.f. The following conditions are equivalent:*

1) *λ may be extended to a Radon measure on E;*
2) *the operator L_∞: $(E', \tau(E', E))_\infty \longrightarrow \widetilde{S}_\infty$ is continuous;*
3) *for every neighbourhood of the origin $U \subset E$, the restriction of L_∞ to $(U^0)_\infty$ is continuous at 0 as the mapping from $((U^0)_\infty, (\tau(E', E))_\infty)$ into $\widetilde{S}_\infty$.*

Proof. The equivalence of 1) and 2) follows from the Prokhorov theorem, Lemma 1 and the fact that the weak and initial topologies on E are Radon-equivalent (see (Schwartz, 1973, p. 162)). Thus it is sufficient to prove that (3)⇒(2). Let $U_1 \supset U_2 \supset \dots$ be a fundamental system of neighbourhoods of the origin in E, and let $V_1 \supset V_2 \supset \dots$ be a fundamental system of neighbourhoods of 0 on $\widetilde{S}_\infty$ such that $V_{n+1} + V_{n+1} \subset V_n$ $(\forall n)$. Using condition (3), we choose a sequence of disked weakly compact sets $C_n \subset E$ such that

$$L_\infty((C_n^0 \cap U_n^0)_\infty) \subset 2^{1-n} V_n,$$

for every $n \in \mathbf{N}$. Let

$$Q_n = \bigcup_{m=1}^{\infty} \sum_{k=1}^{m} 2^{k-1} L\Big(C_{n+k}^0 \bigcap U_{n+k}^0\Big),$$

where $\sum\limits_{k=1}^{m} A_k$ denotes the set

$$\Big\{\sum_{k=1}^{m} a_k\colon\ a_k \in A_k \quad (\forall k)\Big\}.$$

It is obvious that the sets Q_n form a fundamental system of neighbourhoods of the origin for some locally convex topology t on ImL, and, moreover, $(Q_n)_\infty \subset$ $\subset 2^{-n} V_n$ $(\forall n)$. Let us decompose L as

$$\begin{array}{ccc} E' & \xrightarrow{\quad L \quad} & S(\Omega) \\ & {\scriptstyle T}\searrow \quad \nearrow{\scriptstyle j} & \\ & (\operatorname{Im} L, t) & \end{array}$$

where T acts as L and j is the identical embedding. Since $(Q_n)_\infty \subset 2^{-n}V_n \quad (\forall n)$, the operator j_∞: $(\mathrm{Im}L, t)_\infty \longrightarrow \widetilde{S}_\infty$ is continuous. If $m > n$, then

$$T(C_m^0 \cap U_n^0) \subset Q_{m-1}.$$

Hence, the restriction of T to every U_n^0 is continuous at 0 with respect to $\tau(E', E)$ and t. Since E is complete, by the Grothendieck theorem (see (Schaefer, 1966, Chapter IV, Theorem 6.2)), the operator T is continuous with respect to $\tau(E', E)$ and t. Hence, 2) is valid and this completes the proof.

Consider an interpretation of Theorem 2 in terms of measurable seminorms in the sense of (Dudley et al., 1971). Let E be a l.c.s., λ a c.m. on $\langle E, E'\rangle$, $p(x)$ a seminorm on E, $U = \{x \in E\colon p(x) < 1\}$. The seminorm p is called λ-measurable if for all $\varepsilon > 0$ there exists a finite $K \subset E$ such that

$$\lambda^*(\varepsilon U + K) > \lambda(E) - \varepsilon.$$

COROLLARY 3. *Let E be any l.c.s., λ a c.m. on $\langle E, E'\rangle$ and let p be a Mackey-continuous seminorm on E, $U = \{x \in E\colon p(x) < 1\}$. Then p is λ-measurable iff, for every $\varepsilon > 0$, there exists a weakly compact $K \subset E$ such that $\lambda^*(\varepsilon U + K) > \lambda(E) - \varepsilon$.*

Proof. It is enough to combine Theorem 2 from (Dudley et al., 1971) and Theorem 2 of this paper.

Let (E, t) be a l.c.s., λ a c.m. on $\langle E, E'\rangle$, L: $E' \to S(\Omega, \Sigma, \mathbf{P})$ the corresponding l.r.f. Consider the following conditions:

(3.1) λ may be extended to a Radon measure on (E, t);

(3.2) λ may be extended to a Radon measure on $(E, \sigma(E, E'))$;

(3.3) λ is σ-additive;

(3.4) $L(x'_n) \to 0$ a.s., for every sequence (x'_n) in E' such that $x'_n \to 0$ in $(E', \sigma(E', E))$;

(3.5) c.m. $T(\lambda)$ on c_0 is σ-additive for every continuous operator T: $E \to c_0$;

(3.6) every continuous seminorm on E is λ-measurable;

(3.7) $\sup |Lx'_n| < \infty$ a.s., for every sequence (x'_n) in E' such that $x'_n \to 0$ uniformly on some neighbourhood of the origin in E.

LEMMA 3. *For every l.c.s. E and every c.m. λ on E the following implications are fulfilled:*

$$\begin{array}{ccccccc} (3.1) & \Rightarrow & (3.2) & \Rightarrow & (3.3) & \Rightarrow & (3.4) \\ & & \Downarrow & & & & \Downarrow \\ & & (3.6) & & \Longrightarrow & & (3.5) \Rightarrow (3.7) \end{array}$$

Proof. Implications (3.1)⇒(3.2)⇒(3.3) ⇒(3.4)⇒(3.5)⇒(3.7) are obvious. Implications (3.2)⇒(3.6)⇒(3.5) follow from Theorem 2 in (Dudley et al., 1971) and the fact that the weak and initial topologies of a Banach space are Radon-equivalent.

Consider now the question when the above implications may be converted. We start with the following evident statement.

LEMMA 4. *If E is a barrel l.c.s. then* (3.5)⇒(3.4), *for every c.m. λ on E.*

THEOREM 3. *Let E be closed subspace of a Banach WCG-space and let λ be any c.m. on $\langle E, E'\rangle$. Then* (3.5)⇒(3.1).

Proof. It is sufficient to consider the case when E is a Banach WCG-space. According to Lemma 4 it is enough to show that (3.4)⇒(3.1). Let K be a disked weakly compact subset of E such that its linear envelope is dense in E. Let t be the topology on E' of uniform convergence on the set K, and let B be the unit ball of E'. It is obvious that $t \leqslant \tau(E', E)$, and on B the topology t majoryzes the topology $\sigma(E', E)$. Let L: $E' \to S(\Omega, \Sigma, \mathbf{P})$ be a l.r.f. for λ. It follows from (3.4) and Lemma 1 that the operator L_∞: $(E', \sigma(E', E))_\infty \to \widetilde{S}_\infty$ is sequentially continuous. Hence, the restriction of L_∞ to B_∞ is continuous for $(\tau(E', E))_\infty$. So (3.1) follows from Theorem 2.

Now we want to generalize Theorem 3 to l.c.s.

Definition 2. A l.c.s. E is called a quasi-WCG-space (briefly, $E \in q$WCG) if E may be isomorphically imbedded in a product of an arbitrary family of Banach WCG-spaces.

We omit the proof of the following proposition. We shall use the term *quasi-normable space* in the sense of (Grothendieck, 1973, p. 176). Recall that every DF-space is quasi-normable (see (Kac, 1973)).

PROPOSITION 3. *The class of $qWCG$-spaces contains all separable l.c.s. all semi-reflexive quasi-normable spaces, all Schwartz spaces (in particular, all nuclear spaces). The class of $qWCG$-spaces is stable with respect to the following operations: product (of an arbitrary power), taking a subspace, countable inductive limit (in the wide sense). If there exists a continuous operator with a dense image from a l.c.s. E into a l.c.s. F and $E \in qWCG$, then $F \in qWCG$.*

LEMMA 5. *If a l.c.s. E is a $qWCG$-space then the implication* (3.5) ⇒(3.6) *is valid for every c.m. λ on $\langle E, E'\rangle$.*

Proof. Let U be any disked neighbourhood of the origin in E and let E_U be the Banach space associated with U, i.e., E_U is the Hausdorff completion of E by the seminorm $p(x) = \inf\{r \geqslant 0\colon x \in rU\}$. Denote φ: $E \to E_U$ the natural mapping.

It is easily seen that E_U is a subspace of a Banach WCG–space. Suppose that λ satisfies (3.5). Then the c.m. $T(\varphi(\lambda))$ on c_0 is σ–additive for any continuous operator T: $E_U \to c_0$. By Theorem 3, $\varphi(\lambda)$ may be extended to a Radon measure on E_U. Thus, (3.6) follows from Theorem 2 in (Dudley et al., 1971).

Definition 3. **a)** A l.c.s. E is said to belong to the class M if (3.6)$\Rightarrow$(3.1) for every c.m. λ on E; **b)** A l.c.s. E is said to belong to the class V if (3.4) $\Rightarrow$(3.1) for every c.m. λ on E.
It is clear that the classes M and V are closed under the formation of closed subspaces. It follows from the Prokhorov theorem that both these classes are closed also under the formation of countable projective limits and countable products.
In particular, the class M contains all Frechet spaces, see Corollary 2.1 in (Dudley et al., 1971).

PROPOSITION 4. *The class M is closed under the formation of strict inductive limits of such spaces.*

Proof. Let E be the strict inductive limit space of an increasing sequence of subspaces E_n, where $E_n \in M$ $(\forall n)$. Let λ be a c.m. on $\langle E, E'\rangle$ satisfying (3.6) and let L: $E' \to S(\Omega, \Sigma, \mathbf{P})$ be corresponding l.r.f.
It is easily seen that for every neighbourhood of the origin $U \subset E$ the restriction of L to U^0 is continuous for $\sigma(E', E)$, and thus $L(U^0)$ is compact in $S(\Omega, \Sigma, \mathbf{P})$. Hence, $\forall \varepsilon > 0$ $\exists n \in \mathbf{N}$ such that $\mathbf{P}(L(x') \neq 0) \leqslant \varepsilon$ for all $x' \in E'$ such that $x'(E_n) = 0$.
By Lemma 2, $\exists \Omega_\varepsilon \in \Sigma$ with $\mathbf{P}(\Omega_\varepsilon) > 1 - \varepsilon$ such that there is the following commutative diagram

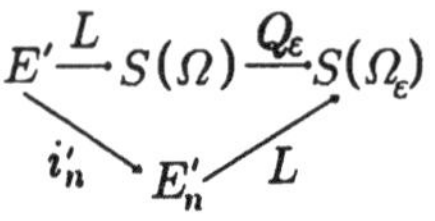

where Q_ε is the natural projection, i_n: $E_n \to E$ is the natural isomorphic embedding. Let η be the c.m. on E_n corresponding to L_ε. Then $i_n(\eta)$ satisfies (3.6) because $i_n(\eta)(A) \leqslant \lambda(A)$ for every Borelian cylinder $A \subset E$. Hence, η also satisfies (3.6) and thus may be extended to a Radon measure on E_n.
Therefore, λ may be extended to a Radon measure on E, and this completes the proof.

COROLLARY 4. *The class M contains all LF–spaces.*

THEOREM 4. *Let E be a l.c.s.*

a) *If $E \in qWCG$ and $E \in M$, then $E \in V$ and conditions* (3.1)–(3.6) *are equivalent for every c.m. λ on $\langle E, E'\rangle$.*

b) *If E is reflexive and E'_β is C–bornological, then conditions* (3.2)–(3.5) *and* (3.7) *are equivalent for every c.m. λ on $\langle E, E'\rangle$.*

Proof. a) Follows from Lemmas 3 and 5.
b) Since E is a barrel space, for any sequence (x'_n) in E' the condition "(x'_n) tends to 0 in the sense of Mackey with respect to $\beta(E',E)$" is equivalent to "$x'_n \to 0$ uniformly on some neighbourhood of 0 in E". To complete the proof we apply Lemma 3, Proposition 1, the definition of C–bornological space and the Prokhorov theorem.

COROLLARY 5. *Let E be a Frechet space or a LF–space. If $E \in qWCG$ or E is reflexive, then $E \in V$ and conditions* (3.1)–(3.6) *are equivalent for every c.m. λ on $\langle E, E'\rangle$. Moreover, if E is reflexive, then conditions* (3.1)–(3.7) *are equivalent.*

Proof follows from Theorem 4, Corollary 1, Corollary 4 and the fact that the weak and the initial topologies on LF–spaces are Radon–equivalent (see (Okada, 1975)).
Let us consider the relation between the class M and the property of Radon–equivalence of the weak and the initial topologies.

COROLLARY 6. *Let E be a l.c.s. If $E \in M$, then the weak and the initial topologies on E are Radon–equivalent. The converse is also true if E is reflexive and E'_β is C–bornological.*

Proof follows from Lemma 3 and Theorem 4.

Remarks and Examples. 1) The implication (3.3)$\Rightarrow$(3.1) of Theorem 3 is known (see (Edgar, 1977)). In (Fremlin and Talagrand, 1980) an example of a σ–additive Gaussian cylindrical measure μ on $\langle l_\infty, l'_\infty\rangle$ is given such that μ can not be extended to a Radon measure. Let us note that its l.r.f. L: $l'_\infty \to S(\Omega)$ satisfies the following condition: the restriction of L to the unite ball of l'_∞ is continuous with respect to $\sigma(l'_\infty, l_\infty)$.
2) Let us give an example of a c.m. on $\langle l_\infty, l'_\infty\rangle$ for which (3.4) is true but (3.3) is not true. Let μ_n be the sequence of the measures on $\mathbf{R}$ defined by

$$\mu_n\{-1\} = \mu_n\{+1\} = 1/2, \quad \mu = \bigotimes_{n=1}^{\infty} \mu_k$$

be the product measure of the μ_n on $\mathbf{R}^\infty$, η be the corresponding c.m. on $\langle l_\infty, l_1\rangle$. Let us extend η to the c.m. λ on $\langle l_\infty, l'_\infty\rangle$ by putting $\lambda(\operatorname{Ker}\varphi) = 1$ for every functional $\varphi \in l'_\infty$ such that $\varphi(c_0) = 0$.
Let us show that λ is not σ–additive. Indeed, let $\mathcal{F}$ be some ultrafilter on N majorizing the Frechet filter. Let us define the functional $f \in (l_\infty)'$ by

$$f(x) = \lim_{\mathcal{F}} x_n, \text{ where } \quad x = (x_n)_{n=1}^{\infty} \in l_\infty.$$

Since $f(c_0) = 0$, we have $\lambda(\operatorname{Ker} f) = 1$. Let us consider the decreasing sequence of Borel cylinders

$$A_n = \{x = (x_k)_{k=1}^{\infty} \in l_\infty \colon |x_k| = 1 \quad \text{for} \quad 1 \leqslant k \leqslant n\}.$$

It is obvious that $\lambda(A_n \cap \operatorname{Ker} f) = 1$ for every n. Since

$$\bigcap_{n=1}^{\infty}\left(A_n \bigcap \operatorname{Ker} f\right) = \emptyset,$$

λ is not σ–additive. On the other hand by the Corollary to Theorem 9 in (Grothendieck, 1953) every continuous operator T: $l_\infty \to c_0$ is weakly compact. Therefore, by Proposition 2, λ satisfies (3.5) and, by Lemma 4, satisfies (3.4).

3) If E is barrel l.c.s. then for any sequence (x'_n) in E' the conditions "$x'_n \to 0$ in $\sigma(E',E)$" and "$x'_n \to 0$ in $c(E',E)$" are equivalent, where $c(E',E)$ is the topology of uniform convergence on the compact disks of E. But if we change condition (3.4) to condition

(3.4)′ $L(x_n)' \to 0$ a.s. for every sequence (x'_n) in E' such that $x'_n \to 0$ in $\tau(E',E)$

then under the assumptions of Theorem 3 the implication (3.4)′ ⇒(3.3) is not true in general.

Indeed, in (Linde and Pietsch, 1974) an example of a Gaussian cylindrical measure λ on $\langle c_0, l_1\rangle$ is given such that λ is not σ–additive but the corresponding c.m. $\tilde{\lambda}$ on $\langle l_\infty, l_1\rangle$ is σ–additive (see also (Vakhania, 1971, *S* 2.4)). Let L: $l_1 \to S(\Omega,\Sigma,\mathbf{P})$ be the l.r.f. for λ. If a sequence (x'_n) in l_1 is such that $x'_n \to 0$ in $\tau(l_1, c_0)$, then $\|x'_n\| \to 0$ and so $L(x'_n) \to 0$ a.s.

Note that if we consider λ as a c.m. on $\langle E', E\rangle$, where $E = (l_1, c(l_1, c_0))$, then λ satisfies (2.4) but does not satisfy (2.3).

4) It follows from Lemmas 3 and 4 that if a l.c.s. E is barrel and $E \in V$, then $E \in M$.

Acknowledgement

The author would like to express his deep gratitude to Prokhorov and Sazonov for their attention and support.

REFERENCES

Dudley, R., Feldman, J. and Le Cam, L. (1971). On seminorms and probabilities, and abstract Wiener spaces. *Ann. Math.* **93**, 390–408.

Edgar, G. (1977). Measurability in Banach space. *Indiana Univ. Math. J.* **26**, 663–677.

Fremlin, D. and Talagrand, M. (1980). A Gaussian measure on l^∞. *Ann. Probab.* **8**, 1192–1193.

Grothendieck, A. (1953). Sur les applications lineaires faiblement compacts d'espaces de type *C(K)*. *Canad. J. Math.* **5**, 129–173.

Grothendieck, A. (1973). *Topological Vector Spaces.* Gordon and Breach, New York etc.

Kac, M. (1973). Every DF–space is quasi–normable. *Funct. Anal. Appl.* **7**, 85–86.

Linde, W. and Pietsch, A. (1974). Abbildungen von Gaußschen Zylindermaßen in Banachräumen. *Theory Probab. Appl.* **19**, 472–487.

Maurey, B. (1973). Theoremes de Nikishin: theoremes de factorisation pour les applications lineaires a valeurs dans un espace $L^0(\Omega,\mu)$. In: *Seminaire Maurey–Schwartz. Espaces L^p et applications radonifiantes.* Paris, exp. 10–11.

Muštari, D. (1970). On convergence almost surely in linear spaces of random variables. *Theory Probab. Appl.* **15**, 351–357.

Okada, S. (1975). Radon–equivalent topologies on (LF)–spaces. *Mem. Fac. Sci., Kyushu Univ.* **A 29**, 223–227.

Okazaki, Y. and Takahashi, Y. (1987). Nuclear subspace of L^0 and the kernel of a linear measure. *J. Multivar. Anal.* **22**, 65–73.

Schaefer, H. (1966). *Topological Vector Spaces.* Macmillan, New York.

Schwartz, L. (1973). *Radon Measures on Arbitrary Topological Spaces and Cylindrical Measures.* Oxford Univ. Press., Oxford.

Vakhania, N. (1971). *Probability Distributions on Linear Spaces.* Metzniereba, Tbilisi (in Russian).

Vakhania, N., Tarieladze, V. and Chobanyan, S. (1987). *Probability Distributions on Banach Spaces.* D. Reidel, Dordrecht etc.

Vershik, A. and Sudakov, V. (1969). Probability measures in infinitedimensional spaces. *Notes Sci. Seminars Leningrad Branch of Math. Inst.* **12**, 7–67.

Vladimirskii, Yu. (1979). On some topological properties of countably additive cylindrical measures. *Theory Probab. Appl.* **24**, 211–215.

Vladimirskii, Yu. (1981). Cylindrical measures and p–summing operators. *Theory Probab. Appl.* **26**, 59–72.

New Trends in Probab. and Statist., pp. 276 – 285
V. Sazonov and T. Shervashidze (Eds.)
1991 VSP/Mokslas

A FAMILY OF MEASURES ADMITTING CONSISTENT ESTIMATES

Z.S. ZERAKIDZE
Tbilisi I. Javakhishvili State University, 380043 Tbilisi, Georgia, USSR

ABSTRACT

This paper continues the investigation of the structure of families of probability measures started in (Zerakidze, 1984a). A maximal family of divisible probability measures and consistent estimates of unknown parameters are constructed; necessary and sufficient conditions are obtained for the existence of consistent estimates.

1. A DIVISIBLE FAMILY OF PROBABILITY MEASURES

Let $(X, \mathcal{B})$ be a measurable space with a given family of measures $\{\mu_\alpha, \alpha \in \mathcal{A}\}$. Let us recall some definitions.

Definition 1.1. The family of measures $\{\mu_\alpha, \alpha \in \mathcal{A}\}$ is called singular if $\alpha \neq \beta \Rightarrow \mu_\alpha$ and μ_β are orthogonal for each $(\alpha, \beta) \in \mathcal{A}^2$.

Definition 1.2. The family of measures $\{\mu_\alpha, \alpha \in \mathcal{A}\}$ is called weakly divisible if there exists a family of $\mathcal{B}$-measurable sets $\{X_\alpha, \alpha \in \mathcal{A}\}$ such that the relations

$$\mu_\alpha(X_\beta) = 0 \quad \forall \alpha \neq \beta,$$
$$\mu_\alpha(X - X_\alpha) = 0 \quad \forall \alpha$$

are fulfilled.

Definition 1.3. The family of measures $\{\mu_\alpha, \alpha \in \mathcal{A}\}$ is called divisible if there exists a family of $\mathcal{B}$-measurable sets $\{X_\alpha, \alpha \in \mathcal{A}\}$ such that

$$\alpha \neq \alpha' \Longrightarrow \operatorname{card}(X_\alpha \cap X_{\alpha'}) < c$$

for each $(\alpha, \alpha') \in \mathcal{A}^2$,where c denotes the power of the continuum and

$$\mu_\alpha(X_\beta) = 0 \quad \forall \alpha \neq \beta,$$
$$\mu_\alpha(X - X_\alpha) = 0 \quad \forall \alpha.$$

Definition 1.4. The family of measures $\{\mu_\alpha, \alpha \in \mathcal{A}\}$ is called strongly divisible if there exist pairwise disjoint $\mathcal{B}$-measurable sets $\{X_\alpha, \alpha \in \mathcal{A}\}$ such that

$$\mu_\alpha(X - X_\alpha) = 0$$

for each $\alpha \in \mathcal{A}$. It is assumed that the parametric set $\mathcal{A}$ has a σ-algebra $\mathcal{F}$ containing all finite subsets of $\mathcal{A}$.

Definition 1.5. The family of measures $\{\mu_\alpha, \alpha \in \mathcal{A}\}$ will be said to admit a consistent estimate of parameters if there exists a measurable map g of the space $(X, \mathcal{B})$ in $(\mathcal{A}, \mathcal{F})$ such that

$$\mu_\alpha\big(X - \{x\colon\ x \in X,\ g(x) = \alpha\}\big) = 0$$

for each $\alpha \in \mathcal{A}$.
The following earlier results were obtained in (Zerakidze,1984b): On any set X with power of continuum we can define

1) a family of pairwise orthogonal probability measures that has the maximal possible power equal to 2^{2^c};
2) a family of weakly divisible probability measures that has the maximal possible power equal to 2^c;
3) a family of strongly divisible probability measures that has the maximal possible power equal to c,

We will prove

THEOREM 1.1. *On any set X with continuum power we can define a divisible family of nonatomic probability measures that has the maximal power greater than c.*

Proof. Let $(Y_i,\ i \in I)$ denote an injective family of all nonempty perfect subsets of $\mathbf{R}^n$. It is obvious that

$$\operatorname{card} I = c, \qquad \operatorname{card} Y_i = c \quad \forall i \in I.$$

By virtue of Theorem 2 from (Kharazishvili, 1984), we find a family $(X_i,\ i \in I)$ such that

$$\operatorname{card} I > c, \qquad X_i \in \mathbf{R}^n \quad \forall i \in I,$$

$$\operatorname{card}(X_i \cap X_{i'}) < c$$

for each $(i, i') \in I^2$, $i \neq i'$, and

$$\operatorname{card}(X_i \cap X_j) = c$$

for each $(i, j) \in I^2$.
Besides, a family $(X_i,\ i \in I)$ can be regarded as a covering of $\mathbf{R}^n$. Let us fix some index $i \in I$ and set

$$X_i^0 = X_i, \qquad X_i^1 = \mathbf{R}^n \backslash X_i.$$

Next, denote by $\mathcal{E}_i$ the σ-algebra generated by $\mathcal{L} \cup \{X_i^0, X_i^1\}$, where $\mathcal{L}$ is the class of all Lebesgue-measurable parts of $\mathbf{R}^n$. It is easy to verify that elements of the σ-algebra $\mathcal{E}_i$ are sets of the form

$$(X_i^0 \cap Z') \cup (X_i^1 \cap Z''),$$

where $Z',\ Z'' \in \mathcal{L}$.

Let $\bar{\mu}_i$ be the measure defined on the σ-algebra $\mathcal{E}_i$ by

$$\bar{\mu}_i((X_i^0 \cap Z') \cup (X_i^1 \cap Z')) = \mathbf{P}(Z'),$$

where $\mathbf{P}$ is a nonatomic probability measure defined on the class $\mathcal{L}$. The relation $\bar{\mu}_i(X_i^1) = 0$ is fulfilled for the measure $\bar{\mu}_i$. Of course, the measure $\bar{\mu}_i$ can be completed and extended to the class $\mathcal{F}(\mathbf{R}^n)$ of all noncontinuum subsets of $\mathbf{R}^n$ by putting $\bar{\mu}_i(Z) = 0$ for any set $Z \in \mathcal{F}(\mathbf{R}^n)$. The measure completed and continued in this manner will also be denoted by the symbol $\bar{\mu}_i$.
In particular, now we have

$$\bar{\mu}_i(X_{i'}) = 0 \quad \text{foreach} \quad (i, i') \in I^2,\ i \neq i'.$$

Finally, denote by $\mathcal{E}$ the σ-algebra generated by

$$\mathcal{L} \cup \mathcal{F}(\mathbf{R}^n) \cup \Big(\bigcup_{i \in I} \{X_i\}\Big).$$

It is clear that regardless of the index $i \in I$ from the set I the σ-algebra $\mathcal{E}$ is contained in the domain of definition of $\bar{\mu}_i$ (since $\bar{\mu}_i$ is a complete measure defined on all elements of the class $\mathcal{F}(\mathbf{R}^n)$). Setting

$$\mu_i = \bar{\mu}_i / \mathcal{E},$$

we obtain the following relations for the σ-algebra $\mathcal{E}_i$ and the families $(\bar{\mu}_i,\ i \in I)$ and $(X_i,\ i \in I)$:

$$\operatorname{card} I > c, \quad X_i \in \mathcal{E} \quad \forall i \in I;$$

$$\operatorname{card}(X_i \cap X_j) > c \quad \text{for each} \quad (i, j) \in I^2,\ i \neq j;$$

for each $i \in I$ the measure μ_i is concentrated on the set X_i.
Let X be any set of the power of continuum, f be a map performing the bijection of the set X on $\mathbf{R}^n$. It is clear that $S = f^{-1}(\mathcal{E})$ is a σ-algebra in the space X. For any element $i \in I$ it is assumed that

$$\lambda_i(A) = \mu_i(f(A)) \quad \text{foreach} \quad A \in S.$$

Obviously, $(\lambda_i,\ i \in I)$ is a divisible family of nonatomic probability measures defined on X.
The theorem is proved.
The following theorem is proved similarly.

THEOREM 1.2. *On any set X with the power of continuum we can define a divisible family of atomic probability measures that has the maximal power greater than c.*

2. CONSISTENT ESTIMATES

THEOREM 2.1. *Let the parametric set $\mathcal{A}$ be a complete metric space, $\mathcal{F}$ be a σ-algebra of Borel subsets of $\mathcal{A}$ containing all finite subsets of $\mathcal{A}$, $\mu_\alpha(A)$ be a measurable function on $\mathcal{A}$ for any $A \in \mathcal{B}$.*
If ν is a measure on $\mathcal{F}$ such that $\nu(U) > 0$ for any open set U and the measures μ_1 and μ_2 defined by

$$\mu_1(A) = \int\limits_U \mu_\alpha(A)\,\nu(d\alpha), \qquad \mu_2(A) = \int\limits_{\mathcal{A}-U} \mu_\alpha(A)\,\nu(d\alpha)$$

are orthogonal, then there exists a measurable function g: $(X, \mathcal{B}) \to (\mathcal{A}, \mathcal{F})$ such that

$$\mu_\alpha(\{x\colon\ g(x) = \alpha\}) = 1$$

for almost all α with respect to the measure ν.

Proof. It is easy to see that the family of sets for which the measures μ_1 and μ_2 are orthogonal is a monotone class: if $C_n \uparrow C$, then

$$\int\limits_{C_n} \mu_\alpha(A)\,\nu(d\alpha) \uparrow \int\limits_{C} \mu_\alpha(A)\,\nu(d\alpha),$$

$$\int\limits_{\mathcal{A}-C_n} \mu_\alpha(A)\,\nu(d\alpha) \downarrow \int\limits_{\mathcal{A}-C} \mu_\alpha(A)\,\nu(d\alpha).$$

Therefore, choosing $D_n \uparrow D$ such that

$$\int\limits_{\mathcal{A}-C_n} \mu_\alpha(D_n)\,\nu(d\alpha) = 0, \qquad \int\limits_{C_n} \mu_\alpha(X - D_n)\,\nu(d\alpha) = 0,$$

we see that

$$\int\limits_{\mathcal{A}-C_n} \mu_\alpha(D)\,\nu(d\alpha) = 0, \qquad \int\limits_{C_n} \mu_\alpha(X - D)\,\nu(d\alpha) = 0,$$

$$\int\limits_{\mathcal{A}-C_m} \mu_\alpha(D_n)\,\nu(d\alpha) = 0 \quad (m > n), \qquad \int\limits_{\mathcal{A}-C} \mu_\alpha(D_n)\,\nu(d\alpha) = 0,$$

$$\int\limits_{\mathcal{A}-C} \mu_\alpha(D)\,\nu(d\alpha) = 0, \qquad \int\limits_{C} \mu_\alpha(X - D)\,\nu(d\alpha) = 0,$$

and hence,

$$\int\limits_{C} \mu_\alpha(X - D)\,\nu(d\alpha) = 0.$$

However, the monotone class containing all open sets coincides with the σ-algebra of Borel sets. This means that under the conditions of the theorem the measures μ_1 and μ_2 will be orthogonal for all Borel sets $C \in \mathcal{F}$.
Next we suppose that $\mathcal{A}$ is separable and construct a sequence of partitions of $\mathcal{A}$ into pairwise disjoint sets

$$\mathcal{A} = \bigcup_{k_1,\dots,k_n} C^{(n)}_{k_1,\dots,k_n}$$

such that the conditions below are fulfilled:

1) Sets $C^{(n)}_{k_1,\dots,k_n}$ are defined for all positive integers n, $k_1,\dots,k_n$ and

$$C^{(n+1)}_{k_1,\dots,k_n,k_{n+1}} \subset C^{(n)}_{k_1,\dots,k_n}, \qquad \bigcup_{k_{n+1}} C^{(n+1)}_{k_1,\dots,k_{n+1}} = \bigcup C^{(n)}_{k_1,\dots,k_n};$$

2) The measure of the boundary of $C^{(n)}_{k_1,\dots,k_n}$ is zero and $\nu(C^{(n)}_{k_1,\dots,k_n}) > 0$;

3) The diameter of $C^{(n)}_{k_1,\dots,k_n}$ is not greater than 2^{-n}.

Since, for various $k_1,\dots,k_n$, the measures

$$\int_{C^{(n)}_{k_1,\dots,k_n}} \mu_\alpha(\cdot)\,\nu(d\alpha)$$

are pairwise orthogonal, there exist Borel sets $X^{(n)}_{k_1,\dots,k_n}$ such that

$$\int_{C^{(n)}_{k_1,\dots,k_n}} \mu_\alpha(X^{(n)}_{k_1,\dots,k_n})\,\nu(d\alpha) = \begin{cases} 0 & \text{if } \sum_{i=1}^{n}(K_i - K_{i'})^2 > 0, \\ \nu(C^{(n)}_{k_1,\dots,k_n}) & \text{if } \sum_{i=1}^{n}(K_i - K_{i'})^2 = 0, \end{cases}$$

where $X^{(n)}_{k_1,\dots,k_n}$ can be chosen pairwise disjoint. Indeed, since this family of measures is countable and pairwise orthogonal, this family is also divisible.
Since $C_{k_1,\dots,k_{n+1}} \subset C_{k_1,\dots,k_n}$, $X^{(n+1)}_{k_1,\dots,k_{n+1}}$ can be chosen so that $X^{(n+1)}_{k_1,\dots,k_{n+1}} \subset$ $\subset X^{(n)}_{k_1,\dots,k_n}$. Denote by $\widehat{C}^{(n)}_{k_1,\dots,k_n}$ the set of those $\alpha \in C^{(n)}_{k_1,\dots,k_n}$ for which

$$\mu_\alpha(X^{(n)}_{k_1,\dots,k_n}) = 1.$$

This is a closed set because $\mathcal{A}$ is separable. Since

$$0 = \int_{C^{(n)}_{k_1,\dots,k_n}} \left(1 - \mu_\alpha(X^{(n)}_{k_1,\dots,k_n})\right)\nu(d\alpha)$$

and the integrand is non-negative, it is equal to zero almost everywhere, i.e.,

$$\nu(C^{(n)}_{k_1,\dots,k_n}) = \nu(\widehat{C}^{(n)}_{k_1,\dots,k_n}).$$

Setting

$$C = \bigcap_n \bigcup_{k_1,\dots,k_n} C^{(n)}_{k_1,\dots,k_n},$$

we have $\nu(C) = \nu(\mathcal{A})$. If $\alpha \in C$, then the equality $\mu_\alpha(X^{(n)}_{k_1,\dots,k_n}) = 1$ will be fulfilled for all n. The functions $g_n(x) = \alpha^{(n)}_{k_1,\dots,k_n}$ on the set

$$\bigcap_n \bigcup_{k_1,\dots,k_n} X^{(n)}_{k_1,\dots,k_n} = \widetilde{X}$$

converge to some limit $g(x)$, since $\rho(g_n, g_{n+1}) < 2^{-n}$. This follows from the fact that, for $x \in X^{(n+1)}_{k_1,\dots,k_{n+1}}$, we have

$$g_n(x) = \alpha^{(n)}_{k_1,\dots,k_n}, \qquad g_{n+1}(x) = \theta^{(n)}_{k_1,\dots,k_n}$$

and these points both belong to $C^{(n)}_{k_1,\dots,k_n}$ whose diameter is not greater than 2^{-n}.
The equality $\mu_\alpha(\tilde{X}) = 1$ is satisfied for $\alpha \in C$.
If $\alpha \in \bigcap \hat{C}^{(n)}_{k_1,\dots,k_n}$, then $\mu_\alpha(\bigcap X^{(n)}_{k_1,\dots,k_n}) = 1$ and, for $x \in \bigcap X^{(n)}_{k_1,\dots,k_n}$, we get $g_n(x) \in C^{(n)}_{k_1,\dots,k_n}$.
Since $\bigcap C^{(n)}_{k_1,\dots,k_n}$ also contains the unique point α, we have $\rho(g_n(x), \alpha) \leqslant 2^{-n}$ and hence, $g(x) = \alpha$.
The theorem is proved in the case of separable $\mathcal{A}$. If $\mathcal{A}$ is not separable this theorem remains true because a Borel measure given on a nonseparable complete metric space has a separable support (Kharazishvili, 1984).
We will recall some other definitions and notation.

Definition 2.1. The family of probability measures $\{\mu_\alpha, \alpha \in \mathcal{A}\}$ will be said to admit a consistent estimate of any parametric function if for any real bounded measurable function $g(\alpha)$ on $(\mathcal{A}, \mathcal{F})$ there exists a real bounded measurable function $f(x)$ on $(X, \mathcal{B})$ such that

$$\mu_\alpha\Big(\{x: f(x) = g(\alpha)\}\Big) = 1$$

for all $\alpha \in \mathcal{A}$.

Definition 2.2. The family of probability measures $\{\mu_\alpha, \alpha \in \mathcal{A}\}$ will be said to admit a unbiased estimate of any parametric function if, for any real bounded function $g(\alpha)$ on $(\mathcal{A}, \mathcal{F})$, there exists a real bounded measurable function $f(x)$ on $(X, \mathcal{B})$ such that

$$\int f(x)\,\mu_\alpha(dx) = g(\alpha)$$

for all $\alpha \in \mathcal{A}$.
Denote by K_1 (K_2 and K_3, respectively) the set of all families of probability measures $\{\mu_\alpha, \alpha \in \mathcal{A}\}$ on $(X, \mathcal{B})$ admitting an unbiased estimate of any parametric function g (a consistent estimate of any g and a consistent estimate of parameters $\alpha \in \mathcal{A}$, respectively).

It is easy to show that $K_3 \subset K_2 \subset K_1$.
In what follows $\mathbf{B}(X, \mathcal{B})$ will always denote a vector space formed by all real bounded measurable functions on $(X, \mathcal{B})$ having the natural order and "$\sup|\cdot|$" norm. It is an (AM)-space with identity which is a function e identically equal to one on X (see (Schaefer, 1966)). Let $\mathbf{B}'(X, \mathcal{B})$ denote the topological conjugate space of $\mathbf{B}(X, \mathcal{B})$; it is an order-complete Banach lattice. The elements of $\mathbf{B}'(X, \mathcal{B})$ are called finitely-additive measures on $(X, \mathcal{B})$, and the canonical bilinear form which puts $\mathbf{B}(X, \mathcal{B})$ and $\mathbf{B}'(X, \mathcal{B})$ in duality is denoted by

$$(f, \mu) = \mu(f) = \int f(x)\, \mu(dx), \quad f \in \mathbf{B}(X, \mathcal{B}), \quad \mu \in \mathbf{B}'(X, \mathcal{B})$$

and called the integral of f with respect to μ.
The construction of a Hilbert space of measures is studied in (Zerakidze, 1986), where it is shown that if M_H is a Hilbert space of measures, then in M_H there exists a family of pairwise orthogonal probability measures $\{\mu_\alpha, \alpha \in \mathcal{A}\}$ such that

$$M_H = \bigoplus_{\alpha \in \mathcal{A}} H_2(\mu_\alpha),$$

where $H_2(\mu_\alpha)$ is the set of measures of the form $\nu(B) = \int f(x)\, \mu_\alpha(dx)$ with $\int f^2(x)\, \mu_\alpha(dx) < \infty$ and the scalar product of two measures ν_1 and ν_2 from this set with the corresponding functions f_1 and f_2 equals

$$\langle \nu_1, \nu_2 \rangle = \int f_1(x) f_2(x)\, \mu_\alpha(dx).$$

Denote by $F = F(M_H)$ the set of f for which $\int f(x)\, \nu(dx)$ is defined.

THEOREM 2.2. *Let the Hilbert space $M_H = \bigoplus_{\alpha \in \mathcal{A}} H_2(\mu_\alpha)$. The family $\{\mu_\alpha, \alpha \in \mathcal{A}\}$ admits a consistent estimate of any parametric function g if and only if it admits a unbiased estimate of any parametric function g and the correspondence $f \to \psi_f$ given by $\int f(x)\, \nu(dx) = \langle \psi_f, \nu \rangle$, $\forall \nu \in M_H$, is one-to-one.*

Proof. *Necessity.* It is clear that if the family $\{\mu_\alpha, \alpha \in \mathcal{A}\}$ admits a consistent estimate of any parametric function g, then it obviously admits an unbiased estimate of any parametric function g.
Let ν_f for $f \in F$ denote the measure for which $\int f(x)\, \psi(dx) = \langle \nu_f, \psi \rangle$. For $f \in F$ denote by A_f the countable subset of $\mathcal{A}$ for which $\int f^2(x)\, \mu_\alpha(dx) = 0$ for $\alpha \in A_f$. Consider the measures $\psi \in H_2(\mu_\alpha)$. Let $f_\psi \in F$ be the function corresponding to a measure ψ.
Then, for $\psi_1, \psi_2 \in H_2(\mu_\alpha)$, we have

$$\int f_{\psi_1}(x)\, \psi_2(dx) = \langle \psi_1, \psi_2 \rangle =$$

$$= \int f_1(x) f_2(x)\, \mu_\alpha(dx) = \int f_{\psi_1}(x) f_2(x)\, \mu_\alpha(dx).$$

Therefore, $f_{\psi_1} = f$ almost everywhere with respect to μ_α. Let $f_\alpha(x) > 0$ almost everywhere with respect to μ_α,

$$\int f_\alpha^2(x)\,\mu_\alpha(dx) < \infty, \qquad \mu_\alpha^* = \int f_\alpha(x)\,\mu_\alpha(dx).$$

Then

$$\int f_{\mu_\alpha^*}(x)\,\mu_\beta(dx) = \langle \mu_\alpha^*, \mu_\beta \rangle = 0 \quad \forall \beta \neq \alpha.$$

Let $C_\alpha = \{x: f_{\mu_\alpha^*}(x) > 0\}$. From the last equality it follows that $\mu_\beta(C_\alpha) = 0$ $(\forall \beta \neq \alpha)$. On the other hand,

$$\mu_\alpha^*(X - C_\alpha) = \int\limits_{X-C_\alpha} f_\alpha(x)\,\mu_\alpha(dx) = \int f_\alpha(x) I_{X-C_\alpha}(x)\,\mu_\alpha(dx) =$$

$$= \int f_{\mu_\alpha^*}(x) I_{X-C_\alpha}(x)\,\mu_\alpha(dx) = 0,$$

since $f_{\mu_\alpha^*} = f_\alpha$ almost everywhere with respect to μ and $f_{\mu_\alpha^*}(x) I_{X-C_\alpha}(x) = 0$. Therefore, the family of measures $\{\mu_\alpha, \alpha \in \mathcal{A}\}$ is weakly divisible.

Sufficiency. From (Schaefer, 1966; Maleev et al., 1978; Zerakidze, 1986) it follows that the family $\{\mu_\alpha, \alpha \in \mathcal{A}\}$ is weakly divisible and the zero-space $\mu_\alpha^{-1}(0)$ of every measure μ_α contains, together with any function $f \in G$, also $|f|$. Here, G is a subspace of the space $\mathbf{B}(X, \mathcal{B})$ containing the identity e and such that $\mathbf{B}(X, \mathcal{B})$ is the direct topological sum of G and the polar with respect to the duality $\langle \mathbf{B}(X, \mathcal{B}), \mathbf{B}'(X, \mathcal{B})\rangle$ of the subspace closed with respect to the topology $\sigma(\mathbf{B}(X, \mathcal{B}), \mathbf{B}'(X, \mathcal{B}))$ generated by the family $\{\mu_\alpha, \alpha \in \mathcal{A}\}$.
Let U be the operator defined for each $v \in \mathbf{B}(X, \mathcal{B})$ by $Uv(\alpha) = \{\langle v, \mu_\alpha\rangle, \alpha \in \mathcal{A}\}$. The fulfillment of the relations $f - e(Uf)(\alpha) \in G$ and

$$0 = \langle f - e(Uf)(\alpha), \mu_\alpha \rangle = \langle |f - e(Uf)(\alpha)|, \mu_\alpha \rangle,$$

for $f \in G$ and each $\alpha \in \mathcal{A}$ means that the function $f - e(Uf)(\alpha)$ is μ_α-equivalent to zero, i.e., f is a consistent estimate for its own image $(Uf)(\alpha)$, since the family $\{\mu_\alpha, \alpha \in \mathcal{A}\}$ admits an unbiased estimate of each parametric function. The theorem is proved.
Let X be a complete separable metric space and $\{\mu_\alpha, \alpha \in \mathcal{A}\}$ be Borel probability measures on $\mathcal{B}$.

THEOREM 2.3. *If*

1) *the Martin axiom is true,*

2) $\{\mu_\alpha, \alpha \in \mathcal{A}\}$ *is a weakly divisible family of Borel probability measures given on the complete separable metric space,*

and

3) *this family admits an unbiased estimate of any parametric function* g,

then this family admits a consistent estimate of parameters $\alpha \in \mathcal{A}$.

Proof. In (Zerakidze, 1984b) it is proved that if the Martin axiom is true, then a weakly divisible family of the Borel probability measures given on the complete separable metric space is strongly divisible.
Let $\mathcal{B}_0$ be the minimal σ-subalgebra of the σ-algebra $\mathcal{B}$ with respect to which all functions from G are measurable. It is clear that $G \subset \mathbf{B}(X, \mathcal{B}_0) \subset \mathbf{B}(X, \mathcal{B})$.
Since the subspace G contains e and is a lattice, we conclude that $G \supset \mathbf{B}(X, \mathcal{B}_0)$ and therefore $G = \mathbf{B}(X, \mathcal{B}_0)$. The ideal I in $\mathcal{B}_0$ of sets of the μ_α-measure zero for all $\alpha \in \mathcal{A}$ consists only of $\{\varnothing\}$. It follows that in X there exists the unique system of sets $\{A_\alpha, \alpha \in \mathcal{A}\}$ dividing the family $\{\mu_\alpha, \alpha \in \mathcal{A}\}$ and being such that $X = \cup A_\alpha, \alpha \in \mathcal{A}$.
Under the conditions of the theorem for each T in G there exists a function f_T which is the consistent estimate for $f_T(\alpha)$. Assume that $A = \{x: f_T(x) \neq 0\}$. Then $\cup_{\alpha \in T} A_\alpha \subset A$, $A_\alpha \cup A = \varnothing$ for all $\alpha \in T$ and hence it follows that $A = \cup_{\alpha \in T} A_\alpha$. This means that $\cup_{\alpha \in T} A_\alpha$ for each T is an element of the σ-algebra $\mathcal{B}_0 \subset \mathcal{B}$. Now it is clear that the mapping $\eta(x) = \alpha,\ x \in A_\alpha$, is a consistent estimate of parameters $\alpha \in \mathcal{A}$.
The theorem is proved.

THEOREM 2.4. *If*

1) the continuum hypothesis is true,

2) a continuum family of probability measures is weakly divisible,

and

3) this family admits an unbiased estimate of any parametric function $g \in \mathcal{B}_L$,

then this family admits a consistent estimate of parameters $\alpha \in \mathcal{A}$.

Proof. Let $\{\mu_\alpha, \alpha \in \mathcal{A}\}$ be a continuum measurable family of probability measures given on the measurable space $(X, \mathcal{B})$. Let us represent $\mathcal{A}$ as a transfinite sequence $\{\alpha_\xi\}_{\xi<\omega_1}$ and put $\mu_{\alpha_\xi} = \mu_\xi$ for an ordinal number ξ. There exists a family of measurable parts $(X_\xi)_{\xi<\omega_1}$ of the set X such that, for all $\xi, \nu < \omega_1$,

$$\mu_\xi(X_\mu) = \delta(\xi, \nu),$$

where $\delta(\xi, \nu)$ is the Kronecker symbol.
Assume that for all $\xi < \omega_1$

$$Y_\xi = X_\xi - \bigcup_{\nu<\xi} X_\nu.$$

Evidently, the family (Y_ξ) of weakly divisible parts of X is disjunctive and, for all $\xi < \omega_1$,

$$\mu_\xi(Y_\xi) = 1,$$

i.e., $\{\mu_\alpha, \alpha \in \mathcal{A}\}$ is a strongly divisible family and now consistent estimates are constructed just like in Theorem 2.3.
The theorem is proved.

REFERENCES

Kharazishvili, A.B. (1984). *Topological Aspects of Measure Theory*. Naukova Dumka, Kiev (in Russian).

Maleev, A.B., Serechenko, A.A. and Tarashchanskii, M.T. (1978). Families of probability measures admitting consistent estimates. In: *Probability Distributions in Infinite-Dimensional Spaces.*, Institute of Mathematics of Ukrainian SSR, Kiev, pp. 125–134 (in Russian).

Schaefer, H.H. (1966). *Topological Vector Spaces*. Collier-Macmillan Limited, New York.

Zerakidze, Z.S. (1984a). Structure of the family of probability measures. *Bull. Acad. Sci. Georgian SSR* **113**, 37–39 (in Russian).

Zerakidze, Z.S. (1984b). On weakly divisible and divisible families of probability measures. *Bull. Acad. Sci. Georgian SSR* **113**, 273–275 (in Russian).

Zerakidze, Z.S. (1986). A Hilbert space of measures. *Ukranian Math. J.* **38**, 148–154. (in Russian).

Part Five

STOCHASTIC ANALYSIS AND CONTROL

New Trends in Probab. and Statist., pp. 289 – 300
V. Sazonov and T. Shervashidze (Eds.)
1991 VSP/Mokslas

ASYMPTOTIC BEHAVIOUR OF SOLUTIONS OF STOCHASTIC EQUATIONS

V. BULDYGIN and V. KOVAL'
Kiev Polytechnical Institute, Pr. Pobedy, 37, 252056 Kiev, USSR
Institute of Mathematics of the Ukrainian Academy of Sciences,
Repin str. 3, 252601 Kiev, USSR

ABSTRACT

Asymptotic behaviour of solutions of linear stochastic recurrent equations is considered. The law of large numbers and the law of iterated logarithm are established.

INTRODUCTION

Consider a stochastic linear recurrent equation of the m-th order

$$\xi_n = a_{n1}\xi_{n-1} + \ldots + a_{mn}\xi_{n-m} + b_n\zeta_n, \quad n \geqslant 1, \tag{1}$$

with zero initial conditions

$$\xi_0 = \ldots = \xi_{1-m} = 0.$$

In this equation $(a_{nk};\ n \geqslant 1,\ 1 \leqslant k \leqslant n)$, $(b_n,\ n \geqslant 1)$ are sequences of real numbers, $(\zeta_n,\ n \geqslant 1)$ is a sequence of independent random variables. For $m \geqslant 2$ it is more convenient to study the asymptotic behaviour of the solution of equation (1) by means of a linear recurrent equation of the first order in $\mathbf{R}^m$ with matrix coefficients. For this purpose put

$$Y_n^T = (\xi_n,\ \ldots,\ \xi_{n-m+1}), \qquad Z_n^T = (b_n\zeta_n,\ 0,\ \ldots,\ 0),$$

$$A_n = \begin{pmatrix} a_{n1} & a_{n2} & \cdots & a_{nm} \\ & & & 0 \\ & I_{m-1} & & \vdots \\ & & & 0 \end{pmatrix}, \quad n \geqslant 1,$$

where the symbol T denotes the transposition of a matrix, I_{m-1} stands for the unit matrix of the $(m-1)$-th order, then one obtains

$$Y_n = A_nY_{n-1} + Z_n, \quad n \geqslant 1, \qquad Y_0 = 0. \tag{2}$$

The random vectors Y_n, $n \geqslant 1$, which are the solutions of equation (2), can be represented as

$$Y_n = A_n \dots A_2 Z_1 + A_n \dots A_3 Z_2 + \dots + Z_n, \quad n \geqslant 2.$$

If, for example, all matrices A_n, $n \geqslant 2$, are nondegenerate then the above representation can be transformed to the growing sums of independent random vectors with matrix normalizations:

$$Y_n = D_n(V_1 + \dots + V_n), \quad n \geqslant 1,$$

where

$$D_1 = I, \qquad D_n = \prod_{k=2}^{n} A_k, \quad n \geqslant 2;$$

$$V_1 = Z_1, \qquad V_k = \left(\prod_{j=2}^{k} A_j\right)^{-1} Z_k, \quad k \geqslant 2.$$

So, the study of the asymptotic behaviour of the sequence of random variables which is the solution of equation (1) can be reduced to limit theorems for sums of independent random vectors with matrix normalizations. Here a special form of the strong law of large numbers (SLLN) which goes back to (Prokhorov, 1950) appears to play an important role. Generalizations of this form of SLLN were obtained in (Loéve, 1955; Martikainen and Petrov 1977; Buldygin, 1978; Martikainen, 1979). For the case of sums of independent random vectors with matrix normalizations the corresponding form of SLLN was studied in (Buldygin and Solntsev, 1985; 1986; 1987). Finally, it was exactly this form of SLLN which led (see (Buldygin and Solntsev, 1987; 1988; 1989)) to the general criteria of convergence and boundedness almost surely (a.s.) of a sequence of random vectors which is the solution of stochastic equation (2) under the additional assumption that the sequence is Gaussian.
In this paper the asymptotic behaviour of the solution of stochastic equation (1) with constant coefficients is studied. The above mentioned criteria together with a special comparison technique and matrix analysis allows to find the asymptotics of the said solution up to a constant for Gaussian and sub-Gaussian perturbations $(\zeta_n,\ n \geqslant 1)$. Besides, the law of large numbers for the solution of stochastic equation (2) with a constant matrix is studied for the same type of perturbation $(Z_n,\ n \geqslant 1)$.

1. LINEAR RECURRENT EQUATIONS WITH CONSTANT COEFFICIENTS

1.1. *Statement of results*

Let a sequence of random variables $(\xi_n,\ n \geqslant 1)$ be the solution of a stochastic linear equation of the m-th order with constant coefficients

$$\xi_n = a_1 \xi_{n-1} + \dots + a_m \xi_{n-m} + b_n \zeta_n, \quad n \geqslant 1, \tag{3}$$

and initial conditions

$$\xi_0 = \xi_1^0, \ldots \xi_{1-m} = \xi_{m-1}^0.$$

The coefficients $a_1, \ldots, a_m$, $(b_n,\ n \geqslant 1)$ are supposed to be real numbers, $a_m \neq 0$; $(\zeta_n,\ n \geqslant 1)$ is a standard Gaussian or sub-Gaussian sequence (the precise definition of a standard sub-Gaussian sequence is given below); the initial values $(\xi_1^0, \ldots, \xi_{m-1}^0)$ form a random vector independent of the sequence $(\zeta_n,\ n \geqslant 1)$. Let us consider a normalizing sequence of real numbers $(c_n,\ n \geqslant 1)$ and suppose that $c_n \neq 0$, $n \geqslant 1$, and that one of the following conditions is fulfilled

$$\overline{\lim_{n\to\infty}} \left|\frac{c_{n+1}}{c_n}\right| < \infty, \qquad \underline{\lim_{n\to\infty}} \left|\frac{c_{n+1}}{c_n}\right| > 0. \tag{4}$$

Note that any monotone sequence satisfies one of these conditions.
Our goal is to establish conditions for $\overline{\lim\limits_{n\to\infty}}\, c_n\xi_n = 0$ a.s.,

$$0 < \overline{\lim_{n\to\infty}} \left|c_n\xi_n\right| < \infty \quad \text{a.s.}, \qquad 0 < \underline{\lim_{n\to\infty}} \left|c_n\xi_n\right| < \infty \quad \text{a.s.}$$

The first of these conditions is an analogue of the law of large numbers, the third one is an analogue of the law of iterated logarithm for the sequence $(\xi_n,\ n \geqslant 1)$. Naturally, the law of iterated logarithm is of greatest interest. But in fact, it is implied by the first relation which will be the central object for the investigation. Every relation indicated above take place under some joint restrictions on the sequences $(b_n,\ n \geqslant 1)$, $(c_n,\ n \geqslant 1)$ and coefficients $a_1, \ldots, a_m$. It is convenient to introduce restrictions on the coefficients $a_1, \ldots, a_m$ in terms of the parameters of the characteristic equation. Consider the characteristic equation

$$\lambda^m - (a_1\lambda^{m-1} + \ldots + a_m) = 0 \tag{5}$$

for equation (3). Let r be the maximal value of the moduli of the roots of equation (5), p be the maximum multiplicity of those roots the moduli of which are equal to r. It is clear that $1 \leqslant p \leqslant m$. Introduce the following notations:

$$f = \big(f(n),\, n \geqslant 1\big), \qquad g = \big(g(n),\, n \geqslant 1\big),$$

where

$$f(n) = \sum_{j=1}^{n} b_j^2 r^{-2j}\left(1 - \frac{j-1}{n}\right)^{2(p-1)}, \qquad g(n) = \sum_{j=1}^{n} b_j^2 r^{-2j}.$$

Further, let $\varphi(n) = c_n^2 r^{2n} n^{2(p-1)}$, $n \geqslant 1$, if the sequence $(c_n,\ n \geqslant 1)$ satisfies the first relation from condition (4), and let $\varphi(n) = c_{n-m+1}^2 r^{2n} n^{2(p-1)}$, $n \geqslant m$, if $(c_n,\ n \geqslant 1)$ satisfies the second relation of (4). Besides, let us fix an arbitrary constant $M > 0$ and put

$$\mathcal{L}_n\{f\} = \log\left(e + \sum_{k=1}^{n-1} \min\left\{M,\ \log\frac{f(k+1)}{f(k)}\right\}\right), \quad n \geqslant 2.$$

The sequence $\mathcal{L}_n\ \{g\}$, $n \geqslant 2$, can be defined in the same way. Note that the sequences $(f(n),\ n \geqslant 1)$ and $(g(n),\ n \geqslant 1)$ are nondecreasing. A sequence of independent Gaussian $N(0,1)$-distributed random variables will be called, as usual, a standard Gaussian sequence.

THEOREM 1. *Let the sequence $(\zeta_n,\ n \geqslant 1)$ in equation (3) be a standard Gaussian sequence.*

A) *If*

$$\lim_{n\to\infty} f(n) < \infty, \tag{6}$$

then

$$\lim_{n\to\infty} c_n \xi_n = 0 \qquad \text{a.s.} \tag{7}$$

holds if and only if

$$\lim_{n\to\infty} \varphi(n) = 0. \tag{8}$$

B) *If*

$$\lim_{n\to\infty} f(n) = \infty \tag{9}$$

and

$$\lim_{n\to\infty} \varphi(n) f(n) \mathcal{L}_n\{f\} = 0, \tag{10}$$

then relation (7) is valid.

C) *If relation (9) is fulfilled, the sequence $(\varphi(n) f(n),\ n \geqslant 1)$ tends decreasing to zero and there exist constants $L_1 \in (0,1]$ and $L_2 \geqslant 1$ such that one of the conditions*

$$L_1 g(n) \leqslant f(n), \quad n \geqslant 1, \tag{11}$$

$$f(n) \leqslant L_2 n^{2(1-p)} g(n), \quad n \geqslant 1, \tag{12}$$

is true, then condition (10) is necessary for relation (7) to be valid.

THEOREM 2. *Let the sequence $(\zeta_n,\ n \geqslant 1)$ in equation (3) be a standard Gaussian sequence.*

A) *If*

$$\lim_{n\to\infty} f(n) < \infty,$$

then

$$\sup_{n\geqslant 1} \frac{|\xi_n|}{r^n n^{p-1}} < \infty \qquad \text{a.s.} \tag{13}$$

B) *If*

$$\lim_{n\to\infty} f(n) = \infty$$

and one of relations (11) and (12) holds, then

$$\overline{\lim_{n\to\infty}} \frac{|\xi_n|}{r^n n^{(p-1)} (f(n)\mathcal{L}_n\{f\})^{1/2}} = \alpha \qquad \text{a.s.,} \tag{14}$$

where α is a nonrandom constant from $(0,\infty)$.

To illustrate the statement of Theorem 2 let us give some its direct corollaries.

COROLLARY 1. *If $p = 1$ and*

$$\lim_{n\to\infty} g(n) = \infty,$$

then there exists a nonrandom constant $\alpha \in (0, \infty)$ such that

$$\overline{\lim_{n\to\infty}} \frac{|\xi_n|}{r^n (g(n)\mathcal{L}_n\{g\})^{1/2}} = \alpha \qquad a.s.$$

COROLLARY 2. *If $r < 1$ and*

$$b_{n+1}^2 \geqslant r b_n^2, \quad n \geqslant n_0 \geqslant 1,$$

then there exists a nonrandom constant $\alpha \in (0, \infty)$ such that

$$\overline{\lim_{n\to\infty}} \frac{|\xi_n|}{|b_n|\sqrt{\log n}} = \alpha \qquad a.s.$$

COROLLARY 3. *If $b_n = b \neq 0$, $n \geqslant 1$, and $r = 1$, then there exists a nonrandom constant $\alpha \in (0, \infty)$ such that*

$$\overline{\lim_{n\to\infty}} \frac{|\xi_n|}{(n^{2p-1} \log\log n)^{1/2}} = \alpha \qquad a.s.$$

COROLLARY 4. *Let $b_n^2 r^{-2n} \sim q n^t$ as $n \to \infty$ with $t \geqslant -1$ and $q > 0$. If $t = -1$, then there exists a nonrandom constant $\alpha \in (0, \infty)$ such that*

$$\overline{\lim_{n\to\infty}} \frac{|\xi_n|}{r^n n^{p-1} \sqrt{\log n \log\log\log n}} = \alpha \qquad a.s.,$$

and, if $t > -1$, then

$$\overline{\lim_{n\to\infty}} \frac{|\xi_n|}{r^n \sqrt{n^{2p+t-1} \log\log n}} = \alpha \qquad a.s.$$

In Theorems 1, 2 and Corollaries 1 to 4 the perturbation sequences are Gaussian. To consider a more general case let us introduce the notion of a standard sub-Gaussian sequence.

Definition. A sequence of independent random variables $(\eta_n,\ n \geqslant 1)$ is called standard sub-Gaussian if

1) $\mathbf{E}\eta_n = 0,\ n \geqslant 1;$
2) $\mathbf{E}\eta_n^2 = 1,\ n \geqslant 1;$

3) $\sup_{n\geqslant 1}\tau(\eta_n)<\infty$, where

$$\tau(\eta_n)=\sup_{|\lambda|>0}\left(\frac{2\log \mathbf{E}\exp(\lambda\eta_n)}{\lambda^2}\right)^{1/2}.$$

The class of sub-Gaussian sequences contains standard Gaussian sequences and a great deal of other sequences. For example, a sequence of independent random variables $(\eta_n,\ n\geqslant 1)$ satisfying condition 1), 2) and such that $\sup_{n\geqslant 1}|\eta_n|\leqslant C<\infty$ a.s. is a standard sub-Gaussian sequence. Sufficient conditions of convergence to zero and boundedness of the solutions of equation (3) turn out to be the same if the sequence $(\zeta_n,\ n\geqslant 1)$ is supposed to be only standard sub-Gaussian.

THEOREM 3. *Let $(\zeta_n,\ n\geqslant 1)$ in equation (3) be a standard sub-Gaussian sequence. Then, under condition (6), relation (8) is sufficient for (7) to be valid. If (9) takes place, then relation (10) is sufficient for (7).*

THEOREM 4. *Let $(\zeta_n,\ n\geqslant 1)$ in equation (3) be a standard sub-Gaussian sequence.*

A) *If $\lim\limits_{n\to\infty} f(n)<\infty$, then (13) takes place.*

B) *If $\lim\limits_{n\to\infty} f(n)=\infty$, then a nonrandom constant α exists such that (14) holds.*

Remark 1. If in Corollaries 1 to 4 one chooses $(\zeta_n,\ n\geqslant 1)$ to be a standard sub-Gaussian sequence, then the statements of these corollaries still hold with $\alpha\in[0,\infty)$.

1.2. *Proofs of theorems*

The proof of Theorem 1 is rather tedious. So let us give a sketch of the proof.
Let $(Y_n,\ n\geqslant 1)$ be a Gaussian Markov sequence in $\mathbf{R}^m$ with zero means. Recall that every Gaussian Markov sequence in $\mathbf{R}^m$ is the solution of equation (2) with a Gaussian perturbation $(Z_n,\ n\geqslant 1)$.
Let $\Phi(l,n+1)=E\|Y_{n+1}-\mathbf{E}(Y_{n+1}\mid Y_l)\|^2$ be the conditional variance of the vector Y_{n+1} with respect to the vector Y_l, $0\leqslant l\leqslant n$, where $\|\cdot\|$ stands for the Euclidean norm of a $\mathbf{R}^m$-vector. Suppose Y_0 to be an arbitrary constant vector in $\mathbf{R}^m$. Then $\Phi(0,n)=E\|Y_n\|^2$. Let $\tilde f=(\tilde f(n),\ n\geqslant 0)$ and $(\tilde\varphi(n),\ n\geqslant 1)$ be sequences of positive numbers such that $\tilde f$ is nondecreasing and $\tilde f(0)=0$.

LEMMA 1 (Comparison principle). *Let $(Y_n,\ n\geqslant 1)$ be a Gaussian Markov sequence in $\mathbf{R}^m$ with zero means. If there exist sequences $(\tilde f(n),\ n\geqslant 1)$ and $(\tilde\varphi(n),\ n\geqslant 1)$ satisfying the above conditions and constants $C_2\geqslant C_1>0$, $L\geqslant 1$ such that for $0\leqslant l\leqslant n$ the inequalities*

$$C_1\tilde\varphi(n+1)\left(1-L\frac{\tilde f(l)}{\tilde f(n+1)}\right)\leqslant\Phi(l,n+1)\leqslant C_2\tilde\varphi(n+1)\left(1-\frac{\tilde f(l)}{\tilde f(n+1)}\right)$$

hold, then:

A) *if* $\lim_{n\to\infty} \tilde{f}(n) < \infty$, *then*

$$\lim_{n\to\infty} Y_n = 0 \quad \text{a.s.} \tag{15}$$

takes place if and only if $\lim_{n\to\infty} \tilde{\varphi}(n) = 0$;

B) *if*

$$\lim_{n\to\infty} \tilde{f}(n) = \infty \tag{16}$$

and

$$\lim_{n\to\infty} \tilde{\varphi}(n) \log \left(e + \sum_{k=1}^{n-1} \min \left\{ M,\ \log \frac{\tilde{f}(k+1)}{\tilde{f}(k)} \right\} \right) = 0 \tag{17}$$

with a fixed constant $M > 0$, *then relation (15) holds;*

C) *if condition* (16) *is fulfilled and the sequence* $(\tilde{\varphi}(n),\ n \geqslant 1)$ *decreases to zero, then condition* (17) *is necessary for* (15).

The proof is based on the following lemmas (see Buldygin and Solntsev (1989)).

LEMMA 2. *Let* $(Y_n, n \geqslant 1)$ *be an arbitrary Gaussian Markov sequence in* $\mathbf{R}^m$ *with zero means. Then the relation* $\lim_{n\to\infty} Y_n = 0$ *a.s. holds if and only if*

(i) $\lim_{n\to\infty} \Phi(0, n) = 0$;

(ii) *for all* $\varepsilon > 0$ *and for all nondecreasing subsequences* $(n_k, k \geqslant 1)$ *of the sequence of natural numbers*

$$\sum_{k=1}^{\infty} \exp\{ -\varepsilon / \Phi(n_k, n_{k+1}\} < \infty.$$

LEMMA 3. *Let* $(\eta_n,\ n \geqslant 1)$ *be a Gaussian Markov sequence in* $\mathbf{R}^1$ *with zero means and*

$$\sigma_n^2 = \mathbf{E}\eta_n^2, \qquad r_{n,n+1} = \mathbf{E}\eta_n\eta_{n+1}/(\sigma_n\sigma_{n+1}), \qquad n \geqslant 1.$$

A) *If* $\prod_{k=l}^{\infty} |r_{k,k+1}| > 0$, *for some* $l \geqslant 1$, *then*

$$\lim_{n\to\infty} \eta_n = 0 \qquad \text{a.s.}, \tag{18}$$

holds if and only if

$$\lim_{n\to\infty} \sigma_n^2 = 0.$$

B) *If, for any* $l \geqslant 1$,

$$\prod_{k=l}^{\infty} |r_{k,k+1}| = 0 \tag{19}$$

and

$$\lim_{n\to\infty} \sigma_n^2 \log \left(e + \sum_{k=1}^{n-1} \min \left\{ M, \log \frac{1}{|r_{k,k+1}|} \right\} \right) = 0, \tag{20}$$

where $M > 0$ is a fixed number, then relation (18) holds.

C) *If condition (19) is valid and the sequence $(\sigma_n^2, n \geqslant 1)$ decreases to zero as $n \to \infty$, then condition (20) is sufficient for (18).*

The proof of Lemma 1 can be obtained from Lemmas 2 and 3 by introducing an auxiliary Gaussian Markov sequence $(\tau_n,\ n \geqslant 1)$ in $\mathbf{R}^1$ with zero means and parameters $\sigma_n^2 = \tilde{\varphi}(n)$ and $r_{n,n+1} = \sqrt{\tilde{f}(n)/\tilde{f}(n+1)}$, $n \geqslant 1$.

LEMMA 4. *Let*

$$A = \begin{pmatrix} a_1 & a_2 & \dots & a_m \\ & & & 0 \\ & I_{m-1} & & \vdots \\ & & & 0 \end{pmatrix}, \qquad B = \begin{pmatrix} 1 \\ 0 \\ \vdots \\ 0 \end{pmatrix},$$

where $a_1, a_2, \dots, a_m$ are the coefficients of equation (3). Then the inequalities

$$Q_1 r^n (n+1)^{p-1} \leqslant \|A^n B\| \leqslant Q_2 r^n (n+1)^{p-1}, \quad n \geqslant 0,$$

hold, where Q_1, Q_2 are some positive constants.

Sketch of the proof of Theorem 1.
For simplicity let us suppose the initial conditions of equation (3) to be equal to zero. Let us distinguish between the cases when $m = 1$ and $m \geqslant 2$, where m is the order of equation (3).
If $m = 1$, then the statement of the theorem is implied by Lemma 3.
Suppose now that $m \geqslant 2$. Introduce the following vectors in $\mathbf{R}^m$

$$Y_n^T = (\xi_n, \dots, \xi_{n-m+1}), \qquad Z_n^T = (\zeta_n, 0, \dots, 0),$$

where T denotes the transposition of a matrix. Using condition (4), it is easy to see that $\lim_{n\to\infty} c_n \xi_n = 0$ a.s. is equivalent to $\lim_{n\to\infty} c_n Y_n = 0$ a.s. or $\lim_{n\to\infty} c_{n-m+1} Y_n = 0$ a.s., where $(Y_n, n \geqslant 1)$ is the solution of the equation

$$Y_n = AY_{n-1} + b_n Z_n, \quad n \geqslant 1, \qquad Y_0 = 0$$

in $\mathbf{R}^m$. For clarity, let us investigate the case of $\lim\limits_{n\to\infty} c_n Y_n = 0$ a.s. One can see that

$$\Phi(l, n+1) = \mathbf{E}\|c_{n+1}Y_{n+1} - \mathbf{E}(c_{n+1}Y_{n+1}|c_l Y_l)\|^2 =$$

$$= c_{n+1}^2 \sum_{i=l+1}^{n+1} b_i^2 \|A^{n+1-i}B\|^2, \qquad 0 \leqslant l \leqslant n.$$

By Lemma 4, the following equalities hold

$$Q_1 \Phi_0(l, n+1) \leqslant \Phi(l, n+1) \leqslant Q_2 \Phi_0(l, n+1), \quad 0 \leqslant l \leqslant n.$$

and $\Phi_0(0, n) = \varphi(n) f(n)$. Hence, using Lemma 1, one can obtain the statement of Theorem 1.

LEMMA 5.

A) *If* $\lim_{n\to\infty} f(n) < \infty$, *then*

$$\overline{\lim_{n\to\infty}} |c_n\xi_n| < \infty \quad \text{a.s.} \tag{21}$$

holds if and only if $\overline{\lim}_{n\to\infty} \varphi(n) < \infty$.

B) *If*

$$\lim_{n\to\infty} f(n) = \infty \tag{22}$$

and

$$\overline{\lim_{n\to\infty}} \varphi(n)f(n)\mathcal{L}_n\{f\} < \infty, \tag{23}$$

then relation (21) *holds.*

C) *If condition* (22) *is valid, the sequence* $(\varphi(n)f(n),\ n \geqslant 1)$ *decreases to zero as* $n \to \infty$ *and one of the conditions* (11) *and* (12) *takes place, then condition* (23) *is necessary for* (21).

The proof of Lemma 5 is immediately implied by Theorem 1 and the following

LEMMA 6 (Egorov, 1969). *Let* $(\xi_n,\ n \geqslant 1)$ *be a sequence of random variables. Then* $\sup_{n\geqslant 1} |\xi_n| < \infty$ a.s. *if and only if, for every sequence* $(v_n,\ n \geqslant 1)$ *decreasing to zero as* $n \to \infty$, *the sequence* $(v_n\xi_n,\ n \geqslant 1)$ *converges to zero almost surely.*

Proof of Theorem 2. Statement **A**) of Theorem 2 is implied by statement **A**) of Lemma 5 and Lemma 6.
Consider statement **B**). Put

$$c_n = \left(r^{2n}n^{2(p-1)}f(n)\mathcal{L}_n\{f\}\right)^{-1/2}, \quad n \geqslant 2.$$

Obviously, $(c_n,\ n \geqslant 2)$ satisfies the first of conditions (4). The condition $\lim_{n\to\infty} f(n) = \infty$ implies that the sequence $(\varphi(n)f(n) = (\mathcal{L}_n\{f\})^{-1}, n \geqslant 1)$ decreases to zero as $n \to \infty$. Using the inequality $\Phi(0,n) \leqslant Q_2\varphi(n)f(n),\ n \geqslant 1$, we conclude that $\mathbf{E}(c_n\xi_n)^2 \to 0,\ n \to \infty$. According to the oscillation theorem for Gaussian sequences, there exists a nonrandom constant $\alpha \in [0,\infty]$ such that

$$\overline{\lim_{n\to\infty}} |c_n\xi_n| = \alpha \quad \text{a.s.}$$

On the other hand, $\overline{\lim}_{n\to\infty} \varphi(n)f(n)\mathcal{L}_n\{f\} = 1$. Then, by statement **B**) of Lemma 5, $d < \infty$, and by statement **C**) of Theorem 1, $\alpha > 0$.
Theorem 2 is proved.
The transition from the Gaussian to the sub-Gaussian perturbation (Theorems 3, 4) is justified in details in (Buldygin and Solntsev, 1989).

2. LINEAR EQUATIONS OF THE FIRST ORDER WITH A CONSTANT MATRIX OF COEFFICIENTS IN $\mathbf{R}^m$

Let a random sequence $(Y_n,\ n \geqslant 1)$ in $\mathbf{R}^m$ be a solution of the equation

$$Y_n = AY_{n-1} + B_n\Gamma_n, \qquad n \geqslant 1, \quad Y_0 = 0,$$

where A, $(B_n,\ n \geqslant 1)$ are arbitrary matrices of order m and $(\Gamma_n,\ n \geqslant 1)$ is a sequence of independent Gaussian $\mathcal{N}(0, I)$-distributed random vectors.

Introduce the following notation: $\rho(A)$ is the spectral radius of the matrix A; $\|\cdot\|$ and $\|\cdot\|_E$ are the vector and matrix Euclidean norms in $\mathbf{R}^m$, respectively; $f = (f(n),\ n \geqslant 1)$, where

$$f(n) = \sum_{i=1}^{n} \|B_i\|_E^2.$$

THEOREM 5. *Let* $\rho(A) < 1$.

A) *If* $\lim_{n\to\infty} f(n) < \infty$, *then*

$$\lim_{n\to\infty} c_n \sum_{i=1}^{n} Y_i = 0 \quad \text{a.s.} \tag{24}$$

holds if and only if

$$\lim_{n\to\infty} c_n = 0.$$

B) *If*

$$\lim_{n\to\infty} f(n) = \infty \tag{25}$$

and

$$\lim_{n\to\infty} c_n^2 f(n) \mathcal{L}_n\{f\} = 0, \tag{26}$$

then relation (24) *holds.*

C) *If condition* (25) *is valid and the sequence* $(c_n^2 f(n),\ n \geqslant 1)$ *decreases to zero as* $n \to \infty$ *then condition* (26) *is necessary for* (24).

Proof can be carried out in the same way as that of Theorem 1.

THEOREM 6. *Let* $\rho(A) < 1$ *and* $\lim_{n\to\infty} f(n) = \infty$. *Then the following analogue of the law of iterated logarithm holds*

$$\lim_{n\to\infty} \frac{\|\sum_{i=1}^{n} Y_i\|}{\sqrt{f(n)\mathcal{L}_n\{f\}}} = \alpha \quad \text{a.s.,}$$

where $\alpha \in (0, \infty)$ *is a nonrandom constant.*

Proof is analogous to the proof of Theorem 2.

COROLLARY 5. *If* $\rho(A) < 1$, $\lim_{n\to\infty} f(n) = \infty$ *and the sequence* $(B_n,\ n \geqslant 1)$ *satisfies the inequality*

$$\|B_{n+1}\|_E^2 \leqslant L \sum_{i=1}^{n} \|B_i\|_E^2$$

with some constant $L > 0$, *then there exists a nonrandom constant* $\alpha \in (0, \infty)$ *such that*

$$\overline{\lim_{n\to\infty}} \frac{\|\sum_{i=1}^{n} Y_i\|}{\sqrt{f(n) \log\log f(n)}} = \alpha \quad \text{a.s.}$$

Remark 2. Similarly to Remark 1 in all the statements of Section 2 the standard Gaussian sequence $(\Gamma_n,\ n \geqslant 1)$ in $\mathbf{R}^m$ can be replaced by a standard sub-Gaussian sequence, the latter one can be understood as a sequence of i.i.d. random vectors in $\mathbf{R}^m$ the components of which are independent sub-Gaussian vectors with zero means and unit variances. Besides, the sufficient conditions of convergence to zero stated in Theorem 5 remain the same and the constant α in Theorem 6 belongs to $[0, \infty)$. As application of the obtained results, consider the problem of finding a solution x_* of a system of linear algebraic equations $Ax = b$. Rewrite this system in an equivalent form $x = (I + \tau A)x - \tau b$, where the number τ is such that $\rho(I + \tau A) < 1$. The method of simple iteration for the determination of the solution x_* under random perturbations (e.g., calculation error) can be represented as follows

$$Y_n = (I + \tau A)Y_{n-1} - \tau b + B_n \Gamma_n, \quad n \geqslant 1,$$

where $Y_0 \in \mathbf{R}^m$ and $(\Gamma_n,\ n \geqslant 1)$ stands for the random sequence defined above. For the approximate determination of the root x_* the averaging procedure is used: $x_* \approx (1/n) \sum_{i=1}^{n} Y_i$, and then the convergence

$$\lim_{n \to \infty} \frac{1}{n} \sum_{i=1}^{n} Y_i = x_* \quad \text{a.s.} \tag{27}$$

is investigated for any $Y_0 \in \mathbf{R}^m$.

THEOREM 7. A) *If* $\lim_{n \to \infty} f(n) < \infty$, *then relation* (27) *is valid.*

B) *If* $\lim_{n \to \infty} f(n) = \infty$, *then the relation*

$$\lim_{n \to \infty} \frac{1}{n^2} f(n) \mathcal{L}_n\{f\} = 0 \tag{28}$$

is sufficient for (27).

C) *If* $\lim_{n \to \infty} f(n) = \infty$ *and the sequence* $((1/n^2) f(n),\ n \geqslant 1)$ *is decreasing to zero as* $n \to \infty$, *then condition* (28) *is necessary for* (27).

Proof is obvious.

Now let us estimate the rate of convergence of this method.

THEOREM 8. *Let* $\lim_{n \to \infty} f(n) = \infty$. *Then*

$$\overline{\lim_{n \to \infty}} \frac{\|(1/n) \sum_{i=1}^{n} Y_i - x_*\|}{n^{-1} \sqrt{f(n) \mathcal{L}_n\{f\}}} = \alpha \qquad \text{a.s.,}$$

where α, $0 < \alpha < \infty$, *is a nonrandom constant.*

REFERENCES

Buldygin, V.V. (1978). Strong law of large numbers and convergence to zero of Gaussian sequences. *Theory Probab. Math. Statist.* **19**, 33-41.

Buldygin, V.V. and Solntsev, S.A. (1985). The contraction principle for cosumming operators and SLLN for weighted sums. In: *Proc. of the Fourth Vilnius Conference on Probability Theory and Mathematical Statistics,*, VNU Science Press, Utrecht, pp. 87-102.

Buldygin, V.V. and Solntsev, S.A. (1986). The contraction principle and strong law of large numbers for weighted sums. *Theory Probab. Appl.* **31**, 452-462.

Buldygin, V.V. and Solntsev, S.A. (1987). The strong law of large numbers for sums of independent random vector with operator normalizations and null convergence of Gaussian sequences. *Theory Probab. Appl.* **32**, 243-256.

Buldygin, V.V. and Solntsev, S.A. (1988). Equivalence of sample and sequential continuity of Gaussian processes and the continuity of Gaussian Markov processes. *Theory Probab. Appl.* **33**, 624-637.

Buldygin, V.V. and Solntsev, S.A. (1989). *The Functional Methods in Problems of Summation of Random Variables.* Naukova Dumka, Kiev (in Russian).

Loéve, M. (1955). *Probability Theory.* Van Nostrand, Princeton.

Martikainen, A.I. and Petrov, V.V. (1977). On necessary and sufficient conditions for the law of iterated logarithm. *Theory Probab. Appl.* **22**, 16-23.

Martikainen, A.I. (1979). On necessary and sufficient conditions for the strong law of large numbers. *Theory Probab. Appl.* **24**, 813-819.

Prokhorov, Yu.V. (1950). *On the Strong Law of Large Numbers.* Izvestija AN SSSR, ser. Math., **14**, pp. 523-536 (in Russian).

New Trends in Probab. and Statist., pp. 301 - 350
V. Sazonov and T. Shervashidze (Eds.)
1991 VSP/Mokslas

DECOMPOSITION OF THE MAXIMUM OF SEMIMARTINGALES AND GENERALIZED ITÔ'S FORMULA

R.J. CHITASHVILI and M.G. MANIA
A. Razmadze Mathematical Institute,
Rukhadze str. 1, 380093 Tbilisi, Georgia, USSR

ABSTRACT

The generalized Itô's formula is derived for the composition $\Phi(\omega, t, \xi_t(\omega))$, where ξ is a semimartingale and the function $\Phi(\omega, t, x)$ is a semimartingale in (ω, t) for every $x \in \mathbf{R}$ and convex in x for every pair (ω, t). This formula is then applied to the uniqueness problem for a stochastic differential equation. As a by-product of the derivation, a decomposition formula for the maximum of semimartingales is obtained which is shown to have interesting applications to the stochastic maximum principle.

Contents

Introduction
1. Decomposition of the maximum of semimartingales
1.1 Finite case
1.2 On convergence of semimartingales
1.3 General case
2. Generalized Itô's formula
2.1 Itô's effect in an implicit form
2.2 Continuity of decomposition terms
2.3 Smooth case. Weakly twice differentiability
2.4 Integration by parts
3. Applications
3.1 Uniqueness problem for a stochastic differential equation
3.2 The stochastic maximum principle
3.3 The variations of Itô's formula
4. Summary of results

INTRODUCTION

In this paper we state conditions under which the maximum of semimartingales is a semimartingale and represent the quantities involved in the decomposition in terms of stochastic line integrals along the maximizing processes. Further, based on the possibility to represent a convex function in terms of maximization, we derive Itô–Ventzel's formula for parametrized semimartingales which are convex in the parameter. We indicate several cases in which explicit expressions for Itô's effect can be given. As applications, we consider the problem of uniqueness of a solution of a stochastic differential equation with a drift expressed in terms of local times and the derivation of an adjoint system of equations of the stochastic maximum principle. Finally, we give a short summary of existing approaches to the generalization of Itô's formula and state some problems.

1. DECOMPOSITION OF THE MAXIMUM OF SEMIMARTINGALES

1.1 *Finite case*

Assume that, on a probability basis $(\Omega, \mathcal{F}, F = (\mathcal{F}_t,\ t \geqslant 0), \mathbf{P})$, a finite family of semimartingales $X = \big(X(t,k),\ t \geqslant 0\big)$, $1 \leqslant k \leqslant n$, is given.
Let us recall the definition of local times. If $\xi = (\xi_t,\ t \geqslant 0)$ is a semimartingale, then, by Tanaka's formula, for every $a \in \mathbf{R}$, we have

$$|\xi_t - a| = |\xi_0 - a| + \int_0^t \operatorname{sign}(\xi_{s-} - a) d\xi_s + L^\xi(t,a) +$$

$$+ \sum_{s \leqslant t} \big(|\xi_s - a| - |\xi_{s-} - a| - \operatorname{sign}(\xi_{s-} - a)\Delta\xi_s \big), \tag{1}$$

where $\operatorname{sign} = -\mathbf{1}\{(-\infty, 0]\} + \mathbf{1}\{(0,\infty)\}$ and L^ξ is a continuous increasing process called the local time of ξ at a point $a \in \mathbf{R}$. We shall consider a measurable modification of L^ξ which is continuous in t and right-continuous having left limits in a. As has been proved in (Yor, 1978), under the assumption that $\sum_{s \leqslant t} |\Delta\xi_s| < \infty$ a.s. for all $t \geqslant 0$, such a modification exists. Moreover, this modification is a continuous at (t,a) if and only if

$$\int_0^t \mathbf{1}\{(\xi_s = a)\} d\alpha_s = 0 \quad \text{a.s.}, \tag{2}$$

where $\alpha = (\alpha_t,\ t \geqslant 0)$ is the finite-variational part in the decomposition of the continuous semimartingale $\xi - \Sigma\Delta\xi$. It should be mentioned that under the latter condition the local time can be defined as a limit depending on (t,a) (Yor, 1978):

$$L^\xi(t,a) = \lim_{\varepsilon \to 0} \frac{1}{\varepsilon} \int_0^t \mathbf{1}\{(a \leqslant \xi_t \leqslant a + \varepsilon)\} d\langle \xi^c, \xi^c \rangle_s \quad \text{a.s.}, \tag{3}$$

where ξ^c is a continuous martingale of ξ and $\langle \xi^c, \xi^c \rangle$ is its square characteristic.

LEMMA 1. *The process*

$$\overline{X}_t^n = \max_{k \leqslant n} X(t,k)$$

is a semimartingale and, for any predictable process $u = (u_t,\ t \geqslant 0)\colon \Omega \times [0,\infty[\to \{1,2,\ldots,n\}$ *with the property*

$$\overline{X}_{t-}^n = \Big(\max_{k \leqslant n} X(\cdot,k)\Big)_{t-} = X\big(t-, u(t)\big), \quad t \geqslant 0, \quad \text{a.s.}, \tag{4}$$

the following decomposition is valid:

$$\overline{X}_t^n = \overline{X}_0^n + \sum_{k=1}^{n} \int_0^t 1\{u_s = k\} X(ds,k) + V_t^n, \tag{5}$$

where V^n *is an increasing process defined by*

$$V_t^n = \sum_{k=1}^{n} \int_0^t 1\{u_s = k\} \mathcal{L}_{(ds,0)}^{\overline{X}^n - X(\cdot,k)}$$

and

$$\mathcal{L}^{\xi}(t,a) = \frac{1}{2} L^{\xi}(t,a) + \sum_{s \leqslant t} 1\{(\xi_{s-} = a)\} \Delta \xi_s$$

for a semimartingale ξ *such that* $\xi - a \geqslant 0$.

Proof. Evidently, $\overline{X}^n$ is a semimartingale (it follows from (1) and the equality $\max(x,y) = (|x-y| + x + y)/2$).
Since, for almost all $\omega \in \Omega$, the left limit of the maximum $\overline{X}^n$ at any point t is equal to the left limit of one of the functions $X(t,k)$, the function u with property (4) exists, $\sum_{k=1}^n 1\{u_s = k\} = 1$, and

$$\{u_s = k\} \subset \big\{\overline{X}^n(s-) = X(s-,k)\big\}. \tag{6}$$

Therefore, decomposition (5) is easily derived from the formula

$$\mathcal{L}^{\xi}(t,0) = \int_0^t 1\{(\xi_{s-} = 0)\} d\xi_s \tag{7}$$

which is valid for non-negative semimartingales ξ.
Indeed, by applying (7) to $\overline{X}_t^n - X(t,k)$, $t \geqslant 0$, we obtain

$$\overline{X}_t^n = \sum_{k=1}^{n} \int_0^t 1\{u_s = k\} X(ds,k) + \sum_{k=1}^{n} \int_0^t 1\{u_s = k\} \big(\overline{X}^n(ds) - X(ds,k)\big) =$$

$$= \sum_{k=1}^{n} \int_0^t 1\{u_s = k\} X(ds,k) + \sum_{k=1}^{n} \int_0^t 1\{u_s = k\} \mathcal{L}_{(ds,0)}^{\overline{X}^n - X(\cdot,k)}.$$

Using the notation

$$\int_0^t \mathcal{Y}(ds, u_s) = \sum_{k=1}^{n} \int_0^t 1\{u_s = k\} \mathcal{Y}(ds,k),$$

for a semimartingale $\mathcal{Y}$, depending on the parameter k and a parameter-valued predictable process u, we can write (5) in terms of line integrals over the path u:

$$\overline{X}_t^n = \overline{X}_0^n + \int_0^t X(ds, u_s) + \int_0^t \overline{\mathcal{L}}(ds, u_s), \tag{5'}$$

where $\overline{\mathcal{L}}(t,k) = \mathcal{L}_{(t,0)}^{\overline{X}^n(\cdot) - X(\cdot,k)}$.

1.2. *On convergence of semimartingales*

For the sake of convenience, we begin this section with listing the basic objects and the notation used in what follows.

Assume that

$(\Omega, \mathcal{F}, \mathbf{P})$ is the main probability space with filtration $F = (\mathcal{F}_t,\ t \geqslant 0)$ and the symbol $\mathcal{P}$ denotes the σ-algebra of predictable subsets of $\Omega \times [0, \infty[$;

$\mathcal{A}^+$ $(\mathcal{A})$ is the set of integrable increasing processes (processes with integrable variation);

$\mathcal{M}$ $(\mathcal{M}^c,\ \mathcal{M}^d,\ \mathcal{M}^2)$ is the set of all (continuous, purely discontinuous, square integrable) martingales;

$\mathcal{M}_{\text{loc}}$ $(\mathcal{M}^2_{\text{loc}},\ \mathcal{A}_{\text{loc}}$, etc.) is the class of local martingales (locally square integrable martingales, processes with locally integrable variation etc.);

$S(S_p)$ is the set of semimartingales (special semimartingales);

$\langle X, m\rangle$ is the mutual characteristic (angle bracket) of the martingale part M of a special semimartingale X which admits the decomposition $X = M + A$, $M \in \mathcal{M}^2_{\text{loc}}$, $A \in \mathcal{A}_{\text{loc}}$, and an element $m \in \mathcal{M}^2_{\text{loc}}$ with $\langle X, m\rangle = \langle M, m\rangle$ which is the unique process from $\mathcal{A} \cap \mathcal{P}$ such that $Mm - \langle M, m\rangle \in \mathcal{M}_{\text{loc}}$;

[X,Y] is the characteristic (square bracket) of $X, Y \in S$;

D denotes the space of the right continuous functions;

$H \cdot X = (\int_0^t H_s dX_s,\ t \geqslant 0)$ is the stochastic integral with respect to $X \in S$;

$X^u = (\int_0^t X(ds, u_s),\ t \geqslant 0)$ is the stochastic line integral with respect to the semimartingale $X = (X_t,\ t \geqslant 0)$ over the path u.

Let $(X_n,\ n \geqslant 1)$ be a sequence of semimartingales and $X = (X_t,\ t \geqslant 0)$ be some adapted process. We are interested in finding out when the convergence in probability (in Skorokhod's or uniform metrics) of the sequence $(X_n,\ n \geqslant 1)$ to the process X implies that X is also a semimartingale and, furthermore, when their martingale parts converge. To handle this problem, we have to introduce the following

Condition **UT** (uniform tightness).

For any $\varepsilon > 0$ there exists $K > 0$ such that

$$\mathbf{P}(|H^n \cdot X_t^n| > k) < \varepsilon$$

for all n and H^n, where $H^n \in \mathcal{H}_t$, $\mathcal{H}_t$ is a set of elementary predictable processes of the form $H_s^n = \sum_i \mathcal{J}_{t_i}^n \mathbf{1}\{]t_i, t_{i+1}]\}$, where $\mathcal{J}_{t_i}^n$ are $\mathcal{F}_{t_i}$-measurable and $|\mathcal{J}_{t_i}^n| \leqslant 1$.

From (Jakubowski et al., 1989) it follows that if condition **UT** is satisfied, then the convergence in probability (in Skorokhod's metric) of the sequence $(X^n,\ n \geqslant 1)$ to the process X is sufficient for the process X to be a semimartingale.

The following lemma states that under conditions 1) – 4) the limit process is a semimartingale and these conditions ensure convergence in probability of the martingale parts $(M^n,\ n \geqslant 1)$ of the sequence $(X^n,\ n \geqslant 1)$ to the martingale M in the decomposition of X. Note that if condition 4) is satisfied, then 2) and 3) together are equivalent to condition **UT**.

LEMMA 2. *Let $(X^n,\ n \geqslant 1)$ be a sequence of semimartingales admitting the decomposition $X^n = M^n + A^n$, $M^n \in \mathcal{M}^2_{\text{loc}}$, $A^n \in \mathcal{A}_{\text{loc}}$, and let X be a process. If the conditions*

1) $\sup_{s\leqslant t} |X_s^n - X_s| \xrightarrow{\mathbf{P}} 0$ *for all* $t \geqslant 0$,

2) $\lim_{N\to\infty} \overline{\lim}_{n\to\infty} \mathbf{P}(\langle M^n\rangle_t \geqslant N) = 0$,

3) $\lim_{N\to\infty} \overline{\lim}_{n\to\infty} \mathbf{P}(\mathrm{Var}_{[0,t]}(A^n) \geqslant N) = 0$,

4) $\sup_n \mathbf{E} \sup_{s\leqslant t} (\Delta X_s^n)^2 < \infty$,

are satisfied, then the process $X = (X_t,\ t \geqslant 0)$ also is a semimartingale and

$$\langle M^n - M\rangle_t \xrightarrow{\mathbf{P}} 0, \qquad \sup_{s\leqslant t} |A_s^n - A_s| \xrightarrow{\mathbf{P}} 0 \tag{8}$$

for all $t \geqslant 0$, where M (respectively, A) is the martingale part (respectively, the part having bounded variation) of the semimartingale X.

Proof. By Itô's formula, we have

$$(X_t^n - X_t^m)^2 = 2\int_0^t (X_{s-}^n - X_{s-}^m)\, d(X_s^n - X_s^m) + [X^n - X^m, X^n - X^m]. \tag{9}$$

Applying Lenglart's inequality (see, e.g., (Liptser and Shiryayev, 1986)) and taking into account condition 2), we have

$$\mathbf{P}\left(\left|\int_0^t (X_{s-}^n - X_{s-}^m)\, d(M_s^n - M_s^m)\right| \geqslant \varepsilon\right) \leqslant$$

$$\leqslant \frac{c}{\varepsilon^2} + \mathbf{P}\left(\int_0^t (X^N_{s-} - X^m_{s-}) d\langle M^n - M^m \rangle_s \leqslant c \right) \leqslant$$

$$\leqslant \frac{c}{\varepsilon^2} + \mathbf{P}\left(\sup_{s \leqslant t} (X^n_s - X^m_s) \langle M^n - M^m \rangle_t > c \right) \longrightarrow 0, \quad n \to \infty, \quad m \to \infty,$$

for all $c > 0$, $\varepsilon > 0$. On the other hand, conditions 1) and 3) imply

$$\mathbf{P}\left(\left| \int_0^t (X^n_{s-} - X^m_{s-})\, d(A^n_s - A^m_s) \right| \geqslant \varepsilon \right) \leqslant \mathbf{P}\Big(\Big(\mathrm{Var}_{[0,t]}(A^n) +$$

$$+ \mathrm{Var}_{[0,t]}(A^m) \Big) \sup_{s \leqslant t} |X^n_s - X^m_s| \geqslant \varepsilon \Big) \longrightarrow 0, \quad n \to \infty, \quad m \to \infty,$$

and from equality (9) we obtain the convergence in probability of the square brackets

$$[X^n - X^m, X^n - X^m]_t \xrightarrow{\mathbf{P}} 0, \quad n \to \infty, \quad m \to \infty. \tag{10}$$

Since, for any stopping time τ, we have the martingale inequalities (see (Dellasherie and Meyer, 1980))

$$\mathbf{E}[M^n - M^m, M^n - M^m]^{1/2}_\tau \leqslant \mathbf{E}[X^n - X^m, X^n - X^m]_\tau, \tag{11}$$

and Davis's inequality

$$\mathbf{E} \sup_{s \leqslant \tau} |M^n_s - M^m_s| \leqslant 4\mathbf{E}[M^n - M^m, M^n - M^m]^{1/2}_\tau,$$

Lenglart's inequality (with an unpredictable dominating process) is applicable:

$$\mathbf{P}\left(\sup_{s \leqslant t} |M^n_s - M^m_s| \geqslant \varepsilon \right) \leqslant$$

$$\leqslant \frac{1}{\varepsilon}\left(b + 4\mathbf{E} \sup_{s \leqslant t} \Delta[X^n - X^m]^{1/2}_s \right) + \mathbf{P}\left([X^n - X^m]^{1/2}_t \geqslant b \right). \tag{12}$$

Since $\Delta[X^n - X^m]^{1/2}_s \leqslant |\Delta X^n_s - \Delta X^m_s|$, conditions 1) and 4) imply that $\mathbf{E}\sup_{s \leqslant t} \Delta[X^n - X^m]^{1/2}_s \to 0$ (for all $t \geqslant 0$ as $n \to \infty$, $m \to \infty$) and from relations (10), (12) we have the convergence

$$\sup_{s \leqslant t} |M^n_s - M^m_s| \xrightarrow{\mathbf{P}} 0 \quad \text{for all } t \geqslant 0 \quad (n \to \infty, \quad m \to \infty).$$

Hence there exists a process $M = (M_t,\ t \geqslant 0)$ such that $M \in \mathcal{M}^2_{\mathrm{loc}}$ and

$$\sup_{s \leqslant t} |M^n_t - M_t| \xrightarrow{\mathbf{P}} 0 \quad (n \to \infty) \quad \text{for all } t \geqslant 0. \tag{13}$$

Note that from condition (4) it follows that (13) is equivalent to

$$\langle M^n - M\rangle_t \xrightarrow{\mathbf{P}} 0 \quad (n \to \infty) \quad \text{for all } t \geqslant 0.$$

Now we can show that $A = X - M$ is a process with locally bounded variation. Relation (13) and condition 1) imply that

$$\sup_{s\leqslant t} |A_s^n - A_s| \xrightarrow{\mathbf{P}} 0 \quad (n \to \infty) \quad \text{for all } t \geqslant 0. \tag{14}$$

Consequently, for some subsequence $(A^{n'}, n')$, we have

$$\sum_{i=1}^{k} |A_{t_{i+1}}^{n'} - A_{t_i}^{n'}| \to \sum_{i=1}^{k} |A_{t_{i+1}} - A_{t_i}| \quad \text{a.s.} \quad (n' \to \infty) \tag{15}$$

for any finite partition $0{=}t_0 < t_1 < \ldots < t_k = t$ of the interval $[0,t]$. On the other hand, by condition 3), for any $\varepsilon > 0$, there exist N and Ω^N such that $\mathbf{P}(\Omega^N) < \varepsilon$ and $\mathrm{Var}_{[0,t]}(A^n) \leqslant k$ for all n on the set $\omega \in \Omega^N$. Therefore, (15) implies that the variation of the process $A = (A_t,\ t \geqslant 0)$ is bounded by a number N when $\omega \in \Omega$ and, since ε is arbitrary, we have $\mathrm{Var}_{[0,t]}(A) < \infty$ a.s. for all $t \geqslant 0$.

COROLLARY 1. *Assume that $(X^n,\ n \geqslant 1)$ is a sequence of semimartingales such that for some process X we have $\sup_{s\leqslant t}|X_s^n - X_s| \xrightarrow{\mathbf{P}} 0$ for all $t \geqslant 0$ as $n \to \infty$. If there exists a sequence $(\tau_k,\ k \geqslant 1)$ of stopping times with $\tau_k \uparrow \infty$ a.s. such that*

$$\sup_n \|X^n\|_{H^2(\tau_k)} < \infty \quad \textit{for every } k \geqslant 1,$$

then the process x also is a semimartingale and, for every $k \geqslant 1$, we have

$$\|M^n - M\|_{H^2(\tau_k)} \to 0, \qquad \|A^n - A\|_{S^2(\tau_k)} \to 0, \quad n \to \infty,$$

where the semimartingale X admits the decomposition $X = M + A$ and, for some stopping time τ, satisfies (see (Emery, 1979))

$$\|X\|_{H^p(\tau)} = \inf_{X=M+A} \left\| [M]_\tau^{1/2} + \int_0^\tau |dA_s| \right\|_{L^p},$$

$$\|X\|_{S^p(\tau)} = \left\| \sup_{s\leqslant\tau} |X_s| \right\|_{L^p}.$$

1.3. *The General Case*

Let a family $\{F(t,x) = M(t,x) + A(t,x),\ t \geqslant 0,\ x \in \mathbf{R}\}$, $M(\cdot,x) \in \mathcal{M}^2_{\mathrm{loc}}$, $A(\cdot,x) \in \mathcal{A}_{\mathrm{loc}}$, of special semimartingales be given. Assume that there exists an increasing predictable process $K = (K_t,\ t \geqslant 0)$ dominating the characteristics of F, i.e.,

$$\langle M(\cdot,x), M(\cdot,x)\rangle \ll K, \quad \mathrm{Var}\, A(\cdot,x) \ll K \quad \text{for each } x \in \mathbf{R}.$$

Let $(\mathcal{X}^n,\ n \geqslant 1)$ be some increasing sequence of finite subsets of $\mathbf{R}$ with $\overline{\bigcup_n \mathcal{X}^n} = \mathbf{R}$. For $x, y \in \mathbf{R}$ we define

$$\overline{F}^n(t) = \sup_{x\in\mathcal{X}^n} F(t,x), \quad \overline{F}(t) = \sup_{x\in\mathbf{R}} F(t,x),$$

$$\begin{aligned} \varphi_t(x,y) &= d\langle M(\cdot,x), M(\cdot,y)\rangle_t / dK_t, \\ \psi_t(x,y) &= \varphi_t(x,x) - 2\varphi_t(x,y) + \varphi_t(y,y), \\ H_t(x,m) &= d\langle M(\cdot,x), m\rangle_t / dK_t, \\ g_t(x) &= dA(t,x)/dK_t. \end{aligned} \tag{16}$$

Let μ^K be the Dolean measure associated with the increasing process $K = (K_t,\ t \geqslant 0)$. It means that, for every bounded predictable process Y, we have

$$\mathbf{E}\int_0^\infty Y_s\, dK_s = \iint_{\Omega\times[0,\infty[} Y \mu^K(d\omega, ds).$$

Besides, we shall use the following notation for integrals:

$$Y \cdot K = (Y \cdot K_t)_{t\geqslant 0} = \left(\int_0^t Y_s\, dK_s,\ t \geqslant 0\right).$$

Denote by U^- the class of predictable processes u^- such that, for all $t \geqslant 0$,

$$\overline{F}(t-) = \left(\sup_{x\in\mathbf{R}} F(\cdot,x)\right)_{t-} = F(t-, u_t^-) \quad \text{a.s.} \tag{17}$$

and let U^+ be the class of optional process u^+ such that, for all $t \geqslant 0$,

$$\overline{F}(t) = F(t, u_t^+) \quad \text{a.s.}$$

Introduce also the class U of predictable process $\widehat{u}$ taking values in the space $Q(\mathbf{R})$ of all probabilities on $(\mathbf{R}, \mathcal{B}(\mathbf{R}))$ such that, for all $t \geqslant 0$,

$$\overline{F}(t-) = \int_{\mathbf{R}} F(t-,x)\widehat{u}_t(dx) \quad \text{a.s.} \tag{18}$$

Suppose that the following conditions are satisfied.
Continuity conditions:

a) the function $F = (F(t,x),\ t \geqslant 0,\ x \in \mathbf{R})$, for every $x \in \mathbf{R}$, is $\mathbf{P}$-a.s. right-continuous in t and continuous in x uniformly with respect to t $\mathbf{P}$-a.s. on every compact from $\mathbf{R}_+ \times \mathbf{R}$.

b) the function $\varphi(x,y)$ ($g(x)$, respectively) a.e. with respect the measure μ^K is continuous in (x,y) at points $x = y$ (respectively, in x).

Boundedness conditions:

c) for every $t \geqslant 0$

$$\sup_{x\in\mathbf{R}} |F(t,x)| < \infty \quad \text{a.s.},$$

for every $n \geqslant 1$ the processes

$$\int_0^t \sup_{x\in[-n,n]} \varphi_s(x,x)\, dK_s, \qquad \int_0^t \sup_{x\in[-n,n]} |g_s(x)|\, dK_s,$$

$$\sup_{s\leqslant t} \sup_{x\in[-n,n]} \big(\Delta F(s,x)\big)^2, \quad t \geqslant 0,$$

are locally integrable.

Conditions of the behavior at the infinity:

d) there exists a sequence of stopping times $(\tau_i,\ i \geqslant 1)$ with $\tau_i \uparrow \infty$ a.s. such that, for every $i \geqslant 1$,

$$\lim_{N\to\infty} \mathbf{P}\Big\{ \sup_{s\leqslant\tau_i} \big(\overline{F}(s-) - \sup_{|x|\geqslant N} F(s-,x)\big) = 0 \Big\} = 0,$$

and

$$\lim_{N\to\infty} \mathbf{P}\Big\{ \sup_{s\leqslant\tau_i} \big(\overline{F}(s) - \sup_{|x|\geqslant N} F(s,x)\big) = 0 \Big\} = 0.$$

Remark 1. Condition **d**) is needed for the function F not to attain its maximum value at the infinity. More precisely, if $u^- = (u_s^-(\omega),\ s \geqslant 0)$ is a maximizing function from U^- (i.e., $\overline{F}(s-) = F(s-,u_s^-)$ a.s. for all $s \geqslant 0$), then

$$\Big(\sup_{s\leqslant\tau} |u_s^-| \geqslant n\Big) \subseteq \Big(\sup_{s\leqslant\tau} \big(\overline{F}(s-) - \sup_{|x|\geqslant n} F(s-,x)\big) = 0\Big)$$

and condition **d**) implies that $\mathbf{P}(\sup_{s\leqslant\tau} |u_s^-| \geqslant n) \to 0,\ n \to \infty$. In the case where there exists a unique pair (u^-,u^+) of maximizing functions (which means that if, for some function $\widetilde{u}^-(\widetilde{u}^+)$, we have

$$\overline{F}(s-) = F(s-,\widetilde{u}_s^-)\big(\overline{F}(s) = F(s,\widetilde{u}_s^+)\big) \quad \text{a.s.}$$

for all $s \geqslant 0$, then $\widetilde{u}^- = u^-$ ($\widetilde{u}^+ = u^+$) a.s. for all $s \geqslant 0$), conditions **a**), **b**), **c**), and **d**) can be replaced by somewhat weaker conditions **a***), **b***), **c***), and **d***). We shall assume that the functions F, φ, and g are separable with respect to x and measurable with respect to (ω, s, x).

a*) For every $x \in \mathbf{R}$ the function F is $\mathbf{P}$-a.s. right-continuous in t and continuous in probability uniformly with respect to t along the functions u^- and u^+ on every compact $[0,t] \times [-n,n]$, i.e., for each $t \geqslant 0,\ n \geqslant 1$, we have

$$\sup_{s\leqslant t} \sup_{|x|\leqslant n,\ |x-u_s^-|\leqslant\varepsilon} \big(F(s-,u_s^-) - F(s-,x)\big) \xrightarrow{\mathbf{P}} 0, \quad \varepsilon \to 0,$$

$$\sup_{s\leqslant t}\ \sup_{|x|\leqslant n,\ |x-u_s^+|\leqslant\varepsilon}(F(s,u_s^+)-F(s,x))\xrightarrow{\mathbf{P}}0,\quad \varepsilon\to 0.$$

b*) The functions $\varphi(x,y)$, $g(x)$ are continuous μ^K-a.s. along the function u^-, i.e., μ^K-a.s.

$$\lim_{\varepsilon\to 0}\ \sup_{|x-u_s^-|\leqslant\varepsilon}\left(\varphi_s(u_s^-,u_s^-)-2\varphi_s(x,u_s^-)+\varphi_s(x,x)\right)=0$$

$$\lim_{\varepsilon\to 0}\ \sup_{|x-u_s^-|\leqslant\varepsilon}\left|g(s,u_s^-)-g(s,x)\right|=0.$$

c*) For every $t\geqslant 0$ we have $|F(t,u_t^+)|<\infty$ $\mathbf{P}$-a.s. The processes

$$\int_0^t\varphi_s(u_s^-,u_s^-)\,dK_s,\qquad \int_0^t\left|g_s(u_s^-)\right|dK_s$$

are locally integrable and there exists some $\varepsilon>0$ such that the process

$$\sup_{s\leqslant t}\ \sup_{|x-u_s^+|\leqslant\varepsilon}\left(\Delta F(s,x)\right)^2$$

also is locally integrable.

d*) For each $t\geqslant 0$

$$\mathbf{P}\left(\sup_{s\leqslant t}|u_s^-|\geqslant n\right)\longrightarrow 0,\qquad \mathbf{P}\left(\sup_{s\leqslant t}|u_s^+|\geqslant n\right)\longrightarrow 0,\qquad n\to\infty.$$

Denote by $\overline{V}=\overline{\max}_a V^a$, the maximum of the set of increasing process $(V^a,\ a\in A)$ with $V_0^a=0$, i.e., the least increasing process such that, for each $a\in A$, the difference $\overline{V}-V^a$ is an increasing process and $\overline{V}_0=0$.

THEOREM 1.

(i) *If conditions* **a**)–**d**) *are satisfied, the process* $\overline{F}$ *is a semimartingale and there exist* $u\in U^-$ *and* $\hat{u}\in\widehat{U}$ *such that*

$$\overline{F}(t)=\overline{F}(0)+\int_0^t M(ds,u_s)+\int_0^t\int_{\mathbf{R}}g_s(x)\hat{u}_s(dx)dK_s+V_t,\tag{19}$$

where $V=(V_t,\ t\geqslant 0)$ *is an increasing process* $(V\in\mathcal{A}_{\mathrm{loc}}^+)$.

(ii) *Under conditions* **a**)–**d**) *the martingale part* M *of the process* $\overline{F}$ *is independent of the choice of* u, *i.e.,*

$$\overline{M}=\int M(ds,u_s)$$

for any function $u\in U^-$.

(iii) *If there exists a unique pair of maximizing processes* (u^-,u^+), $u^-\in U^-$, $u^+\in U^+$ *and conditions* **a***)–**d***) *are satisfied, then* $\overline{F}$ *can be expressed as*

$$\overline{F}(t)=\overline{F}(0)+\int_0^t M(ds,u_s^-)+\int_0^t g(s,u_s^-)\,dK_s+\overline{\max_x}\,\mathcal{L}^{\overline{F}-F(\cdot,x)}(t,0).\tag{20}$$

Remark 2. The notation

$$M_t^u = \int_0^t M(ds, u_s)$$

is used here for a stochastic line integral which is uniquely determined (Chitashvili and Mania, 1986) as an element of the stable space of martingales generated by the family $(M(\cdot, x),\ x \in \mathbf{R})$ for which

$$\langle M^u, m\rangle_t = \int_0^t H_s(u_s, m)\, dKs$$

for every $m \in \mathcal{M}^2$, where $H_s(x, m) = d\langle M(\cdot, x), m\rangle_s / dK_s$. If conditions **b***), **c***) (respectively, **b**), **c**)) are satisfied for $u \in U^-$, then M^u is uniquely defined (respectively, M^u is uniquely defined for all predictable functions u: $\Omega \times [0, \infty[\to \mathbf{R})$.

Proof of Theorem 1. i) Let $\mathcal{X}^n = \{k/2^n\colon -n2^n \leqslant k \leqslant n2^n\}$. By Lemma 1, the process $\overline{F}^n(t) = \sup_{x \in \mathcal{X}^n} F(t, x)$ for each $n \geqslant 1$ is a semimartingale and for every u^n: $\Omega \times [0, \infty[\to \mathcal{X}^n$, in addition to $\overline{F}^n(t-) = F(t-, u_t^n)$, we have

$$\overline{F}^n(t) = \overline{F}^n(0) + \overline{M}_t^n + \overline{B}_t^n + V_t^n, \tag{21}$$

where

$$\overline{M}_t^n = \sum_{x \in \mathcal{X}^n} \int_0^t \mathbf{1}\{(u_s^n = x)\}\, M(ds, x),$$

$$\overline{B}_t^n = \sum_{x \in \mathcal{X}^n} \int_0^t \mathbf{1}\{(u_s^n = x)\}\, A(ds, x) = \int_0^t g_s(u_s^n)\, dK_s,$$

$$V_t^n = \sum_{x \in \mathcal{X}^n} \int_0^t \mathbf{1}\{(u_s^n = x)\}\, \overline{\mathcal{L}}^n(ds, x) \in \mathcal{A}_{\text{loc}}^+.$$

We can show that, for the semimartingale $(\overline{F}^n,\ n \geqslant 1)$ and for the process $\overline{F}$, conditions of Lemma 2 are satisfied. First, we verify that $\sup_{s \leqslant t} |\overline{F}^n(s) - \overline{F}(s)| \to 0$ for any $t \geqslant 0$. Let u^+ be some maximizing function from U^+. Since,

$$\overline{F}(t) - \overline{F}^n(t) \leqslant \min_{x \in \mathcal{X}^n} \left(\overline{F}(t) - F(t, x)\right) =$$

$$= \min_{x \in \mathcal{X}^n} \Big[\mathbf{1}\{(|u_t^+| < n)\}\big(F(t, u_t^+) - F(t, x)\big) + \mathbf{1}\{(|u_t^+| \geqslant n)\}\big(F(t, u_t^+) - F(t, x)\big)\Big] \leqslant$$

$$\leqslant \sup_{x\colon |x - u_t^+| \leqslant 1/2^n} \mathbf{1}\{(|u_t^+| < n)\}\big(F(t, u_t^+) - F(t, x)\big) + \mathbf{1}\{(|u_t^+| \geqslant n)\}\big(\overline{F}(t) - F(t, n)\big) \leqslant$$

$$\leqslant \sup_{\substack{|x|\leqslant n,\ |y|\leqslant n \\ |x-y|\leqslant 1/2^n}} \big|F(t,x)-F(t,y)\big| + \mathbf{1}\{(|u_t^+|\geqslant n)\}\big(\overline{F}(t)-F(t,n)\big) \tag{22}$$

and

$$\mathbf{P}(\sup_{s\leqslant t}\sup_{|x-y|\leqslant 1/2^n}\big|F(t,x)-F(t,y)\big|\geqslant\varepsilon)\to 0$$

as $n\to\infty$ by condition **a**) of the theorem, the convergence of $\sup_{s\leqslant t}|\overline{F}^n(s)-\overline{F}(s)|$ follows from condition **d**) because

$$\mathbf{P}\Big(\sup_{s\leqslant t}\mathbf{1}\{(|u_s^+|\geqslant s)\}>0\Big)=\mathbf{P}\Big(\sup_{s\geqslant t}|u_s^+|\geqslant n\Big)\leqslant$$

$$\leqslant\mathbf{P}\Big\{\sup_{s\leqslant t}(\overline{F}(s)-\sup_{|x|\geqslant n}F(s,x))=0\Big\}. \tag{23}$$

Now we can show that condition 2) of Lemma 2 is satisfied. Let

$$\Omega_j=\Big\{\omega\colon \sup_{s\geqslant t}(\overline{F}(s-)-\sup_{|x|\geqslant j}F(s-,x))>0\Big\}.$$

Evidently, $\Omega_j\subseteq\{\omega\colon \sup_{s\leqslant t}|u_s^n|\geqslant j \text{ for all } n\geqslant 1,\}$ and hence, for all n, $\varphi(u^n,u^n)\leqslant\sup_{x\in[-j,j]}\varphi(x,x)$ on the set $[0,t]\times\Omega_j$. Therefore (note that $\langle\overline{M}^n\rangle = = \int\varphi(u^n,u^n)\,dK$),

$$\lim_N\overline{\lim_n}\mathbf{P}\Big(\int_0^t\varphi_s(u_s^n,u_s^n)\,dK_s>N\Big)=$$

$$=\lim_N\overline{\lim_n}\mathbf{P}\Big[\Big(\int_0^t\varphi_s(u_s^n,u_s^n)\,dK_s>N,\ \Omega_j\Big)\Big]+$$

$$+\mathbf{P}\Big[\Big(\int_0^t\varphi_s(u_s^n,u_s^n)\,dK_s>N,\ \Omega_j^c\Big)\Big]\leqslant$$

$$\leqslant\lim_N\mathbf{P}\Big(\int_0^t\sup_{x\in[-j,j]}\varphi_s(x,x)\,dK_s>N\Big)+\mathbf{P}(\Omega_j^c) \tag{24}$$

and condition 2) follows from **c**) and **d**), since j is arbitrary and $\lim_{j\to\infty}\mathbf{P}(\Omega_j^c)=0$ by condition **d**).

Equality (21) implies that

$$\mathrm{Var}_{[0,t]}(\overline{B}^n+\overline{V}^n)=\int_0^t|g(s,u_s^n)|\,dK_s+V_t^n\leqslant$$

$$\leqslant 2 \int_0^t |g(s,u_s^n)|\, dK_s + |\overline{M}_t^n| + |\overline{F}_t^n|. \tag{25}$$

Since the sequence $(\langle \overline{M}^n \rangle_t,\ n \geqslant 1)$ satisfies condition 2 of Lemma 2, i.e., is tight, the tightness of the sequence $\mathrm{Var}_{[0,t]}(\overline{B}^n + \overline{V}^n)$ follows from (25), the tightness of the sequence $(|g(u^n)| \cdot K,\ n \geqslant 1)$ (which in turn can be deduced from conditions **c**) and **d**) by following the same line of argument as in the proof of tightness of $(\langle \overline{M}^n \rangle_t,\ n \geqslant 1)$, and the fact that $\sup_x |F(t,x)| < \infty$ a.s. for all $t \geqslant 0$.
The validity of condition 4) of Lemma 2 is implied by condition **c**) and the inequality

$$(\Delta \overline{F}^n(s))^2 \leqslant \sup_{x \in [-n,n]} (\Delta F(s,x))^2. \tag{26}$$

Now, by Lemma 2, we obtain that the process $(\overline{F}(t),\ t \geqslant 0)$ is a special semimartingale and, for any $t \geqslant 0$, we have

$$\langle \overline{M}^n - \overline{M} \rangle_t \xrightarrow{\mathbf{P}} 0, \quad n \to \infty, \tag{27}$$

where $\overline{M}$ is the semimartingale part of the semimartingale $\overline{F}$. We need to show the existence of $u \in U^-$ such that

$$\overline{M}_t = M_t^u = \int_0^t M(ds, u_s), \quad t \geqslant 0.$$

Since

$$\langle \overline{M}^i - \overline{M}^j \rangle_t = \int_0^t \psi_s(u_s^i, u_s^j)\, dK_s \xrightarrow{\mathbf{P}} 0, \quad i \to \infty, \quad j \to \infty,$$

and the inequality

$$\psi(u^i, u^j) \leqslant 4 \sup_{x \in [-N,N]} \varphi(x,x) \quad \mu^K\text{-a.e.}$$

holds on the set

$$\Omega_N = (\omega\colon \sup_{s \leqslant t} (\overline{F}(s-) - \sup_{|x| \geqslant N} F(s-,x)) > 0)$$

for all i and j, we have, by condition **c**),

$$\mathbf{E}\langle \overline{M}^j - \overline{M}^i \rangle_{t \wedge \tau_m} = \iint \psi_s(u_s^i,\, u_s^j) \mathbf{1}\{[\![0, \tau_m, \wedge t]\!]\} \mathbf{1}\{\Omega_N\} \mu^K(d\omega, ds) \to 0, \quad i,j \to \infty$$

where $(\tau_m,\ m \geqslant 1)$ is a common localizing sequence for the processes $(\sup_{x \in [-n,n]} \psi(x,x) \cdot K,\ n \geqslant 1)$ (the existence of such sequence of stopping times follows from (Dellasherie, 1978)). Consequently, for every m and N a subsequence $i(m,N)$ can be chosen for which

$$\Psi_s(u_s^i,u_s^j)1\{\Omega_N\}1\{s \leqslant \tau_m \wedge t\} \to 0 \quad \mu^K - \text{a.e.}, \quad i \to \infty, \quad j \to \infty,$$

and since $\mathbf{P}(\Omega_N) \to 1$, $\tau_m \uparrow \infty$ a.s., $m, N \to \infty$, for diagonal subsequence (for convenience we preserve the earlier notation of i,j), we can state that

$$\psi_s(u_s^i,u_s^j)I_{[0,t]} \to 0, \quad \mu^K\text{-a.e.} \quad i \to \infty,\ j \to \infty. \tag{28}$$

We need the statement of a general character presented in (Chitashvili and Toronjadze, 1981).

LEMMA 3. *Let $(f_n(x),\ n \geqslant 1)$ be some sequence of measurable mappings $f_n\colon (X,\mathcal{B}) \to Y$ with some metric space Y and let, for each $x \in X$, the set $\{f_n(x),\ n \geqslant 1\}$ be a subset of some compact . Then some subsequence $(n_j(x),\ j \geqslant 1)$ of measurable functions $n_j\colon (X,\mathcal{B}) \to \mathbf{N} = (0,1,\dots)$ can be found such that $n_j(x) \leqslant n_{j+1}(x)$, $\lim_{j\to\infty} n_j(x) = \infty$ and the sequence $f_{n_j(x)}(x)$ is convergent for each $x \in X$.*

Thus, the sequence $\big(n_j(\omega,\, t),\ j \geqslant 1\big)$ of predictable indices can be found such that the sequence $\tilde{u}_t^j(\omega) = (u_t^{n_j(\omega,t)}(\omega),\ j \geqslant 1)$ converges to some predictable process u. Since, for each n, we have $\overline{F}^n(t-) = F(t-,u_t^n)$ a.s. for all $t \geqslant 0$, it follows that $u \in U^-$, i.e.,

$$\overline{F}(t-) = F(t-,u_t) \quad \text{a.s. for all } t \geqslant 0.$$

Let us show that the processes $\overline{M}$ and $M^u = \int M(ds,u_s)$ are indistinguishable. Clearly, the sequence $(\overline{M}^i = M^{u^i},\ i \geqslant 1)$ converges weakly (in $\mathcal{H}^2_{\text{loc}}$) in probability to the martingale $\overline{M}$, i.e., for each $m \in \mathcal{M}^2$, we have

$$\mathbf{P}\{|\langle M^{u^i} - \overline{M},\ m\rangle_t| \geqslant \varepsilon\} \longrightarrow 0, \quad n \to \infty. \tag{29}$$

It is not difficult to check the inequality (see (Chitashvili and Mania, 1986))

$$\big[H(x,m) - H(y,m)\big]^2 \leqslant \psi(x,y)\frac{d\langle m\rangle}{dK} \quad \mu^K \ - \ \text{a.e.} \tag{30}$$

(We recall that $H(x,m) = d\langle M(\cdot\,,x),m\rangle/dK$.)
From (30) and (28) it follows that the sequence $\big(H(u^i,m),\ i \geqslant 1\big)$ converges μ^K-a.e. to a predictable process $\big(H_s(m),\ s \geqslant 0\big)$. Therefore the sequence $\big(H(\tilde{u}^j),\ j \geqslant 1\big)$ converges to the same limit. But, by the continuity of the function $H(x,m)$ μ^K-a.e. in x (which follows from inequality (30) and condition **b**)), the construction of the sequence $(\tilde{u}^j,\ j \geqslant 1)$ implies that the seqence $H(\tilde{u}^j,m)$ converges μ^K-a.e. to the function $H(u,m)$. So, for any $m \in \mathcal{M}^2$, we have

$$\mathbf{P}\Big\{|\langle M^{u^i} - M^u,m\rangle_t| \geqslant \varepsilon\Big\} = \mathbf{P}\Bigg\{\Bigg|\int_0^t (H_s(u_s^i,m) - H_s(u_s,m))dK_s\Bigg| \geqslant \varepsilon\Bigg\} \to 0$$

as $i \to \infty$, which together with (29) gives

$$\mathbf{P}\Big\{\big|\langle M^u - \overline{M}, m\rangle_t\big| \geqslant \varepsilon\Big\} = 0$$

for any $m \in \mathcal{M}^2$, $t \geqslant 0$, $\varepsilon \geqslant 0$. Hence, $\langle M^u - \overline{M}, m\rangle = 0$ for every $m \in \mathcal{M}^2$ and we obtain that M^u and $\overline{M}$ are indistinguishable.
Now consider the sequence

$$\overline{B}_t^n = \int_0^t g(s, u_s^n)dK_s, \quad n \geqslant 1.$$

Provisionally assume that the sequence $\big(g(u^n),\ n \geqslant 1\big)$ is weakly compact in $L_1\big([0,t]\times\Omega, \mathcal{P}, \mu^K\big)$ (Banach space of predictable processes integrable with respect to the measure μ^K).
Then there exists a subsequence $g(u^{n'})$ weakly convergent in L_1 to some $h \in L_1(\mu^K)$, i.e., for each $f \in L_\infty(\mu^K)$ (we do not change the notation for $g(u^n)$),

$$\mathbf{E}\int_0^t \big(g(s, u_s^n) - h(s)\big)f(s)dK_s \to 0, \quad n \to \infty. \tag{31}$$

On the other hand, the element h can be represented as a strong limit of convex combinations of elements $g(u^n)$ (see, e.g., (Yosida, 1967)). Let α_i^n, $1 \leqslant i \leqslant n$, with $\alpha_i^n \geqslant 0$, $\sum_{i=1}^n \alpha_i^n = 1$, be such that

$$\mathbf{E}\int_0^t \Big|\sum_{i=1}^n \alpha_i^n g(s, u_s^i) - h(s)\Big| dK_s \to 0, \quad n \to \infty. \tag{32}$$

Let $\hat{u}_s^n(dx)$ be a distribution concentrated for each (ω, t) at points u_s^j

$$\hat{u}_s^n(u_s^j) = \alpha_j^n, \quad 1 \leqslant j \leqslant n.$$

Evidently,

$$\sum_{i=1}^n \alpha_i^n g(s, u_s^i) = \int_{\mathbf{R}} g(s, x)\hat{u}_s^n(dx), \tag{33}$$

and we can choose a subsequence for which

$$\int g(s, x)\hat{u}_s^n(dx) \longrightarrow h(s), \quad \mu^K\text{-a.e.}, \quad n \to \infty. \tag{34}$$

Applying Lemma 3, a sequence $n_j(\omega, t) \uparrow \infty$ can be found such that the sequence of measures $\hat{u}_t^{n_j(\omega,t)}$ at each (ω, t) converges weakly to some predictable distribution $\hat{u}$. By the continuity of the function g in x and by (34), we have

$$h_t = \int g(t, x)\hat{u}_t(dx) \quad \mu^K\text{-a.e.} \tag{35}$$

To prove that $\hat{u} \in \widehat{U}$, it is sufficient to show that, for every fixed $m \geqslant 1$, the inequality

$$\left(\sup_{x \in X^m} F(\cdot, x)\right)_{t-} \leqslant \int_{\mathbf{R}} F(t-, x)\hat{u}_t(dx)$$

is correct.
Obviously, we can state that, for each fixed $i = 1, 2, \ldots,$

$$\alpha_i^n \to 0, \qquad n \to \infty.$$

Indeed, for each $k \geqslant 1$, we can choose a sequence $\alpha_{j,k}^n \geqslant 0$, $\sum_{j=k}^n \alpha_{j,k}^n = 1$, $k \leqslant j \leqslant n$, $k \geqslant 1$, such that

$$\varepsilon_{n,k} = \mathbf{E} \int_0^t \left| \sum_{j=k}^n \alpha_{j,k}^n g(s, u_s^j) - h(s) \right| dK_s \longrightarrow 0, \quad n \to \infty.$$

It is necessary only to choose k_n such that $k_n \to \infty$, $\varepsilon_{n,k_n} \to 0$ and set

$$\alpha_i^n = \alpha_{i,k_n}^n, \quad k_n \leqslant i \leqslant n.$$

Since $F(t-, u_t^j) \geqslant \overline{F}^m(t-)$, for $j > m$, we have

$$\int F(t-, x)\hat{u}_t^n(dx) = \sum_{j=1}^m \alpha_j^n F(t-, u_t^j) + \sum_{j=m+1}^n \alpha_j^n F(t-, u_t^j) \geqslant$$

$$\geqslant \sum_{j=1}^m \alpha_j^n F(t-, u_t^j) + \overline{F}^m(t-)$$

and, passing to the limit as $n \to \infty$, we obtain the desired inequality. Now we must get rid of the assumption concerning the weak compactness in $L_1(\mu^K)$ of the sequence $(g(u^n), n \geqslant 1)$. Let

$$\Omega_j = \left\{\omega : \sup_{s \leqslant t}(\overline{F}(s-) - \sup_{|x| \geqslant j} F(s-, x)) > 0\right\}.$$

Since

$$g(u^n) \leqslant \sup_{x \in [-j,j]} |g(x)|$$

on the set Ω_j and

$$\mathbf{E} \int 1\{s \leqslant \tau_m\} \sup_{x \in [-j,j]} |d_s(x)| K_s < \infty$$

the sequence $(g_s(u_s^n) 1\{\Omega_j\} 1\{\tau_m \geqslant s\}, n \geqslant 1)$ is uniformly integrable for every j, m and hence (see, e.g., (Dunford and Schwartz, 1962)) is weakly compact in $L_1(\Omega \times [0, t], \mu^K)$. Therefore, by (31), (35), there exists a predictable random measure $\hat{u}_s^{j,m}(dx)$ such that

$$\mathbf{E}\int_0^t \mathbf{1}\{s \leqslant \tau_m\}\mathbf{1}\{\Omega_j\}\Big(g(s,u_s^n) - \int_{\mathbf{R}} g(s,x)\hat{u}_s^{j,m}(dx)\Big)f(s)dK_s \longrightarrow 0 \qquad (36)$$

as $n \to \infty$, for each $j \geqslant 1$, $m \geqslant 1$, and every bounded function f. Hence, $\hat{u}_s^{i,m}(dx) = \hat{u}_s^{i',m'}(dx)$ on the set $\Omega_i \mathbf{1}\{s \leqslant \tau_m\}$ for all $i' > i$, $m' > m$, $\mathbf{P}(\Omega_i \cap (\tau_m \geqslant \geqslant t)) \to 1$ as $i \to \infty$, $m \to \infty$, and the limit

$$\hat{u}_s(dx) = \lim_{\substack{i\to\infty \\ m\to\infty}} \hat{u}_s^{m,i}(dx) \qquad (37)$$

exists.

By virtue of the fact that, for every m and j, $\hat{u}_s^{m,j}(dx)$ belongs to the class $\widehat{U}$ (see (37)) and by the continuity of F in x, we have that the predictable random measure $\hat{u}_s(dx)$ also belongs to the class $\widehat{U}$. Finally, we can show that the process

$$\overline{V}(t) = \overline{F}(0) - \int_0^t M(ds,u_s) - \int_0^t\int_{\mathbf{R}} g_s(x)\hat{u}_s(dx)dk_s \qquad (38)$$

is increasing.

By Lemma 1,

$$V^n(t) = \overline{F}^n(t) - \overline{F}^n(0) - \int_0^t M(ds,u_s^n) - \int_0^t g_s(u_s^n)dK_s \qquad (39)$$

is an increasing process. Besides, we have proved that

$$\sup_{s\leqslant t}\left|\overline{F}^n(s) - \overline{F}(s)\right| \xrightarrow{\mathbf{P}} 0, \quad n\to\infty, \quad t\geqslant 0, \qquad (40)$$

$$\Big\langle \int M(ds,u_s^n) - \int M(ds,u_s)\Big\rangle_t \xrightarrow{\mathbf{P}} 0, \quad n\to\infty, \quad t\geqslant 0, \qquad (41)$$

and, for some subsequence (it can be assumed that the sequence $(V^n,\ n\geqslant 1)$ itself possesses this property),

$$\mathbf{E}\int_0^t \mathbf{1}\{\Omega_j \cap (s\leqslant \tau_m)\}\Big[g(s,u_s^n) - \int_{\mathbf{R}} g_s(x)\hat{u}_s(dx)\Big]f(s)\,dK_s \to 0 \qquad (42)$$

for every bounded function f and for any fixed pair (j,m). Relations (38) and (42) imply that for any bounded $\mathcal{F}_t$-measurable random variable ι we have (recall that the sequences $\overline{F}^n$ and $\overline{M}^n$ also converge in the sense of (42) to $\overline{F}$ and $\overline{M}$, respectively)

$$\mathbf{E}\,\mathbf{1}\{\Omega_j\cap(t\leqslant\tau_m)\}\iota(V_s - V_{s'}) = \lim_{n\to\infty}\mathbf{E}\,\mathbf{1}\{\Omega_j\cap(\tau_m\geqslant t)\}\iota(V_s^n - V_{s'}^n)\geqslant 0,$$

if $s \geqslant s'$, for every j and m. Hence, $V_s \geqslant V_{s'}$ a.s. on the set $\Omega_j \cap (\tau_m \geqslant t)$. Since $\tau_m \uparrow \infty$, $m \to \infty$, and a.s. $\mathbf{P}(\Omega_j) \to 1$, $j \to \infty$, passing to the limit, we obtain $V_s \geqslant V_{s'}$ a.s.
ii) We prove that the martingale part $\overline{M} = M^u$ of the process $\overline{F}$ is independent of $u \in U^-$. For any $(x, y) \in \mathbf{R}^2$

$$\int_0^t \mathbf{1}\{F(s-,x) = F(s-,y)\} d\big\langle M(\cdot,x) - M(\cdot,y)\big\rangle_s = 0 \qquad \text{a.s.}$$

which is a direct consequence of (1), (7). Since $\psi(x,y) = d\langle M(x) - M(y)\rangle / dK$, this means that, for each $(s,y) \in \mathbf{R}^2$,

$$\mathbf{1}\big\{F(s-,x) = F(s-,y)\big\}\psi_s(x,y) = 0 \quad \mu^K\text{-a.e.} \tag{43}$$

To prove that

$$\langle M^u - M^v \rangle = 0$$

for each u, $v \in U$, we have to show that there exists a predictable set $\mathcal{D} \subset \Omega \times [0, \infty[$ such that $\mu^K(\mathcal{D}^c) = 0$ and, for each $(\omega, t) \in \mathcal{D}$,

$$\mathbf{1}\big\{F(s-,x) = F(s-,y)\big\}\psi_s(x,y) = 0$$

for any pair $(x,y) \in \mathbf{R}^2$ satisfying

$$F(t-,x) = F(t-,y) = \overline{F}(t-).$$

Consider the set

$$\mathcal{D} = \bigcap_{(x,y)\in Q^2} \mathcal{D}(x,y),$$

where $\mathcal{D}(x,y) = \Big\{(\omega,t)\colon \mathbf{1}\big\{F(t-,x) = F(t-,y)\big\}\psi_t(x,y) = 0\Big\}$ and $Q \subset \mathbf{R}$ is a countable dense set in $\mathbf{R}$. Obviously, $\mu^K(\mathcal{D}^c) = 0$.
Let $(\omega,t) \in \mathcal{D}$ and fix a pair (x,y) such that

$$F(t-,x) = F(t-,y) = \overline{F}(t-).$$

Consider the following three cases.
1°. In some neighbourhoods $\mathcal{O}(x)$, $\mathcal{O}(y)$ of the points x, y, we have

$$F(t-,a) = F(t-,b) = \overline{F}(t-), \quad a \in \mathcal{O}(x),\ b \in \mathcal{O}(y).$$

Evidently, there exist sequences $a_n \to x$, $b_n \to y$ (as $n \to \infty$), $a_n \in \mathcal{O}(x) \bigcap Q$, $b_n \in \mathcal{O}(y) \bigcap Q$, and, by continuity of ψ, we have (43).
2°. There exist sequences $a_n \to x$, $b_n \to y$ $(n \to \infty)$ such that $F(t-,a_n) < \overline{F}(t-)$, $F(t-,b_n) < \overline{F}(t-)$. Since $F(t,x)$ is continuous in x, it is easy to find sequences $\widetilde{a}_n$, $\widetilde{b}_n$ $(n \geqslant 1)$ such that $\widetilde{a}_n \to x$, $\widetilde{b}_n \to y$, $\widetilde{a}_n \in Q$, $\widetilde{b}_n \in Q$, $F(t-,\widetilde{a}_n) = F(t-,\widetilde{b}_n)$, and we have (43) again.

3°. In a neighbourhood $\mathcal{O}(x)$

$$F(t-,a) = \overline{F}(t-), \quad a \in \mathcal{O}(x),$$

and there exists a sequence $b_n \in Q$, $b_n \to y$, for which $F(t-,b_n) < \overline{F}(t-)$.
We find the maximal r_0 such that

$$F(t-,a) = \overline{F}(t-), \quad |x-a| \leqslant r_0.$$

There exists $\bar{a}$ with $F(t-,a) = \overline{F}(t-)$ for which a sequence can be found such that $a_n \in Q$, $F(t-,a_n) < \overline{F}(t-)$, $a_n \to \bar{a}$, and, moreover, there exist sequence a'_n, $a''_n \in Q$, $a_n \to x$, $a''_n \to \bar{a}$, for which

$$F(t-,a'_n) = F(t-,a''_n).$$

Thus, (43) holds for the pair $(x,\bar{a})$. As for $(\bar{a},y)$, the question reduces to case 2°.
iii) Unlike the martingale part $\overline{M}$, the bounded-variation terms $\overline{B}$ and V in the decomposition of the process $\overline{F}$ generally depend (by the construction) on the chosen sequence of maximizing finite-valued processes $(u^n,\ n \geqslant 1)$. However, if there exists a unique maximizing process $u^- \in U^-$: $\overline{F}(t-) = F(t-,u_t^-)$ a.s. for all $t \geqslant 0$, then the increasing process can be directly characterized in terms of the process $F(t,x)$. In the case where there exists a unique maximizing function u^- the measure $\hat{u}$ for every couple (ω,t) is concentrated at the point $u^-(\omega,t)$ and hence

$$\overline{B}_t = \int_0^t \int_{\mathbf{R}} g(s,x)\hat{u}_s(dx)dK_s = \int_0^t g(s,u_s^-)dK_s. \tag{44}$$

The proof of the fact that

$$V_t = \overline{F}(t) - \overline{F}(0) - \int_0^t M(ds,u_s^-) - \int_0^t g(s,u_s^-)dK_s$$

is an increasing process follows the same line of argument as the proof of the first part of the theorem.
Let us show, for example, the tightness of the family $(\langle \overline{M}^n\rangle_t,\ n \geqslant 1)$.
By conditions **a***) and **b***), we have that

$$\lim_{\varepsilon\to 0} \mathbf{P}\Big\{ \inf_{|x-u_s^-|\leqslant\varepsilon} F(s-,x) - \sup_{|x-u_s^-|>\varepsilon} F(s-,x) > 0 \quad \text{for all } s \leqslant t \Big\} = 0. \tag{45}$$

Let

$$\Omega_\varepsilon = \Big\{\omega\colon \inf_{s\leqslant t}\big(\inf_{|x-u_s^-|\leqslant\varepsilon} F(s-,x) - \sup_{|x-u_s^-|>\varepsilon} F(s-,x)\big) > 0\Big\}.$$

Since on the set Ω_ε we have

$$\varphi(u^n,u^n) \leqslant 4 \sup_{|x-u_s^-|\leqslant\varepsilon} \varphi(x,x) \qquad \text{a.e.}$$

for all $s \leqslant t$, and (45) implies that $\mathbf{P}(\Omega_\varepsilon^c) \to 0,\ \varepsilon \to 0$, we obtain

$$\lim_{N\to\infty} \overline{\lim_n} \mathbf{P}\Big(\int_0^t \varphi(u^n, u^n) dK \geqslant N \Big) =$$

$$= \lim_{N\to\infty} \overline{\lim_{\varepsilon\to 0}} \overline{\lim_n} \mathbf{P}\Big[\Big(\int_0^t \varphi(u^n, u^n) dK \geqslant N,\ \Omega_\varepsilon \Big) \bigcup$$

$$\bigcup \Big(\int_0^t \varphi(u^n, u^n) dK \geqslant N,\ \Omega_\varepsilon^c \Big) \Big] \leqslant$$

$$\leqslant \lim_{N\to\infty} \overline{\lim_{\varepsilon\to 0}} \mathbf{P}\Big(\int_0^t \sup_{|x-u_s^-|\leqslant\varepsilon} \varphi_s(x,x) dK_s \geqslant N \Big) \leqslant$$

$$\leqslant \lim_{N\to\infty} \mathbf{P}\Big(\overline{\lim_\varepsilon} \int_0^t \sup_{|x-u_s^-|\leqslant\varepsilon} \varphi_s(x,x) dK_s \geqslant N \Big) \leqslant$$

$$\leqslant \lim_{N\to\infty} \mathbf{P}\Big(\int_0^t \varphi(u_s^-, u_s^-) dK_s \geqslant N \Big).$$

But condition **c)** implies that $\int_0^t \varphi(u_s^-, u_s^-) dK_s < \infty$ a.s. for any $t \geqslant 0$ and hence the sequence $\big(\langle \overline{M}^n \rangle_t,\ n \geqslant 1\big)$ is tight. Now we can show that

$$V = \overline{\max_{x\in\mathbf{R}}}\, \mathcal{L}^{\overline{F}-F(x)}(\cdot, 0). \tag{46}$$

Since $\{u_s^- = x\} = \big\{\overline{F}(s-) = F(s-, x)\big\}$ for any $x \in \mathbf{R}$, and

$$\begin{aligned} \int_0^t \mathbf{1}\{u_z^- = x\} d\Big(\int_0^z M(ds, u_s^-) \Big) &= M(t,x), \\ \int_0^t \mathbf{1}\{u_s^- = x\} d\overline{B}_s &= \int_0^t g(s,x) dK_s, \end{aligned} \tag{47}$$

the equality

$$\overline{F}(t) - \overline{F}(0) = F(t,x) - F(0,x) + \int_0^t M(ds, u_s^-) - M(t,x) +$$

$$+ \int_0^t g(s, u_s^-) dK_s - \int_0^t g(s,x) dK_s + V_t$$

and the definition of local time (see also equality (7)) imply

$$\mathcal{L}^{\overline{F}-F(x)}(t,0) = \int_0^t 1\{\overline{F}(s-)-F(s-,x) = 0\}d(\overline{F}(s)-F(s,x)) = \int_0^t 1\{u_s^- = x\}dV_s,$$

which results in (46).

COROLLARY 2. *Suppose that there exists a semimartingale ξ such that the processes $\xi = (\xi_t,\ t \geqslant 0)$ and $\xi_- = (\xi_{t-},\ t \geqslant 0)$ are maximizing functions of the family F (i.e., $\overline{F}(t) = F(t,\xi_t)$ and $\overline{F}(t-) = F(t-,\xi_{t-})$ a.s. for all $t \geqslant 0$).*
If, for the processes $(\xi, \xi_-, F(x),\ x \in \mathbf{R})$, conditions $\mathbf{a}^*$), $\mathbf{b}^*$), and $\mathbf{c}^*$) are satisfied, then the process $\overline{F} = (\overline{F}(t),\ t \geqslant 0)$ is a semimartingale representable as

$$\overline{F}(t) = \overline{F}(0) + \int_0^t M(ds, \xi_{s-}) + \int_0^t g(s, \xi_{s-})dK_s + V_t, \tag{48}$$

where $V = (V_t,\ t \geqslant 0)$ is an increasing process.

Proof. Let $F^n(t,x) = F(t,x) - (\xi_t - x)^2/n$. Obviously, ξ_- (ξ) is a unique maximizing function of the family $(F^n(x), x \in \mathbf{R})$ for every $n \geqslant 1$. Clearly, $F^n(x)$ is a semimartingale for every $n \geqslant 1$, $x \in \mathbf{R}$, and the martingale part and the part of bounded variation of $F^n(x)$ are equal to

$$\overline{M}^n(x) = M(t,x) - \frac{2}{n}\int_0^t (\xi_{s-} - x)dm_s,$$

$$\overline{A}^n(x) = A(t,x) - \frac{2}{n}\int_0^t (\xi_{s-} - x)d\alpha_s - \frac{1}{n}[\xi]_t,$$

respectively, where m and α are canonical decomposition terms of ξ. It can be easily verified that conditions $\mathbf{a}^*$), $\mathbf{b}^*$), and $\mathbf{c}^*$) are also satisfied for processes (ξ_-, F^n) for every $n \geqslant 1$ (condition $\mathbf{d}$) follows from the left (right) continuity of the process ξ_- (ξ)).
Hence, part iii) of Theorem 1 implies that $\overline{F}^n(t) = \sup_{x\in R} F^n(t, x)$ is a semimartingale and

$$\overline{F}^n(t) = \overline{F}(0) + \int_0^t M(ds, \xi_{s-}) + \int_0^t g(s, \xi_{s-})dK_s - \frac{1}{n}[\xi]_t + V_t^n, \tag{48'}$$

where $V^n = (V_t^n, t \geqslant 0)$ is an increasing process. It should be taken into account that a stochastic line integral with respect to martingales $(\xi_- - x)\cdot m$, $x \in \mathbf{R}$, along the curve ξ_- is equal to zero. Since $\overline{F}^n(t-) = \overline{F}(t)$ for every n, (48′) follows immediately from (40).
Note that the jump of the line integral

$$\int \Phi(ds,\, \xi_{s-}) = \int M(ds,\, \xi_{s-}) + \int A(ds,\, \xi_{s-})$$

at point t is equal to $\Phi(t,\, \xi_{t-}) - \Phi(t-,\, \xi_{t-})$. Since every increasing process is the sum of its jumps and of a continuous increasing process, and inasmuch as

$$\Delta V_t = \Delta \overline{F}(t) - \big(F(t,\, \xi_{t-}) - F(t-,\, \xi_{t-})\big) = F(t,\, \xi_t) - F(t,\, \xi_{t-})$$

in the case where ξ and ξ_- are unique maximizing functions of the family $\mathcal{F} = (F(\cdot,\, x),\, x \in R)$, we obtain

COROLLARY 3. *If the conditions of Corollary 1 are satisfied, then*

$$\overline{F}(t) = \overline{F}(0) + \int_0^t F(ds,\, \xi_{s-}) + \sum_{s \leqslant t} \big((s,\, \xi_s) - F(s,\, \xi_{s-})\big) + V_t^c,$$

where $(V_t^c,\, t \geqslant 0)$ *is a continuous increasing process and it can be expressed as*

$$V_t^c = \overline{\max_{x \in R}}\, L^{\overline{F} - F(x)}(t,\, 0).$$

2. GENERALIZED ITÔ'S FORMULA

2.1. *Itô's effect in an implicit form*

Now we shall apply decomposition formula (48) to the derivation of Itô's formula. We restrict ourselves to continuous processes. Let the family $(\Phi(t,\, x),\ t \geqslant 0,\ x \in \mathbf{R})$ of continuous semimartingales be given for which the decomposition

$$\Phi(t,\, x) = N(t,\, x) + B(t,\, x), \qquad N(\cdot,\, x) \in \mathcal{M}_{\text{loc}}, \qquad B(\cdot,\, x) \in \mathcal{A}_{\text{loc}}$$

holds and let $\xi = (\xi_t,\, t \geqslant 0)$ be a continuous semimartingale:

$$\xi_t = m_t + \alpha_t, \qquad m \in \mathcal{M}_{\text{loc}}, \qquad \alpha \in \mathcal{A}_{\text{loc}}.$$

Assume that Φ is convex in x for any $(\omega,\, t)$, the left derivative $\Phi'_-(t,\, x)$ is a semimartingale for every $x \in \mathbf{R}$ admitting the decomposition

$$\Phi'(t,\, x) = N'(t,\, x) + B'(t,\, x), \qquad N'(\cdot,\, x) \in \mathcal{M}_{\text{loc}}, \qquad B'(\cdot,\, x) \in \mathcal{A}_{\text{loc}}$$

(here, N', B' are not assumed to be derivatives of N and B, respectively, whose existence does not necessarily follow from the convexity in x of the function Φ).

Furthermore, assume that there exists an increasing process $K = (K_t,\ t \geqslant 0)$ dominating characteristics of ξ, Φ and Φ', i.e., such that, for each $t \geqslant 0$, we have

$$\langle N(\cdot,\, x),\, N(\cdot,\, x)\rangle_t \ll K_t, \qquad \langle N'(\cdot,\, x),\, N'(\cdot,\, x)\rangle_t \ll K_t, \qquad \langle m \rangle \ll K_t,$$

$$\operatorname*{Var}_{[0,\,t]} B(\cdot,\, x) \ll K_t, \qquad \operatorname*{Var}_{[0,\,t]} B'(\cdot,\, x) \ll K_t, \qquad \operatorname*{Var}_{[0,\,t]} \alpha \ll K_t.$$

Define

$$f^N(x,\, x) = d\langle N(\cdot,\, x),\, N(\cdot,\, y)\rangle/dK, \quad f'(x,\, y) = d\langle N'(\cdot,\, x),\, N'(\cdot,\, y)\rangle/dK,$$

$$f^{N',\,N}(x,\, y) = d\langle N'(\cdot,\, x),\, N(\cdot,\, y)\rangle/dK, \quad f^{N',\,\xi}(x) = d\langle N'(\cdot,\, x),\, \xi\rangle/dK,$$

$$f^{N,\,\xi}(x) = d\langle N(\cdot,\, x),\, \xi\rangle/dK, \qquad f^{\xi} = d\langle m\rangle/dK,] \tag{49}$$

$$f^{B'}(x) = dB'(\cdot,\, x)/dK, \qquad f^{\alpha} = d\alpha/dK,$$

$$f^B(x) = dB(\cdot,\, x)/dK.$$

Clearly, $\Phi(t,\, \xi_t)$ is representable as a maximum of semimartingales. Indeed, it follows from the well-known properties of convex functions that

$$\Phi(t,\, \xi_t) = \sup_{x\in\mathbf{R}} F(t,\, x),$$

where

$$F(t,\, x) = \Phi'_-(t,\, x)(\xi_t - x) + \Phi(t,\, x). \tag{50}$$

Below we use the notation μ^K for the Dolean measure of K.

THEOREM 2. *Let the following conditions be valid:*

A) *for each $(\omega,\, t)$ the function Φ is convex in x,*

B) *for each $x \in R$ the derivative $\Phi_- = (\Phi'_-(t,\, x),\, t \geqslant 0)$ is a semimartingale;*

C) *the functions Φ'_-, f^N, $f^{N',\xi}$, f^B are continuous μ^K-a.e. along the path ξ;*

D) *the processes*

$$f^N(\xi,\, \xi)\cdot K, \quad f^{N',\xi}(\xi)\cdot K, \quad f^B(\xi)\cdot K$$

are locally integrable and the process $(\Phi'_-(s,\, \xi_s),\, s \geqslant 0)$ is locally bounded. *Then the process $(\Phi(t,\, \xi_t),\, t \geqslant 0)$ is a semimartingale admitting the decomposition*

$$\Phi(t,\, \xi_t) = \Phi(0,\, \xi_0) + \int_0^t \Phi(ds,\, \xi_s) + \int_0^t \Phi'_-(s,\, \xi_s)d\xi_s +$$

$$+\left\langle \int_0^{\cdot} \Phi'_-(ds,\, \xi_s),\, \xi.\right\rangle_t + V_t, \tag{51}$$

where $V = (V_t,\, t \geqslant 0)$ is an increasing process.

Proof. Since the function Φ is convex with respect to x for any $(\omega,\, t)$, the derivative Φ'_- is an increasing function with respect to x for all $(\omega,\, t)$. Therefore, the function $\Phi(t,\, \xi_t)$ can be represented as

$$\Phi(t,\, \xi_t) = \sup_{x \in \mathbf{R}} \left[\Phi'_-(t,\, x)\,(\xi_t - x) + \Phi(t,\, x)\right]$$

Moreover, the process $\xi = (\xi_t,\, t \geqslant 0)$ is a maximizing function for $F(t,\, x) = \Phi'_-(t,\, x)\,(\xi_t - x) + \Phi(t,\, x)$ (i. e., $\Phi(t,\, \xi_t) = F(t,\, \xi_t)$). So, to apply Theorem 1 (more precisely, Corollary 2), we must verify that conditions **a***) – **c***) of Theorem 1 can be satisfied under the assumptions **A**) – **D**) of the theorem. Obviously, for every $x \in \mathbf{R}$, the process $F(t,\, x)$ from (50) is a semimartingale, and Itô's formula yields representations for its martingale part and finite-variation part in the decomposition

$$F(t,\, x) = M(t,\, x) + A(t,\, x),$$

$$M(t,\, x) = N(t,\, x) + \int_0^t \Phi'_-(s,\, x) dm_s + \int_0^t (\xi_s - x)\, N'(ds,\, x),$$

$$A(t,\, x) = B(t,\, x) + \int_0^t \Phi'_-(s,\, x)\, d\alpha_s + \int_0^t (\xi_s - x)\, B'(ds,\, x) +$$

$$+ \langle N'(\cdot,\, x),\, \xi\rangle_t. \tag{52}$$

From this and (49), we obtain expressions for the functions $\varphi(x,\, y)$, $g(x)$ which lead to conditions **a***), **b***) and **c***):

$$g(x) = dA(\cdot,\, x)/dK = f^B(x) + \Phi'_-(x)\, f^\alpha + (\xi - x)\, f^{B'}(x) + f^{N',\xi}(x), \tag{53}$$

$$\varphi(x,\, y) = d\langle M(\cdot,\, x),\, M(\cdot,\, y)\rangle/dK =$$

$$= f^N(x,\, y) + (\xi - x)\,(\xi - y)\, f^{N'}(x,\, y) + \Phi'_-(x)\, \Phi'_-(y)\, f^\xi +$$

$$+ (\xi - x)\, \Phi'_-(y)\, f^{N',\xi}(x) + (\xi - y)\, \Phi'_-(x)\, f^{N',\xi}(y) +$$

$$+ (\xi - x)\, f^{N,N'}(x,\, y) + (\xi - y)\, f^{N',N}(x,\, y) +$$

$$+ \Phi'_-(x)\, f^{N,\xi}(y) + \Phi'_-(y)\, f^{N,\xi}(x). \tag{54}$$

The continuity of the function $\Phi'_-(x)$ along the process ξ and the local boundedness of the process $(\Phi'_-(s,\, \xi_s),\; c \geqslant 0)$ imply

$$\sup_{s\leqslant t}\ \sup_{|x|\leqslant n,\ |x-\xi_s|\leqslant\varepsilon}\big(F(s,\,\xi_s)-F(s,\,x)\big)=$$

$$=\sup_{s\leqslant t}\ \sup_{|x|\leqslant n,\ |x-\xi_s|\leqslant\varepsilon}\big(\Phi(s,\,\xi_s)-\Phi(s,\,x)-(\xi_s-x)\,\Phi'_-(s,\,x)\big)\leqslant$$

$$\leqslant 2\varepsilon\sup_{s\leqslant t}\ \sup_{|x|\leqslant n,\ |x-\xi_s|\leqslant\varepsilon}|\Phi'_-(s,\,x)|\xrightarrow{\mathbf{P}}0,\quad \varepsilon\to 0.$$

Therefore, condition **a***) of Theorem 1 is satisfied. Expressions (53) and (54) show that condition **c**) implies condition **b***).

It can be easily verified that the processes $\varphi(\xi,\,\xi)\cdot K$ and $|g(\xi)|\cdot K$ are locally integrable (see condition **c***)). As a localizing sequence for the process

$$\varphi(\xi,\,\xi)\cdot K=f^N(\xi,\,\xi)\cdot K+\big(\Phi'_-(\xi)\big)^2 f^{\xi}\cdot K+2\Phi'_-(\xi)\,f^{N,\xi}(\xi)\cdot K$$

one can take the sequence $(t_j\wedge\sigma_j\wedge\gamma_j,\ j\geqslant 1$, where $(\tau_j,\,j\geqslant 1)$ is a localizing sequence of stopping times for the process $f^N(\xi,\,\xi)$, i.e., $\tau_j\uparrow\infty$ and, for every $j\geqslant 1$,

$$\mathbf{E}\int_0^{\tau_j} f^N(\xi,\,\xi)\,dK<\infty,\qquad \sigma_j=\inf\langle t\colon\,|\xi_t|\geqslant j\rangle$$

and $\gamma_j\uparrow\infty$ is a sequence of stopping times for which there exist constants $c_j,\ j\geqslant 1$, such that

$$|\Phi'_-(t,\,\xi_t)|\mathbf{1}\{t\leqslant\gamma_j\}\leqslant c_j\quad\text{a.s.}$$

Since, for $\tau=\tau_j\wedge\sigma_j\wedge\gamma_j$, we have

$$\mathbf{E}\int_0^{\tau}\big(\Phi'_-(s,\,\xi_s)\big)^2 f^{\xi}(s)\,dK_s\leqslant\text{const}\cdot\mathbf{E}\langle\xi\rangle_\tau<\infty$$

and

$$\mathbf{E}\int_0^{\tau}\Phi'_-(s,\,\xi_s)\,f^{N,\xi}(s,\xi_s)\,dK_s=\mathbf{E}\int_0^{\tau}\Phi'_-(s,\,\xi_s)\,d\left\langle\int_0^{\cdot} N(du,\,\xi_u),\,\xi.\right\rangle_s\leqslant$$

$$\leqslant\left(\mathbf{E}\int_0^{\tau}\varphi_s(\xi_s,\,\xi_s)\,dK\right)^{1/2}\left(\mathbf{E}\int_0^{\tau}\big(\Phi'_-(s,\,\xi_s)\big)^2 d\langle\xi\rangle_s\right)^{1/2}<\infty,$$

we clearly obtain $\varphi(\xi,\,\xi)\cdot K\in\mathcal{A}^+_{\text{loc}}$. The fact that

$$|g(\xi)|\cdot K=|f^B(\xi)+\Phi'_-(\xi)\,f^{\alpha}+f^{N',\xi}(\xi)|\cdot K\in\mathcal{A}^+_{\text{loc}}$$

is proved in a similar way. Hence, for the family of semimartingales $(F(x),\, x \in \mathbf{R})$ and the semimartingale ξ, all conditions of Corollary 2 are satisfied and therefore we are led to formula (51):

$$\Phi(t,\, \xi_t) = \sup_{x\in\mathbf{R}} F(t,\, x) = \int_0^t M(ds,\, \xi_s) + \int_0^t g(s,\, \xi_s)\, dK_s =$$

$$= \int_0^t N(ds,\, \xi_s) + \int_0^t f^B(s,\, \xi_s)\, dK_s + \int_0^t \Phi'_-(s,\, \xi_s)\, dm_s + \int_0^t \Phi'_-(s,\, \xi_s)\, d\alpha_s +$$

$$+ \int_0^t f^{N',\xi}(s,\, xi_s)\, dK_s + V_t =$$

$$= \int_0^t \Phi(ds,\, \xi_s) + \int_0^t \Phi'_-(s,\, \xi_s)\, d\xi_s + \left\langle \int_0^{\cdot} \Phi'_-(ds,\, \xi_s),\, \xi_{\cdot} \right\rangle_t + V_t,$$

since $\int B(ds,\, \xi_s) = \int f^B(\xi)\, dK$, $\langle \int_0^{\cdot} \Phi'_-(ds,\, \xi_s),\, \xi_{\cdot}\rangle_t = \int_0^t f^{N',\xi}(s, \xi_s)\, dK_s$ and $\int Y(ds,\, \xi_s) = 0$ for the family $Y(x) = (\xi - x)\cdot \Phi'_-(x)$. $x \in \mathbf{R}$.

COROLLARY 4. *If, at each point $(\omega,\, t)$, the function Φ is the difference of two convex functions (i.e., if the left-hand derivative of the function Φ is a function with bounded variation with respect to x on every compact set for every pair $(\omega,\, t)$), then formula* (51) *remains true with the only distinction that the process V in the decomposition of $\Phi\,(t,\, \xi_t)$ is now process with locally bounded variation.*

Remark 3. If we consider the family $(\Phi\,(t,\, x),\; t \geqslant 0)$, $x \in \mathbf{R}$, of right-continuous special semimartingales with locally square-integrable martingale parts and if $\xi = (\xi_t,\; t \geqslant 0)$ is some locally bounded right-continuous semimartingale, then under some extra condition on jumps (additional to A) – D)) (namely, $\sup_{s\leqslant t} \sup_{x\in[-n,n]} (\Delta\Phi\,(s,\, \xi))^2 \in \mathcal{A}^+_{\text{loc}}$), we can apply Corollary 3, to prove that the process $(\Phi\,(t,\, \xi_t),\; t \geqslant 0)$ is a semimartingale having the decomposition

$$\Phi\,(t,\, \xi_t) = \Phi\,(0,\, \xi_0) + \int_0^t \Phi\,(ds,\, \xi_s) + \int_0^t \Phi'_-(s^-,\, \xi_{s-}\, d\xi_s + \left[\int_0^{\cdot} \Phi'_-(ds,\, \xi_-),\, \xi_{\cdot} \right]_t +$$

$$+ \sum_{s\leqslant t} \left[\Phi\,(s,\, \xi_s) - \Phi\,(s, \xi_{s-}) - \Phi\,(s,\, \xi_{s-})\Delta\xi_s\right] + V_t, \tag{55}$$

the sum in this formula being representable as

$$\sum_{s\leqslant t} \Big(\Phi(s-,\, \xi\,(s)) - \Phi(s-,\, \xi\,(s-)) - \Phi'(s-,\, \xi\,(s-))\Delta\xi_s\Big) +$$

$$+\sum_{s\leqslant t}\Big(\Phi\big(s,\,\xi(s)\big)-\Phi\big(s,\,\xi(s-)\big)+\Phi\big(s-,\,\xi(s)\big)\Big)-$$

$$-\Big(\Phi'\big(s,\,\xi(s)\big)-\Phi'\big(s-,\,\xi(s)\big)\Big)\Delta\xi_s\Big)\Big),$$

where $V=(V_t,\ t\geqslant 0)$ is a continuous increasing process.

2.2. *Continuity property of decomposition terms*

By Theorem 2, the expression (Itô's effect)

$$J_t^{\xi}(\Phi)=V_t=\Phi(t,\,\xi_t)-\Phi(0,\,\xi_0)-$$

$$-\int_0^t \Phi(ds,\,\xi_s)-\int_0^t \Phi'_-(s,\,\xi_s)\,d\xi_s-\Big\langle\int_0^{\cdot}\Phi'_-(ds,\,\xi_s),\,\xi\Big\rangle_t$$

is a continuous increasing process. To derive explicit expressions for $J^{\xi}(\Phi)$, some continuity properties of this functional with respect to Φ (i.e., continuity properties of the terms on the right-hand side of (56)) should be established.

LEMMA 4. *Assume that, for every element of the sequence* $(\Phi_n,\ n\geqslant 1)$ *from a family of semimartingales, the conditions of Theorem 2 are satisfied and there exists a sequence of stopping times* $(\tau_j,\ j\geqslant 1)$, $\tau_j\uparrow\infty$, *such that for every* $\tau=\tau_j$ *and* $t\geqslant 0$ *as* $n\to\infty$

$$\Phi_n(t\wedge\tau,\,\xi_{t\wedge\tau})\xrightarrow{\text{w}}\Phi(t\wedge\tau,\,\xi_{t\wedge\tau})\quad in\quad L_1(\Omega,\,\mathcal{F}_t,\,\mathbf{P}),\tag{57}$$

$$\Phi'_n(t,\,\xi_t)\xrightarrow{\text{w}}\Phi'(t,\,\xi_t)\quad in\quad L_1([\![0,\,\tau\wedge t]\!],\,\mathcal{P},\,\mu^K),\tag{58}$$

$$f^{N'_n,\xi}(t,\,\xi_t)\xrightarrow{\text{w}}f^{N',\xi}(t,\,\xi_t)\quad in\quad L_1([\![0,\,\tau\wedge t]\!],\,\mathcal{P},\,\mu^K),\tag{59}$$

$$f^{B_n}(t,\,\xi_t)\xrightarrow{\text{w}}f^{B}(t,\,\xi_t)\quad in\quad L_1([\![0,\,\tau\wedge t]\!],\,\mathcal{P},\,\mu^K),\tag{60}$$

$$f^{N_n,m}(t,\,\xi_t)\xrightarrow{\text{w}}f^{N,m}(t,\,\xi_t)\quad in\quad L_2([\![0,\,\tau\wedge t]\!],\,\mathcal{P},\,\mu^K)\tag{61}$$

for all $m\in\mathcal{M}^2$ *(the symbol* $\xrightarrow{\text{w}}$ *denotes the weak convergence). Then, for every* $t\geqslant 0$ *and* $\tau=\tau_j$, $j\geqslant 1$, *we have*

$$J^{\xi}_{t\wedge\tau}(\Phi_n)\xrightarrow{\text{w}}J^{\xi}_{t\wedge\tau}(\Phi)\quad in\quad L_1(\Omega,\,\mathcal{F}_t,\,\mathbf{P}).\tag{62}$$

Proof. Let ι be an $\mathcal{F}$-measurable random variable with $\mathbf{E}\iota^2<\infty$ and m^{ι} be a right-continuous modification of the martingale $\mathbf{E}(\iota/\mathcal{F}_t)$. Consider the weak convergence in $L_2([\![0,\,\tau\wedge t]\!],\mathcal{P},\,\mu^K)$ of the third term on the right-hand side of equality (56). By the properties of stochastic line integrals and conditions (60), (61), we have

$$\mathbf{E}\Big(\int_0^t\Phi_n(ds,\,\xi_s)-\int_0^t\Phi(ds,\,\xi_s)\Big)\iota=$$

$$= \mathbf{E}\Big(\int_0^t N^n(ds,\, \xi_s) - \int_0^t N(ds,\, \xi_s)\Big)\iota + \mathbf{E}\Big(\int_0^t B^n(ds,\, \xi_s) - \int_0^t B(ds,\, \xi_s)\Big)\iota =$$

$$= \mathbf{E}\Big\langle \int_0^{\cdot} N^n(ds,\, \xi_s) - \int_0^{\cdot} N(ds,\, \xi_s),\, m^{\iota}\Big\rangle_t + \mathbf{E}\int_0^t m_s^{\iota}(f^{B^n}(s,\, \xi_s) - f^B(s,\, \xi_s))\, dK_s =$$

$$= \mathbf{E}\int_0^t (f^{N_n,\, m^{\iota}}(s,\, \xi_s) - f^{N,m^{\iota}}(s,\, \xi_s))\, dK_s +$$

$$+ \mathbf{E}\int_0^t m_s^{\iota}(f^{B^n}(s,\, \xi_s) - f^B(s,\, \xi_s))\, dK_s \longrightarrow 0, \quad n \to \infty.$$

The convergence of the other terms in (56) is proved in a similar way.

2.3. *Smooth case. Weak two-fold differentiability*

LEMMA 5. *Let $\Phi(\omega,\, t,\, x) = f_t(\omega)\, \varphi(x)$, where $f = (f_t,\, t \geqslant 0)$ is a semimartingale and φ is a difference of two convex functions. Then*

$$J_t^{\xi}(\Phi) = \frac{1}{2}\int_{\mathbf{R}}\int_0^t f(s)\, L^{\xi}(ds,\, x)\, \varphi'_-(dx). \tag{63}$$

Proof. In this case decomposition (51) and formula (63) can be obtained by a direct application of Itô's formula for nonrandom convex functions (see Meyer 1976) and Itô's formula for the product

$$\Phi(t,\, \xi_t) = f(t)\, \varphi(\xi_t) = \int_0^t \varphi(\xi_s)\, df_s + \int_0^t f_s \varphi'_-(\xi_s)\, d\xi_s +$$

$$+ \int_0^t \varphi'_-(\xi_s)\, d\langle \xi,\, f\rangle_s + \frac{1}{2}\int_{\mathbf{R}}\int_0^t f_s L^{\xi}(ds,\, x)\, \varphi'_-(dx). \tag{64}$$

Note that the conditions of Theorem 2 are satisfied because the function $\varphi'_-(x)$ has a countable number of break points and hence the function $\Phi'_-(t,\, x) = f_t \varphi(x)$ is continuous μ^K-a.e. along ξ. Since in this case

$$\int \Phi(ds,\, \xi_s) = \int \varphi(\xi_s)\, df_s, \qquad \Big\langle \int_0^{\cdot} \Phi'_-(ds,\, \xi_s),\, \xi\Big\rangle = \int \varphi(\xi_s)\, d\langle \xi,\, f\rangle_s,$$

by comparing formulas (51) and (64), we see that

$$V_t = \frac{1}{2} \int_{\mathbf{R}} \int_0^t f_s L^{\xi}(ds, x)\, \varphi'_-(dx).$$

Therefore, if the function Φ has the form

$$\Phi(\omega, t, x) = \sum_{i=1}^{n} f_t^i(\omega)\, \varphi^i(x), \tag{64}$$

where f^i, $1 \leqslant i \leqslant n$, are semimartingales and φ^i, $1 \leqslant i \leqslant n$, are convex functions the left derivatives of which are absolutely continuous with respect to some measure $G(dx)$ on $(\mathbf{R}, \mathcal{B}(\mathbf{R}))$, then the process $V = J^{\xi}(\Phi)$ can be represented as

$$V_t = \frac{1}{2} \int_{\mathbf{R}} \int_0^t \left(\sum_{i=1}^{n} f_i(s) \frac{d\varphi'_i(x)}{dG(x)} \right) L^{\xi}(ds, x)\, G(dx) =$$

$$= \frac{1}{2} \int_{\mathbf{R}} \int_0^t \frac{d\Phi'_-(s, x)}{dG(x)} L^{\xi}(ds, x)\, G(dx), \tag{65}$$

Let $(G(x), x \in \mathbf{R})$ be a function with locally bounded variation. By $\Phi_-^{v, G}$ we denote the Radon–Nikodym derivative of the left derivative Φ'_- with respect to the measure $G(dx)$. Denote by μ^G Dolean's measure of the process $\int_{\mathbf{R}} L^{\xi}(t, x)\, G(dx)$. We say that the function $\Phi_-^{\prime, G}$ is continuous μ^G-a.e. along the curve ξ if

$$\lim_{\varepsilon \to 0} \sum_{|x - \xi_s| \leqslant \varepsilon} \left| \Phi_-^{\prime, G}(s, \xi_s) - \Phi_-^{\prime, G}(s, x) \right| = 0 \quad \mu^G\text{-a.e.}$$

THEOREM 3. *Let conditions* A) – C) *of Theorem* 2 *be satisfied and assume that there exists* $\varepsilon > 0$ *such that the processes*

$$\sup_{|\xi_s - x| \leqslant \varepsilon} f^N(x, x) \cdot K, \qquad \sup_{|\xi_s - x| \leqslant \varepsilon} f^{N'}(x, x) \cdot K, \qquad \sup_{|\xi_s - x| \leqslant \varepsilon} |f^B(x)| \cdot K$$

are locally integrable. Then

i) *weakly (locally) in* $L_1(\Omega, \mathcal{F}, \mathbf{P})$ *we have*

$$V_t = J_t^{\xi}(\Phi) = \lim_{\varepsilon \to 0} \frac{1}{\varepsilon} \int_0^t \left(\Phi'_-(s, \xi_s) - \Phi'_-(s, \xi_s - \varepsilon) \right) d\langle \xi \rangle_s; \tag{66}$$

ii) *If the left derivative* Φ'_- *of the function* Φ *at every point* (ω, t) *is absolutely continuous with respect to the measure* $G(dx)$, *the Radon–Nikodym* *derivative $\Phi'^{,G}_{-} = d\Phi'_{-}(x)/dG(x)$ is continuous μ^G-a.e. along ξ and the processes $\sup_{x\in[-n,n]} \Phi'^{,G}(t, x)$, $n \geqslant 1$, are locally bounded for each n, then the increasing process V in (51) has the form*

$$V_t = J^\xi_t(\Phi) = \frac{1}{2}\int_{\mathbf{R}}\int_0^t \Phi'^{,G}_{-}(s,x)\, L^\xi(ds,x)\, G(dx). \tag{67}$$

Proof. First we prove assertion ii). By Lemma 2 from (Emery, 1979), there exists a sequence of stopping times $(\tau_m,\ m \geqslant 1)$ such that $\tau_m \uparrow \infty$, $m \to \infty$, and the processes $\sup_{x\in[n,-n]} \Phi'^{,G}_{-}(t \wedge \tau_{m,x})$ are bounded for every $n \geqslant 1$. Therefore, using a localizing sequence of stopping times $(\tau_m \wedge \sigma_m,\ m \geqslant 1)$, where $\sigma_m = \inf(t$: $|\xi_t| \geqslant m)$, we can regard the processes $\xi, \Phi'^{,G}_{-}$ to be bounded.
Without loss of generality we can assume that $|\xi_t| \leqslant 1$.
Let

$$\Phi_n(t,x) = \Phi(t,-1) + \Phi'_{-}(t,-1)(x+1)+$$

$$+\sum_{k=1}^{n} \frac{\varphi'_{-}(t,2(k+1)/n) - \Phi'_{-}(t,2k/n)}{G(2(k+1)/n) - G(2k/n)} \int_{-1}^{x} \Big(G\big(y\wedge \frac{2(k+1)}{n}\big) - G\big(y\wedge \frac{2k}{n}\big)\Big)dy. \tag{68}$$

Evidently, the functions $(\Phi_n, n \geqslant 1)$ are of the form (64) and, by Lemma 5, we have

$$V^n_t = J^\xi_t(\Phi_n) = \frac{1}{2}\int_{\mathbf{R}}\int_0^t \Phi'^{,G}_{-}(s,x)\, L^\xi(ds,x)\, G(dx). \tag{69}$$

Since the function $\Phi'^{,G}_{-}$ is continuous μ^G-a.e. along ξ, we have

$$\Phi'^{,G}_{n}(s,\xi_s) - \Phi'^{,G}(s,\xi_s) =$$

$$= \sum_{k=1}^{n} \frac{(\Phi'_{-}(s,2(k+1)/n) - \Phi'_{-}(s,2k/n))}{G(2(k+1)/n) - G(2k/n)} \times$$

$$\times \mathbf{1}\Big\{\frac{2k}{n} < \xi_s \leqslant \frac{2(k+1)}{n}\Big\} - \Phi'^{,G}_{-}(s,\xi_s) =$$

$$= \sum_{k=1}^{n} \mathbf{1}\Big\{\frac{2k}{n} < \xi_s \leqslant \frac{2(k+1)}{n}\Big\}\Big(G\Big(\frac{2(k+1)}{n}\Big) - G\Big(\frac{2k}{n}\Big)\Big)^{-1} \times$$

$$\times \int_{2k/n}^{2(k+1)/n} (\Phi'^{,G}_{-}(s,x) - \Phi'^{,G}_{-}(s,\xi_s))G(dx) \leqslant$$

$$\leqslant \sup_{|x-\xi_s|\leqslant 1/n} |\Phi'^{,G}_{-}(s,x) - \Phi'^{,G}_{-}(s,\xi_s)| \to 0, \quad \mu^G - \text{a.e.}, \quad n \to \infty,$$

and from boundedness of $\Phi'^{,G}_{-}$(implied by $\mathbf{1}\{\xi \neq x\} \cdot L(x) = 0$) it follows that

$$\mathbf{E}\iota \int_{\mathbf{R}} \int_0^t (\Phi'^{,G}_n(s,x) - \Phi'^{,G}(s,x)) L^{\xi}(ds,x) G(dx) \to 0, \quad n \to \infty, \tag{70}$$

for each bounded random variable ι.
It is easy to show that relations (57) − −(61) of Lemma 4 hold true. Thus, we can verify the validity of (59).
By μ^K-a.e. continuity of the function $f^{N',\xi}(x)$ along the path ξ, we have

$$\mathbf{E}\int_0^t \left| \left(\frac{d\langle \Phi'_n(\cdot,x), m\rangle_s}{dK_s} - \frac{d\langle \Phi'(\cdot,x), m\rangle_s}{dK_s} \right)_{\xi_s} \right| dK_s =$$

$$= \mathbf{E}\int_0^t \left| \left(\frac{d\langle \Phi'(\cdot,x), m\rangle_s}{dK_s} + \sum_{k=1}^{n} \frac{d\langle \Phi'(\cdot, 2(k+1)/n) - \Phi'(i, 2k/n), m\rangle_s}{dK_s} \times \right.\right.$$

$$\left.\left. \times \frac{G(x \wedge 2(k+1)/n) - G(x \wedge 2k/n)}{G(2(k+1)/n) - G(2k/n)} - \frac{d\langle \Phi'(\cdot,x), m\rangle_s}{dK_s} \right)_{\xi_s} \right| dK_s =$$

$$= \mathbf{E}\int_0^t \left| \sum_{k=1}^{n} \mathbf{1}\left\{ \frac{2k}{n} < \xi_s \leqslant \frac{2(k+1)}{n} \right\} \left[f^{N',m}\left(s, \frac{2(k+1)}{n}\right) + \right.\right.$$

$$+ \frac{G(\xi_s) - G(2k/n)}{G(2(k+1)/n) - G(2k/n)} \left(f^{N',m}\left(s, \frac{2(k+1)}{n}\right) - \right.$$

$$\left.\left. \left. - f^{N',m}(s, 2k/n) \right) \right] - f^{N',m}(s,\xi_s) \right| dK_s \leqslant$$

$$\leqslant 2\mathbf{E}\int_0^t \sup_{|\xi_s - x| \leqslant 1/n} \left| f^{N',m}(s,\xi_s) - f^{N',m}(s,x) \right| dK_s \longrightarrow 0, \quad n \to \infty.$$

The other relations are verified in a similar way.
By Theorem 2,

$$V_t = \Phi(t,\xi_t) - \Phi(0,\xi_0) - \int_0^t \Phi(ds,\xi_s) - \int_0^t \Phi'_-(s,\xi_s) d\xi_s - \left\langle \int_0^{\cdot} \Phi'_-(ds,\xi_s), \xi \right\rangle_t$$

is an increasing process and Lemma 4 implies that the increasing process

$$V_t^n = \frac{1}{2} \int_{\mathbf{R}} \int_0^t \Phi'^{,G}_n(s,x)\, L^{\xi}(ds,x)\, G(dx)$$

converges to the process V, as $n \to \infty$, weakly in L_1. But by (70) (since there is a unique week limit), we have

$$V_t = \frac{1}{2}\int_{\mathbf{R}}\int_0^t \Phi'^{,G}_{-}(s,x)\, L^{\xi}(ds,x)\, G(dx)$$

P-a.s. for every $t \geqslant 0$.
Because V is a process with **P**-a.s. continuous paths, it is indistinguishable from the regular modification of the integral $(1/2)\int_{\mathbf{R}}\int_0^t \Phi'^{,G}_{-}(s,x)\, L^{\xi}(ds,x)\, G(dx)$.
To prove the assertion i), we consider a sequence of functions $(\Phi_n,\ n \geqslant 1)$ such that

$$\Phi_n(\omega,t,x) = \Phi(t,-1) + \int_{-1}^{x} n\big((\Phi(t,y) - \Phi(t,y-1/n)\big)dy.$$

Obviously, for every n, the function Φ_n satisfies all the conditions of the second part of Theorem 3. Hence,

$$V_t^n = \Phi_n(t,\xi_t) - \Phi_n(0,\xi_0) - \int_0^t \Phi_n(ds,\xi_s) - \int_0^t \Phi'_n(s,\xi_s)d\xi_s - \Big\langle \int_0^{\cdot} \Phi'_n(ds,\xi_s),\xi_s),\xi \Big\rangle_t$$

is an increasing process and, by (67),

$$V_t^n = \int_{\mathbf{R}}\int_0^t \frac{d\Phi'(s,x)}{dx} L^{\xi}(ds,x)dx =$$

$$= \int_{\mathbf{R}}\int_0^t n(\Phi'_{-}(s,x) - \Phi'_{-}(s,x-1/n)L^{\xi}(ds,x)dx. \tag{71}$$

Now we can show that the conditions of Lemma 4 are satisfied.
Since

$$\frac{d\langle\Phi(\cdot,x),m\rangle}{dK} = \frac{d\langle\Phi(\cdot,-1),m\rangle}{dK} + \int_{-1}^{x}\frac{d\langle\Phi'(\cdot,y),m\rangle}{dK}dy \qquad \mu^K\text{-a.e.}$$

and

$$n\int_{x-1/n}^{x} \frac{d\langle\Phi'(\cdot,y),m\rangle}{dK}dy \longrightarrow \frac{d\langle\Phi'(\cdot,x),m\rangle}{dK} \qquad d\mu^K\,dx - \text{a.e.}$$

as $n \to \infty$, by Lebesgue's theorem, we have

$$\mathbf{E}\Big(\int_0^t \frac{d\langle\Phi_n(\cdot,x),m\rangle_s}{dK_s} - \frac{d\langle\Phi(\cdot,x),m\rangle_s}{dK_s}\Big)_{\xi_s} dK_s =$$

$$= \mathbf{E}\int_0^t \int_{-1}^{\xi_s} \left[n\left(\frac{d\langle\Phi(\cdot,y),m\rangle_s}{dK_s} - \frac{d\langle\Phi(\cdot,y-1/n),m\rangle_s}{dK_s} \right) - \frac{d\langle\Phi'(\cdot,y),m\rangle}{dK_s} \right] dy\, dK_s \leqslant$$

$$\leqslant \iiint\limits_{\Omega\times[0,t]\times[-1,1]} \left| n \int_{x-1/n}^{x} \frac{d\langle\Phi'(\cdot,y),m\rangle_s}{dK_s} dy - \frac{d\langle\Phi'(\cdot,x),m\rangle}{dK_s} \right| \mu^K(d\omega,ds)dx \longrightarrow 0,$$

as $n \to \infty$, for every $m \in \mathcal{M}^2$ and, hence, relation (61) is true.
Since the function $f^{N',\xi}(x)$ is μ^K-a.e. continuous along the path ξ, the validity of relation (59) follows from

$$\mathbf{E}\int_0^t \left| \left(\frac{d\langle\Phi_n'(\cdot,x),m\rangle_s}{dK_s} - \frac{d\langle\Phi'(\cdot,x),m\rangle_s}{dK_s} \right)_{\xi_s} \right| dK_s =$$

$$= \mathbf{E}\int_0^t \left| n \int_{\xi_s-1/n}^{\xi_s} \left(f^{N',m}(s,y) - f^{N',m}(s,\xi_s) \right) dy \right| dK_s \leqslant$$

$$\leqslant \mathbf{E}\int_0^t \sup_{\xi_s-1/n<y\leqslant\xi_s} \left| f^{N',m}(s,y) - f^{N',m}(s,\xi_s) \right| dK_s.$$

It can also be proved that the remaining relations of Lemma 4 are true.
Therefore, the increasing process $V^n = (V_t^n,\ t \geqslant 0)$ weakly converges in L_1, for every $t \geqslant 0$, to the increasing process $V = (V_t,\ t \geqslant 0)$, and it follows from (71) that the increasing process V can be defined, for every $t \geqslant 0$, as the weak limit in $L_1(\Omega,\ \mathcal{F},\ \mathbf{P})$ of the expression

$$\int_{\mathbf{R}} \int_0^t n(\Phi_-'(s,x) - \Phi_-'(s,x-1/n)) L^\xi(ds,x)dx.$$

Since the equality

$$\int_{\mathbf{R}} \int_0^t G(s,x)L^\xi(ds,x)dx = \int_0^t G(s,\xi_s)d\langle\xi\rangle_s$$

is true for every $\mathcal{P}\otimes\mathcal{B}(R)$-measurable function $G(\omega,s,x)$, we have

$$V_t = \lim_{\varepsilon\to 0} \int_0^t \frac{1}{\varepsilon}(\Phi_-'(s,\xi_s) - \Phi_-'(s,\xi_s-\varepsilon))d\langle\xi\rangle_s.$$

This means that, in a sense, the functional $J^\xi(\Phi)$ can be viewed as an integral of the second derivative along the path ξ, i.e., for every bounded random variable ι, we have

$$\mathbf{E}J_t^\xi(\Phi)\iota = \lim_{\varepsilon\to 0} \mathbf{E}\frac{1}{\varepsilon} \int_0^t (\Phi'(s,\xi_s) - \Phi'(s,\xi_s-\varepsilon))d\langle\xi\rangle_s. \qquad (72)$$

COROLLARY 6 (Lazrieva and Torondjadze, 1986). *If the conditions of part* i) *of Theorem 3 are satisfied, and the function* Φ *is twice continuously differentiable, then*

$$V_t = J_t^\xi(\Phi) = \int_0^t \Phi''(s, \xi_s) d\langle\xi\rangle_s.$$

Proof follows from part i) of Theorem 3 and the occupation formula. In this case, $G(x) = x$, $\Phi'^{,G}_{-} = \Phi''$ and $\mu^G = \mu^{\langle\xi\rangle}$.

2.4. *Integration by parts*

Extending formula (67) to smooth functions under the assumption that the function Φ is convex (or it is the difference of two convex functions), we introduce the following formal expression

$$J_t^\xi(\Phi) = \frac{1}{2}\int_0^t \int_{\mathbf{R}} \Phi'_-(s, dx) L^\xi(ds, x) \tag{73}$$

for the increasing process V which enters the decomposition (51) (Φ'_- denotes the left derivative of Φ). The meaning of (73) can be specified in different ways. Some of them (at least formally) are as follows.

(a) An integral of the weak second derivative considered in the previous section:

$$J_t^\xi(\Phi) = \lim_{\varepsilon\to 0} \frac{1}{2\varepsilon} \int_0^t (\Phi'_-(s, \xi_s) - \Phi'(s, \xi_s - \varepsilon)) d\langle\xi\rangle_s$$

weakly in $L_1(\Omega, \mathcal{F}, \mathbf{P})$.

(b) Double integral (or the mixed variation):

$$J_t^\xi(\Phi) = \lim_{\Delta\to 0,\, \delta\to 0} \frac{1}{2} \sum_{i,j} (\Phi'_-(s_i, x_{j+1}) - \Phi'_-(s_i, x_j))(L^\xi(s_{i+1}, x_j) - L^\xi(s_i, x_j)),$$

where $\Delta = \max_j |x_{j+1} - x_j|$, $\delta = \max_i |s_{i+1} - s_i|$.

(c) Integration by parts (I):

$$J_t^\xi(\Phi) = \int_{\mathbf{R}} L^\xi(t, x)\Phi'_-(t, dx) - \int_{\mathbf{R}}\int_0^t L^\xi(s, x)\Phi'_-(ds, dx).$$

(d) Integration by parts (II):

$$J_t^\xi(\Phi) = \int_{\mathbf{R}} \Phi'(t, x) L^\xi(t, dx) - \int_{\mathbf{R}}\int_0^t (\Phi'(s, x) - \Phi'(0, x)) L^\xi(ds, dx).$$

In what follows we use (c) and (b).
If the function Φ is twice continuously differentiable, then by Theorem 3, we have

$$J_t^\xi(\Phi) = \frac{1}{2}\int_{\mathbf{R}}\int_0^t \Phi''(s,x)L^\xi(ds,x)dx.$$

If, in addition, Φ'' is a semimartingale for every $x \in \mathbf{R}$, then integration by parts (with respect to s) of this expression yields

$$J_t^\xi(\Phi) = \frac{1}{2}\int_{\mathbf{R}} L^\xi(t,x)\Phi''(t,x)dx - \frac{1}{2}\int_{\mathbf{R}}\int_0^t L^\xi(s,x)\Phi''(ds,x)dx. \tag{74}$$

(Since L^ξ is locally bounded, the stochastic integral $\int L^\xi(\cdot,x)\Phi''(ds,x)$ is defined for every $x \in \mathbf{R}$ and $\mathcal{O} \otimes \mathcal{B}(\mathbf{R})$-measurable modification of this integral can be chosen (Jacod, 1979).) So, if the conditions of the second part of Theorem 3 are satisfied and Φ'' is a continuous semimartingale, then the process $V = J^\xi(\Phi)$ can be also defined as the difference of integrals (74).
In case Φ is a convex function, the integral $J^\xi(\Phi)$ given by (73) is defined as a weak limit in L_1 of the second derivative of Φ along the path ξ. This definition is, however, impracticable since it is difficult to find such a limit. Under the additional assumption **D'**) below, we define $J_t^\xi(\Phi)$ as the difference

$$J_t^\xi(\Phi) = \frac{1}{2}\int_{\mathbf{R}} L^\xi(t,x)\Phi'(t,dx) - \frac{1}{2}\int_{\mathbf{R}}\int_0^t L^\xi(s,x)\Phi'(ds,dx). \tag{75}$$

D') The functions $(B'(s,x),\ s \geqslant 0,\ x \in R), (N'(s,x),\ s \geqslant 0,\ x \in R)$ are $\mathbf{P}$-a.s. jointly left–continuous in (s,x), right–limited in x and continuous in s. The function $B'(s,x)$ has bounded variation on every compact $[0,t] \times [-n,n]$ ($\mathbf{P}$-a.e.). The function $\langle N'(\cdot,x),\ N'(\cdot,x)\rangle_t$, for every $t \geqslant 0$, has $\mathbf{P}$-a.e. bounded variation along the diagonal on every compact $[-n,n]$, i.e.,

$$\lim_{\Delta_n\to 0}\sum_j \langle N'(\cdot,x_{j+1}) - N'(\cdot,x_j)\rangle = \int_{-n}^{n} \langle N'(\cdot,|dx|)\rangle_t,$$

where $\Delta = \max_j|x_{j+1} - x_j|, \quad -n = x_0 < x_1 < \ldots < x_k = n.$

Suppose also that, for each $n \geqslant 0$, the processes

$$\int_0^t\int_{-n}^{n} L^\xi(s,x)|B'(ds,dx)|, \qquad \int_0^t\int_{-n}^{n} L^\xi(s,x)\langle N'(\cdot,|dx|)\rangle(ds) \tag{76}$$

are locally integrable.

LEMMA 6. *If condition* **D'**) *is satisfied, then the function* $(\langle N'(\cdot,x),m\rangle_s,$ $s \geqslant 0,\ x \in \mathbf{R})$ *is of bounded variation on every compact* $[0,t]\times[-n,n]$ *for each* $m \in \mu^2$ *and*

$$\mathbf{E}\int_{\mathbf{R}}\int_0^t L^\xi(s,x)\langle N'(\cdot,x),m\rangle(ds,dx) \leqslant$$

$$\leqslant \left(\mathbf{E}\langle m\rangle_t\right)^{1/2}\left(\mathbf{E}\int_{\mathbf{R}}\int_0^t L^\xi(s,x)\langle N'(\cdot,|dx|)(ds)\right)^{1/2}. \tag{77}$$

Proof. For $x<y$ and $s<t$, the Kunita–Watanabe inequality implies that

$$\langle N'(\cdot,y),m\rangle_t - \langle N'(\cdot,y),m\rangle_s - \langle N'(\cdot,x),m\rangle_t +$$

$$+\langle N'(\cdot,x),m\rangle_s = \langle N'(\cdot,y)-N'(\cdot,x),m\rangle_t -$$

$$-\langle N'(\cdot,y)-N'(\cdot,x),m\rangle_s \leqslant$$

$$\leqslant (\langle m\rangle_s^t)^{1/2}(\langle N'(\cdot,y)-N'(\cdot,x)\rangle_s^t)^{1/2}, \tag{78}$$

where $\langle m\rangle_s^t = \langle m\rangle_t - \langle m\rangle_s$. In view of that, inequality (77) for the elementary functions $(L_i,\ i\geqslant 1)$ approximating L^ξ follows from Schwartz's inequality for sums, passing to the limit with respect to i, we extend (77) to the general case. Consequently, the integral $\iint L^\xi(s,x)\Phi'(ds,dx)$ in equality (75) can be defined as the sum

$$\int_{\mathbf{R}}\int_0^t L^\xi(s,x)B'(ds,dx) + \widetilde{m},$$

where $\widetilde{m}$ is a martingale such that

$$\langle \widetilde{m},m\rangle_t = \int_0^t\int_{\mathbf{R}} L^\xi(s,x)\big\langle N'(\cdot,x),m\big\rangle(ds,dx) \tag{79}$$

for every $m\in\mathcal{M}^2$. (It follows from (77) that such $\widetilde{m}$ exists and is uniquely defined by (79).)

THEOREM 4. *Assume that, in addition to* **A**) $--$ **C**) *and* **D'**), *the following condition is also satisfied:*

E) *For almost all* $\omega\in\Omega$ *and every* $t\geqslant 0$,

$$\int_{\mathbf{R}}\int_0^t \mathbf{1}\Big\{(s,x)\colon \int_0^s \mathbf{1}\{\xi=x\}d\alpha \neq 0\Big\}\Big(B'(ds,dx)+\big\langle N'(\cdot,|dx|)\big\rangle(ds)\Big)=0.$$

Then the increasing process $V = (J_t^\xi(\Phi),\, t \geqslant 0)$ *can be represented as*

$$V_t = \frac{1}{2}\int_{\mathbf{R}} L^\xi(t,x)\Phi'(t,dx) - \frac{1}{2}\int_0^t\int_{\mathbf{R}} L^\xi(s,x)\Phi'(ds,dx). \tag{80}$$

Remark 4. Condition **E**) is satisfied, for instance, when $\alpha \ll \langle m\rangle$, because $\int \mathbf{1}\{\xi = x\}d\langle m\rangle = 0$ (a.s.).

Proof. Let $(g(y),\, y \in \mathbf{R})$ be an infinitely differentiable non-negative function such that $g(y) = 0$ for $|y| \geqslant 1$ and $\int g(y)dy = 1$.
Consider the following sequence of functions infinitely differentiable for any (ω, t):

$$\Phi_n(\omega,t,x) = n\int \Phi(\omega,t,x+y)g(ny)dy = n\int \Phi(\omega,t,y)g(n(y-x))dy.$$

Since, for every pair (ω, t), the set of points x where the function Φ has no derivatives is countable, we have

$$\Phi_n'(\omega,t,x) = n\int \Phi_-'(\omega,t,x+y)g(ny)dy = n\int \Phi_-'(\omega,t,y)g(n(y-x))dy,$$

where Φ_-' is the left derivative of the function Φ.
By localizing, we can suppose that ξ is bounded. Obviously, the function Φ_n is convex for all (ω, t), the functions Φ_n, Φ_n' are semimartingales for every x (this is implied by conditions **B**) and **D**)), and all the conditions of part ii) of Theorem 3 are satisfied. Applying Theorem 3 and equality (74), we conclude that the processes $(J_t^\xi(\Phi),\, t \geqslant 0)$ and

$$\frac{1}{2}\int_{\mathbf{R}} L^\xi(t,x)\Phi_n'(t,dx) - \frac{1}{2}\int_{\mathbf{R}}\int_0^t L^\xi(s,x)\Phi_n'(ds,dx) \tag{81}$$

are indistinguishable.
It can be easily shown that the functions $(\Phi_n,\, n \geqslant 1)$ satisfy relations (57)–(61) of Lemma 4. In particular, we can show the validity of relation (61).
Furthermore, we have

$$\frac{d\langle\Phi_n(\cdot,x),m\rangle_s}{dK_s} = n\int f^{N_n,m}(s,x+y)g(ny)dy =$$

$$= \int_{-1}^{1} f^{N_n,m}\Big(S, x + \frac{z}{n}\Big)g(z)dz \quad \mu^k\text{-a.e.}$$

Hence, using inequality (30), condition **D'**), the boundedness of ξ, and the fact that $f^N(x,y)$ is continuous μ^k-a.e., we obtain

$$\mathbf{E}\int_0^t (f^{N_n,m}(s,\xi_s) - f^{N,m}(s,\xi_s)dK_s =$$

$$= \mathbf{E}\int_0^t\int_{-1}^1 (f^{N_n,m}(s,\xi_s+\frac{z}{n}) - f^{N,m}(s,\xi_s))g(z)dzdK_s \leqslant$$

$$\leqslant \mathbf{E}\int_0^t \sup_{|z|\leqslant 1/n} |f^{N,m}(s,\xi+Z) - f^{N,m}(s,\xi_s)|dK_s \leqslant$$

$$\leqslant \left(\mathbf{E}\int_0^t \sup_{|z|\leqslant 1/n} \Psi(\xi_s,\xi_s+z)dk_s)^{1/2}(\mathbf{E} < m >\right)^{1/2} \longrightarrow 0, \quad n\to\infty,$$

for each $m \in \mathcal{M}^2$.

Relations (57)–(60) can be proved in a similar manner. Hence, the sequence of increasing processes $(V_t^n,\ t \geqslant 0)$ converges weakly in L_1 (for each $t \geqslant 0$) to the process

$$V_t = \Phi(t,\xi_t) - \Phi(0,\xi_0) - \int_0^t \Phi(ds,\xi_s) - \int_0^t \Phi'_-(s,\xi_s)d\xi_s - \left\langle \int_0^{\cdot} \Phi'(ds,\xi_s),\xi \right\rangle_t$$

which is an increasing process according to Theorem 2, and in order to prove the theorem it is sufficient to show that the two terms appearing in (81) converge to the integrals

$$\int L^\xi(t,x)\Phi'_-(t,dx), \qquad \iint L^\xi(t,x)\Phi'_-(dt,dx),$$

respectively. Since, $\mathbf{P}$-a.e. for every $s \geqslant 0$, we have

$$\Phi'_n = \int_{-1}^1 \Phi'_-\left(s,x+\frac{z}{n}\right)g(z)dz,$$

$$\langle N'_n(\cdot,x),m\rangle_s = \int_{-1}^1 \langle N'\left(\cdot,x+\frac{z}{n}\right),m\rangle_s g(z)dz,$$

$$B''_n = \int_{-1}^1 B'\left(s,x+\frac{z}{n}\right)g(z)dz$$

and

$$\langle N_n'(x), m\rangle_s - \langle N'(x), m\rangle_s = \int_{-1}^{1} \langle N'\Big(x+\frac{z}{n}\Big) - N'(x), m >_s g(z)dz \leqslant$$

$$\leqslant \sup_{|z|\leqslant 1/n} \langle N'(x+z) - N'(x), m\rangle_s \leqslant (\langle m\rangle_s)^{1/2} \sup_{|z|\leqslant 1/n} \langle N'(x+z) - N'(x)\rangle_s,$$

the sequence of functions $\langle N_n'(x), m\rangle_s$ ($B_n'(s,x)$, $\Phi_n'(s,x)$, respectively) converges to the function $\langle N'(x), m\rangle_s$ ($B'(s,x)$, $\Phi_-(s,x)$, respectively) **P**-a.s. at all continuity points (s,x) of the function $\langle N'(x), m\rangle_s$ ($B'(s,x)$, $\Phi_-(s,x)$, respectively) and therefore at all points (s,x) for which $\langle N'(x+) - N'(x)\rangle_s = 0$.
Since the function $L^\xi(s,x)$ is continuous in s and has jumps only at points (s,x) for which $\int_0^s \mathbf{1}\{\xi = x\}d\alpha = 0$, it follows from conditions **D′**), **E**) that

$$\int_{\mathbf{R}}\int_0^t L^\xi(s,x)B_n'(ds,dx) \longrightarrow \int_{\mathbf{R}}\int_0^t L^\xi(s,x)B'(ds,dx) \quad \text{a.s.}$$

as $n \to \infty$, and, for every bounded random variable ι, we have

$$\mathbf{E}\,\iota(\tilde{m}_n - \tilde{m}) = \mathbf{E}\langle \mathbf{E}\,(\iota|\mathcal{F}_\cdot), \tilde{m}_n - \tilde{m}\rangle =$$

$$= \mathbf{E}\int_0^t\int_{\mathbf{R}} L^\xi(s,x)\langle N_n'(\cdot), E(\iota|(\mathcal{F}_\cdot))\rangle(ds,\, dx) -$$

$$-\mathbf{E}\int_0\int_{\mathbf{R}} L^\xi(s,x)\langle N'(\cdot), E(\iota|\mathcal{F}_\cdot)\rangle(ds,\, dx) \longrightarrow 0, \quad n \to \infty.$$

Hence, the integral

$$\int\int L^\xi(s,x)\Phi_n'(ds,\, dx)$$

converges weakly in L_1 (for every $t \geqslant 0$) to $\int\int L^\xi(s,x)\Phi_-'(ds,\, dx)$. Evidently, it follows from condition **E**) that $\int L^\xi(t,x)\,\Phi_n'(t,dx) \to \int L^\xi(t,x)\,\Phi_-'(t,dx)$, $n \to \infty$, and, in view of the uniqueness of the weak limit, we have

$$V_t = \frac{1}{2}\int_{\mathbf{R}} L^\xi(t,x)\Phi'(t,dx) - \frac{1}{2}\int_{\mathbf{R}}\int_0^t L^\xi(s,x)\Phi'(ds,dx). \tag{82}$$

Now we define the integral $J^\xi(\Phi)$ as the mixed variation of the functions L^ξ and Φ_-'. We say that this integral exists if, **P**-a.s. for all $t \geqslant 0$,

$$V_t(L^\xi, \Phi_-') = \sup_{\substack{0<s_1<\dots<s_n=t\\ -\infty<x_1<\dots<x_m<\infty}} \sum_{i,j} \big(L^\xi(s_{i+1}, x_j) - L^\xi(s_i x_j)\big)\times$$

$$\times\big(\Phi_-'(x_{j+1}, s_i) - \Phi_-'(x_j, s_i)\big) < \infty,$$

where the maximum is taken over all finite partitions of the set $[0,t] \times \mathbf{R}$. Evidently, $V_t(L^\xi, \Phi'_-) \geqslant V_s(L^\xi, \Phi'_-)$ a.s. if $s \leqslant t$, and we can consider the right-continuous modification of $(V_t(L^\xi, \Phi'_-),\ t \geqslant 0)$. Note that if the conditions of Theorem 3, part ii), are satisfied, then $J_t(L^\xi,\ \Phi'_-)$ exists and coincides with the integral

$$\int_{\mathbf{R}}\int_0^t \Phi'^{,G}_-(s,x) L^\xi(ds,\ x)\, dG(x),$$

and, if the conditions of Theorem 4 are satisfied, then the mixed variation is equal to

$$\int_{\mathbf{R}} L^\xi(t,x)\, \Phi'_-(t,dx) - \int_{\mathbf{R}}\int_0^t L^\xi(s,x)\, \Phi'_-(ds,dx).$$

Finally, we give one simple property of the process $V(L^\xi, \Phi'_-)$. If for semimartingales ξ and ι there exist mixed variations $V(L^\xi, \Phi'_-), V(L^\iota, \Phi'_-)$, and if f, φ are bounded optional processes, then, $\mathbf{P}$-a.e. for all $t \geqslant 0$, we have

$$V(f \cdot L^\xi + \varphi \cdot L^\iota, \varphi'_-) = f \cdot V(L^\xi, \Phi'_-) + \varphi \cdot V(L^\iota, \Phi'). \tag{83}$$

3. APPLICATIONS

3.1. *Uniqueness of a stochastic differential equation*

Consider the stochastic equation (involving local times)

$$X_t = X_0 + \int_0^t \int_{\mathbf{R}} L^X(ds,y) G(s,dy) + m_t, \tag{84}$$

where m is a continuous local martingale, $G(t,y)$ is a semimartingale having bounded variation in y for each pair (ω,t).

We say that the process $X = (X_t,\ t \geqslant 0)$ is a solution of (84) if and only if the statement of Theorem 2 is applicable to the pair $(\Phi(t,x) = \int_{x_0}^x G(t,y)\,dy,\ \xi = X)$ and the expression of the double integral on the right–hand side of (84) mean the finite-variation part V of (51). (Thus, one of the definitions of the integral

$$V_t = J_t^X(G) = \int_0^t \int_{\mathbf{R}} L^X(ds,y) G(s,dx)$$

considered above can be applied.)

LEMMA 7. *If X^1 and X^2 are two solutions of* (84), *then $X = \max(X^1, X^2)$ is also a solutions of equation* (84).

Proof. Since $X^1 - X^2$ is of finite variation, we obtain $L^{X^1-X^2} = 0$ and according to (Weinryb, 1983)

$$L^{X^1 \vee X^2}(t,y) = \int_0^t 1\{X_s^1 \leqslant Y\} L^{X^2}(ds,y) + \int_0^t 1\{X_s^2 \leqslant Y\} L^{X^1}(ds,y). \quad (85)$$

By Tanaka's formula, property (83) of the double integrals $V(L^X, G)$, and (85), we have

$$X_t^1 \vee X_t^2 = (X_t^1 - X_t^2)^+ + X_t^2 = \int_0^t 1\{X_s^1 > X_s^2\} dX_s^1 + \int_0^t 1\{X_s^1 \leqslant X_s^2\} dX_s^2 =$$

$$= \int_0^t 1\{X_s^1 > X_s^2\} d\Big(\int_0^s \int_{\mathbf{R}} L^{X^1}(du,y) G(u,dy)\Big) +$$

$$+ \int_0^t 1\{X_s^1 > X_s^2\} d\Big(\int_0^s \int_{\mathbf{R}} L^{X^2}(du,y) G(u,dy)\Big) + m_t =$$

$$= \int_0^t \int_{\mathbf{R}} (1\{X_s^1 > X_s^2\} L^{X^1}(ds,y) + 1\{X_s^1 \leqslant X_s^2\} L^{X^2}(ds,y)) G(s,dy) + m_t =$$

$$= \int_0^t \int_{\mathbf{R}} L^{X^1 \vee X^2}(ds,y) G(s,dy) + m_t.$$

As a direct consequence of this statement we have

COROLLARY 7. *For equation* (84) *the unique strong solvability is equivalent to the unique weak solvability.*

In general, the conclusion of Lemma 7 is stated using the definition of the double integral from Itô's formula.
In (Chitashvili and Toronjadze, 1981) this approach was used to prove the strong uniqueness of equation (84) of a special form. The general equation of the form (84) with $G(\omega, s, dy) = G(dy)$ was investigated in (Strook and Yor, 1981; Le Gall, 1982; Engelbert and Schmidt, 1985).

3.2. *The stochastic maximum principle*

Here we show how the statement ii) of Theorem 1 can be used to derive the adjoint system of equations of the stochastic maximum principle.

The general problem of choosing an optimal absolutely continuous measure reduces (see (Chitashvili and Mania, 1986; 1987)) the maximization problem

$$\mathbf{E}\mathcal{E}_T(M^u)\iota \to \max_u,$$

where $M^a = (M(\cdot,a), a \in A)$ is a set of martingales defining the exponential densities $\mathcal{E}(M^a)$ for each decision $a \in A$, u is a predictable process with values in A defining the controls, ι is an $\mathcal{F}$-measurable random variable, and M^u denotes a stochastic line integral along the curve u, i.e., $M^u = \int M(ds, u_s)$ (see Remark 2). The dynamic programming approach to the construction of optimal controls involves the stochastic differential equation for the value process

$$S_t = \sup_u \mathbf{E}\Big[(\mathcal{E}(M^u)/\mathcal{E}_t(M^u))\iota \mid \mathcal{F}_t\Big],$$

$$dS_t = -\max_{a \in A} H(t,a)dK_t + d\overline{m}_t, \quad S_T = \iota, \tag{86}$$

where $\overline{m}$ is the martingale part of the value process uniquely defined by the boundary condition at the right-hand side of the interval $[0,T]$ and the Hamiltonian is expressed as

$$H(t,a) = d\langle M^a, m\rangle_t / dK_t$$

with some increasing process $K = (K_t,\ t \geqslant 0)$ dominating the square characteristics of martingales M^a, $a \in A$. The problem of deriving an adjoint system of equations is motivated by the following argument.

Assume that there exist a continuous system $(N^i,\ 1 \leqslant i \leqslant n)$ of martingales with deterministic characteristics and suppose (without loss of generality) that $\langle N^i\rangle = K$, $1 \leqslant i \leqslant n$. Assume that any $(\mathcal{F}, \mathbf{P})$-martingale can be expressed as an integral with respect to the system $(N^i,\ i \leqslant n)$. In particular, let

$$M_t^a = \sum_i \int_0^t f_i(s,a)dN_s^i, \qquad \overline{m}_t = \int_0^t \Psi_i(s)dN_s^i.$$

Now the Hamiltonian can be expressed in terms of adjoint processes $(\Psi_i,\ 1 \leqslant i \leqslant n)$

$$H(t,a) = \sum_i f_i(t,a)\Psi_i(t),$$

and equation (86) can be viewed as an implicit equation for the system $(\Psi_i,\ 1 \leqslant i \leqslant n)$.

Denote by Z the class of processes X representable as

$$X_t = X_0 + \sum_i \int_0^t X_s^{(i)} dN_s^i + \int_0^t dK_s.$$

Note that for the value process $S \in Z$ with "derivatives" we have

$$S^{(i)} = \Psi_i, \quad 1 \leqslant i \leqslant n, \qquad S^{(0)} = -\max_a \sum_i f_i(a)\Psi_i, \tag{87}$$

and by the representation property there exists a predictable process $\iota^{(i)} = (\iota_t^{(i)}, t \geqslant 0)$ such that

$$\mathbf{E}(\iota \mid \mathcal{F}_t) = \mathbf{E}\iota + \sum_i \int_0^t \iota_s^{(i)} dN_s^i. \tag{88}$$

We shall use the notation $S^{(i,j)} = (S^{(i)})^{(j)}$ for the second "derivatives" $(i, j = 0, 1, \dots, n)$

THEOREM 5. *Let* $\Psi_i \in Z$, $f_i(a) \in Z$ (i.e., $S^{(i)} \in Z$, $S^{(0)} \in Z$), $i \leqslant n$, $\iota^{(i)}$ *be continuous at* T *and the condition*

$$S^{(0,i)} = S^{(i,j)}, \qquad S^{(i,j)} = S^{(i,j)}, \qquad 1 \leqslant i \leqslant n, \quad 1 \leqslant j \leqslant n, \tag{89}$$

be satisfied. Then for any optimal control u^* *the following system of (backward) equations*

$$d\Psi_j(t) = -\sum_i \Psi_i(t) f_i^{(j)}(t, u_t^*) dK_t + dm_t^*, \qquad \Psi_i(T) = \iota_T^{(i)}, \qquad 1 \leqslant j \leqslant u,$$

where m^* *is a martingale with respect to the measure* $d\mathbf{P}^{u^*} = \mathcal{E}(M^{u^*})d\mathbf{P}$, *and the maximum principle*

$$\max_a \sum_i \Psi_i(t) f_i(t, a) = \sum_i \Psi_i(t) f_i(t, u_t^*)$$

are satisfied.

Proof. Using statement ii) of Theorem 1, for any optimal control u^* from equation (86), for the value process we have

$$S^{(0,j)} = -\sum_i f_i^{(j)}(t, u_t^*)\Psi_i(t) - \sum_i f_i(t, u_t^*)\Psi_i^{(j)}(t).$$

By relation (89),

$$\Psi_j^{(0)}(t) = S^{(j,0)} = S^{(0,j)} = -\sum_i f_i^{(j)}(t, u_t^*)\Psi_i(t) - \sum_i f_i(t, u_t^*)\Psi_j^{(i)}(t),$$

and, since the last summand can be rewritten as

$$d\langle \Psi_j, M^{u^*}\rangle_t \, dK_t = \left\langle \sum_i \int_0^{\cdot} \Psi_j^{(i)}(s) dN_s^i, \sum_i \int_0^{\cdot} f_i(s, u^*) dN_s^i \right\rangle_t / dK_t,$$

we obtain that

$$d\Psi_j(t) = -\sum_i f_i^{(j)}(t, u^*)\Psi_i(t) dK_t + dm^{*j},$$

where

$$m_t^{*j} = \int_0^t \sum_i \Psi_i^{(i)}(s)dN_s^i - \langle \Psi_j, M^{u^*} \rangle_t$$

is a martingale with respect to the measure $d\mathbf{P}^{u^*} = \mathcal{E}(M^{u^*})d\mathbf{P}$.
Now we can show that the boundary condition $\Psi_i(T) = \iota_T^{(i)}$ holds.
According to Girsanov's theorem, for any control u^* for which

$$\max_a H(t,a) = H(t, u_t^*),$$

the value process S is a martingale with respect to the measure $d\mathbf{P}^{u^*} = \mathcal{E}(M^{u^*})d\mathbf{P}$ and, by (86),

$$S_t = \mathbf{E}^{u^*}(\iota \mid \mathcal{F}_t) = \mathbf{E}\Big[\iota(\mathcal{E}_T(M^{u^*})/\mathcal{E}_t(M^{u^*})) \mid \mathcal{F}_t\Big] =$$

$$= -\int_0^t \sum_i \Psi_i(s) f_i(s,u^*)dK_s + \sum_i \int_0^t \Psi_i(s)dN_s^i.$$

From this relation and (88) it follows that

$$\mathbf{E}\Big(\Big[(\mathcal{E}_T(M^{u^*})/\mathcal{E}_t(M^{u^*}))\iota\Big] \mid \mathcal{F}_t\Big) - \mathbf{E}(\iota \mid \mathcal{F}_t) =$$

$$\int_t^T \sum_i \Psi_i f_i(u^*)dK + \int_t^T \sum_i (\iota^{(i)} - \Psi_i)dN^i. \tag{90}$$

On the other hand, using the expression for the differential of an exponent it can be easily seen that the left-hand side of this equality can be written as

$$\mathbf{E}\Big(\int_t^T \sum_i [\mathcal{E}_s(M^{u^*})/\mathcal{E}_t(M^{u^*})]\iota_s^{(i)} f_i(s,u^*)dK_s \mid \mathcal{F}_t\Big), \tag{91}$$

and Schwartz's inequality yields

$$\frac{1}{K_T - K_t}\mathbf{E}\int_t^T \sum_i (\Psi_i(s) - \iota_s^{(i)})^2 dK_s \leqslant$$

$$\leqslant 2\mathbf{E}\int_t^T [\sum_i \Psi_i(s) f_i(s,u^*)]^2 dK_s +$$

$$+2\mathbf{E}\int_t^T [(\mathcal{E}_s(M^{u^*})/\mathcal{E}_t(M^{u^*})) \sum_i \iota_s^{(i)} f_i(s,u^*)]^2 dK_s.$$

Thus taking $t \to T$, we obtain $\Psi_i(T) = \iota_T^{(i)}$.
Condition (89) is satisfied for controlled Markov diffusion process, i.e., when "controlled" measures $d\mathbf{P}^u = \mathcal{E}(M^u)d\mathbf{P}$ represent the distribution of a solution of the stochastic differential equation

$$dX_t = f(t, X_t, u_t)dt + dW_t$$

with a vector function $f = (f_i,\ 1 \leqslant i \leqslant n)$, a Wiener process $W = (W_i,\ 1 \leqslant i \leqslant n)$ and the reward $\iota = \Phi(X_T)$ with some smooth function Φ.
The value process can be given as $S_t = S(t, X_t)$ for some (deterministic) sufficiently smooth function $S(t,x),\ t \leqslant T,\ x \in \mathbf{R}^n$. In this "classical" case the adjoint system is derived under less restrictive conditions (see, e.g., (Haussman, 1978)). The above way derivation was suggested in (Chitashvili, 1986).

3.3. *Variations of Itô's formula*
The formula

$$\Phi(t, \xi_t) = \Phi(0, \xi_0) + \int_0^t \Phi(ds, \xi_s) + \int_0^t \Phi'(s, \xi_s)d\xi_s +$$

$$+ \left\langle \int_0^{\cdot} \Phi'(ds, \xi_s), \xi \right\rangle_t + J_t^{\xi}(\Phi) \tag{92}$$

with the increasing process $V = (J_t^{\xi}(\Phi),\ t \geqslant 0)$, formally given as

$$J_t^{\xi}(\Phi) = \int_{\mathbf{R}} \int_0^t L^{\xi}(ds, x)\Phi'_{-}(s, dx), \tag{93}$$

enables us to give a rough but rather useful summary of different versions of Itô's formula. All the decomposition formulas for the substitution of a semimartingale ξ for a one-dimensional variable x in the (random) function $\Phi(t,x)$ can be classified according to the requirements imposed on the two items involved in the substitution procedure, i.e., on the semimartingale ξ and the function Φ.
The table below shows the relationship between different results (which actually concern the continuous processes).
Except for the last two cases, where the process $(J_t^{\xi}(\Phi),\ t \geqslant 0)$ is not of bounded variation, all cases presented in the table are concerned with the semimartingale Φ having at least the first derivative in x of bounded variation (i.e., represented as a difference of two convex functions in x).
The only "stable" term which remains unchanged in all the versions of Itô's formula is the classical term $\int \Phi'(s, \xi_s)d\xi_s$, while the line integral $\int \Phi(ds, \xi_s)$ takes the stochastic integral form if the martingale part of Φ is provided with the finite integral representation of the form

$$N(t,x)=\sum_{i=1}^{n}\int_{0}^{t}\varphi_i(s,x)dN_s^i$$

and reduces to the classical term $\int_0^t(\partial/\partial s)\Phi(s,\xi_s)ds$ for the differentiable in t function Φ not containing the martingale part ($N\equiv 0$). Ventzel's effect

$$\left\langle \int_0^{\cdot}\Phi'_-(ds,\xi_s),\ \xi\right\rangle$$

takes an integral form if both martingale parts $(N'(t,x),m)$ of (Φ',ξ) are provided with a finite integral representation.

Generally, for Itô's effect $J^\xi(\Phi)$, under the rather weak conditions **A**) – **D**) we can give only a nonexplicit expression (by applying part iii) of Theorem 1)

$$J_t^\xi(\Phi)=\overline{\max_x}\,L^{\Phi(\cdot,\xi)-\Phi(\cdot,x)(\xi.-x)}(\cdot,0)$$

and the nonconstructive definition of $J_t^\xi(\Phi)$ as a weak limit (in L_1) of the expression

$$\frac{1}{2\varepsilon}\int_0^t(\Phi'_-(s,\xi_s)-\Phi'_-(s,\xi_s-\varepsilon))d\langle\xi\rangle_s.$$

Under the stronger conditions **A**) – **C**), **D**′), **E**) we define the process $(J_t^\xi(\Phi),\ t\geqslant 0)$ as a difference of the following integrals

$$\frac{1}{2}\int_{\mathbf{R}}L^\xi(t,x)\Phi'_-(t,dx)-\frac{1}{2}\int_{\mathbf{R}}\int_0^t L^\xi(s,x)\Phi'_-(ds,dx) \qquad (94)$$

and prove that the processes $(J_t^\xi(\Phi),\ t\geqslant 0)$ and $(V_t,\ t\geqslant 0)$ from (51) are indistinguishable.

Note that if Φ is a convex function independent of (ω,t), then all the conditions **A**) – **C**), **D**′), **E**) are satisfied, the last integral in (94) is equal to zero, and we get Meyer's formula (Meyer, 1976), i.e., Itô's formula for the semimartingale ξ and the convex function Φ independent of (ω,t).

We conclude this short summary with the statement of some problems which seem natural to investigate.

1) For a fixed semimartingale ξ, find a description of the class of semimartingales $(\Phi(\cdot,x),\ x\in\mathbf{R})$ for which the substitution $\Phi(t,\xi_t)$ is a semimartingale.

2) For a fixed increasing predictable process $K=(K_t,\ t\geqslant 0)$, find a description of the class of semimartingales $(\Phi(\cdot,x),\ x\in\mathbf{R})$ for which the substitution $\Phi(t,\xi_t)$ is a semimartingale characteristics of which are dominated by K (i.e., $\Phi(t,\xi_t)=M_t+A_t$, $\operatorname{Var}A\ll K$, $\langle M,M\rangle\ll K$) for any semimartingale ξ dominated by K (generalization of Krylov's results).

3) Find a description of the class of functions Φ which preserve the domination property (as in 2) for any increasing process K.
4) Find the description of the class of semimartingales $(\Phi(\cdot, x),\ x \in \mathbf{R})$ for which $(F(t, \xi_t),\ t \geqslant 0)$ is a semimartingale for any semimartingale ξ (generalization of Wang's result). Is in this case boundedness of variation of the first derivative (for each (ω, t)) also necessary?

Remark 5. In the recent paper (Menin and Slominski, 1990) it was proved that the condition **UT** guarantees the convergence of the martingale parts M^n of the semimartingales X^n from Lemma 2.

4. SUMMARY OF RESULTS

Classes of ξ	Requirements on $\Phi\left(\omega, t, \xi_t(\omega)\right)$	Condition on $\Phi(\omega, t, x)$	Authors
Itô process	Itô process	$\Phi \in C^{1,2}$, deterministic	Itô (1953)
Itô process	Itô process	$\Phi \in W^{1,2}$ (Sobolev space), deterministic	Krylov (1969)
Wiener process	Submartingale	$\Phi=\|x\|$	Tanaka (1963)
Wiener process	Semimartingale	Convex in x, not depending on (ω, t)	Brosamler (1970) Wang (1977)
Semimartingale	Semimartingale	Convex in x, not depending on (ω, t)	Meyer (1976) Jacod (1978)
Itô process	Itô process	Twice differentiable in x, Itô process in (ω, t)	Ventzel (1965) Rozovski (1973)
Semimartingale	Semimartingale	Twice differentiable in x, semimartingale in (ω, t) with integral representation	Kunita (1981) Bismut (1981) Mikulevičius (1983) Trofimov (1983)
Semimartingale	Semimartingale	Twice differentiable with Hőlder condition on Φ'' martingale in (ω, t)	Sznitman (1982)
Semimartingale	Semimartingale	Twice differentiable in x, semimartingale in (ω, t)	Lazrieva and Toronjadze (1986)
Semimartingale	Semimartingale	Convex in x, bounded variation in t, not depending on ω	Ouknine (1989)
Semimartingale	Semimartingale	Convex in x, semimartingale in (ω, t)	The present paper
Wiener process	Zero square variant for nonmartingale part	Absolutely continuous in x, deterministic	Yor (1982)
Semimartingale	Zero square variation for nonmartingale part	Absolutely continuous in x, differentiable in t, not depending on ω	Al-Hussaini and Elliott (1987)
Semimartingale	Zero square variation for nonmartingale	Differentiable in t, Φ and $d\Phi/dt$ absolutely continuous in x, with integral representation of Φ	Al-Hussaini (1988)

REFERENCES

Al-Hussaini, A.N. (1988). A generalization of Itô's formula. *Lect. Notes Control. Inf. Sci.* **126**, 46–54.

Al-Hussaini, A.N. and Elliot, R.J. (1987). An extension of Itô's differential formula. *Nagoya Math. J.* **105**, 9–18.

Bismut, J.M. (1981). A generalized formula of Itô and some other properties of stochastic flows *Z.Wahrsch. verw. Gebiete* **55**, 174–211.

Brosamler, G.A. (1970). Quadratic variation of potentials and harmonic functions. *Trans. Amer. Math. Soc.* **159**, 243–257.

Chitashvili, R.J. (1986). Stochastic maximum principle in the problem of optimal absolutely continuous change of measure. *Lect. Notes Control Inf. Sci.* **78**, 111–120.

Chitashvili, R.J. and Mania, M.G. (1987). On optimal controls in the problem of locally absolutely continuous change of measure. In: *Probability Theory and Mathematical Statistics. Proc. Fourth Vilnius Conf.,* VNU Science Press, Utrecht **1**, pp. 331–356.

Chitashvili, R.J. and Mania, M.G. (1987). Optimal locally absolutely continuous change of measures. *Stochastics* **21**, 131–229.

Chitashvili, R.J. and Toronjadze, T.A. (1981). On one-dimensional stochastic differential equations with unit diffusion coefficient. Structure of solutions. *Stochastics* **4**, 281–315.

Dellasherie, C. (1978). Quelques applications du lemme Borel–Cantelli a la theorie de semimartingales. *Lect. Notes Math.* **649**, 742–746.

Dellasherie, C. and Meyer, P.A. (1980). *Probabilités et potentiel.* **2**, Herman, Paris

Emery, M. (1979). Une topologie sur l'espace des semimartingales. *Lect. Notes Math.* **721**, 260–281.

Engelbert, H.J. and Schmidt, W. (1985). On one-dimensional stochastic differential equations with generalized drift. *Lect. Notes Control Inf. Sci.* **69**, 143–155.

Haussman, U.G. (1978). On the stochastic maximum principle. *SIAM J.Control and Optimization* **16**, 236–251.

Jacod, J. (1978). Calcul stochastique et problemes de martingales. *Lect. Notes Math.* **714**,

Jakubowski, A., Memin, J. and Pages, G. (1989). Convergence en loi des suites d'integrales stochastiques sur i'espace D^1 de Skorokhod. *Probab. Theory Rél. Fields* **81**, 111–137.

Kunita, H. (1981). On the decomposition of solutions of stochastic differential equations. *Lect. Notes Math.* **851**, 213–255.

Lazrieva, N.L. and Toronjadze, T.A. (1986). Itô–Ventzel's formula for semimartingales, asymptotic properties of MLE and recursive estimation. *Lect. Notes Control Inf. Sci.* **96**, 346–355.

Le Gall, J.F. (1982). *Temps locaux et equations differentielles stochastiques..* These 3e cycle Universite de Paris VI.

Liptser, R.Sh. and Shiryaev, A.N. (1986). *Theory of martingales.* Nauka, Moscow (in Russian)

Memin, J and Slominski, B. (1990). Condition UT et stabilite en loi des solutions equations differentielles stochastiques. *Preprint.* , .

Meyer, P.A. (1976). Un cours sur les integrales stochastiques Seminare de Probabilites. *Lect. Notes Math.* **511**, 245-511.

Mikulevichius, R. (1983). On some properties of solutions of stochastic differential equations *Lithuanian Math. J.* **23**, 18–31.

Ouknine, J. (1989). Functions de semimartingales et applications aux equations differentielles stochastiques. *Stochastics* **28**, 115–122.

Rozovsky B.L. (1973). On Itô-Ventzel's formula. *Vestnik Moscow Univ.* **1**, 26–32 (in Russian).

Stroock, D.W. and Yor, M. (1981). Some remarkable martingales. *Lect. Notes Math.* **850**, 590–603.

Sznitman A. (1982). Martingales dependant d'un parametre: une formule d'Itô. *Z. Wahrsch. verw. Greb.* **60**, 41–70.

Tanaka, H. (1963). Note on continuous additive functionals of the 1-dimensional Brown motion path. *Z. Wahrsch. verw. Greb.* **1**, 251–257.

Trofimov, E.J. A theorem on the implicit function for semimartingales and its application. *Russian Math. Surveys* **38**, 215–216.

Ventzel' A.D. (1965). On equations of the theory of conditionally Markov processes. *Theory Probab. Appl.* **10**, 358–361.

Wang, A.T. (1977). Generalized Itô's formula and additive functionals of Brownian motion. *Z. Wahrsch. verw. Greb.* **41**, 153–159.

Weinryb, S. (1983). Etude d'une equation differentielle stochastique avec temps local. *Lect. Notes Math.* **986**, 72–78.

Yor, M. (1978). Sur la continuite des temps locaux associes a certaines semimartingales. *Asterisque* **52–53**, 23–35.

Yor, M. (1982). Sur la transformation de Hilbert de temps locaux Browniens, et une extension de la formule d'Itô. *Lect. Notes Math.* **920**, 338–347.

New Trends in Probab. and Statist., pp. 351 – 374
V. Sazonov and T. Shervashidze (Eds.)
1991 VSP/Mokslas

OPTIMAL TRANSMISSION OF GAUSSIAN SIGNALS INVOLVING COST FOR TRANSMISSION AND FEEDBACK

O.A. GLONTI and L.G. JAMBURIA

Tbilisi I. Javakhishvili State University, Laboratory of Probabilistic Statistical Methods, David Agmashenebeli Ave. 150[a], 380012 Tbilisi, Georgia, USSR

ABSTRACT

Optimal coding and decoding functionals and an optimal switching strategy for Gaussian signals transmitted through a noiseless feedback channel involving cost for the transmission and feedback are constructed.

INTRODUCTION

The aim of the investigation is to construct optimal schemes for Gaussian signal transmission through a channel with additive "white" noise and noiseless feedback involving cost for the transmission and feedback. As opposed to traditional optimal schemes (see, e.g., (Liptser and Shiryayev, 1977; 1978)) the existence of cost makes the optimization problem more sophisticated and along with optimal coding and decoding functionals one has also search for optimal transmission cut–in and cut–off strategy or feedback or both using the square criterion with additive loss for the cut–in periods. The obtained results are based on the results of filtering of conditionally Gaussian diffusion type processes (Liptser and Shiryayev, 1978) and a lemma, which is given in the Appendix together with its proof. It should be noted that we construct optimal transmission schemes of both Gaussian random variables and Gaussian processes which are solutions of linear stochastic differential equations. Since it is impossible to find out whether there is a unique strong solution of the stochastic equation defining the transmission scheme after the substitution of codings, decodings and strategies claiming to be optimal, no optimal strategy can be constructed when the transmission of such processes involving cost for the feedback is studied and one has to look for ε–optimal ones. This problem will be the subject of a special paper by the authors. It should be also noted that a part of the results of this work has been announced in (Glonti and Jamburia, 1989).

1. OPTIMAL TRANSMISSION OF GAUSSIAN SIGNALS INVOLVING COST FOR THE TRANSMISSION

Let on a complete probability space $(\Omega, \mathcal{F}, (\mathcal{F}_t), \mathbf{P})$ a Gaussian process $\theta = (\theta_t, \mathcal{F}_t)$, $t \in [0,T]$ with a differential

$$d\theta_t = a(t)\theta_t\,dt + b(t)\,d\widetilde{W}_t, \quad \theta_0 = \theta,$$

be given, where $\widetilde{W} = (\widetilde{W}_t, \mathcal{F}_t)$ is a Wiener process independent of the Gaussian random variable θ with $\mathbf{E}\theta = m$ and $\mathbf{E}(\theta - m)^2 = \gamma$; $a(t)$ and $b(t)$ are nonrandom functions and $|a(t)| \leqslant K$, $|b(t)| \leqslant K$.
Let the process $\theta = (\theta_t, \mathcal{F}_t)$ be transmitted according to the following scheme

$$d\xi_t = \alpha_t\big[A_0(t,\xi) + A_1(t,\xi)\theta_t\big]dt + dW_t, \quad \xi_0 = 0, \tag{1}$$

where the Wiener process $W = (W_t, \mathcal{F}_t)$ is independent of $\widetilde{W}$ and $\theta = (\theta_t, \mathcal{F}_t)$, and nonanticipative coding functionals A_0 and A_1 satisfy the following conditions

$$\mathbf{P}\left\{ \int_0^T A_0^2(t,\xi)dt < \infty \right\} = 1, \qquad \sup_{x\in C,\ t\leqslant T} \big|A_1(t,x)\big| < \infty.$$

The process $\alpha = (\alpha_t, \mathcal{F}_t^\xi)$ is a strategy taking only two values: 0 and 1. It determines the moments of signal transmission. Its appearance depends on the presence of the cost for the transmission. Let the strategy α and the coding functionals A_0 and A_1 be such that equation (1) has a unique strong solution. According to the received messages, we must construct decoding $\hat{\theta}_t(\xi)$ which reproduces the signal θ_t. Let

$$\delta(t) = \inf_{\alpha, A_0, A_1, \hat{\theta}} \mathbf{E}\left[\left(\theta_t - \hat{\theta}_t(\xi)\right)^2 + c\int_0^t \alpha_s\,ds\right], \quad c > 0,$$

where c is the cost for the transmission per unit of time. Let the following power type condition

$$\mathbf{E}\left[\left(A_0(t,\xi) + A_1(t,\xi)\theta_t\right)^2 \Big| \mathcal{F}_t^\xi\right] \leqslant P \tag{2}$$

be satisfied, where P is some constant.
Denote

$$m_t = \mathbf{E}\left(\theta_t | \mathcal{F}_t^\xi\right), \quad \gamma_t = \mathbf{E}\left[(\theta_t - m_t)^2 | \mathcal{F}_t^\xi\right].$$

It can be easily seen that the optimal decoding is m_t and, therefore,

$$\delta(t) = \inf_{\alpha, A_0, A_1} \mathbf{E}\left[\gamma_t + c\int_0^t \alpha_s\,ds\right].$$

It is known (Liptser and Shiryayev, 1978) that $m = (m_t, \mathcal{F}_t^\xi)$ and $\gamma = (\gamma_t, \mathcal{F}_t^\xi)$, $t \in [0, T]$ satisfy the following equations:

$$dm_t = a(t)m_t\,dt + \alpha_t\gamma_t A_1(t,\xi)\Big[d\xi_t - (A_0(t,\xi) + A_1(t,\xi)m_t)\alpha_t\,dt\Big], \quad m_0 = m, \tag{3}$$

$$d\gamma_t = \Big[2a(t)\gamma_t - \alpha_t\gamma_t^2 A_1^2(t,\xi) + b^2(t)\Big]dt, \quad \gamma_0 = \gamma. \tag{4}$$

The second equation can be rewritten in the following form

$$\gamma_t = \gamma \exp\left\{2\int_0^t a(s)\,ds - \int_0^t \alpha_s\gamma_s A_1^2(s,\xi)\,ds\right\} +$$

$$+\int_0^t b^2(s)\exp\left\{2\int_s^t a(u)\,du - \int_s^t \alpha_u\gamma_u A_1^2(u,\xi)\,du\right\}ds.$$

From condition (2),

$$\mathbf{E}\big[A_0(t,\xi) + A_1(t,\xi)m_t\big]^2 + \gamma_t A_1^2(t,\xi) \leqslant P$$

and, therefore, $\gamma_t A_1^2(t,\xi) \leqslant P$. Using this fact, we have

$$\gamma_t \geqslant \gamma \exp\left\{\int_0^t \Big[2a(s) - P\alpha_s\Big]\,ds\right\} +$$

$$+\int_0^t b^2(s)\exp\left\{\int_s^t \Big[2a(u) - P\alpha_u\Big]\,du\right\}ds.$$

If we take $A_1^*(t,\xi) = \sqrt{P/\gamma_t^*}$, where γ_t^* is the solution of (4) for a given A_1^*, then

$$\gamma_t^* = \gamma \exp\left\{\int_0^t \Big[2a(s) - P\alpha_s\Big]\,ds\right\} +$$

$$+\int_0^t b^2(s)\exp\left\{\int_s^t \Big[2a(u) - P\alpha_u\Big]\,du\right\}ds.$$

For $A_1^*(t,\xi) = \sqrt{P/\gamma_t^*}$, condition (2) is fulfilled if $A_0^*(t,\xi) = -A_1^*(t,\xi)m_t^*$, where m_t^* is the solution of (3) for given A_0^*, A_1^* and γ_t^*.
This implies that

$$\delta(t) = \inf_\alpha \mathbf{E}\left[\gamma \exp\left\{2\int_0^t a(s)\,ds - P\int_0^t \alpha_s\,ds\right\} +\right.$$

$$+\int_0^t b^2(s)\exp\left\{2\int_s^t a(u)\,du - P\int_s^t \alpha_u\,du\right\}ds + c\int_0^t \alpha_s\,ds\Bigg] \tag{5}$$

and, for a fixed strategy α, the optimal codings and decoding are A_0^*, A_1^* and m_t^*. Now we construct the optimal strategy. Denote the expression in square brackets in (5) by $\delta_\alpha(t)$ and rewrite it in the following form

$$\delta_\alpha(t) = \exp\left\{2\int_0^t a(s)\,ds - P\int_0^t \alpha_s\,ds\right\}\times$$

$$\times\left[\gamma + \int_0^t b^2(s)\exp\left\{\int_0^s \alpha_u\,du - 2\int_0^s a(u)\,du\right\}ds\right] + c\int_0^t \alpha_s\,ds. \tag{6}$$

Using inequality (A.2) from the Appendix of this paper, we obtain

$$\delta_\alpha(t) \geqslant R_t\exp\left\{-P\int_0^t \alpha_s\,ds\right\}\left[\gamma + \int_0^{t-\varkappa_t} b^2(s)R_s^{-1}\,ds + \right.$$

$$\left. + \exp\left\{-P\left(t - \int_0^t \alpha_s\,ds\right)\right\}\int_{t-\varkappa_t}^t b^2(s)R_s^{-1}e^{Ps}\,ds\right] + c\int_0^t \alpha_s\,ds, \tag{7}$$

where $R_t = \exp\left\{2\int_0^t a(s)\,ds\right\}$, $\varkappa_t = \int_0^t \alpha_s\,ds$.
Consider the function

$$f(x) = R_t e^{-Px}\left[\gamma + \int_0^{t-x} b^2(s)R_s^{-1}\,ds + \right.$$

$$\left. + e^{-P(t-x)}\int_{t-x}^t b^2(s)R_s^{-1}e^{Ps}ds\right] + cx, \quad x \in [0,t].$$

Then

$$f'(x) \geqslant 0 \Longleftrightarrow ce^{Px} \geqslant PR_t\left(\int_0^{t-x} b^2(s)R_s^{-1}ds + \gamma\right).$$

Investigating the functions ce^{Px} and $PR_t\left(\int_0^{t-x} b^2(s)R_s^{-1}ds + \gamma\right)$ one can easily arrive at the following conclusions:

1) If $c \geqslant PR_t\left(\gamma + \int_0^t b^2(s)R_s^{-1}ds\right)$ then $f'(x) \geqslant 0$, $x \in [0,t]$, and, therefore, $f(x) \geqslant f(0)$.

2) If $\gamma P R_t e^{-Pt} < c < P R_t\left(\gamma + \int_0^t b^2(s) R_s^{-1}\, ds\right)$, then there exists a unique number $x_0 \in (0,t)$ such that $f'(x) \leqslant 0$ if $x \leqslant x_0$, and $f'(x) \geqslant 0$ if $x \geqslant x_0$, i.e., $f(x) \geqslant f(x_0)$, $x \in [0,t]$ and x_0 is the unique solution of the equation

$$ce^{Px} = PR_t\left(\gamma + \int_0^{t-x} b^2(s) R_s^{-1} ds\right). \tag{8}$$

3) If $c \leqslant \gamma P R_t e^{-Pt}$, then $f'(x) \leqslant 0$ and $f(x) \leqslant f(t)$, $x \in [0,t]$.

Using the properties of the function f and inequality (7) we obtain that if $c \geqslant$ $\geqslant P R_t\left(\gamma + \int_0^t b^2(s) R_s^{-1}\, ds\right)$, then, for any strategy α,

$$\delta_\alpha(t) \geqslant R_t\left(\gamma + \int_0^t b^2(s) R_s^{-1}\, ds\right).$$

But for $\alpha_s^* = 0$, $s \leqslant t$, from (6) we obtain

$$\delta_{\alpha^*}(t) = R_t\left(\gamma + \int_0^t b^2(s) R_s^{-1}\, ds\right).$$

Therefore, in that case the optimal strategy is $\alpha_s^* = 0$, $s \leqslant t$, and

$$\delta(t) = R_t\left(\gamma + \int_0^t b^2(s) R_s^{-1}\, ds\right).$$

Now, let

$$\gamma P R_t e^{-Pt} < c < P R_t\left(\gamma + \int_0^t b^2(s) R_s^{-1}\, ds\right).$$

Then from (7) we have that, for any strategy α,

$$\delta_\alpha(t) \geqslant R_t e^{-Px_0}\left[\gamma + \int_0^{t-x_0} b^2(s) R_s^{-1}\, ds + \right.$$

$$\left. + e^{-P(t-x_0)} \int_{t-x_0}^{t} b^2(s) R_s^{-1} e^{Ps}\, ds\right] + cx_0,$$

where x_0 is the unique solution of equation (8).

Consider the strategy

$$\alpha_s^* = I(t - x_0 \leqslant s \leqslant t), \quad s \in [0,t].$$

From the second proposition of Lemma 1 from the Appendix, using $\int_0^t \alpha_s^* \, ds = x_0$, we obtain

$$\delta_{\alpha^*}(t) = R_t e^{-Px_0}\Bigg[\gamma + \int_0^{t-x_0} b^2(s) R_s^{-1} \, ds + e^{-P(t-x_0)} \int_{t-x_0}^{t} b^2(s) R_s^{-1} e^{Ps} \, ds\Bigg] + cx_0.$$

Therefore, $\delta(t) = \delta_{\alpha^*}(t)$ and the constructed strategy is optimal.
Finally, when $c \leqslant \gamma P R_t e^{-Pt}$, in a similar way we obtain that the optimal strategy is $\alpha_s^* = 1, \ s \leqslant t$, and

$$\delta(t) = R_t e^{-Pt}\left(\gamma + \int_0^t b^2(s) R_s^{-1} e^{Ps} \, ds\right) + ct.$$

Hence the following theorem is proved.

THEOREM 1. *Let a Gaussian process $\theta = (\theta_t, \mathcal{F}_t)$ with a differential*

$$d\theta_t = a(t)\theta_t \, dt + d\widetilde{W}_t$$

be transmitted according to scheme (1). *Then the optimal coding is*

$$A_1^*(t,\xi) = \sqrt{P/\gamma_t^*}, \quad A_0(t,\xi) = -A_1^*(t,\xi) m_t^*,$$

where the optimal decoding m_t^* and the process γ_t^* are determined by equations (3) *and* (4) *for given A_0^*, A_1^* and for the optimal strategy α^* which is determined in the following way.*

If

$$c \geqslant P \exp\left\{2\int_0^t a(s) \, ds\right\}\left[\gamma + \int_0^t b^2(s) \exp\left\{-2\int_0^s a(u) \, du\right\} ds\right],$$

then $\alpha_s^* = 0, \ s \leqslant t$. In this case

$$\delta(t) = \exp\left\{2\int_0^t a(s) \, ds\right\}\left[\gamma + \int_0^t b^2(s) \exp\left\{-2\int_0^s a(u) \, du\right\} ds\right].$$

If

$$\gamma P \exp\left\{2\int_0^t \left[a(s) - \frac{P}{2}\right] ds\right\} < c < P \exp\left\{2\int_0^t a(s) \, ds\right\} \times$$

$$\times\left[\gamma + \int_0^t b^2(s)\exp\left\{-2\int_0^s a(u)\,du\right\}ds\right],$$

then the optimal strategy is

$$\alpha_s^* = I(t - x_0 \leqslant s \leqslant t), \quad s \in [0,t],$$

where x_0 *is the unique solution of equation* (8) *and*

$$\delta(t) = \exp\left\{2\int_0^t a(s)\,ds - Px_0\right\}\left[\gamma + \right.$$

$$+\int_0^{t-x_0} b^2(s)\exp\left\{-2\int_0^s a(u)\,du\right\}ds +$$

$$\left. + e^{-P(t-x_0)}\int_{t-x_0}^t b^2(s)\exp\left\{-2\int_0^s\left[a(u)+\frac{P}{2}\right]du\right\}ds\right] + cx_0.$$

If

$$c \leqslant \gamma P\exp\left\{2\int_0^t\left[a(s)-\frac{P}{2}\right]ds\right\},$$

then $\alpha_s^* = 1,\ s \leqslant t$, *and*

$$\delta(t) = \exp\left\{2\int_0^t\left[a(s)-\frac{P}{2}\right]ds\right\}\left[\gamma + \right.$$

$$\left. + \int_0^t b^2(s)\exp\left\{-2\int_0^s\left[a(u)+\frac{P}{2}\right]du\right\}ds\right] + ct.$$

Now consider the case when the transmitted signal is a Gaussian random variable θ with $\mathbf{E}\theta = m$ and $\mathbf{E}(\theta - m)^2 = \gamma$.
Let the transmission be according to scheme (1) and let all the conditions which were satisfied for the transmission of the Gaussian process be fulfilled. Then from Theorem 1 the following theorem $(a(t) = 0,\ b(t) = 0)$ can be obtained as a special case.

THEOREM 2. *Let* $c \geqslant \gamma P$. *Then* $\delta(t) = \gamma$, $\alpha_s^* = 0$, $s \leqslant t$, *i.e., there is no transmission. Let* $c < \gamma P$.
If $t < t_0$, *where* $t_0 = (1/P)\log(\gamma P/c)$, *then*

$$\delta(t) = \gamma e^{-Pt} + ct.$$

The optimal strategy is $\alpha_s^* = 1$, $s \in [0,t]$, *the optimal coding functionals have the forms*

$$A_1^*(s) = \sqrt{P/\gamma}e^{Ps/2}, \quad A_0^*(s,\xi) = -A_1^*(s)m_s^*,$$

where the optimal decoding is a solution of equation (3) for given A_0^* and A_1^*.
If $t \geqslant t_0$ then

$$\delta(t) = \gamma e^{-Pt_0} + ct_0.$$

The optimal strategy is $\alpha_s^* = I(t - t_0 \leqslant s \leqslant t)$,

$$A_1^*(s) = \begin{cases} \sqrt{\frac{P}{\gamma}}e^{Ps/2}, & s < t_0; \\ \sqrt{\frac{P}{\gamma}}e^{Pt_0/2}, & t_0 \leqslant s \leqslant t; \end{cases}$$

$A_0^*(t,\xi)$ and m_t^* are determined as in the previous case.

Remark. In the case when $c < \gamma P$ and $t \geqslant t_0$, any strategy α_s, for which $\int_0^t \alpha_s ds = t_0$, is optimal.

Now consider the case when the transmission of Gaussian random variables is without feedback according to the scheme

$$d\xi_t = \alpha_t\big[B_0(t) + B_1(t)\theta\big]dt + dW_t, \quad \xi_0 = 0, \quad t \in [0,T].$$

Let the following power condition be satisfied

$$\mathbf{E}\big[B_0(t) + B_1(t)\theta\big]^2 \leqslant P.$$

THEOREM 3. *Let $c \geqslant \gamma P$. Then the optimal strategy is $\alpha_s^* = 0$, $s \leqslant t$, and $\delta(t) = \gamma$.*
If $c < \gamma P$ and $t < t_1$, where $t_1 = (\sqrt{c\gamma P} - c)(cP)^{-1}$, then

$$\delta(t) = \frac{\gamma}{1 + Pt} + ct, \quad \alpha_s^* = 1, \quad s \in [0,t],$$

$$B_1^*(s) = \sqrt{\frac{P}{\gamma}}, \quad B_0^*(s) = -m\sqrt{\frac{P}{\gamma}}.$$

If $c < \gamma P$ and $t \geqslant t_1$ then

$$\delta(t) = \frac{\gamma}{1 + Pt_1} + ct_1, \quad \alpha_s^* = I(s < t_1),$$

$$B_1^*(s) = \sqrt{\frac{P}{\gamma}}, \quad B_0^*(s) = -m\sqrt{\frac{P}{\gamma}}.$$

Proof. As in Theorem 1, let

$$\delta(t) = \inf_{\alpha, B_0, B_1} \mathbf{E}\left[\gamma_t + c\int_0^t \alpha_s\, ds\right].$$

In this case

$$\gamma_t = \frac{\gamma}{1+\gamma\int_0^t \alpha_s B_1^2(s)\, ds}, \qquad t \in [0,T].$$

From the power condition it follows easily, that $B_1^2(s) \leqslant P/\gamma$.
Therefore,

$$\delta(t) = \inf_{\alpha} \mathbf{E}\left[\frac{\gamma}{1+P\int_0^t \alpha_s\, ds} + c\int_0^t \alpha_s\, ds\right]$$

and the optimal coding is $B_1^*(s) = \sqrt{P/\gamma}$, $B_0^*(s) = -m\sqrt{P/\gamma}$. Consider the following function for the optimal strategy construction

$$f(x) = \frac{\gamma}{1+Px} + cx, \quad x \in [0,t].$$

If $c \geqslant \gamma P$ then $f'(x) \geqslant 0$, $x \in [0,t]$, and $f(x) \geqslant f(0) = \gamma$. If $c < \gamma P$ then $f(x) \geqslant f(t_1)$, $x \in [0,t]$, where $t_1 = (\sqrt{c\gamma P} - c)(cP)^{-1}$. Consequently, when $c \geqslant \gamma P$,

$$\delta_\alpha(t) \geqslant \frac{\gamma}{1+P\int_0^t \alpha_s\, ds} + c\int_0^t \alpha_s\, ds \geqslant \gamma.$$

If $\alpha_s^* = 0$, $s \in [0,t]$, then $\delta_{\alpha^*}(t) = \gamma$ and, hence, $\delta(t) = \gamma$.
If $c < \gamma P$ and $t < t_1$ then $\delta_\alpha(t)$ decreases with respect to $\int_0^t \alpha_s\, ds$ and

$$\delta_\alpha(t) \geqslant \frac{\gamma}{1+Pt} + ct.$$

When $\alpha_s^* = 1$, $s \in [0,t]$,

$$\delta_{\alpha^*}(t) = \frac{\gamma}{1+Pt} + ct.$$

If $c < \gamma P$ and $t \geqslant t_1$ then

$$\delta_\alpha(t) \geqslant \frac{\gamma}{1+Pt_1} + ct_1$$

and when $\alpha_s^* = I(s < t_1)$,

$$\delta_{\alpha^*}(t) = \frac{\gamma}{1+Pt_1} + ct_1.$$

Note that $\delta_\alpha(t) = \gamma/(1+Pt_1) + ct_1$ for any strategy α for which $\int_0^t \alpha_s\, ds = t_1$. The theorem is proved.

2. OPTIMAL TRANSMISSION OF GAUSSIAN SIGNALS INVOLVING COST FOR FEEDBACK

Suppose that the transmitted message is a Gaussian random variable θ with $\mathbf{E}\theta = m$ and $\mathbf{E}(\theta - m)^2 = \gamma$ and the transmission of θ is carried out according to the following scheme

$$d\xi_t = \Big\{\beta_t\big[A_0(t,\xi) + A_1(t,\xi)\theta\big] + \\ +(1-\beta_t)\big[B_0(t) + B_1(t)\theta\big]\Big\}dt + dW_t, \ \xi_0 = 0, \tag{9}$$

where $t \in [0,T]$, $\beta = (\beta_t)_{t\leqslant T}$ is a $(\mathcal{F}_t^\xi)_{t\leqslant T}$-adapted stochastic process taking two values: 0 and 1. If $\beta_t = 1$ the transmission is with feedback and if $\beta_t = 0$ it is without feedback. Let

$$\delta(t) = \inf_{A_0, A_1, B_0, B_1, \beta, \hat{\theta}} \mathbf{E}\left[\left(\theta - \hat{\theta}_t\right)^2 + c\int_0^t \beta_s\, ds\right], \quad c > 0,$$

where c is the cost for the feedback.

Let the following power conditions

$$\mathbf{E}\Big\{\big[A_0(t,\xi) + A_1(t,\xi)\theta\big]^2 \,|\, \mathcal{F}_t^\xi\Big\} \leqslant P, \quad \mathbf{E}\big[B_0(t) + B_1(t)\theta\big]^2 \leqslant P \tag{10}$$

be satisfied.

The functionals A_0, A_1, B_0, B_1, β are such that equation (9) has a unique strong solution. Denote

$$m_t = \mathbf{E}(\theta|\,\mathcal{F}_t^\xi), \quad \gamma_t = \mathbf{E}\big[(\theta - m_t)^2|\,\mathcal{F}_t^\xi\big].$$

The equations for m_t and γ_t have the following forms

$$dm_t = \gamma_t\Big[\beta_t A_1(t,\xi) + (1-\beta_t)B_1(t)\Big]\Big\{d\xi_t - \Big[\beta_t A_0(t,\xi) + \\ +(1-\beta_t)B_0(t) + \Big(\beta_t A_1(t,\xi) + (1-\beta_t)B_1(t)\Big)m_t\Big]dt\Big\}, \quad m_0 = m, \tag{11}$$

$$d\gamma_t = -\gamma_t^2\Big[\beta_t A_1(t,\xi) + (1-\beta_t)B_1(t)\Big]^2 dt, \quad \gamma_0 = \gamma. \tag{12}$$

THEOREM 4. *Let the transmission of the Gaussian variable θ take place according to scheme (9).*
If $c \geqslant \gamma P/4$, then the optimal strategy $\beta_s^* = 0$, $s \in [0,t]$, i.e., the transmission is without feedback. The optimal coding is

$$B_1^*(s) = \sqrt{\frac{P}{\gamma}}, \quad B_0^*(s) = -m\sqrt{\frac{P}{\gamma}}$$

and the optimal decoding m_t^* is determined from (10) and (11) for given β^*, B_0^*, B_1^*

In that case,

$$\delta(t) = \frac{\gamma}{1+Pt}.$$

Exactly the same statement is true if $c < \gamma P/4$ *and* $t \leqslant a$ *or* $t \geqslant b$, *where*

$$a = \frac{\gamma P - 2c - \sqrt{\gamma^2 P^2 - 4c\gamma P}}{2cP}, \quad b = \frac{\gamma P - 2c + \sqrt{\gamma^2 P^2 - 4c\gamma P}}{2cP}.$$

If $c < \gamma P/4$ *and* $t \in (a, b)$ *then the optimal strategy has the following form* $\beta_s^* = I(t - x_0 \leqslant s \leqslant t)$, *where* x_0 *is the unique solution of the equation*

$$ce^{Px}\left[1 + P(t-x)\right]^2 = \gamma P^2 (t - x)$$

in the interval $(0, t)$.
The optimal codings are determined in the following way

$$B_1^*(s) = \sqrt{\frac{P}{\gamma}}, \quad B_0^*(s) = -m\sqrt{\frac{P}{\gamma}},$$

$$A_1^*(s,\xi) = \sqrt{\frac{P}{\gamma}} \exp\left\{\frac{P}{2}\int_0^s \beta_u^*\,du\right\}\Bigg[1+$$

$$+\gamma\int_0^s \left(1-\beta_u^*\right)\left(B_1^*(u)\right)^2 \exp\left\{-P\int_0^u \beta_r^* dr\right\} du\Bigg],$$

$$A_0^*(t,\xi) = -A_1^*(t,\xi)m_t^*,$$

where m_t^* *and* γ_t^* *are determined by equations* (11) *and* (12). *In this case*

$$\delta(t) = \frac{\gamma e^{-Px_0}}{1 + P(t - x_0)} + cx_0.$$

Proof. For fixed coding functionals A_0, A_1, B_0, B_1 and strategy β, the optimal decoding is m_t. Hence,

$$\delta(t) = \inf_{A_0, A_1, B_0, B_1, \beta} \mathbf{E}\left[\gamma_t + c\int_0^t \beta_s\,ds\right].$$

First we construct the optimal coding functionals for a fixed strategy β. Rewrite the power condition in the following way

$$\left[A_0(t,\xi) + A_1(t,\xi)m_t\right]^2 + \gamma_t A_1^2(t,\xi) \leqslant P,$$

$$\left[B_0(t) + B_1(t)m\right]^2_+ \gamma B_1^2(t) \leqslant P. \tag{13}$$

The equation for γ_t can be written in the following form

$$\gamma_t = \gamma \exp\left\{ -\int_0^t \beta_s \gamma_s A_1^2(s,\xi)\, ds - \int_0^t (1-\beta_s)\gamma_s B_1^2(s)\, ds \right\}.$$

From (13) it follows that $\gamma_t A_1^2(t,\xi) \leqslant P$. Therefore,

$$\gamma_t \geqslant \gamma \exp\left\{ -P\int_0^t \beta_s\, ds - \int_0^t (1-\beta_s) B_1^2(s)\, ds \right\}. \tag{14}$$

If we take $A_1^*(t,\xi) = \sqrt{P/\gamma_t^*}$, where γ_t^* is a solution of (12) for a given functional A_1^* and a fixed B_1, then in (14) we obtain an equality. Hence A_1^* is the optimal coding. For the realization of the power condition we can take $A_0^*(t,\xi) = -A_1^*(t,\xi)m_t^*$, where m_t^* is a solution of (11) for given A_0^*, A_1^*, γ_t^*. If we solve equation (12) for $A_1^*(t,\xi) = \sqrt{P/\gamma_t^*}$, then we obtain

$$\gamma_t^* = \frac{\gamma \varkappa(t)}{1+\gamma\int_0^t (1-\beta_s) B_1^2(s)\varkappa(s)\, ds}, \quad \varkappa(t) = \exp\left\{ -P\int_0^t \beta_s\, ds \right\},$$

and, therefore, the optimal coding has the following form

$$A_1^*(t,\xi) = \sqrt{\frac{P}{\gamma}}\Big(\varkappa(t)\Big)^{-1/2}\left[1+\gamma\int_0^t \Big(1-\beta_s\Big) B_1^2(s)\varkappa(s)\, ds\right].$$

Thus,

$$\delta(t) = \inf_{B_0,B_1,\beta} \mathbf{E}\left[\frac{\gamma\varkappa(t)}{1+\gamma\int_0^t \Big(1-\beta_s\Big) B_1^2(s)\varkappa(s)\, ds} + c\int_0^t \beta_s\, ds\right].$$

From (13), $B_1^2(t) \leqslant P/\gamma$ and, if we put $B_1^*(t) = \sqrt{P/\gamma}$ and $B_0^*(t) = -m\sqrt{P/\gamma}$, then condition (13) is fulfilled and

$$\delta(t) = \inf_{\beta} \mathbf{E}\left[\frac{\gamma\varkappa(t)}{1+P\int_0^t \Big(1-\beta_s\Big)\varkappa(s)\, ds} + c\int_0^t \beta_s\, ds\right].$$

Thus, for a fixed strategy β the optimal coding functionals and the optimal decoding are constructed.

Now we can construct the optimal strategy. It is easy to see that

$$\delta(t) = \inf_{\beta} \mathbf{E}\left[\frac{\gamma}{1+P\Big(\varkappa(t)\Big)^{-1}\int_0^t \varkappa(s)\, ds} + c\int_0^t \beta_s\, ds\right].$$

We denote the expression in the brackets by $\delta_\beta(t)$. Using inequality (A.1) from the Appendix, we have

$$\delta_\beta(t) \geqslant \frac{\gamma \varkappa(t)}{1+P\Big(t-\int_0^t \beta_s\, ds\Big)} + c\int\limits_0^t \beta_s\, ds.$$

Consider the function

$$f(x) = \frac{\gamma e^{-Px}}{1+P(t-x)} + cx, \quad x \in [0,t].$$

Investigating the derivatives of this function, we easily obtain that if $c \geqslant \gamma P/4$ then this function increases for all $x \in [0,t]$ and, therefore, $f(x) \geqslant f(0) = \gamma/(1+Pt)$. If $c < \gamma P/4$ and $t \leqslant a$ or $t \geqslant b$, where a and b are defined in the formulation of the theorem, then again $f(x) \geqslant \gamma/(1+Pt)$, $x \in [0,t]$.
If $c < \gamma P/4$ and $t \in (a,b)$ then $f'(x) < 0$ for $x \in [0,x_0)$ and $f'(x) > 0$ for $x \in (x_0,t]$, i.e., $f(x) \geqslant f(x_0)$, $x \in [0,t]$, where x_0 is the unique solution of the equation

$$ce^{Px}\big[1+P(t-x)\big]^2 = \gamma P^2(t-x).$$

Using the properties of the function f, we can obtain the following results.
If $c \geqslant \gamma P/4$, then (note that $0 \leqslant \int_0^t \beta_s\, ds \leqslant t$), for any strategy β, we have

$$\delta_\beta(t) \geqslant \frac{\gamma}{1+Pt}$$

and, for $\beta^*_s = 0$, $s \in [0,t]$,

$$\delta_{\beta^*}(t) = \frac{\gamma}{1+Pt},$$

i.e., β^* is the optimal strategy.
If $c < \gamma P/4$ and $t \leqslant a$ or $t \geqslant b$ then we obtain the same.
If $c < \gamma P/4$ and $t \in (a,b)$ then

$$\delta_\beta(t) \geqslant \frac{\gamma e^{-Px_0}}{1+P(t-x_0)} + cx_0.$$

Let $\beta^*_s = I(t-x_0 \leqslant s \leqslant t)$. Then, according to the second assertion of Lemma 1 from the Appendix, we obtain

$$\delta_{\beta^*}(t) = \frac{\gamma \exp\Big\{-P\int_0^t \beta^*_s\, ds\Big\}}{1+P\Big(t-\int_0^t \beta^*_s\, ds\Big)} + c\int\limits_0^t \beta^*_s\, ds = \frac{\gamma e^{-Px_0}}{1+P(t-x_0)} + cx_0,$$

which completes the proof of the theorem.

3. OPTIMAL TRANSMISSION OF GAUSSIAN SIGNALS INVOLVING COST FOR TRANSMISSION AND FEEDBACK

Consider the problem of optimal transmission of the Gaussian random variable θ under the condition that there is a cost both for the transmission and feedback. Let the transmission be carried out according to the following scheme

$$d\xi_t = \alpha_t\Big\{\beta_t\big[A_0(t,\xi)+A_1(t,\xi)\theta\big]+(1-\beta_t)\big[B_0(t)+B_1(t)\theta\big]\Big\}\,dt+dW_t,$$

$$\xi_0 = 0, \quad t\in[0,T]. \tag{15}$$

The strategy α defines whether the transmission occurs or not. The strategy β defines whether the transmission is with feedback or not. It is assumed that α and β take two values 0 and 1 and are such that equation (15) has a unique strong solution. Let again, power condition (10) be satisfied and

$$\delta(t) = \inf_{A_0,A_1,B_0,B_1,\alpha,\beta,\hat\theta} \mathbf{E}\left[\left(\theta-\hat\theta_t\right)^2+c_1\int_0^t \alpha_s\,ds+c_2\int_0^t \alpha_s\beta_s\,ds\right], \quad c_1>0, \quad c_2>0,$$

where c_1 is the cost for the transmission and c_2 is the cost for the feedback. The equations for m_t and γ_t have the following forms

$$dm_t = \alpha_t\gamma_t\Big[\beta_tA_1(t,\xi)+(1-\beta_t)B_1(t)\Big]\Big\{d\xi_t-\Big[\beta_tA_0(t,\xi)+$$

$$+(1-\beta_t)B_0(t)+\big(\beta_tA_1(t,\xi)+(1-\beta_t)B_1(t)\big)m_t\Big]dt\Big\}, \quad m_0=m, \tag{16}$$

$$d\gamma_t = -\alpha_t\gamma_t^2\Big[\beta_tA_1(t,\xi)+(1-\beta_t)B_1(t)\Big]^2dt, \quad \gamma_0=\gamma.$$

Using the fact that the optimal decoding is m_t and exactly repeating computations of Section 2, we obtain that, for fixed strategies α and β, the following coding functionals

$$B_1^*(t)=\sqrt{\frac{P}{\gamma}}, \quad B_0^*(t)=-m\sqrt{\frac{P}{\gamma}},$$

$$A_1^*(t,\xi)=\sqrt{\frac{P}{\gamma_t^*}}, \quad A_0^*(t,\xi)=-A_1^*(t,\xi)m_t^*$$

are optimal, where

$$\gamma_t^* = \frac{\gamma e^{-P\nu(t)}}{1+P\int_0^t\alpha_s(1-\beta_s)e^{-P\nu(s)}\,ds}, \quad \nu(t)=\int_0^t\alpha_s\beta_s\,ds,$$

and m_t^* is determined from (16). Taking all this into account, we have

$$\delta(t) = \inf_{\alpha,\beta} \mathbf{E}\left[\frac{\gamma e^{-P\nu(t)}}{1 + P\int_0^t \alpha_s(1-\beta_s)e^{-P\nu(s)}\,ds} + c_1\int_0^t \alpha_s\,ds + c_2\nu(t)\right]. \tag{17}$$

Now we can construct the optimal strategy β under a fixed strategy α. Let, for fixed $t \in [0,T]$,

$$\int_0^t \alpha_s\,ds = v.$$

Denote the expression in brackets in (17) by $\delta_{\alpha\beta}(t)$ and write it in the following form

$$\delta_{\alpha\beta}(t) = \frac{\gamma}{1 + Pe^{P\nu(t)}\int_0^t \alpha_s e^{-P\nu(s)}\,ds} + c_1\int_0^t \alpha_s\,ds + c_2\nu(t). \tag{18}$$

Using inequality (A.3) from the Appendix, after simple transformations we obtain

$$\delta_{\alpha\beta}(t) \geqslant \frac{\gamma e^{-P\nu(t)}}{1 + P(v - \nu(t))} + c_1 v + c_2\nu(t). \tag{19}$$

Consider the function

$$f(x) = \frac{\gamma e^{-Px}}{1 + P(v-x)} + c_1 v + c_2 x, \quad x \in [0,v].$$

Examining the derivatives of f, we easily obtain that for $c_2 \geqslant \gamma P/4$, $f'(x) \geqslant 0$, $x \in [0,v]$, and, therefore, $f(x) \geqslant f(0)$, i.e., $f(x) \geqslant \gamma/(1+Pv) + c_1 v$, $x \in [0,v]$. If $c_2 < \gamma P/4$ and $v \notin (a,b)$, where

$$a = \frac{\gamma P - 2c_2 - \sqrt{\gamma^2P^2 - 4c_2\gamma P}}{2c_2 P}, \qquad b = \frac{\gamma P - 2c_2 + \sqrt{\gamma^2P^2 - 4c_2\gamma P}}{2c_2 P},$$

then we again have $f(x) \geqslant f(0)$, $x \in [0,v]$. If $c_2 < \gamma P/4$ and $v \in (a,b)$ then $f(x) \geqslant f(x_0)$, $x \in [0,v]$, where $x_0 \in [0,v]$ is the unique solution of the equation

$$c_2\left[1 + P(v-x)\right]^2 e^{Px} = \gamma P^2(v-x). \tag{20}$$

Using the properties of the function f, from inequality (19) we obtain that if $c_2 \geqslant \gamma P/4$,

$$\delta_{\alpha\beta}(t) \geqslant \frac{\gamma}{1+Pv} + c_1 v,$$

but if $\beta_s^* = 0$, $s \in [0,t]$, from (18) we have that

$$\delta_{\alpha\beta^*}(t) = \frac{\gamma}{1+Pv} + c_1 v,$$

i.e., when $c_2 \geqslant \gamma P/4$,

$$\delta(t) = \inf_{\alpha} \mathbf{E}\left[\frac{\gamma}{1 + P\int_0^t \alpha_s\, ds} + c\int_0^t \alpha_s\, ds\right].$$

The same is true if $c_2 < \gamma P/4$ and $v \notin (a, b)$.
Now let $c_2 < \gamma P/4$ and $v \in (a, b)$. Then

$$\delta_{\alpha\beta}(t) \geqslant \frac{\gamma e^{-Px_0}}{1 + P(v - x_0)} + c_1 v + c_2 x_0,$$

where x_0 is the unique solution of equation (20). Taking into account that if $c_2 < \gamma P/4$,

$$c_2(1 + Pa)^2 = \gamma P^2 a \quad \text{and} \quad c_2(1 + Pb)^2 = \gamma P^2 b,$$

from (20) we can easily see that if $v = a$ and $v = b$ then $x_0 = 0$. In the sequel denote by x_0 or $x_0(v)$ or $x_0(\alpha)$ the solution of equation (20) if $v \in (a, b)$ and zero if $v \notin (a, b)$.
Then, if $c_2 < \gamma P/4$,

$$\delta_{\alpha\beta}(t) \geqslant \frac{\gamma e^{-Px_0(v)}}{1 + P(v - x_0(v))} + c_1 v + c_2 x_0(v).$$

Let $\beta_s^* = I(s_{x_0} \leqslant s \leqslant t)$, where $s_{x_0} = \sup\left\{s \geqslant 0\colon \int_s^t \alpha_u\, du = x_0\right\}$.
Then, from (18), we obtain

$$\delta_{\alpha\beta^*}(t) = \frac{\gamma e^{-Px_0(v)}}{1 + P(v - x_0(v))} + c_1 v + c_2 x_0(v).$$

Thus, if $c_2 < \gamma P/4$, the optimal strategy is $\beta_s^* = \alpha_s I(s_{x_0} \leqslant s \leqslant t)$ and

$$\delta(t) = \inf_{\alpha} \mathbf{E}\left[\frac{\gamma e^{-Px_0(\alpha)}}{1 + P\left(\int_0^t \alpha_s\, ds - x_0(\alpha)\right)} + c_1\int_0^t \alpha_s\, ds + c_2 x_0(\alpha)\right].$$

Now construct the optimal strategy α. First consider the case $c_2 \geqslant \gamma P/4$. Then

$$\delta(t) = \inf_{\alpha} \mathbf{E}\left[\frac{\gamma}{1 + P\int_0^t \alpha_s\, ds} + c_1\int_0^t \alpha_s\, ds\right].$$

Consider the function

$$g(v) = \frac{\gamma}{1 + Pv} + c_1 v, \quad v \in [0, t].$$

If $c_1 \geqslant \gamma P$ then $g'(v) \geqslant 0$ and, therefore, $g(v) \geqslant g(0) = \gamma$.
If $c_1 < \gamma P$ then $g'(v) \leqslant 0$ for $v \leqslant v_1$, where $v_1 = (\sqrt{c_1\gamma P} - c_1)(c_1 P)^{-1}$ and $g'(v) \geqslant 0$ for $v \geqslant v_1$. Therefore, if $v_1 \geqslant t$ then $g(v) \geqslant g(t)$, and if $v_1 < t$ then $g(v) \geqslant g(v_1)$.
Denote

$$\delta_\alpha(t) = \frac{\gamma}{1 + P\int_0^t \alpha_s\,ds} + c_1 \int_0^t \alpha_s\,ds.$$

Using the properties of the function g, we have:
If $c_1 \geqslant \gamma P$ then $\delta_\alpha(t) \geqslant \gamma$. If $\alpha_s^* = 0,\ s \leqslant t$, then $\delta_{\alpha^*}(t) = \gamma$.
Thus, if $c_2 \geqslant \gamma P/4$ and $c_1 \geqslant \gamma P$, the optimal strategy is $\alpha_s^* = 0,\ s \leqslant t$, and $\delta(t) = \gamma$.
If $c_1 < \gamma P$ then

$$\delta_\alpha(t) \geqslant \begin{cases} \gamma/(1+Pt) + c_1 t, & t < v_1; \\ \gamma/(1+Pv_1) + c_1 v_1, & t \geqslant v_1. \end{cases}$$

If $\alpha_s^* = I(0 \leqslant s < v_1)$,

$$\delta_{\alpha^*}(t) = \begin{cases} \gamma/(1+Pt) + c_1 t, & t < v_1; \\ \gamma/(1+Pv_1) + c_1 v_1, & t \geqslant v_1. \end{cases}$$

Thus, if $c_2 \geqslant \gamma P/4$ and $c_1 < \gamma P$,

$$\delta(t) = \begin{cases} \gamma/(1+Pt) + c_1 t, & t < v_1; \\ \gamma/(1+Pv_1) + c_1 v_1, & t \geqslant v_1, \end{cases}$$

and the optimal strategy is $\alpha_s^* = I(0 \leqslant s < v_1)$.
Now consider the case when $c_2 < \gamma P/4$. Then

$$\delta(t) = \inf_\alpha \mathbf{E}\left[\frac{\gamma e^{-Px_0(\alpha)}}{1 + P\Big(\int_0^t \alpha_s\,ds - x_0(\alpha)\Big)} + c_1 \int_0^t \alpha_s\,ds + c_2 x_0(\alpha)\right]. \tag{21}$$

Denote the expression in the brackets by $\delta_\alpha(t)$. Consider the function

$$F(v) = \frac{\gamma e^{-Px_0(v)}}{1 + P\big(v - x_0(v)\big)} + c_1 v + c_2 x_0(v), \quad v \in [0,t].$$

When $v \in [a,b]$, $x_0(v)$ is a solution of equation (20) and, taking this into account, we have

$$F'(v) = c_1 - \frac{c_2}{P\big(v - x_0(v)\big)}.$$

Using the rule of differentiation of an implicit function, from (20) we can obtain

$$x_0'(v) = \frac{1 - P\big(v - x_0(v)\big)}{1 + P^2\big(v - x_0(v)\big)^2}$$

and so $d(v - x_0(v))/dx \geqslant 0$. Therefore, if v increases from a to b, then $v - x_0(v)$ increases from a to b too.
Let $c_2 < \gamma P/4$ and $c_1 \geqslant \gamma P$. Then $c_1 > c_2 Pb$, because $\gamma P > c_2 Pb$ (this can be shown by direct verification).
Obviously, $Pb = 1/Pa$, and, therefore,

$$\frac{c_2}{P(v - x_0(v))} \leqslant \frac{c_2}{Pa} = c_2 Pb < c_1$$

and, hence, $F'(v) \geqslant 0$ for $v \in [a, b]$. Consequently, $F(v) \geqslant F(a)$ for $v \in [a, b]$.
If $v \notin [a, b]$ then $x_0(v) = 0$ and the function $F(v)$ has the form

$$F(v) = \frac{\gamma}{1 + Pv} + c_1 v.$$

When $c_1 \geqslant \gamma P$, $F(v)$ increases if $v \notin [a, b]$ and $F(v) \geqslant \gamma$.
Thus, when $c_2 < \gamma P/4$ and $c_1 \geqslant P$

$$F(v) \geqslant \gamma, \quad v \in [0, t].$$

Hence, from (21), $\delta_\alpha(t) \geqslant \gamma$ and if $\alpha_s^* = 0$, $s \leqslant t$, then $\delta_{\alpha^*}(t) = \gamma$ and, therefore, $\delta(t) = \gamma$.
Now let $c_2 < \gamma P/4$ and $c_1 < \gamma P$.
Consider separately three cases.

1) $c_2 < \gamma P/4$, $c_1 \leqslant c_2 Pa$. In this case

$$\frac{c_2}{P(v - x_0(v))} \geqslant \frac{c_2}{Pb} = c_2 Pa \geqslant c_1$$

and since

$$F'(v) = c_1 - \frac{c_2}{P(v - x_0(v))}\,.$$

for $v \in [a, b]$, then $F'(v) \leqslant 0$, and $F(v) \geqslant F(b)$ for $v \in [a, b]$.
If $v \notin [a, b]$ then $F'(v) \leqslant 0$ for $v \leqslant v_1$, and $F'(v) \geqslant 0$ for $v \geqslant v_1$, where $v_1 = (\sqrt{c_1 \gamma P} - c_1)(c_1 P)^{-1}$. Under the condition $c_1 \leqslant c_2 Pa$, we have $v_1 \geqslant b$. Indeed,

$$F'(b) = \frac{c_1(1 + Pb)^2 - \gamma P}{(1 + Pb)^2},$$

$$c_1(1 + Pb)^2 \leqslant \frac{c_2}{Pb}(1 + Pb)^2 = \frac{\gamma P^2 b}{Pb} = \gamma P$$

and $F'(b) \leqslant 0$, hence, $v_1 \leqslant b$.
It follows that, for all $v \in [0, t]$,

$$F(v) \geqslant F(v_1) = \frac{\gamma}{1 + Pv_1} + c_1 v_1$$

and $F(v)$ decreases if $v \leqslant v_1$ and increases if $v \geqslant v_1$.
Therefore, for $c_2 < \gamma P/4$ and $c_1 \leqslant c_2 Pa$,

$$\delta_\alpha(t) \geqslant \begin{cases} \gamma/(1 + Pt) + c_1 t, & t < v_1; \\ \gamma/(1 + Pv_1) + c_1 v_1, & t \geqslant v_1. \end{cases}$$

If $\alpha_s^* = I(s < v_1)$ then we obtain

$$\delta(t) = \delta_{\alpha^*}(t) = \begin{cases} \gamma/(1 + Pt) + c_1 t, & t < v_1; \\ \gamma/(1 + Pv_1) + c_1 v_1, & t \geqslant v_1. \end{cases}$$

2) $c_2 < \gamma P/4,\ c_1 \geqslant c_2 Pb$.

We assume that $c_1 < \gamma P$. The case $c_1 \geqslant \gamma P$ has been considered earlier. In this case $F'(v) \geqslant 0,\ v \in [a,b]$, and $F'(v) \leqslant 0,\ v \leqslant v_1$, and $F'(v) \geqslant 0,\ v \geqslant v_1$ for $v \notin [a,b]$. If $c_1 \geqslant c_2 Pb$ then $v_1 \leqslant a$.
Indeed,

$$F'(a) = \frac{c_1(1+Pa)^2 - \gamma P}{(1+Pa)^2},$$

$$c_1(1+Pa)^2 \geqslant \frac{c_2}{Pa}(1+Pa)^2 = \frac{\gamma P^2 a}{Pa} = \gamma P$$

and, therefore, $F'(a) \geqslant 0$. So $v_1 \leqslant a$. Therefore, as in 1), we obtain that in this case the optimal strategy is $\alpha_s^* = I(s < v_1)$ and

$$\delta(t) = \begin{cases} \gamma/(1+Pt) + c_1 t, & t < v_1; \\ \gamma/(1+Pv_1) + c_1 v_1, & t \geqslant v_1. \end{cases}$$

3) $c_2 Pa < c_1 < c_2 Pb$. Let $v \in [a,b]$. Then $F'(v) \leqslant 0 \Leftrightarrow v - x_0(v) \leqslant c_2/c_1 P$.
From equation (20) it follows that if $v - x_0(v) = c_2/c_1 P$, then

$$x_0 = \frac{1}{P}\log\frac{c_1\gamma P}{(c_1+c_2)^2} \quad \text{and} \quad v = v_0 \equiv \frac{c_2}{c_1 P} + \frac{1}{P}\log\frac{c_1\gamma P}{(c_1+c_2)^2}.$$

Thus, $F'(v) \leqslant 0 \Leftrightarrow v \leqslant v_0$ and $F(v) \geqslant F(v_0)$ for $v \in [a,b]$.
If $v \notin [a,b]$ then $F'(v) \leqslant 0 \Leftrightarrow v \leqslant v_1$. We can prove that $v_1 \in [a,b]$. It is sufficient to show that $F'(a) < 0$ and $F'(b) > 0$. Since the condition $c_2 Pa < c_1 < c_2 Pb$ is the same as $c_2/Pb < c_1 < c_2/Pa$, we have

$$c_1(1+Pa)^2 < \frac{c_2}{Pa}(1+Pa)^2 = \frac{\gamma P^2 a}{Pa} = \gamma P \Rightarrow F'(a) < 0,$$

$$c_1(1+Pb)^2 < \frac{c_2}{Pb}(1+Pb)^2 = \frac{\gamma P^2 b}{Pb} = \gamma P \Rightarrow F'(b) < 0.$$

It follows that, for $v \notin [a,b]$,

$$F(v) \geqslant \min\left[F(a), F(b)\right]$$

and, since $F(v_0) \leqslant \min\left[F(a), F(b)\right]$, we obtain $F'(v) \leqslant 0 \Leftrightarrow v \leqslant v_0$ and $F(v) \geqslant$ $\geqslant F(v_0)$ for all $v \in [0,t]$.
Hence, using (21), we obtain

$$\delta_\alpha(t) \geqslant \begin{cases} \dfrac{\gamma e^{-Px_0}}{1+P(t-x_0)} + c_1 t + c_2 x_0, & t < v_0; \\ \dfrac{\gamma e^{-Px_0}}{1+P(v_0-x_0)} + c_1 v_0 + c_2 x_0, & t \geqslant v_0. \end{cases}$$

If we take $\alpha_s^* = I(s < v_0)$ then this inequality becomes an equality and $\delta(t) =$ $= \delta_{\alpha^*}(t)$. Thus, the following theorem is proved.

THEOREM 5. *Let the transmission of a Gaussian random variable θ be carried out according to scheme (15), and power condition (10) be satisfied.*

1) *Let $c_1 \geqslant \gamma P$. Then the optimal strategy α^* is $\alpha^*_s = 0$, $s \leqslant t$, and $\delta(t) = \gamma$, i.e., the transmission does not occur.*
2) *Let $c_1 < \gamma P$, $c_2 \geqslant \gamma P/4$. Then*

$$\alpha^*_s = I(s < v_1), \quad v_1 = \left(\sqrt{c_1\gamma P} - c_1\right)\left(c_1 P\right)^{-1},$$

$$\beta^*_s = 0, \quad A^*_1(s,\xi) = \sqrt{\frac{P}{\gamma^*_s}}, \quad A^*_0(s,\xi) = -A^*_1(s,\xi)m^*_s,$$

$$B^*_1(s) = \sqrt{\frac{P}{\gamma}}, \quad B^*_0(s) = -m\sqrt{\frac{P}{\gamma}},$$

where m^*_t and γ^*_t are solutions of equation (16) for optimal α^*, β^*, A^*_0, A^*_1, B^*_0, B^*_1. Since the optimal codings and decoding are identical in all cases, in the sequel we shall give only the form of optimal strategies and the form of $\delta(t)$:

$$\delta(t) = \begin{cases} \gamma/(1+Pt) + c_1 t, & t < v_1; \\ \gamma/(1+Pv_1) + c_1 v_1, & t \geqslant v_1. \end{cases}$$

3) *Let $c_1 < \gamma P$, $c_2 < \gamma P/4$, $c_1 \notin \left(c_2 P a_1 c_2 P b\right)$, where*

$$a = \frac{\gamma P - 2c_2 - \sqrt{\gamma^2 P^2 - 4c_2\gamma P}}{2c_2 P}, \quad b = \frac{\gamma P - 2c_2 + \sqrt{\gamma^2 P^2 - 4c_2\gamma P}}{2c_2 P}.$$

Then $\alpha^*_s = I(s < v_1)$, $\beta^*_s = \alpha^*_s I(s \geqslant s_{x_0})$, where

$$s_{x_0} = \sup\left\{s \geqslant 0\colon \int_s^t \alpha^*_u \, du = x_0\right\},$$

x_0 is the unique solution of the equation

$$c_2 e^{Px}\left[1 + P(v - x)\right]^2 = \gamma P^2(v - x),$$

where $v = \int_0^t \alpha^*_s \, ds$. So,

$$\delta(t) = \begin{cases} \gamma/(1+Pt) + c_1 t, & t < v_1; \\ \gamma/(1+Pv_1) + c_1 v_1, & t \geqslant v_1. \end{cases}$$

Note that if $t \geqslant v_1$ then $\int_0^t \alpha^*_s \, ds = v_1$. Under the condition $c_1 \leqslant c_2 P a$, $v_1 \geqslant b$ and, if $c_1 \geqslant c_2 P b$, then $v_1 \leqslant a$, i.e., the equation (22) has no solution and $x_0 = 0$. Then $s_{x_0} = t$ and $\beta^*_s = 0$, $s \leqslant t$.

4) *Let $c_1 \in (c_2 P a, c_2 P b)$. Then*

$$\alpha^*_s = I(s < v_0), \quad v_0 = \frac{c_2}{c_1 P} + \frac{1}{P}\log\frac{c_1\gamma P}{\left(c_1 + c_2\right)^2},$$

$$\beta_s^* = \alpha_s^* I(s \geqslant s_{x_0}),$$

where x_0 is the unique solution of the equation (22) with v_0 instead of v and

$$x_0 = \frac{1}{P} \log \frac{c_1 \gamma P}{(c_1 + c_2)^2}, \quad s_{x_0} = \sup \left\{ s \geqslant 0 : \int_s^t \alpha_u^* \, du = x_0 \right\}.$$

Thus

$$\delta(t) = \begin{cases} \dfrac{\gamma e^{-P x_0}}{1 + P(t - x_0)} + c_1 t + c_2 x_0, & t < v_0; \\ \dfrac{\gamma e^{-P x_0}}{1 + P(v_0 - x_0)} + c_1 v_0 + c_2 x_0, & t \geqslant v_0. \end{cases}$$

APPENDIX

LEMMA 1. *Let $\alpha = (\alpha_t)_{t \geqslant 0}$ be a real integrable function taking values in the interval $[a, b]$, and $f = (f_t)_{t \geqslant 0}$ be a real intagrable non-negative function. Then the following inequalities are valid*

$$\int_0^t f_s e^{-P \varkappa_s} \, ds \leqslant \int_0^{t_0} f_s e^{-P a s} \, ds + e^{P(bt - \varkappa_t)} \int_{t_0}^t f_s e^{-P b s} \, ds,$$

$$\int_0^t f_s e^{P \varkappa_s} \, ds \geqslant \int_0^{t_0} f_s e^{P a s} \, ds + e^{-P(bt - \varkappa_t)} \int_{t_0}^t f_s e^{P b s} \, ds,$$

where $t_0 = (bt - \varkappa_t)(b - a)^{-1}$, $\varkappa_t = \int_0^t \alpha_s \, ds$.
If the function $\alpha = (\alpha_t)_{t \geqslant 0}$ has the form $\alpha_s = aI(s < x) + bI(s \geqslant x)$, where $x \in [0, t]$, then both inequalities become equalities.

Proof. Take $\alpha_s^* = aI(s < t_0) + bI(s \geqslant t_0)$. Then $\int_0^t \alpha_s^* \, ds = \varkappa_t$. We can show that $\int_0^s \alpha_u^* \, du \leqslant \int_0^s \alpha_u \, du$, for any $s \in [0, t]$. It is evident if $s < t_0$. Assume that, for some $t_1 \geqslant t_0$, the opposite inequality takes place. Then

$$\varkappa_t = \int_0^t \alpha_u^* \, du = \int_0^{t_1} \alpha_u^* \, du + \int_{t_1}^t \alpha_u^* \, du >$$

$$> \int_0^{t_1} \alpha_u \, du + \int_{t_1}^t \alpha_u \, du = \int_0^t \alpha_u \, du,$$

which is not true. It follows that

$$\int_0^t f_s e^{-P\varkappa_s}\, ds \leqslant \int_0^t f_s e^{-P\varkappa_s^*}\, ds, \quad \int_0^t f_s e^{P\varkappa_s}\, ds \geqslant \int_0^t f_s e^{P\varkappa_s^*}\, ds,$$

where $\varkappa_s^* = \int_0^s \alpha_u^*\, du$.

Using the definition of $\alpha^* = (\alpha_t^*)_{t\geqslant 0}$, one can easily see that

$$\int_0^t f_s e^{-P\varkappa_s^*}\, ds = \int_0^{t_0} f_s e^{-Pas}\, ds + e^{P(bt-\varkappa_t)} \int_{t_0}^t f_s e^{-Pbs}\, ds,$$

$$\int_0^t f_s e^{P\varkappa_s^*}\, ds = \int_0^{t_0} f_s e^{Pas}\, ds + e^{-P(bt-\varkappa_t)} \int_{t_0}^t f_s e^{Pbs}\, ds$$

and we obtain the required inequality. The validity of the second assertion follows by a direct verification.

COROLLARY. *If $\alpha = (\alpha_t)_{t\geqslant 0}$ is a real function with values from $[0,1]$ then*

$$\int_0^t e^{-P\varkappa_s}\, ds \leqslant t - \varkappa_t + \frac{1}{P}\left(1 - e^{-P\varkappa_t}\right). \tag{A.1}$$

This follows from the first inequality of Lemma 1 with $a = 0,\ b = 0,\ f_s = 1$. From the second inequality with $a = 0,\ b = 1$, we obtain

$$\int_0^t f_s e^{P\varkappa_s}\, ds \geqslant \int_0^{t-\varkappa_t} f_s ds + e^{-P(t-\varkappa_t)} \int_0^t f_s e^{Ps}\, ds. \tag{A.2}$$

Let $\alpha = (\alpha_t)_{t\geqslant 0}$ and $\beta = (\beta_t)_{t\geqslant 0}$ be real integrable functions taking two values 0 and 1. Put $\nu(t) = \int_0^t \alpha_u \beta_u\, du$ and $s_0 = \sup\{s \geqslant 0\colon \int_0^t \alpha_u\, du = \nu(t)\}$.

LEMMA 2. *The inequality*

$$\int_0^t \alpha_s e^{-P\nu(s)}\, ds \leqslant \varkappa_t - \nu(t) + \frac{1}{P}\left(1 - e^{-P\nu(t)}\right) \tag{A.3}$$

is valid. If $\beta_s = \alpha_s I(s \geqslant x)$, where $x \in [0,t]$, then the inequality becomes an equality.

Proof. Let $\beta^*_s = \alpha_s I(s \geqslant s_0)$. We can show that, for any $s \in [0,t]$,

$$\int_0^s \alpha_u \beta^*_u\, du \leqslant \int_0^s \alpha_u \beta_u\, du. \tag{A.4}$$

If $s < s_0$ then

$$\int_0^s \alpha_u \beta^*_u\, du = 0 \leqslant \int_0^s \alpha_u \beta_u\, du.$$

If, for some $s_1 \in (s_0, t]$, the opposite inequality

$$\int_0^{s_1} \alpha_u \beta^*_u\, du > \int_0^{s_1} \alpha_u \beta_u\, du$$

takes place, then we have

$$\nu(t) = \int_0^t \alpha_s \beta^*_s\, ds = \int_0^{s_1} \alpha_s \beta^*_s\, ds + \int_{s_1}^t \alpha_s \beta^*_s\, ds > \int_0^{s_1} \alpha_s \beta_s\, ds + \int_{s_1}^t \alpha_s\, ds \geqslant$$

$$\geqslant \int_0^{s_1} \alpha_s \beta_s\, ds + \int_{s_1}^t \alpha_s \beta_s\, ds = \int_0^t \alpha_s \beta_s\, ds.$$

That contradicts the assumption that $\nu(t) = \int_0^t \alpha_u \beta_u\, du$.
It follows that

$$\int_0^t \alpha_s e^{-P\nu(s)}\, ds \leqslant \int_0^t \alpha_s e^{-P\nu^*(s)}\, ds, \tag{A.5}$$

where $\nu^*(t) = \int_0^t \alpha_s \beta^*_s\, ds$. Let $w_s = \varkappa_s - \varkappa_{s_0}$. It can be easily seen that

$$\int_0^t \alpha_s e^{-P\nu^*(s)}\, ds = \int_0^{s_0} \alpha_s e^{-P\nu^*(s)}\, ds + \int_{s_0}^t \alpha_s e^{-P\nu^*(s)}\, ds =$$

$$= \int_0^{s_0} \alpha_s\, ds + \int_{s_0}^{t} \alpha_s e^{-P w_s}\, ds = \int_0^{t} \alpha_s\, ds - \int_0^{t} \alpha_s \beta_s\, ds + \frac{1}{P}\left(1 - e^{-P\nu(t)}\right),$$

which together with (A.4) gives (A.3).

REFERENCES

Glonti, O.A. and Jamburia, L.G. (1989). On optimal transmission of signals involving the cost for transmission. In: *The Fifth Intern. Vilnius Conf. on Probab. Theory and Math. Statist., Abstracts of Comm.* **III.** IMC. Vilnius, pp. 142–143.

Liptser, R.S. and Shiryayev, A.N. (1977). *Statistics of Random Processes* I. *General. Theory.* Springer, Berlin etc.

Liptser, R.S. and Shiryayev, A.N. (1978). *Statistics of Random Processes* II. *Applications.* Springer, Berlin etc.

New Trends in Probab. and Statist., pp. 375 – 390
V. Sazonov and T. Shervashidze (Eds.)
1991 VSP/Mokslas

STATIONARY AND PERIODIC SOLUTIONS OF STOCHASTIC DIFFERENCE AND DIFFERENTIAL EQUATIONS IN BANACH SPACE

A. Ya. DOROGOVTSEV
Kiev State University, Vladimirskaya str. 64, 252001 Kiev, USSR

ABSTRACT

Criteria of existence of stationary and periodic solutions of stochastic difference and differential linear equations with operator coefficients are given. Here a periodic solution is a random process whose finite dimensional distributions are periodic with the same period with respect to shift of time. Sufficient conditions for existence of periodic solutions of nonlinear equations are obtained for close to linear nonlinearities.

INTRODUCTION

Stochastic difference and differential equations are convenient for applications and mathematical investigations models of stationary processes. These models are well studied in the finite dimensional case (see (Arato, 1982)) and references there in. Stationary and periodic solutions of stochastic difference and differential equations in Banach spaces are also of interest especially in connection with different modern applications. Here some new difficulties arise. Although the study of random processes with abstract values has started comparatively long ago (see (Chobanyan and Weron, 1975; Bharucha-Reid, 1972)) and references therein, the available theory is poorer than in the finite dimensional case. The difficulties related to the infinite dimensionality of the phase space are also considerable. For instance, in a Banach space there is no simple analogue to the elementary finite dimensional theory of linear difference and differential equations with constant coefficients. In this paper we give criteria of existence of stationary and periodic solutions of linear difference and differential equations in a Banach space. Here a periodic solution is a random process whose finite dimensional distributions are periodic with the same period with respect to shift of time (Dorogovtsev, 1962). The considered linear models are related to some applied problems (Dorogovtsev, 1990b) and obtained results are also of interest in those nonlinear situations where investigated nonlinearities are close to linear. Periodic solutions of stochastic equations were studied by several authors, for references see (Dorogovtsev, 1988; 1990a).

1. DEFINITIONS

Let $(B, \|\cdot\|)$ be a complex separable Banach space, $\mathcal{L}(B)$ be the Banach space of linear bounded operators acting in B with operator norm denoted by $\|\cdot\|$. Let $\bar{0}$ and E be the null element in B and the identity operator in $\mathcal{L}(B)$, respectively. For an operator A the sets $\sigma(A)$ and $\rho(A)$ are its spectrum and resolvent set, respectively. The value of the functional $f \in B^*$ at $x \in B$ will be denoted by $\langle f, x \rangle$.
Let $\mathcal{B}(B)$ be the Borel σ–field in B and

$$S: = \{z \in \mathbf{C} \big| \, |z| = 1\}.$$

All random elements considered below are defined on some complete probability space $(\Phi, \mathcal{F}, \mathbf{P})$.

Definition 1.1. Let $\tau \in \mathbf{N}$. A B–valued random process $\{x(n)\colon n \in \mathbf{Z}\}$ is called periodical with period τ if

$$\forall\, n \in \mathbf{N}, \quad \forall\, \{m_1, \ldots, m_n\} \subset \mathbf{Z}, \quad \forall\, \{A_1, \ldots, A_n\} \subset \mathcal{B}(B):$$

$$\mathbf{P}\Big\{ \bigcap_{j=1}^{n} \big\{\omega \big| \, x(m_j + \tau, \omega) \in A_j\big\}\Big\} = \mathbf{P}\Big\{ \bigcap_{j=1}^{n} \big\{\omega \big| \, x(m_j, \omega) \in A_j\big\}\Big\}.$$

A periodic with period $\tau = 1$ process is stationary (in the narrow sense).

Definition 1.2. Let $\tau > 0$. A process $\{x(t)\colon t \in \mathbf{R}\}$ in B is called periodic with period τ if

$$\forall\, n \in \mathbf{N}, \quad \forall\, \{t_1, \ldots, t_n\} \subset \mathbf{R}, \quad \forall\, \{A_1, \ldots, A_n\} \subset \mathcal{B}(B):$$

$$\mathbf{P}\Big\{ \bigcap_{j=1}^{n} \big\{\omega \big| \, x(t_j + \tau, \omega) \in A_j\big\}\Big\} = \mathbf{P}\Big\{ \bigcap_{j=1}^{n} \big\{\omega \big| \, x(t_j, \omega) \in A_j\big\}\Big\}.$$

We should mention that trajectories of a periodic process are not necessarily periodic functions, only its mean characteristics are periodic.
The uniqueness of the process satisfying some equation means its uniqueness up to stochastic equivalence.

2. DIFFERENCE EQUATIONS

Let $A \in \mathcal{L}(B)$ be a fixed operator.

THEOREM 2.1. *The equation*

$$x(n+1) = Ax(n) + \varepsilon(n), \quad n \in \mathbf{Z}, \tag{2.1}$$

has a unique stationary solution $\{x(n)\colon n \in \mathbf{Z}\}$ *with* $\mathbf{E}\|x(0)\| < \infty$, *for every stationary process* $\{\varepsilon(n)\colon n \in \mathbf{Z}\}$ *in* B *with* $\mathbf{E}\|\varepsilon(0)\| < \infty$ *iff* $\sigma(A) \cap S = \varnothing$.

Proof. Necessity . Let an arbitrary element $z \in B$, $z \neq \bar{0}$, and an arbitrary number $\lambda_0 \in S$ be fixed, the random variable θ be uniformly distributed on $[0, 2\pi]$. The process

$$\{e^{i\theta}\lambda_0^n\colon\ n \in \mathbf{Z}\}$$

is stationary in $\mathbf{C}$. Therefore, the process

$$\{-e^{i\theta}\lambda_0^n z\colon\ n \in \mathbf{Z}\} \tag{2.2}$$

is stationary in B. Let $\{x(n)\colon n \in \mathbf{Z}\}$ be the unique stationary solution of equation (2.1) with process (2.2) instead of $\{\varepsilon(n)\colon\ n \in \mathbf{Z}\}$. Since in this case, for every $n \in \mathbf{Z}$

$$e^{i\theta}\lambda_0^n = \langle f_0,\ Ax(n) - x(n+1)\rangle,$$

where the functional $f_0 \in B^*$ satisfies $\langle f_0, z\rangle = 1$, the process in $B \times \mathbf{C}$

$$\{(x(n),\ \lambda_0^n e^{i\theta})\colon\ n \in \mathbf{Z}\}$$

is stationary. The functional f_0 exists by the Hahn–Banach theorem. Here the condition $z \neq \bar{0}$ is used. Hence, the process

$$\{x(n)\lambda_0^{-n}e^{-i\theta}\colon\ n \in \mathbf{Z}\}$$

is also stationary and

$$\mathbf{E}\|x(n)\lambda_0^{-n}e^{-i\theta}\| = \mathbf{E}\|x(0)\| < \infty.$$

Let $u := \mathbf{E}(x(n)\lambda_0^{-n}e^{-i\theta})$, $n \in \mathbf{Z}$. Taking mathematical expectations in the equality

$$x(n+1)\lambda_0^{-n}e^{-i\theta} = Ax(n)\lambda_0^{-n}e^{-i\theta} - z, \quad n \in \mathbf{Z},$$

we obtain

$$\lambda_0 u = Au - z. \tag{2.3}$$

Thus, for every $z \in B$, equation (2.3) has a solution $u \in B$. This solution is unique. Indeed, if an element $v \in B$, $v \neq u$, is also a solution of equation (2.3), then the process

$$\left\{x(n) + e^{i\theta}\lambda_0^n(u - v)\colon\ n \in \mathbf{Z}\right\}$$

is a stationary solution of equation (2.1) which differs from $\{x(n)\colon\ n \in \mathbf{Z}\}$.
So the operator $A - \lambda_0 E$ is a one–to–one mapping onto B and, therefore, it has a bounded inverse operator defined on B. Consequently, $\lambda_0 \in \rho(A)$.

Sufficiency. Let $\sigma(A) \cap S = \varnothing$ and $\{\varepsilon(n)\colon\ n \in \mathbf{Z}\}$ be a stationary process with $\mathbf{E}\|\varepsilon(0)\| < \infty$. Now let P_- (P_+) be the Riesz projectors corresponding to the part of the spectrum $\sigma(A)$ which is inside S (outside S). The space B is partitioned in direct sum of subspaces invariant for the operator A (Dunford and Schwartz, 1958). For every $n \in \mathbf{Z}$ define the random element

$$x(n) := \sum_{j=0}^{\infty}(AP_-)^j\varepsilon(n-1-j) - \sum_{j=-\infty}^{-1}(AP_+)^j\varepsilon(n-1-j).$$

The series in the formula for $x(n)$ are convergent in norm in B with probability 1, since

$$\sum_{j=0}^{\infty} \|(AP_-)^j\| + \sum_{j=-\infty}^{-1} \|(AP_+)^j\| < \infty.$$

The process $\{x(n)\colon n \in \mathbf{Z}\}$ is stationary and $\mathbf{E}\|x(0)\| < \infty$. This process satisfies equation (2.1).

Let $\{x(n)\colon n \in Z\}$ and $\{u(n)\colon n \in \mathbf{Z}\}$ be stationary solutions of equation (2.1) with $\mathbf{E}\|x(0)\| < \infty$, $\mathbf{E}\|u(0)\| < \infty$. Then, for every $n \in \mathbf{Z}$, we have

$$\mathbf{E}\|P_-(x(n)-u(n))\| \leqslant \|(AP_-)^m\| \sup_{k\in\mathbf{Z}} \mathbf{E}\|P_-(x(k)-u(k))\| \leqslant$$

$$\leqslant \|P_-\| \Big(\mathbf{E}\|x(0)\| + \mathbf{E}\|u(0)\|\Big) \|(AP_-)^m\|, \quad m \geqslant 1;$$

$$\mathbf{E}\|P_+(x(n)-u(n))\| \leqslant$$

$$\leqslant \|P_+\| \Big(\mathbf{E}\|x(0)\| + \mathbf{E}\|u(0)\|\Big) \|(AP_+)^{-m}\|, \quad m \geqslant 1.$$

From these inequalities it follows that $x(n) = u(n)$ for every $n \in \mathbf{Z}$ with probability 1. Theorem 2.1 is proved.

Let $A(n) \in \mathcal{L}(B)$, $n \in \mathbf{Z}$, and

$$A(n+\tau) = A(n), \quad n \in \mathbf{Z}.$$

Consider equation

$$x(n+1) = A(n)x(n) + \varepsilon(n), \quad n \in \mathbf{Z}. \tag{2.4}$$

Put

$$\begin{aligned} B_0 &:= A(\tau-1)A(\tau-2)\cdots A(1)A(0),\\ B_1 &:= A(0)A(\tau-1)\cdots A(2)A(1),\\ B_2 &:= A(1)A(0)A(\tau-1)\cdots A(3)A(2),\\ &\cdots\cdots\cdots\cdots\cdots\cdots\\ B_{\tau-1} &:= A(\tau-2)A(\tau-3)\cdots A(0)A(\tau-1). \end{aligned}$$

THEOREM 2.2. *Equation* (2.4) *has a unique periodic with period* τ *solution* $\{x(n)\colon n \in \mathbf{Z}\}$ *with* $\mathbf{E}\|x(k)\| < \infty$, $k = 1,2,\dots,\tau$, *for every periodic with period* τ *process* $\{\varepsilon(n)\colon n \in \mathbf{Z}\}$ *in* B *with* $\mathbf{E}\|\varepsilon(k)\| < \infty$, $k = 1,2,\dots,\tau$, *if* $\sigma(B_k) \cap S = \varnothing$ *for some* k *and only if* $\sigma(B_k) \cap S = \varnothing$ *for all* $k = 1,2,\dots,\tau$.

Proof. Necessity. Let $\{z(k)\colon k \in \mathbf{Z}\}$ be an arbitrary stationary process in B with $\mathbf{E}\|z(0)\| < \infty$, and the process $\{\varepsilon(n)\colon n \in \mathbf{Z}\}$ be defined by the following relations, for $\nu \in \mathbf{Z}$,

$$\varepsilon(\nu\tau + j - 1) := z(\nu),$$

$$\varepsilon(\nu\tau + s) := \bar{0}, \quad s \in \{0,1,2,\dots,\tau-1\}\setminus\{j-1\}.$$

Here $j \in \{1,2,\dots,\tau\}$ is fixed. The process $\{\varepsilon(n)\colon n \in \mathbf{Z}\}$ is periodic with period τ and $\mathbf{E}\|\varepsilon(k)\| < \infty$, $k = 1,2,\dots,\tau$. Let $\{x(n)\colon n \in \mathbf{Z}\}$ be the unique periodic with period τ solution of equation (2.4) for the process $\{\varepsilon(n)\colon n \in \mathbf{Z}\}$. Then the process $\{x(\nu\tau + j)\colon \nu \in \mathbf{Z}\}$ is stationary and satisfies the following equation

$$x((\nu+1)\tau + j) = B_j x(\nu\tau + j) + z(\nu), \quad \nu \in \mathbf{Z}, \quad B_\tau := B_0.$$

This solution is unique. So, by Theorem 2.1, $\sigma(B_j) \cap S = \varnothing$.

Sufficiency. Suppose that $\sigma(B_0) \cap S = \varnothing$ and let $\{\varepsilon(n)\colon\ n \in \mathbf{Z}\}$ be a periodic with period τ process in B with $\mathbf{E}\|\varepsilon(k)\| < \infty,\ k = 1, 2, \ldots, \tau$. Let

$$z_0(\nu) := \sum_{t=1}^{\tau-1} A(\tau-1)A(\tau-2)\ldots A(\tau-t)\varepsilon((\nu+1)\tau - t - 1) +$$

$$+\varepsilon((\nu+1)\tau - 1), \quad \nu \in \mathbf{Z}.$$

The process $\{z_0(\nu)\colon\ \nu \in \mathbf{Z}\}$ is stationary with $\mathbf{E}\|z_0(0)\| < \infty$. By Theorem 2.1, there is a unique stationary process $\{x_0(\nu)\colon\ \nu \in \mathbf{Z}\}$ with $\mathbf{E}\|x_0(0)\| < \infty$ satisfying the following equation

$$x_0(\nu+1) = B_0 x_0(\nu) + z_0(\nu), \quad \nu \in \mathbf{Z}.$$

Put now, for $\nu \in \mathbf{Z}$,

$$x(\nu\tau) := x_0(\nu),$$

$$x(\nu\tau+1) := A(0)x_0(\nu) + \varepsilon(\nu\tau),$$

$$x(\nu\tau+2) := A(1)x(\nu\tau+1) + \varepsilon(\nu\tau+1) = A(1)A(0)x_0(\nu) + A(1)\varepsilon(\nu\tau) + \varepsilon(\nu\tau+1),$$

$$\cdots\cdots\cdots\cdots\cdots\cdots\cdots\cdots\cdots\cdots\cdots\cdots$$

$$\begin{aligned} x(\nu\tau+\tau-1) := {} & A(\tau-2)A(\tau-3)\ldots A(0)x_0(\nu) + \\ & +A(\tau-2)A(\tau-3)\ldots A(1)\varepsilon(\nu\tau) + \\ & +A(\tau-2)A(\tau-3)\ldots A(2)\varepsilon(\nu\tau+1) + \\ & +A(\tau-2)\ldots A(3)\varepsilon(\nu\tau+2) + \ldots + \varepsilon(\nu\tau+\tau-2). \end{aligned} \tag{2.5}$$

The process in B^τ

$$\Big\{(x(\nu\tau), x(\nu\tau+1), \ldots, x(\nu\tau+\tau-1))\colon\ \nu \in \mathbf{Z}\Big\}$$

is stationary, so the process $\{x(n)\colon\ n \in \mathbf{Z}\}$ is a periodic with period τ process in B. By (2.5), the equality

$$x(n+1) = A(n)x(n) + \varepsilon(n) \tag{2.6}$$

holds for n of the form $n = \nu\tau + j,\ j = 0, 1, \ldots, \tau-2,\ \nu \in \mathbf{Z}$. Thus, the correctness of equality (2.6), for all $n \in \mathbf{Z}$, is justified by the equality

$$x(\nu\tau+\tau) = A(\tau-1)x(\nu\tau+\tau-1) + \varepsilon(\nu\tau+\tau-1)$$

for all $\nu \in \mathbf{Z}$. Indeed, taking into consideration (2.6), we have

$$A(\tau-1)x(\nu\tau+\tau-1) = B_0 x_0(\nu) + A(\tau-1)\ldots A(1)\varepsilon(\nu\tau) +$$

$$+A(\tau-1)\ldots A(2)\varepsilon(\nu\tau+1) + A(\tau-1)\ldots A(3)\varepsilon(\nu\tau+2) + \ldots$$

$$\ldots + A(\tau-1)\varepsilon(\nu\tau+\tau-2) =$$

$$= B_0 x_0(\nu) + z_0(\nu) - \varepsilon\big((\nu+1)\tau - 1\big) = x_0(\nu+1) - \varepsilon\big((\nu+1)\tau - 1\big).$$

Theorem 2.2 is proved.

Remark. By Theorem 2.2 the validity of condition $\sigma(B_k) \cap S = \emptyset$ for all $k = 0, 1, \ldots, \tau - 1$ is equivalent to the validity of this condition for any one arbitrary value of k. This statement is also a consequence of the following fact. For any two operators $\{A, C\} \subset \mathcal{L}(B)$ the equality

$$\sigma(AC) \setminus \{0\} = \sigma(CA) \setminus \{0\}$$

is true. Indeed, if for $\lambda \neq 0$ the operator $AC - \lambda E$ has bounded inverse D defined on B, then the operator $(CDA - E)/\lambda$ is inverse to $CA - \lambda E$.

The method of proof of the existence of stationary and periodic solutions in Theorems 2.1 and 2.2 is not always applicable to the difference equations of more complicated structure than equation (2.1). Another method is proposed in (Dorogovtsev, 1990b) for general situation with one operator coefficient. Let

$$\mathfrak{M} := \Big\{ \{c(\nu)\colon\ \nu \in \mathbf{Z}\} \subset \mathcal{L}(B) \Big|\ \exists\, t_0 \in (0,1) \quad \exists t_1 \in (1, +\infty)\colon$$

$$\sum_{\nu=-\infty}^{-1} \|c(\nu)\| t_0^{\nu} + \sum_{\nu=0}^{\infty} \|c(\nu)\| t_1^{\nu} < \infty;\ Ac(\nu) = c(\nu)A,\ \nu \in \mathbf{Z} \Big\}.$$

THEOREM 2.3. *The set of operators* $\{c(\nu)\colon\ \nu \in \mathbf{Z}\} \in \mathfrak{M}$ *exists and the equation*

$$Ax(n) = x(n+1) - 2x(n) + x(n+1) + \varepsilon(n), \quad n \in \mathbf{Z},$$

has a unique stationary solution of the form

$$a + \sum_{\nu \in \mathbf{Z}} c(n-\nu)\varepsilon(\nu), \quad n \in \mathbf{Z}; \quad a \in B, \tag{2.7}$$

for every stationary B-valued process

$$\{\varepsilon(n)\colon\ n \in \mathbf{Z}\},\ \mathbf{E}\|\varepsilon(0)\| < \infty,$$

iff $\sigma(A) \cap [-4, 0] = \emptyset$.

For a proof of this theorem see (Dorogovtsev, 1990b), where a generalization of it, for the case of closed operator A, is also given. The same method is used in the proof of the following statements.

Let ω be analytic in some neighbourhood of the unit circle S function with the Laurent expansion

$$\omega(\lambda) = \sum_{n \in \mathbf{Z}} a_n \lambda^n, \quad \{a_n\colon\ n \in \mathbf{Z}\} \subset \mathbf{C}.$$

THEOREM 2.4. *The set of operators* $\{c(\nu)\colon\ \nu \in \mathbf{Z}\} \in \mathfrak{M}$ *exists and the equation*

$$Ax(n) = \sum_{k \in \mathbf{Z}} a_k x(n+k) + \varepsilon(n), \quad n \in \mathbf{Z},$$

has a unique stationary solution of the form (2.7), for every stationary process in B,

$$\{\varepsilon(n)\colon\ n \in \mathbf{Z}\}, \quad \mathbf{E}\|\varepsilon(0)\| < \infty,$$

iff $\sigma(A) \cap \omega(S) = \varnothing$.

THEOREM 2.5. *Let $p \in \{0\} \cup \mathbf{N}$ and $\{A_0, A_1, \ldots, A_p\} \subset \mathcal{L}(B)$. The equation*

$$A_0 x(n) + A_1 x(n-1) + \ldots + A_p x(n-p) = \varepsilon(n), \quad n \in \mathbf{Z},$$

has a unique stationary solution

$$\{x(n)\colon\ n \in \mathbf{Z}\}, \quad \mathbf{E}\|x(0)\| < \infty,$$

for every stationary process $\{\varepsilon(n)\colon\ n \in \mathbf{Z}\}$ in B with $\mathbf{E}\|\varepsilon(0)\| < \infty$ iff, for every $\lambda \in S$, the operator

$$A_0 \lambda^p + A_1 \lambda^{p-1} + \ldots + A_p$$

has an inverse operator defined on B.

Now we shall present one of the theorems related to the nonlinear equation.

THEOREM 2.6. *Let a function $G\colon\ \mathbf{Z} \times B \to B$ and an operator $A \in \mathcal{L}(B)$ satisfy conditions:*

1) $\forall\, n \in \mathbf{Z},\ \forall\, x \in B\colon\ G(n+\tau, x) = G(n, x)$;
2) $\sigma(A) \cap S = \varnothing$;
3) $\forall\, 1 \leqslant k \leqslant \tau,\ \forall\, \{x, y\} \subset B$:

$$\|G(k,x) - G(k,y) - A(x-y)\| \leqslant L\|x-y\|;$$

4) $L\Big(\sum\limits_{j=-\infty}^{-1} \|(AP_+)^j\| + \sum\limits_{j=0}^{\infty} \|(AP_-)^j\|\Big) < 1,$

where P_-, P_+ are the Riesz projectors for operator A from Theorem 2.1. Then, for every periodic with period τ process in B,

$$\{\varepsilon(n)\colon\ n \in \mathbf{Z}\}; \quad \mathbf{E}\|\varepsilon(k)\| < \infty, \quad k = 1, 2, \ldots, \tau,$$

there is unique periodic with period τ solution

$$\{x(n)\colon\ n \in \mathbf{Z}\}; \quad \mathbf{E}\|x(k)\| < \infty, \quad k = 1, 2, \ldots, \tau,$$

of the equation

$$x(n+1) = G(n, x(n)) + \varepsilon(n), \quad n \in \mathbf{Z}.$$

A special case of Theorem 2.6 was proved in (Dorogovtsev and Le Vigne Thuang, 1988), Theorem 2.6 is proved in similar fashion.

3. LINEAR DIFFERENTIAL EQUATIONS DISTURBED BY PERIODIC PROCESSES

A solution of the equation considered in this section is B-valued process on $\mathbf{R}$ whose almost all trajectories are continuously differentiable on $\mathbf{R}$ in norm and satisfy the equation. Below number $\tau > 0$ is fixed.
Suppose that a function

$$A \in C(\mathbf{R}, \mathcal{L}(B))\colon\ A(t+\tau) = A(t), \quad t \in \mathbf{R},$$

is fixed and $U\colon\ \mathbf{R} \to \mathcal{L}(B)$ is a solution of the following problem

$$\frac{dU(t)}{dt} = A(t)U(t), \quad t \in \mathbf{R}, \quad U(0) = E. \tag{3.1}$$

For every $t \in \mathbf{R}$ the operator $U(t)$ is invertible and

$$U(t) = U(t - n\tau)U(\tau)^n, \quad t \in \mathbf{R}, \quad n \in \mathbf{Z}. \tag{3.2}$$

Besides,

$$\|U(t)U(t_0)^{-1}\| \leqslant \exp\left(\int_{t_0}^{t} \|A(s)\|\, ds\right), \quad t_0 < t. \tag{3.3}$$

The properties of the function U are well known (Massera and Schaffer, 1966).
Let $\mathcal{P}$ be the class of all periodic with period τ random processes $\{\xi(t)\colon\ t \in \mathbf{R}\}$ in B for which

$$\int_0^\tau \mathbf{E}\|\xi(t)\|\, dt < \infty.$$

THEOREM 3.1. *The equation*

$$\frac{dx(t)}{dt} = A(t)x(t) + \xi(t), \quad t \in \mathbf{R}, \tag{3.4}$$

has a unique periodic with period τ solution $\{x(t)\colon\ t \in \mathbf{R}\}$ in B with

$$\sup_{0 \leqslant t \leqslant \tau} \mathbf{E}\|x(t)\| < \infty$$

for every process $\xi \in \mathcal{P}$ if and only if

$$\sigma(U(\tau)) \cap S = \varnothing. \tag{3.5}$$

Proof. Necessity. Let the process $\{x(t)\colon\ t \in \mathbf{R}\}$ be a periodic with period τ solution of the equation (3.4) for the process $\xi \in \mathcal{P}$. It is easy to verify that the processes in B

$$\{x(n\tau)\colon\ n \in \mathbf{Z}\}, \quad \left\{\xi(n) := \int_0^\tau U(\tau)U(s)^{-1}\xi(n\tau + s)\, ds\colon\ n \in \mathbf{Z}\right\} \tag{3.6}$$

are stationary and

$$\mathbf{E}\|\xi(0)\| \leqslant \exp\left(\int_0^\tau \left\|A(s)\right\| ds\right) \int_0^\tau \mathbf{E}\|\xi(s)\|\, ds.$$

Since for the solution $\{x(t)\colon\ t \in \mathbf{R}\}$ of the equation (3.4) the following representation holds with probability 1

$$x(t) = U(t)U(t_0)^{-1}x(t_0) + \int_{t_0}^t U(t)U(s)^{-1}\xi(s)\, ds$$

for $\{t_0, t\} \subset \mathbf{R}$ (see (Massera and Schaffer, 1966)), the stationary process $\{x(n\tau)\colon$ $n \in \mathbf{Z}\}$ satisfies the equation

$$x((n+1)\tau) = U(\tau)x(n\tau) + \xi(n), \quad n \in \mathbf{Z}. \tag{3.7}$$

Let us mention now that for every stationary process $\{\tilde{\xi}(n)\colon\ n \in \mathbf{Z}\}$ with $\mathbf{E}\|\tilde{\xi}(0)\| <$ $< \infty$ defined on $\mathbf{R}$ the process

$$\tilde{\xi}(n\tau + s) := 6\tau^{-3}s(\tau - s)U(s)U(\tau)^{-1}\tilde{\xi}(n),$$

where $n \in \mathbf{Z}$ and $s \in [0, \tau]$ is periodic with period τ and belongs to $\mathcal{P}$. Besides, for $n \in \mathbf{Z}$,

$$\int_0^\tau U(\tau)U(s)^{-1}\tilde{\xi}(n\tau + s)\, ds =$$

$$= \int_0^\tau U(\tau)U(s)^{-1}U(s)U(\tau)^{-1}6\tau^{-3}s(\tau - s)\tilde{\xi}(n)\, ds = \tilde{\xi}(n).$$

Thus, for every stationary process

$$\{\xi(n)\colon\ n \in \mathbf{Z}\}, \quad \mathbf{E}\|\xi(0)\| < \infty,$$

equation (3.7) has a unique stationary solution $\{x(n\tau)\colon\ n \in \mathbf{Z}\}$, $\mathbf{E}\|x(0)\| < \infty$. By Theorem 2.1, condition (3.5) is satisfied.

Sufficiency. Let the process $\{\xi(t)\colon\ t \in \mathbf{R}\} \in \mathcal{P}$ and $\{\xi(n)\colon\ n \in \mathbf{Z}\}$ be a stationary process defined by formula (3.6) and $\mathbf{E}\|\xi(0)\| < \infty$. By Theorem 2.1, under condition (3.5), there is a unique stationary solution $\{\widetilde{x}(n\tau)\colon\ n \in \mathbf{Z}\}, \mathbf{E}\|\widetilde{x}(0)\| < < \infty$, of equation (3.7). Define now the process $\{x(t)\colon\ t \in \mathbf{R}\}$ in the following way

$$x(n\tau + s) := U(s)\widetilde{x}(n\tau) + U(s)\int_0^s U(t)^{-1}\xi(n\tau + t)\,dt, \quad n \in \mathbf{Z}, \quad s \in [0, \tau].$$

The process $\{x(t)\colon\ t \in \mathbf{R}\}$ is periodic with period τ.
Using properties of the function U, for every $n \in \mathbf{Z}$ and $s \in (0, \tau)$ we have with probability 1

$$\frac{d}{ds}x(n\tau + s) = \frac{dU(s)}{ds}\widetilde{x}(n\tau) +$$

$$+\frac{dU(s)}{ds}\int_0^s U(t)^{-1}\xi(n\tau + t)\,dt + U(s)U(s)^{-1}\xi(n\tau + s) =$$

$$= A(s)U(s)\widetilde{x}(ns) + A(s)U(s)\int_0^s U(t)^{-1}\xi(n\tau + t)\,dt +$$

$$+\xi(n\tau + s) = A(s)x(n\tau + s) + \xi(n\tau + s).$$

Analogously it is checked that with probability 1

$$D_+x(n\tau) = A(0)x(n\tau) + \xi(n\tau), \quad D_-x(n\tau) = A(\tau)x(n\tau) + \xi(n\tau),$$

where $D_+f(s)$ ($D_-f(s)$) is the right (left) derivative (in norm) of a function f at the point s. Thus the process $\{x(t)\colon\ t \in \mathbf{R}\}$ satisfies the equation (3.4) and

$$\sup_{0 \leqslant t \leqslant \tau} \mathbf{E}\|x(t)\| < \infty.$$

Now we shall prove uniqueness. Let $\{x_i(t)\colon\ t \in \mathbf{R}\}$, $i = 1, 2$, be periodic with period τ solutions of equation (3.4) and

$$a := \sup_{0 \leqslant t \leqslant \tau}\Big(\mathbf{E}\|x_1(t)\| + \mathbf{E}\|x_2(t)\|\Big) < \infty.$$

Then for $z(t) := x_1(t) - x_2(t)$, $t \in \mathbf{R}$, we have with probability 1

$$\frac{dz(t)}{dt} = A(t)z(t), \quad t \in \mathbf{R},$$

so

$$z(t) = U(t)U(t_0)^{-1}z(t_0), \quad \{t_0,\ t\} \subset \mathbf{R},$$

whence

$$U(t)^{-1}z(t) = U(\tau)^{-n}z(n\tau), \quad t \in \mathbf{R}, \quad n \in \mathbf{Z}.$$

Now let P_- and P_+ be spectral projectors constructed according to the operator $U(\tau)$ like in the proof of Theorem 2.1. Then with probability 1

$$P_-U(t)^{-1}z(t) = \big(U(\tau)P_-\big)^{-n}P_-z(n\tau),$$

$$P_+U(t)^{-1}z(t) = \big(U(\tau)P_+\big)^{-n}P_+z(n\tau),$$

whence

$$\mathbf{E}\big\|P_-U(t)^{-1}z(t)\big\| \leqslant \big\|(U(\tau)P_-)^{-n}\big\|\|P_-\|a,$$

$$\mathbf{E}\big\|P_+U(t)^{-1}z(t)\big\| \leqslant \big\|(U(\tau)P_+)^{-n}\big\|\|P_+\|a. \tag{3.8}$$

Now let $n \to -\infty$ in the first inequality of (3.8) and $n \to +\infty$ in the second one. We obtain $U(t)^{-1}z(t) = \bar{0}$ with probability 1 for every $t \in \mathbf{R}$.
Theorem 3.1 is proved.

Remark. Suppose that A satisfies the additional condition $A(t)A(s) = A(s)A(t)$, $\{s,t\} \subset [0,\tau]$. In particular, it is so if $A(t) \equiv A \in \mathcal{L}(B)$, $t \in [0,\tau]$. Then

$$U(\tau) = \exp\left(\int_0^\tau A(s)\,ds\right)$$

and by the spectrum transformation theorem the condition $\sigma(U(\tau)) \cap S = \varnothing$ is equivalent to the condition that operators $\int_0^\tau A(s)\,ds$ spectrum and the imaginary axis are disjoint.
In some cases Theorem 3.1 allows to obtain conditions for existence of periodic solutions of the equations of higher orders. Here is an example of such result.

THEOREM 3.2. *Let $T \in \mathcal{L}(B)$. The equation*

$$\frac{d^2x(t)}{dt^2} = Tx(t) + \xi(t), \quad t \in \mathbf{R}, \tag{3.9}$$

has a unique, periodic with period τ solution $\{x(t)\colon\ t \in \mathbf{R}\}$ with

$$\sup_{0\leqslant t\leqslant\tau} \mathbf{E}\|x(t)\| + \sup_{0\leqslant t\leqslant\tau} \mathbf{E}\|x'(t)\| < \infty$$

for every process $\xi \in \mathcal{P}$ iff $\sigma(T) \cap (-\infty,0] = \varnothing$.

4. STOCHASTIC EQUATIONS IN HILBERT SPACE

In this section $B = H$ is a complex Hilbert space with inner product $(\cdot,\cdot)$, $\{e_n\colon\ n \geqslant 1\}$ is an orthonormal basis in H.

Definition 4.1. A complex-valued process $\{w(t)\colon\ t \in \mathbf{R}\}$ is called a Wiener process if $\operatorname{Re} w$ and $\operatorname{Im} w$ are independent real Wiener processes on $\mathbf{R}$. Let $\{w_n(t)\colon\ t \in \mathbf{R}\}, n \geqslant 1$, be a sequence of independent complex-valued Wiener processes with parameters $\{\lambda_n\colon\ n \geqslant 1\}$ respectively and

$$\sum_{n=1}^{\infty} \lambda_n < \infty.$$

The process

$$w(t) := \sum_{n=1}^{\infty} w_n(t) e_n$$

is called a Wiener process on $\mathbf{R}$ with values in H.
The series for the value of $w(t)$ in Definition 4.1 is convergent with probability 1 for every $t \in \mathbf{R}$, besides $\mathbf{E}w(t) = \bar{0},\ t \in \mathbf{R}$, and the mutual covariance operator of the elements $w(t)$ and $w(s)$ when $s < t$ is of the form $(t-s)W$, where

$$Wz := \sum_{n=1}^{\infty} \lambda_n (z, e_n) e_n, \quad z \in H.$$

Concerning properties of the Wiener process in H and stochastic equations with this process (see (Curtain and Falb, 1971)).
For $t \in \mathbf{R}$, let

$$\mathcal{F}_t := \sigma\Big(w_n(v) - w_n(u) \Big|\ u \leqslant v \leqslant t,\ n \geqslant 1 \Big).$$

Let $A \in \mathcal{L}(H)$ be fixed.

Definition 4.2. An H-valued process $\{x(t)\colon\ t \in \mathbf{R}\}$ is called a solution of the equation

$$\frac{dx(t)}{dt} = Ax(t) + \frac{dw(t)}{dt}, \quad t \in \mathbf{R}, \tag{4.1}$$

if:

a) for every $t \in \mathbf{R}$ the random element $x(t)$ is $\mathcal{F}_t$-measurable;

b) for every $-\infty < t_0 < t_1 < \infty$,

$$\sup \left\{ \mathbf{E}\|x(t)\|^2 \,|\, t_0 \leqslant t \leqslant t_1 \right\} < \infty;$$

c) the process $\{x(t)\colon\ t \in \mathbf{R}\}$ is continuous on $\mathbf{R}$ with respect to the norm in H;

d) for every $-\infty < t_0 < t < \infty$ with probability 1 the following equality is true

$$x(t) = x(t_0) + \int_{t_0}^{t} Ax(s)\, ds + w(t) - w(t_0). \tag{4.2}$$

THEOREM 4.1. *For every H-valued Wiener process $\{w(t)\colon\ t \in \mathbf{R}\}$ stochastic equation* (4.1) *has unique stationary solution in H*

$$\{x(t)\colon\ t \in \mathbf{R}\}, \quad \mathbf{E}\|x(0)\|^2 < \infty,$$

such that for every $t \in \mathbf{R}$ the element $x(t)$ is $\mathcal{F}_t$-measurable, iff A the spectrum of the operator belongs to the left half-plane.

Proof. Necessity. First we observe that $\mathbf{E}x(0) = \bar{0}$, since due to (4.2), $A\mathbf{E}x(0) = \bar{0}$. If in this case $\mathbf{E}x(0) \neq \bar{0}$, then equation (4.1) has two different solutions $\{x(t)\colon t \in \mathbf{R}\}$ and $\{x(t) - \mathbf{E}x(0)\colon t \in \mathbf{R}\}$. Consequently, the operator A should be one–to–one. Let $S(t)$ be the mutual correlation operator of the elements $x(t)$ and $x(0)$ for $t \geqslant 0$. By elementary arguments we obtain the following formulae

$$S(t - t_0) = S(0)\exp\left((t - t_0)A^*\right), \quad t \geqslant t_0;$$

$$S(0)A^* + AS(0) = -W. \tag{4.3}$$

By Krein's results (Krein, 1964) it follows from (4.3) that the spectrum of the operator A belongs to the left half–plane. In this case

$$S(0) = \int_0^{+\infty} e^{At}We^{A^*t}dt.$$

Sufficiency. Suppose that the spectrum of the operator A belongs to the left half–plane. Put $G(t) := e^{tA}$, $t \geqslant 0$, so

$$\|G(t)\| \leqslant Le^{ta_-}, \quad t \geqslant 0,$$

with some numbers $L \geqslant 0$ and $a_- > 0$. For every $t \in \mathbf{R}$ the stochastic integral

$$x(t) := \int_{-\infty}^{t} G(t - s)\, dw(s)$$

is defined. Consider properties of the process $\{x(t)\colon t \in \mathbf{R}\}$. This process is continuous on $\mathbf{R}$. Indeed, every summand in the representation

$$x(t) = \sum_{n=1}^{\infty} \int_{-\infty}^{t} G(t - s)e_n\, dw_n(s) \tag{4.4}$$

is continuous in norm in H process on $\mathbf{R}$. Besides, for every $m \geqslant 1$, $\{t', t''\} \subset \mathbf{R}$, we have

$$\Delta(t', t'') := \mathbf{E}\left\| \sum_{n=1}^{m} \int_{-\infty}^{t'} G(t' - s)e_n\, dw_n(s) - \sum_{n=1}^{m} \int_{-\infty}^{t''} G(t'' - s)e_n\, dw_n(s)\right\|^4 \leqslant$$

$$\leqslant 8\mathbf{E}\left\| \sum_{n=1}^{m} \int_{-\infty}^{t'} \Big(G(t' - s) - G(t'' - s)\Big)e_n\, dw_n(s)\right\|^4 +$$

$$+8\mathbf{E}\left\| -\sum_{n=1}^{m} \int_{t'}^{t''} G(t'' - s)e_n\, dw_n(s)\right\|^4. \tag{4.5}$$

To estimate the right-hand side in (4.5) we observe that for a Gaussian random element η with mean value $\bar{0}$ and correlation operator T

$$\mathbf{E}\|\eta\|^4 \leqslant 3(\operatorname{tr} T)^2. \tag{4.6}$$

A simple calculation shows that the correlation operators of the Gaussian elements

$$\sum_{n=1}^{m} \int_{-\infty}^{t'} (G(t'-s) - G(t''-s))e_n \, dw_n(s), \quad \sum_{n=1}^{m} \int_{t'}^{t''} G(t''-s)e_n \, dw_n(s)$$

are equal to

$$\int_{-\infty}^{t'} \Big(G(t'-s) - G(t''-s)\Big) W_m \Big(G(t'-s) - G(t''-s)\Big)^* ds,$$

$$\int_{t'}^{t''} G(t''-s) W_m G(t''-s)^* ds,$$

where

$$W_m z := \sum_{n=1}^{m} \lambda_n (z, e_n) e_n, \quad z \in H.$$

With the help of (4.6) we obtain from (4.5) that for $t' < t''$,

$$\Delta(t', t'') \leqslant 3\Big(\frac{L^2}{2a_-} \operatorname{tr} W\Big)^2 \|G(t''-t') - E\|^4 + \big(\operatorname{tr} W L^2 (t''-t')\big)^2. \tag{4.7}$$

Inequality (4.7) allows us to apply Kolmogorov's theorem about continuity of a random process on every finite segment. Thus process (4.4) is continuous on $\mathbf{R}$. By definition for every $t \in \mathbf{R}$ the element $x(t)$ is $\mathcal{F}_t$-measurable. Moreover, for every $t \in \mathbf{R}$ we have

$$\mathbf{E}x(t) = \bar{0}, \quad \mathbf{E}\|x(t)\|^2 \leqslant \operatorname{tr} W \int_{-\infty}^{t} \|G(t-s)\| \, ds \leqslant L a_-^{-1} \operatorname{tr} W.$$

Process (4.4) is stationary.

Now we shall prove that the process $\{x(t)\colon t \in \mathbf{R}\}$ satisfies the equation in Theorem 4.1. Let numbers $t_0 < t$ be fixed. First we have

$$x(t) - x(t_0) - \int_{t_0}^{t} Ax(s) \, ds =$$

$$= \int_{-\infty}^{u} G(u-s) \, dw(s) \Big|_{u=t_0}^{u=t} - \int_{t_0}^{t} A\Big(\int_{-\infty}^{s} G(s-v) \, dw(v) \Big) \, ds. \tag{4.8}$$

Then

$$\int_{t_0}^{t} A\left(\int_{-\infty}^{s} G(s-v)\,dw(v)\right) ds = \int_{t_0}^{t}\left(\int_{-\infty}^{s} AG(s-v)\,dw(v)\right) ds =$$

$$= \int_{t_0}^{t}\left(\int_{-\infty}^{s} \frac{dG(s-v)}{ds}\,dw(v)\right) ds =$$

$$= \int_{-\infty}^{t_0}\left(\int_{t_0}^{t} \frac{dG(s-v)}{ds}\,ds\right) dw(v) + \int_{t_0}^{t}\left(\int_{s}^{t} \frac{dG(s-v)}{ds}\,ds\right) dw(v) =$$

$$= \int_{-\infty}^{t_0} G(s-v)\,dw(v)\Big|_{s=t_0}^{s=t} + \int_{t_0}^{t} G(t-v)\,dw(v) - \int_{t_0}^{t} dw(v). \qquad (4.9)$$

From (4.8) and (4.9), we have

$$x(t) - x(t_0) - \int_{t_0}^{t} Ax(s)\,ds = w(t) - w(t_0).$$

Now let $\{y(t)\colon\ t \in \mathbf{R}\}$ be another stationary solution of the equation in Theorem 4.1. For $z(t) := y(t) - x(t),\ t \in \mathbf{R}$, we have

$$\frac{dz(t)}{dt} = Az(t), \quad t \in \mathbf{R}.$$

Let t_0 be a fixed arbitrary number. Since

$$z(t_0) = G(t-t_0)z(t), \quad t < t_0,$$

we have

$$\mathbf{E}\|z(t_0)\| \leqslant L\exp\big(-a_-(t_0-t)\big)\mathbf{E}\|z(t)\| \leqslant L\exp(-a_-(t_0-t))C, \quad t < t_0,$$

where

$$C := \sup_{t\in\mathbf{R}} \mathbf{E}\|z(t)\| < \infty.$$

Thus $\mathbf{E}\|z(t_0)\| = 0$. Theorem 4.1 is proved.

REFERENCES

Arato, A. (1982). *Linear Stochastic Systems with Constant Coefficients. A Statistical Approach.* Springer, New York.

Bharucha-Reid, A. (1972). *Random Integral Equations.* Academic Press, New York.

Chobanyan, S. and Weron, A. (1975). Banach space valued stationary processes and their linear prediction. *Diss. Math.* **125.**

Curtain, R.F. and Falb, P.L. (1971). Stochastic differential equations in Hilbert space. *J. Different. Equat.* **10**, 412–430.

Krein, M.G. (1964). *Lectures on the Stability Theory of Solutions of Differential Equations in Banach Space.* Inst. Techn. Inf., Academy Sci. Ukrainian SSR, Kiev.

Massera, L.J. and Schaffer, J.J. (1966). *Linear Differential Equations and Functions Spaces.* Academic Press, New York.

Dorogovtsev, A.Ya. (1962). Some remarks on differential equations perturbed by periodic stochastic processes. *Ukrainian Math. J.* **14**, 119–128.

Dorogovtsev, A.Ya. (1988). Existence of periodic solutions of abstract stochastic equation. Asymptotic periodicity of Cauchy problem solutions. *Theory Probab. Math. Statist.* **39**, 47–52 (in Russian).

Dorogovtsev, A.Ya. and Le Vigne Thuang. (1988). On the existence of periodic and stationary regimes of discrete dynamic systems in Banach space. *Cybernetics.* No. 6, 121–123 (in Russian).

Dorogovtsev, A.Ya. (1990a). Necessary and sufficient conditions for existence of stationary and periodic solutions of a stochastic difference equation in Hilbert space. *Comp. Math. Appl.* **19**, 31–37.

Dorogovtsev, A.Ya. (1990b). Stationary and periodic solutions of stochastic difference equation in Banach space. *Theory Probab. Math. Statist.* **42**, 35–42 (in Russian).

Dunford, N. and Schwartz, J.T. (1958). *Linear Operators. Part I: General Theory.* Interscience Publishers, New York.

New Trends in Probab. and Statist., pp. 391 – 412
V. Sazonov and T. Shervashidze (Eds.)
1991 VSP/Mokslas

CONTROLLED RANDOM FIELDS ON GRAPHS AND STOCHASTIC MODELS OF ECONOMIC EQUILIBRIUM

I.V. EVSTIGNEEV
CEMI, USSR Academy of Sciences, Krasikova 32, 117418 Moscow, USSR

ABSTRACT

The paper examines a new scheme of stochastic control: an "equilibrium" scheme of controlling a random vector field on a directed graph. The problems and the results are motivated by mathematical economics.

1. The aim of this paper is to construct a stochastic dynamic model of economic equilibrium which takes into account the structure of commodity flows in the economic system. Studying this model, we make an attempt to integrate the theory of spatial economic equilibrium (Samuelson, 1952; Takayama and Judge, 1971; Nagurney, 1987) and the analysis of equilibrium trajectories of economic growth (Polterovich, 1978; Bewley, 1981; Evstigneev and Katyshev, 1982). The model under consideration is based on a scheme of controlling a random vector field on a directed graph. This scheme represents a modification of that studied in (Evstigneev, 1984; 1988).
The main result contained in the paper is an existence theorem for equilibrium states. Its proof uses the method of monotone operators (see, e.g., (Vainberg, 1972)).

2. We list the basic data which are assumed to be fixed in what follows. Let T be a finite set, and $(\Omega, \mathcal{F}, \mathbf{P})$ a probability space (Ω is a set, $\mathcal{F}$ a σ–algebra of subsets of Ω, and $\mathbf{P}$ a probability measure on $\mathcal{F}$). Suppose that to each $t \in T$ correspond a σ–algebra $\mathcal{F}_t$ generated by a countable family of sets and a positive integer m_t. Suppose that for each $t \in T$ a set $K(t) \subseteq T$ containing t is given. Thus, T is endowed with a structure of a directed graph: the inclusion $s \in K(t)$ means that there is an "arrow" of the graph leading from t to s. It is assumed everywhere that

$$s \in K(t) \Longrightarrow \mathcal{F}_s \supseteq \mathcal{F}_t \tag{1}$$

(when we move along the arrows of the graph, the σ–algebras $\mathcal{F}_t$ do not decrease). It is convenient to introduce the following notation:

$$M(t) = \{s \in T\colon\ t \in K(s)\}; \quad K(t+) = K(t)\backslash\{t\}; \quad M(t-) = M(t)\backslash\{t\}.$$

Points $t \in T$, are interpreted as economic units (plants, firms, regions etc.). The economic units consume and produce commodities. The structure of commodity flows between them is described by the graph $(T, K(\cdot))$: if there is an arrow leading from t to s, this means that t supplies some kinds of commodities to s. The number m_t fixed above for each $t \in T$ represents the quantity of different kinds of commodities that may be received by t. We denote by $\mathcal{F}$ the collection of all random events influencing the economic system. The σ–algebra $\mathcal{F}_t$ consists of the events that influence the economic unit t. Condition (1) means that the random factors which influence the economic unit t supplying commodities may thus influence the economic unit s receiving them.

3. For each $t \in T$, we consider the space $\mathcal{X}_t$ whose elements are random $\mathcal{F}_t$–measurable m_t–dimensional vectors $x(\omega)$ $(\omega \in \Omega)$ with

$$\|x(\cdot)\|_\infty = \operatorname{ess\,sup} |x(\cdot)| < \infty.$$

(For a finite–dimensional vector $a = (a^1, a^2, \ldots, a^l)$, we denote by $|a|$ the sum $|a^1| + |a^2| + \ldots + |a^l|$.) We write $p(\cdot) \in \mathcal{P}_t$ if $p(\omega)$ $(\omega \in \Omega)$ is a $\mathcal{F}_t$–measurable m_t–dimensional random vector such that $p(\omega) \geqslant 0$ almost surely (a.s.) and the expectation $\|p(\cdot)\|_1 = \mathbf{E}|p(\omega)|$ is finite. We will identify $\mathcal{F}_t$–measurable random vectors coinciding a.s. Then, using the standard notation for the spaces of measurable vector functions, we can write

$$\mathcal{X}_t = \mathcal{L}^\infty(\Omega, \mathcal{F}_t, \mathbf{P}, \mathbf{R}^{m_t}); \quad \mathcal{P}_t = \mathcal{L}^1(\Omega, \mathcal{F}_t, \mathbf{P}, \mathbf{R}^{m_t}) \cap \{p\colon\ p \geqslant 0 \text{ a.s.}\}.$$

The space $\mathcal{L}^1(\Omega, \mathcal{F}_t, \mathbf{P}, \mathbf{R}^{m_t})$ will be also denoted for shortness by $\mathcal{L}^1(\mathcal{F}_t)$ or $\mathcal{L}^1_t$ and the space $\mathcal{L}^\infty(\Omega, \mathcal{F}_t, \mathbf{P}, \mathbf{R}^{m_t})$ by $\mathcal{L}^\infty(\mathcal{F}_t)$ or $\mathcal{L}^\infty_t$.

Non–negative vectors $x = (x^1, \ldots, x^{m_t}) \in \mathcal{X}_t$ will be interpreted as commodity bundles (x^i is the quantity of the i–th commodity in the bundle x) and $p = (p^1, \ldots, p^{m_t})$ as price vectors (p^i is the price of a unit of the i–th commodity). For $x = (x^1, \ldots, x^{m_t})$ and $p = (p^1, \ldots, p^{m_t})$, the scalar product

$$p(\omega)x(\omega) = p^1(\omega)x^1(\omega) + \ldots + p^{m_t}(\omega)x^{m_t}(\omega)$$

(computed for each $\omega \in \Omega$) is the cost of the commodity bundle x in the price system p.

A family of vectors $v = (v_s)_{s \in K(t)}$, where $v_s \in \mathcal{X}_s$, will be called a program of the economic unit t. Here $-v_t$ may be interpreted as the input vector and v_s, $s \in K(t+)$, as the output vectors destined for the economic units $s \in K(t+)$ (provided $K(t+) \neq \emptyset$).

From the point of view of interpretation, it is natural to assume that for each program $v = (v_s)_{s \in K(t)}$, the vectors $-v_t$ and v_s, $s \in K(t+)$ are non–negative a.s. However, this assumption is not necessary for the results formulated below to be true.

A collection $z = (z_t)_{t \in T}$ of programs $z_t = (z_{ts})_{s \in K(t)}$ of economic units $t \in T$ will be called a program of the economic system.

4. For $t \in T$, we define

$$\mathcal{Z}_t = \prod_{s\in K(t)} \mathcal{X}_s, \quad \mathcal{Q}_t = \prod_{s\in K(t)} \mathcal{P}_s. \tag{2}$$

Here $\mathcal{Z}_t$ is the set of programs of the economic unit t. The elements of $\mathcal{Q}_t$ are collections of price vectors $q = (q_s)_{s\in K(t)}$, where q_t describes the prices corresponding to the input vector of the economic unit t and q_s, $s \in K(t+)$, the prices corresponding to the output vectors of t.
In the model we are given:

a) multivalued mappings

$$\mathcal{Q}_t \ni q \mapsto Z_t(q) \subset \mathcal{Z}_t, \quad t \in T, \tag{3}$$

b) random vectors

$$h_t \in \mathcal{X}_t, \quad t \in T. \tag{4}$$

The mappings $q \mapsto Z_t(q)$ describe the behaviour of economic units, their reaction on introducing a price system $q = (q_s)_{s\in K(t)} \in \mathcal{Q}_t$. Namely, $Z_t(q)$ is the set of preferable programs of the economic unit t for the price system q. The vectors h_t (for $h_t \geqslant 0$ a.s.) may be regarded as vectors of exogenous resources received by the system from without. Negative coordinates of h_t mean that the corresponding goods are withdrawn from the system. The role of the vectors h_t is clear from the definition of equilibrium states given below.

5. For $p = (p_s)_{s\in T}$, $p_s \in \mathcal{P}_s$, let

$$q^t(p) = (p_s)_{s\in K(t)} \tag{5}$$

and

$$Z(p) = \left\{z = (z_t)_{t\in T}\colon\ z_t \in Z_t\big(q^t(p)\big),\ t \in T\right\} \tag{6}$$

($Z(p)$ is the set of preferable programs of the economic system for the prices $p = (p_s)$).
Let $z = (z_t)_{t\in T}$ $[z_t = (z_{ts})_{s\in K(t)} \in \mathcal{Z}_t]$ be a program of the economic system under consideration and $p = (p_s)_{s\in T}$ $[p_s \in \mathcal{P}_s]$ a collection of price vectors. We say that (z,p) is an equilibrium state if the following conditions are satisfied:

(i) the inclusion

$$z \in Z(p) \tag{7}$$

is valid;

(ii) for each $s \in T$

$$h_s + \sum_{t\in M(s)} z_{ts} \geqslant 0 \quad \text{(a.s.)}; \tag{8}$$

(iii) for all $s \in T$

$$p_s\left(h_s + \sum_{t\in M(s)} z_{ts}\right) = 0 \quad \text{(a.s.).} \tag{9}$$

Condition (i) states that in the equilibrium every economic unit chooses a preferable program. Inequality (8), which may be rewritten in the form

$$-z_{ss} \leqslant h_s + \sum_{t\in M(s-)} z_{ts} \quad \text{(a.s.),} \tag{10}$$

expresses a balance constraint: the input vector of an economic unit s a.s. does not exceed the sum of outputs of economic units $t \in M(s-)$ and h_s. The equality (9) means that the cost of unused resources equals zero.

6. We list the conditions to be imposed on the model to guarantee the existence of equilibrium states.

(**A**) There exists a program

$$\overset{\circ}{z} = (\overset{\circ}{z}_t) \quad [\overset{\circ}{z}_t = (\overset{\circ}{z}_{ts})_{s\in K(t)} \in \mathcal{Z}_t] \tag{11}$$

satisfying the following requirements:

(**A**.1) For all $t \in T$, $q \in \mathcal{Q}_t$ and each program $v = (v_s)_{s\in K(t)} \in Z_t(q)$, we have

$$v_s \geqslant \overset{\circ}{z}_{ts}, \quad s \in K(t+). \tag{12}$$

(Equalities and inequalities for random vectors considered here and in what follows are assumed to hold almost surely.)

(**A**.2) There exists a positive constant δ such that

$$h_s + \sum_{t\in M(s)} \overset{\circ}{z}_{ts} \geqslant \delta e_s \tag{13}$$

for all $s \in T$. Here $e_s = (1, 1, \ldots, 1) \in \mathbf{R}^{m_s}$.

(**A**.3) A constant $M_0 > 0$ exists for which

$$\mathbf{E}(qv|\,\mathcal{F}_t) \geqslant \mathbf{E}(q\overset{\circ}{z}_t|\,\mathcal{F}_t) - M_0 \tag{14}$$

for all $t \in T$, $q \in \mathcal{Q}_t$ and $v \in Z_t(q)$.

(**B**) For every $t \in T$, the mapping $Z_t(\cdot)$ has the following properties:

(**B**.1) The set $Z_t(q)$ (for any $q \in \mathcal{Q}_t$) is nonempty, convex and bounded in the norm $\|\cdot\|_\infty$ uniformly with respect to $q \in \mathcal{Q}_t$.

(**B**.2) The mapping $q \mapsto Z_t(q)$ is closed in the following sense. If

$$q^k \in \mathcal{Q}_t, \quad v^k \in Z_t(q^k) \quad (k = 1, 2, \ldots), \quad q \in \mathcal{Q}_t, \quad v \in \mathcal{Z}_t, \tag{15}$$

$$\|q^k - q\|_1 \to 0 \quad \text{and} \quad \mathbf{E}wv^k \to \mathbf{E}wv \tag{16}$$

for all $w \in \prod_{s\in K(t)} \mathcal{L}_s^1$, then $v \in Z_t(q)$.

(**B**.3) For any $H \in [0,\infty)$ there exists a non-negative, nondecreasing $\mathcal{F}\times\mathcal{B}[0,\infty)$-measurable function ($\mathcal{B}(\cdot)$ is the Borel σ-algebra) $\gamma_t^H(\omega,r)$ ($\omega \in \Omega$, $r \in [0,\infty)$) such that $\gamma_t^H(\omega,r) > 0$ for $r > 0$ and

$$\mathbf{E}(q-q')\,(v-v') \geqslant \mathbf{E}\gamma_t^H(\omega, |\, q_t(\omega) - q_t'(\omega)|) \tag{17}$$

for all $v \in Z_t(q)$, $v' \in Z_t(q')$, $q = (q_s)_{s\in K(t)} \in \mathcal{Q}_t$, $q' = (q'_s)_{s\in K(t)} \in \mathcal{Q}_t$ with $\|q\|_\infty \leqslant H$ and $\|q'\|_\infty \leqslant H$.

THEOREM 1. *Conditions* (**A**) *and* (**B**) *are sufficient for the existence of equilibrium states.*

7. Let us discuss the conditions of Theorem 1. The family of vectors $\overset{\circ}{z}_{ts}$, $t \in T$, $s \in K(t+)$, described in (**A**) may be interpreted as a "minimal output plan". Namely, if $v = (v_s)_{s\in K(t)}$ belongs to the set $Z_t(q)$ of preferable programs of an economic unit t (for some price system $q = (q_s)_{s\in K(t)}$), then, according to (**A**.1), each of the output vectors v_s, $s \in K(t+)$, of the program v majorizes the corresponding vector $\overset{\circ}{z}_{ts}$. By (**A**.2) the program $\overset{\circ}{z} = (\overset{\circ}{z}_t)$ makes it possible to realize the minimal output plan "with a surplus".
We note that, for any program $v = (v_s)_{s\in K(t)}$ of an economic unit t and any price system $q = (q_s)_{s\in K(t)}$, the scalar product

$$qv = \sum_{s\in K(t+)} q_s v_s - q_t(-v_t)$$

may be interpreted as the profit yielded by the program v (the difference between the total output cost and the input cost). Thus, (**A**.3) has the following meaning: if v is a preferable program for some price system q, then the conditional expectation of the profit qv of this program cannot be essentially less than the conditional expectation of the profit $q\overset{\circ}{z}_t$ of the program $\overset{\circ}{z}_t$.
The assumption (**B**.1) is standard. In (**B**.2) it is stated that the mapping $q \mapsto Z_t(q)$ is closed with respect to the norm $\|\cdot\|_1$ in the space $\mathcal{Q}_t$ and the $*$-weak convergence in the space

$$\mathcal{Z}_t = \prod_{s\in K(t)} \mathcal{L}_s^\infty.$$

(If v^k, $k = 1,2,\ldots$, and v are elements of $\prod_{t\in S}\mathcal{L}_t^\infty$, where S is a subset of T, then the $*$-weak convergence of v^k to v means that $\mathbf{E}wv^k \to \mathbf{E}wv$ for all $w \in \prod_{t\in S}\mathcal{L}_t^1$.) It follows from (**B**.3) that $\mathbf{E}(q-q')(v-v') \geqslant 0$ for $v \in Z_t(q)$, $v' \in Z_t(q')$ and $q, q' \in \mathcal{Q}_t$ with $\|q\|_\infty$, $\|q'\|_\infty < \infty$. This is the monotonicity condition for the multivalued operator $Z_t(q)$ considered for $q \in \mathcal{Q}_t$ with $\|q\|_\infty < \infty$. In (**B**.3) a stronger requirement is contained: a strict monotonicity condition for $Z_t(q)$ with respect to q_t (which belongs to the set of variables $q = (q_s)_{s\in K(t)}$). See, e.g. (Polterovich, 1978) for conditions of this type for deterministic models.
We note that the model under consideration has the following essential features.

1) Different economic units are influenced by different random factors and have different information.

2) The structure of commodity flows between economic units is given by an arbitrary finite directed graph. This structure and the "stochastic basis" of the model, specified by the family of σ-algebras $\mathcal{F}_t$, are compatible with each other in the sense of condition (1).
3) The incentives of economic units have a more general nature than maximizing utilities or preferences under budget constraints. They are described in terms of abstract multivalued mappings $q \mapsto Z_t(q)$ satisfying a certain system of axioms.

8. We establish some consequences of assumptions (**A**) and (**B**) which are formulated in terms of the mapping $Z(\cdot)$. Let

$$\mathcal{P} = \prod_{t \in T} \mathcal{P}_t.$$

PROPOSITION 1. *Conditions* (**A**) *and* (**B**) *imply that the following conditions* (**A**$'$) *and* (**B**$'$) *hold.*

(**A**$'$) *There exists a program* $\overset{\circ}{z}$ *of the form* (11) *which satisfies the requirements listed below.*

(**A**$'$.1) *For every* $p \in \mathcal{P}$ *and each program* $z = (z_t) \in Z(p)$ $[z_t = (z_{ts})_{s \in K(t)}]$, *we have*

$$z_{ts} \geqslant \overset{\circ}{z}_{ts}, \quad t \in T, \quad s \in K(t+). \tag{18}$$

(**A**$'$.2) *For some* $\delta > 0$ *inequality* (13) *is valid.*

(**A**$'$.3) *There is* $M_0 > 0$ *such that*

$$\mathbf{E}\Big(q^t(p) z_t \Big| \mathcal{F}_t\Big) \geqslant \mathbf{E}\Big(q^t(p) \overset{\circ}{z}_t \Big| \mathcal{F}_t\Big) - M_0 \tag{19}$$

for all $p \in \mathcal{P}$, $t \in T$ *and* $z = (z_t) \in Z(p)$.

(**B**$'$) *The mapping* $Z(\cdot)$ *possesses the following properties.*

(**B**$'$.1) *For any* $p \in \mathcal{P}$, *the set* $Z(p)$ *is nonempty, convex and, for all* $p \in \mathcal{P}$, $z = (z_t) \in Z(p)$ *and* $t \in T$, *we have* $\|z_t\|_\infty \leqslant M_1$, *where* M_1 *is a constant independent of* p, z *and* t.

(**B**$'$.2) *If*

$$p^k \in \mathcal{P}, \quad z^k = (z_t^k) \in Z(p^k), \quad p \in \mathcal{P}, \quad z = (z_t), \quad z_t \in \mathcal{Z}_t, \tag{20}$$

$$\|p^k - p\|_1 \longrightarrow 0 \quad \text{and} \quad \mathbf{E} w_t z_t^k \longrightarrow \mathbf{E} w_t z_t \tag{21}$$

for all $t \in T$ *and* $w_t \in \prod_{s \in K(t)} \mathcal{L}_s^1$, *then* $z \in Z(p)$.

(**B**$'$.3) *For every* $H \in [0, \infty)$, *there exists a non-negative nondecreasing* $\mathcal{F} \times \mathcal{B}[0,\infty)$*-measurable function* $\gamma^H(\omega, r)$ $(\omega \in \Omega,\ r \in [0,\infty))$ *such that* $\gamma^H(\omega, r) > 0$ *for* $r > 0$ *and*

$$\sum_{t \in T} \mathbf{E}\Big(q^t(p) - q^t(p')\Big)(z_t - z_t') \geqslant \mathbf{E}\gamma^H\Big(\omega, |p(\omega) - p'(\omega)|\Big) \tag{22}$$

for all $(z_t) \in Z(p)$, $(z'_t) \in Z(p')$ *and* $p, p' \in \mathcal{P}$ *with* $\|p\|_\infty, \|p'\|_\infty \leqslant H$.

9. *Proof.* The properties (**A**′) and (**B**′.**1**) immediately follow from (**A**), (**B**.**1**) and the definitions of $Z(\cdot)$ and $q^t(\cdot)$ given in (5) and (6). Let us verify (**B**′.**2**). It follows from (20) that $z_t^k \in Z_t(q^t(p^k))$, and from (21) that $\|q^t(p^k) - q^t(p)\|_1 \to 0$ and $z_t^k \to z_t$ in the $*$-weak topology. Using (**B**.**1**), we derive from this that $z_t \in Z_t(q^t(p))$, $t \in T$, i.e., $z \in Z(p)$.

It remains to check (**B**′.**3**). Let $\gamma_t^H(\omega, r)$ $(t \in T,\ q \in \mathcal{Q}_t)$ be the function described in (**B**.**3**). Fix some $p = (p_t) \in \mathcal{P}$ and define

$$\bar{\gamma}^H(\omega, r) = \min_{t \in T} \gamma_t^H(\omega, r), \quad \gamma^H(\omega, r) = \bar{\gamma}^H(\omega, r/N), \tag{23}$$

where N is the number of elements in T. The function $\gamma^H(\omega, r)$ is non-negative, nondecreasing, $\mathcal{F} \times \mathcal{B}[0, \infty)$-measurable and positive for $r > 0$ because, by (**B**.**3**), each of the functions $\gamma_t^H(\omega, r)$ has these properties. The inequality (22) follows from the chain of relations:

$$\sum_{t \in T} \mathbf{E}\Big(q^t(p) - q^t(p')\Big)(z_t - z'_t) \geqslant \mathbf{E} \sum_{t \in T} \gamma_t^H\Big(\omega, |p_t(\omega) - p'_t(\omega)|\Big) \geqslant$$

$$\geqslant \mathbf{E} \sum_{t \in T} \bar{\gamma}^H\Big(\omega, |p_t(\omega) - p'_t(\omega)|\Big) \geqslant \mathbf{E}\bar{\gamma}^H\Big(\omega, \max_{t \in T} |p_t(\omega) - p'_t(\omega)|\Big) \geqslant$$

$$\geqslant \mathbf{E}\bar{\gamma}^H\Big(\omega, N^{-1} \sum_{t \in T} |p_t(\omega) - p'_t(\omega)|\Big) =$$

$$= \mathbf{E}\bar{\gamma}^H\Big(\omega, N^{-1}|p(\omega) - p'(\omega)|\Big) = \mathbf{E}\gamma^H\Big(\omega, |p(\omega) - p'(\omega)|\Big),$$

where the first inequality follows from (17), the second one from the definition of $\bar{\gamma}^H$, the third one from the non-negativity of $\bar{\gamma}^H$ and the fourth one from the monotonicity of $\bar{\gamma}^H(\omega, r)$ with respect to r. Thus, (22) holds with γ^H defined by (23).

10. Theorem 1 is a direct consequence of Proposition 1 and the following result.

THEOREM 2. *Let*

$$\mathcal{P} \ni p \mapsto Z(p) \subseteq \prod_{t\in T} \mathcal{Z}_t$$

be a multivalued mapping satisfying (**A′**) *and* (**B′**). *Then there exist* $z = (z_t)$ $[z_t = (z_{ts})_{s\in K(t)} \in \mathcal{Z}_t]$ *and* $p = (p_s) \in \mathcal{P}$ *with the properties* (7)–(9).

The pair (z, p) described in Theorem 2 may be regarded as an equilibrium state in a general (non–local) model specified by the family of vectors $h_s \in \mathcal{X}_s$, $s \in T$, and by the multivalued mapping $p \mapsto Z(p)$ which characterizes preferable programs of the economic system depending on existing prices. In contrast to the model defined in Section 4, the set $Z(p)$ in Theorem 2 does not necessarily admit a representation in the form

$$Z(p) = \prod_{t\in T} Z_t(q^t(p)),$$

where $Z_t(\cdot)$ depends only on $q^t(p)$ and not on the whole collection of price vectors $p = (p_s)_{s\in T}$. If the set $Z(p)$ admits such a representation (as it does in Section 4), it is natural to say that the model under consideration is local.

11. Theorem 2 will be derived from a general theorem which gives sufficient conditions for the existence of equilibrium states of monotone operators in the spaces of random vector functions on graphs. Define

$$\mathcal{L}^\infty = \prod_{t\in T} \mathcal{L}_t^\infty = \Big(\prod_{t\in T} \mathcal{X}_t\Big), \quad \mathcal{L}^1 = \prod_{t\in T} \mathcal{L}_t^1,$$

and consider the multivalued operator

$$\mathcal{P} \ni p \mapsto D(p) \subseteq \mathcal{L}^\infty,$$

which associates with every

$$p = (p_t)_{t\in T} \in \mathcal{P}$$

the set $D(p)$ consisting of vectors

$$d = (d_t)_{t\in T} \in \mathcal{L}^\infty$$

representable in the form

$$d_t = h_t + \sum_{s\in M(t)} z_{st}, \quad t \in T, \tag{24}$$

where

$$z_s = (z_{st})_{t\in K(s)}, \quad (z_s)_{s\in T} \in Z(p). \tag{25}$$

We say that a pair of vectors (d, p), $(d \in \mathcal{L}^\infty,\ p \in \mathcal{P})$ is an equilibrium state for the operator $p \mapsto D(p)$ if

$$d \in D(p), \quad d \geqslant 0, \tag{26}$$

and

$$pd = 0. \tag{27}$$

(In another words, (d, p) is a solution of the complementarity problem for the operator $p \mapsto D(p)$.)
It is clear from the above definitions that equilibrium states in the economic model under consideration exist if and only if there exist equilibrium states for the operator $p \mapsto D(p)$. As it will be shown, the following conditions guarantee the existence of equilibrium states of such an operator.

(D.1) For each $p \in \mathcal{P}$, the set $D(p)$ is nonempty, convex and bounded in $\mathcal{L}^\infty$ uniformly with respect to $p \in \mathcal{P}$.

(D.2) If $d^k \in D(p^k)$, $\|p^k - p\|_1 \to 0$ and $\mathbf{E}d^k b \to \mathbf{E}db$ for all $b \in \mathcal{L}^1$, then $d \in D(p)$.

(D.3) For each $H \in [0, \infty)$, there exists a non-negative, nondecreasing $\mathcal{F} \times \mathcal{B}[0,\infty)$-measurable function $\gamma^H(\omega, r)$ ($\omega \in \Omega$, $r \geqslant 0$) taking positive values for $r > 0$ and such that

$$\mathbf{E}(d - d')(p - p') \geqslant \mathbf{E}\gamma^H\Big(\omega, \big|p(\omega) - p'(\omega)\big|\Big) \tag{28}$$

for $d \in D(p)$, $d' \in D(p')$ and $p, p' \in \mathcal{P}$ with $\|p\|_\infty, \|p'\|_\infty \leqslant H$.

(D.4) There exist positive constants A and B such that, for all

$$d = (d_t)_{t \in T} \in D(p), \quad p = (p_t)_{t \in T} \in \mathcal{P} \tag{29}$$

and each $t_0 \in T$, we have the inequality

$$|p_{t_0}| \leqslant A \sum_{t \in T(t_0)} \mathbf{E}(p_t d_t | \mathcal{F}_{t_0}) + B, \tag{30}$$

where $T(t_0) = \{t : t \succcurlyeq t_0\}$.

We write $t \succcurlyeq t_0$ $(t, t_0 \in T)$ and we say that t *majorizes* t_0 if there is a sequence $t_1, t_2, \ldots, t_k$ such that $t_k = t$ and $t_{i+1} \in K(t_i)$ for $i = 0, 1, \ldots, k-1$.

THEOREM 3. *Suppose that a set $D(p) \subseteq \mathcal{L}^\infty$ is associated with each $p \in \mathcal{P}$. For a multivalued operator $p \mapsto D(p)$ to have an equilibrium state, it is sufficient that it satisfies conditions* (**D**.1)–(**D**.4).

This theorem was formulated in (Evstigneev, 1984). Its proof will be given in Sections 15–20. Before proving the theorem, we shall deduce Theorem 2 from it.

12. *Proof of Theorem* 2. It is sufficient to verify that properties (**D**.1)–(**D**.4) of the operator $D(\cdot)$ defined in Section 11 follow from (**A**$'$) and (**B**$'$). It is easily seen that (**B**$'$.1) implies (**D**.1). Let us consider sequences d^k, p^k and vectors d and p satisfying the conditions listed in (**D**.2). For each k, the vector d^k is representable in the form

$$d^k = (d_t^k), \quad d_t^k = h_t + \sum_{s \in M(t)} z_{st}^k, \quad t \in T,$$

where

$$(z_{st}^k)_{t \in K(s)} = z_s^k, \quad (z_s^k)_{s \in T} \in Z(p^k).$$

It is clear that $\|q^s(p^k) - q^s(p)\|_1 \to 0$, since $\|p^k - p\|_1 \to 0$. Furthermore, the vectors z^k_{st} $(k = 1, 2, \dots,\ s \in T,\ t \in K(s))$ are uniformly bounded in the norm $\|\cdot\|_\infty$, and hence there exist vectors $z_{st} \in \mathcal{X}_t$ $(s \in T,\ t \in K(s))$ such that, for some subsequence $k(n)$, $n = 1, 2, \dots$,

$$\mathbf{E}wz_s^{k(n)} \longrightarrow \mathbf{E}wz_s \quad [z_s = (z_{st})_{t\in K(s)}]$$

for all $s \in T$ and $w \in \prod_{t\in K(s)} \mathcal{L}^1_t$. (This follows, e.g., from the $*$–weak sequential compactness of a ball in the dual space to the separable Banach space $\mathcal{L}^1_t$.) Passing to the limit and using (**B**$'$.**2**), we conclude

$$(z_s)_{s\in T} \in Z(p), \quad d = (d_t), \quad d_t = h_t + \sum_{s\in M(t)} z_{st},$$

which means that $d \in D(p)$.

13. Let us show that the operator $p \mapsto D(p)$ satisfies (**D.3**). Fix some $p = (p_t) \in \mathcal{P}$ and take

$$p' = (p'_t) \in \mathcal{P}, \quad d = (d_t) \in D(p), \quad d' = (d'_t) \in D(p').$$

By the definition of $D(\cdot)$, we can represent d_t and d'_t in the form

$$d_t = h_t + \sum_{s\in M(t)} z_{st}, \quad d'_t = h_t + \sum_{s\in M(t)} z'_{st}, \tag{31}$$

where

$$(z_{st})_{t\in K(s)} = z_s, \quad (z'_{st})_{t\in K(s)} = z'_s, \quad (z_s)_{s\in T} \in Z(p), \quad (z'_s)_{s\in T} \in Z(p').$$

The property (**D.3**) follows from (**B**$'$.**3**) and the chain of relations

$$\mathbf{E}(d-d')(p-p') = \sum_{t\in T} \mathbf{E}(d_t - d'_t)(p_t - p'_t) = \mathbf{E}\sum_{t\in T}\sum_{s\in M(t)} (z_{st} - z'_{st})(p_t - p'_t) =$$

$$= \mathbf{E}\sum_{s\in T}\sum_{t\in K(s)} (z_{st} - z'_{st})(p_t - p'_t) = \mathbf{E}\sum_{s\in T} (z_s - z'_s)\big(q^s(p) - q^s(p')\big).$$

14. Now let us verify (**D.4**). Fix some $t_0 \in T$ and consider the sets

$$T_0 = \{t\colon\ t \succcurlyeq t_0\}, \quad T_1 = T \backslash T_0.$$

Before proving (30), we make the following remark. Let S be the subset of $T \times T$ defined by

$$S = \Big\{(s,t) \in T \times T\colon\ s \succcurlyeq t_0,\ t \in K(s)\Big\}.$$

Then

$$S = S' \backslash S'', \ S'' \subseteq S',$$

where

$$S' = \Big\{(s,t):\ t \in T_0,\ s \in M(t)\Big\},$$

$$S'' = \Big\{(s,t):\ t \in T_0,\ s \in M(t) \cap T_1\Big\}.$$

Indeed, if $(s,t) \in S$, then $t \in T_0$ and $s \in M(t)$. Therefore, $(s,t) \in S'$. Suppose that $(s,t) \in S''$. Then $s \in T_1$, which contradicts the assumption $(s,t) \in S$. Consequently, $(s,t) \in S' \backslash S''$. Suppose now that $(s,t) \in S' \backslash S''$. Then $t \in K(s)$, and, since $t \in T_0$ and $(s,t) \notin S''$, we have $s \notin M(t) \cap T_1$. Thus, $s \in T_1$ (because $s \in M(t)$). Therefore, $s \in T_0$ and, consequently, $(s,t) \in S$.

Let us take d and p of the form (29). Since $d \in D(p)$, there exist vectors $z_s = (z_{st})_{t \in K(s)}$, $s \in T$, satisfying (24) and (25).

Let us write for shortness $\mathbf{E}_s(\cdot)$ instead of $\mathbf{E}(\cdot \mid \mathcal{F}_s)$. From $(\mathbf{A}'.\mathbf{3})$ and the relation $(z_s) \in Z(p)$, it follows that

$$\mathbf{E}_s\Big(q^s(p)z_s\Big) \geqslant \mathbf{E}_s\Big(q^s(p)\overset{\circ}{z}_s\Big) - M_0, \quad s \in T. \tag{32}$$

Since

$$q^s(p)z_s = \sum_{t \in K(s)} p_t z_{st}, \quad q^s(p)\overset{\circ}{z}_s = \sum_{t \in K(s)} p_t \overset{\circ}{z}_{st},$$

we obtain from (32)

$$\mathbf{E}_s\left(\sum_{t \in K(s)} p_t z_{st}\right) \geqslant \mathbf{E}_s\left(\sum_{t \in K(s)} p_t \overset{\circ}{z}_{st}\right) - M_0. \tag{33}$$

For each $s \in T_0$, we have $\mathcal{F}_s \supseteq \mathcal{F}_{t_0}$. Thus, taking the conditional expectation $\mathbf{E}_{t_0}(\cdot)$ of both sides of inequality (33) with $s \succcurlyeq t_0$ and summing, we obtain

$$\mathbf{E}_{t_0}\left(\sum_{s \succcurlyeq t_0} \sum_{t \in K(s)} p_t z_{st}\right) \geqslant \mathbf{E}_{t_0}\left(\sum_{s \succcurlyeq t_0} \sum_{t \in K(s)} p_t \overset{\circ}{z}_{st}\right) - M_1, \tag{34}$$

where M_1 is a positive constant (M_1 is equal to the product of M_0 and the number of elements of $\{s:\ s \succcurlyeq t_0\}$). By (34) and the definitions of S' and S'', we have

$$-M_1 \leqslant \mathbf{E}_{t_0} \sum_{s \succcurlyeq t_0} \sum_{t \in K(s)} p_t(z_{st} - \overset{\circ}{z}_{st}) = \mathbf{E}_{t_0} \sum_{(s,t) \in S} p_t(z_{st} - \overset{\circ}{z}_{st}) =$$

$$= \mathbf{E}_{t_0} \sum_{(s,t) \in S'} p_t(z_{st} - \overset{\circ}{z}_{st}) - \mathbf{E}_{t_0} \sum_{(s,t) \in S''} p_t(z_{st} - \overset{\circ}{z}_{st}). \tag{35}$$

If $(s,t) \in S''$, then $t \in T_0$ and $s \in M(t) \cap T_1$. Thus, $t \in K(s+)$, and, consequently, $z_{st} \geqslant \overset{\circ}{z}_{st}$ by $(\mathbf{A}'.1)$. Applying this to (35) and using the definition of S', we get

$$-M_1 \leqslant \mathbf{E}_{t_0} \sum_{t \succcurlyeq t_0} \sum_{s \in M(t)} p_t(z_{st} - \overset{\circ}{z}_{st}) = \mathbf{E}_{t_0} \sum_{t \succcurlyeq t_0} p_t \sum_{s \in M(t)} (z_{st} - \overset{\circ}{z}_{st}) =$$

$$= \mathbf{E}_{t_0} \sum_{t \succcurlyeq t_0} p_t \left[\left(h_t + \sum_{s \in M(t)} z_{st} \right) - \left(h_t + \sum_{s \in M(t)} \overset{\circ}{z}_{st} \right) \right] \leqslant$$

$$\leqslant \mathbf{E}_{t_0} \sum_{t \succcurlyeq t_0} p_t (d_t - \delta e_t) = \mathbf{E}_{t_0} \sum_{t \succcurlyeq t_0} p_t d_t - \mathbf{E}_{t_0} \sum_{t \succcurlyeq t_0} \delta |p_t| \leqslant$$

$$\leqslant \mathbf{E}_{t_0} \sum_{t \succcurlyeq t_0} p_t d_t - \delta |p_{t_0}|. \tag{36}$$

The second inequality in this chain of relations follows from (13), and the third one is valid since

$$\mathbf{E}_{t_0} \sum_{t \succcurlyeq t_0} \delta |p_t| \geqslant \mathbf{E}_{t_0} \delta |p_{t_0}| = \delta |p_{t_0}|.$$

From (36), we derive

$$|p_{t_0}| \leqslant \frac{1}{\delta} \mathbf{E}_{t_0} \sum_{t \succcurlyeq t_0} p_t d_t + \frac{M_1}{\delta},$$

which yields (**D**.**4**).

15. Our aim now is to prove Theorem 3. First we shall prove it under the additional assumption that each of the σ-algebras $\mathcal{F}_t$, $t \in T$, consists of a finite number of sets. After that, we shall establish it in the general case by passing to the limit.

Proof of Theorem 3 in the case of finite algebras $\mathcal{F}_t$. In this situation the space $\mathcal{L}^\infty$ is finite-dimensional and coincides with $\mathcal{L}^1$. Let us fix in the space $\mathcal{L} \equiv \mathcal{L}^\infty = \mathcal{L}^1$ the topology corresponding to the convergence in norm $\|\cdot\|_1$ (and the $*$-weak convergence).
We note that (**D**.**4**) implies

$$\mathbf{E}|p_{t_0}| \leqslant A \sum_{t \succcurlyeq t_0} \mathbf{E} p_t d_t + B \tag{37}$$

for $p = (p_t) \in \mathcal{P}$, $d = (d_t) \in D(p)$ and $t_0 \in T$. From (37) we get

$$\mathbf{E}|p| = \sum_{t_0 \in T} \mathbf{E}|p_{t_0}| \leqslant \sum_{t_0 \in T} \sum_{t \succcurlyeq t_0} A \mathbf{E} p_t d_t + \sum_{t_0 \in T} B =$$

$$= \sum_{t \in T} \left(\sum_{t_0 \preccurlyeq t} A \right) \mathbf{E} p_t d_t + \overline{B} = \sum_{t \in T} A_t \mathbf{E} p_t d_t + \overline{B}, \tag{38}$$

where

$$A_t = A \# \{t_0 \colon\ t_0 \preccurlyeq t\}, \quad \overline{B} = B \# T$$

(the symbol # denotes the number of elements of a finite set).
We define: $\overline{A} = A \# T$;

$$\Pi = \left\{ p \in \mathcal{P} \colon \sum_{t \in T} A_t \mathbf{E}|p_t| \leqslant 2\overline{AB} \right\}; \quad \Delta = \left\{ d \in \mathcal{L} \colon\ \|d\|_\infty \leqslant M \right\},$$

where M is a constant such that $\|d\|_\infty \leqslant M$ for all $d \in D(p)$ and $p \in \mathcal{P}$ (see (**D**.1)). With each $d \in \Delta$, we associate the set $U(d)$ of those $p = (p_t) \in \Pi$ for which

$$\sum_{t\in T} A_t \mathbf{E} p_t d_t = \min_{q\in\Pi} \sum_{t\in T} A_t \mathbf{E} q_t d_t, \tag{39}$$

where $q = (q_t)$. The sets Π and Δ are closed and bounded in the finite-dimensional space $\mathcal{L}$ (the boundedness of Π follows from the positivity of the constants A_t). Consequently, these sets are compact, and thus the minimum in (39) is achieved, i.e., the set $U(d)$ is nonempty. Furthermore, Π, Δ and $U(d)$ $(d \in \Delta)$ are convex.

16. Let us consider the multivalued mapping

$$(p, d) \mapsto U(d) \times D(p). \tag{40}$$

Here (p, d) belongs to the convex compact set $\Pi \times \Delta$, and $U(d) \times D(p)$ is a convex compact set contained in $\Pi \times \Delta$. Since $\mathcal{L}$ is finite-dimensional, it follows from (**D**.2) that the graph of mapping (40) is closed. By using Kakutani's theorem (see, e.g., (Nikaido, 1968, Theorem 4.4)), we conclude that mapping (40) has a fixed point, i.e., there exist $p \in \Pi$ and $d \in \Delta$ for which

$$p \in U(d) \quad \text{and} \quad d \in D(p). \tag{41}$$

Now we shall deduce from (41) that $d \geqslant 0$ (a.s.) and $pd = 0$ (a.s), and thus the desired assertion will be proved. Since $p = (p_t)$ minimizes the functional defined in (39) on the set Π and $0 \in \Pi$, we have

$$\sum_{t\in T} A_t \mathbf{E} p_t d_t \leqslant 0, \tag{42}$$

which implies

$$\mathbf{E}|p| \leqslant \overline{B} \tag{43}$$

(see (38)). From this it follows that

$$\sum_{t\in T} A_t \mathbf{E}|p_t| \leqslant \overline{A} \sum_{t\in T} \mathbf{E}|p_t| \leqslant \overline{A}\,\overline{B}. \tag{44}$$

Suppose that the inequality $d \geqslant 0$ (a.s) does not hold. This means that, for some $t = s$, some coordinate $d_s^j(\omega)$ of the vector $d_s(\omega)$ is negative with positive probability. Let Γ be an $\mathcal{F}_s$-measurable event such that $d_s^j(\omega) < 0$ for $\omega \in \Gamma$ and $\mathbf{P}(\Gamma) > 0$. Consider the random vector $q(\omega) = (q_t(\omega))$, where $q_t(\omega) = p_t(\omega)$, for $t \neq s$ and

$$q_s^l(\omega) = \begin{cases} p_s^l(\omega), & l \neq j, \\ p_s^j(\omega) + \varepsilon \chi_\Gamma(\omega), & l = j. \end{cases}$$

If $\varepsilon > 0$ is sufficiently small then, by (44),

$$\sum_{t\in T} A_t \mathbf{E}|q_t| = \sum_{t\in T} A_t \mathbf{E}|p_t| + \varepsilon A_s \mathbf{P}(\Gamma) \leqslant 2\overline{A}\,\overline{B},$$

i.e., $q \in \Pi$. Furthermore we have

$$\sum_{t\in T} A_t \mathbf{E} p_t d_t \leqslant \sum_{t\in T} A_t \mathbf{E} q_t d_t,$$

because $p \in U(d)$. On the other hand,

$$\sum_{t\in T} A_t \mathbf{E} q_t d_t = \sum_{t\in T} A_t \mathbf{E} p_t d_t + A_s \varepsilon \mathbf{E} \chi_\Gamma d_s^j < \sum_{t\in T} A_t \mathbf{E} p_t d_t.$$

Hence, we are led to a contradiction.

Thus, $d \geqslant 0$ (a.s.). This and (42) imply that $pd = 0$ (a.s.). So the desired result is obtained in the case of finite algebras $\mathcal{F}_t$.

The idea used in the proof is typical for various applications of Kakutani's theorem in the theory of economic equilibrium (cf., in particular (Polterovich, 1978)). It should be noted also that we have not yet used the condition (**D.3**); we need it only for the case of infinite σ-algebras $\mathcal{F}_t$.

17. *Proof of Theorem* 3 *in the general case.* According to the assumption made in Section 2, each of the σ-algebras $\mathcal{F}_t$ is generated by a countable family of sets $\Gamma_t^1, \Gamma_t^2, \ldots$. Hence, there exist finite algebras $\mathcal{F}_t^k$ $(t \in T,\ k = 1, 2, \ldots)$ such that for any k,

$$t' \in K(t) \Rightarrow \mathcal{F}_{t'}^k \supseteq \mathcal{F}_t^k, \tag{45}$$

and for any t,

$$\mathcal{F}_t^1 \subseteq \mathcal{F}_t^2 \subseteq \ldots \subseteq \mathcal{F}_t^k \subseteq \ldots \subseteq \mathcal{F}_t \tag{46}$$

and

$$\mathcal{F}_t = \bigvee_{k=1}^{\infty} \mathcal{F}_t^k \tag{47}$$

(the latter notation means that $\mathcal{F}_t$ is the smallest σ-algebra containing all the σ-algebras $\mathcal{F}_t^k$, $k = 1, 2, \ldots$). As $\mathcal{F}_t^k$ we may take the σ-algebra generated by the sets

$$\Gamma_s^i, \quad s \preccurlyeq t, \quad i = 1, 2, \ldots, k. \tag{48}$$

Indeed, (45) is true because $\{s\colon\ s \preccurlyeq t'\} \supseteq \{s\colon\ s \preccurlyeq t\}$ for $t' \in K(t)$. We have $\mathcal{F}_t^k \subseteq \mathcal{F}_t^{k+1}$, since the family of sets (48) is smaller than the corresponding family for $k+1$. Every $\mathcal{F}_t^k$ is contained in $\mathcal{F}_t$ because $\Gamma_s^i \in \mathcal{F}_s \subseteq \mathcal{F}_t$ if $s \preccurlyeq t$. Finally, (47) follows from the inclusion $\Gamma_t^1, \Gamma_t^2, \ldots \in \bigvee_{k=1}^{\infty} \mathcal{F}_t^k$.

Thus, the σ-algebras $\{\mathcal{F}_t\}$ are approximated by the finite algebras $\{\mathcal{F}_t^k\}$ with properties (45)–(47). Now consider the spaces

$$\mathcal{L}^{\infty,k} = \prod_{t\in T} \mathcal{L}^\infty(\mathcal{F}_t^k), \quad \mathcal{L}^{1,k} = \prod_{t\in T} \mathcal{L}^1(\mathcal{F}_t^k),$$

and define

$$\mathcal{P}^k = \{p \in \mathcal{L}^{1,k}\colon\ p \geqslant 0\}.$$

With each vector

$$d = (d_t)_{t\in T} \in \mathcal{L}^1,$$

we associate the vector

$$\mathbf{E}^k d = \Big(\mathbf{E}(d_t \,|\, \mathcal{F}_t^k)\Big)_{t\in T} \in \mathcal{L}^{1,k},$$

and for $p \in \mathcal{P}$, we put $D^k(p) = \mathbf{E}^k D(p)$, where

$$\mathbf{E}^k D(p) = \{\mathbf{E}^k d\colon\ d \in D(p)\} \subseteq \mathcal{L}^{\infty,k}.$$

18. Let us show that the multivalued operator

$$\mathcal{P}^k \ni p \mapsto D^k(p) \subseteq \mathcal{L}^{\infty,k},$$

defined above, satisfies conditions (**D**.1)–(**D**.4) (formulated in terms of the σ-algebras $\mathcal{F}_t^k$ and the spaces $\mathcal{L}^{1,k}$, $\mathcal{L}^{\infty,k}$ and $\mathcal{P}^k$).

1) From property (**D**.1) of the operator $D(\cdot)$, it follows that for each $p \in \mathcal{P}^k$, the set $D^k(p)$ is nonempty, convex and bounded in the norm $\|\cdot\|_\infty$ uniformly with respect to $p \in \mathcal{P}^k$.

2) Suppose that we are given random vectors $d^m, d \in \mathcal{L}^{\infty,k}$, $p^m, p \in \mathcal{P}^k$ ($m = 1, 2, \ldots$) satisfying the following conditions:

$$d^m \in D^k(p^m), \quad \|p^m - p\|_1 \to 0$$

and $\mathbf{E} d^m b \to \mathbf{E} d b$ for all $b \in \mathcal{L}^{1,k}$. Let us verify the relation $d \in D^k(p)$. We have $d^m = \mathbf{E}^k d^{(m)}$, where $d^{(m)}$ is an element of $D(p^m)$. Since the sets $D(p^m)$ for $m = 1, 2, \ldots$ are uniformly bounded in the space $\mathcal{L}^\infty$, it may be assumed without loss of generality that $d^{(m)}$ converges to some $d^{(0)} \in \mathcal{L}^\infty$ in the $*$-weak topology of $\mathcal{L}^\infty$ (we use the $*$-weak compactness of a ball in $\mathcal{L}^\infty$). By virtue of property (**D**.2) of the operator $D(\cdot)$, $d^{(0)} \in D(p)$, and, consequently, $\mathbf{E}^k d^{(0)} \in D^k(p)$. Since $d^{(m)} \to d^{(0)}$ in the $*$-weak topology of the space $\mathcal{L}^\infty$, $d^m = \mathbf{E}^k d^{(m)} \to \mathbf{E}^k d^{(0)}$ in the $*$-weak topology of the space $\mathcal{L}^{\infty,k}$. Indeed, for each $a \in \mathcal{L}^{1,k}$,

$$\mathbf{E}\big(\mathbf{E}^k d^{(m)}\big)a = \mathbf{E} d^{(m)} a \longrightarrow \mathbf{E} d^{(0)} a = \mathbf{E}\big(\mathbf{E}^k d^{(0)}\big)a.$$

Therefore,

$$d = \mathbf{E}^k d^{(0)} \in \mathbf{E}^k D(p) = D^k(p).$$

3) Consider vectors $p, p' \in \mathcal{P}^k$ with $\|p\|_\infty, \|p'\|_\infty \leqslant H$ and vectors $d \in D^k(p)$ and $d' \in D^k(p')$. Take $b \in D(p)$ and $b' \in D(p')$ such that $d = \mathbf{E}^k b$ and $d' = \mathbf{E}^k b'$. By virtue of property (**D**.3) of the operator $D(\cdot)$, we obtain

$$\mathbf{E}(d - d')(p - p') = \mathbf{E}(\mathbf{E}^k b - \mathbf{E}^k b')(p - p') =$$

$$= \mathbf{E}(b - b')(p - p') \geqslant \mathbf{E}\gamma^H\Big(\omega, \big|p(\omega) - p'(\omega)\big|\Big), \tag{49}$$

where the second equality follows from the fact that $p, p' \in \mathcal{P}^k$. Thus, the operator $D^k(\cdot)$ possesses property (**D.3**) (with the same function $\gamma^H(\omega, r)$ whose existence is postulated for the operator $D(\cdot)$).

4) Let us now establish an estimate analogous to (**D.4**) for the operator $p \mapsto D^k(p)$ $(p \in \mathcal{P}^k)$. Take

$$p = (p_t)_{t\in T} \in \mathcal{P}^k, \quad d = (d_t)_{t\in T} \in D^k(p).$$

Since $d \in D^k(p)$, there exists $b = (b_t)_{t\in T} \in D(p)$ such that $d = \mathbf{E}^k b$. Applying (30) to p and b, we obtain

$$|p_{t_0}| \leqslant A \sum_{t \geqslant t_0} \mathbf{E}(p_t b_t | \mathcal{F}_{t_0}) + B. \tag{50}$$

Taking the conditional expectation with respect to $\mathcal{F}^k_{t_0}$ of both sides of inequality (50), we find that

$$|p_{t_0}| \leqslant A \sum_{t \geqslant t_0} \mathbf{E}(p_t b_t | \mathcal{F}^k_{t_0}) + B \tag{51}$$

(because $\mathcal{F}^k_{t_0} \subseteq \mathcal{F}_{t_0}$ and $|p_{t_0}|$ is $\mathcal{F}^k_{t_0}$-measurable). Further, for $t \succcurlyeq t_0$, we have $\mathcal{F}^k_t \supseteq \mathcal{F}^k_{t_0}$, consequently,

$$\mathbf{E}(p_t b_t | \mathcal{F}^k_{t_0}) = \mathbf{E}\Big(\mathbf{E}(p_t b_t | \mathcal{F}^k_t) | \mathcal{F}^k_{t_0}\Big) =$$

$$= \mathbf{E}\Big(p_t \mathbf{E}(b_t | \mathcal{F}^k_t) | \mathcal{F}^k_{t_0}\Big) = \mathbf{E}(p_t d_t | \mathcal{F}^k_{t_0}).$$

From this and (51), the required inequality follows:

$$|p_{t_0}| \leqslant A \sum_{t \geqslant t_0} \mathbf{E}(p_t d_t | \mathcal{F}^k_{t_0}) + B. \tag{52}$$

19. Thus, the operator

$$\mathcal{P}^k \ni p \mapsto D^k(p) \subseteq \mathcal{L}^{\infty,k}$$

satisfies conditions (**D.1**)–(**D.4**) (in the scheme described in terms of the σ-algebras $\mathcal{F}^k_t$). According to the result established in Sections 15 and 16, the operator $D^k(\cdot)$ possesses an equilibrium state, i.e., there exist random vectors $d^k = (d^k_t)$ and $p^k = (p^k_t)$ such that

$$p^k \in \mathcal{P}^k, \quad d^k \in D^k(p^k), \tag{53}$$

$$d^k \geqslant 0, \quad p^k d^k = 0. \tag{54}$$

In order to complete the proof of Theorem 3, it now suffices to prove the following

PROPOSITION 2. *Suppose that, for each $k = 1, 2, \ldots$, there exist random vectors p^k and d^k with the properties* (53) *and* (54). *Then the operator $D(\cdot)$ has an equilibrium state.*

(We recall that $D(\cdot)$ is supposed to satisfy (**D.1**)–(**D.4**).)

Proof. For all $k = 1, 2, \ldots$ and $t \in T$, we have

$$\|p_t^k\|_\infty \leqslant B. \tag{55}$$

Indeed, by (54),

$$0 = p^k d^k = \sum_{t \in T} p_t^k d_t^k,$$

and thus $p_t^k d_t^k = 0$ since $d_t^k \geqslant 0$ and $p_t^k \geqslant 0$. Using this fact and applying (52) to $p = p^k$ and $d = d^k$, we obtain (55).
Let $d^{(k)} (k = 1, 2, \ldots)$ be a random vector for which $d^{(k)} \in D(p^k)$ and $d^k = \mathbf{E}^k d^{(k)}$. Such a random vector exists because of (53) and the definition of $D^k(\cdot)$. Inequality (55) implies that $\{p^k\}$ is a bounded sequence in $\mathcal{L}^\infty$. The same is true for $\{d^{(k)}\}$ (since the sets $D(p)$, $p \in \mathcal{P}$, are uniformly bounded in $\mathcal{L}^\infty$). Thus, for some subsequence, p^k and $d^{(k)}$ $*$–weakly converge in $\mathcal{L}^\infty$:

$$p^k \xrightarrow{*w} p, \quad d^{(k)} \xrightarrow{*w} d \tag{56}$$

where p and d are some elements of $\mathcal{L}^\infty$. Since $p^k \geqslant 0$, it follows from (56) that $p \in \mathcal{P}$.

20. Let us show that

$$d^k = \mathbf{E}^k d^{(k)} \xrightarrow{*w} d. \tag{57}$$

For this, we take an arbitrary $a \in \mathcal{L}^1$ and write

$$\left|\mathbf{E}d^k a - \mathbf{E}da\right| = \left|\mathbf{E}(\mathbf{E}^k d^{(k)})a - \mathbf{E}da\right| = \left|\mathbf{E}d^{(k)}\mathbf{E}^k a - \mathbf{E}da\right| \leqslant$$

$$\leqslant \left|\mathbf{E}(\mathbf{E}^k a)d^{(k)} - \mathbf{E}ad^{(k)}\right| + \left|\mathbf{E}ad^{(k)} - \mathbf{E}ad\right|.$$

Here the second summand tends to zero by (56). The first one is not greater than

$$\mathbf{E}\left|\mathbf{E}^k a - a\right|\left|d^{(k)}\right| \leqslant \sup_k \|d^{(k)}\|_\infty \mathbf{E}\left|\mathbf{E}^k a - a\right|,$$

where $\sup_k \|d^{(k)}\|_\infty < \infty$, because the sets $D(p^k)$ are bounded in $\mathcal{L}^\infty$. Furthermore, it follows from properties (46) and (47) of the σ–algebras $\mathcal{F}_t^k$ and from the convergence theorem for conditional expectations (see, e.g., (Meyer, 1966, Theorem V. 18)) that $\mathbf{E}|\mathbf{E}^k a - a| \to 0$. Thus, $d^k \xrightarrow{*w} d$.
Now we prove that, for some subsequence, p^k converges to p in the norm $\|\cdot\|_1$, and not only in the $*$–weak topology of $\mathcal{L}^\infty$ (cf., (Evstigneev and Katyshev, 1982, Lemma 4)). This is the part of the proof of Proposition 2, where (**D.3**) is used. Fix some $d_0 \in D(p)$. By (**D.3**), (55) and (56),

$$\mathbf{E}(d^{(k)} - d_0)(p^k - p) \geqslant \mathbf{E}\gamma^H\left(\omega, \left|p^k(\omega) - p(\omega)\right|\right) \tag{58}$$

for $H = B(\#T)$. The left–hand side of (58) is equal to

$$\mathbf{E}d_0(p - p^k) + \mathbf{E}d^{(k)}(p^k - p),$$

where the first summand tends to zero since $p^k \xrightarrow{*w} p$. Further, by (54), $\mathbf{E}d^{(k)}p^k = \mathbf{E}d^k p^k = 0$, and

$$-\mathbf{E}d^{(k)}p = -\mathbf{E}d^{(k)}\mathbf{E}^k p + \mathbf{E}d^{(k)}(\mathbf{E}^k p - p) =$$

$$= -\mathbf{E}d^{(k)}p + \mathbf{E}d^{(k)}(\mathbf{E}^k p - p) \leqslant \|d^{(k)}\|_\infty \mathbf{E}\big|\mathbf{E}^k p - p\big| \to 0,$$

which follows from the fact that $d^k \geqslant 0$ and from the convergence theorem for conditional expectations.

Thus, the left-hand side of (58) tends to zero and, consequently,

$$\lim_{k\to\infty} \mathbf{E}\gamma^H\Big(\omega, |p(\omega) - p^k(\omega)|\Big) = 0.$$

The sequence of random variables $\gamma^H(\omega, |p(\omega) - p^k(\omega)|) \geqslant 0$ converging in the norm $\mathbf{E}|\cdot|$ contains a subsequence converging a.s. Therefore,

$$\gamma^H\Big(\omega, |p(\omega) - p^k(\omega)|\Big) \to 0 \quad \text{a.s.} \tag{59}$$

for some subsequence.

It follows from (59) that

$$r_k(\omega) \equiv \big|p(\omega) - p^k(\omega)\big| \to 0 \quad \text{a.s.} \tag{60}$$

Indeed, suppose it is false. Then there exists $\Gamma \in \mathcal{F}$ with $\mathbf{P}(\Gamma) > 0$ such that $r_k(\omega)$ does not converge to zero for $\omega \in \Gamma$, i.e., for each $\omega \in \Gamma$, there is $r(\omega) > 0$ such that $r_k(\omega) \geqslant r(\omega)$ for infinitely many values of k. Consequently, if $\omega \in \Gamma$, then

$$\gamma^H\Big(\omega, |p(\omega) - p^k(\omega)|\Big) \geqslant \gamma^H(\omega, r(\omega)) > 0$$

for some infinite sequence of k's (we use the monotonicity and positivity properties of $\gamma^H(\omega, r)$). Thus, with a positive probability, $\gamma^H(\omega, |p(\omega) - p^k(\omega)|)$ does not converge to zero, which is a contradiction.

According to (55), $\sup \|p^k\|_\infty < \infty$. This, together with (60), for the selected subsequence yields $\|p^k - p\|_1 \to 0$. Furthermore, by (56), $d^{(k)} \xrightarrow{*w} d$. Since $d^{(k)} \in D(p^k)$, it follows from (**D**.2) that $d \in D(p)$.

By (54) and (57), $d^k \geqslant 0$ and $d^k \xrightarrow{*w} d$. Therefore, $d \geqslant 0$. Finally, the relations

$$d^k \xrightarrow{*w} d, \quad \sup_k \|d^k\|_\infty < \infty, \quad \mathbf{E}p^k d^k = 0, \quad \|p^k - p\|_1 \to 0,$$

imply the equality $\mathbf{E}pd = 0$. Thus, $pd = 0$ (because $p \geqslant 0$ and $d \geqslant 0$) and, consequently, (d, p) is an equilibrium state for the operator $D(\cdot)$.

Proposition 2 and Theorem 3 are proved.

21. In conclusion, we make several remarks.

Remark 1. In Sections 2–4, the description of the model was given in a static form (no time parameter was explicitly introduced). Dynamics can be included in the above scheme as follows. Assume that the elements of the set T are pairs

$$t = (n, b), \quad n \in \{1, 2, \ldots, N\}, \quad b \in B(n),$$

where $\{1, 2, \ldots, N\}$ is the set of values of the time parameter n and $B(n)$ is the set of economic units functioning at time n. With each pair $t = (n, b)$ $(b \in B(n))$ a set $K(n, b) \subseteq T$ is associated such that $(n', b') \in K(n, b)$ implies $n' \geqslant n$. Here, $K(n, b)$ describes the set of economic units depending (via direct commodity supplies) on b and functioning at times $n' \geqslant n$. As before, we suppose that, for all $(n, b) \in T$, σ-algebras $\mathcal{F}_{(n,b)}$ are given, satisfying the condition

$$\mathcal{F}_{(n',b')} \supseteq \mathcal{F}_{(n,b)} \quad \text{for} \quad (n', b') \in K(n, b)$$

(the economic units depending on supplies of goods thus depend on random factors influencing suppliers). We note that in this scheme the moment of shipping off goods and the moment of their receipt are indicated by n and n'.
There is also another way of introducing time in the model. We may simply suppose that an integer-valued function $\nu(t)$, $t \in T$, is given on the graph T, and its values $\nu(t)$ are interpreted as moments of functioning of economic units $t \in T$. It is natural to assume here that $\nu(s) \geqslant \nu(t)$ for $s \in K(t)$.

Remark 2. The model under consideration differs from the standard stochastic control schemes in the following respects. a) The set T and the σ-algebras $\mathcal{F}_t$, $t \in T$, are not linearly ordered. b) We do not maximize or minimize any explicitly given functional: our aim is to find a random vector field (z_t, p_t) on the graph T such that $z = (z_t)$, $p = (p_t)$ is an equilibrium state. In this sense, we say that the above scheme is an equilibrium scheme of controlling a random vector field on a directed graph.
The following comments are also in order. The above description of the model is given in terms of the mappings $q \mapsto Z_t(q)$ associating with each family of price vectors the set of preferable programs. However, we do not specify the procedure of decision making which determines the choice of preferable programs. In order to define such a procedure, we need to introduce some additional elements into the model.
For example, suppose that, to each $t \in T$, there correspond: a measurable space A_t (the space of decisions), a set U_t of $\mathcal{F}_t$-measurable functions $u\colon \Omega \to A_t$ (the set of controls), an operator $\Phi_t\colon U_t \to \mathcal{Z}_t$ from the space of controls into the space of programs, and a set $U_t(q) \subseteq U_t$ including all the preferable controls for the price system $q \in \mathcal{Q}_t$. By definition, the set $Z_t(q)$ consists of programs $v \in \mathcal{Z}_t$ such that $v \in \Phi_t(U_t(q))$.
This "parametric" way of describing the model (cf., e.g., (Arkin and Evstigneev, 1987, Chapter 2, Section 4.7)) is close to the conventional control theory, where two groups of parameters are usually distinguished: controls and states of the system. Such an approach was applied to optimization problems for random fields on graphs in (Evstigneev, 1988) and to equilibrium problems in (Evstigneev, 1989). In the present work, we follow another, "functional", approach (cf. (Arkin and Evstigneev, 1987, Chapter 2, Section 4.1)), which, on one hand, yields some gain in generality and, on the other hand, simplifies the model description.

Remark 3. If each of the σ-algebras $\mathcal{F}_t$ is finite, Theorems 1, 2 and 3 are valid, respectively, without the assumptions (**B**.3), (**B**$'$.3) and (**D**.3). This follows from the results of Sections 15 and 16.

Remark 4. We point to an important special case of the considered model (an analogue of Polterovich's model (Polterovich, 1978)). Suppose that each economic unit $t \in T$ with $K(t+) \neq \varnothing$ is described as follows. For each

$$q = (q_s)_{s \in K(t)} \in Q_t,$$

the set $Z_t(q)$ consists of programs $v = (v_s)_{s \in K(t)}$ which can be represented in the form

$$-v_t = x + c, \quad v_s = y_s, \quad s \in K(t+),$$

where

$$c \in C_t(q_t), \quad \left(x, (y_s)_{s \in K(t+)}\right) \in W_t$$

and

$$\sum_{s \in K(t+)} \mathbf{E} q_s y_s - \mathbf{E} q_t x \geqslant \sum_{s \in K(t+)} \mathbf{E} q_s y'_s - \mathbf{E} q_t x' \tag{61}$$

for all

$$\left(x', (y'_s)_{s \in K(t+)}\right) \in W_t.$$

Here $C_t(\cdot)$ is a given multivalued mapping that associates with each $a \in \mathcal{P}_t$ a set of non-negative vectors $c \in \mathcal{X}_t$. We denote by W_t a set consisting of collections of non-negative vectors

$$\left(x, (y_s)_{s \in K(t+)}\right), \tag{62}$$

where $x \in \mathcal{X}_t$, $y_s \in \mathcal{X}_s$, $s \in K(t+)$.
In this scheme, we distinguish the production sector and the consumption sector of an economic unit $t \in T$. The first one is described by the set W_t (the technological set) which consists of all admissible technological processes of the form (62). Condition (61) means that the production sector maximizes the expected profit. The consumption of the economic unit t is described by the multivalued demand function $C_t(\cdot)$. Some natural assumptions concerning $C_t(\cdot)$ and W_t (in particular, the strict monotonicity of $-C_t(\cdot)$) guarantee the fulfillment of conditions (**A**) and (**B**) and thus the existence of equilibria. This model will be studied in detail in a subsequent paper, where a special attention will be devoted to turnpike properties of equilibrium states (similar to those in (Evstigneev, 1988)).

Remark 5. As a background, this work has the studies of stochastic analogues of the classical von Neumann, Gale, Arrow and Debreu models (Dynkin, 1971; 1972; 1976; Radner, 1971; 1982; Bewley, 1972; 1981; Brock and Mirman, 1972; Evstigneev, 1972; Bhattacharya and Majumdar, 1973; Arkin and Evstigneev, 1987). The following topics are also closely related to the subject of this paper: the theory of team problems (Radner, 1962; Jordan, 1975; Ho, 1980; Witsenhausen, 1980) and the analysis of convex multistage problems of stochastic optimization (Dynkin, 1972; Rockafellar and Wets, 1975; 1976; Evstigneev, 1976). Economic

models involving random fields on some undirected graphs (lattices Z^d) were considered in (Föllmer, 1976).
The present paper is a continuation of article (Evstigneev, 1984) that was kindly recommended for publication in the journal "Doklady AN SSSR" by Yu.V. Prokhorov. I take the opportunity to thank Yu.V. Prokhorov for his encouragement and attention to my work in probabilistic mathematical economics.

REFERENCES

Arkin, V.I. and Evstigneev, I.V. (1987). *Stochastic Models of Control and Economic Dynamics.* Academic Press, London.

Bewley, T. (1972). Existence of equilibria in economies with infinitely many commodities. *J. Econ. Theory.* **4**, 514–540.

Bewley, T. (1981). Stationary equilibrium. *J. Econ. Theory.* **24**, 265–295.

Bhattacharya, R.N. and Majumdar, M. (1973). Random exchange economies. *J. Econ. Theory.* 6, 37–67.

Brock, W.A. and Mirman, L.J. (1972). Optimal economic growth and uncertainty: The discounted case. *J. Econ. Theory.* **4**, 479–513.

Dynkin, E.B. (1971). Some probability models for a developing economy. *Soviet Math. Dokl.* **12**, 1422–1425.

Dynkin, E.B. (1972). Concave stochastic dynamic programming. *Math. USSR Sbornik.* **16**, 501–515.

Dynkin, E.B. (1976). Economic equilibrium under uncertainty. In: *Computing Equilibria: How and Why.* Ed. by M.W. Loś and J. Loś. Nort Holland, Amsterdam, pp. 207–218.

Evstigneev, I.V. (1972). Optimal economic planning taking account of stationary random factors. *Soviet Math. Dokl.* **13**, 1357–1359.

Evstigneev, I.V. (1976). Lagrange multipliers for the problems of stochastic programming. In: *Warsaw Fall Seminars in Mathematical Economics.* Ed. by M.W. Loś, J. Loś and A. Wieczorek. Springer, Berlin, pp. 34–48.

Evstigneev, I.V. and Katyshev P.K. (1982). Equilibrium paths in stochastic models of economic dynamics. *Theory Probab. Appl.* **27**, 127–135.

Evstigneev, I.V. (1984). An optimality principle and an equilibrium theorem for controlled random fields on a directed graph. *Soviet Math. Dokl.* **29**, 89–93.

Evstigneev, I.V. (1988). Controlled random fields on an oriented graph. *Theory Probab. Appl.* **33**, 433–445.

Evstigneev, I.V. (1989). Controlled random fields on graphs and stochastic economic models. In: *Fifth Intern. Vilnius Conf. on Probab. Theory and Math. Statist., Abstracts of Comm.* 1, IMC, Vilnius, pp. 140–141.

Föllmer, H. (1974). Random economies with many interacting agents. *J. Math. Econ.* **1**, 51–62.

Ho, Y.C. (1980). Team decision theory and information structures. *Proc. IEEE.* **68**, 654–666.

Jordan, J.C. (1975). Information and shadow prices for the constrained concave team problem. *J. Math. Econ.* **2**, 371–393.

Meyer, P.A. (1966). *Probability and Potentials.* Ginn (Blaisdell), Boston.

Nagurney, A. (1987). Competitive equilibrium problems, variational inequalities and regional science. *J. Regional Sci.* **27**, 503–517.

Nikaido, H. (1968). *Convex Structures and Economic Theory.* Academic Press, New York.

Polterovich, V.M. (1978). Equilibrium trajectories of economic growth. In: *Methods of Functional Analysis in Mathematical Economics.* Ed. by B.A. Efimov. Nauka, Moscow, pp. 56–97 (In Russian. English transl. In: *Econometrica* (1983). **51**, 693–729).

Radner, R. (1962). Team decision problems. *Ann. Math. Statist.* **33**, 857–881.

Radner, R. (1971). Balanced stochastic growth at the maximum rate. In: *Contributions to the von Neumann Growth Model.* Ed. by G. Bruckmann and W. Weber. Springer, Vienna, pp. 39–52.

Radner, R. (1982). Equilibrium under uncertainty. In: *Handbook of Mathematical Economics.* Ed. by K.J. Arrow and M. Intrilligator. **1**, North Holland, Amsterdam, pp. 923–1006.

Rockafellar, R.T. and Wets, R.J.-B. (1975). Stochastic convex programming: Kuhn — Tucker conditions. *J. Math. Econ.* **2**, 349–370.

Rockafellar, R.T. and Wets, R.J.-B. (1976). Non-anticipativity and L^1-martingales in stochastic optimization problems. *Math. Programming Study.* **6**, 170–187.

Samuelson, P.A. (1952). Spatial price equilibrium and linear programming. *Amer. Econ. Rev.* **42**, 283–303.

Takayama, T. and Judge, G.G. (1971). *Spatial and Temporal Price and Allocation Models.* North Holland, Amsterdam.

Vainberg, M.M. (1973). *The Variational Method and the Method of Monotone Operators in the Theory of Nonlinear Equations.* Wiley, New York.

Witsenhausen, H. (1980). Informational aspects of stochastic control. In: *Analysis and Optimization of Stochastic Systems.* Ed. by O.L.R. Jacobs et al. Academic Press, London, pp. 273–284.

New Trends in Probab. and Statist., pp. 413 - 435
V. Sazonov and T. Shervashidze (Eds.)
1991 VSP/Mokslas

ASYMPTOTIC EXPANSIONS FOR SINGULARY PERTURBED STOCHASTIC DIFFERENTIAL EQUATIONS

Yu.M. KABANOV, S.M. PERGAMENSHCHIKOV and J.M. STOYANOV
Central Economics and Mathematical Institute of the USSR Academy of Sciences, Krasikova str., 32, 117418 Moscow, USSR
Department of Theoretical Cybernetics, Tomsk State University, Tomsk, USSR
Institute of Mathematics of the Bulgarian Acad. of Sci., Sofia, Bulgaria

ABSTRACT

An asymptotic expansions for the SDE

$$\varepsilon\, dx_t^\varepsilon = F(x_t^\varepsilon)\, dt + \varepsilon^{1/2+\delta} G(x_t^\varepsilon)\, dW_t$$

is obtained under the assumption that the real parts of the eigenvalues of the matrix $F'(x_\infty)$ are negative, where x_∞ is a root of the equation $F(x) = 0$. The proofs are based on the moment bounds for solutions of linear SDE's.

INTRODUCTION

Analogously to the classical theory, we use the term "singularly perturbed" for a stochastic differential equation (SDE) with a small parameter in the left-hand side. Such SDE's arise in models of two-scale stochastic systems containing "fast" and "slow" variables. In (Kabanov and Pergamenshchikov, 1990 a,b) for such systems it was established a stochastic version of the well-known Tikhonov theorem about asymptotic behavior of solutions when the parameter tends to zero (see, e.g., (Vasilyeva and Butuzov, 1973)). The present paper deals with homogeneous systems containing only "fast" variables. For this particular case we present results on the rate of convergence in the stochastic analogue of the Tikhonov theorem including asymptotic expansions of the solution of a singularly perturbed SDE.
We consider the following SDE in $\mathbf{R}^n$:

$$\varepsilon\, dx_t^\varepsilon = F(x_t^\varepsilon)\, dt + \beta\sqrt{\varepsilon} G(x_t^\varepsilon)\, dW_t, \quad x_0^\varepsilon = \zeta, \tag{0.1}$$

where $W = (W_t)_{t\geqslant 0}$ is a standard one-dimensional Wiener process, $\zeta \in \mathbf{R}^n$, $\varepsilon \in]0,1[$.
Assume that the following three conditions are satisfied:

(A) The functions F and G with values in $\mathbf{R}^n$ are Lipschitz on each compact subset of $\mathbf{R}^n$ and have linear growth. The matrix $F' = \partial F/\partial x$ is continuous.

(B) There exists a root x_∞ of the equation $F(x) = 0$ such that the real parts of all eigenvalues of the matrix $A_\infty := F'(x_\infty)$ are negative, and

$$\lim_{r\to\infty} \tilde{x}_r = x_\infty, \tag{0.2}$$

where $\tilde{x}$ is the solution of the ordinary differential equation

$$d\tilde{x}_r = F(\tilde{x}_r)dr, \quad \tilde{x}_0 = \zeta \tag{0.3}$$

(this means that the initial value ζ belongs to the domain of the influence of the root x_∞).

(C) $\beta := \beta_\varepsilon = O(\varepsilon^\delta)$ as $\varepsilon \to 0$ for a fixed $\delta > 0$. Without loss of generality we assume that $\beta \in [0,1]$.

Recall that the results of (Kabanov and Pergamenshchikov, 1990 a) imply, in particular, that

$$\mathbf{P}\text{-}\lim_{\varepsilon\to 0} \sup_I \left|x_t^\varepsilon - x_\infty\right| = 0, \tag{0.4}$$

where $I = [T_0, T]$ is any interval with $0 < T_0 < T$. Our goal is to find a rate of convergence in (0.4) and obtain an asymptotic expansion for x^ε.

The paper is organized as follows. Section 1 contains results about the first order approximations of x^ε in β. In Section 2 we give asymptotic expansions of x^ε in power series of β. The results of Sections 1 and 2 are based on the moment bounds for the solutions of linear SDE's with negative Lyapunov exponents proved in Section 3. These moment bounds seem to be of an independent interest, and Section 3 can be read separately.

It is worth to note that in the time-scale $r := t/\varepsilon$, the asymptotic expansion of Theorem 2.1 coincides with a known expansion for "regularly" perturbed SDE's on the fixed interval $[0,T]$ (see (Freidlin and Wentzell, 1984, Theorem 2.2)). However, in our case condition (B) allows to prove the uniform character of the approximation on the increasing time intervals $[0, T/\varepsilon]$.

1. FIRST ORDER APPROXIMATIONS

The following notations will be used in the sequel:

$$||y||_{s,t} := \sup\left\{|y_u|,\ u \in [s,t]\right\}, \quad ||y||_t := ||y||_{0,t}.$$

THEOREM 1.1. *Suppose that the conditions* (A)–(C) *are satisfied and, moreover, F' is locally Lipschitz. Let $\tilde{x}$ be the solution of* (0.1), *$\tilde{x}^1$ be the solution of* SDE:

$$d\tilde{x}_r^1 = F'(\tilde{x}_r)\tilde{x}_r^1\,dr + G(\tilde{x}_r)\,d\widetilde{W}_r, \quad \tilde{x}_0^1 = 0, \tag{1.1}$$

with the Wiener process $\widetilde{W}_r = W_{r\varepsilon}/\sqrt{\varepsilon}$, $r > 0$.
Then the following expansion holds

$$x_t^\varepsilon = \tilde{x}_{t/\varepsilon} + \beta_\varepsilon \tilde{x}_{t/\varepsilon}^1 + \beta_\varepsilon \Delta_t^1, \tag{1.2}$$

where

$$\mathbf{P}\text{-}\lim_{\varepsilon\to 0} \|\Delta^1\|_T = 0. \tag{1.3}$$

Proof. Let us change the time parameter in (0.1) putting $r := t/\varepsilon$ and $\tilde{x}^\varepsilon_r :=$ $:= x_{r\varepsilon}$. Then $\tilde{x}^\varepsilon_r$ satisfies the equation

$$d\tilde{x}^\varepsilon_r = F(\tilde{x}^\varepsilon_r)\,dr + \beta G(\tilde{x}^\varepsilon_r)\,d\widetilde{W}_r, \quad \tilde{x}^\varepsilon_0 = \zeta. \tag{1.4}$$

Put

$$\tilde{z}_r := \tilde{x}_r + \beta\tilde{x}^1_r, \tag{1.5}$$

$$\widetilde{\Delta}^1_r := \beta^{-1}(\tilde{x}^\varepsilon_r - \tilde{z}_r). \tag{1.6}$$

Obviously $\Delta^1_t = \widetilde{\Delta}^1_{t/\varepsilon}$, and the relation (1.3) is equivalent to the uniform convergence in probability of the process $\widetilde{\Delta}^1$ on the increasing time–intervals $[0, T/\varepsilon]$. In other words, we have to check that

$$\mathbf{P}\text{-}\lim_{\varepsilon\to 0} \|\widetilde{\Delta}^1\|_{T/\varepsilon} = 0. \tag{1.7}$$

Indeed, according to the finite increments formula, we have

$$F(\tilde{x}^\varepsilon_r) = F(\tilde{z}_r + \beta\widetilde{\Delta}^1_r) = F(\tilde{z}_r) + \beta A^\varepsilon_r \widetilde{\Delta}^1_r, \tag{1.8}$$

where A^ε_r is a matrix whose i–th column has the form

$$\frac{\partial F}{\partial x_1}\Big(\tilde{z}_r + \vartheta^1_r \beta\widetilde{\Delta}^1_r\Big), \tag{1.9}$$

$0 \leqslant \vartheta^i_r \leqslant 1$ (due to Filippov's lemma, the functions ϑ^i can be chosen predictable). Put $A_r := F'(\tilde{x}_r)$. It follows from (0.1), (1.4)–(1.6), (1.8) that $\widetilde{\Delta}^1_r$ is the solution of the following SDE with zero initial condition

$$d\widetilde{\Delta}^1_r = A\widetilde{\Delta}^1_r\,dr + (A^\varepsilon_r - A_r)\widetilde{\Delta}^1_r\,dr +$$

$$+\beta^{-1}\Big(F(\tilde{z}_r) - F(\tilde{x}_r) - \beta F'(\tilde{x}_r)\tilde{x}^1_r\Big)\,dr + \Big(G(\tilde{x}^\varepsilon_r) - G(\tilde{x}_r)\Big)\,d\widetilde{W}_r. \tag{1.10}$$

Let introduce the stopping time $\rho := \tau_{\mu,\varepsilon} \wedge \sigma_\varepsilon$ depending on ε and $\mu \in]0,1[$, where

$$\tau_{\mu,\varepsilon} := \inf\big\{r\colon\ |\widetilde{\Delta}^1_r| \geqslant \mu\big\} \wedge (T/\varepsilon), \quad \sigma_\varepsilon := \inf\big\{r\colon\ |\tilde{x}^1_r| \geqslant \beta_\varepsilon^{-1/4}\big\} \wedge (T/\varepsilon).$$

Put $N = \|\tilde{x}\|_\infty$. Then, for $r \leqslant \rho$, we have

$$|\tilde{z}_r| \leqslant |\tilde{x}_r| + \beta|\tilde{x}^1_r| \leqslant N + 1, \quad |\tilde{x}^\varepsilon_r| \leqslant |\tilde{z}_r| + \beta|\widetilde{\Delta}^1_r| \leqslant N + 2.$$

The local Lipschitz condition for F' and G imply that there exists a constant K such that, for all $r \leqslant \rho$,

$$\left|(A_r^\varepsilon - A_r)\widetilde{\Delta}_r^1\right| \leqslant K\big(|\tilde{z}_r - \tilde{x}_r| + \beta|\widetilde{\Delta}_r^1|\big) \leqslant K\big(\beta|\tilde{x}_r^1| + \beta\big) \leqslant 2K\beta^{3/4},$$

$$\beta^{-1}\left|F(\tilde{z}_r) - F(\tilde{x}_r) - \beta F(\tilde{x}_r)\tilde{x}_r^1\right| \leqslant K\beta|\tilde{x}_r^1|^2 \leqslant K\beta^{1/2},$$

$$\left|G(\tilde{x}_r^\varepsilon) - G(\tilde{x}_r)\right| \leqslant K|\tilde{x}_r^\varepsilon - \tilde{x}_r| \leqslant K\big(\beta|\tilde{x}_r^1| + \beta|\widetilde{\Delta}_r^1|\big) \leqslant 2K\beta^{3/4}.$$

Since A_r converges to the constant matrix $A_\infty := F'(x_\infty)$ satisfying (B), the matrix-valued function A_r satisfies the condition H_2 of Section 3 (this fact can be deduced from Corollary 10.2.2 in (Bylov et al., 1974)). Applying Theorem 3.2 to the equation (1.10) and taking into account (C), we find that

$$\mathbf{P}\text{-}\lim_{\varepsilon\to 0} \|\widetilde{\Delta}^1\|_\rho = 0. \tag{1.11}$$

The next step is to show that, for an arbitrary $a > 0$,

$$\lim_{\varepsilon\to 0} \mathbf{P}\Big(\|\tilde{x}^1\|_{T/\varepsilon} \geqslant a\beta^{-1/4}\Big) = 0. \tag{1.12}$$

Indeed, the Chebyshev inequality, Theorem 3.1 and (C) imply that

$$\mathbf{P}\big(\|\tilde{x}^1\|_{T/\varepsilon} \geqslant a\beta^{-1/4}\big) \leqslant a^{-4m}\beta^m \mathbf{E}\|\tilde{x}^1\|_{T/\varepsilon}^{4m} \leqslant \text{const}\,\beta^m T/\varepsilon \longrightarrow 0$$

as $\varepsilon \to 0$, for sufficiently large m.
Evidently,

$$\mathbf{P}\big(\|\widetilde{\Delta}^1\|_{T/\varepsilon} \geqslant \mu\big) \leqslant \mathbf{P}\big(\|\widetilde{\Delta}^1\|_\rho \geqslant \mu\big) + \mathbf{P}\big(\|\tilde{x}^1\|_{T/\varepsilon} \geqslant \beta^{-1/4}\big),$$

and (1.7) follows from (1.11), (1.12). Thus, (1.3) holds. The theorem is proved.
Consider now the process $\tilde{y}^1$ satisfying the linear SDE

$$d\tilde{y}_r^1 = A_\infty \tilde{y}_r^1\,dr + G_\infty\,d\widetilde{W}_r, \quad \tilde{y}_0^1 = 0, \tag{1.13}$$

where $A_\infty := F'(x_\infty)$, $G_\infty := G(x_\infty)$.

THEOREM 1.2. *Suppose that the conditions* (A)–(C) *are satisfied and, moreover, F' is locally Lipschitz.*
Then

$$x_t^\varepsilon = x_\infty + \beta_\varepsilon \tilde{y}_{t/\varepsilon}^1 + \beta_\varepsilon \widehat{\Delta}_t^1, \tag{1.14}$$

where $\tilde{y}^1$ is the solution of (1.13) *and for any $T_0 \in]0,T]$*

$$\mathbf{P}\text{-}\lim_{\varepsilon\to 0} \|\widehat{\Delta}^1\|_{T_0,T} = 0. \tag{1.15}$$

Proof. It is well known that (B) implies an exponential rate of convergence in (0.2). A more formal statement is given in

LEMMA 1.1. *Let $\widetilde{R}_r := \tilde{x}_r - x_\infty$. Then, for any $T_0 \in]0,T]$, there are constants c_1, c_2 such that, for all $\varepsilon > 0$,*

$$\|\widetilde{R}\|_{T_0/\varepsilon,\infty} \leqslant c_1 \exp\{-c_2/\varepsilon\}. \tag{1.16}$$

Proof. It follows from (B) and the finite increments formula that $d\widetilde{R}_r = \widehat{A}_r \widetilde{R}_r\, dr$, where $\widehat{A}_r$ is a matrix with i–th column of the form

$$\frac{\partial F}{\partial x_1}(x_\infty + \vartheta_r^i \widetilde{R}_r), \quad 0 \leqslant \vartheta_r^i \leqslant 1.$$

Using (B) and continuity of F', we can conclude that $\widehat{A}$ satisfies the condition H_2. Hence, for all $r \geqslant r_0$,

$$|\widetilde{R}_r| \leqslant L|\widetilde{R}_{r_0}| \exp\big\{-\varkappa(r - r_0)\big\}.$$

Thus, if $r_0 \in]0, T_0/\varepsilon]$ we have, for all $r \geqslant T_0/\varepsilon$, that

$$|\widetilde{R}_r| \leqslant |\widetilde{R}_{r_0}| \exp\{\varkappa r_0\} \exp\{-\varkappa T_0/\varepsilon\},$$

i.e., (1.16) holds for sufficiently small ε. Since $\widetilde{R}$ is bounded, we can choose such constant c_1 that (1.16) will be satisfied for all $\varepsilon \in]0,1[$.
Now we establish the exponential bound for the deviation $\tilde{y}^1$ of $\tilde{x}$.

LEMMA 1.2. *Put $\widetilde{R}_r^1 := \tilde{x}_r^1 - \tilde{y}_r^1$. Then, for any $m \in \mathbf{N}$, there exist constants γ_1, γ_2 such that, for all $\varepsilon \in]0,1[$,*

$$\mathbf{E}\|\widetilde{R}^1\|^m_{T_0/\varepsilon, T/\varepsilon} \leqslant \gamma_1 \exp\{-\gamma_2/\varepsilon\}. \tag{1.17}$$

Proof. It follows from (1.2) and (1.13) that

$$d\widetilde{R}_r^1 = A_\infty \widetilde{R}_r^1\, dr + V_r \tilde{x}_r^1\, dr + g_r\, d\widetilde{W}_r, \quad \widetilde{R}_0^1 = 0, \tag{1.18}$$

where $A_\infty := F'(x_\infty)$, $V_r := F'(\tilde{x}_r) - F'(x_\infty)$, $g_r := G(\tilde{x}_r) - G(x_\infty)$.
Since the functions F' and G are locally Lipschitz, $\|\tilde{x}\|_\infty < \infty$, there exists a constant K such that

$$|V_r| + |g_r| \leqslant K|\widetilde{R}_r|. \tag{1.19}$$

Fix $S \in]0, T_0[$. It follows from (1.18) that, for $r \geqslant T_0/\varepsilon$, we have a representation $\widetilde{R}_r^1 = \zeta_r^1 + \zeta_r^2 + \zeta_r^3$, where

$$\zeta_r^1 := \exp\big\{(r - S/\varepsilon)A_\infty\big\}\widetilde{R}_S^1,$$

$$\zeta_r^2 := \int_{S/\varepsilon}^{r} \exp\big\{(r-u)A_\infty\big\} V_u \tilde{x}_u^1\, du, \quad \zeta_r^3 := \int_{S/\varepsilon}^{r} \exp\big\{(r-u)A_\infty\big\} g_u\, d\widetilde{W}_u.$$

The condition (B) implies that there are constants L and $\varkappa$ such that, for $r \geqslant u$, the following inequality holds:

$$\left|\exp\{(r-u)A_\infty\}\right| \leqslant L\exp\{-\varkappa(r-u)\}. \tag{1.20}$$

Therefore, we evidently have

$$\mathbf{E}\|\zeta^1\|^{4m}_{T_0/\varepsilon} \leqslant \text{const}\exp\{-4m(T_0-S)\varkappa/\varepsilon\}. \tag{1.21}$$

Further, (1.19) and Lemma 1.1 imply that V can be bounded on $[S,\infty[$ as in the left-hand side of (1.17). It follows easily, from Theorem 3.1, that

$$\mathbf{E}\|\tilde{x}^1\|^{4m}_{T/\varepsilon} \leqslant \text{const}\,\varepsilon^{-1}.$$

Thus,

$$\mathbf{E}\|\zeta^2\|^{4m}_{T_0/\varepsilon,T/\varepsilon} \leqslant \text{const}\,\mathbf{E}\|\tilde{x}^1\|^{4m}_{T/\varepsilon}\exp\{-4mc_2/\varepsilon\} \leqslant$$
$$\leqslant \text{const}\exp\{-4mc_2/\varepsilon\}/\varepsilon. \tag{1.22}$$

It is clear that ζ^3 on $[S/\varepsilon,\infty]$ satisfies the linear SDE:

$$d\zeta^3_r = A_\infty\zeta^3_r\,dr + g_r\,d\widetilde{W}_r, \quad \zeta^3_{S/\varepsilon} = 0.$$

Due to (1.19) and Lemma 1.1, the diffusion coefficient g_r satisfies on $[S/\varepsilon, T/\varepsilon]$ the exponential inequality like (1.17).
Applying again Theorem 3.1, we obtain that

$$\mathbf{E}\|\zeta^3\|^{4m}_{T_0/\varepsilon,T/\varepsilon} \leqslant \text{const}\exp\{-c/\varepsilon\}/\varepsilon \leqslant \gamma_1\exp\{-\gamma_2/\varepsilon\}, \tag{1.23}$$

for suitably chosen constants γ_1 and γ_2.
Thus, the statement of Lemma 2.1 for $m = 4k$ follows from (1.21)–(1.23). The extension for all $m \in \mathbf{N}$ is obvious.
Theorem 1.2 is an evident corollary of Theorem 1.1 and exponentional bounds (1.16), (1.17).

THEOREM 1.3. *Suppose that conditions* (A)–(C) *are satisfied and* F' *satisfies the local Lipschitz condition.*
Then, for any $m \in \mathbf{N}$ *and* $t_1,\ldots,t_m$: $0 < t_1 < \ldots < t_m$, *the distribution of the random vector* $\beta_\varepsilon^{-1}(x^\varepsilon_{t_1} - x_\infty,\ldots,x^\varepsilon_{t_m} - x_\infty)$ *converges weakly as* $\varepsilon \to 0$ *to the distribution of the random vector* $(\xi_1,\ldots,\xi_m)$ *whose components are independent and have* n*-dimensional normal distribution with zero mean and the covariance matrix* $G_\infty\Gamma G_\infty^T$, *where* Γ *is the unique solution of the algebraic equation*

$$A_\infty\Gamma + \Gamma A_\infty^T + G_\infty G_\infty^T = 0. \tag{1.24}$$

Proof. Put, for convenience, $y_r := \tilde{y}^1_r$, where $\tilde{y}^1_r$ is the solution of (1.13). Due to Theorem 1.2, it is sufficient to study the weak convergence of $(y_{t_1/\varepsilon},\ldots,y_{t_m/\varepsilon})$. We have

$$y_r = \int_0^r \exp\left\{(r-s)A_\infty\right\} G_\infty \, d\widetilde{W}_s,$$

and, hence $\mathbf{E}y_r = 0$,

$$\Gamma_r := \mathbf{E}y_r y_r^T = \int_0^r \exp\left\{(r-s)A_\infty\right\} G_\infty G_\infty^T \exp\left\{(r-s)A^T\right\} ds =$$

$$= \int_{-r}^0 \exp\{-sA_\infty\} G_\infty G_\infty^T \exp\{-sA^T\} \, ds.$$

Since the eigenvalues of A_∞ are located in left semi-plane, there exist the limit Γ of Γ_r as $r \to \infty$ and Γ is the unique solution of (1.24) (see (Liptser and Shiryaev, 1977; 1978) for details).
As y is a Gaussian process, the limiting distribution of $y_{t/\varepsilon}$ (as $\varepsilon \to 0$) is the n-dimensional Gaussian distribution with zero mean and covariance matrix Γ. Finally, the independence of the components of the limiting distribution for $(y_{t_1/\varepsilon}, \ldots, y_{t_m/\varepsilon})$ can be easily derived from the expression for the covariance of y,

$$\mathbf{E}y_t y_s^T = \exp\left\{(t-s)A_\infty\right\}\Gamma_s,$$

where $s \leqslant t$.

2. HIGHER ORDER APPROXIMATIONS

In this section we construct asymptotic expansions of the solution of (0.1) with respect to the small parameter $\beta = \beta_\varepsilon$. At first, we will be interested in an expansion of the type

$$x_t^\varepsilon = \tilde{x}_r + \beta \tilde{x}_r^{(1)} + \ldots + \beta^k \tilde{x}_r^{(k)} + \ldots, \tag{2.1}$$

where $r := t/\varepsilon$, $\tilde{x}$ is the solution of (0.3) and $\tilde{x}^{(1)} = \tilde{x}^1$ is the solution of (1.1). In order to explain the structure of the terms of this expansion we present some formal manipulations, following (Freidlin and Wentzell, 1984, Chapter 2).
Let

$$Z(\beta) := \sum_{q \geqslant 0} c_q \beta^q. \tag{2.2}$$

Assuming that F and G in (0.1) are infinitely differentiable, we can expand $F(Z(\beta))$ and $G(Z(\beta))$ in power series of β. The higher order derivatives of the function $f(\beta) := F(Z(\beta))$ are given by

$$D^q f(\beta) = \sum A^q_{i_1,\ldots,i_j} F^{(j)}(Z(\beta))\left(D^{i_1} Z(\beta), \ldots, D^{i_j} Z(\beta)\right),$$

where the summation is taken over all natural numbers $i_1, \ldots, i_j$ such that $i_1 + +i_2 + \ldots + i_j = q$, $j \leqslant q$. Here the derivative $F^{(j)}(Z(\beta))$ is a j–linear form on $\mathbf{R}^n$ with values in $\mathbf{R}^n$. It follows that

$$F(Z(\beta)) = \sum_{q \geqslant 0} F_q(c_0, \ldots, c_q)\beta^q \tag{2.3}$$

with $F_0(c_0) = F(c_0)$, $F_1(c_0, c_1) = F'(c_0)c_1$ and, for $q \geqslant 1$,

$$F_q(c_0, \ldots, c_q) = F'(c_0)c_q + f_q(c_0, \ldots, c_{q-1}), \tag{2.4}$$

$$f_q(c_0, \ldots, c_{q-1}) := \sum B^q_{i_1, \ldots, i_j} F^{(j)}(c_0)(c_{i_1}, \ldots, c_{i_j}), \tag{2.5}$$

where $2 \leqslant j \leqslant q$. Notice that F_q is a polynomial function (of degree not greater then q) with coefficients depending on c_0.
In the one–dimensional case we can use the well–known De Bruno formula (see, e.g., (Kaufmann, 1968, Section 10)) to get a more explicit expression for F_q:

$$F_q(c_0, \ldots, c_q) = \sum \frac{1}{i_1! \ldots i_q!} F^{(j)}(c_0) c_1^{i_1} \ldots c_q^{i_q},$$

where the sum is taken over all j from 1 to q and all non–negative integers $i_1, \ldots, i_q$ such that $i_1 + i_2 + \ldots + i_q = j$, $i_1 + 2i_2 + \ldots + qi_q = q$.
Analogously,

$$G(Z(\beta)) = \sum_{q \geqslant 0} G_q(c_0, \ldots, c_q)\beta^q. \tag{2.6}$$

Let us substitute the decomposition (2.1) into (1.4) and make use of (2.3) and (2.6). We have

$$\sum_{q \geqslant 0} \beta^q d\tilde{x}_r^{(q)} = \sum_{q \geqslant 0} \beta^q F_q(\tilde{x}_r^{(0)}, \ldots, \tilde{x}_r^{(q)})\, dr + \sum_{q \geqslant 0} \beta^{q+1} G_q(\tilde{x}_r^{(0)}, \ldots, \tilde{x}_r^{(q)})\, d\widetilde{W}_r.$$

The coefficients at the same power of β should be equal. This implies that, for $q = 0$, $d\tilde{x}_r^{(0)} = F(\tilde{x}_r^{(0)})dr$, $\tilde{x}_0^{(0)} = \zeta$, i.e., $\tilde{x}^{(0)} = \tilde{x}$.
For $q \geqslant 1$, we get the following SDE:

$$d\tilde{x}_r^{(q)} = F_q(\tilde{x}_r, \tilde{x}_r^{(1)}, \ldots, \tilde{x}_r^{(q)})\, dr + G_{q-1}(\tilde{x}_r, \tilde{x}_r^{(1)}, \ldots, \tilde{x}_r^{(q-1)})\, d\widetilde{W}_r, \tag{2.7}$$

with the initial condition $\tilde{x}_0^{(q)} = 0$. In particular, for $q = 1$, we have

$$d\tilde{x}_r^{(1)} = F_1(\tilde{x}_r, \tilde{x}_r^{(1)})dr + G(\tilde{x}_r)\, d\widetilde{W}_r, \quad \tilde{x}_0^{(1)} = 0,$$

and this SDE coincides with (1.2), since $F_1(\tilde{x}_r, \tilde{x}_r^{(1)}) = F'(\tilde{x}_r)\tilde{x}_r^{(1)}$.
The structure of functions F_q, G_q (see (2.4)) enables us to solve the system (2.7) in a recurrent way. When the solutions $\tilde{x}^{(0)}, \tilde{x}^{(1)}, \ldots, \tilde{x}^{(q-1)}$ of the first q equations are known, then the $\tilde{x}^{(q)}$ can be found from the linear SDE

$$d\tilde{x}_r^{(q)} = \big(F'(\tilde{x}_r)\tilde{x}_r^{(q)} + \tilde{f}_q\big)dr + \widetilde{G}_{q-1}\, d\widetilde{W}_r, \quad \tilde{x}_0^{(q)} = 0, \tag{2.8}$$

where due to (B), $F'(\tilde{x}_r)$ tends to the matrix A_∞ which has all its eigenvalues in the left half-plane. The coefficients

$$\tilde{f}_q := f_q\big(\tilde{x}_r, \ldots, \tilde{x}_r^{(q-1)}\big), \quad \widetilde{G}_{q-1} := G_{q-1}\big(\tilde{x}_r, \ldots, \tilde{x}_r^{(q-1)}\big) \tag{2.9}$$

are polynomials in $\tilde{x}_r^{(1)}, \ldots, \tilde{x}_r^{(q-1)}$, whose degrees are q and $q-1$.
The above structure of the system allows to extend the approach of Section 1 to obtain results on higher order approximations.
Now we consider a generalization of Theorem 1.1.

THEOREM 2.1. *Assume that* (A)–(C) *are satisfied and, moreover, that F and G have k and $k-1$ continuous derivatives, respectively, $F^{(k)}$ and $G^{(k-1)}$ are locally Lipschitz.*
Then

$$x_t^\varepsilon = \tilde{x}_r + \beta\tilde{x}_r^{(1)} + \ldots + \beta^k\tilde{x}_r^{(k)} + \beta^k\Delta_t^{(k)}, \tag{2.10}$$

where $r := t/\varepsilon$, $\tilde{x}$ is the solution of (0.3), *$\tilde{x}^{(q)}$ is the solution of* (2.7), $1 \leqslant q \leqslant k$, *with coefficients given by* (2.3), (2.6), *and where the remainder term satisfies*

$$\mathbf{P}\text{-}\lim_{\varepsilon\to 0} \|\Delta^{(k)}\|_T = 0. \tag{2.11}$$

Proof. The reasoning follows the same line as in the proof of Theorem 1.1. However, we need some auxiliary results.

LEMMA 2.1. *Under assumptions of Theorem 2.1, for any $m \in \mathbf{N}$ and $1 \leqslant q \leqslant k$, we have*

$$\sup_{r\geqslant 0} \mathbf{E}\big|\tilde{x}_r^{(q)}\big|^{2m} < \infty. \tag{2.12}$$

Proof. We use induction in q. For $q = 1$ the process $\tilde{x}^{(1)}$ is the solution of an equation of the form (3.32) with $h = 0$ and a deterministic function g (see Section 3). Hence (2.12) holds by virtue of (B) and Lemma 3.4. Suppose that (2.12) holds for all $q \leqslant M$. Since, in general, $\tilde{x}^{(M+1)}$ satisfies an equation of the form (3.32) with polynomial coefficients h and g of the variables $\tilde{x}^{(1)}, \ldots, \tilde{x}^{(M)}$, when, by the induction hypothesis, the conditions of Lemma 3.4 are fulfilled. Thus, (2.12) holds for $q = M + 1$.

LEMMA 2.2. *Under the assumptions of Theorem 2.1, for any $m \in \mathbf{N}$ and $1 \leqslant q \leqslant k$, there are positive constants $C_0(q,m)$, $C_1(q,m)$ such that*

$$\mathbf{E}\big\|\tilde{x}^{(q)}\big\|_T^{2m} \leqslant C_0(q,m) + C_1(q,m)T. \tag{2.13}$$

Proof. Recall that $\tilde{x}^{(q)}$ is the solution of the SDE (2.8). By Lemma 2.1, its coefficients $\tilde{f}_q$ and $\widetilde{G}_{q-1}$ satisfy the condition

$$\sup_{t\geqslant 0}\mathbf{E}\big|\tilde{f}_q(t)\big|^{2m}<\infty,\quad \sup_{t\geqslant 0}\mathbf{E}\big|\tilde{g}_{q-1}(t)\big|^{2m}<\infty\quad (\forall m\in\mathbf{N}).$$

Hence (2.13) follows from Lemma 3.5.

LEMMA 2.3. *Under the assumptions of Theorem 2.1, for any non-negative a, μ and $1\leqslant q\leqslant k$,*

$$\lim_{\varepsilon\to 0}\mathbf{P}\Big(\big\|\tilde{x}^{(q)}\big\|_{T/\varepsilon}\geqslant \mu\varepsilon^{\alpha}\Big)=0. \tag{2.14}$$

Proof. Let m be a natural number such that $m>1/(2\alpha)$. By the Chebyshev inequality and (2.13), we get

$$\mathbf{P}\Big(\big\|\tilde{x}^{(q)}\big\|_{T/\varepsilon}\geqslant \mu\varepsilon^{\alpha}\Big)\leqslant \big(\varepsilon^{2m\alpha}/\mu^{2m}\big)\mathbf{E}\big\|\tilde{x}^{(q)}\big\|_{T/\varepsilon}^{2m}\leqslant$$

$$\leqslant\big(\varepsilon^{2m\alpha}C_0(q,m)+\varepsilon^{2m\alpha-1}C_1(q,m)T\big)/\mu^{2m}\longrightarrow 0$$

as $\varepsilon\to 0$.

Proof of Theorem 2.1. Put

$$\tilde{z}_r:=\tilde{x}_r+\beta\tilde{x}_r^{(1)}+\ldots+\beta^k\tilde{x}_r^{(k)}, \tag{2.15}$$

$$\widetilde{\Delta}_r^{(k)}:=\beta^{-k}\big(\tilde{x}_r^{\varepsilon}-\tilde{z}_r\big). \tag{2.16}$$

Thus, we have to establish that

$$\mathbf{P}\text{-}\lim_{\varepsilon\to 0}\big\|\widetilde{\Delta}^{(k)}\big\|_{T/\varepsilon}=0. \tag{2.17}$$

By finite increments formula, we have

$$d\tilde{x}_r^{\varepsilon}=\big(F(\tilde{x}_r^{\varepsilon})-F(\tilde{z}_r)\big)\,dr+\beta\big(G(\tilde{x}_r^{\varepsilon})-G(\tilde{z}_r)\big)\,d\widetilde{W}_r+$$

$$+F(\tilde{z}_r)\,dr+\beta G(\tilde{z}_r)\,d\widetilde{W}_r=\beta^kA_r^{\varepsilon}\widetilde{\Delta}_r^{(k)}\,dr+\beta^{k+1}B_r^{\varepsilon}\widetilde{\Delta}_r^{(k)}\,d\widetilde{W}_r+$$

$$+F(\tilde{z}_r)\,dr+\beta G(\tilde{z}_r)\,d\widetilde{W}_r,\quad \tilde{x}_0^{\varepsilon}=\zeta, \tag{2.18}$$

where A_r^{ε}, B_r^{ε} are the matrices whose i-th columns are, respectively,

$$\frac{\partial F}{\partial x_1}\Big(\tilde{z}_r+\vartheta_r^i\big(\tilde{x}_r^{\varepsilon}-\tilde{z}_r\big)\Big),\quad \frac{\partial G}{\partial x_1}\Big(\tilde{z}_r+\eta_r^i\big(\tilde{x}_r^{\varepsilon}-\tilde{z}_r\big)\Big),$$

with $0\leqslant\vartheta_r^i\leqslant 1$, $0\leqslant\eta_r^i\leqslant 1$.
From (2.15), (2.18) and (2.7) it follows that

$$d\widetilde{\Delta}_r^{(k)}=A_r\widetilde{\Delta}_r^{(k)}\,dr+\big(A_r^{\varepsilon}-A_r\big)\widetilde{\Delta}_r^{(k)}\,dr+\beta B_r^{\varepsilon}\widetilde{\Delta}_r^{(k)}\,d\widetilde{W}_r+$$

$$+\Phi_r\,dr + \Psi_r\,d\widetilde{W}_r, \quad \widetilde{\Delta}_r^{(k)} = 0, \tag{2.19}$$

where

$$\Phi_r := \beta^{-k}\Big(F(\tilde{z}_r) - \sum_{q=0}^{k} F_q(\tilde{x}_r, \ldots, \tilde{x}_r^{(q)})\beta^q\Big), \tag{2.20}$$

$$\Psi_r := \beta^{-k+1}\Big(G(\tilde{z}_r) - \sum_{q=0}^{k-1} G_q(\tilde{x}_r, \ldots, \tilde{x}_r^{(q)})\beta^q\Big). \tag{2.21}$$

Introduce a stopping time ρ depending on parameters ε, $\mu \in]0,1[$ and $\alpha \in \in]0, 1/(k+2)[$ by the formula $\rho := \tau_{\mu,\varepsilon} \wedge \sigma_{\alpha,\varepsilon}$, where

$$\tau_{\mu,\varepsilon} := \inf\big\{r\colon\ |\widetilde{\Delta}_r^{(k)}| \geqslant \mu\big\} \wedge (T/\varepsilon),$$

$$\sigma_{\alpha,\varepsilon} := \inf\big\{r\colon\ |\tilde{x}_r^{(1)}| + \ldots + |\tilde{x}_r^{(k)}| \geqslant \beta^{-\alpha}\big\} \wedge (T/\varepsilon).$$

Put $N := \|\tilde{x}\|_\infty$. For $r \leqslant \rho$, we have

$$|\tilde{z}_r| \leqslant |\tilde{x}_r| + \beta\Big(|\tilde{x}_r^{(1)}| + \ldots + |\tilde{x}_r^{(k)}|\Big) \leqslant N+1,$$

$$|\tilde{x}_r^\varepsilon| \leqslant |\tilde{z}_r| + \beta|\widetilde{\Delta}_r^{(k)}| \leqslant N+2.$$

The local Lipschitz condition for F' implies that there is a constant K such that, for all $r \leqslant \rho$, we have

$$\Big|(A_r^\varepsilon - A_r)\widetilde{\Delta}_r^{(k)}\Big| \leqslant K\Big(|\tilde{z}_r - \tilde{x}_r| + \beta|\widetilde{\Delta}_r^{(k)}|\Big) \leqslant$$

$$\leqslant K\Big(\beta\Big(|\tilde{x}_r^{(1)}| + \ldots + |\tilde{x}_r^{(k)}|\Big) + \beta\Big) \leqslant 2K\beta^{1-\alpha}.$$

By virtue of Lemma 2.4 given below after the proof of Theorem 2.1, there exist a constant C_N such that, for all $r \leqslant \rho$,

$$|\Phi_r| \leqslant C_N\beta^{1-(k+2)\alpha}, \quad |\Psi_r| \leqslant C_N\beta^{1-(k+1)\alpha}.$$

Applying Theorem 3.2 to the equation (2.19), we get

$$\lim_{\varepsilon\to 0} \mathbf{E}\|\widetilde{\Delta}^{(k)}\|_\rho = 0. \tag{2.22}$$

Further,

$$\mathbf{P}\Big(\|\widetilde{\Delta}^{(k)}\|_{T/\varepsilon} \geqslant \mu\Big) \leqslant \mathbf{P}\Big(\|\widetilde{\Delta}^{(k)}\|_\rho \geqslant \mu\Big) + \mathbf{P}\Big(\sigma_{\alpha,\varepsilon} < T/\varepsilon\Big). \tag{2.23}$$

But, due to Lemma 2.3,

$$\mathbf{P}\Big(\sigma_{\alpha,\varepsilon} < T/\varepsilon\Big) \leqslant \sum_{q=1}^{k} \mathbf{P}\Big(\|\tilde{x}^{(q)}\|_{T/\varepsilon} \geqslant \beta^{-\alpha}/k\Big) \longrightarrow 0 \tag{2.24}$$

as $\varepsilon \to 0$.
The desired relation (2.17) follows from (2.22)–(2.24).

LEMMA 2.4. *Let* $f(\beta) := F(z(\beta))$, F: $\mathbf{R}^n \to \mathbf{R}^n$, $\beta \in [0,1]$,

$$z(\beta) := \sum_{q=0}^{k} c_q \beta^q, \quad f_k(\beta) := \sum_{q=0}^{k} f^{(q)}(0)\beta^q/q!,$$

$|c_0| \leqslant N$, $|c_1| + \ldots + |c_k| \leqslant \beta^{-\alpha}$, *where* $\alpha \leqslant 1/(k+2)$. *Assume that* $F^{(k)}$ *satisfies the Lipschitz condition in the ball* $\{z\colon |z| \leqslant N+1\}$.
Then there exists a constant C *such that*

$$\beta^{-k}|f(\beta) - f_k(\beta)| \leqslant C\beta^{1-(k+2)\alpha}.$$

Proof. Without loss of generality we consider the case $m = 1$. Applying the Taylor formula of order $k-1$ to $f(\beta)$, it is easily seen that we have to estimate $|f^{(k)}(\vartheta\beta) - f^{(k)}(0)|$, where $0 \leqslant \vartheta \leqslant 1$. The formula for the derivative of a composite function reduce the problem to the estimation of expressions of the form

$$F^{(j)}\big(z(\vartheta\beta)\big)\big(D^{i_1}z(\vartheta\beta), \ldots, D^{i_j}z(\beta)\big) - F^{(j)}\big(z(0)\big)\big(D^{i_1}z(0), \ldots, D^{i_j}z(0)\big),$$

where $i_1 + \ldots + i_j = k$, $j \leqslant k$. Notice that $D^i z(\vartheta\beta) = D^i z(0) + R_i(\beta) = i!c_1 + R_i(\beta)$, where $|R_i(\beta)| \leqslant \text{const}\,\beta^{1-\alpha}$. Now, taking into account the Lipschitz condition for $F^{(j)}$, we get that the absolute value of the difference of the derivatives does not exceed const $(c_0 + \ldots + c_k)^{j+1}\beta^{1-\alpha} \leqslant \text{const}\,\beta^{1-(j+2)\alpha}$.
This yields the desired result.
The following result generalizes Theorem 1.2.

THEOREM 2.2. *Assume that conditions* (A)–(C) *are satisfied and, moreover, that* F *and* G *have* k *and* $k-1$ *continuous derivatives, respectively,* $F^{(k)}$ *and* $G^{(k-1)}$ *are locally Lipschitz.*
Then

$$x_t^\varepsilon = x^* + \beta\tilde{y}_r^{(1)} + \ldots + \beta^k\tilde{y}_r^{(k)} + \beta^k\widehat{\Delta}_t^{(k)}, \tag{2.25}$$

where $r := t/\varepsilon$, $\tilde{y}^{(q)}$, $q \leqslant k$, *are solutions of the following* SDE's:

$$d\tilde{y}_r^{(1)} = A_\infty \tilde{y}_r^{(1)}\,dr + G_\infty\,d\widetilde{W}_r, \quad \tilde{y}_0^{(1)} = 0, \tag{2.26}$$

$$d\tilde{y}_r^{(q)} = \big(A_\infty\tilde{y}_r^{(q)} + f_q(x_\infty, \tilde{y}_r^{(1)}, \ldots, \tilde{y}_r^{(q-1)})\big)\,dr +$$

$$+ G_{q-1}\big(x_\infty, \tilde{y}_r^{(1)}, \ldots, \tilde{y}_r^{(q-1)}\big)\,d\widetilde{W}_r, \quad \tilde{y}_0^{(1)} = 0, \tag{2.27}$$

with $A_\infty := F'(x_\infty)$, $G_\infty := G(x_\infty)$, G_{q-1} *and* f_q *are defined by* (2.3)–(2.6), *and where the remainder term satisfies*

$$\mathbf{P}\text{-}\lim_{\varepsilon\to 0} \|\widehat{\Delta}^{(k)}\|_{T_0,T} = 0 \tag{2.28}$$

for each $T_0 \in]0,T]$.

The above theorem is an evident corollary of Theorem 2.1, and the exponential bound for the difference $\tilde{x}^{(q)} - \tilde{y}^{(q)}$ is given in

LEMMA 2.5. *Let* $R_r^{(q)} := \tilde{x}_r^{(q)} - \tilde{y}_r^{(q)}$. *Then, for any* $T_0 \in]0,T]$ *and* $m \in \mathbf{N}$, *there are constants* γ_1, γ_2 *such that, for all* $q \leqslant k$ *and* $\varepsilon \in]0,1[$,

$$\mathbf{E}\|\widetilde{R}^{(q)}\|_{T_0/\varepsilon, T/\varepsilon} \leqslant \gamma_1 \exp\{-\gamma_2/\varepsilon\}. \tag{2.29}$$

Proof. It follows from (2.8), (2.9) and (2.27) that

$$d\widetilde{R}_r^{(q)} = \left(A_\infty \widetilde{R}_r^{(q)} + V_r \tilde{x}_r^{(q)} + h_r^{(q)}\right) dr + g_r^{(q)}\, d\widetilde{W}_r, \quad \widetilde{R}_0^{(q)} = 0, \tag{2.30}$$

where

$$V_r := F'(\tilde{x}_r) - F'(x^*),$$

$$h_r^{(q)} := f_q\left(\tilde{x}_r, \tilde{x}_r^{(1)}, \dots, \tilde{x}_r^{(q-1)}\right) - f_q\left(x^*, \tilde{y}_r^{(1)}, \dots, \tilde{y}_r^{(q-1)}\right),$$

$$g_r^{(q)} := G_{q-1}\left(\tilde{x}_r, \tilde{x}_r^{(1)}, \dots, \tilde{x}_r^{(q-1)}\right) - G_{q-1}\left(x^*, \tilde{y}_r^{(1)}, \dots, \tilde{y}_r^{(q-1)}\right).$$

At first we notice that Lemma 1.3 gives the desired bound for $q = 1$. Further we proceed by induction. Assume that there are constants such that (2.28) holds for all $q \leqslant M - 1 < k$ and $\varepsilon \in [0,1[$. Let us show that the inequality similar to (2.28) holds for $q = M$.

Fix $S \in]0, T_0]$. For $r \geqslant T_0/\varepsilon$ we have $\widetilde{R}_r^{(M)} = \zeta_r^1 + \zeta_r^2 + \zeta_r^3$, where

$$\zeta_r^1 := \exp\left\{(r - S/\varepsilon)A_\infty\right\}\widetilde{R}_S^{(M)},$$

$$\zeta_r^2 := \int_{S/\varepsilon}^{r} \exp\left\{(r-u)A_\infty\right\}\left(V_u \tilde{x}_u^{(M)} + h_u^{(M)}\right) du,$$

$$\zeta_r^3 := \int_{S/\varepsilon}^{r} \exp\left\{(r-u)A_\infty\right\} g_u^{(M)}\, d\widetilde{W}_u.$$

As in the proof of Lemma 1.3, we have

$$\mathbf{E}\|\zeta^1\|_{T_0/\varepsilon}^{4m} \leqslant \text{const}\exp\left\{-4m(T_0 - S)\gamma/\varepsilon\right\}. \tag{2.31}$$

The induction assumption and the structure of f_M and G_{M-1} imply that, for any $p \in \mathbf{N}$,

$$\mathbf{E}\|h^{(M)}\|^p_{S/\varepsilon,T/\varepsilon} \leqslant \gamma_1 \exp\{-\gamma_2/\varepsilon\}, \tag{2.32}$$

$$\mathbf{E}\|g^{(M)}\|^p_{S/\varepsilon,T/\varepsilon} \leqslant \gamma_1 \exp\{-\gamma_2/\varepsilon\}. \tag{2.33}$$

The exponential bounds for V (see (1.17) and (1.19)) and $h^{(M)}$, the inequality $|\exp\{(r-u)A_\infty\}| \leqslant L\exp\{-\varkappa(r-u)\}$ and Lemma 2.2 imply the exponential bound for $\mathbf{E}\|\zeta^2\|^{4m}_{T_0/\varepsilon,T/\varepsilon}$.
As ζ^3 satisfies the linear SDE

$$d\tilde{\zeta}^3_r = A_\infty \tilde{\zeta}^3_r\,dr + g^{(M)}_r\,d\widetilde{W}_r, \quad \tilde{\zeta}^3_{S/\varepsilon} = 0,$$

for $r \geqslant S/\varepsilon$ with the diffusion coefficient admitting exponential bound (2.23), inequality (3.41) yields the exponential bound for $\mathbf{E}\|\zeta^3\|^{4m}_{T_0/\varepsilon,T/\varepsilon}$.
The above reasoning gives the desired exponential bound (for $m = 4k$ and, hence, for all $m \in \mathbf{N}$).

3. MOMENT BOUNDS FOR LINEAR SDE

This section contains bounds for the even moments of the solutions of linear SDE's. We assume that the coefficient $A = (A_t)$ in the drift term is the random process taking values in the set of $n \times n$-matrices.
The main hypothesis about A will be the following.

H$_1$) For any t, all eigenvalues of $(A_t + A_t^T)/2$ do not exceed some number $-\gamma < 0$.

H$_2$) There exist positive constant L and $\varkappa$ such that, for all $s, t \in \mathbf{R}_+$, $s \leqslant t$,

$$|\Phi(t,s)| \leqslant L\exp\{-\varkappa(t-s)\}, \tag{3.1}$$

where $\Phi(t,s)$ is the fundamental matrix, i.e., the solution of the equation

$$\frac{\partial\Phi(t,s)}{\partial t} = A_t\Phi(t,s), \quad \Phi(s,s) = I, \tag{3.2}$$

I is the identity matrix of order n.

Let us note that condition H_2 and more strong condition H_1 are well known in the theory of Lyapunov exponents (see, e.g., (Bylov et al., 1974)).
These conditions imply that the Lyapunov exponents are strictly negative and hence the solutions of the corresponding homogeneous equation tends to zero with exponential rate.
If A is a constant matrix and the real parts of its eigenvalues do not exceed $-\gamma < 0$, then the condition H_2 is satisfied (for any $\varkappa < \gamma$). However, for A depending on t the homogeneous equation may have an exponentially increasing solution even for the case when all eigenvalues of A_t are negative and do not depend on t.
It is important for us, that if $A = (A_t)$ is deterministic and $A_t \to A_\infty$ as $t \to \infty$, where all real parts of the eigenvalues of A_∞ do not exceed $-\gamma < 0$, then the condition H_2 is satisfied for any $\varkappa < \gamma$ (see (Bylov et al., 1974, Corollary 10.2.2)).

LEMMA 3.1. *Let y be the solution of the linear SDE*

$$dy_t = A_t y_t\, dt + g_t\, dW_t, \quad y_0 = 0, \tag{3.3}$$

where A is a bounded predictable process satisfying H_1, g is a bounded predictable process with values in $\mathbf{R}^n$, $\|g\| \leqslant K$, W is a one-dimensional Wiener process. Then, for an arbitrary $m \in \mathbf{N}$,

$$\mathbf{E}|y_t|^{2m} \leqslant k_m, \tag{3.4}$$

for any $t \in \mathbf{R}_+$,

$$\mathbf{E}|y_\tau|^{2m} \leqslant c_m T, \tag{3.5}$$

for any stopping time τ with values in the interval $[0,T]$, $T \in \mathbf{R}_+$,

$$\mathbf{E}\|y\|_T^{4m} \leqslant (2c_m + c_{4m})T, \tag{3.6}$$

where

$$k_m := (2m-1)!!\, K^{2m}/(2\gamma)^m, \tag{3.7}$$

$$c_m := m(2m-1)!!\, K^{2m}/(2\gamma)^{m-1}. \tag{3.8}$$

Proof. By the Itô formula

$$\begin{aligned} |y_t|^{2m} = 2m \int_0^t |y_s|^{2(m-1)} y_s^T A_s y_s\, ds + 2m \int_0^t |y_s|^{2(m-1)} y_s^T g_s\, dW_s + \\ + 2m(m-1) \int_0^t |y_s|^{2(m-2)} \left(y_s^T g_s\right)^2 ds + m \int_0^t |y_s|^{2(m-1)} |g_s|^2 ds. \end{aligned} \tag{3.9}$$

Put $S_t(m) := \mathbf{E}|y_t|^{2m}$ and take the expectation of both sides of the above equality. It is well known (see, e.g., (Liptser and Shiryaev, 1989, Chapter 4, §4)) that the norms of solutions of linear SDE's with bounded A and g have moments of any order, and hence $\mathbf{E}\|y\|_T^{2m} < \infty$.
Thus, the stochastic integral in (3.9) is a martingale with zero mean. Taking into account that $x^T A x \leqslant -\gamma |x|^2$, by virtue of H_1, we get the differential inequality

$$S_T'(m) \leqslant -2m\gamma S_t(m) + m(2m-1)K^2 S_t(m-1), \quad S_0(m) = 0.$$

Applying the Gronwall–Bellman lemma we obtain recurrent relations

$$S_t(m) \leqslant m(2m-1)K^2 \int_0^t S_u(m-1) \exp\left\{ -2m\gamma(t-u) \right\} du. \tag{3.10}$$

In particular, for $m=1$, we have $S_t(1) \leqslant K^2/(2\gamma)$. By induction, we easily derive from (3.10), that

$$S_t(m) \leqslant k_m = (2m-1)!!\, K^{2m}/(2\gamma)^m. \tag{3.11}$$

Therefore, (3.4) holds.
Notice now that the first integral in the right-hand side of (3.9) is negative, while the stochastic integral is a square integrable martingale. Hence it follows from (3.9), (3.11) that, for any stopping time τ with values in $[0,T]$,

$$\mathbf{E}|y_\tau|^{2m} \leqslant m(2m-1)K^2 \int_0^T S_u(m-1)\,du. \tag{3.12}$$

A substitution of (3.11) into (3.12) yields (3.5).
It remains to establish the bound (3.6). For this purpose, introduce the stopping time $\tau := \tau_a := \inf\{t\colon\ |y_t|^{4m} \geqslant a\} \wedge T$. The Chebyshev inequality and (3.5) imply that

$$\mathbf{P}\big(\|y\|_T^{4m} \geqslant a\big) = \mathbf{P}\big(|y_\tau|^{4m} \geqslant a\big) \leqslant$$

$$\leqslant \min\big\{\mathbf{E}|y_\tau|^{8m}/a^2,\ \mathbf{E}|y_\tau|^{2m}/\sqrt{a}\big\} \leqslant \min\big\{c_{4m}/a^2,\ c_m/\sqrt{a}\big\}T.$$

Hence,

$$\mathbf{E}\|y\|_T^{4m} = \int_0^\infty \mathbf{P}\Big(\|y\|_T^{4m} \geqslant a\Big)\,da \leqslant c_m T \int_0^1 da/\sqrt{a} + c_{4m}T\int_1^\infty da/a^2,$$

implying (3.6).

LEMMA 3.2. *Let y be the solution of the following linear differential equation*

$$dy_t = (A_t y_t + \alpha_t)\,dt, \quad y_0 = 0, \tag{3.13}$$

where A is a bounded measurable process satisfying H_1, α is a measurable process,

$$C := C_m := \sup_{t\leqslant T} \mathbf{E}|\alpha_t|^{2m} < \infty.$$

Then

$$\mathbf{E}\|y\|_T^{2m} \leqslant 2m\gamma^{1-2m}CT. \tag{3.14}$$

Proof. Put

$$S_t := S_t(m) := \mathbf{E}|y_t|^{2m}, \quad \overline{S}_T := \sup_{t\leqslant T} S_t, \quad p := 2m, \quad 1/q := 1 - 1/p.$$

The Cauchy formula and boundedness of A imply that $\overline{S}_T < \infty$. Since $y^TAy \leqslant -\gamma|y|^2$ and $y^T\alpha \leqslant |y||\alpha|$, then, from the formula

$$|y_t|^{2m} = 2m \int_0^t |y_s|^{2(m-1)} y_s^T A_s y_s \, ds + 2m \int_0^t |y_s|^{2(m-1)} y_s^T \alpha_s \, ds \tag{3.15}$$

and the Hölder inequality, we get

$$dS_t/dt \leqslant -p\gamma S_t + pC^{1/p} S_t^{1/q}.$$

The Gronwall–Bellman lemma yields

$$S_t \leqslant pC^{1/p} \int_0^t S_u^{1/q} \exp\{-p\gamma(t-u)\} du.$$

Hence, $\overline{S}_T \leqslant (1/\gamma) C^{1/p} \overline{S}_T^{1/q}$ implying $\overline{S}_T \leqslant C/\gamma^p$. Using this bound and taking into account that the first term in the right-hand side of (3.15) is negative, we obtain, by the Hölder inequality, that

$$\mathbf{E}\|y\|_T^{2m} \leqslant pC^{1/p} \overline{S}_T^{1/q} T \leqslant pC/\gamma^{p/q} T = 2mC\gamma^{1-2m} T.$$

This completes the proof of Lemma 3.2.

LEMMA 3.3. *Let z be the solution of the following linear differential equation*

$$dz_t = (A_t z_t + \alpha_t)\, dt, \quad z_0 = 0, \tag{3.16}$$

where A is a bounded measurable process satisfying H_2, $\|A\|_T \leqslant K_A$, α is measurable process,

$$C := C_m := \sup_{t \leqslant T} \mathbf{E}|\alpha_t|^{2m} < \infty.$$

Then

$$\mathbf{E}\|z\|_T^{2m} \leqslant 2K(L, m, K/\varkappa)\varkappa^{1-2m} CT, \tag{3.17}$$

where

$$K(L, m, K_A/\varkappa) = m + (2(1 + K_A/\varkappa)L)^{2m}. \tag{3.18}$$

Proof. The Cauchy formula, (3.1) and the Hölder inequality with $p := 2m$, $1/q := 1 - 1/p$, imply that

$$\mathbf{E}|z_t|^p \leqslant \mathbf{E}\Big(\int_0^t |\Phi(t,s)||\alpha_s|\, ds\Big)^p \leqslant L^p \mathbf{E}\Big(\int_0^t \exp\{-\varkappa(t-s)\}|\alpha_s|\, ds\Big)^p \leqslant$$

$$\leqslant L^p \mathbf{E}\Big(\int_0^t \exp\{-\varkappa p(t-s)/2\}|\alpha_s|^p\,ds\Big)\Big(\int_0^t \exp\{-\varkappa q(t-s)/2\}\,ds\Big)^{p/q} \leqslant$$

$$\leqslant L^p C\varkappa^{-p}(2/p)(2/q)^{p/q} \leqslant (2L)^p\varkappa^{-p}p^{-1}C. \tag{3.19}$$

Consider the equation $dy_t = \big(-\varkappa y_t + \alpha_t\big)\,dt,\ y_0 = 0.$
According to Lemma 3.2, its solution admits the bound

$$\mathbf{E}\|y\|^{2m} \leqslant 2m\varkappa^{1-2m}CT. \tag{3.20}$$

Put $X := y - z$. Then $dX_t = \big(-\varkappa X_t + \nu_t\big)\,dt,\ X_0 = 0$, where $\nu_t := (-\varkappa I - A_t)z_t$. Applying (3.19), we easily get

$$\sup_{t\leqslant T}\mathbf{E}|\nu_t|^{2m} \leqslant \big(2L(\varkappa + K_A)\big)^{2m}(2m)^{-1}\varkappa^{-2m}C.$$

According to Lemma 3.2,

$$\mathbf{E}\|X\|_T^{2m} \leqslant \big(2(1+K_A/\varkappa)L\big)^{2m}\varkappa^{1-2m}CT. \tag{3.21}$$

Now the statement of Lemma 3.3 follows from (3.20) and (3.21).
Lemma 3.2 can be used to obtain moment bounds for SDE's whose drift coefficient includes the matrix A which satisfies only condition H_2. However, they are more complicated than those provided by Lemma 3.1.

THEOREM 3.1. *Let y be the solution of the linear SDE (3.3), where A is a bounded function (with values in the set of $n\times n$–matrices) satisfying H_2, $\|A\|_T \leqslant K_A$, g is a bounded predictable process with values in $\mathbf{R}^n$, $\|g\| \leqslant K$, W is a one–dimensional Wiener process.*
Then, for an arbitrary $m \in \mathbf{N}$,

$$\mathbf{E}|y_t|^{2m} \leqslant C_m L^{2m}K^{2m}\varkappa^{-m}, \tag{3.22}$$

for any $t \in \mathbf{R}_+$,

$$\mathbf{E}|y_\tau|^{2m} \leqslant B_m K^{2m}\varkappa^{1-m}T, \tag{3.23}$$

for any stopping time τ with values in $[0,T]$, $T \in \mathbf{R}_+$,

$$\mathbf{E}\|y\|_T^{4m} \leqslant 2(B_{4m}+B_m)\big(K^{8m}\varkappa^{1-4m} + K^{2m}\varkappa^{1-m}\big)T, \tag{3.24}$$

where

$$B_m := B_m(L, K_A/\varkappa) := C'_m\big(1 + L^{2m}(1+K_A/\varkappa)^{2m}\big), \tag{3.25}$$

C_m, C'_m depending only on m.

Proof. The Burkholder inequality (see, e.g., (Liptser and Shiryaev, 1989)), implies that

$$\mathbf{E}\left|\int_0^t f_s\,dW_s\right|^{2m} \leqslant C_m\mathbf{E}\left(\int_0^t |f_s|^2\,ds\right)^m. \tag{3.26}$$

Therefore, using the Cauchy formula $y_t = \int_0^t \Phi(t,s)g_s\,dW_s$ and H_2, we obtain

$$\mathbf{E}|y_t|^{2m} \leqslant C_m\mathbf{E}\left(\int_0^t \left|\Phi(t,s)\right|^2 |g_s|^2\,ds\right)^m \leqslant C_m L^{2m}K^{2m}\varkappa^{-m}. \tag{3.27}$$

Thus, (3.22) holds.
To prove (3.23) and (3.24), we consider the equation

$$dz_t = -\varkappa z_t\,dt + g_t\,dW_t, \quad z_0 = 0,$$

whose solution z, according to Lemma 3.1, admits the bound

$$\mathbf{E}|z_\tau|^{2m} \leqslant m(2m-1)!!\,2^{1-m}K^{2m}\varkappa^{1-m}T, \tag{3.28}$$

for any stopping time $\tau \leqslant T$.
Put $Z := z - y$. Then $dZ_t = \left(-\varkappa Z_t + \alpha_t\right)dt$, $Z_0 = 0$, where $\alpha_t := (-\varkappa I - A_t)y_t$. Combining Lemma 3.2 and (3.27), we get

$$\mathbf{E}\left\|Z\right\|_T^{2m} \leqslant 2mC_m(LK)^{2m}\left(1 + K_A/\varkappa\right)^{2m}\varkappa^{1-m}. \tag{3.29}$$

The desired inequality (3.23) follows from (3.28) and (3.29). The bound (3.24) can be deduced from (3.23) by the same arguments as in Lemma 3.1.

THEOREM 3.2. *Let Y^ε be the solution of the linear* SDE's

$$dY_t^\varepsilon = \left(A_t Y_t^\varepsilon + \Phi_t^\varepsilon\right)dt + \Psi_t^\varepsilon\,dW_t, \quad Y_0^\varepsilon = 0, \tag{3.30}$$

$\rho = \rho^\varepsilon$ be a stopping time with values in $[0, T/\varepsilon]$.
Assume the following:

1) *the bounded matrix-valued function A satisfies H_2;*
2) *the processes Φ^ε, Ψ^ε are predictable and*

$$\|\Phi^\varepsilon\|_\rho \leqslant C\varepsilon^\nu, \quad \|\Phi^\varepsilon\|_\rho \leqslant C\varepsilon^\nu, \tag{3.31}$$

for any ε and some positive constants C and ν.
Then, for any $m \in \mathbf{N}$,

$$\lim_{\varepsilon\to 0}\mathbf{E}\left\|Y^\varepsilon\right\|_\rho^{2m} = 0.$$

Proof. By the Cauchy formula we easily find that $\|Y^\varepsilon\|_\rho = \|y^\varepsilon\|_{T/\varepsilon} + \|z^\varepsilon\|_{T/\varepsilon}$, where

$$dy_t^\varepsilon = A_t y_t^\varepsilon\,dt + \psi_t^\varepsilon I\{t \leqslant \rho\}\,dW_t, \quad y_0^\varepsilon = 0,$$

$$dz_t^\varepsilon = \big(A_t z_t^\varepsilon + \phi_t^\varepsilon I\{t \leqslant \rho\}\big)\,dt, \quad z_0^\varepsilon = 0.$$

But the inequalities of Theorem 3.1 and Lemma 3.3 imply that

$$\lim_{\varepsilon\to 0} \mathbf{E}\|y^\varepsilon\|_{T/\varepsilon}^{2m} = 0, \quad \lim_{\varepsilon\to 0} \mathbf{E}\|z^\varepsilon\|_{T/\varepsilon}^{2m} = 0,$$

for sufficiently large m. Thus, (3.31) holds.
Now we consider the linear SDE's with nonhomogeneous drift and unbounded coefficients.

LEMMA 3.4. *Let y be the solution of the linear* SDE

$$dy_t = \big(A_t y_t + h_t\big)\,dt + g_t\,dW_t, \quad y_0 = \xi, \tag{3.32}$$

where A is a bounded function (with values in the set of $n\times n$–matrices) satisfying H_2, g and h are predictable processes with values in $\mathbf{R}^n$, W is a one-dimensional Wiener process.
Then there is a constant K such that, for any T,

$$\sup_{t\leqslant T} \mathbf{E}|y_t|^{2m} \leqslant K\Big(\mathbf{E}|\xi|^{2m} + \sup_{t\leqslant T}\mathbf{E}|h_t|^{2m} + \sup_{t\leqslant T}\mathbf{E}|g_t|^{2m}\Big). \tag{3.33}$$

In particular, if

$$\mathbf{E}|\xi|^{2m} < \infty, \quad \sup_t \mathbf{E}|h_t|^{2m} < \infty, \quad \sup_t \mathbf{E}|g_t|^{2m} < \infty, \tag{3.34}$$

then $\sup_{t\leqslant T}\mathbf{E}|y_t|^{2m} < \infty$.

Proof. Without loss of generality we can suppose that the right-hand side of (3.33) is finite. Let $\Phi(t,s)$ be the fundamental matrix satisfying, by assumption, (3.1). We introduce the processes

$$y_t^{(0)} := \Phi(t,0)\xi, \quad y_t^{(1)} := \int_0^t \Phi(t,s)h_s\,ds, \quad y_t^{(2)} := \int_0^t \Phi(t,s)g_s\,dW_s. \tag{3.35}$$

It follows from (3.14) that

$$\sup_t \mathbf{E}\big|y_t^{(0)}\big|^{2m} \leqslant \mathbf{E}\big\|y^{(0)}\big\|_T^{2m} \leqslant L^{2m}\mathbf{E}|\xi|^{2m}. \tag{3.36}$$

Put $p := 2m$, $1/q := 1 - 1/p$. By the Hölder inequality, we get

$$|y_t^{(1)}|^p \leqslant \Big(\int_0^t |\Phi(t,s)|^q \exp\big\{-\varkappa q(t-s)/2\big\}\,ds\Big)^{p/q}\int_0^t \exp\big\{-\varkappa p(t-s)/2\big\}|h_s|^p ds.$$

Hence, taking into account (3.1), we have

$$\sup_{t \leqslant T} \mathbf{E}|y_t^{(1)}|^{2m} \leqslant \mathbf{E}\|y^{(1)}\|_T^{2m} \leqslant \text{const} \sup_{t \leqslant T} \mathbf{E}|h_t|^{2m}. \tag{3.37}$$

By virtue of (3.26),

$$\mathbf{E}|y_t^{(2)}|^{2m} \leqslant C_m \mathbf{E}\Big(\int_0^t |\Phi(t,s)|^2 |g_s|^2 ds \Big)^m,$$

where C_m is some constant. The Hölder inequality together with (3.1) yields

$$\sup_{s \leqslant T} \mathbf{E}|y_s^{(2)}|^{2m} \leqslant \text{const} \sup_{s \leqslant T} \mathbf{E}|g_s|^{2m} < \infty. \tag{3.38}$$

Since $y = y^{(0)} + y^{(1)} + y^{(2)}$, the statement of Lemma 3.4 follows from (3.36)–(3.38).

LEMMA 3.5. *Let y be the solution of the linear* SDE

$$dy_t = (A_t y_t + h_t)\, dt + g_t\, dW_t, \quad y_0 = \xi, \tag{3.39}$$

where A is a bounded function (with values in the set of $n \times n$–matrices) satisfying H_2, $\mathbf{E}|\xi|^{4m} < \infty$, g and h are predictable processes with values in $\mathbf{R}^n$ such that, for any $k \in \mathbf{N}$,

$$\mathbf{E}\|h\|_T^{2k} < \infty, \quad \mathbf{E}\|g\|_T^{2k} < \infty. \tag{3.40}$$

Then there exist constants C_0, C_1 (depending on m, L and $\varkappa$) such that

$$\mathbf{E}\|y\|_T^{4m} \leqslant C_0 \mathbf{E}|\xi|^{4m} + C_1 \big(K_T(m,h) + L_T(m,g)\big)T, \tag{3.41}$$

where

$$K_T(m,h) := \sup_{s \leqslant T} \mathbf{E}|h_s|^{4m},$$

$$L_T(m,g) := \Big(K_T^{1/2}(m-1,g) + K_T^{1/2}(4m-1,g)\Big) K_T^{1/2}(1,g) + K_T(m,g).$$

Proof. As in the proof of Lemma 3.4, we use the representation $y = y^{(0)} + y^{(1)} + y^{(2)}$, where $y_t^{(0)} := \Phi(t,0)\xi$, and the processes $y^{(1)}$ and $y^{(2)}$ are defined by (3.35). The exponential bound (3.1) implies that

$$\mathbf{E}\|y^{(0)}\|_T^{4m} \leqslant L^{4m} \mathbf{E}|\xi|^{4m} < \infty. \tag{3.42}$$

The process $y_t^{(1)}$ is the solution of the equation $dy_t^{(1)} = (A_t y_t^{(1)} + h_t)\, dt$, $y_0^{(1)} = 0$. Thus, by Lemma 3.3, we have

$$\mathbf{E}\|y^{(1)}\|_T^{4m} \leqslant C^{(1)} \sup_{s \leqslant T} \mathbf{E}|h_s|^{4m} T < \infty. \tag{3.43}$$

Put $S_t(m) := \mathbf{E}|y_t^{(2)}|^{2m}$. In accordance with (3.38),

$$S_t(m) \leqslant \text{const} \sup_{s\leqslant T} \mathbf{E}|g_s|^{2m}, \tag{3.44}$$

where the constant depends only on m, L and $\varkappa$.
Suppose additionally that, for any t, the matrix A_t is symmetric and has negative eigenvalues bounded from zero. In such a case we can follow the method of the proof of Lemma 3.1.
In the Itô formula (3.9) written for $|y_t^{(2)}|^{2m}$, the first integral in the right–hand side is negative, while the stochastic integral is a locally square integrable martingale. Hence, for any stopping time $\tau \in [0,T]$, we have

$$\mathbf{E}\left|y_\tau^{(2)}\right|^{2m} \leqslant (2m-1)m \int_0^T \mathbf{E}\left|y_s\right|^{2(m-1)}\left|g_s\right|^2 ds.$$

Applying the Schwarz inequality and taking into account (3.44), we get

$$\mathbf{E}\left|y_t^{(2)}\right|^{2m} \leqslant \text{const} \left(\sup_{s\leqslant T} \mathbf{E}|g_s|^{4(m-1)} \sup_{s\leqslant T} \mathbf{E}|g_s|^4 \right)^{1/2} T. \tag{3.45}$$

The method used to derive (3.6) from (3.5), gives the bound

$$\mathbf{E}\left\|y^{(2)}\right\|_T^{4m} \leqslant \text{const}\, l_T(m,g)T, \tag{3.46}$$

where

$$l_T(m,g) := \left(K_T^{1/2}(m-1,g) + K_T^{1/2}(4m-1,g)\right)K_T^{1/2}(1,g).$$

In order to cover the general case, we consider the process

$$dz_t = -\varkappa z_t\, dt + g_t\, dW_t, \quad z_0 = 0,$$

admitting the bound like (3.46), namely,

$$\mathbf{E}\left\|z\right\|_T^{4m} \leqslant \text{const}\, l_T(m,g)T. \tag{3.47}$$

Then, by Lemma 3.3 and (3.44), we get

$$\mathbf{E}\left\|z - y^{(2)}\right\|_T^{4m} \leqslant \text{const} \sup_{s\leqslant T} \mathbf{E}|g_s|^{4m}T. \tag{3.48}$$

It follows from (3.47) and (3.48), that

$$\mathbf{E}\left\|y^{(2)}\right\|_T^{4m} \leqslant \text{const}\, L_T(m,g)T. \tag{3.49}$$

The desired inequality (3.41) is a consequence of (3.42), (3.43) and (3.49).

Remark. This research was partially supported by the Ministry of Sciences and Higher Education in Bulgaria under Grant No. 1035.

REFERENCES

Bylov, B.F., Vinograd, R.E., Grobman, D.M. and Nemytskii, V.V. (1966). *Theory of the Lyapunov Exponents.* Nauka, Moscow (in Russian).

Freidlin, M.I. and Wentzell, A.D. (1984). *Random Perturbations of Dynamical Systems.* Springer, Berlin etc.

Kabanov, Yu.M. and Pergamenshchikov, S.M. (1990 a). On singularly perturbed stochastic differential equations and partial differential equations. *Soviet. Math. Dokl.* **311**, 1039–1042 (In Russian).

Kabanov, Yu.M. and Pergamenshchikov, S.M. (1990 b). Singular perturbations of stochastic differential equations: the Tikhonov theorem. *Soviet. Math. Sbornik.* **181**, 1170–1182 (In Russian).

Kaufmann, A. (1968). *Introduction à la Combinatorique en vue des Applications.* Dunod, Paris.

Liptser, R.Sh. and Shiryaev, A.N. (1977, 1978). *Statistics of Random Processes.* **1, 2**, Springer, Berlin etc.

Liptser, R.Sh. and Shiryaev, A.N. (1989). *Theory of Martingales.* Kluwer, Dordrecht.

Vasilyeva, A.B. and Butuzov, V.F. (1973). *Asymptotic Expansions of Solutions of Singularly Perturbed Equations.* Nauka, Moscow (in Russian).

New Trends in Probab. and Statist., pp. 436 - 444
V. Sazonov and T. Shervashidze (Eds.)
1991 VSP/Mokslas

ON UNIFORMLY OPTIMAL CODING OF GAUSSIAN MESSAGES IN WHITE GAUSSIAN CHANNELS WITH FEEDBACK

P.K. KATYSHEV
Central Economics and Mathematical Institute, the USSR Academy of Sciences, Krasikova str. 32, 117418 Moscow, USSR

ABSTRACT

The problem of transmission of Gaussian messages through a white Gaussian channel with feedback is considered. It is proved that in the general setting there is no uniformly optimal coding scheme.

1. INTRODUCTION

The white Gaussian channel with feedback is described by the following stochastic differential equation

$$d\xi_t = A(t,\theta,\xi)\,dt + dw_t, \quad \xi_0 = 0, \quad t \in [0,T], \tag{1}$$

where θ is the message to be transmitted, ξ_t is the channel output at a moment t, A is a coding function, w is a standard Wiener process (white noise) independent on θ. Throughout the paper θ will be a Gaussian vector or a process. The coding A is a nonanticipative process with respect to ξ, i.e., A depends on $\xi_0^t = \{\xi_s\colon 0 \leqslant s \leqslant t\}$ at any moment t (the exact definitions will be given later). This means that the feedback is instantaneous and noiseless. Moreover, throughout the paper we shall assume that the mean power constraint

$$\mathbf{E}A^2(t,\theta,\xi) \leqslant P, \quad t \in [0,T], \quad P > 0, \tag{2}$$

holds. Let $\widehat{\theta}(r,\xi_0^r)$ be the reproduction of message θ based on observations ξ_0^r (decoding) and $\Delta_{A,\hat\theta}(r)$ be the error of reproduction at a moment $r \in [0,T]$ (as it will be clear later it is expedient to "separate" the time t in (1), (2) and the time r of reproduction of the message). Let us denote $\Delta(r) = \inf \Delta_{A,\hat\theta}(r)$, where inf is taken over the whole set of admissible coding and decoding functions (the minimal error of reproduction). The general problem of optimal transmission of a message θ may be formulated as follows: to calculate the value $\Delta(r)$ and to construct optimal or ε-optimal coding and decoding functions (if they exist). Let us make two important remarks: 1) an optimal or even an ε-optimal scheme may *apriori* depend on r. If this is not the case then corresponding optimal coding and decoding functions are called *uniformly optimal*; 2) in the problem formulated above the error $\Delta_{A,\hat{\theta}}(r)$, as a rule, is one of the variants of mean-square deviation, therefore the optimal decoding may be found easily: it is the corresponding conditional expectation. So the main problem is to construct an optimal coding function.

Let us make a short review.

For the case when θ is a Gaussian random variable with variance γ and $\Delta_{A,\hat{\theta}}(r) = \mathbf{E}\big(\theta - \widehat{\theta}(r, \xi_0^r)\big)^2$, in (Ihara, 1973) and in (Liptser and Shiryaev, 1978, Chapter 6) it is proved that $\Delta(r) = \gamma e^{-Pr}$ and an optimal transmission scheme is found. This scheme turned out to be uniformly optimal and, moreover, the optimal coding function A^* is linear in θ: $A^*(t, \theta, \xi) = f_0^*(t, \xi) + \theta f_1^*(t, \xi)$, and the system (θ, ξ_0^T) is Gaussian. In (Katyshev, 1975) these results are generalized for the case when $\theta = (\theta_1, \dots, \theta_n)$ is a Gaussian vector and

$$\Delta_{A,\hat{\theta}}(r) = \sum_{i=1}^{n} \mathbf{E}\big(\theta_i - \widehat{\theta}_i(r, \xi_0^r)\big)^2$$

but with the difference that only an ε-optimal coding is found. It should be mentioned the paper (Ovseevich, 1970) which contains similar results.

The problem of transmission of Gaussian processes through the channel, described by (1) and (2), is considered in subsequent papers (Katyshev, 1976; 1978; Ihara, 1974; 1975; Liptser, 1974). In this setting the form of dependence of the coding A on θ is important. In (Katyshev, 1976) it is assumed that the whole trajectory θ_0^T of the process θ may be used in the channel input at any moment t and the results analogous to (Katyshev, 1975) obtained. In (Ihara, 1974; 1975; Liptser, 1974) it is supposed that $A(t, \theta, \xi) = A(t, \theta_t, \xi)$, i.e., at any moment t only the "current" value θ_t of θ may be used, and $\Delta_{A,\hat{\theta}}(r) = \mathbf{E}\big(\theta_r - \widehat{\theta}(r, \xi_0^r)\big)^2$. Moreover, in (Ihara, 1974; 1975) the process θ may be arbitrary but the coding function A has to be linear in θ and ξ, in (Liptser, 1974) it is vice versa: the coding A may be arbitrary but the process θ satisfies a linear stochastic differential equation and is therefore a Gaussian–Markov process. In these papers an optimal scheme of transmission is constructed and the function $\Delta(r)$ is calculated. Finally, in (Katyshev, 1978) it is assumed that θ is an arbitrary Gaussian Markov process and at any moment t the whole history $\theta_0^t = \{\theta_s\colon\ 0 \leqslant s \leqslant t\}$ of the process θ is known and may be used in the channel input. An expression for $\Delta(r)$ is obtained in terms of the covariance function of θ and an optimal coding is found which indeed at any t depends only on θ_t. It should be emphasized that in all cases mentioned above the optimal (or ε-optimal) schemes of transmission possess the following properties:

1) the optimal coding A^* is linear in θ;

2) the system (θ, ξ_0^T) is Gaussian;

3) the optimal scheme is uniformly optimal.

The question is: will the property of uniform optimality be preserved in a more general situation? The main result of this work is that in the linear Gaussian framework the answer is negative.

2. STATEMENT OF THE PROBLEM AND THE MAIN RESULT

We consider the following scheme of transmission. Let a Gaussian random variable θ_0 be given, which cannot be directly used in coding, but there is a piece-wise constant observable stochastic process, constituting with θ_0 a Gaussian system. The problem is to find the coding minimizing the mean-square error of reproduction of θ_0. It is easy to show that this approach is the basis for solution of the problems of transmissions of Gaussian processes considered in (Ihara, 1974; 1975; Katyshev, 1976; 1978; Liptser, 1974). Now we give the exact formulation.

Let $\mathcal{T} = \{0 = t_0 < t_1 < \ldots\}$ be a (finite or infinite) partition of $[0, \infty)$. Let also a Gaussian system $\theta = (\theta_0, \theta_1, \theta_2, \ldots)$ be given where θ_0 is a random variable with $\mathbf{E}\theta_0 = m$, $\operatorname{var}\theta_0 = \gamma$, and θ_i, $i = 1, 2, \ldots$, is a random vector of dimension m_i. Let $w = \{w_t\colon t \geqslant 0\}$ be a Wiener process independent of θ. Denote $\mathbf{C} = \mathbf{C}[0, T]$ the space of continuous functions on $[0, T]$ and let $\mathcal{C}_t = \sigma(x_s\colon 0 \leqslant s \leqslant t)$. The function A in equation (1) is called *admissible coding* if

1) $A(t, \theta, x) = f_0^{(i)}(t, x) + \theta_i f_1^{(i)}(t, x)$ for $t \in [t_{i-1}, t_i]$, where $x \in \mathbf{C}$, $f_0^{(i)}(t, \cdot)$ is a $\mathcal{C}_t$-measurable functional, $f_1^{(i)}(t, \cdot)$ is a m_i-vector with $\mathcal{C}_t$-measurable components and the second term in the right-hand side of the above formula is a scalar product;
2) equation (1) has a strong solution in the following sense:

$$\mathcal{F}_t^\xi \subseteq \mathcal{F}_{t_{i-1}}^\xi \vee \sigma(\theta_i) \vee \sigma(w_s - w_{t_{i-1}},\ t_{i-1} \leqslant s \leqslant t)$$

if $t \in [t_{i-1}, t_i]$, where, as usually, $\mathcal{F}_t^\xi = \sigma(\xi_s,\ 0 \leqslant s \leqslant t)$;
3) for any $t \in [0, T]$ mean power constraint (2) holds;
4) the system $(\theta, \xi) = (\theta, \{\xi_t\colon 0 \leqslant t \leqslant T\})$ is Gaussian.

Remarks. 1. Conditions 1) and 4) mean that we deal with a linear Gaussian scheme.

2. Usually the existence of a strong solution means that $\mathcal{F}_t^\xi \subseteq \mathcal{F}_t^w \vee \sigma(\theta_1, \ldots, \theta_i)$ for $t \in [t_{i-1}, t_i]$. Condition 2) is more restrictive but it may be verified that if the functional A satisfies integral Lipshitz's condition and has linear growth then this condition holds.

Let us denote by $\mathcal{A}$ the class of all admissible coding functions. (Obviously the set $\mathcal{A}$ depends on the partition $\mathcal{T}$ and the system θ.)

Let $\widehat{\theta}(r, x)$ $(r \in [0, T],\ x \in \mathbf{C})$ be some $\mathcal{C}_r$-measurable functional such that for any $A \in \mathcal{A}$ the inequality $\mathbf{E}\widehat{\theta}^2(r, \xi) < \infty$ holds (for the corresponding process ξ). We shall call such functional an *admissible decoding* and denote the set of all such functionals by θ. For $A \in \mathcal{A}$ and $\theta \in \Theta$ denote

$$\Delta_{A,\hat{\theta}}(r) = \mathbf{E}\big(\theta_0 - \widehat{\theta}(r, \xi)\big)^2, \quad \Delta(r) = \inf_{\mathcal{A},\Theta} \Delta_{A,\hat{\theta}}(r).$$

The value $\Delta(r)$ represents the minimal mean-square error of reproduction of the message θ_0 based on the output $\xi_0^r = \{\xi_s\colon 0 \leqslant s \leqslant r\}$. The coding and decoding functions realizing minimum (if they exist) are called, of course, *optimal.*

Let $A \in \mathcal{A}$; denote $m(t) = \mathbf{E}(\theta_0 | \mathcal{F}_t^\xi)$, $t \in [0,T]$. Since, for any fixed A,

$$\Delta_{A,\hat{\theta}}(r) \geqslant \Delta_A(r) \equiv \mathbf{E}\big(\theta_0 - m(r)\big)^2,$$

the optimal decoding may be found easily: it is the conditional expectation, and $\Delta(r) = \inf_{\mathcal{A}} \Delta_A(r)$.

Definition. The functional $A^* \in \mathcal{A}$ is called an *uniformly optimal* coding, if $\Delta_{A^*}(r) = \Delta(r)$ for any $r \in [0,T]$.
The following theorem represents the main result of this paper.

THEOREM. *There exist a partition $\mathcal{T}$ and system θ such that there is no uniformly optimal coding in the corresponding class $\mathcal{A}$.*

3. AUXILIARY RESULTS

Let ζ be a Gaussian random variable independent of the Wiener process w with $\operatorname{var}\zeta = \delta > 0$. Suppose that the process ξ satisfies the equation

$$d\xi_t = \big[f_0(t,\xi) + \zeta f_1(t,\xi)\big]dt + dw_t, \quad \xi_0 = 0, \quad t \in [0,T], \tag{3}$$

where $f_0(t,x)$, $f_1(t,x)$ $(x \in \mathbf{C})$ are measurable nonanticipative functionals and

$$\int_0^T f_1^2(t,x) < \infty,$$

for any $x \in \mathbf{C}$. Denote $n(t) = \mathbf{E}(\zeta | \mathcal{F}_t^\xi)$.

LEMMA 1. *Let the following conditions hold:*

a) *equation* (3) *has a strong solution on* $[0,T]$;

b) $\mathbf{E}[f_0(t,\xi) + \zeta f_1(t,\xi)]^2 \leqslant P$,

c) $\mathbf{E}\big(\zeta - n(t)\big)^2 = \delta e^{-Pt}$,

for any $t \in [0,T]$. Then

1) $f_1^2(t,\xi) = f_1^2(t) = Pe^{Pt}/\delta$ a.s., *for any* $t \in [0,T]$;

2) $f_0(t,\xi) = -f_1(t)n(t)$ a.s., *for almost any* $t \in [0,T]$ (*with respect to the Lebesgue measure*).

Remark. In (Ihara, 1973) and in (Liptser and Shiryaev, 1978, Chapter 15) it is proved that under conditions a), b) the inequality $\mathbf{E}\big(\zeta - n(t)\big)^2 \geqslant \delta e^{-Pt}$ holds for any $t \in [0,T]$ and there exist functions f_0, f_1 with properties a), b) such that the corresponding process defined by (3) satisfies the equation

$$d\xi_t = f_1(t)\big(\zeta - n(t)\big)\,dt + dw_t, \quad \xi_0 = 0, \quad t \in [0,T],$$

and previous inequality is relaxed to a strict equality. So, this lemma asserts that in the class of linear (in ζ) coding functions the optimal scheme is unique.

Proof. Let $\delta_t = \mathbf{E}\big[(\zeta - n(t))^2 | \mathcal{F}_t^\xi\big]$. It is known that δ_t satisfies the following equation

$$\dot{\delta}_t = -f_1^2(t,\xi)\delta_t^2, \quad \delta_0 = \delta, \tag{4}$$

(Lipcer and Shiryaev, 1978, Chapter 12), which has the solution

$$\delta_t = \frac{\delta}{1 + \delta \int_0^t f_1^2(s,\xi)\,ds},$$

therefore, $\inf_{0 \leqslant t \leqslant T} \delta_t > 0$. Then (4) implies that $\dot{\delta}_t/\delta_t = -f_1^2\delta_t$, and

$$\delta_t = \delta \exp\Big(-\int_0^t f_1^2(s,\xi)\delta_s\,ds\Big). \tag{5}$$

Let us rewrite inequality b) omitting the arguments,

$$\mathbf{E}(f_0 + \zeta f_1)^2 = \mathbf{E}\Big(f_0 + f_1 n(t) + f_1\big(\zeta - n(t)\big)\Big)^2 =$$

$$= \mathbf{E}\big(f_0 + f_1 n(t)\big)^2 + \mathbf{E} f_1^2 \delta_t \leqslant P. \tag{6}$$

Then the Jensen inequality ($\mathbf{E}e^\eta \geqslant e^{\mathbf{E}\eta}$) and c), (5), (6) provide the following sequence of inequalities:

$$\delta e^{-Pt} = \mathbf{E}\delta_t \geqslant \delta \exp\Big(-\int_0^t \mathbf{E}\big(f_1^2(s,\xi)\delta_s\big)ds\Big) \geqslant \delta e^{-Pt}. \tag{7}$$

Using the strong convexity of the function e^x, we have

$$\int_0^t f_1^2(s,\xi)\delta_s\,ds = \int_0^t \mathbf{E}(f_1^2(s,\xi)\delta_s)\,ds = Pt \quad \text{a.s.},$$

hence, $\delta_t = \mathbf{E}\delta_t = \delta e^{-Pt}$ a.s. From the continuity of the process δ_t it follows that $\delta_t = \delta e^{-Pt}$ a.s. for any $t \in [0,T]$. Now equation (4) implies statement 1). Finally, 6) and 7) imply $\int_0^T \mathbf{E}\big(f_0 + f_1 n(t)\big)^2 dt = 0$, which is equivalent to 2).
Now we suppose that on some probability space with filtration $\mathbf{F} = (\mathcal{F}_t,\ 0 \leqslant t \leqslant T)$ a Wiener process $w = (w_t,\ 0 \leqslant t \leqslant T)$ (with respect to $\mathbf{F}$) and a measurable F-adapted process $A = (A_t,\ 0 \leqslant t \leqslant T)$ with

$$\mathbf{E}\int_0^T A_t^2\,dt < \infty$$

are given.
The following lemma gives an extension of the well-known result from (Kadota, Zakai and Ziv, 1971).

LEMMA 2 (Katyshev, 1978; Glonti, 1978). *Let y be a $\mathcal{F}_0$–measurable random element taking values in some measurable space and let the process ξ be defined by the equality*

$$\xi_t = \int_0^t A_s\,ds + w_t. \tag{8}$$

Then

$$I(y,\xi_0^t) = \frac{1}{2}\mathbf{E}\int_0^t (\overline{A}_s^2 - \overline{\mathbf{A}}_s^2)ds, \tag{9}$$

where $\xi_0^t = \{\xi_s\colon\ 0 \leqslant s \leqslant t\}$, $I(y,\xi_0^t)$ is the mutual information of y, ξ_0^s, and $\overline{A}_s = \mathbf{E}(A_s|\mathcal{F}_s^\xi, y)$, $\overline{\mathbf{A}}_s = \mathbf{E}(A_s|\mathcal{F}_s^\xi)$.

Assume now that the conditions of Lemma 2 hold for $y = \zeta$. .

LEMMA 3. *Suppose that the system (ζ,ξ) is Gaussian, $\mathbf{E}A_t^2 \leqslant P$ for any $t \in [0,T]$, and $0 \leqslant u \leqslant v \leqslant T$. Then*

1) $\mathbf{E}\big(\zeta - n(v)\big)^2 \geqslant \mathbf{E}\big(\zeta - n(u)\big)^2 e^{-P(v-u)}$;
2) *there exists a process $\widetilde{A} = (\widetilde{A}_t,\ 0 \leqslant t \leqslant T)$ (independent of v) such that*
   - (i) $\widetilde{A}_t = A_t$, *for* $0 \leqslant t \leqslant u$;
   - (ii) *for the process $\widetilde{\xi}$, defined by (8), $\widetilde{A}_t = f_1(t)\big(\zeta - \widetilde{n}(t)\big)$, for $t \geqslant u$, where $f_1(t)$ is the function obtained in Lemma 1;*
   - (iii) *the system $(\zeta,\widetilde{\xi})$ is Gaussian and $\mathbf{E}\widetilde{A}_t^2 = P$, for any $t \geqslant u$;*
   - (iv) $\mathbf{E}\big(\zeta - \widetilde{n}(v)\big)^2 = \mathbf{E}\big(\zeta - \widetilde{n}(u)\big)^2 e^{-P(v-u)}$ *(obviously* (i) *implies $\widetilde{n}(u) = n(u)$).*

Proof. 1) It is known that if the system (ζ,ξ) is Gaussian then $I(\zeta,\xi_0^t) = I\big(\zeta,n(t)\big)$ for any t (see, e.g., (Van de Water, 1980)). Moreover, since $\big(\zeta,n(t)\big)$ is a two–dimensional Gaussian vector, it may be easily calculated that

$$\mathbf{E}\big(\zeta - n(t)\big)^2 = \delta\exp\big(-2I(\zeta,n(t))\big).$$

Using Lemma 2, we have

$$\mathbf{E}\big(\zeta - n(v)\big)^2 = \delta\exp\big(-2I(\zeta,n(v))\big) =$$

$$= \delta\exp\big(-2I(\zeta,\xi_0^v)\big) = \delta\exp\left(-2I(\zeta,\xi_0^u) - \mathbf{E}\int_u^v(\overline{A}_s^2 - \overline{\mathbf{A}}_s^2)\,ds\right) =$$

$$= \mathbf{E}\big(\zeta - n(u)\big)^2\exp\left(-\mathbf{E}\int_u^v(\overline{A}_s^2 - \overline{\mathbf{A}}_s^2)\,ds\right) \geqslant \mathbf{E}\big(\zeta - n(u)\big)^2 e^{-P(v-u)}.$$

The last equality holds since $\mathbf{E}\overline{A}_s^2 \leqslant \mathbf{E}A_s^2 \leqslant P$ and $\mathbf{E}\overline{\mathbf{A}}_s^2 \geqslant 0$.

2) When ζ is a Gaussian process and $u = 0$ in (Ihara, 1974) the coding $\widetilde{A}$ is constructed for which conditions (i)—(iii) hold. It may be verified that the same

approach allows to obtain the desired process $\widetilde{A}$ for any u. Condition (iv) holds, since due to (ii), (iii), we have $\mathbf{E}(\widetilde{A}_s | \mathcal{F}_s^{\bar{\xi}}, \zeta) = \widetilde{A}_s$, $\mathbf{E}(\widetilde{A}_s | \mathcal{F}_s^{\bar{\xi}}) = 0$ and therefore in the previous formula the last inequality becomes an equality.

4. PROOF OF THE THEOREM

In the sequel we shall deal with σ–algebras $\mathcal{F}_t^\xi$ for different moments t. If these moments have sub– or superscripts (e.g., $t = t_i$), we shall write $\mathcal{F}^\xi(t_i)$.
Before proving the theorem we establish a proposition which has an independent interest.
Let $\mathcal{T} = \{0 = t_0 < t_1 < \ldots\}$ be some partition, $A \in \mathcal{A}$ be any admissible coding and ξ be the solution of equation (1). For any random vector η denote $m(\eta, t) = \mathbf{E}(\eta | \mathcal{F}_t^\xi)$ (recall that $m(\theta_0, t) = m(t)$).

PROPOSITION. *Let $t_{i-1} \leqslant t < t_i$. Then*

1) $\Delta_A(t) \geqslant \mathbf{E}\eta_i^2 \exp\big(-P(t - t_{i-1})\big) + \mathbf{E}\lambda_i^2$, *where*

$$\eta_i = \mathbf{E}\big(\theta_0 - m(t_{i-1}) | \theta_i - m(\theta_i, t_{i-1})\big), \quad \lambda_i = \theta_0 - m(t_{i-1}) - \eta_i;$$

2) *there exists admissible coding A^*, coinciding with A on the interval $[0, t_{i-1}]$, such that*

$$\Delta_{A^*}(t) = \mathbf{E}\eta_i^2 \exp\big(-P(t - t_{i-1})\big) + \mathbf{E}\lambda_i^2.$$

Proof. 1) Since the system (θ, ξ) is Gaussian the random variable λ_i does not depend on $\theta_i - m(\theta_i, t_{i-1})$. Moreover, λ_i is independent of the σ–algebra $\mathcal{F}^\xi(t_{i-1})$, because $\mathbf{E}(\eta_i | \mathcal{F}^\xi(t_{i-1})) = 0$, and hence, $\mathbf{E}(\lambda_i | \mathcal{F}^\xi(t_{i-1})) = 0$. Thus λ_i and θ_i are independent. Obviously, λ_i also does not depend on the σ–algebra $\sigma(w_s - -w_{t_{i-1}}, t_{i-1} \leqslant s \leqslant t)$. Hence, due to condition 2) in the definition of admissible coding, the random variable λ_i and the σ–algebra $\mathcal{F}_t^\xi$ are independent. Since $\theta_0 = \eta_i + \lambda_i + m(t_{i-1})$, $m(\lambda_i, t) = 0$ and $m(t_{i-1})$ is $\mathcal{F}_t^\xi$–measurable, we have $m(t) = m(\eta_i, t) + m(t_{i-1})$. It follows that

$$\Delta_A(t) = \mathbf{E}\big(\theta_0 - m(t)\big)^2 = \mathbf{E}\big(\eta_i - m(\eta_i, t)\big)^2 + \mathbf{E}\lambda_i^2, \tag{10}$$

because the random variables $\eta_i - m(\eta_i, t)$ and λ_i are independent. Using statement 1) of Lemma 3 with $\zeta = \eta_i$, $u = t_{i-1}$, $v = t$ and taking into account that $m(\eta_i, t_{i-1}) = 0$, we obtain the required inequality.
2) In order to obtain this result it is sufficient to use in (10) the statement 2) of Lemma 3.
Essentially this proposition means that if some admissible coding were used till the moment t_{i-1} then to minimize the value $\Delta_A(t)$ for $t \in [t_{i-1}, t_i]$, it is necessary to construct the Gaussian random variable η_i and to transmit it on this time interval using the optimal scheme described in 2) of Lemma 3.

Proof of the theorem. Let us consider a partition $\mathcal{T}$ containing only two points: $\mathcal{T} = \{0 = t_0 < t_1\}$ and let $\theta = (\alpha_1, \alpha_2, \alpha_3)$ be a three-dimensional Gaussian vector. For simplicity we assume that $\mathbf{E}\alpha_1 = \mathbf{E}\alpha_2 = \mathbf{E}\theta_2 = 0$. Let $\theta_0 = \alpha_1$, $\theta_1 = (\alpha_1, \alpha_2)$, $\theta_2 = \alpha_3$. Finally, assume that α_1 and α_2 are independent and denote $\mathbf{E}\alpha_1^2 = \gamma_1$, ρ_{ij} is the coefficient of correlation of random variables α_i and α_j, $i, j = 1, 2, 3$. Now suppose that there exists a uniformly optimal coding $A \in \mathcal{A}$. Then the proposition asserts that $\Delta(t) = \mathbf{E}\eta_1^2 e^{-Pt} + \mathbf{E}\lambda_1^2$, for $0 \leqslant t \leqslant t_1$, where $\eta_1 = \mathbf{E}(\theta_0 | \theta_1) = \alpha_1$, $\lambda_1 = \theta_0 - \eta_1 = 0$, i.e., $\Delta(t) = \gamma_1 e^{-Pt}$. If $\Delta_A(t) = \Delta(t)$ for $t \in [0, t_1]$, then, according to Lemma 1, the corresponding process ξ satisfies on the interval $[0, t_1]$ the equation

$$d\xi_t = f(t)\big(\theta_0 - m(t)\big)dt + dw_t, \quad \xi_0 = 0, \tag{11}$$

and $f^2(t)\mathbf{E}\big(\theta_0 - m(t)\big)^2 = P$. Now we may use the filtering equations for conditionally-Gaussian processes (see (Liptser and Shiryaev, 1978, Chapter 12)). Then, for random variables

$$\eta_2 = \mathbf{E}(\theta_0 - m(t_1) | \theta_2 - m(\theta_2, t_1)), \quad \lambda_2 = \theta_0 - m(t_1) - \eta_2,$$

we obtain

$$\mathbf{E}\eta_2^2 = \gamma_1 \exp(-2Pt_1) \frac{\rho_{13}^2}{1 - \rho_{13}^2(1 - \exp(-Pt_1))},$$

$$\mathbf{E}\lambda_2^2 = \gamma_1 \exp(-Pt_1) \frac{1 - \rho_{13}^2}{1 - \rho_{13}^2(1 - \exp(-Pt_1))}.$$

The first statement of the proposition implies the inequality

$$\Delta_A(t) \geqslant \mathbf{E}\eta_2^2 \exp\big(-P(t - t_1)\big) + \mathbf{E}\lambda_2^2, \tag{12}$$

for $t \geqslant t_1$.

Let us consider now such a coding $\widehat{A}$ that the corresponding process $\widehat{\xi}$ satisfies on the interval $[0, t_1]$ the equation

$$d\widehat{\xi}_t = \widehat{f}(t)\big(\alpha_2 - m(\alpha_2, t)\big)dt + dw_t, \quad \widehat{\xi}_0 = 0,$$

and $\widehat{f}^2(t)\mathbf{E}\big(\alpha_2 - m(\alpha_2, t)\big)^2 = P$. This means that we use the optimal scheme for transmission of the random variable α_2 on $[0, t_1]$. Again, using the filtering equations for conditionally Gaussian processes, we have

$$\Delta_{\widehat{A}}(t) = \gamma_1, \quad \mathbf{E}\widehat{\eta}_2^2 = \gamma_1 \frac{\rho_{13}^2}{1 - \rho_{23}^2(1 - \exp(-Pt_1))},$$

$$\mathbf{E}\widehat{\lambda}_2^2 = \gamma_1 \left(1 - \frac{\rho_{13}^2}{1 - \rho_{23}^2(1 - \exp(-Pt_1))}\right),$$

for $t \in [0, t_1]$, where the variables $\widehat{\eta}_2$ and $\widehat{\lambda}_2$ are defined by the process $\widehat{\xi}$, similarly to η_2 and λ_2. Due to statement 2) of the proposition, we can define the coding $\widehat{A}$ for $t \geqslant t_1$ such that

$$\Delta_{\widehat{A}}(t) = \mathbf{E}\widehat{\eta}_2^2 \exp\big(-P(t - t_1)\big) + \mathbf{E}\widehat{\lambda}_2^2. \tag{13}$$

Put now $\rho_{23}^2 = 1 - \rho_{13}^2$, providing the non-negative definiteness of the covariance matrix of θ. Then routine calculations show that if $\rho_{13}^2 > 1/2$, the inequality

$$\mathbf{E}\widehat{\lambda}_2^2 < \mathbf{E}\lambda_2^2 \tag{14}$$

holds. The first terms in the right–hand side of (12) and (13) tend to zero as $t \to \infty$, hence, due to (14), there exists a moment τ such that $\Delta_A(t) > \Delta_{\hat{A}}(t)$ for any $t > \tau$. Therefore, if $T > \tau$ then the coding A is not uniformly optimal on $[0, T]$.

REFERENCES

Glonti, O.A. (1978). On mutual information of signals in the case of transmission over a "noisy" feedback channel. *Theory Probab. Appl.* **23**, 377–379.

Ihara, S. (1973). Optimal coding in white Gaussian channel with feedback. *Lect. Notes Math.* **330**, 120–123.

Ihara, S. (1974). Coding theory in white Gaussian channel with feedback. *J. Mult. Anal.* **4**, 74–87.

Ihara, S. (1975). Coding theory in Gaussian channel with feedback, II: Evaluation of the filtering error. *Nagoya Math. J.* **58**, 127–147.

Kadota, T.T., Zakai, M. and Ziv, J. (1971). Mutual information of the white Gaussian channel with and without feedback. *IEEE Trans. Inf. Theory,* **IT–17**, 368–371.

Katyshev, P.K. (1975). Transmission of Gaussian vector through the channel with noiseless feedback. *Theory Probab. Appl.* **20**, 185–192.

Katyshev, P.K. (1976). Transmission of Gaussian process through the channel with feedback. *Probl. Inf. Transm.* **12**, 112–118 (in Russian).

Katyshev, P.K. (1978). On a problem of control of stochastic process arising in information theory. In: *Stochastic Processes and Control.* Nauka, Moscow, pp. 75–93 (in Russian).

Liptser, R.S. (1974). Optimal coding and decoding in transmission of Gaussian Markov message through the channel with noiseless feedback. *Probl. Inf. Transm.* **10**, 3–15.

Liptser, R.S. and Shiryaev, A.N. (1978). *Statistics of Random Processes.* Springer, Berlin etc.

Ovseevich, I.A. (1970). Optimal transmission of the Gaussian message through the white Gaussian channel with feedback. *Probl. Inf. Trans.* **6**, 3–14.

Van de Water Henk (1980). *The Value of Information in Stochastic Control.* Working paper.

New Trends in Probab. and Statist., pp. 445 - 456
V. Sazonov and T. Shervashidze (Eds.)
1991 VSP/Mokslas

ON THE STRUCTURE OF A RANDOM GRAPH WITH NONUNIFORM DISTRIBUTION

V.I. KHOKHLOV and V.F. KOLCHIN
Steklov Mathematical Institute of the USSR Academy of Sciences,
Vavilov str. 42, 117966 Moscow, USSR

ABSTRACT

We consider a random graph $G_{N,T}$ with N labelled vertices and T edges. These T edges are obtained by T independent trials. In each trial the edge between vertices i and j occurs with probability $2p_ip_j$ and the loop at the vertex i occurs with probability p_i^2, $i, j = 1, 2, \ldots, N$, $p_1, \ldots, p_N \geqslant 0$, $p_1 + \ldots + p_N = 1$. Let $N \to \infty$, $2T/N \to \lambda$, $p_i = a_i/N$, $a_i = a_i(N)$, $i = 1, 2, \ldots, N$, and there exist a limit

$$a^2 = \lim_{N\to\infty} \frac{1}{N} \sum_{i=1}^{N} a_i^2$$

and positive constants ε and E such that $\varepsilon \leqslant a_i \leqslant E$, $i = 1, 2, \ldots, N$. Then, under the additional condition $a^2\lambda < 1$, the graph $G_{N,T}$, with probability approaching one, does not contain components with more than one cycle and tree-components that have more than $c \log N$ vertices, where c is a constant. Moreover, under these conditions the distribution of the total number of cycles converges to the Poisson distribution with parameter $\Lambda = -2^{-1} \log(1 - a^2\lambda)$.

1. MAIN RESULTS

A lot of papers are devoted to investigations of asymptotic characteristics of random graphs that take their values with equal probabilities (see (Karonski, 1982; Bollobas, 1985; Palmer, 1985), where surveys of results may be found, and (Stepanov, 1987; Kolchin, 1986; Takács, 1988)). Only a few of works deal with random graphs that take their values with unequal probabilities (Kovalenko, 1971, 1972; Ivchenko, 1975). However, it is of interest to study zones of stability of limit distributions of graph characteristics as the probabilities become unequal. Without doubt, it is useful to know what deviations of the probabilities are negligible and what more significant deviations result in.

In (Erdös and Rényi, 1960) the random graph which takes any value from the set of all graphs with fixed numbers of labelled vertices and edges with equal probabilities was investigated. Consider a slight modification of the graph. Let $\tilde{G}_{N,T}$ have N

labelled vertices and T edges. These T edges are obtained as outcomes of T independent trials. In each trial the edge between i and j occurs with probability $2N^{-2}$ and the loop at the vertex i occurs with probability N^{-2}, $i, j = 1, \ldots, N$. Thus, a realization of the graph $\widetilde{G}_{N,T}$ may have loops and multiple edges. From the results of (Erdös and Rényi, 1960) it follows that if $N \to \infty$, $2T/N \to \lambda$, $0 < \lambda < 1$, then with probability approaching one all connected components of $\widetilde{G}_{N,T}$ are trees or components with exactly one cycle, and the total number of cycles is distributed in the limit according to the Poisson law with the parameter $\Lambda = -2^{-1} \log(1 - \lambda)$.

In this paper we consider a similar random graph with N labelled vertices, of which T edges are also obtained by T independent trials. But here, in each trial the edge between the vertices i and j occurs with probability $2p_i p_j$ and the loop at the vertex i occurs with probability p_i^2, $i, j = 1, 2, \ldots, N$, $p_1, p_2, \ldots, p_N \geqslant 0$, $p_1 + p_2 + \ldots + p_N = 1$. The equiprobable case $p_1 = p_2 = \ldots = p_N = 1/N$ corresponds to the above mentioned graph $\widetilde{G}_{N,T}$. We shall use the similar notation $G_{N,T}$ for the nonequiprobable graph. It turns out that even rather significant deviations of probabilities $p_1, p_2, \ldots, p_N$ from $1/N$ lead to small changes of the structure of the graph $G_{N,T}$. Let

$$p_i = a_i/N, \quad a_i = a_i(N), \quad 0 < \varepsilon \leqslant a_i \leqslant E < \infty, \quad i = 1, 2, \ldots, n, \tag{1}$$

where ε and E are constants, the limit

$$a^2 = \lim_{N \to \infty} \frac{1}{N} \sum_{i=1}^{N} a_i^2 = \lim_{N \to \infty} N \sum_{i=1}^{N} p_i^2 \tag{2}$$

exists, and

$$2T/N \to \lambda \quad \text{as} \quad N, T \to \infty. \tag{3}$$

We shall assume also that

$$a^2 \lambda < 1. \tag{4}$$

Now we can formulate the main results of the paper.

THEOREM 1. *If conditions* (1)–(4) *hold, then, with probability approaching one, the random graph* $G_{N,T}$ *does not contain components with more than one cycle.*

Denote by α_r the number of cycles of length r and let $\zeta_{N,T} = \alpha_1 + \ldots + \alpha_N$ be the total number of cycles in $G_{N,T}$.

THEOREM 2. *Under the conditions* (1)–(4), *for any fixed non-negative integer number* k,

$$\mathbf{P}\{\zeta_{N,T} = k\} \to \frac{1}{k!} \Lambda^k e^{-\Lambda},$$

where $\Lambda = -2^{-1} \log(1 - a^2 \lambda)$.

THEOREM 3. *If the conditions* (1)–(4) *are satisfied, then, for any fixed* $1 \leqslant$

$\leqslant r_1 < \ldots < r_s$, *the joint distribution of the random variables* $\alpha_{r_1}, \ldots, \alpha_{r_s}$ *converges to the distribution of s independent random variables distributed according to the Poisson laws with parameters* $\lambda_1, \ldots, \lambda_s$, *respectively,* $\lambda_r = \lambda^r a^{2r}/(2r)$.

THEOREM 4. *If the conditions (1)–(4) are satisfied, then there exists a constant c such that the random graph* $G_{N,T}$, *with probability tending to one, does not contain tree-components with more than* $c \log N$ *vertices.*

Theorems 2 and 3 was proved in (Kolchin and Khokhlov, 1990) at great length. The main purpose of this article is to prove Theorems 1 and 4.

2. PROOF OF THEOREM 1

We have to prove that, under the conditions (1)–(4), with probability tending to one the random graph $G_{N,T}$ does not contain some configurations consisting of two connected cycles. Everywhere below, we shall denote by c and c_i some positive constants (may be not the same). We shall use also notation $\mathcal{N} = \{1,2,\ldots,N\}, \mathcal{T} = \{1,2,\ldots,T\}, a^{[b]} = a(a-1)\ldots(a-b+1)$ and, for the given set $\mathcal{M} = \{1,2,\ldots,M\}$, we shall denote by symbols $\{x_1,\ldots,x_r\}_{\mathcal{M}\neq}$ and $\langle x_1,\ldots,x_r\rangle_{\mathcal{M}\neq}$ the set of all $\binom{M}{r}$ unordered r-collections and the set of all $r!\binom{M}{r}$ ordered r-collections of different elements of $\mathcal{M}$, respectively. Note that if, for any $x_1,\ldots,x_r$, a function $f(x_1,\ldots,x_r)$ takes the same values with any permutation of $x_1,\ldots,x_r$, then obviously

$$\sum_{\langle x_1,\ldots,x_r\rangle_{\mathcal{N}\neq}} f(x_1,\ldots,x_r) = r! \sum_{\{x_1,\ldots,x_r\}_{\mathcal{N}\neq}} f(x_1,\ldots,x_r).$$

To reduce expressions it is convenient to use short notation of the type B_r for $(b_1,\ldots,b_r)$; for instance, $\langle B_r\rangle_{\mathcal{N}\neq}$ is $\langle b_1,\ldots,b_r\rangle_{\mathcal{N}\neq}$, and $\{B_r\}_{\mathcal{N}\neq}$ is $\{b_1,\ldots,b_r\}_{\mathcal{N}\neq}$.

LEMMA 1. *Under the conditions (1)–(4), with probability tending to one, the random graph* $G_{N,T}$ *does not contain any subgraph which consists of two cycles with exactly one common vertex.*

Proof. The subgraph which consists of two cycles with exactly one common vertex may be called the eight. For given s_1, s_2, it is natural to associate the introduced symbol $\langle v, X_{s_1}, X_{s_2}\rangle_{N\neq} = \langle v, x_1,\ldots,x_{s_1}, y_1,\ldots,y_{s_2}\rangle_{N\neq}$ with such subgraph consisting of cycles on vertices $v, x_1,\ldots,x_{s_1}$ and $v, y_1,\ldots,y_{s_2}$ and having edges $(v,x_1),(x_1,x_2),\ldots,(x_{s_1},v)$ and $(v,y_1),(y_1,y_2),\ldots,(y_{s_2},v)$, correspondingly. Observe that if one of the numbers s_1 and s_2 equal zero then we have a loop at the vertex v and if both s_1 and s_2 are zeros then we have two loops at v. The expected number Σ_1 of eights in $G_{N,T}$ can be written in the following form

$$\Sigma_1 = \sum_{s_1,s_2=0}^{N} \sum_{\langle v,X_{s_1};Y_{s_2}\rangle_{\mathcal{N}\neq}} \sum_{\langle l_1,\ldots,l_{s_1+s_2+2}\rangle_{\mathcal{T}\neq}} \mathbf{P}\{\xi(v,X_{s_1};Y_{s_2};l_1,\ldots$$

$$\ldots,l_{s_1+s_2+2}) = 1\} =$$

$$= \sum_{s_1,s_2=0}^{N} \binom{T}{s_1+s_2+2} \sum_{\langle v,X_{s_1},Y_{s_2}\rangle_{\mathcal{N}\neq}} \mathbf{P}\{\xi(v,X_{s_1};Y_{s_2};1,\dots,s_1+s_2+2)=1\},$$

where the random variable $\xi(v,X_{s_1};Y_{s_2};l_1,\dots,l_{s_1+s_2+2})$ equals one if the trials with numbers $l_1,\dots,l_{s_1+s_2+2}$ from those T trials that form the edges of $G_{N,T}$ have the outcomes $(v,x_1),(x_1,x_2),\dots,(x_{s_1},v),(v,y_1),(y_1,y_2),\dots,(y_{s_2},v)$ in the indicated order. Otherwise, let the random variable be equal to zero. The probability of occurring an edge (i,j) in a fixed trial does not exceed $2p_ip_j = 2a_ia_jN^{-2}$ for all i,j, therefore,

$$\mathbf{P}\{\xi(v,X_{s_1};Y_{s_2};1,\dots,s_1+s_2+2)=1\} \leqslant$$

$$\leqslant 2^{s_1+s_2+2}a_v^4a_{x_1}^2\dots a_{x_{s_1}}^2a_{y_1}^2\dots a_{y_{s_2}}^2N^{-2(s_1+s_2+2)} \leqslant$$

$$\leqslant E^2 2^{s_1+s_2+2}a_v^2a_{x_1}^2\dots a_{x_{s_1}}^2a_{y_1}^2\dots a_{y_{s_2}}^2N^{-2(s_1+s_2+2)},$$

where the right-hand side does not depend on the order of $v,x_1,\dots,x_{s_1},y_1,\dots,y_{s_2}$. Moreover, under the conditions of Lemma 1, for sufficiently large N,

$$\frac{1}{N}\sum_{i=1}^{N}a_i^2 \leqslant a^2+\delta, \quad \frac{2T}{N} \leqslant \lambda+\delta, \tag{5}$$

where δ is an arbitrary small fixed positive number.

We shall find only an upper bound for Σ_1. Take δ so that $(a^2+\delta)(\lambda+\delta)<1$. Then, for sufficiently large N,

$$\Sigma_1 \leqslant E^2 \sum_{s_1,s_2=0}^{\infty} 2^{s_1+s_2+2}\binom{T}{s_1+s_2+2}\times$$

$$\times \sum_{\langle v,X_{s_1};Y_{s_2}\rangle_{\mathcal{N}\neq}} a_v^2a_{x_1}^2\dots a_{x_{s_1}}^2a_{y_1}^2\dots a_{y_{s_2}}^2N^{-2(s_1+s_2+2)} =$$

$$= E^2 \sum_{s_1,s_2=0}^{\infty} 2^{s_1+s_2+2}\binom{T}{s_1+s_2+2}(s_1+s_2+2)!\times$$

$$\times \sum_{\{v,X_{s_1};Y_{s_2}\}_{\mathcal{N}\neq}} a_v^2a_{x_1}^2\dots a_{x_{s_1}}^2a_{y_1}^2\dots a_{y_{s_2}}^2N^{-2(s_1+s_2+2)} \leqslant$$

$$\leqslant \frac{1}{N}E^2\sum_{s_1,s_2=0}^{\infty}\left(\frac{2T}{N}\right)^{s_1+s_2+2}\left(\frac{1}{N}\sum_{i=1}^{N}a_i^2\right)^{s_1+s_2+1} \leqslant$$

$$\leqslant \frac{E^2(\lambda+\delta)}{N}\sum_{s_1,s_2=0}^{\infty}(\lambda+\delta)^{s_1+s_2+1}(a^2+\delta)^{s_1+s_2+1} =$$

$$= \frac{E^2(\lambda+\delta)^2(a^2+\delta)}{N\big(1-(\lambda+\delta)(a^2+\delta)\big)}.$$

Thus, under the conditions of Lemma, we have $\Sigma_1 \to 0$ as $N \to \infty$. The lemma is proved.

LEMMA 2. *Under the conditions (1)–(4), with probability tending to one, the random graph $G_{N,T}$ does not contain any subgraph which consists of two cycles with a sequence of common edges.*

Proof. A subgraph which consists of two cycles with a sequence of common edges may be called the cycle with a bridge. Let such a bridge consist of two vertices, v and w, that are connected by three paths v, X_{s_1}, w; v, Y_{s_2}, w; v, Z_{s_3}, w. Introduce the random variable $\xi(v, w, X_{s_1}; Y_{s_2}; Z_{s_3}; l_1, \dots, l_{s_1+s_2+s_3+3})$ that is equal to one if the trials with numbers $l_1, \dots, l_{s_1+s_2+s_3+3}$ from those T trials that form the random graph $G_{N,T}$ have the outcomes $(v, x_1), \dots, (x_{s_1}, w)$, $(v, y_1), \dots, (y_{s_2}, w), (v, z_1), \dots, (z_{s_3}, w)$ in the indicated order. Otherwise, the random variable is zero. It is clear that

$$\mathbf{P}\{\xi(v, w, X_{s_1}; Y_{s_2}; Z_{s_3}) = 1\} \leqslant$$

$$\leqslant 2^{s_1+s_2+s_3+3} a_v^3 a_w^3 a_{x_1}^2 \dots a_{x_{s_1}}^2 a_{y_1}^2 \dots a_{y_{s_2}}^2 a_{z_1}^2 \dots a_{z_{s_3}}^2 N^{-2(s_1+s_2+s_3+3)} \leqslant$$

$$\leqslant E^2 2^{s_1+s_2+s_3+3} a_v^2 a_w^2 a_{x_1}^2 \dots a_{x_{s_1}}^2 a_{y_1}^2 \dots a_{y_{s_2}}^2 a_{z_1}^2 \dots a_{z_{s_3}}^2 N^{-2(s_1+s_2+s_3+3)},$$

and the right-hand side of the estimate does not depend on the order of v, w, $X_{s_1}, Y_{s_2}, Z_{s_3}$.

The expectation Σ_2 of the total number of cycles with bridges can be estimated as follows

$$\Sigma_2 \leqslant \sum_{s_1,s_2,s_3=0}^{\infty} \sum_{\langle v,w,X_{s_1},Y_{s_2},Z_{s_3}\rangle_{N\neq}} \sum_{\langle l_1,\dots,l_{s_1+s_2+s_3+3}\rangle_{T\neq}}' \mathbf{P}\{\xi(v, w, x_1, \dots$$

$$\dots, x_{s_1}; y_1, \dots, y_{s_2}; z_1, \dots, z_{s_3}; l_1, \dots, l_{s_1+s_2+s_3+3}) = 1\} =$$

$$= \sum_{s_1,s_2,s_3=0}^{\infty} \binom{T}{s_1 + s_2 + s_3 + 3} \sum_{\langle v,w,X_{s_1},Y_{s_2},Z_{s_3}\rangle_{N\neq}} \mathbf{P}\{\xi(v, w, x_1, \dots$$

$$\dots, x_{s_1}; y_1, \dots, y_{s_2}; z_1, \dots, z_{s_3}; 1, \dots, s_1 + s_2 + s_3 + 3) = 1\} \leqslant$$

$$\leqslant E^2 \sum_{s_1,s_2,s_3=0}^{\infty} \binom{T}{s_1 + s_2 + s_3 + 3} \sum_{\langle v,w,X_{s_1};Y_{s_2};Z_{s_3}\rangle_{N\neq}} a_v^2 a_w^2 a_{x_1}^2 \dots$$

$$\dots a_{x_{s_1}}^2 a_{y_1}^2 \dots a_{y_{s_2}}^2 a_{z_1}^2 \dots a_{z_{s_3}}^2 N^{-2(s_1+s_2+s_3+3)} = E^2 \times$$

$$\times \sum_{s_1,s_2,s_3=0}^{\infty} \binom{T}{s_1 + s_2 + s_3 + 3} (s_1 + s_2 + s_3 + 3)! \sum_{\{v,w,X_{s_1},Y_{s_2},Z_{s_3}\}_{N\neq}} a_v^2 a_w^2 a_{x_1}^2 \dots$$

$$\dots a_{x_{s_1}}^2 a_{y_1}^2 \dots a_{y_{s_2}}^2 a_{z_1}^2 \dots a_{z_{s_3}}^2 N^{-2(s_1+s_2+s_3+3)} \leqslant$$

$$\leqslant \frac{1}{N}E^2 \sum_{s_1,s_2,s_3=0}^{\infty} \left(\frac{2T}{N}\right)^{s_1+s_2+s_3+3} \left(\frac{1}{N}\sum_{i=1}^{N} a_i^2\right)^{s_1+s_2+s_3+2}$$

Taking δ the same as in the proof of Lemma 1, we see that, for sufficiently large N,

$$\Sigma_2 \leqslant \frac{1}{N}E^2(\lambda+\delta)^3(a^2+\delta)^2\frac{1}{1-(\lambda+\delta)(a^2+\delta)}.$$

Thus, under the conditions (1)–(4) we have $\Sigma_2 \to 0$ as $N \to \infty$. Lemma 2 is proved.

LEMMA 3. *Under the conditions* (1)–(4), *with probability tending to one, the random graph* $G_{N,T}$ *does not contain any subgraph which consists of two cycles connected by a sequence of edges.*

Proof. A subgraph which consists of two cycles connected by a sequence of edges may be called the pince-nez. Let

$$\xi\,(X_{s_1};\, Y_{s_2};\, U_{s_3};\, l_1,\dots,l_{s_1+s_2+s_3+1}) = 1$$

if the trials with numbers $l_1,\dots,l_{s_1+s_2+s_3+1}$ from those T trials that form the random graph $G_{N,T}$ have the outcomes $(x_1,x_2),\dots,(x_{s_1},x_1),(y_1,y_2),\dots$ $\dots,(y_{s_2},y_1),(x_1,u_1),(u_1,u_2),\dots,(u_{s_3},y_1)$ in the indicated order. Otherwise, let the random variable be equal to zero. It is clear that

$$\mathbf{P}\,\{\xi\,(X_{s_1};\, Y_{s_2};\, Z_{s_3};\, l_1,\dots,l_{s_1+s_2+s_3+1}) = 1\} \leqslant$$

$$\leqslant 2^{s_1+s_2+s_3+1}a_{x_1}^3 a_{y_1}^3 a_{x_2}^2 \dots a_{x_{s_1}}^2 a_{y_2}^2 \dots a_{y_{s_2}}^2 a_{u_1}^2 \dots a_{u_{s_3}}^2 N^{-2(s_1+s_2+s_3+1)} \leqslant$$

$$\leqslant E^2 2^{s_1+s_2+s_3+1}a_{x_1}^2 \dots a_{x_{s_1}}^2 a_{y_1}^2 \dots a_{y_{s_2}}^2 a_{u_1}^2 \dots a_{u_{s_3}}^2 N^{-2(s_1+s_2+s_3+1)},$$

and the right-hand side does not depend on the order of $x_{s_1},\dots,x_{s_1},y_1,\dots,y_{s_2}$, $u_1,\dots,u_{s_3}$. As in the proof of Lemma 1, using this estimate one can obtain the following upper bound for the expectation Σ_3 of the total number of pince-nezes in the random graph $G_{N,T}$:

$$\Sigma_3 \leqslant \sum_{s_1,s_2,s_3=0}^{\infty} \quad \sum_{\langle X_{s_1},Y_{s_2},Z_{s_3}\rangle_{N\neq}} \quad \sum_{\langle l_1,\dots,l_{s_1+s_2+s_3+1}\rangle_{T\neq}} \mathbf{P}\{\xi(X_{s_1};\, Y_{s_2};\, Z_{s_3};\, l_1,\dots$$

$$\dots,l_{s_1+s_2+s_3+1}) = 1\} \leqslant \sum_{s_1,s_2,s_3=0}^{\infty} \binom{T}{s_1+s_2+s_3+1} \times$$

$$\times \sum_{\langle x_{s_1},Y_{s_2},Z_{s_2}\rangle_{N\neq}} \mathbf{P}\,\{\xi\,(X_{s_1};\, Y_{s_2};\, Z_{s_3};\, l_1,\dots,l_{s_1+s_2+s_3+1}) = 1\} \leqslant$$

$$\leqslant \frac{1}{N}E^2 \sum_{s_1,s_2,s_3=0}^{N} \left(\frac{2T}{N}\right)^{s_1+s_2+s_3+1} \left(\frac{1}{N}\sum_{i=1}^{N} a_i^2\right)^{s_1+s_2+s_3}$$

Taking δ as in the proof of Lemma 1, we obtain, for sufficiently large N,

$$\Sigma_3 \leqslant \frac{1}{N} E^2(\lambda+\delta) \frac{1}{1-(\lambda+\delta)(a^2+\delta)}$$

Thus, under the conditions (1)–4), we have $\Sigma_3 \to 0$ as $N \to \infty$. Lemma 3 is proved.

If there exists a component of $G_{N,T}$ that contains more than one cycle then there exists a subgraph of $G_{N,T}$ which contains an eight, a cycle with a bridge or pince-nez. In accordance with Lemmas 1, 2, and 3, the probability of such events tends to zero. The theorem is proved.

Before the proving of Theorem 4 we need an auxiliary result.

LEMMA 4. *Let the conditions* (1)–(4) *be satisfied. Then, with probability approaching one, the random graph* $G_{N,T}$ *does not contain a component with more than* $cN^{1/4}$ *vertices, where c is a sufficiently large constant.*

Proof. Suppose that the assertion of Lemma 4 is not true. Then there exist subsequences $\{N\}, \{T\}$ and a positive constant γ such that $2T/N \to \lambda$, $\lambda a^2 < 1$, and, with probability not less than γ, the random graph $G_{N,T}$ contains a component K which has $m \geqslant cN^{1/4}$ vertices. Let us choose the parameter θ such that $(\lambda + 2\theta)a^2 < 1$ and add randomly θN edges to the set of edges of the graph $G_{N,T}$. Now let us estimate the probability of occurring at least one new cycle in the component K after the allocation of the θN new edges. The probability that an edge randomly allocated has both its ends in K is equal to

$$\sum_{\langle v_1, v_2 \rangle_{K \neq}} \frac{2a_{v_1}a_{v_2}}{N^2} + \sum_{v \in K} \frac{a_v^2}{N^2} = \frac{(a_{v_1} + \ldots + a_{v_m})^2}{N^2},$$

where $v_1, \ldots, v_m$ are the vertices that form the initial component K. The probability that none of new edge has both its ends in K is equal to

$$\left(1 - \frac{(a_{v_1} + \ldots + a_{v_m})^2}{N^2}\right)^{\theta N}.$$

Therefore, the probability that at least one new cycle will appear in K is equal to

$$1 - \left(1 - \frac{(a_{v_1} + \ldots + a_{v_m})^2}{N^2}\right)^{\theta N}.$$

Since

$$\frac{(a_{v_1} + \ldots + a_{v_m})}{N} > \frac{\varepsilon m}{N},$$

this probability is not less than

$$1 - \exp\{-\varepsilon^2 \theta m^2 / N\}. \tag{6}$$

Now add a new portion of θN edges to the graph obtained after the addition of the first portion. The same quantity (6) is the low bound for the probability that one more cycle occurs in the component K. By virtue of the independence of these two allocations, the probability to obtain two new cycles in K after the allocation of $2\theta N$ edges is not less than

$$\left(1 - \exp\{-\varepsilon^2\theta m^2/N\}\right)^2.$$

Thus the component K in the random graph $G_{N,T+2\theta N}$ has more than one cycle with probability not less than

$$\gamma\left(1 - \exp\{-\varepsilon^2\theta m^2/N\}\right)^2.$$

On the other hand, as it was established in Lemmas 1–3, for the number X of components of the random graph $G_{N,T+2\theta N}$ that contains more than one cycle, we have

$$\mathbf{E}X \leqslant c_1/N,$$

where c_1 is some constant. Therefore, the chain of inequalities

$$\gamma(1 - \exp\{-\varepsilon^2\theta m^2/N\})^2 \leqslant \mathbf{P}\{X > 0\} \leqslant \mathbf{E}X \leqslant \frac{c_1}{N}$$

is true. It was supposed that $m \geqslant cN^{1/4}$. Therefore,

$$\frac{c_1}{N} \geqslant \gamma(1 - \exp\{-\varepsilon^2\theta m^2/N\})^2 \geqslant \gamma(1 - \exp\{-\varepsilon^2\theta c^2/\sqrt{N}\})^2$$

and so the assumption that there exists a component with more than $cN^{1/4}$ vertices leads to a contradiction if c is sufficiently large since for sufficiently large c this inequality is impossible.

3. PROOF OF THEOREM 4

Denote by $\zeta(a,b)$ the number of components in the graph $G_{N,T}$ that are trees for which the number of their vertices lies in the segment $[a,b]$. Let us represent this random variable as a sum of indicators. We denote by $\xi^{d_1\ldots d_s}_{v_1\ldots v_s;\, e_1\ldots e_{s-1}}$ the random variable which equals one if the trials with numbers $e_1,\ldots,e_{s-1}$ from those T trials that form the random graph $G_{N,T}$ have outcomes that form a component-tree on different vertices $v_1,\ldots,v_s$ having degrees $d_1,\ldots,d_s$, respectively. Otherwise, $\xi^{d_1\ldots d_s}_{v_1\ldots v_s;\, e_1\ldots e_{s-1}} = 0$.
It is easy to check that

$$\mathbf{E}\zeta(a,b) =$$

$$= \sum_{a\leqslant s\leqslant b}\ \sum_{\{e_1,\ldots,e_s\}_{T\neq}}\ \sum_{\{v_1,\ldots,v_s\}_{N\neq}}\ \sum_{d_1,\ldots,d_s=1}^{\infty} \mathbf{P}\{\xi^{d_1\ldots d_s}_{v_1\ldots v_s;\, e_1\ldots e_{s-1}} = 1\} =$$

$$= \sum_{a\leqslant s\leqslant b} \binom{T}{s-1} \sum_{\{v_1,\ldots,v_s\}_{N\neq}}\ \sum_{d_1,\ldots,d_s=1}^{\infty} t(s;\, d_1,\ldots,d_s)\times$$

$$\times \frac{a_{v_1}^{d_1} \dots a_{v_s}^{d_s} 2^{s-1}(s-1)!}{N^{2(s-1)}} \left(1 - \frac{a_{v_1} + \dots + a_{v_s}}{N}\right)^{2T-2(s-1)},$$

where $t(s;\, d_1, \dots, d_s)$ is the number of trees that have $s \geqslant 2$ vertices $v_1, \dots, v_s$, the vertex v_j having a degree d_j, $j = 1, 2, \dots, s$. It is known that

$$t(s;\, d_1, \dots, d_s) = \frac{(s-2)!}{(d_1 - 1)! \dots (d_s - 1)!},$$

where

$$d_1 + \dots + d_s = 2(s-1).$$

(This formula was obtained in (Moon, 1967), and it was rediscovered later on and independently in (Rényi, 1970).) Hence,

$$\mathbf{E}\zeta(a, b) =$$

$$= \sum_{a \leqslant s \leqslant b} 2^{s-1} T^{[s-1]} \sum_{\{v_1, \dots, v_s\}_{\mathcal{N}\neq}} \sum_{\mathcal{D}} \frac{(s-2)!}{(d_1 - 1)! \dots (d_s - 1)!\, N^{2(s-1)}} \times$$

$$\times \left(1 - \frac{a_{v_1} + \dots + a_{v_s}}{N}\right)^{2T-2(s-1)} a_{v_1}^{d_1} \dots a_{v_s}^{d_s} \leqslant$$

$$\leqslant \sum_{a \leqslant s \leqslant b} \left(\frac{2T}{N}\right)^{s-1} N \sum_{\{v_1, \dots, v_s\}_{\mathcal{N}\neq}} \frac{a_{v_1} \dots a_{v_s}}{N^s} (a_{v_1} + \dots$$

$$\dots + a_{v_s})^{s-2} \left(1 - \frac{a_{v_1} + \dots + a_{v_s}}{N}\right)^{2T-2(s-1)} \leqslant$$

$$\leqslant \sum_{a \leqslant s \leqslant b} \left(\frac{2T}{N}\right)^{s-1} N s^{s-2} \sum_{\{v_1, \dots, v_s\}_{\mathcal{N}\neq}} \frac{a_{v_1} \dots a_{v_s}}{N^s} \left(\frac{a_{v_1} + \dots + a_{v_s}}{s}\right)^{s-2} \times$$

$$\times \exp\left\{\left(-\frac{2T}{N} - \frac{2(s-1)}{N}\right) s \frac{a_{v_1} + \dots + a_{v_s}}{s}\right\},$$

where $\mathcal{D} = \{d_1 + \dots + d_s = 2s - 2,\ d_1, \dots, d_s \geqslant 1\}$. Now, let $a \to \infty$, $b = cN^{1/4}$, where c is such that Theorem 1 is true. Then there exists c_1 such that

$$\exp\left\{\frac{2s(s-1)}{N} \frac{a_{v_1} + \dots + a_{v_s}}{s}\right\} \leqslant c_1.$$

Therefore,

$$\mathbf{E}\zeta(a,\, cN^{1/4}) \leqslant c_1 N \sum_{a \leqslant s} s^{s-2} \left(\frac{2T}{N}\right)^{s-1} \times$$

$$\times \sum_{\{v_1, \dots, v_s\}_{\mathcal{N}\neq}} \frac{a_{v_1} \dots a_{v_s}}{N^s} \left(\frac{a_{v_1} + \dots + a_{v_s}}{s} \exp\left\{-\frac{2T}{N} \frac{a_{v_1} + \dots + a_{v_s}}{s}\right\}\right)^{s-2}.$$

As it was mentioned above, if summands $f_{v_1,\dots,v_s}$ are invariant with respect to any permutation of their indices, then

$$\sum_{\{v_1,\dots,v_s\}_{\mathcal{N}\neq}} f_{v_1,\dots,v_s} = \frac{1}{s!} \sum_{(v_1,\dots,v_s)_{\mathcal{N}\neq}} f_{v_1,\dots,v_s} \leqslant \cdot$$

$$\leqslant \frac{1}{s!} \sum_{v_1,\dots,v_s\in\mathcal{N}} f_{v_1,\dots,v_s}.$$

Consequently,

$$\mathbf{E}\zeta(a,\, cN^{1/4}) \leqslant c_1 N \sum_{a\leqslant s} \frac{s^{s-2}}{s!} \frac{2T}{N} \times$$

$$\times \sum_{v_1,\dots,v_s\in\mathcal{N}} \frac{a_{v_1}\dots a_{v_s}}{N^s} \left(\frac{2T}{N} \frac{a_{v_1}+\dots+a_{v_s}}{s} \exp\left\{ -\frac{2T}{N} \frac{a_{v_1}+\dots+a_{v_s}}{s} \right\} \right)^{s-2}. \tag{6}$$

Let us consider s independent identically distributed random variables $\xi_1,\dots,\xi_s$ with outcomes $a_{v_1},\dots,a_{v_s}$ having probabilities

$$\frac{a_1}{N},\ \frac{a_2}{N},\ \dots,\ \frac{a_N}{N},$$

correspondingly. Setting

$$\eta_s = \frac{2T}{N} \frac{\xi_1+\dots+\xi_s}{s},$$

we can rewrite the inequality (6) in the form

$$\zeta(a,\, cN^{1/4}) \leqslant 2c_1 T \sum_{s\geqslant a} \frac{s^{s-2}}{s!} \mathbf{E}\left(\eta_s e^{-\eta_s}\right)^{s-2}.$$

Using Stirling's formula, we find the inequality

$$\mathbf{E}\zeta(a,\, cN^{1/4}) \leqslant c_2 T \sum_{s\geqslant a} \frac{1}{s^{5/2}} \mathbf{E} f^{s-2}(\eta_s),$$

where $f(x) = xe^{1-x}$. It is not hard to see that, under the conditions of the theorem,

$$\mathbf{E}\eta_s = \frac{2T}{N} \sum_{i=1}^{N} a_i^2 \to \lambda a^2.$$

Therefore, starting with some N and T, for sufficiently small positive Δ, we have

$$\Delta + \mathbf{E}\eta_s \leqslant q_1 < 1,$$

and, by monotonicity of f,

$$f(\mathbf{E}\eta_s + \Delta) \leqslant f(q_1) = q_2 < 1,$$

where q_1 and q_2 are some constants. The function $f(x)$ attains its maximum which is equal to one at the point $x = 1$. Hence, for sufficiently small positive Δ,

$$\mathbf{E}f^{s-2}(\eta_s) \leqslant f^{s-2}(\mathbf{E}\eta_s + \Delta)\mathbf{P}\{\eta_s < \mathbf{E}\eta_s + \Delta\} +$$

$$+ f^{s-2}(1)\mathbf{P}\{\eta_s \geqslant \mathbf{E}\eta_s + \Delta\} \leqslant$$

$$\leqslant q_2^{s-2} + \mathbf{P}\{\eta_s \geqslant \mathbf{E}\eta_s + \Delta\}. \tag{7}$$

Let us apply to the second summand in (7) the exponential estimate of this probability given in (Hoeffding, 1963): if $\xi_1, \ldots, \xi_s$ are independent identically distributed random variables, $0 < \varepsilon \leqslant \xi_i \leqslant E$, $i = 1, 2, \ldots, s$, then, for any positive Δ,

$$\mathbf{P}\{\xi_1 + \ldots + \xi_s - s\mathbf{E}\xi_1 \geqslant s\Delta\} =$$

$$= \mathbf{P}\left\{\frac{\xi_1 + \ldots + \xi_s}{s} \geqslant \mathbf{E}\xi_1 + \Delta\right\} \leqslant$$

$$\leqslant \exp\left\{-\frac{2s\Delta^2}{(E-\varepsilon)^2}\right\} = \exp\left\{-2s\left(\frac{\Delta}{E-\varepsilon}\right)^2\right\}.$$

Hence, for any fixed positive Δ,

$$\mathbf{P}\{\eta_s \geqslant \mathbf{E}\eta_s + \Delta\} \leqslant \left(\exp\left\{-\frac{2\Delta^2}{(E-\varepsilon)^2}\left(\frac{N}{2T}\right)^2\right\}\right)^s,$$

and there exists $q_3 < 1$ such that, starting with some N and T,

$$\mathbf{P}\{\eta_s \geqslant \mathbf{E}\eta_s + \Delta\} \leqslant q_3^s.$$

Taking into account that $q_2 = f(q_1) < 1$, $q_3 < 1$, we obtain the inequality

$$\mathbf{E}f^{s-2}(\eta_s) \leqslant q_2^{s-2} + q_3^s,$$

where $q = \max(q_2, q_3) < 1$. Thus,

$$\mathbf{E}\zeta(a, cN^{1/4}) \leqslant 2c_2T\sum_{s\geqslant a}\frac{1}{s^{5/2}}q_s \leqslant 2c_2T\frac{q^a}{1-q}.$$

If $a > c_3 \log N$, where c_3 is sufficiently large, then under the conditions of the theorem the right-hand side of the last expression tends to zero. It means that the graph $G_{N,T}$ does not contain trees that have the size $s \in [c_3 \log N, cN^{1/4}]$ with probability approaching one. Taking into account the assertion of Lemma 4, we obtain that the graph $G_{N,T}$ does not contain the components that are trees of the size larger than $c_3 \log N$ with probability approaching one.

REFERENCES

Bollobas, B. (1985). *Random Graphs*. Academic Press, London.

Erdös, P. and Rényi, A. (1960). On the evolution of random graphs. *Publ. Math. Inst. Hungaria Acad. Sci.* **5**, 17–61.

Hoeffding, W. (1963). Probability inequalities for sums of bounded random variables. *J. Amer. Statist. Assoc.* **58**, 13–30.

Ivchenko, G. I. (1975). Nonequiprobable random graphs. *Proc. MIEE* **44**, 34–66. (in Russian).

Karonski, M. (1982). A review of random graphs. *J. Graph Theory* **6**, 349–389.

Kolchin, V.F. (1986). On the behaviour of a random graph near a critical point. *Theory Probab. Appl.* **31**, 439–451.

Kolchin, V.F. and Khokhlov, V.I. (1990). On the number of cycles in nonequiprobable random graphs. *Discrete Math. Appl.* **2**, 137–145. (in Russian).

Kovalenko, I.N. (1971). On the theory of random graphs. *Kibernetika* **4**, 1–4. (in Russian).

Kovalenko, I.N. (1972). On the structure of a random digraph. *Theory Probab. and Math. Statist.* **6**, 83–91.

Palmer, E.M. (1985). *Graphical Evolution*. Wiley, New York.

Stepanov, V.E. (1987). On some features of the structure of a random graph near a critical point. *Theory Probab. Appl.* **32**, 573–594.

Takács, L. (1988). On the limit distribution of the number of cycles in a random graph. *J. Appl. Probab.* **25A**, 359–376.

New Trends in Probab. and Statist., pp. 457 - 473
V. Sazonov and T. Shervashidze (Eds.)
1991 VSP/Mokslas

IDENTITIES FOR LADDER FUNCTIONALS OF HOMOGENEOUS RANDOM PROCESSES AND WALKS WITH INDEPENDENT INCREMENTS

V.S. LUGAVOV and B.A. ROGOZIN
Pedagogical Institute, Kurgan, USSR;
Computing Centre of the Siberian Branch of USSR Acad. Sci., Omsk, USSR

ABSTRACT

Integral transformations of the joint distributions of random variables, connected with fluctuations of a process or a walk, are obtained for homogeneous processes and walks with independent increments.

1. LADDER FUNCTIONALS FOR WALKS

Let $\xi_1, \dots, \xi_n, \dots$ be a sequence of independent identically distributed random variables

$$S(n) = \sum_{k=1}^{n} \xi_k, \quad \overline{S}(n) = \max_{0 \leqslant k \leqslant n} S(k), \quad S(0) = \overline{S}(0) = 0, \quad n = 1, \dots$$

Define the random sets

$$T = \{n \geqslant 1: S(n) > \overline{S}(n-1)\} \cup \{0\}, \quad \widetilde{T} = \{n \geqslant 0: S(n) = \overline{S}(n)\}.$$

Random variables, determined by the sets $T, \widetilde{T}$ and by t e values of the walk S on T and $\widetilde{T}$ we shall call upper ladder functionals of the walk S. The following variables are upper ladder functionals

$$p_k = \sup \big\{[0,k] \cap T\big\} = \inf \big\{l: S(l) = \overline{S}(k)\big\},$$
$$q_k = \inf \big\{[k+1, \infty) \cap T\big\} = \inf \big\{l: S(l) > \overline{S}(k)\big\},$$
$$l_k = \sup \big\{p_l - p_{l-1}: l \leqslant k\big\}, \quad m_1(k) = N\big([0,k] \cap T\big), \tag{1}$$

for $k = 0, 1, \dots$, and

$$\gamma_x = \sup \{k: k \in T, \overline{S}(k) < x\},$$
$$\theta_x = \inf \{k: k \in T, \overline{S}(k) \geqslant x\} = \inf \{k: S(k) \geqslant x\},$$
$$h_x = \sup \{\overline{S}(k) - \overline{S}(k-1);\ k \in T, \overline{S}(k) < x\},$$

$$m_2(x) = N(\{k\colon k \in T,\ \overline{S}(k) < x\}) \tag{2}$$

for $x > 0$, where $N(A)$ denotes the number of elements in A, $\inf A = \infty$, $\sup A = 0$ if A is an empty set. If we replace T by $\widetilde{T}$ in (1), (2), we shall obtain the upper ladder functionals $\tilde{p}_k$, $\tilde{q}_k$, $\tilde{l}_k$, $\widetilde{m}_1(k)$ for $k = 0, 1, \ldots$ and $\widetilde{\gamma}_x$, $\widetilde{\theta}_x$, $\tilde{h}_x$, $\widetilde{m}_2(x)$ for $x > 0$, for which with probability 1

$$\overline{S}(k) = S(p_k) = S(\tilde{p}_k) = \overline{S}(p_k) = \overline{S}(\tilde{p}_k), \quad \theta_x = \widetilde{\theta}_x,$$

$$h_x = \tilde{h}_x, \quad S(\gamma_x) = S(\widetilde{\gamma}_x) = \overline{S}(\gamma_x) = \overline{S}(\widetilde{\gamma}_x).$$

The first ladder moments and heights:

$$t_1 = \inf\{n > 0\colon n \in T\}, \quad \tilde{t}_1 = \inf\{n > 0\colon n \in \widetilde{T}\},$$

$S(t_1), S(\tilde{t}_1)$ are also ladder functionals. Put for $|z| \leqslant 1$, $\operatorname{Re}\lambda \leqslant 0$

$$\Psi_+(z,\lambda) = 1 - \mathbf{E}\big[z^{t_1} e^{\lambda S(t_1)};\ t_1 \neq \infty\big],$$

$$\widetilde{\Psi}_+(z,\lambda) = 1 - \mathbf{E}\big[z^{\tilde{t}_1} e^{\lambda S(\tilde{t}_1)};\ \tilde{t}_1 \neq \infty\big].$$

For $|z| < 1$, the functions $\Psi_+(z,\lambda)$, $\widetilde{\Psi}_+(z,\lambda)$ are positive components of the canonical factorization of the function $1 - z\mathbf{E}\exp\{\lambda\xi_1\}$ for $\operatorname{Re}\lambda = 0$ (Borovkov, 1972). In this paper, using results of the renewal theory, we shall obtain the integral transformations of the following random vectors

$$(m_1(k),\ p_k,\ \overline{S}(k),\ q_k,\ \overline{S}(q_k),\ l_k), \quad (\widetilde{m}_1(k),\ \tilde{p}_k,\ \overline{S}(k),\ \tilde{q}_k,\ \overline{S}(\tilde{q}_k),\ \tilde{l}_k) \tag{3}$$

for $k = 0, 1, \ldots,$ and

$$(m_2(x),\ \gamma_x,\ \overline{S}(\gamma_x),\ \theta_x,\ \overline{S}(\theta_x),\ h_x), \quad (\widetilde{m}_2(x),\ \widetilde{\gamma}_x,\ \overline{S}(\widetilde{\gamma}_x),\ \theta_x,\ \overline{S}(\theta_x),\ h_x) \tag{4}$$

for $x > 0$, in terms of Ψ_+, $\widetilde{\Psi}_+$.

The book (Borovkov, 1972) contains a review of early results on such identities, which are called factorization identities. Factorization identities for functionals (3), (4) are obtained in this paper in a similar way.

2. LADDER FUNCTIONALS FOR PROCESSES

Let $\xi = \{\xi(t)\colon t \geqslant 0\}$, $\xi(0) = 0$, be a homogeneous process with independent increments having right continuous paths with left limits. Then

$$\mathbf{E}\exp\{\lambda\xi(t)\} = \exp\{t\Psi(\lambda)\},$$

for $\operatorname{Re}\lambda = 0$. Put

$$\bar{\xi}(t) = \sup_{0 \leqslant u \leqslant t} \xi(u), \quad \bar{\xi}(0) = 0, \quad \widehat{\xi}(t) = \bar{\xi}(t-0),$$

$$\widetilde{Z} = \{t \geqslant 0\colon \xi(t) = \bar{\xi}(t)\}, \quad Z = \{t > 0\colon \xi(t) > \bar{\xi}(u),\ 0 \leqslant u < t\} \cup \{0\},$$

for $t > 0$. Random variables, determined by the sets Z, $\widetilde{Z}$ and by the values of the process ξ on Z and $\widetilde{Z}$, we shall call upper ladder functionals of the process ξ. The following variables are upper ladder functionals of ξ:

$$p(t) = \sup \{Z \cap [0,t)\} = \inf \{u\colon \widehat{\xi}(u) = \widehat{\xi}(t)\},$$

$$q(t) = \inf \{Z \cap [t,\infty)\} = \inf \{u \geqslant t\colon \xi(u) > \widehat{\xi}(t)\},$$

$$l(t) = \sup \{q(u) - p(u)\colon q(u) < t\} \tag{5}$$

for $t > 0$ and

$$\gamma(x) = \sup \{t\colon t \in Z,\ \xi(t) < x\},$$

$$\theta(x) = \inf \{t\colon t \in Z,\ \xi(t) \geqslant x\} = \inf \{t\colon \xi(t) \geqslant x\},$$

$$h(x) = \sup \{\bar{\xi}(t) - \widehat{\xi}(t)\colon t \in Z,\ \bar{\xi}(t) < x\} \tag{6}$$

for $x > 0$. If we replace Z by $\widetilde{Z}$ in (5), (6), we shall obtain the following upper ladder functionals: $\tilde{p}(t)$, $\tilde{q}(t)$, $\tilde{l}(t)$, $\tilde{\gamma}(x)$, $\widetilde{\theta}(x)$, $\tilde{h}(x)$. Almost surely

$$\widehat{\xi}(x) = \bar{\xi}(t) = \bar{\xi}(p(t)) = \bar{\xi}(\tilde{p}(t)), \quad h(x) = \tilde{h}(x), \quad \theta(x) = \widetilde{\theta}(x)$$

for $t > 0$, $x > 0$. The ladder subordinators, introduced in (Fristedt, 1973), are ladder functionals, see also (Bingham, 1975; Lugavov and Rogozin, 1987b).
A two dimensional homogeneous process with independent increments

$$\mathcal{Z} = \{(\tau(l), \eta(l))\colon l \geqslant 0\} \quad \text{with } \mathbf{R}^2_+ = [0,\infty) \times [0,\infty)$$

as the state space and the terminal state (∞, ∞) is called a strictly increasing ladder subordinator for the process ξ if $\tau(0) = \eta(0) = 0$ and

1) $C(Z) = C(Q(\tau))$, where $Q(\tau) = \{\tau(l)\colon \tau(l) \neq \infty,\ l \geqslant 0\}$ and $C(A)$ is the closure of A,
2) $\mathbf{P}(\tau(1) = 0) < 1$,
3) $\eta(l) = \xi(\tau(l))$ if $\tau(l) \neq \infty$, and $\eta(l) = \infty$ if $\tau(l) = \infty$.

In the same way, using $\widetilde{Z}$ instead of Z, one can define a weakly increasing ladder subordinator

$$\widetilde{\mathcal{Z}} = \left\{(\tilde{\tau}(l), \tilde{\eta}(l))\colon l \geqslant 0\right\}.$$

Below we shall suppose that the path–functions of ladder subordinators are right–continuous with left–limits.
The characteristic exponents $f_+(\lambda_1, \lambda_2)$, $\tilde{f}_+(\lambda_1, \lambda_2)$ of the processes $\mathcal{Z}$, $\widetilde{\mathcal{Z}}$ are defined by

$$f_+(\lambda_1, \lambda_2) = -\log \mathbf{E}\big[\exp\{\lambda_1 \tau(1) + \lambda_2 \eta(1)\};\ \tau(1) \neq \infty\big],$$

$$\tilde{f}_+(\lambda_1, \lambda_2) = -\log \mathbf{E}\big[\exp\{\lambda_1 \tilde{\tau}(1) + \lambda_2 \tilde{\eta}(1)\};\ \tilde{\tau}(1) \neq \infty\big],$$

for $\operatorname{Re}\lambda_i \leqslant 0$, $i = 1, 2$.

If $\{(\tau(l),\ \eta(l))\colon l \geqslant 0\}$ is a ladder subordinator, then, for an arbitrary constant $c > 0$, the process $\{(\tau(cl),\ \eta(cl))\colon l \geqslant 0\}$ is a ladder subordinator too. In the sequel we shall assume that the characteristic exponents of the processes

$$\mathcal{Z} = \{(\tau(l),\ \eta(l))\colon l \geqslant 0\} \quad \text{and} \quad \widetilde{\mathcal{Z}} = \{(\tilde{\tau}(l),\ \tilde{\eta}(l))\colon l \geqslant 0\}$$

satisfy the condition $f_+(-1,0) = \tilde{f}_+(-1,0) = 1$. Under this condition the characteristic exponents of ladder subordinators are determined uniquely by the distribution of the process ξ (Fristedt, 1973). In (Lugavov and Rogozin, 1987b) a factorization method to determine these exponents was proposed.
Introduce ladder functionals of the local time type for the random sets Z, $\widetilde{Z}$, $\mathcal{X} = \{\bar{\xi}(t)\colon t \geqslant 0\}$,

$$n_1(t) = \inf\{l\colon \tau(l) \geqslant t\}, \quad n_2(x) = \inf\{l\colon \eta(l) \geqslant x\},$$

$$\tilde{n}_1(t) = \inf\{l\colon \tilde{\tau}(l) \geqslant t\}, \quad \tilde{n}_2(x) = \inf\{l\colon \tilde{\eta}(l) \geqslant x\},$$

for $t > 0$ and $x > 0$. Furthermore, in terms of f_+, $\tilde{f}_+$ we shall obtain the integral transformations of the joint distributions of the following random variables

$$\begin{gathered}
\big(n_1(t),\ p(t),\ \bar{\xi}(t),\ q(t),\ \bar{\xi}(q(t)),\ l(t)\big),\\
\big(\tilde{n}_1(t),\ \tilde{p}(t),\ \bar{\xi}(t),\ \tilde{q}(t),\ \bar{\xi}(\tilde{q}(t)),\ \tilde{l}(t)\big),\\
\big(N_1(t),\ p(t),\ \bar{\xi}(t),\ q(t),\ \bar{\xi}(q(t)),\ l(t)\big),
\end{gathered} \tag{7}$$

$$\begin{gathered}
\big(n_2(x),\ \gamma(x),\ \hat{\xi}(\theta(x)),\ \theta(x),\ \bar{\xi}(\theta(x)),\ h(x)\big),\\
\big(\tilde{n}_2(x),\ \tilde{\gamma}(x),\ \hat{\xi}(\theta(x)),\ \theta(x),\ \bar{\xi}(\theta(x)),\ \tilde{h}(x)\big),\\
\big(N_2(x),\ \gamma(x),\ \hat{\xi}(\theta(x)),\ \theta(x),\ \bar{\xi}(\theta(x)),\ h(x)\big),
\end{gathered} \tag{8}$$

where

$$N_1(t) = N(Z \cap [0,t)), \quad N_2(x) = N(\mathcal{X} \cap [0,x)).$$

If

$$\lim_{\lambda_1 = \operatorname{Re}\lambda_1 \to -\infty} f_+(\lambda_1, 0) = c_0 < \infty$$

then $N_1(t) < \infty$ for $t > 0$, and $N_2(x) < \infty$ for $x > 0$. In this case

$$\mathbf{P}\big(z_1 = \inf\{t > 0\colon t \in Z\} = 0\big) = 0$$

and

$$f_+(\lambda_1, \lambda_2) = c_0\big(1 - \mathbf{E}\big[e^{\lambda_1 z_1 + \lambda_2 \xi(z_1)};\ z_1 \neq \infty\big]\big),$$

for $\operatorname{Re}\lambda_i \leqslant 0$, $i = 1, 2$.
Reviews of the results on factorization identities for stochastic processes are contained in (Bingham, 1975; Fristedt, 1973; Gikhman and Skorokhod, 1975; Korolyuk, 1975; Pecherskii and Rogozin, 1969). In the present paper, contrary to the preceding works, factorization identities for the functionals (7) and (8) are obtained by a single method and include also the functionals $n_1(t)$, $\tilde{n}_1(t)$, $n_2(x)$, $\tilde{n}_2(x)$, $N_1(t)$, $N_2(x)$, $l(t)$, $\tilde{l}(t)$, $h(x)$.

For one class of processes in (Csáki and Földes, 1987) a factorization identity, containing the integral transformation of the distribution of the vector $(p(t),$ $\bar{\xi}(t),\ \max\{l(t),\ t-p(t)\})$, was obtained.

3. LAPLACE TRANSFORMS OF THE DISTRIBUTIONS OF THE RENEWAL FUNCTIONALS FOR TWO-DIMENSIONAL WALKS

Let

$$D_1(0)=D_2(0)=0,\quad \overline{D}(0)=\big(D_1(0),D_2(0)\big),$$
$$\overline{D}(1)=\big(D_1(1),D_2(1)\big),\ldots,\quad \overline{D}(n)=\big(D_1(n),D_2(n)\big),\ldots$$

be successive states of a two-dimensional homogeneous walk $\overline{D}$ with independent jumps with $\mathbf{R}_+^2$ as the state space and the terminal state (∞,∞). Denote for $\operatorname{Re}\lambda_i \leqslant 0,\ i=1,2,$

$$f(\lambda_1,\lambda_2)=\mathbf{E}\Big[\exp\big\{\bar{\lambda}\overline{D}(1)\big\};\ \overline{D}(1)\neq(\infty,\infty)\Big],$$

where $\mathbf{E}[\eta;A]=\mathbf{E}[\eta\chi_A]$, χ_A is the indicator of the event A and, for $\bar{\lambda}=(\lambda_1,\lambda_2)$ and $\bar{x}=(x_1,x_2)$, $\bar{\lambda}\bar{x}=\lambda_1x_1+\lambda_2x_2$. Put $g(0)=0$ and, for $k=1,\ldots,$

$$g(k)=\sup_{1\leqslant i\leqslant k}\big(D_2(i)-D_2(i-1)\big)$$

and

$$m(x)=\inf\{l\colon D_2(l)\geqslant x\}\quad \text{for } x>0.$$

LEMMA 1. *For* $\mathbf{P}(D_2(1)=0)<1,\ \nu\leqslant 0,\ \lambda_1\leqslant 0,\ \lambda_1+\mu_1\leqslant 0,\ \lambda_2\leqslant 0,$ $\lambda_2+\omega<0,\ \lambda_2+\mu_2+\omega<0,\ s>0,$

$$-\omega\int_0^\infty e^{\omega x}\mathbf{E}\Big[\exp\Big\{\nu m(x)+\bar{\mu}\overline{D}(m(x)-1)+\bar{\lambda}\overline{D}(m(x))\Big\};$$
$$g(m(x)-1)<s,\ \overline{D}(m(x))\neq(\infty,\infty)\Big]\,dx=$$
$$=\frac{e^{\nu}\big(f(\lambda_1,\lambda_2)-f(\lambda_1,\lambda_2+\omega)\big)}{1-e^{\nu}\mathbf{E}\big[\exp\big\{(\lambda_1+\mu_1)D_1(1)+(\lambda_2+\mu_2+\omega)D_2(1)\big\};\ D_2(1)<s\big]}. \tag{9}$$

For $\mathbf{P}(D_2(1)=0)<1,\ \omega<0,\ \nu\leqslant 0,\ \mu_1\leqslant 0,\ \mu_2+\omega<0,\ s>0,$

$$-\omega\int_0^\infty e^{\omega x}\mathbf{E}\Big[\exp\Big\{\nu m(x)+\bar{\mu}\overline{D}(m(x)-1)\Big\};\ g(m(x)-1)<s\Big]\,dx=$$
$$=\frac{e^{\nu}[1-f(0,\omega)]}{1-e^{\nu}\mathbf{E}\Big[\exp\big\{\mu_1D_1(1)+(\mu_2+\omega)D_2(1)\big\};\ D_2(1)<s\Big]}. \tag{10}$$

Proof. The condition $\mathbf{P}(D_2(1)=0)<1$ implies $\mathbf{P}(m(x)<+\infty)=1$. Therefore, the integral in relation (9) equals

$$\sum_{k=1}^{\infty} e^{k\nu}(-\omega)\int_0^{\infty} e^{\omega x}\mathbf{E}\Big[\exp\big\{\bar\mu\overline{D}(k-1)+\bar\lambda\overline{D}(k)\big\};\ g(k-1)<s,$$

$$m(x)=k,\ \overline{D}(k)\neq(\infty,\infty)\Big]\,dx=\sum_{k=1}^{\infty} e^{k\nu}H_k(\omega,\bar\mu,\bar\lambda,s). \tag{11}$$

For $k\geqslant 2,\ \bar y=(y_1,y_2),\ \bar z=(z_1,z_2),\ \bar\mu=(\mu_1,\mu_2)$,

$$H_k(\omega,\bar\mu,\bar\lambda,s)=-\omega\int_0^{\infty} e^{\omega x}\int_{y_1\geqslant 0}\int_{y_2\geqslant 0}\int_{z_1\geqslant 0}\int_{z_2\geqslant 0}\exp\{\bar\mu\bar y+\bar\lambda\bar z\}\times$$

$$\times\mathbf{P}\Big(D_2(k-1)<x,\ D_2(k)\geqslant x,\ \overline{D}(k-1)\in d\bar y,\ \overline{D}(k)\in d\bar z,\ g(k-1)<s\Big)\,dx=$$

$$=\iiiint\limits_{\substack{0\leqslant y_1\leqslant z_1\\ 0\leqslant y_2<z_2}}\exp\{\bar\mu\bar y+\bar\lambda\bar z\}\Big(\int_{(y_2,z_2)}(-\omega)e^{\omega x}dx\Big)\times$$

$$\times\mathbf{P}\Big(\overline{D}(k-1)\in d\bar y,\ \overline{D}(k)\in d\bar z,\ g(k-1)<s\Big)=$$

$$=\iiiint\limits_{\substack{0\leqslant y_1\leqslant z_1\\ 0\leqslant y_2<z_2}}\Big(\exp\big\{\mu_1y_1+(\mu_2+\omega)y_2+\lambda_1z_1+\lambda_2z_2\big\}-\exp\big\{\mu_1y_1+\mu_2y_2+\lambda_1z_1+$$

$$+(\lambda_2+\omega)z_2\big\}\Big)\mathbf{P}\Big(\overline{D}(k-1)\in d\bar y,\ \overline{D}(k)\in d\bar z,\ g(k-1)<s\Big).$$

Making the change of variables $y_1=y_1',\ y_2=y_2',\ z_1=z_1'+y_1',\ z_2=z_2'+y_2'$, we obtain

$$\iiiint\limits_{\substack{z_1'\geqslant 0,\ z_2'\geqslant 0,\\ y_1'\geqslant 0,\ y_2'\geqslant 0}}\exp\big\{(\lambda_1+\mu_1)y_1'+(\lambda_2+\mu_2+\omega)y_2'\big\}\Big[\exp\{\lambda_1z_1'+\lambda_2z_2'\}-$$

$$-\exp\{\lambda_1z_1'+(\lambda_2+\omega)z_2'\}\Big]\mathbf{P}\Big(\overline{D}(k-1)\in d\bar y',g(k-1)<s\Big)\mathbf{P}(\overline{D}(1)\in d\bar z')=$$

$$=\Big(f(\lambda_1,\lambda_2)-f(\lambda_1,\lambda_2+\omega)\Big)\int_{y_1'\geqslant 0}\int_{y_2'\geqslant 0}\exp\big\{(\lambda_1+\mu_1)y_1'+$$

$$+(\lambda_2+\mu_2+\omega)y_2'\big\}\mathbf{P}\Big(D_1(k-1)\in dy_1',\ D_2(k-1)\in dy_2',\ g(k-1)<s\Big)=$$

$$=\big(f(\lambda_1,\lambda_2)-f(\lambda_1,\lambda_2+\omega)\big)\Big(\mathbf{E}\Big[\exp\{(\lambda_1+\mu_1)+D_1(1)+$$

$$+(\lambda_2+\mu_2+\omega)D_2(1)\}; \; g(1)<s\big]\Big)^{k-1}.$$

Further, for $k=1$, we have

$$H_1(\omega,\bar{\mu},\bar{\lambda},s)=-\omega\int\limits_0^\infty e^{\omega x}\int\limits_{y_1\geqslant 0}\int\limits_{y_2\geqslant x}\exp\{\bar{\lambda}\bar{y}\}\mathbf{P}\big(\overline{D}(1)\in d\bar{y}\big)\,dx=$$

$$=\int\limits_0^\infty\int\limits_0^\infty \exp\{\bar{\lambda}\bar{y}\}\big(1-\exp\{\omega y_2\}\big)\mathbf{P}\big(\overline{D}(1)\in d\bar{y}\big)=f(\lambda_1,\lambda_2)-f(\lambda_1,\lambda_2+\omega). \quad (12)$$

Using the expressions for $H_k(\omega,\bar{\mu},\bar{\lambda},s)$ from (11), we obtain identity (9). Similar arguments prove identity (10).
From Lemma 1 we have

COROLLARY 1. *For* $\mathbf{P}(D_2(1)=0)<1$, $\nu\leqslant 0$, $\lambda_1\leqslant 0$, $\lambda_1+\mu_1\leqslant 0$, $\lambda_2\leqslant 0$, $\lambda_2+\omega<0$, $\lambda_2+\mu_2+\omega<0$, $u>0$, $v>0$,

$$\int\limits_0^\infty e^{\omega x}\mathbf{E}\Big[\exp\Big\{\nu m(x)+\bar{\mu}\overline{D}\big(m(x)-1\big)+\bar{\lambda}\overline{D}\big(m(x)\big)\Big\};$$

$$g\big(m(x)-1\big)<u,\; x-D_2\big(m(x)-1\big)<v,\; \overline{D}\big(m(x)\big)\neq(\infty,\infty)\Big]\,dx=$$

$$=e^{\nu}\Big(\int\limits_0^v e^{\omega s}\big(f(\lambda_1,\lambda_2)-f_2(\lambda_1,\lambda_2,s)\big)\,ds\Big)\times$$

$$\times\Big(1-e^{\nu}f_2(\mu_1+\lambda_1,\mu_2+\lambda_2+\omega,u)\Big)^{-1}. \quad (13)$$

For $\mathbf{P}(D_2(1)=0)<1$, $\omega<0$, $\nu\leqslant 0$, $\mu_1\leqslant 0$, $\mu_2+\omega<0$, $u>0$, $v>0$,

$$\omega\int\limits_0^\infty e^{\omega x}\mathbf{E}\Big[\exp\Big\{\nu m(x)+\bar{\mu}\overline{D}\big(m(x)-1\big)\Big\};\; g\big(m(x)-1\big)<u,$$

$$x-D_2\big(m(x)-1\big)<v,\overline{D}\big(m(x)\big)=(\infty,\infty)\Big]\,dx=$$

$$=e^{\nu}\big(1-f(0,0)\big)\big(e^{\omega v}-1\big)\big(1-f_2(\mu_1,\mu_2+\omega,u)\big)^{-1}, \quad (14)$$

where

$$f_2(\lambda_1,\lambda_2,s)=\mathbf{E}[\exp\{\bar{\lambda}\overline{D}(1)\};\; D_2(1)<s].$$

Proof. Considering (9) with ω replaced by $\omega+\omega'$ and μ_2 replaced by $\mu_2-\omega'$, we obtain

$$-(\omega+\omega')^{-1}e^{\nu}\big(f(\lambda_1,\lambda_2)-f(\lambda_1,\lambda_2+\omega+\omega')\big)\times$$

$$\times\Big(1 - e^{\nu}\mathbf{E}\big[\exp\{(\lambda_1+\mu_1)D_1(1) + (\lambda_2+\mu_2+\omega)D_2(1)\};\ D_2(1) < u\big]\Big)^{-1} =$$

$$= \int\limits_0^\infty e^{\omega x}\mathbf{E}\Big[\exp\Big\{\nu m(x) + \bar{\mu}\overline{D}\big(m(x)-1\big) + \bar{\lambda}\overline{D}\big(m(x)\big) +$$

$$+\omega'(x) - D_2\big(m(x)-1\big)\Big\};\ g\big(m(x)-1\big) < u,\ \overline{D}(m(x)) \neq (\infty,\infty)\Big]\,dx =$$

$$= \int\limits_0^\infty e^{\omega' v}d_v \int\limits_0^\infty e^{\omega x}\mathbf{E}\Big[\exp\Big\{\nu m(x) + \bar{\mu}\overline{D}\big(m(x)-1\big) + \bar{\lambda}\overline{D}\big(m(x)\big)\Big\};$$

$$g\big(m(x)-1\big) < u,\ \overline{D}\big(m(x)\big) \neq (\infty,\infty),\ x - D_2\big(m(x)-1\big) < v\Big]\,dx.$$

Using relation (12), we obtain

$$-(\omega+\omega')^{-1}\big(f(\lambda_1,\lambda_2) - f(\lambda_1,\lambda_2+\omega-\omega')\big) =$$

$$= \int\limits_0^\infty e^{(\omega'+\omega)v}\mathbf{E}\Big[\exp\{\lambda_1 D_1(1) + \lambda_2 D_2(1)\};\ v \leqslant D_2(1) < \infty\Big]\,dv =$$

$$= \int\limits_0^\infty e^{\omega' v}d_v\Big(\int\limits_0^v e^{\omega s}\mathbf{E}\Big[\exp\{\bar{\lambda}\overline{D}(1)\};\ s \leqslant D_2(1) < \infty\Big]\,ds\Big)$$

and this implies (13). In the same way we prove (14).

4. LAPLACE TRANSFORMS OF THE DISTRIBUTIONS OF THE RENEWAL FUNCTIONALS FOR TWO-DIMENSIONAL SUBORDINATORS

Let $\overline{T} = \{\overline{T}(t) = (T_1(t), T_2(t))\colon t \geqslant 0\}$; $\overline{T}(0,0) = (0,0)$ be a homogeneous two-dimensional process with independent increments, with $\mathbf{R}_+^2$ as the state space and the terminal state (∞,∞) with $\mathbf{P}(\overline{T}(1) = (\infty,\infty)) < 1$. The following Lévy representation holds

$$\mathbf{E}\Big[\exp\{\lambda_1 T_1(t) + \lambda_2 T_2(t)\};\ \overline{T}(t) \neq (\infty,\infty)\Big] = \exp\{-t\varphi(\lambda_1,\lambda_2)\},$$

$$-\varphi(\lambda_1,\lambda_2) = -a_0 + a_1\lambda_1 + a_2\lambda_2 + \iint\limits_{\mathbf{R}_+^2}\big(\exp\{\lambda_1 y_1 + \lambda_2 y_2\} - 1\big)\,d\Pi, \qquad (15)$$

for $\operatorname{Re}\lambda_i \leqslant 0,\ i = 1,2$, where $a_i \geqslant 0,\ i = 0,1,2$, and Π is a measure on $\mathbf{R}_+^2$ such that

$$\iint\limits_{\mathbf{R}_+^2} \min\{y_1 + y_2,\ 1\}\,d\Pi < \infty.$$

We shall call the process $\overline{T}$ a two-dimensional subordinator and the function $\varphi(\lambda_1, \lambda_2)$, for $\operatorname{Re}\lambda_i \leqslant 0$, $i = 1,2$, the characteristic exponent of $\overline{T}$. We shall assume the path-functions of $\overline{T}$ to be right-continuous with left-limits. Put, for $x > 0$,

$$n(x) = \inf\{t\colon T_2(t) \geqslant x\}, \quad d(x) = \sup_{0 \leqslant t < x} \big(T_2(t) - T_2(t-0)\big).$$

LEMMA 2. *For* $\mathbf{P}(T_2(1) = 0) < 1$, $\nu \leqslant 0$, $\lambda_1 \leqslant 0$, $\lambda_1 + \mu_1 \leqslant 0$, $\lambda_2 \leqslant 0$, $\lambda_2 + \omega < 0$, $\mu_2 + \lambda_2 + \omega < 0$, $u > 0$,

$$-\omega \int_0^\infty e^{\omega x} \mathbf{E}\Big[\exp\big\{\nu n(x) + \bar{\mu}\overline{T}(n(x) - 0) + \bar{\lambda}\overline{T}(n(x))\big\};$$

$$d(n(x)) < u,\ \overline{T}(n(x)) \neq (\infty, \infty)\Big]\, dx = \big(\varphi(\lambda_1, \lambda_2 + \omega) - \varphi(\lambda_1, \lambda_2)\big) \times$$

$$\times\big(-\nu + \varphi(\lambda_1 + \mu_1, \lambda_2 + \mu_2 + \omega) + \varphi_2(\lambda_1 + \mu_1, \lambda_2 + \mu_2 + \omega, u)\big)^{-1}. \tag{16}$$

For $\mathbf{P}(T_2(1) = 0) < 1$, $\nu \leqslant 0$, $\omega < 0$, $\mu_1 \leqslant 0$, $\mu_2 + \omega < 0$, $u > 0$,

$$-\omega \int_0^\infty e^{\omega x} \mathbf{E}\Big[\exp\big\{\nu n(x) + \bar{\mu}\overline{T}(n(x) - 0)\big\};\ d(n(x)) < u\Big]\, dx =$$

$$= \varphi(0, \omega)\big(-\nu + \varphi(\mu_1, \mu_2 + \omega) + \varphi_2(\mu_1, \mu_2 + \omega, u)\big)^{-1}, \tag{17}$$

where

$$\varphi_2(\lambda_1, \lambda_2, u) = \int_{y_1 \geqslant 0} \int_{y_2 \geqslant u} \exp\{\lambda_1 y_1 + \lambda_2 y_2\}\, d\Pi.$$

Proof. Consider the sequence of processes

$$\overline{T}^{(k)} = \{\overline{T}^{(k)}(t)\colon t \geqslant 0\}, \quad k = 1, \dots,$$

where

$$\overline{T}^{(k)}(t) = \overline{T}(n/2^k) \quad \text{for } n/2^k \leqslant t < (n+1)/2^k, \quad n = 0, 1, \dots.$$

In the same way as we have defined $n(x)$, $\overline{T}(n(x) - 0)$, $\overline{T}(n(x))$ for $\overline{T}$, we define $n^{(k)}(x)$, $\overline{T}^{(k)}(n^{(k)}(x) - 0)$, $\overline{T}^{(k)}(n^{(k)}(x))$ for the process $\overline{T}^{(k)}$. Consider the walk $\overline{D}^{(k)} = \{\overline{D}^{(k)}(n) = \overline{T}(n2^{-k})\colon n = 0, 1, \dots\}$. From relation (9), we obtain

$$-\omega \int_0^\infty e^{\omega x} \mathbf{E}\Big[\exp\Big\{\frac{\nu}{2^k} n^{(k)}(x) 2^k + \bar{\mu}\overline{T}\Big(\frac{2^k n^{(k)}(x) - 1}{2^k}\Big) + \bar{\lambda}\overline{T}\Big(\frac{2^k n^{(k)}(x)}{2^k}\Big)\Big\};$$

$$\max\Big\{0, T_2\Big(\frac{1}{2^k}\Big), T_2\Big(\frac{2}{2^k}\Big) - T_2\Big(\frac{1}{2^k}\Big), \ldots, T_2\Big(\frac{2^k n^{(k)}(x) - 1}{2^k}\Big) -$$

$$-T_2\Big(\frac{2^k n^{(k)}(x) - 2}{2^k}\Big)\Big\} < u,\ \overline{T}\Big(\frac{2^k n^{(k)}(x)}{2^k}\Big) \neq (\infty, \infty)\Big]\, dx =$$

$$= e^{\nu/2^k}\Big(\mathbf{E}\Big[\exp\{\bar{\lambda}\overline{T}(2^{-k})\};\ \overline{T}(2^{-k}) \neq (\infty,\infty)\Big] - \tag{18}$$

$$-\mathbf{E}\Big[\exp\{\lambda_1 T_1(2^{-k}) + (\lambda_2 + \omega) T_2(2^{-k})\};\ \overline{T}(2^{-k}) \neq (\infty,\infty)\Big]\Big)\times$$

$$\times\Big(1 - e^{\nu/2^k}\mathbf{E}\Big[\exp\{(\lambda_1 + \mu_1)T_1(2^{-k}) + (\lambda_2 + \mu_2 + \omega)T_2(2^{-k})\};\ T_2(2^{-k}) < u\Big]\Big)^{-1}.$$

The limit of the integral in (18) as $k \to \infty$ coincides with the integral in the left side of (16). Moreover

$$\lim_{k\to\infty} 2^k\Big(1 - \mathbf{E}\Big[\exp\{\bar{\lambda}\overline{T}(2^{-k})\};\ \overline{T}(2^{-k}) \neq (\infty,\infty)\Big]\Big) = \varphi(\lambda_1, \lambda_2),$$

for $\operatorname{Re}\lambda_i \leqslant 0$, $i = 1, 2$, and, for point u such that $\Pi(A_u) = 0$,

$$\lim_{k\to\infty} 2^k\mathbf{E}\Big[\exp\Big\{(\lambda_1 + \mu_1)T_1(2^{-k}) + (\lambda_2 + \mu_2 + \omega)T_2(2^{-k})\Big\};\ u \leqslant T_2(2^{-k}) < \infty\Big] =$$

$$= \int\limits_{y_1 \geqslant 0}\ \int\limits_{y_2 \geqslant u} \exp\Big\{(\lambda_1 + \mu_1)y_1 + (\lambda_2 + \mu_2 + \omega)y_2\Big\}\, d\Pi,$$

where $A_u = \{(y_1, y_2)\colon 0 \leqslant y_1 < \infty,\ y_2 = u\}$. Therefore, the limit of the right-hand side of (18) as $k \to \infty$ coincides with the right side of (16) for each $u > 0$ such that $\Pi(A_u) = 0$. Using the left continuity of both sides of (16), we prove (16) for all $u > 0$. One can similarly prove (17).
From Lemma 2 we have

COROLLARY 2. For $\mathbf{P}(T_2(1) = 0) < 1$, $\nu \leqslant 0$, $\lambda_1 \leqslant 0$, $\mu_1 + \lambda_1 \leqslant 0$, $\lambda_2 \leqslant 0$, $\lambda_2 + \omega < 0$, $\mu_2 + \lambda_2 + \omega < 0$, $u > 0$, $v > 0$,

$$\int\limits_0^\infty e^{\omega x}\mathbf{E}\Big[\exp\Big\{\nu n(x) + \bar{\mu}\overline{T}(n(x) - 0) + \bar{\lambda}\overline{T}(n(x))\Big\};$$

$$x - T_2(n(x) - 0) < v,\ d(n(x)) < u,\ \overline{T}(n(x)) \neq (\infty,\ \infty)\Big]\, dx =$$

$$= \Big(a_2 + \int\limits_0^v e^{\omega s}\varphi_2(\lambda_1, \lambda_2, s)ds\Big)\Big(-\nu + \varphi(\lambda_1 + \mu_1, \lambda_2 + \mu_2 + \omega) +$$

$$+\varphi_2(\lambda_1 + \mu_1, \lambda_2 + \mu_2 + \omega, u)\Big)^{-1}. \tag{19}$$

For $\mathbf{P}(T_2(1) = 0) < 1,\ \nu \leqslant 0,\ \omega < 0,\ \mu_1 \leqslant 0,\ \mu_2 + \omega < 0,\ u > 0,\ v > 0,$

$$\omega \int_0^\infty e^{\omega x} \mathbf{E}\Big[\exp\Big\{ \nu n(x) + \bar{\mu} \overline{T}\big(n(x) - 0\big) \Big\};\ x - T_2\big(n(x) - 0\big) < v,$$

$$d\big(n(x)\big) < u,\ \overline{T}\big(n(x)\big) = (\infty, \infty) \Big]\, dx =$$

$$= \varphi(0,0)\Big(\exp\{\omega v\} - 1 \Big) \Big[-\nu + \varphi(\mu_1, \mu_2 + \omega) + \varphi_2(\mu_1, \mu_2 + \omega, u) \Big]^{-1}. \tag{20}$$

Proof of Corollary 2 is similar to the proof of Corollary 1.

Remark 1. Putting $\lambda_1 = \lambda_2 = \mu_1 = \mu_2 = \nu = 0,\ u = v$ in (19), we get an identity for the distribution of $\max(x - T_2(n(x) - 0),\ d(n(x)))$. An equivalent formula for this distribution was given in (Krylov and Juškevič, 1965) and in (Horowitz, 1972).
Putting $\mu_1 = \lambda_1 = 0,\ u = \infty$ in (16), we get the identity which is known from (Fristedt, 1973).

Remark 2. The reasoning from Lemma 1 is applicable also under some other restrictions on the jumps of the walk $\overline{D}$ before the first-passage across a level x. This remark concerns the other statements of this section too.

5. IDENTITIES FOR WALKS

Consider the sequences

$$\big\{ (t_k, \overline{S}(t_k)) \colon k = 0, 1, \dots \big\}, \quad \big\{ (\tilde{t}_k, \overline{S}(\tilde{t}_k)) \colon k = 0, 1, \dots \big\},$$

where

$$t_0 = \tilde{t}_0 = 0, \quad t_k = \inf\{[t_{k-1} + 1, \infty) \cap T\},$$
$$\tilde{t}_k = \inf\{[\tilde{t}_{k-1} + 1, \infty) \cap \widetilde{T}\},$$

$k = 1, \dots;\ \overline{S}(\infty) = \infty$. These sequences are homogeneous two-dimensional walks with independent jumps, determining the ladder moments and heights for the walk S. Putting $\overline{D}(k) = (t_k, \overline{S}(t_k)),\ k = 0, 1, \dots$, we obtain

$$m(x) = m_2(x), \quad D_1\big(m(x)\big) = \theta_x, \quad D_1\big(m(x) - 1\big) = \gamma_x, \quad g\big(m(x) - 1\big) = h_x,$$

for $x > 0$. Furthermore, putting $\lambda = \lambda_2,\ \mu = \mu_2,\ \alpha = e^{\lambda_1},\ \beta = e^{\mu_1},\ p = e^{\nu}$ in (13) and (14), we get

THEOREM 1. *For* $|\rho| \leqslant 1,\ |\alpha| \leqslant 1,\ |\alpha\beta| \leqslant 1,\ \mathrm{Re}\,\lambda \leqslant 0,\ \mathrm{Re}(\lambda + \omega) < 0,$ $\mathrm{Re}(\mu + \lambda + \omega) < 0,\ u > 0,\ v > 0,\ \widehat{S} = \lim_{n \to \infty} \overline{S}(n)$

$$\int_0^\infty e^{\omega x} \mathbf{E}\Big[\rho^{m_2(x)} \beta^{\gamma_x} \alpha^{\theta_x} \exp\Big\{ \mu \overline{S}(\gamma_x) + \lambda \overline{S}(\theta_x) \Big\};$$

$$h_x < u,\ x - \overline{S}(\gamma_x) < v,\ \widehat{S} \geqslant x\Big] dx = \rho \int_0^v e^{\omega s} \mathbf{E}\Big[\alpha^{t_1} e^{\lambda \overline{S}(t_1)};\ s \leqslant \overline{S}(t_1) < \infty\Big] ds \times$$

$$\times \Big(1 - \rho \mathbf{E}\Big[(\alpha\beta)^{t_1} e^{(\lambda+\mu+\omega)\overline{S}(t_1)};\ \overline{S}(t_1) < u\Big]\Big)^{-1}. \tag{21}$$

For $\operatorname{Re}\omega < 0,\ |\rho| \leqslant 1,\ |\beta| \leqslant 1,\ \operatorname{Re}(\mu+\omega) < 0,\ u > 0,\ v > 0,$

$$\omega \int_0^\infty e^{\omega x} \mathbf{E}\Big[\rho^{m_2(x)} \beta^{\gamma_x} \exp\{\mu \overline{S}(\gamma_x)\};\ h_x < u,\ x - \overline{S}(\gamma_x) < v,\ \widehat{S} < x\Big] dx =$$

$$= \rho \mathbf{P}(t_1 = \infty)(e^{\omega v} - 1)\Big(1 - \rho \mathbf{E}\Big[\beta^{t_1} e^{(\mu+\omega)\overline{S}(t_1)};\ \overline{S}(t_1) < u\Big]\Big)^{-1}. \tag{22}$$

In the same way, using the walk $\{(\tilde{t}_k, \overline{S}(\tilde{t}_k)) \colon k = 0, 1, \ldots\}$, we obtain

THEOREM 2. *If* $\mathbf{P}(\xi_1 = 0) < 1$ *and if the conditions of Theorem* 1 *are satisfied, then in* (21), (22) *one may replace* $m_2(x),\ \gamma_x,\ t_1$ *by* $\widetilde{m}_2(x),\ \widetilde{\gamma}_x,\ \tilde{t}_1$, *respectively.*

Put $\overline{D}(k) = (\overline{S}(t_k), t_k),\ k = 0, 1, \ldots$ In this case we get

$$m(x) = m_1(k), \quad D_2(m(x)) = q_k, \quad D_2(m(x) - 1) = p_k, \quad g(m(x) - 1) = l_k$$

for $k < x < k+1,\ k = 0, 1, \ldots$ Putting $\lambda = \lambda_1,\ \mu = \mu_1,\ \alpha = e^{\lambda_2},\ \beta = e^{\mu_2},\ \gamma = e^{\omega},\ \rho = e^{\nu}$ in (13) and (14), we obtain

THEOREM 3. *For* $|\rho| \leqslant 1,\ |\alpha| \leqslant 1,\ |\alpha\gamma| < 1,\ |\alpha\beta\gamma| < 1,\ \operatorname{Re}\lambda \leqslant 0,\ \operatorname{Re}(\lambda+\mu) \leqslant 0,\ u > 0,\ v > 0,\ q = \lim_{k\to\infty} p_k$

$$\sum_{k=0}^\infty \gamma^k \mathbf{E}\Big[\rho^{m_1(k)} \beta^{p_k} \alpha^{q_k} \exp\Big\{\mu \overline{S}(k) + \lambda \overline{S}(q_k)\Big\};\ l_k < u,\ k - p_k < v,\ q > k\Big] =$$

$$= \rho \sum_{0 \leqslant k < v} \gamma^k \mathbf{E}\Big[\alpha^{t_1} e^{\lambda \overline{S}(t_1)};\ k \leqslant t_1 < \infty\Big] \Big(1 - \rho \mathbf{E}\Big[(\alpha\beta\gamma)^{t_1} e^{(\lambda+\mu)\overline{S}(t_1)};\ t_1 < u\Big]\Big)^{-1}.$$

For $|\rho| \leqslant 1,\ |\gamma| < 1,\ |\beta\gamma| < 1,\ \operatorname{Re}\mu \leqslant 0,\ u > 0,\ v > 0,$

$$\sum_{k=0}^\infty \gamma^k \mathbf{E}\Big[\rho^{m_1(k)} \beta^{p_k} \exp\Big\{\mu \overline{S}(k)\Big\};\ l_k < u,\ k - p_k < v,\ q \leqslant k\Big] =$$

$$= \rho \mathbf{P}(t_1 = \infty) \sum_{0 \leqslant k < v} \gamma^k \Big(1 - \rho \mathbf{E}\Big[(\beta\gamma)^{t_1} \exp\Big\{\mu \overline{S}(t_1)\Big\};\ t_1 < u\Big]\Big)^{-1}.$$

Similar statements are true for the distributions of the vectors

$$(\widetilde{m}_1(k),\ \tilde{p}_k, \overline{S}(k),\ \tilde{q}_k,\ \overline{S}(\tilde{q}_k),\ \tilde{l}_k).$$

6. IDENTITIES FOR PROCESSES

Express the ladder functionals introduced in Section 2, in terms of the ladder subordinators $\mathcal{Z} = \{(\tau(l), \eta(l)) \colon l \geqslant 0\}$, $\widetilde{\mathcal{Z}} = \{(\tilde{\tau}(l), \tilde{\tau}(l)) \colon l \geqslant 0\}$ and the functionals $n_1(t)$, $\tilde{n}_1(t)$, $n_2(x)$, $\tilde{n}_2(x)$, for $t > 0$, $x > 0$.

LEMMA 3. *With probability* 1,

$$p(t) = \tau(n_1(t) - 0), \quad q(t) = \tau(n_1(t)),$$

$$l(t) = \sup \{\tau(l) - \tau(l - 0) \colon l < n_1(t)\},$$

$$\tilde{p}(t) = \tilde{\tau}(\tilde{n}_1(t) - 0), \quad \tilde{q}(t) = \tilde{\tau}(\tilde{n}_1(t)),$$

$$\tilde{l}(t) = \sup \{\tilde{\tau}(l) - \tilde{\tau}(l - 0) \colon l < \tilde{n}_1(t)\},$$

for $t > 0$ *and*

$$\gamma(x) = \tau(n_2(x) - 0), \quad \theta(x) = \tau(n_2(x)),$$

$$h(x) = \sup \{\eta(l) - \eta(l - 0) \colon l < n_2(x)\},$$

for $x > 0$.
If $\mathbf{P}(\xi(1) = 0) < 1$ *and* $x > 0$, *then with probability* 1,

$$\tilde{\gamma}(x) = \tilde{\tau}(\tilde{n}_2(x) - 0), \quad \tilde{\theta}(x) = \tilde{\tau}(\tilde{n}_2(x)),$$

$$\tilde{h}(x) = \sup \{\tilde{\eta}(l) - \tilde{\eta}(l - 0) \colon l < \tilde{n}_2(x)\}.$$

Proof. With probability 1,

$$p(t) = \sup \{Z \cap [0, t)\} = \sup \{C(Z) \cap [0, t)\} =$$

$$= \sup \{C(Q(\tau)) \cap [0, t)\} = \sup \{Q(\tau) \cap [0, t)\} = \tau(n_1(t) - 0),$$

$$l(t) = \sup \{\tau(n_1(u)) - \tau(n_1(u) - 0) \colon \tau(n_1(u)) < t\} =$$

$$= \sup \{\tau(n_1(u)) - \tau(n_1(u) - 0) \colon n_1(u) < n_1(t)\} = \sup \{\tau(l) - \tau(l - 0) \colon l < n_1(t)\}.$$

Denote

$$\widehat{\xi} = \widehat{\xi}(\infty) = \lim_{t \to \infty} \bar{\xi}(t), \quad q = \lim_{t \to \infty} p(t), \quad \tilde{q} = \lim_{t \to \infty} \tilde{p}(t).$$

We shall use the Lévy representation (15) for the characteristic exponent $f_+(\lambda_1, \lambda_2)$ of the process $\mathcal{Z}$.

THEOREM 4. *For* $\operatorname{Re} \nu \leqslant 0$, $\operatorname{Re} \lambda_1 \leqslant 0$, $\operatorname{Re}(\mu_1 + \lambda_1) \leqslant 0$, $\operatorname{Re} \lambda_2 \leqslant 0$, $\operatorname{Re}(\lambda_2 + \omega) < 0$, $\operatorname{Re}(\lambda_2 + \mu_2 + \omega) < 0$, $u > 0$, $v > 0$,

$$\int_0^\infty e^{\omega x} \mathbf{E}\Big[\exp \Big\{\nu n_2(x) + \mu_1 \gamma(x) + \mu_2 \widehat{\xi}(\theta(x)) + \lambda_1 \theta(x) +$$

$$+\lambda_2\bar{\xi}(\theta(x))\Big\}; \ x-\widehat{\xi}(\theta(x))<v, \ h(x)<u, \ \widehat{\xi}\geqslant x\Big]\,dx=$$

$$=\Big(a_2+\int_0^v e^{\omega s}f_{2+}(\lambda_1,\lambda_2,s)\,ds\Big)\Big(-\nu+f_+(\lambda_1+\mu_1,\lambda_2+\mu_2+\omega)+$$

$$+f_{2+}(\lambda_1+\mu_1,\lambda_2+\mu_2+\omega,u)\Big)^{-1},$$

where

$$f_{2+}(\lambda_1,\lambda_2,u)=\int_{y_1\geqslant 0}\int_{y_2\geqslant u}\exp\{\lambda_1y_1+\lambda_2y_2\}\,d\Pi.$$

For $\operatorname{Re}\nu\leqslant 0,\ \operatorname{Re}\mu_1\leqslant 0,\ \operatorname{Re}\omega<0,\ \operatorname{Re}(\mu_2+\omega)<0,\ u>0,\ v>0,$

$$\omega\int_0^\infty e^{\omega x}\mathbf{E}\Big[\exp\Big\{\nu n_2(x)+\mu_1\gamma(x)+\mu_2\widehat{\xi}(\theta(x))\Big\}; \ h(x)<u, \ x-\widehat{\xi}(\theta(x))<v,$$

$$\widehat{\xi}<x\Big]\,dx=f_+(0,0)(e^{\omega v}-1)\Big[-\nu+f_+(\mu_1,\mu_2+\omega)+f_{2+}(\mu_1,\mu_2+\omega,u)\Big]^{-1}.$$

For $\operatorname{Re}\nu\leqslant 0,\ \operatorname{Re}\lambda_1\leqslant 0,\ \operatorname{Re}(\lambda_1+\omega)<0,\ \operatorname{Re}(\lambda_1+\mu_1+\omega)<0,\ \operatorname{Re}\lambda_2\leqslant 0,$ $\operatorname{Re}(\lambda_2+\mu_2)\leqslant 0,\ u>0,\ v>0,$

$$\int_0^\infty e^{\omega t}\mathbf{E}\Big[\exp\Big\{\nu n_1(t)+\mu_1p(t)+\mu_2\widehat{\xi}(t)+\lambda_1q(t)+\lambda_2\bar{\xi}(q(t))\Big\};$$

$$t-p(t)<v, \ l(t)<u, \ q\geqslant t\Big]\,dt=\Big(a_1+\int_0^v e^{\omega s}f_{1+}(\lambda_1,\lambda_2,s)\,ds\Big)\times$$

$$\times\Big(-\nu+f_+(\lambda_1+\mu_1+\omega,\lambda_2+\mu_2)+f_{1+}(\lambda_1+\mu_1+\omega,\lambda_2+\mu_2),u)\Big)^{-1}, \quad (23)$$

where

$$f_{1+}(\lambda_1,\lambda_2,u)=\int_{y_1\geqslant u}\int_{y_2\geqslant 0}\exp\{\lambda_1y_1+\lambda_2y_2\}\,d\Pi.$$

For $\operatorname{Re}\nu\leqslant 0,\ \operatorname{Re}\mu_2\leqslant 0,\ \operatorname{Re}\omega<0,\ \operatorname{Re}(\mu_1+\omega)<0,\ u>0,\ v>0,$

$$\omega\int_0^\infty e^{\omega t}\mathbf{E}\Big[\exp\Big\{\nu n_1(t)+\mu_1p(t)+\mu_2\widehat{\xi}(t)\}; \ l(t)<u, \ t-p(t)<v, \ q<t\Big]\,dt=$$

$$=f_+(0,0)(e^{\omega v}-1)\Big(-\nu+f_+(\mu_1+\omega,\mu_2)+f_{1+}(\mu_1+\omega,\mu_2,u)\Big)^{-1}.$$

Proof. The statement of this theorem follows from Corollary 2, Lemma 3 and from the following relations:

$$\eta(n_2(x)-0)=\sup\left\{\eta(l)\colon l<n_2(x)\right\}=\sup\left\{\bar\xi(v)\colon v<\theta(x)\right\}=\widehat\xi(\theta(x)),$$

$$\eta(n_1(t)-0)=\sup\left\{\eta(l)\colon l<n_1(t)\right\}=\sup\left\{\bar\xi(v)\colon v<t\right\}=\widehat\xi(t)=\bar\xi(t),$$

$$\left\{\tau(n_1(t))<\infty\right\}=\{q\geqslant t\},\quad \left\{\tau(n_2(x))<\infty\right\}=\{\widehat\xi\geqslant x\}$$

with probability 1. In the same way we may obtain the distributions of the vectors $(\tilde n_1(t),\ \tilde p(t),\ \bar\xi(t),\ \tilde q(t),\ \bar\xi(\tilde q(t)),\ \tilde l(t))$ for $t>0$, and under the condition $\mathbf{P}(\xi(1)= =0)<1$ of the vectors $(\tilde n_2(x),\ \tilde\gamma(x),\ \widehat\xi(\theta(x)),\ \theta(x),\ \bar\xi(\theta(x)),\ h(x))$ for $x>0$.
Put $z_0=0$. If $\mathbf{P}(z_1=\inf\{t>0\colon t\in Z\}=0)=0$ then we shall define $z_k=\inf\{t> >z_{k-1}\colon t\in Z\}$ for $k=1,2,\ldots$ Note that if $\mathbf{P}(z_1=0)=1$, then $\mathbf{P}(\tau(1)=0)=0$ and, hence,

$$\lim_{\lambda_1=\mathrm{Re}\,\lambda_1\to-\infty} f_+(\lambda_1,0)=\infty.$$

If $\mathbf{P}(z_1=0)=0$ then $\mathbf{P}(\tau(1)\mp 0)>0$ and, hence,

$$\lim_{\lambda_1=\mathrm{Re}\,\lambda_1\to-\infty} f_+(\lambda_1,0)=c_0, \tag{24}$$

where c_0 is a finite positive constant. If $\mathbf{P}(z_1=0)=0$ then the ladder subordinator $\mathcal{Z}\{(\tau(l),\ \eta(l))\colon l\geqslant 0\}$ is a generalized Poisson process (Fristedt, 1973) and, hence,

$$f_+(\lambda_1,\lambda_2)=a_0+\int\limits_0^\infty\int\limits_0^\infty\left(1-\exp\left\{\lambda_1y_1+\lambda_2y_2\right\}\right)d\Pi=$$

$$=c_0\Big(1-\mathbf{E}\Big[\exp\{\lambda_1z_1+\lambda_2\bar\xi(z_1)\};\ z_1\neq\infty\Big]\Big),$$

for $\mathrm{Re}\,\lambda_i\leqslant 0,\ i=1,2$.
If we put, using the notation from Corollary 1 and Lemma 1, $\overline{D}(k)=(z_k,\bar\xi(z_k))$, $k=0,1,\ldots$, then we shall obtain the homogeneous two-dimensional random walk with independent increments, for which

$$m(x)=N_2(x),\quad D_1(m(x)-1)=\gamma(x),\quad D_1(m(x))=\theta(x),$$

$$D_2(m(x)-1)=\widehat\xi(\theta(x)),\quad D_2(m(x))=\bar\xi(\theta(x)),\quad g(m(x)-1)=h(x).$$

From Corollary 1 we obtain the following statement.

THEOREM 5. *If* (24) *is valid for* ξ, *then for* $\mathrm{Re}\,\nu\leqslant 0$, $\mathrm{Re}\,\lambda_1\leqslant 0$, $\mathrm{Re}(\lambda_1+ +\mu_1)\leqslant 0$, $\mathrm{Re}\,\lambda_2\leqslant 0$, $\mathrm{Re}(\lambda_2+\omega)<0$, $\mathrm{Re}(\lambda_2+\mu_2+\omega)<0$, $u>0$, $v>0$

$$\int\limits_0^\infty e^{\omega x}\mathbf{E}\Big[\exp\left\{\nu N_2(x)+\mu_1\gamma(x)+\mu_2\widehat\xi(\theta(x))+\lambda_1\theta(x)+\lambda_2\bar\xi(\theta(x))\right\};$$

$$x - \widehat{\xi}(\theta(x)) < v,\ h(x) < u,\ \widehat{\xi} \geqslant x\Big] dx =$$

$$= \left(e^{\nu} \int_0^v e^{\omega s} \left(\int_{y_1 \geqslant 0} \int_{y_2 \geqslant s} \exp\{\lambda_1 y_1 + \lambda_2 y_2\}\, d\Pi \right) ds \right) \times$$

$$\times \left(c_0 - e^{\nu} \int_{y_1 \geqslant 0} \int_{y_2 < u} \exp\left\{(\lambda_1 + \mu_1) y_1 + (\lambda_2 + \mu_2 + \omega) y_2\right\} d\Pi \right)^{-1}.$$

For $\operatorname{Re}\nu \leqslant 0,\ \operatorname{Re}\mu_1 \leqslant 0,\ \operatorname{Re}\omega < 0,\ \operatorname{Re}(\mu_2 + \omega) < 0,\ u > 0,\ v > 0,$

$$\omega \int_0^\infty e^{\omega x} \mathbf{E}\Big[\exp\left\{\nu N_2(x) + \mu_1 \gamma(x) + \mu_2 \widehat{\xi}(\theta(x))\right\};$$

$$x - \widehat{\xi}(\theta(x)) < v,\ h(x) < u,\ \widehat{\xi} < x\Big] dx =$$

$$= e^{\nu} f_+(0,0)(e^{\omega v} - 1)\left(c_0 - e^{\nu} \int_{y_1 \geqslant 0} \int_{y_2 < u} \exp\left\{\mu_1 y_1 + (\mu_2 + \omega) y_2\right\} d\Pi \right)^{-1}.$$

A similar statement is true for the distributions of the vectors

$$\left(N_1(t),\ p(t),\ \bar{\xi}(t),\ q(t),\ \bar{\xi}(q(t)),\ l(t)\right).$$

Remark 3. Theorems 1–4 are proved in (Lugavov and Rogozin, 1987a; 1987b) for $\nu = 0,\ u = v = \infty$. Identity (23) is proved in (Csáki and Földes, 1988) for $\nu = \lambda_1 = \lambda_2 = 0,\ u = v$ for one class of homogeneous processes with independent increments.

REFERENCES

Bingham, N.H. (1975). Fluctuation theory in continuous time. *Adv. Appl. Probab.* **7**, 705–766.

Borovkov, A.A. (1976). *Stochastic Processes in Queueing Theory.* Springer, Berlin etc.

Csáki, E. and Földes, A. (1988). On the length of the longest flat interval. In: *Probab. Theory and Math. Statist. Appl., Proc. of the 5th Pannonian Symp. on Math. Stat.*, Visegrad, 1985. Ed. by Grossmann W. Budapest, pp. 23–33.

Fristedt, B. (1973). Sample functions of stochastic processes with stationary independent increments. *Adv. Probab.* **3**, 241–396.

Gikhman, I.I. and Skorokhod, A.V. (1975). *Theory of Stochastic Processes.* **2**, Springer, Berlin etc.

Horowitz, J. (1972). Semilinear Markov processes, subordinators and renewal theory. *Z. Wahrsch. verw. Geb.* **24**, 167–193.

Korolyuk, V.S. (1975). *Boundary Problems for Compound Poisson Processes.* Naukowa Dumka, Kiev (in Russian).

Krylov, N.V. and Jushkevich, A.A. (1965). Markov random sets. *Trans. Moscow Math. Soc.* **13**, 114–135.

Lugavov, V.S. and Rogozin, B.A. (1987a). Renewal theory and factorization identities for random walks. In: *Stochastic Models and Information Systems.* Ed. by Topchii V.A., Computing Centre of the Siberian Branch of USSR Ac. Sci., Novosibirsk, pp. 74–85.

Lugavov, V.S. and Rogozin, B.A. (1987b). Ladder subordinators and factorization identities for processes with independent increments. In: *Stochastic Models and Information Systems.* Ed. by Topchii V.A., Computing Centre of the Siberian Branch of USSR Academy Sci., Novosibirsk, pp. 86–107.

Pecherskii, E.A. and Rogozin, B.A. (1969). On joint distributions of random variables associated with fluctuations of a process with independent increments. *Theory Probab. Appl.* **14**, 410–423.

New Trends in Probab. and Statist., pp. 474 - 484
V. Sazonov and T. Shervashidze (Eds.)
1991 VSP/Mokslas

ON CONVERGENCE OF SOLUTIONS OF STOCHASTIC EQUATIONS

S. MAKHNO

Institute of Applied Mathematics and Mechanics of the Ukrainian Academy of Sciences, R. Luxemburg str., 74, 340114 Donetsk, USSR

ABSTRACT

We consider the problems of weak convergence of solutions of stochastic equations of general type to the solution of the stochastic diffusion equation.

INTRODUCTION

Let $\xi^\varepsilon(t)$ be a solution of the stochastic differential equation

$$\xi^\varepsilon(t) = x^\varepsilon + \int_0^t b^\varepsilon\big(s, \xi^\varepsilon(s), \omega\big)\, ds + \int_0^t \sigma^\varepsilon\big(s, \xi^\varepsilon(s), \omega\big)\, dw^\varepsilon(s) +$$

$$+ \int_0^t \int c^\varepsilon\big(s, \xi^\varepsilon(s), \omega, \theta\big)\, \mu^\varepsilon(d\theta \times ds). \tag{1}$$

The coefficients of this equation are random and depend on the parameter ε irregularly. This means that they have no limit when $\varepsilon \to 0$. We shall study the convergence of the measures on $(D_T, \mathcal{M}_t)$ associated with the process ξ^ε as $\varepsilon \to 0$. Under the conditions of Theorem 1 the random process ξ^ε converges weakly (in the sense of distributions) to the solution of the stochastic equation

$$\xi(t) = x + \int_0^t f\big(s, \xi(s)\big)\, ds + \int_0^t g^{1/2}\big(s, \xi(s)\big)\, dw(s). \tag{2}$$

In Theorems 2–5, as an application of this result, we shall discuss the problem of weak convergence for special classes of stochastic equations with coefficients irregularly depending on ε.

---

1. DEFINITIONS

Let $(\Omega, \mathfrak{F}, \mathfrak{F}_t, \mathbf{P})$ be a filtered probability space, E_d be the Euclidean space of dimension d and $(\cdot,\cdot)$ be the scalar product of its elements. The symbols $L_p([0,T]\times D)$, $W_p^{1,2}([0,T]\times D)$ (the Sobolev space), $L_{p,\mathrm{loc}}$, $W^{1,2}_{p,\mathrm{loc}}$, C_0^∞ are used in the usual sense. The norm in $L_{p,\mathrm{loc}}$ will be denoted by $\|\cdot\|_{p,\mathrm{loc}}$. The symbol $\rightharpoonup$ is used to denote the weak convergence in the space $L_{2,\mathrm{loc}}$, $'$ means conjugation, ∇ stands for gradient.

Let $(D_T, \mathcal{M}_t)$, $t\in[0,T]$, be the space of functions $x(t)$ defined on $[0,T]$, right continuous and having left limits with values in E_d, $\mathcal{M}_t = \sigma\{x(s),\ s\leqslant t\}$.

Denote the convergence of measures on this space by $\Rightarrow$, $\mathfrak{F}_t^\varepsilon = \sigma\{\xi^\varepsilon(s),\ s\leqslant t\}$.

In equation (1) $w^\varepsilon(t)$, for every ε, is a standard $\mathfrak{F}_t$-measurable Wiener process with values in E_m, $\mu^\varepsilon(d\theta\times dt)$ is a Poisson martingale measure with $\mathbf{E}[\mu^\varepsilon(d\theta\times dt)]^2 = \Pi^\varepsilon(d\theta)dt$. Assume that w^ε and μ^ε are mutually independent. Let $b^\varepsilon(t,x,\omega)$ and $c^\varepsilon(t,x,\omega,\theta)$ be d-vectors and $\sigma^\varepsilon(t,x,\omega)$ be a $d\times m$-matrix. These functions are defined on $[0,T]\times E_d\times\Omega$, $\theta\in\Theta$ and supposed to be Borel $\mathfrak{F}_t$-adopted functions. Let $a^\varepsilon = \sigma^\varepsilon(\sigma^\varepsilon)'$ and let the following condition be satisfied: there exist constants $K>0$ and $\alpha>0$ such that

$$|b_i^\varepsilon(t,x,\omega)|^2 + |a_{ij}^\varepsilon(t,x,\omega)| + \int |c^\varepsilon(t,x,\omega,\theta)|^2\Pi^\varepsilon(d\theta) \leqslant K(1+|x|^2), \qquad (3)$$

$$(a^\varepsilon\lambda,\lambda)\geqslant \alpha|\lambda|^2. \qquad (4)$$

Suppose that a solution of (1) exists.

The coefficients of the limit equation (2) may be obtained from conditions (N), (V) given below. Note that these coefficients are uniquely determined by these conditions (Makhno, 1988).

Let, for each value of the parameter $\varepsilon>0$, a d-vector $f^\varepsilon(t,x)$ and a $d\times d$-matrix $g^\varepsilon(t,x)$, $t\in[0,T]$, $x\in E_d$, be given.

Condition (N). There exists a sequence of the functions $N_{kl}^\varepsilon(t,x)\in W^{1,2}_{d+1,\mathrm{loc}}$, $k,l=1,\dots,d$, such that $N_{kl}^\varepsilon = N_{lk}^\varepsilon$ and

1) $\lim_{\varepsilon\to0}\sup_{x\in D}|N_{kl}^\varepsilon(t,x)| = 0$ for any bounded domain $D\subset E_d$ and for each $t\in[0,T]$;

2) $\lim_{\varepsilon\to0}\|N_{kl}^\varepsilon\|_{d+1,\mathrm{loc}} = \lim_{\varepsilon\to0}\|\partial N_{kl}^\varepsilon/\partial x_i\|_{d+1,\mathrm{loc}} = 0$, $i=1,\dots,d$;

3) $\hat g_{kl}^\varepsilon := g_{kl}^\varepsilon + \sum_{i,j} g_{i,j}^\varepsilon \partial^2 N_{kl}^\varepsilon/\partial x_i\partial x_j \rightharpoonup g_{kl}$, $\varepsilon\to0$;

4) $\lim_{\varepsilon\to0}\|\partial N_{kl}^\varepsilon/\partial t + (1/2)(\hat g_{kl}^\varepsilon - g_{kl})\|_{d+1,\mathrm{loc}} = 0$.

Condition (V). There exists a sequence of functions $V_k^\varepsilon(t,x)\in W^{1,2}_{d+1,\mathrm{loc}}$, $k=1,\dots,d$, such that

1) $\lim_{\varepsilon\to0}\sup_{x\in D}|V_k^\varepsilon(t,x)| = 0$ for any bounded domain $D\subset E_d$ and for each $t\in[0,T]$;

2) $\lim_{\varepsilon\to0}\|V_k^\varepsilon\|_{d+1,\mathrm{loc}} = \lim_{\varepsilon\to0}\|\partial V_k^\varepsilon/\partial x_i\|_{d+1,\mathrm{loc}} = 0$, $i=1,\dots,d$;

3) $\hat f_k^\varepsilon := f_k^\varepsilon + (1/2)\sum_{i,j} g_{i,j}^\varepsilon \partial^2 V_k^\varepsilon/\partial x_i\partial x_j \rightharpoonup f_k$, $\varepsilon\to0$;

4) $\lim_{\varepsilon\to 0} \left\| \partial V_k^\varepsilon/\partial t + \hat{f}_k^\varepsilon - f_k \right\|_{d+1,\mathrm{loc}} = 0.$

We shall say that condition (M) is satisfied if:

(M$_1$) There exist a d-vector $f^\varepsilon(t,x)$ and a $d\times d$-matrix $g^\varepsilon(t,x)$ which satisfy conditions (3), (4). The functions $f_k^\varepsilon(t,x)$ are measurable, the functions $g_{kl}^\varepsilon(t,x)$ are continuous and, for every $\Phi \in \mathbf{C}_0^\infty$,

$$\lim_{\varepsilon\to 0} \mathbf{E}\left\{ \int_s^t \left[L_\Phi^\varepsilon\big(v,\xi^\varepsilon(v)\big) - \mathcal{L}^\varepsilon\Phi\big(v,\xi^\varepsilon(v)\big)\right] dv \,|\, \mathfrak{F}_s^\varepsilon \right\} = 0,$$

where

$$L_\Phi^\varepsilon = \frac{\partial\Phi}{\partial t} + (b^\varepsilon, \nabla\Phi) + \frac{1}{2}(a^\varepsilon\nabla,\nabla)\Phi +$$

$$+ \int \left[\Phi\big(t, x + c^\varepsilon(t,x,\omega,\theta)\big) - \Phi - \big(c^\varepsilon,\nabla\Phi\big)\right]\Pi^\varepsilon(d\theta),$$

$$\mathcal{L}^\varepsilon\Phi = \frac{\partial\Phi}{\partial t} + (f^\varepsilon,\nabla\Phi) + \frac{1}{2}(g^\varepsilon\nabla,\nabla)\Phi.$$

(M$_2$) The functions $f^\varepsilon, g^\varepsilon$ satisfy conditions (V), (N), the limit function f is measurable, the limit function g is continuous and

$$|f_i|^2 + |g_{ij}| \leqslant K(1+|x|^2), \quad (g\lambda,\lambda) \geqslant \alpha|\lambda|^2, \quad \alpha > 0.$$

(M$_3$) $\forall\Phi \in \mathbf{C}_0^\infty$

$$\lim_{\varepsilon\to 0} \mathbf{E}\left\{ \int_s^t \left[\frac{\partial V_k^\varepsilon}{\partial v} + \frac{1}{2}(g^\varepsilon\nabla,\nabla)V_k^\varepsilon \right] \Phi\big(v,\xi^\varepsilon(v)\big) dv \,|\, \mathfrak{F}_s^\varepsilon \right\} = 0,$$

$$\lim_{\varepsilon\to 0} \mathbf{E}\left\{ \int_s^t \left[\frac{\partial N_{kl}^\varepsilon}{\partial v} + \frac{1}{2}(g^\varepsilon\nabla,\nabla)N_{kl}^\varepsilon \right] \Phi\big(v,\xi^\varepsilon(v)\big) dv \,|\, \mathfrak{F}_s^\varepsilon \right\} = 0.$$

2. GENERAL THEOREM

Let us formulate our general limit theorem.

THEOREM 1. *Let $x^\varepsilon \to x$ and condition* (M) *be satisfied. Then $\xi^\varepsilon \Rightarrow \xi$.*

Proof. Under condition (3) the set of measures Q^ε associated with solutions ξ^ε on $(D_T, \mathcal{M}_t)$ is weakly compact. Let Q be a limit point of this sequence. We must show that $\forall\Phi \in \mathbf{C}_0^\infty$

$$\Phi\big(t,x(t)\big) - \Phi\big(s,x(s)\big) - \int_s^t \left[\frac{\partial\Phi}{\partial v} + (f,\nabla\Phi) + \frac{1}{2}(g\nabla,\nabla)\Phi\right]\Big(v,x(v)\Big)dv$$

is an $\mathcal{M}_t$–martingale. By (M_2) the solution of (2) is weakly unique (Strook and Varadhan, 1979). Therefore, the sequence Q^ε possesses a unique limit point. Let $\Psi_s(x)$ be bounded, continuous and $\mathcal{M}_s$–measurable functional. Applying Ito's formula and condition (M_1), we have

$$\lim_{\varepsilon\to 0} \mathbf{E}\Psi_s(\xi^\varepsilon)\left\{\Phi\big(t,\xi^\varepsilon(t)\big) - \Phi\big(s,\xi^\varepsilon(s)\big) - \int_s^t \mathcal{L}^\varepsilon\Phi\big(v,\xi^\varepsilon(v)\big)dv\right\} = 0. \tag{5}$$

Note that

$$\Psi_s(\xi^\varepsilon)\int_s^t \left[f_k^\varepsilon\big(v,\xi^\varepsilon(v)\big) - f_k\big(v,\xi^\varepsilon(v)\big)\right]\frac{\partial\Phi}{\partial x_k}\big(v,\xi^\varepsilon(v)\big)dv \doteq I_1 + I_2,$$

where

$$I_1 = \Psi_s(\xi^\varepsilon)\int_s^t \left(\frac{\partial V_k^\varepsilon}{\partial v} + \hat{f}_k^\varepsilon - f_k\right)\frac{\partial\Phi}{\partial x_k}\big(v,\xi^\varepsilon(v)\big)dv,$$

$$I_2 = -\Psi_s(\xi^\varepsilon)\int_s^t \left(\frac{\partial V_k^\varepsilon}{\partial v} + \frac{1}{2}\sum_{i,j} g_{ij}^\varepsilon \frac{\partial^2 V_k^\varepsilon}{\partial x_i \partial x_j}\right)\frac{\partial\Phi}{\partial x_k}\big(v,\xi^\varepsilon(v)\big)dv.$$

By Krylov's estimate (Anulova and Pragarauskas, 1977),

$$\mathbf{E}|I_1| \leqslant C\left\|\left(\frac{\partial V_k^\varepsilon}{\partial t} + \hat{f}_k^\varepsilon - f_k\right)\frac{\partial\Phi}{\partial x_k}\right\|_{d+1}.$$

Consequently, $\lim_{\varepsilon\to 0}\mathbf{E}I_1 = 0$. From condition ($M_3$) we have that $\lim_{\varepsilon\to 0}\mathbf{E}I_2 = 0$. Therefore, for any continuous function $f_k(t,x)$

$$\lim_{\varepsilon\to 0}\mathbf{E}\Psi_s(\xi^\varepsilon)\int_s^t f_k^\varepsilon \frac{\partial\Phi}{\partial x_k}\big(v,\xi^\varepsilon(v)\big)dv =$$

$$= \mathbf{E}^Q\Psi_s(x)\int_s^t f_k\frac{\partial\Phi}{\partial x_k}\big(v,x(v)\big)dv. \tag{6}$$

This formula is valid also for the measurable functions f_k from condition (M) (Makhno, 1988). Analogously,

$$\lim_{\varepsilon\to 0}\mathbf{E}\Psi_s(\xi^\varepsilon)\int_s^t g_{kl}^\varepsilon \frac{\partial^2\Phi}{\partial x_k\partial x_l}\big(v,\xi^\varepsilon(v)\big)dv =$$

$$= \mathbf{E}^Q\Psi_s(x)\int_s^t g_{kl}\frac{\partial^2\Phi}{\partial x_k\partial x_l}\big(v,x(v)\big)dv. \tag{7}$$

Now (5)–(7) imply the theorem.

3. EQUATIONS WITH PERIODIC COEFFICIENTS

Let the random process $\left(x^\varepsilon(t), y^\varepsilon(t)\right)$ be a solution of the system

$$x^\varepsilon(t) = x^\varepsilon + \frac{1}{\varepsilon}\int_0^t F\Big(\frac{s}{\varepsilon^2}, \frac{x^\varepsilon(s)}{\varepsilon}, y^\varepsilon(s), x^\varepsilon(s)\Big)ds +$$

$$+\int_0^t G\Big(\frac{s}{\varepsilon^2}, \frac{x^\varepsilon(s)}{\varepsilon}, y^\varepsilon(s), x^\varepsilon(s)\Big)dw_1(s), \tag{8}$$

$$y^\varepsilon(t) = y^\varepsilon + \frac{1}{\varepsilon^2}\int_0^t \alpha\Big(\frac{s}{\varepsilon^2}, \frac{x^\varepsilon(s)}{\varepsilon}, y^\varepsilon(s), x^\varepsilon(s)\Big)ds +$$

$$+\frac{1}{\varepsilon}\int_0^t \beta\Big(\frac{s}{\varepsilon^2}, \frac{x^\varepsilon(s)}{\varepsilon}, y^\varepsilon(s), x^\varepsilon(s)\Big)dw_2(s) +$$

$$+\int_0^t\int \delta\Big(\frac{s}{\varepsilon^2}, \frac{x^\varepsilon(s)}{\varepsilon}, y^\varepsilon(s), x^\varepsilon(s), \theta\Big)\mu^\varepsilon(d\theta \times ds). \tag{9}$$

We shall assume that the following condition is fulfilled:

(A_1) The coefficients $F_i(s,z,y,x)$, $H_{ij} = (GG')_{ij}(s,z,y,x)$, $i,j = 1,\dots,d$; $\alpha_r(s,z,y,x)$, $\delta_r(s,z,y,x,\theta)$, $h_{rl} = (\beta\beta')_{rl}(s,z,y,x)$, $r,l = 1,\dots,n$; $s \in E_1$, $x,z \in E_d$, $y \in E_n$, $\theta \in \Theta$, are bounded and periodic in (s,z,y) functions, with period 1 in all components.

We shall discuss the convergence of the random process $x^\varepsilon(t)$. Sometimes in addition to condition (A_1) the following condition is imposed:

(A_2) The functions H_{ij}, h_{rl} are twice continuously differentiable in all arguments; the matrices H and h are uniformly nondegenerate.

(A_3) The functions α_r, F_i are continuously differentiable in (z,y,x) and α_r, F_i are twice continuously differentiable in s.

(A_4) The functions δ_r are continuously differentiable in (s,z,y,x).

(A_5) $w_1(s)$ is an m_1-dimensional Wiener process, $w_2(s)$ an m_2-dimensional Wiener process, $\mu^\varepsilon(d\theta \times dt)$ is a Poisson martingale and

$$\mathbf{E}\left[\mu^\varepsilon(d\theta \times dt)\right]^2 = \frac{1}{\varepsilon}\Pi(d\theta)dt, \qquad \int |\delta(s,z,y,x,\theta)|^2\Pi(d\theta) \leqslant K.$$

Assume that w_1, w_2, and μ^ε are mutually independent.

Denote by $p(t,z,y,x)$ the periodic in (t,z,y) solution of the equation (Bensoussan et al., 1978)

$$\frac{\partial}{\partial t}p-\frac{1}{2}\sum_{i,j}\frac{\partial^2}{\partial z_i\partial z_j}(H_{ij}p)-\frac{1}{2}\sum_{r,l}\frac{\partial^2}{\partial y_r\partial y_l}(h_{rl}p)+$$

$$+\sum_i\frac{\partial}{\partial z_i}(F_ip)+\sum_r\frac{\partial}{\partial z_r}(\alpha_r p)=0,\quad \langle p(\cdot,\cdot,\cdot,x)\rangle=1.$$

We consider x as a parameter. Here $\langle f(\cdot)\rangle=\int_0^1 f(z)dz$. Throughout this section we assume that

$$\langle F_i(\cdot,\cdot,\cdot,x)p(\cdot,\cdot,\cdot,x)\rangle=0. \tag{10}$$

Then (Bensoussan et al., 1978) there exist sufficiently smooth solutions $u_i(t,z,y,x)$ of the equation

$$\frac{\partial}{\partial t}u_i+(F,\nabla_z,u_i)+(\alpha,\nabla_y u_i)+\frac{1}{2}(H\nabla_z,\nabla_z)u_i+\frac{1}{2}(h\nabla_y,\nabla_y)u_i=-F_i,$$

$$\langle u_i(\cdot,\cdot,\cdot,x)\rangle=0,\quad i=1,\dots,d. \tag{11}$$

Put

$$b_i=(F,\nabla_z u_i)+(H\nabla_z,\nabla_z)u_i+$$

$$+\int\left[u_i\big(t,z,y+\delta(t,z,y,x,\theta),x\big)-u_i-(\nabla_y u_i,c)\right]\Pi(d\theta),$$

$$\sigma_{ir}=\begin{cases}G_{ir}+\sum_k \partial u_i/\partial z_k G_{kr}, & r=1,\dots,m_1,\quad i=1,\dots,d.\\ \sum_l \partial u_i/\partial y_l\beta_{lr}, & r=m_1+1,\dots,m_1+m_2.\end{cases}$$

THEOREM 2. *Let $x^\varepsilon\to x$ and conditions* (A_1)–(A_5), (10) *be valid. Then $x^\varepsilon\Rightarrow$ $\Rightarrow\xi$ and*

$$f_i(x)=\langle b_i(\cdot,\cdot,\cdot,x)p(\cdot,\cdot,\cdot,x)\rangle,\qquad g_{ij}(x)=\langle a_{ij}(\cdot,\cdot,\cdot,x)p(\cdot,\cdot,\cdot,x)\rangle.$$

Proof. Denote

$$\zeta_i^\varepsilon(t)=\varepsilon\Big[u_i\Big(\frac{t}{\varepsilon^2},\frac{x^\varepsilon(t)}{\varepsilon},y^\varepsilon(t),x^\varepsilon(t)\Big)-u_i\Big(0,\frac{x^\varepsilon}{\varepsilon},y^\varepsilon,x^\varepsilon\Big)-$$

$$-\frac{1}{2}\int_0^t(H\nabla_x,\nabla_x)u_i ds-\int_0^t(\nabla_x u_i,Gdw_1)\Big],$$

$$b_i^\varepsilon(t,x,\omega)=b_i\Big(\frac{t}{\varepsilon^2},\frac{x-\zeta^\varepsilon(t)}{\varepsilon},y^\varepsilon(t),x-\zeta^\varepsilon(t)\Big),$$

$$\sigma_{ir}^\varepsilon(t,x,\omega)=\sigma_{ir}\Big(\frac{t}{\varepsilon^2},\frac{x-\zeta^\varepsilon(t)}{\varepsilon},y^\varepsilon(t),x-\zeta^\varepsilon(t)\Big),$$

$$c_i^\varepsilon(t,x,\omega,\theta)=\varepsilon\Big[u_i\Big(\frac{t}{\varepsilon^2},\frac{x-\zeta^\varepsilon(t)}{\varepsilon},y^\varepsilon(t)+\delta\Big(\frac{t}{\varepsilon^2},\frac{x-\zeta^\varepsilon(t)}{\varepsilon},$$

$$y^\varepsilon(t),x-\zeta^\varepsilon(t),\theta\Big),x-\zeta^\varepsilon(t)\Big)-u_i\Big(\frac{t}{\varepsilon^2},\frac{x-\zeta^\varepsilon(t)}{\varepsilon},y^\varepsilon(t),x-\zeta^\varepsilon(t)\Big)\Big],$$

$$w(t)=(w_1,w_2)'.$$

Applying Ito's formula with $\varepsilon u_i(t/\varepsilon^2,x/\varepsilon,y,x)$, $x^\varepsilon(t)$, $y^\varepsilon(t)$ and combining with (8), we obtain equation (1) for the process $\xi^\varepsilon(t)=x^\varepsilon(t)+\zeta^\varepsilon(t)$ with coefficients $b^\varepsilon(t,x,\omega)$, $\sigma^\varepsilon(t,x,\omega)$, $c^\varepsilon(t,x,\omega,\theta)$. If we prove that condition (M) is satisfied with the limit functions f,g, then the assertion of Theorem 2 will be a direct consequence of Theorem 1.

Note that for "sufficiently" smooth periodic in (t,z,y) function $\chi(t,z,y,x)$ such that $\langle\chi(\cdot,\cdot,\cdot,x)p(\cdot,\cdot,\cdot,x)\rangle=0$ we have

$$\lim_{\varepsilon\to0}\mathbf{E}\Big\{\Big(\int_s^t\chi\big(\frac{v}{\varepsilon^2}\frac{x^\varepsilon(v)}{\varepsilon},y^\varepsilon(v),x^\varepsilon(v)\big)dv\Big)^2\mid\mathfrak{F}_s^\varepsilon\Big\}=0. \tag{12}$$

We define Ψ as a solution of equation (11) with F replaced by χ. Applying Ito's formula with $\varepsilon^2\Psi(t/\varepsilon^2,x/\varepsilon,y,x)$, $x^\varepsilon(t),y^\varepsilon(t)$, we obtain (12).

Furthermore,

$$L_\Phi^\varepsilon(t,x)=\frac{\partial\Phi}{\partial t}(t,x)+\big(b^\varepsilon(t,x,\omega),\nabla\Phi\big)+$$

$$+\frac{1}{2}\big(a^\varepsilon(t,x,\omega)\nabla,\nabla\big)\Phi+O(\varepsilon),\quad\varepsilon\to0. \tag{13}$$

So, condition (M_1) $\big(f^\varepsilon(t,x)=f(x),g^\varepsilon(t,x)=g(x)\big)$ follows from (13), (12). Conditions (M_2), (M_3) are clearly satisfied $(N_{kl}^\varepsilon\equiv V_k^\varepsilon\equiv0)$.

4. SECOND ORDER DIFFERENTIAL EQUATIONS

We shall discuss the problem of weak convergence of solutions of the stochastic equations:

$$\varepsilon^p\ddot{x}^\varepsilon+\mu\dot{x}^\varepsilon=B^\varepsilon(t,x^\varepsilon)+A^\varepsilon(t,x^\varepsilon)\dot{w},$$

$$x^\varepsilon(0)=y^\varepsilon,\quad p>0,\quad\mu>0. \tag{14}$$

Equations (14) should be interpreted as a system of two stochastic equations of the form

$$x^\varepsilon(t)=x^\varepsilon+\int_0^t y^\varepsilon(s)ds,$$

$$y^\varepsilon(t)=y^\varepsilon+\frac{1}{\varepsilon^p}\int_0^t\big[-\mu y^\varepsilon(s)+B^\varepsilon\big(s,x^\varepsilon(s)\big)\big]ds+$$

$$+\frac{1}{\varepsilon^p}\int_0^t A^\varepsilon\big(s,x^\varepsilon(s)\big)dw(s). \tag{15}$$

From (15) we obtain

$$x^\varepsilon(t)+\eta^\varepsilon(t)=x^\varepsilon+\frac{1}{\mu}\int_0^t B^\varepsilon\big(s,x^\varepsilon(s)\big)ds+\frac{1}{\mu}\int_0^t A^\varepsilon\big(s,x^\varepsilon(s)\big)dw(s),$$

where

$$\eta^\varepsilon(t)=e^{-t\mu/\varepsilon^p}\Big[y^\varepsilon+\int_0^t e^{s\mu/\varepsilon^p}B^\varepsilon\big(s,x^\varepsilon(s)\big)ds+$$

$$+\int_0^t e^{s\mu/\varepsilon^p}A^\varepsilon\big(s,x^\varepsilon(s)\big)dw(s)\Big].$$

Hence, $\xi^\varepsilon(t)=x^\varepsilon(t)+\eta^\varepsilon(t)$ is a solution of (1), where

$$b^\varepsilon(t,x,w)=\frac{1}{\mu}B^\varepsilon\big(t,x-\eta^\varepsilon(t)\big),\qquad \sigma^\varepsilon(t,x,w)=\frac{1}{\mu}A^\varepsilon\big(t,x-\eta^\varepsilon(t)\big).$$

Let the following conditions be satisfied:

(B_1) There exists a constant $K>0$ such that $|A^\varepsilon|+|B^\varepsilon|\leqslant K$.

(B_2) There exists Ψ_ε such that for every x,y

$$|A^\varepsilon(t,x)-A^\varepsilon(t,y)|+|B^\varepsilon(t,x)-B^\varepsilon(t,y)|\leqslant\Psi_\varepsilon|x-y|.$$

(B_ε) $\lim_{\varepsilon\to0}\Psi_3\sup_{t\in[0,T]}\mathbf{E}|\eta^\varepsilon(t)|=0$; $\lim_{\varepsilon\to0}y^\varepsilon=0$.

THEOREM 3. *Let $x^\varepsilon\to x$, conditions (B_1)–(B_3) be valid and the functions B^ε, $(A^\varepsilon)^2$ satisfy conditions (M_2), (M_3). Then $x^\varepsilon\Rightarrow\xi$.*

Proof. It follows from the assumptions of Theorem 3 that condition (M_1) is satisfied for the functions

$$f^\varepsilon=\frac{1}{\mu}B^\varepsilon,\qquad g^\varepsilon=\frac{1}{\mu^2}(A^\varepsilon)^2.$$

Hence the assertion of Theorem 3 is a direct consequence of Theorem 1.

We shall consider solutions of equation (14). Let the following conditions be satisfied

$$B^\varepsilon(t,x)=B\Big(\frac{t}{\varepsilon^K},\frac{x}{\varepsilon}\Big),\qquad A^\varepsilon(t,x)=A\Big(\frac{t}{\varepsilon^K},\frac{x}{\varepsilon}\Big), \tag{16}$$

where the functions $B(t,x)$, $A(t,x)$ are periodic in t,x with period 1 in all components.

There are four cases for description of the limit process: $K=0$, $K\in(0,2)$,

$K = 2$, $K > 2$. We shall discuss, for example, the convergence of solutions of stochastic equation (14) when $K = 2$ and $K > 2$. Let the following conditions be satisfied:

(B_4) The function $A(t,x)$ is twice continuously differentiable and $A^2(t,x) \geqslant \alpha > 0$.

(B_5) The function $B(t,x)$ is continuously differentiable. Denote by $m(t,x)$ the periodic solution of the equation

$$\frac{\partial}{\partial t}m - \frac{1}{2\mu^2}\frac{\partial^2}{\partial x^2}(A^2 m) = 0, \quad \langle m\rangle = 1.$$

THEOREM 4. *Let the coefficients of equation* (14) *satisfy condition* (16), $K = 2$, $p > 1$, $y^\varepsilon = o(\varepsilon)$, $x^\varepsilon \to x$, $\varepsilon \to 0$, *and let conditions* (B_4), (B_5) *be valid. Then* $x^\varepsilon \Rightarrow \xi$ *and*

$$f = \frac{1}{\mu}\langle Bm\rangle, \qquad g = \frac{1}{\mu^2}\langle A^2 m\rangle.$$

Proof. Clearly, conditions (B_1)–(B_3) hold with $\Psi_\varepsilon = \text{const}/\varepsilon$. There exist periodical "sufficiently" smooth functions $N(t,x), V(t,x)$ satisfying the equations (Bensoussan et al., 1978)

$$\frac{\partial N}{\partial t} + \frac{1}{2\mu^2}A^2\frac{\partial^2 N}{\partial x^2} = \frac{1}{\mu^2}\left[A^2 - \langle A^2 m\rangle\right], \quad \langle N\rangle = 0,$$

$$\frac{\partial V}{\partial t} + \frac{1}{2\mu^2}A^2\frac{\partial^2 V}{\partial x^2} = \frac{1}{\mu}\left[B - \langle Bm\rangle\right], \quad \langle V\rangle = 0.$$

Set $N^\varepsilon(t,x) = \varepsilon^2 N(t/\varepsilon^2, x/\varepsilon)$, $V^\varepsilon(t,x) = \varepsilon^2 V(t/\varepsilon^2, x/\varepsilon)$. It can be easily verified that the functions $N^\varepsilon(t,x)$, $V^\varepsilon(t,x)$ satisfy conditions (N), (V), and

$$f = \frac{1}{\mu}\langle Bm\rangle, \qquad g^2 = \frac{1}{\mu^2}\langle A^2 m\rangle.$$

We shall now verify the validity of condition M_3. Applying Ito's formula to ξ^ε and to the function $N^\varepsilon(t,x)$, we obtain

$$\mathbf{E}\Psi_s(\xi^\varepsilon)\int_s^t\left[\frac{\partial N^\varepsilon}{\partial v}(v,\xi^\varepsilon(v)) + \frac{1}{2\mu^2}A^2\left(\frac{v}{\varepsilon^2},\frac{x^\varepsilon(v)}{\varepsilon}\right)\times\right.$$

$$\left.\times\frac{\partial^2 N^\varepsilon}{\partial x^2}(v,\xi^\varepsilon(v))\right]\Phi(v,\xi^\varepsilon(v))dv = O(\varepsilon), \quad \varepsilon \to 0. \tag{17}$$

Analogously,

$$\mathbf{E}\Psi_s(\xi^\varepsilon)\int_s^t\left[\frac{\partial V^\varepsilon}{\partial v}(v,\xi^\varepsilon(v)) + \frac{1}{2\mu^2}A^2\left(\frac{v}{\varepsilon^2},\frac{x^\varepsilon(v)}{\varepsilon}\right)\times\right.$$

$$\times \frac{\partial^2 V^\varepsilon}{\partial x^2}(v, \xi^\varepsilon(v))\Big]\Phi(v, \xi^\varepsilon(v))dv = O(\varepsilon), \quad \varepsilon \to 0. \tag{18}$$

Consequently, we can conclude from (17), (18), (B_3) that condition (M_3) is satisfied. Thus, the conditions of Theorem 3 are fulfilled. Theorem 4 is proved.
Let us introduce the condition

(B_6) The functions $A(t,x)$, $B(t,x)$ are four times continuously differentiable in x. Denote $\widetilde{a}(x) = \int_0^1 A^2(t,x)dt$, $\widetilde{B}(x) = \int_0^1 B(t,x)dt$.

THEOREM 5. *Let the coefficients of equation* (14) *satisfy condition* (16), $K > 2$, $p > 1$, $y^\varepsilon = o(\varepsilon)$, $x^\varepsilon \to x$, $\varepsilon \to 0$, *and conditions* (B_4)–(B_6) *be valid. Then* $x^\varepsilon \Rightarrow \xi$ *and*

$$f = \frac{1}{\mu}\left\langle \frac{\widetilde{B}}{\widetilde{a}} \right\rangle \langle \widetilde{a}^{-1} \rangle^{-1}, \qquad g^2 = \frac{1}{\mu^2}\langle \widetilde{a}^{-1} \rangle^{-1}.$$

Proof of this theorem is essentially the same as the proof of Theorem 4 and we only indicate how to define the functions N^ε, V^ε. It is clear that the function $p(x) = \widetilde{a}^{-1}(x)\langle \widetilde{a}^{-1} \rangle^{-1}$ is a solution of the equation

$$\frac{d^2}{dx^2}(\widetilde{a}p) = 0, \quad \langle p \rangle = 1.$$

This implies that the following equations

$$\widetilde{a}(x)\frac{d^2\psi}{dx^2} = \langle \widetilde{a}^{-1} \rangle^{-1} - \widetilde{a}(x), \quad \langle \psi \rangle = 0,$$

$$\frac{1}{2}\widetilde{a}(x)\frac{d^2\varphi}{dx^2} = \left\langle \frac{\widetilde{B}}{\widetilde{a}} \right\rangle \langle \widetilde{a}^{-1} \rangle^{-1} - \widetilde{B}(x), \quad \langle \varphi \rangle = 0,$$

possess unique solutions (Bensoussan et al., 1978).
Put

$$\psi_1(t,x) = -\frac{1}{2}\int_0^t \left[A^2(s,x)\frac{d^2\psi(x)}{dx^2} + A^2(s,x) - \langle \widetilde{a}^{-1} \rangle^{-1}\right]ds,$$

$$\varphi_1(t,x) = -\int_0^t \frac{1}{2}\left[A^2(s,x)\frac{d^2\varphi(x)}{dx^2} + B(s,x) - \left\langle \frac{\widetilde{B}}{\widetilde{a}} \right\rangle \langle \widetilde{a}^{-1} \rangle^{-1}\right]ds.$$

It can be easily verified that the functions

$$N^\varepsilon(t,x) = \varepsilon^2\psi\left(\frac{x}{\varepsilon}\right) + \varepsilon^k\psi_1\left(\frac{t}{\varepsilon^k}, \frac{x}{\varepsilon}\right),$$

$$V^\varepsilon(t,x) = \varepsilon^2\varphi\left(\frac{x}{\varepsilon}\right) + \varepsilon^k\varphi_1\left(\frac{t}{\varepsilon^k}, \frac{x}{\varepsilon}\right),$$

satisfy conditions (N), (V), and the limit functions are

$$f = \frac{1}{\mu}\left\langle \frac{\widetilde{B}}{\widetilde{a}} \right\rangle \langle \widetilde{a}^{-1} \rangle^{-1} \quad \text{and} \quad g^2 = \frac{1}{\mu^2}\langle \widetilde{a}^{-1} \rangle^{-1}.$$

Theorem 5 is proved.

REFERENCES

Anulova, S. and Pragarauskas, H. (1977). On strong Markov weak solutions of stochastic equations. *Lithuanian Math. J.* **17**, 5–26.

Bensoussan, A., Lions, J.-L. and Papanicolau, G. (1978). *Asymptotic Analysis for Periodic Structures.* North-Holland, Amsterdam.

Makhno, S. (1988). Sufficient conditions for convergence of stochastic differential equations. *Theory Random Proc.* **16**, 66–73. (in Russian).

Strook, D. and Varadhan, S. (1979). *Multidimensional Diffusion Processes.* Springer, Berlin etc.

New Trends in Probab. and Statist., pp. 485 - 502
V. Sazonov and T. Shervashidze (Eds.)
1991 VSP/Mokslas

STOCHASTIC INTEGRALS AND STOCHASTIC DIFFERENTIAL EQUATIONS ON THE PLANE INVOLVING STRONG SEMIMARTINGALES

Yu.S. MISHURA
Kiev State University, Vladimirskaya str. 64, 252601 Kiev, USSR

ABSTRACT

This paper contains a local version of Gundy's decomposition of a two-parameter strong martingale and a construction of stochastic integral with respect to a local strong martingale and semimartingale. The existence and the uniqueness of a solution of a stochastic differential equation that contains a two-parameter strong semimartingale is established.

1. DEFINITIONS

Let $t = (t_1, t_2) \in \mathbf{R}^2_+$, $s = (s_1, s_2) \in \mathbf{R}^2_+$. We write $s \leqslant t$ if $s_i \leqslant t_i$ and $s < t$ if $s_i < t_i$ $i = 1,2$. Let $(\Omega, \mathfrak{F}, \mathfrak{F}_t, P)$ be a complete stochastic basis, where $\mathfrak{F}_t$, $t \in \mathbf{R}^2_+$, $\mathfrak{F}_t \subset \mathfrak{F}$, satisfy conditions (F1)–(F4) from (Cairoli and Walsh, 1975),

$$\mathfrak{F}^1_{t_1} = \bigvee_{t_2 \geqslant 0} \mathfrak{F}_t, \qquad \mathfrak{F}^2_{t_2} = \bigvee_{t_1 \geqslant 0} \mathfrak{F}_t,$$

$$\mathfrak{F}^*_t = \mathfrak{F}^1_{t_1} \vee \mathfrak{F}^2_{t_2}, \qquad \widehat{\mathfrak{F}}_t = \mathfrak{F}_t \cap \{ \bigvee_{s<t} \mathfrak{F}^*_s \}.$$

We consider the integrable stochastic D-processes

$$\left(x = x_t(\omega),\ \mathfrak{F}_t,\ t \in \mathbf{R}^2_+\right)$$

which are fixed on G, $G = (\{0\} \times [0, \infty[) \cup ([0, \infty[\times \{0\})$, and have trajectories that a.s. belong to the space D of functions with four limits $x^{\pm\pm}_t$ for all $t \in \mathbf{R}^2_+ \backslash G$; $x_t = x^{++}_t$. Let

$$x]s,\ t] = x_t - x_{s_1 t_2} - x_{t_1 s_2} + x_s, \quad s \leqslant t.$$

Definition 1. The process $(x_t, \mathfrak{F}_t)$ is a strong martingale (s.m.) if

$$\mathbf{E}\,(x]s,\ t] \mid \mathfrak{F}^*_s) = 0, \quad \forall s \leqslant t.$$

According to (Fouque and Millet, 1980) and (Millet, 1981) a strong martingale has a D–modification.
Let X^* be the class of processes with integrable supremum, $\mathfrak{M}_s$ be the class of s.m., $\mathfrak{M}_s^* = X^* \cap \mathfrak{M}_s$, $\mathfrak{M}_s^p$, $p \geqslant 1$, be the class of s.m. which are in L^p, $\mathcal{A}$ be the class of increasing integrable processes, and let

$$\mathcal{B} = \mathcal{A} - \mathcal{A}.$$

We shall use the notions of a weakly predictable process and an optional $(\mathbf{O}x)$, a weakly predictable $(\widehat{\Pi}x)$, and a dual weakly predictable $(x^{\widehat{\Pi}})$ projections of x (Gushchin, 1982).
The set $F \subset \mathbf{R}_+^2 \times \Omega$ is said to be weakly predictable if its indicator is a weakly predictable process.

Definition 2. A weakly predictable set $F = F(\omega)$ is called a stopping set if $(\{t \in F\} \Rightarrow \{[0,t] \subset F\})$.
Let E be some class of stochastic processes. We say that $x \in lE$, if there exist sequences of $x^n \in E$ and those of stopping sets F^n, $n \geqslant 1$, such that

$$F^n \subset F^{n+1}, \quad \bigcup_n F^n = \mathbf{R}_+^2 \times \Omega,$$

and

$$(x_t - x_t^n) I_{F^n}(t) = 0.$$

The sequence (x^n, F^n) is called a localizing sequence for x.
If x is a D-process, then its jump is defined as

$$\Box x_t = x_t - x_t^{-+} - x_t^{+-} + x_t^{--}, \quad t \in \mathbf{R}_+^2 \backslash G.$$

Let $T \in \mathbf{R}_+^2$ be fixed, and let

$$\lambda_{n,n}^T = \lambda_{1,n}^{T_1} \times \lambda_{2,n}^{T_2}$$

be an increasing sequence of grids on $[0,T]$, where

$$\lambda_{i,n}^{T_i} = \{0 = t_{i,n}^0 < t_{i,n}^1 < \ldots < t_{i,n}^n = T_i\}, \quad i = 1,2.$$

Define

$$|\lambda^T| = \max_{i=1,2,\ 0 \leqslant j \leqslant n-1} \left(t_{i,n}^{j+1} - t_{i,n}^j\right),$$

$$t_{jk} = (t_{1,n}^j,\ t_{2,n}^k), \qquad t_{jk}(t) = t_{jk} \wedge t,$$

$$\Box_{jk} x(t) = x\big]t_{jk}(t),\ t_{j+1\,k+1}(t)\big].$$

Let (x, y, z) be some stochastic processes. Consider the following stochastic functionals (under the hypothesis that the limits exist and do not depend on λ_{nn}^T):

$$[x]_t = \mathbf{P}\text{-}\lim_{n\to\infty} \sum_{j,k} (\Box_{jk} x(t))^2$$

(quadratic variation of x);

$$\big((x,\ z)\cdot y\big)_t = \mathbf{P}\text{-}\lim_{n\to\infty} \sum_{j,k} x_{j\,k+1}(t) z_{j+1\,k}(t) \Box_{jk} y(t).$$

2. GUNDY'S DECOMPOSITION OF LOCAL STRONG MARTINGALES

LEMMA 1. *If $x \in lX^*$ is a D-process then*

$$A_t = \sum_{s\leqslant t} \Box x_s I\{|\Box x_s| > a\} \in l\mathcal{A}$$

for all $a > 0$.

Proof. Let $(x^n,\ E^n)$, $n \geqslant 1$ be a localizing sequence for x. Since x is a D-process, $A_t < \infty$ a.s. $\forall t \in \mathbf{R}^2_+$. Assume that $A_\infty < \infty$ a.s. (otherwise consider the sequence of increasing squares $[0,m]^2$, $m \to \infty$ and $A_{(m,m)} < \infty$ a.s.).
Let

$$\tau_1^n = \inf\{t_1\colon\ A_{t_1\infty} > n\}, \qquad \tau_2^n = \inf\{t_2\colon\ A_{\infty t_2} > n\},$$

$$\tau^n = (\tau_1^n,\ \tau_2^n),$$

$$\widehat{A}_t^n = \int_{[0,t]} I\Big\{\widehat{\Pi} I_{[0,\tau^n]}(s) > \frac{1}{2}\Big\} I_{E^n}(s)\, dA_s,$$

$$F^{i,n} = \Big\{t\colon\ \widehat{\Pi} I_{[0,\tau^n]}(s) > \frac{1}{2} \quad \forall s \leqslant t\Big\}, \quad i = 1, 2,$$

$$F^n = F^{1,n} \cap F^{2,n}.$$

Then F^n are stopping sets, $F^n \subset F^{n+1}$, $A^n = A$ on $F^n \cap E^n$; $\widehat{\Pi} I_{[0,\tau^n]} = \mathrm{O} I_{[0,\tau^n]}$ according to Theorem 5 in (Gushchin, 1982);

$$\mathbf{E}\,\widehat{A}_t^n \leqslant 2\mathbf{E} \int_{[0,t]} \mathrm{O} I_{[0,\tau^n]}(s) I_{E^n}(s)\, dA_s = 2\mathbf{E} \int_{[0,t]} \mathrm{O} I_{[0,\tau^n]\cap E^n}(s)\, dA_s \leqslant$$

$$\leqslant 4n + 2\mathbf{E} \sup_{s\in E^n\cap[0,\tau^n\wedge t]} \Box A_s \leqslant 4n + 8\mathbf{E}\,(x^n)_t^* < \infty,$$

$$x_t^* = \sup_{s\leqslant t} |x_s|.$$

Using Doob's maximal inequality for usual martingales, we obtain

$$\mathbf{P}\{t \notin F^{1,n}\} = \mathbf{P}\Big\{ \sup_{s_2\leqslant t_2} \mathbf{E}\{I_{]\tau_n^1,\infty[}(t_1) \mid \mathfrak{F}^2_{s_2}\} > \frac{1}{2}\Big\} \leqslant$$

$$\leqslant 16\mathbf{E}\left((I_{]\tau_n^1,\infty[}(t_1) \mid \mathfrak{F}^2_{t_2})^2\right) \leqslant 16\mathbf{P}\{\tau_n^1 < t_1\} \to 0.$$

Similarly,

$$\mathbf{P}\{t \notin F^{2,n}\} \to 0, \quad n \to \infty.$$

Therefore

$$\bigcup_n F^n = \mathbf{R}_+^2 \times \Omega,$$

and $(\widehat{A}^n,\ F^n \cap E^n)$, $n \geqslant 1$, is a localizing sequence for A.

THEOREM 1 (Gundy's decomposition). *Let $M \in l\mathfrak{M}_s^*$. Then $M = M^1 + M^2$, where*

$$M^1 \in l\mathfrak{M}_s^2, \quad M^2 \in l\mathfrak{M}_s^* \cap l\mathcal{B}.$$

Proof. Consider the process

$$A_t = \sum_{s \leqslant t} \square M_s I\{|\square M_s| > a\}.$$

According to Lemma 1, $A \in l\mathcal{A}$. So there exists $A^{\widehat{\Pi}}$, and $A - A^{\widehat{\Pi}} \in l\mathfrak{M}_s^* \cap l\mathcal{B}$. We have $M - A = -A^{\widehat{\Pi}} + L$, where $L \in l\mathfrak{M}_s^*$ and $|\square(M - A)_t| \leqslant a$, therefore $|\square(-A^{\widehat{\Pi}} + L)_t| \leqslant a$. According to (Mishura and Gushchin, 1990), $\widehat{\Pi}\square L I_{E^n} = 0$, where $(L^n,\ E^n)$, $n \geqslant 1$ is a localizing sequence for L. So,

$$a \geqslant \left|\widehat{\Pi}\big(-\square(A^{\widehat{\Pi}} + L)\big)_t\right| = |\square A_t^{\widehat{\Pi}}| \quad \text{and} \quad |\square L_t| \leqslant 2a.$$

If follows from Theorem 2 in (Mishura and Gushchin, 1990) that there exists quadratic variation $[L]_t$, $\square[L]_t \leqslant 4a^2$, so $[L] \in l\mathcal{A}$ and $[L]\widehat{\Pi} \in l\mathcal{A}$. Same as before, suppose that $[L]_\infty^{\widehat{\Pi}} < \infty$ a.s. Let

$$\tau_i^n = \inf\{t_i\colon\ [L]_{\infty t_i}^{\widehat{\Pi}} > n\}, \quad i = 1, 2,$$

$$\xi_t^{i,n} = \widehat{\Pi} I_{[0,\tau_i^n]}(t), \qquad F^{i,n} = \left\{t\colon\ \xi_s^{i,n} > \frac{1}{2} \quad \forall s \leqslant t\right\}$$

$$\alpha_t^i(\tau_i^n) = I\left\{\xi_t^{i,n} > \frac{1}{2}\right\} + 2\xi_t^{i,n} I\left\{\xi_t^{i,n} \leqslant \frac{1}{2}\right\}, \quad n \geqslant 1.$$

According to (Mishura and Gushchin, 1990), there exists stochastic integral

$$\gamma_t^n = \Big(\big(\alpha^1(\tau_1^n),\ \alpha^2(\tau_2^n)\big) \cdot L\Big) \in l\mathfrak{M}_s^*,$$

$$|\square\gamma_t^n| \leqslant 8a, \qquad \mathbf{E}\,[\gamma^n]_t \leqslant 2n + 64a^2,$$

$$(\gamma_t^n - L_t) I_{F^n}(t) = 0, \qquad F^n = F^{1,n} \cap F^{2,n}.$$

Let us prove that $\gamma^n \in \mathfrak{M}_s^2$. The index n will be omitted, $\gamma^n = \gamma$. Let $(\bar{\gamma}^m,\ D^m)$ be a localizing sequence for γ, $\bar{\gamma}^m \in \mathfrak{M}_s^*$. Fix $T \in \mathbf{R}_+^2$ and consider

$$\sigma_1^m = \inf\{t_1\colon\ (t_1,\ T_2) \notin D^m\} \wedge T_1,$$

$$\sigma_2^m = \inf\left\{t_2\colon\ (T_1,\, t_2) \notin D^m\right\} \wedge T_2,$$

$$\sigma^m = (\sigma_1^m,\, \sigma_2^m).$$

Obviously, $\sigma_i^{m+1} \geqslant \sigma_i^m$, and $\sigma_i^m = T_i$ for all $m > m(\omega)$. Consider

$$A_t^m = \sum_{s \leqslant t} \Box\bar{\gamma}_s^m I\left\{|\Box\bar{\gamma}_s^m| > 8a\right\} \in I\!B.$$

Note that $(\bar{\gamma}^m - \gamma)I_{D^m} = 0$, whence $A^m I_{D^m} = 0$ and therefore $A^{m,\widehat{\Pi}} I_{D^m} = 0$. Let

$$\tau_i^{m,l} = \inf\{t_i\colon\ A_{t_i T_j}^m > l\} \wedge T_i, \qquad i = 1, 2, \quad j = 3 - i.$$

Then

$$A^{m,l} = ((\alpha^1(\tau_1^{m,l}),\, \alpha^2(\tau_2^{m,l})) \cdot A^m) \in B.$$

Note that $|\Box A^{m,\widehat{\Pi}}| \leqslant 8a$. Now we construct the integral

$$\zeta^{m,l} = \left((\alpha^1(\tau_1^{m,l})\alpha^1(\sigma_1^m),\, \alpha^2(\tau_2^{m,l})\alpha^2(\sigma_2^m)) \cdot \left(\bar{\gamma}^m - A^m + A^{m,\widehat{\Pi}}\right)\right) \in \mathfrak{M}_s^*;$$

$$\mathbf{E}\,[\zeta^{m,l}]_t \leqslant 2^4 \mathbf{E}\left((\alpha^1(\tau_1^{m,l})\alpha^1(\sigma_1^m),\, \alpha^2(\tau_2^{m,l})\alpha^2(\sigma_2^m)) \cdot \left[\bar{\gamma}^m - A^m + A^{m,\widehat{\Pi}}\right]\right)_t \leqslant$$

$$\leqslant 2^8 \mathbf{E}\left((\alpha^1(\tau_1^{m,l}),\, \alpha^2(\tau_2^{m,l})) \cdot \left[\bar{\gamma}^m - A^m + A^{m,\widehat{\Pi}}\right]_{\sigma^m}^{\widehat{\Pi}}\right) \leqslant$$

$$\leqslant 2^{10} \mathbf{E}\left(\left(\left[\bar{\gamma}^m - A^m + A^{m,\widehat{\Pi}}\right]_{\sigma^m}^{\widehat{\Pi}}\right)^{-+} +\right.$$

$$\left. + \left(\left[\bar{\gamma}^m - A^m + A^{m,\widehat{\Pi}}\right]_{\sigma^m}^{\widehat{\Pi}}\right)^{+-} + \Box\left[\bar{\gamma}^m - A^m + A^{m,\widehat{\Pi}}\right]_{\sigma^m}^{\widehat{\Pi}}\right).$$

Since $(s_1,\, \sigma_2^m) \in D^m$ if $s_1 < \sigma_1^m$, we have

$$\mathbf{E}\left(\left[\bar{\gamma}^m - A^m + A^{m,\widehat{\Pi}}\right]_{\sigma^m}^{\widehat{\Pi}}\right)^{-+} = \mathbf{E}\left(\left[\bar{\gamma}^m\right]_{\sigma^m}^{\widehat{\Pi}}\right)^{-+} =$$

$$= \mathbf{E}\left([\gamma]_{\sigma^m}^{\widehat{\Pi}}\right)^{-+} \leqslant 2n + 64a^2.$$

Similarly,

$$\mathbf{E}\left(\left[\bar{\gamma}^m - A^m + A^{m,\widehat{\Pi}}\right]_{\sigma^m}^{\widehat{\Pi}}\right)^{+-} \leqslant 2n + 64a^2.$$

Note that

$$|\Box(\bar{\gamma}^m - A^m)_t| \leqslant 8a, \qquad |\Box A_t^{m,\widehat{\Pi}}| \leqslant 8a,$$

so,

$$\mathbf{E}\,\Box_{\sigma^m}[\bar{\gamma}^m - A^m + A^{m,\widehat{\Pi}}]^{\widehat{\Pi}} \leqslant 2^8 a^2.$$

Therefore, $\mathbf{E}\,[\zeta^{m,l}]_t \leqslant C$, where C is a constant. By Theorem 2 in (Mishura and Gushchin, 1990), $\zeta^{m,l} \in \mathfrak{M}^2_s$ for all $m \geqslant 1$, $l \geqslant 1$. Consider the stopping sets

$$K^{i,m} = \left\{t\colon\ \widehat{\Pi}\ I_{[0,\sigma_i^m]}(s) > \frac{1}{2} \qquad \forall s \leqslant t\right\}, \qquad K^m = K^{1,m} \cap K^{2,m}.$$

Obviously,

$$\bigcup_m K^m = [0,T] \times \Omega.$$

Since $\tau_i^{m,l} \geqslant \sigma_i^m$, we have $\alpha^i(\tau_i^{m,l})\alpha^i(\sigma_i^m) = 1$ on $K^{i,m}$. Therefore

$$(\zeta_t^{m,l} - \gamma) I_{K^m \cap D^m \cap [0,T]} = 0,$$

and $\zeta_t^{m,l} \to \gamma_t$ for all $l \geqslant 1$ as $m \to \infty$, $t \in [0,T]$.
Using Theorem 2 in (Mishura and Gushchin, 1990), we obtain

$$\mathbf{E}\,\gamma_t^2 \leqslant \lim_{m\to\infty} \mathbf{E}\,(\zeta_t^{m,l})^2 \leqslant C_1 \lim_{m\to\infty} \mathbf{E}[\zeta^{m,l}]_t \leqslant C_2.$$

Thus, $\gamma = \gamma^n \in \mathfrak{M}^2_s$; $(\gamma^n,\ F^n)$ is localizing sequence for L.

3. STOCHASTIC INTEGRATION WITH RESPECT TO STRONG MARTINGALES

Let $M \in \mathfrak{M}^2_s$, $\langle M\rangle = [M]^{\widehat{\Pi}}$.
The construction of the stochastic integral $(\varphi \cdot M)$ for a weakly predictable φ such that $\mathbf{E}\int \varphi_s^2\, d\langle M\rangle_s < \infty$ was described in (Gushchin, 1988). We extend it to $M \in l\mathfrak{M}^*_s$. Recall that, for $M \in l\mathfrak{M}^*_s$, there exists $\{[M]_t,\ t \in [0,T]\}$ for all $T \in \mathbf{R}^2_+$.

LEMMA 2. *Let $M \in l\mathfrak{M}^*_s$, $M = M^1 + M^2$ according to Theorem* 1, *and* $[M^i] = 0$, $i = 1, 2$.
Then $M = 0$.

Proof. Let $M^1 \in l\mathfrak{M}^2_s$, $M^2 \in l\mathfrak{M}^*_s \cap l\mathcal{B}$, and let $(M^{1,n},\ M^{2,n},\ D^n)$, $n \geqslant 1$, be a localizing sequence for M^1 and M^2.
Since $[M^2]_t = \sum_{s \leqslant t} (\Box M^2_s)^2 = 0$, we have $\Box M^2 = 0$. Thus,

$$\Box M^{2,n} I_{D^n} = \Box M^2 I_{D^n} = 0.$$

According to Lemma 17 in (Gushchin, 1982),

$$M_t^{2,n} = \sum_{s \leqslant t} \Box M_s^{2,n} - \left(\sum_{s \leqslant t} \Box M_s^{2,n}\right)^{\widehat{\Pi}}.$$

Consequently, $M^{2,n} I_{D^n} = 0$. Furthermore,

$$[M^{1,n}] I_{D^n} = \langle M^{1,n}\rangle I_{D^n} = 0$$

and $\mathbf{E}(\int I_{D^n}\, dM^{1,n})^2 = 0$. This means that $M^{1,n} I_{D^n} = 0$ and, therefore,

$$MI_{D^n} = (M^{1,n} + M^{2,n})I_{D^n} = 0 \quad \text{and} \quad M = 0.$$

Let $M \in \mathfrak{M}_s^2 \cap \mathcal{B}$, and let φ be a weakly predictable process such that

$$\mathbf{E} \int\limits_{[0,t]} \varphi_s^2 \, d\langle M \rangle_s < \infty,$$

and

$$\mathbf{E} \int\limits_{[0,t]} |\varphi_s| \, d|M|_s < \infty, \qquad |M|_t = \operatorname*{var}_{[0,t]} M.$$

Then the stochastic integral $(\varphi \cdot M)_t$ and the Stieltjes integral $\int_{[0,t]} \varphi_s \, dM_s$ coincide (Gushchin, 1988).

Remark 1. Let $M^1 \in l\mathfrak{M}_s^2$, $(M^{1,n}, D^n)$, $n \geqslant 1$, be a localizing sequence for M, $\langle M^1 \rangle_t = \lim\limits_{n\to\infty} \langle M^{1,n} \rangle_t$, and let φ be a weakly predictable process, $\varphi_s^n = (\varphi_s \wedge n) \vee (-n)$,

$$\mathbf{P}\Bigg\{ \int\limits_{[0,t]} \varphi_s^2 \, d\langle M^1 \rangle_s < \infty \Bigg\} = 1, \quad t \in \mathbf{R}_+^2.$$

Since

$$\mathbf{E} \int\limits_{[0,t]} (\varphi_s^n)^2 \, d\langle M^{1,n} \rangle_s < \infty \quad \forall t \in \mathbf{R}_+^2,$$

the stochastic integral $\zeta^{1,n} = (\varphi^n \cdot M^{1,n})$ exists.
It follows from Theorem 2 in (Gushchin, 1988) that

$$(\zeta^{1,n} - \zeta^{1,k}) I_{D^n \cap E^n} = 0 \quad \forall k \geqslant n,$$

where $E^n = \{t \colon \varphi_t^* \leqslant n\}$.
So, there exists $\zeta_t^1 = \lim\limits_{n\to\infty} \zeta_t^{1,n}$ a.s., $\zeta^1 \in l\mathfrak{M}_s^2$,

$$\langle \zeta^1 \rangle_t = \int\limits_{[0,t]} \varphi_s^2 \, d\langle M^1 \rangle_s;$$

ζ^1 is proposed as a definition of the stochastic integral $(\varphi \cdot M^1)$.
Furthermore, let $M^2 \in l\mathfrak{M}_s^* \cap l\mathcal{B}$, $(M^{2,n}, F^n)$, $n \geqslant 1$, be a localizing sequence for M^2, φ be a weakly predictable process,

$$\mathbf{P}\Bigg\{ \int\limits_{[0,t]} |\varphi_s| \, d|M^2|_s < \infty \Bigg\} = 1.$$

Set $\zeta_t^{2,n} = \int_{[0,t]} \varphi_s^n \, dM_s^{2,n}$ (the Stieltjes integral). Obviously,

$$(\zeta^{2,n} - \zeta^{2,k}) I_{F^n \cap E^n} = 0 \quad \forall k \geqslant n$$

and there exists $\zeta_t^2 = \lim_{n\to\infty} \zeta_t^{2,n}$ a.s., $\zeta^2 \in l\mathfrak{M}_s^* \cap l\mathcal{B}$, ζ^2 is proposed as a definition of the stochastic integral $(\varphi \cdot M^2)$. If $M \in l\mathfrak{M}_s^2 \cap l\mathcal{B}$ and

$$\mathbf{P}\Big\{ \int\limits_{[0,t]} \varphi_s^2 \, d\langle M\rangle_s < \infty, \int\limits_{[0,t]} |\varphi_s| \, d|M|_s < \infty \Big\} = 1, \quad \forall t \in \mathbf{R}_+^2,$$

then ζ^1 and ζ^2 coincide.
If $\zeta \in l\mathfrak{M}_s^*$, then $\zeta = \zeta^1 + \zeta^2$ is Gundy's decomposition (Theorem 1).

THEOREM 2. *Let $M \in l\mathfrak{M}_s^*$, φ be a weakly predictable and locally bounded process. Then*

1) *there exists a unique process ζ such that, for any decomposition $M = M^1 + M^2$, there exists a decomposition $\zeta = \zeta^1 + \zeta^2$ with the property that, for all $N \in l\mathfrak{M}_s^*$,*

$$[\zeta^i, N]_t = \int\limits_{[0,t]} \varphi_s \, d[M^i, N]_s, \quad i = 1,2; \tag{1}$$

2) *the processes $\Box\zeta$ and $\varphi\Box M$ are indistinguishable;*

3) *if $M \in l\mathfrak{M}_s^2$ and*

$$\mathbf{P}\{ \int_{[0,t]} \varphi_s^2 \, d\langle M\rangle_s < \infty\} = 1$$

or $M \in l\mathfrak{M}_s^* \cap l\mathcal{B}$ and

$$\mathbf{P}\Big\{ \int\limits_{[0,t]} |\varphi_s| \, d|M|_s < \infty \Big\} = 1, \quad \forall t \in \mathbf{R}_+^2,$$

then ζ coincides with the stochastic integral introduced in Remark 1.

Proof. 1) *Existence.* Let $(M^{i,n}, N^{i,n}, i = 1,2, \varphi^n, D^n)$, $n \geqslant 1$, be localizing sequences for the corresponding processes.
Construct the stochastic integral

$$(\varphi^n \cdot M^n) := (\varphi^n \cdot M^{1,n}) + \int\limits_{[0,t]} \varphi_s^n \, dM_s^{2,n}.$$

Using the properties of stochastic integrals (Gushchin, 1988), we have

$$[(\varphi^n \cdot M^{1,n}), N^n]_t = \int\limits_{[0,t]} \varphi_s^n \, d[M^{1,n}, N^{1,n}]_s + \sum_{s \leqslant t} \varphi_s^n \Box M_s^{1,n} \Box N_s^{2,n} =$$

$$= \int_{[0,t]} \varphi_s^n \, d[M^{1,n}, N^n]_s,$$

$$[(\varphi^n \cdot M^{2,n}), N^n]_t = \sum_{s\leqslant t} \varphi_s^n \square M_s^{2,n} \square N^n = \int_{[0,t]} \varphi_s^n \, d[M^{2,n}, N^n]_s.$$

Passing to the limit as $n \to \infty$, we obtain (1). Obviously, $\zeta \in l\mathfrak{M}_s^*$.
Uniqueness. If $[\bar{\zeta}^i, N]_t = \int_{[0,t]} \varphi_s \, d[M^i, N]_s$, $i = 1,2$, then $[\zeta^i - \bar{\zeta}^i, N] = 0$. Set $N = \zeta^i - \bar{\zeta}^i$. It follows that $[\zeta^i - \bar{\zeta}^i] = 0$, $i = 1,2$. By Lemma 2, $\zeta = \bar{\zeta}$.
2) Since

$$\square\zeta I_{D^n} = \square\zeta^n I_{D^n} = \square(\varphi^n \cdot M^{1,n}) I_{D^n} + \square(\varphi^n \cdot M^{2,n}) I_{D^n} =$$

$$= \varphi^n \square M^n I_{D^n} = \varphi \square M I_{D^n},$$

the processes $\square\zeta$ and $\varphi\square M$ are indistinguishable.
Assertion 3) follows from the method of construction of ζ.

Definition 3. The stochastic integral $(\varphi \cdot M)$, where φ is weakly predictable and locally bounded; $M \in l\mathfrak{M}_s^*$, is the unique process $\zeta \in l\mathfrak{M}_s^*$ such that, for arbitrary Gundy's decomposition $M = M^1 + M^2$, there exists a decomposition $\zeta = \zeta^1 + \zeta^2$ with the property that, for all $N \in l\mathfrak{M}_s^*$,

$$[\zeta^i, N]_t = \int_{[0,t]} \varphi_s \, d[M^i, N]_s.$$

The correctness of Definition 3 follows from Theorem 2.

Definition 4. A process $(x_t, \mathfrak{F}_t)$ is a strong semimartingale if $x = M + A$, where $M \in l\mathfrak{M}_s^*$, $A \in l\mathcal{B}$.

Remark 2. According to Theorem 1, a strong semimartingale x has a representation $x = M^1 + M^2 + A$, where $M^1 \in l\mathfrak{M}_s^2$, $M^2 \in l\mathfrak{M}_s^* \cap l\mathcal{B}$. Define $B = M^2 + A$. Let φ be a weakly predictable and locally bounded process.
Let the stochastic integral $(\varphi \cdot x)$ be defined as the sum

$$(\varphi \cdot M^1) + \int \varphi \, dB.$$

LEMMA 3. *The integral $(\varphi \cdot x)$ does not depend on the decomposition $x = M^1 + b$, $M^1 \in l\mathfrak{M}_s^2$, $B \in l\mathcal{B}$.*

Proof. Let $x = \overline{M}^1 + \bar{b}$, $\overline{M}^1 \in l\mathfrak{M}_s^2$, $\overline{B} \in l\mathcal{B}$. Then $M^1 - \overline{M}^1 = \overline{B} - B \in l\mathfrak{M}_s^2 \cap l\mathcal{B}$. Let $(M^{1,n} - \overline{M}^{1,n}, \overline{B}^n - B^n, \varphi^n, D^n)$, $n \geqslant 1$, be a localizing sequence. According to Lemma 2, the stochastic integral $(\varphi^n \cdot (M^{1,n} - \overline{M}^{1,n}))$ and the Stieltjes integral $\int \varphi^n \, d(\overline{B}^n - B^n)$ are indistinguishable, i.e.,

$$(\varphi^n \cdot M^{1,n}) + \int \varphi^n \, dB^n = \int \varphi^n \, d\overline{B}^n + (\varphi^n \cdot \overline{M}^{1,n}).$$

Letting $n \to \infty$, we complete the proof.

Remark 3. Obviously, $(\varphi \cdot x) = (\varphi \cdot M) + \int \varphi\, dA$, where $(\varphi \cdot M)$ is a stochastic integral (Definition 3), $\int \varphi\, dA$ is a Stieltjes integral.

4. STOCHASTIC DIFFERENTIAL EQUATIONS ON THE PLANE WITH RESPECT TO STRONG SEMIMARTINGALES. EXISTENCE AND UNIQUENESS OF SOLUTIONS

Let f: $\mathbf{R}_+^2 \times \Omega \times D \to \mathbf{R}$ satisfy the condition

(a) for arbitrary D-process $(x_t, \mathfrak{F}_t)$, the mapping

$$(f \circ x)\colon\ \mathbf{R}_+^2 \times \Omega \to \mathbf{R},$$

where $(f \circ x)(t, \omega) = f(t, \omega, x_\circ(\omega))$, is weakly predictable and locally bounded.

Let $z = M + A$ be a strong semimartingale, $M \in l\mathfrak{M}_s^*$, $A \in l\mathcal{B}$; H is also a strong semimartingale. Construct the stochastic integral $((f \circ x)\cdot z)$ according to Remark 2.

Definition 5. Suppose $(x_t, \mathfrak{F}_t)$ is a D-process such that

$$x_t = H_t + \big((f \circ x) \cdot z\big)_t \quad \forall t \in \mathbf{R}_+^2. \tag{2}$$

Then x is said to be a strong solution of (2).

Suppose that f satisfies also the conditions

(b) there exists an increasing predictable process K_t such that for all $\mathfrak{F}_t$-adapted D-processes x and y,

$$\big|(f \circ x)_t - (f \circ y)_t\big| \leqslant K_t |x - y|_t^*.$$

(c) $|f(s, \omega, 0)| \leqslant C$ a.s., where $C > 0$ is a constant.

Equation (2) may be written in an equivalent form

$$x_t = H_t + \big((f \circ x) \cdot M^1\big)_t + \int\limits_{[0,t]} (f \circ x)_s\, dB_s, \quad B = M^2 + A. \tag{3}$$

Suppose, for simplicity, that $K_t \leqslant K$, where $K > 0$ is some constant.

LEMMA 4. *Assume that, for all $K > 0$, there exists $\alpha = \alpha(K)$ such that equation* (3) *has a unique strong solution x provided that M^1 and B satisfy the condition*

(d) $|\Box M_s^1| \leqslant \alpha, \quad |\Box B_s| \leqslant \alpha.$

Then equation (3) *has a unique strong solution regardless of condition* (d).

Proof. Let $(M^{1,n}, D^n)$ be a localizing sequence for M^1.

Set

$$N_t^n = \sum_{s \leqslant t} \Box M_s^{1,n} I\Big\{ |\Box M_s^{1,n}| > \frac{\alpha}{2} \Big\}.$$

Then

$$\mathbf{E}\,|N_t^n| \leqslant 2\alpha^{-1}\mathbf{E}\sum_{s\leqslant t}(\square M_s^{1,n})^2 \leqslant$$

$$\leqslant 2\alpha^{-1}\mathbf{E}\,[M^{1,n}]_t \leqslant 2C_1\alpha^{-1}\mathbf{E}\,(M_t^{1,n})^2 \leqslant \infty,$$

where $C_1 > 0$ is a constant. So there exists a dual weakly predictable projection $N^{n,\widehat{\Pi}}$, $N^n - N^{n,\widehat{\Pi}} \in \mathfrak{M}_s^*$, and

$$M^{1,n} - N^n = -N^{n,\widehat{\Pi}} + L^n, \quad L^n \in \mathfrak{M}_s^*.$$

Note that $|\square(M^{1,n} - N^n)_t| \leqslant \alpha/2$. Therefore,

$$\left|\square(-N^{n,\widehat{\Pi}} + L^n)_t\right| \leqslant \alpha/2.$$

Since $\widehat{\Pi}(\square L^n)_t = 0$ (Mishura and Gushchin, 1990), we have $|\square N_t^{n,\widehat{\Pi}}| \leqslant \alpha/2$ and $|\square L_t^n| \leqslant \alpha$. So, $M^{1,n} = L^n + P^n$, where $P^n \in \mathfrak{M}_s^* \cap \mathcal{B}$.
Furthermore,

$$[P^n]_t \leqslant 2[N^n]_t + 2[N^{n,\widehat{\Pi}}]_t \leqslant 2\sum_{s\leqslant t}(\square M_s^{1,n})^2 + \alpha \operatorname*{var}_{[0,t]} N^{n,\widehat{\Pi}}$$

and, by Theorem 2 in (Mishura and Gushchin, 1990), $P^N \in \mathfrak{M}_s^2$. Consequently, $L^n \in \mathfrak{M}_s^2$.
Consider the equation

$$x_t = H_t + \left((f \circ x) \cdot M^{1,n}\right)_t + \int\limits_{[0,t]} (f \circ x)_s \, dB_s,$$

where $B = M^2 + A$, and rewrite it in the following form:

$$x_t = H_t + \left((f \circ x) \cdot L^n\right)_t + \int\limits_{[0,t]} (f \circ x)_s \, dC_s^n, \tag{4}$$

where $C^n = P^n + B$.
Let $E_1 = \{t\colon\ |\square C^n|_t^* < \alpha\}$. The set $\overline{E}_1$ is closed, so it is predictable. Consider the auxiliary equation

$$x_t^1 = H_t + \left((f \circ x^1) I_{\overline{E}_1} \cdot L^n\right)_t + \int\limits_{[0,t]} (f \circ x^1)_s I_{\overline{E}_1}(s) \, dC_s^n. \tag{5}$$

It follows from the condition of Lemma 4 that (5) has a unique strong solution. This statement means that equation (3) has a unique strong solution on E_1.
Let us extend the solution to $\overline{E}_1$. Let $\partial E_1 = \overline{E}_1 \backslash E_1$. Then ∂E_1 a.s. consists of a finite number of horizontal and vertical segments.

Let $(t, \omega) \in \partial E_1$. If $[0,t]\backslash\{t\} \in E_1$, we set

$$\bar{x}_t^1 = (x_t^1)^{+-} + (x_t^1)^{-+} - (x_t^1)^{--} + (f \circ x^1)_t \Box C_t^n.$$

If

$$([0,t]\backslash[0,t[) \cap \partial E_1 \neq \varnothing,$$

then there exists, for example, $s_2 \in [0,t_2]$ such that $(t_1, z_2) \in E_1$ for all $z_2 < s_2$ and $(t_1, z_2) \in \partial E_1$ for all $s_2 \leqslant z_2 \leqslant t_2$. In this case, we set

$$\bar{x}_{t_1 s_2}^1 = (x_{t_1 s_2}^1)^{+-} + (x_{t_1 s_2}^1)^{-+} - (x_{t_1 s_2}^1)^{--} + (f \circ x^1)_{t_1 s_2} \Box C_{t_1 s_2}^n,$$

$$\bar{x}_{t_1 z_2}^1 = (x_{t_1 z_2}^1)^{-+} + x_{t_1 s_2}^1 - (x_{t_1 s_2}^1)^{-+} +$$

$$+ \int_{s_2}^{z_2} (f \circ \bar{x}^1)_{t_1 u_2}\, d_2\big(C_{t_1 u_2}^n - (C_{t_1 u_2}^n)^{-+}\big), \quad s_2 < z_2 \leqslant t_2. \tag{6}$$

Note that (6) is a "one-parameter" equation having a unique solution (Elliott, 1982, Theorem 14.6). Finally, we set

$$\bar{x}^1 I_{E_1} = x^1 I_{E_1}.$$

Then $\overline{X}^1$ is the unique solution of equation (4) on the set $\overline{E}_1$.
Now let

$$E_2 = \{(t, \omega)\colon\ t \notin \overline{E}_1,\ |\Box C_s^n| \leqslant \alpha \quad \forall s \in [0,t]\backslash\overline{E}_1\}.$$

Consider the equation

$$x_t^2 = H_t + ((f \circ \bar{x}^1) I_{\overline{E}_1} \cdot L^n)_t + \int_{[0,t]} (f \circ \bar{x}^1)_s I_{\overline{E}_1}(s)\, dC_s^n +$$

$$+ ((f \circ x^2) I_{\overline{E}_2 \backslash \overline{E}_1} \cdot L^n)_t + \int_{[0,t]} (f \circ x^2)_s I_{E_2 \backslash \overline{E}_1}(s)\, dC_s^n. \tag{7}$$

The sum of H_t and the terms containing $\bar{x}^1$ is a strong semimartingale, therefore it follows from the hypothesis that (7) has a unique solution.
But

$$x_t^2 I_{\overline{E}_1(t)} = \Big(H_t + ((f \circ \bar{x}^1) \cdot L^n)_t + \int_{[0,t]} (f \circ \bar{x}^1)_s\, dC_s^n\Big) I_{\overline{E}_1}(t) = \bar{x}_t^1 I_{\overline{E}_1}(t).$$

So

$$x_t^2 = H_t + ((f \circ x^2) I_{\overline{E}_2} \cdot L^n)_t + \int_{[0,t]} (f \circ x^2)_s I_{E_2}(s)\, dC_s^n.$$

Since this equation has a unique solution, equation (4) has a unique solution on $\overline{E}_1 \cup E_2$. Extending the solution to $\overline{E}_1 \cup \overline{E}_2 \cup \ldots \cup \overline{E}_k \cup E_{k+1}$ for all $k \geqslant 1$, we finally obtain that the equation

$$x_t^{(n)} = H_t + \left((f \circ x^{(n)}) \cdot L^n\right)_t + \int\limits_{[0,t]} (f \circ x^{(n)})_s \, dC_s^n =$$

$$= H_t + \left((f \circ x^n) \cdot M^{1,n}\right)_t + \int\limits_{[0,t]} (f \circ x^{(n)})_s \, dB_s$$

has a unique solution $x_t^{(n)}$. Note that

$$(x_t^k - x_t^n) I_{D^n}(t) = 0 \quad \text{for all} \quad k \geqslant n,$$

so there exists $x_t = \lim_{n\to\infty} x_t^{(n)}$. The process x_t is the unique solution of (3).

LEMMA 5. *Assume that equation* (3) *has a unique strong solution when* $H_t = 0$. *Then* (3) *has a unique strong solution for an arbitrary strong semimartingale* H_t.

Proof is similar to the one given for the "one-parameter" case in (Elliott, 1982, Lemma 14.11) and therefore is omitted.

THEOREM 3. *Under conditions* (**a**), (**b**) *and* (**c**) *equation* (2) *has a unique strong solution.*

Proof. By Lemmas 4–5 we may assume that, for some $\alpha > 0$,

$$|\Box M_t^1| \leqslant \alpha, \qquad |\Box B_t| \leqslant \alpha; \qquad H_t = 0.$$

For simplicity, assume also that $K_t \leqslant K$, $B \in \mathcal{B}$, $M^1 \in \mathfrak{M}_s^2$. Set $\alpha = (8\sqrt{582}K)^{-1}$. Let $[M^1]$ be the quadratic variation of M^1. From Theorem 1 in (Mishura and Gushchin, 1990) it follows that $\Box[M^1]_t \leqslant \alpha^2$, therefore, $\Box\langle M^1\rangle_t \leqslant \alpha^2$. Fix $[0,T]$ and consider a sequence of increasing grids λ_{nn}^T on $[0,T]$, $|\lambda_{nn}^T| \to 0$ as $n \to \infty$. Define

$$\Box_{jk} =]t_{1,n}^j,\, t_{1,n}^{j+1}] \times]t_{2,n}^k,\, t_{2,n}^{k+1}],$$

$$C_s^{jk} = C_s - C_{(s_1, t_{2,n}^k)} - C_{(t_{1,n}^j, s_2)} + C_{t_{jk}}, \quad s \geqslant t_{jk}; \tag{8}$$

$$\tau_j^1 = \inf\left\{s_1\colon \left(\langle M^1\rangle^{jk}_{(s_1, t_{2,n}^{k+1})}\right)^{1/2} \vee \left(\operatorname{var} B\right)^{jk}_{(s_1, t_{2,n}^{k+1})} > \alpha\right\} \wedge t_{1,n}^{j+1},$$

$$\tau_k^2 = \inf\left\{s_2\colon \left(\langle M^1\rangle^{jk}_{(t_{1,n}^{j+1}, s_2)}\right)^{1/2} \vee \left(\operatorname{var} B\right)^{jk}_{(t_{1,n}^{j+1}, s_2)} > \alpha\right\} \wedge t_{2,n}^{k+1}, \tag{9}$$

$$\tau_{jk} = (\tau_j^1,\, \tau_k^2), \qquad \Delta_{jk} = [0,\, \tau_{jk}] \cap \Box_{jk},$$

$$I_s^{jk} = I\{\widehat{\Pi} I_{\Delta_{jk}}(s) > 1/2\}.$$

Note that $\langle M^1\rangle^{jk}_{\tau_{jk}} \leqslant 3\alpha^2$, $(\operatorname{var} B)^{jk}_{\tau_{jk}} \leqslant 3\alpha$.

First consider the equation

$$x_t^{11} = H_t + \left((f \circ x^{11}) I^{00} \cdot M^1\right)_t + \int_{[0,t]} (f \circ x^{00})_s I_s^{00}\, dB_s, \tag{10}$$

where $\Delta_{00} = [0, \tau_{00}]$.
Consider the space

$$\mathfrak{L}^2_{[0,T]} = \{x = x_t,\ t \in [0,T]:\ \mathbf{E}(x_T^*)^2 < \infty\}$$

with the usual norm and let a linear operator T be defined on this space by

$$(Tx)_t = \left((f \circ x) I^{00} \cdot M^1\right)_t + \int_{[0,t]} (f \circ x)_s I_s^{00}\, dB_s.$$

It is easily seen that T is a contracting operator. Indeed, from Doob's maximal inequality for martingales and from Lemma 16 in (Gushchin, 1982), we obtain $(\widehat{\Pi} I_{\Delta_{00}} = \mathbf{O} I_{\Delta_{00}})$

$$\begin{aligned}
\mathbf{E}\Big(((Tx)-(Ty))_T^*\Big)^2 &\leqslant 2\mathbf{E}\Big(\Big(\big(((f\circ x)-(f\circ y))I^{00}\cdot M^1\big)_T^*\Big)^2 + \\
&+2\mathbf{E}\Big(\Big(\int I_s^{00}((f\circ x)_s-(f\circ y)_s)\,dB_s\Big)_T^*\Big)^2 \leqslant \\
&\leqslant 64\mathbf{E}\int_{[0,T]} \widehat{\Pi} I_{[0,\tau_{00}]}(s)\big((f\circ x)_s-(f\circ y)_s\big)^2\, d\langle M^1\rangle_s + \\
&+8\mathbf{E}\Big(\int_{[0,T]} \mathbf{O} I_{[0,\tau_{00}]}(s)\big|(f\circ x)_s-(f\circ y)_s\big|\, d|B|_s\Big)^2 \leqslant \\
&\leqslant 64\mathbf{E}\int_{[0,T]} I_{[0,\tau_{00}]}(s)\big((f\circ x)_s-(f\circ y)_s\big)^2\, d\langle M^1\rangle_s + \\
&+2^{11}\mathbf{E}\Big(\int_{[0,T]} I_{[0,\tau_{00}]}(s)\big|(f\circ x)_s-(f\circ y)_s\big|\, d|B|_s\Big)^2 \leqslant \\
&\leqslant 8^2\cdot 582 K^2\alpha^2 \mathbf{E}\big((x-y)_T^*\big)^2 \leqslant \mathbf{E}\big((x-y)_T^*\big)^2/2.
\end{aligned}$$

Thus, equation (10) has a unique strong solution which can be obtained by successive approximations:

$$\mathbf{E}\big((x^{11}-x^{11,r})_T^*\big)^2 \longrightarrow 0, \quad r \to \infty.$$

Set

$$x^{11,0} = 0, \qquad x_t^{11,r} = \left((f \circ x^{11,r-1}) I^{00} \cdot M^1\right)_t + \int_{[0,t]} (f \circ x^{11,r-1})_s I_s^{00}\, dB_s.$$

By the same estimations as above we obtain

$$\mathbf{E}\big((x^{11,r})^*_T\big)^2 \leqslant \frac{1}{2}C^2 + \mathbf{E}\big((x^{11,r-1})^*_T\big)^2/2$$

(C is taken from (**c**)), therefore,

$$\mathbf{E}\big((x^{11,r})^*_T\big)^2 \leqslant C^2 \quad \text{and} \quad \mathbf{E}\big((x^{11})^*_T\big)^2 \leqslant C^2.$$

Note that x^{11}_t is a strong semimartingale.
Now consider the equation

$$x^{12}_t = x^{11}_t + \big((f \circ x^{12})I^{01} \cdot M^1\big)_t + \int\limits_{[0,t]} (f \circ x^{12})_s I^{01}_s \, dB_s. \tag{11}$$

Following the same line of argument, we conclude that (11) has a unique strong solution x^{12}_t and $\mathbf{E}\,((x^{12})^*_T)^2 < \infty$. But

$$x^{12}_t I_{\square_{00}}(t) = x^{11}_t I_{\square_{00}}(t) + \big((f \circ x^{12})I^{01} I_{\square_{00}} \cdot M^1\big)_t \cdot I_{\square_{00}}(t) +$$
$$+ \int\limits_{[0,t]} (f \circ x^{12})_s I^{01}_s I_{\square_{00}}(s) \, dB_s \cdot I_{\square_{00}}(t) = x^{00}_t I_{\square_{00}}(t),$$

because

$$I^{01}_s I_{\square_{00}}(s) = I\Big\{\widehat{\Pi} I_{\Delta_{01}}(s) > 1/2,\ s \in \square_{00} \cap \square_{11}\Big\} = 0.$$

So

$$x^{12}_t = \sum_{(j,k)=(0,0),(0,1)} \Big(\big((f \circ x^{12})I^{jk} \cdot M^1\big)_t +$$
$$+ \int\limits_{[0,t]} (f \circ x^{12})_s I^{jk}_s \, dB_s\Big), \quad t \in [0,T].$$

Proceeding in this way, we obtain a stochastic process x^{nn}_t which is the unique strong solution of the equation

$$x^{nn}_t = \sum_{j,k=0}^{n-1} \Big(\big((f \circ x^{nn})I^{jk} \cdot M^1\big)_t + \int\limits_{[0,t]} (f \circ x^{nn})_s I^{jk}_s \, dB_s\Big), \tag{12}$$

$$\mathbf{E}\big((x^{nn})^*_T\big)^2 < \infty. \tag{13}$$

Let

$$F^n = \Big\{(t,\omega)\colon \sum_{j,k=0}^{n-1} I^{jk}_s = 1 \quad \forall s \leqslant t\Big\}.$$

Then

$$\mathbf{P}\{T \in F^n\} = \mathbf{P}\Big\{\bigcap_{j,k=0}^{n-1} \Big\{\sum_{j,k=0}^{n-1} I^{jk}_s = 1 \quad \forall s \in \square_{jk}\Big\}\Big\} =$$

$$= \mathbf{P}\Big\{\bigcap_{j,k=0}^{n-1}\Big\{\widehat{\Pi}I_{\bigcup_{j,k=0}^{n-1}\Delta_{jk}}(s) > \frac{1}{2} \quad \forall s \in \square_{jk}\Big\}\Big\} =$$

$$= \mathbf{P}\{\widehat{\Pi}I_{\bigcup_{j,k=0}^{n-1}\Delta_{jk}}(s) > \frac{1}{2} \quad \forall s \in [0,T]\} =$$

$$= 1 - \mathbf{P}\Big\{\sup_{s \leqslant T} I_{[0,T]\setminus(\bigcup_{j,k=0}^{n-1}\Delta_{jk})}(s) \geqslant \frac{1}{2}\Big\} = 1 - p.$$

According to Lemma 9 in (Gushchin, 1982),

$$\widehat{\Pi}I_{[0,T]\setminus\bigcup_{j,k=0}^{n-1}\Delta_{jk}}(s) \leqslant \overline{M}_s + \widetilde{M}_s,$$

where $\overline{M} \in \mathfrak{M}_s^2$, $\widetilde{M} \in \mathfrak{M}^2$. Applying Doob's inequality for two-parameter martingales to $\overline{M}$ and $\widetilde{M}$, we obtain, after some calculation, that

$$p \leqslant 2^8\mathbf{P}\Big\{[0,T] \neq \bigcup_{j,k=0}^{n-1}\Delta_{jk}\Big\} =$$

$$= 2^8\mathbf{P}\Big\{\sup_{j,k}\big((\square_{jk}\langle M^1\rangle)^{1/2} \vee \square_{jk}(\mathrm{var}\, A)\big) > \alpha\Big\} \longrightarrow$$

$$\longrightarrow 2^8\mathbf{P}\Big\{\sup_{t \leqslant T}\big((\square\langle M^1\rangle_t)^{1/2} \vee |\square A_t|\big) > \alpha\Big\} = 0.$$

So $\mathbf{P}\{T \in F^n\} \to 1$ as $n \to \infty$. Furthermore,

$$\Big(\big((f \circ x^{nn}) \cdot M^1\big)_t + \int_{[0,t]} (f \circ x^{nn})_s \, dB_s\Big) I_{F^n}(t) =$$

$$= \sum_{j,k=0}^{n-1}\Big(\big((f \circ x^{nn})I^{jk} \cdot M^1\big)_t + \int_{[0,t]} (f \circ x^{nn})_s I_s^{jk} \, dB_s\Big) I_{F^n}(t) = x_t^{nn} I_{F^n}(t),$$

i.e., x_t^{nn} is a solution of equation (3) on F^n. Let also

$$\Big[y_t - \big((f \circ y) \cdot M^1\big)_t - \int_{[0,t]} (f \circ y)_s \, dB_s\Big] I_{F^n}(t) = 0.$$

Then y_t satisfies on F^n equation (12). Therefore, $y_t = x_t^{nn}$, $t \in F^n$. Let $r > n$. Then $F^r \supset F^n$ and

$$\Big(x_t^{rr} - \big((f \circ x^{rr}) \cdot M^1\big)_t - \int_{[0,t]} (f \circ x^{rr})_s \, dB_s\Big) I_{F^n}(t) =$$

$$= \Big(x_t^{rr} - \big((f \circ x^{rr}) \cdot M^1\big)_t - \int_{[0,t]} (f \circ x^{rr})_s \, dB_s\Big) I_{F^r}(t) I_{F^n}(t) = 0.$$

It follows from the uniqueness of the solution that

$$(x_t^{nn} - x_t^{rr}) I_{F^n}(t) = 0.$$

So there exists $x_t = \lim_{n\to\infty} x_t^{nn}$ and x_t is the unique solution of (3) and, consequently, of (2) on $[0,T]$. Since T is arbitrary, equation (2) has a unique solution on $\mathbf{R}_+^2$.

It follows from (13) that x and $(f \circ x)$ are locally square integrable, so $(f \circ x)$ is locally bounded and the right-hand side of (2) is a strong semimartingale.

The next theorem is proved in a similar way.

THEOREM 4. *Let* $(\bar{z}_t = (z_t^i)_{i=1}^k, \mathfrak{F}_t)$ *be a vector-valued strong semimartingale,*

$$(\bar{f} = (f^i)_{i=1}^k)$$

be a vector-valued function with components satisfying (**a**), (**b**) *and* (**c**), *and* H_t *be a strong semimartingale. Then the equation*

$$x_t = H_t + \big((\bar{f} \circ x), \bar{z}\big)_t = H_t + \sum_{i=1}^{k} \big((f^i \circ x) \cdot z^i\big)_t$$

has a unique strong solution.

COROLLARY. *Let* $M \in l\mathfrak{M}_s^2$, $A \in lB$, f_i, $i = 1,2$, *satisfy* (**a**), (**b**) *and* (**c**), *and* H_t *be a strong semimartingale. Then the equation*

$$x_t = H_t + \big((f_1 \circ x) \cdot M\big)_t + \int_{[0,t]} (f_2 \circ x)_s \, dA_s \tag{14}$$

has a unique strong solution.

Remark 4. The papers (Yeh, 1981; Tudor, 1981; Dozzi, 1986) provide conditions for existence and uniqueness of solutions of equations similar to (14) under rigid restrictions on f and $\langle M \rangle$ (f is assumed to be predictable and $\langle M \rangle$ is assumed either to be continuous or to satisfy $\langle M \rangle_t = \alpha_{t_1} \beta_{t_2}$ or some other equivalent assumptions). Our methods enable us to widen the class of functions f and to avoid any restrictions on $\langle M \rangle$. The results obtained correspond to the most general theorems for the one-parameter case (Elliott, 1982).

REFERENCES

Cairoli, R. and Walsh, J.B. (1975). Stochastic integrals in the plane. *Acta Math.* **134**, 111–183.

Dozzi, M. (1986). Stochastic processes with multidimensional parameter. *Preprint.* Bern Univ. Press, Bern.

Elliott, R.J. (1982). *Stochastic Calculus and Applications.* Springer, Berlin etc.

Fouque, J.-P. and Millet, A. (1980). Regularité à gauche des martingales fortes a plusieurs indices. *C. R. Acad. Sci.* **290**, 773–776.

Gushchin, A. (1982). On the general theory of random fields on the plane. *Russian Math. Surveys* **37**, 55–80.

Gushchin, A. (1988). Stochastic integration with respect to strong martingales on the plane. In: *Reports on Mathematics and its Applications.* 2. Moscow–Tula, pp. 242–256 (in Russian).

Millet, A. (1981). Convergence and regularity of strong martingales. *Lect. Notes Math.* **863**, 50–58.

Mishura, Ju. and Gushchin, A. (1990). Strong martingales inequalities for quadratic variation and some decompositions. In: *Probab. Theory and Math. Statist. Proc. Fifth Vilnius Conf.* **2**, Mokslas, Vilnius.

Tudor, C. (1981). Stochastic integral equations in the plane. *Rev. Roum. Math. Pures et Appl.* **26**, 507–538.

Yeh, J. (1981). Existence of strong solutions for stochastic integral equations in the plane. *Pacific Math. J.* **97**, 217–247.

New Trends in Probab. and Statist., pp. 503 – 527
V. Sazonov and T. Shervashidze (Eds.)
1991 VSP/Mokslas

ITÔ–VENTSEL' FORMULA FOR ANTICIPATIVE PROCESSES

O. PURTUKHIA
Razmadze Mathematical Institute of the Georgian Academy of Sciences,
Z. Rukhadze str. 1, 380093 Tbilisi, Georgia, USSR

ABSTRACT

Some properties of the Skorokhod integral and of the stochastic derivative are studied and the Itô–Ventsel' formula is generalized to the case of anticipative processes.

INTRODUCTION

The aim of this study is to derive a rule for the change of variables generalizing the Itô–Ventsel' formula to the anticipative case.
A formula of this type in the Wiener space was proved in (Ustunel, 1988) in the nonadapted case or, more precisely, an Itô–type formula was derived there for a random field $F(x,\omega)$ which is fast decreasing function of x when in F the argument x is replaced by the so–called anticipative Itô process. The formula obtained in (Ustunel, 1988) contains, in its turn, the relevant results from (Hitsuda, 1979).
Furthermore, in (Martias, 1988) was considered the case when for each x function $F_t(x)$ is an adapted process of diffusion type

$$F_t(x) = x + \int_0^t \sigma(F_s(x))\,dW_s + \int_0^t b(F_s(x))\,ds,$$

where the anticipative process

$$U_t = U_0 + \int_0^t h_s\,\delta W_s + \int_0^t K_s\,ds$$

is substituted; the integral $\delta_t\varphi = \int_0^t \dot{\varphi}_s\,\delta W_s$ is defined as a divergence operator conjugated with the Sobolev derivative in the direction of the Cameron–Martin space.
We consider the case when $F_t(x,\omega)$ (for each x) and U_t are both anticipative processes and admit a representation as a sum of an integral in the Skorokhod sense and of the Lebesgue integral.

As it is well known, the proof of the classical Itô formula is based on the following main fact: if u_t is a continuous adapted bounded process and v_t is a process adapted to $\mathbf{E}\int_0^1 v_s^2\,ds < \infty$ then, as $n \to \infty$,

$$\sum_{k=0}^{n-1} u_{t_{k,n}} \left(\int_{t_{k,n}}^{t_{k+1,n}} v_s dW_s \right)^2 \longrightarrow \int_0^1 u_s v_s^2 ds$$

in probability, where $\{0 = t_{0,n} < t_{1,n} < \dots < t_{n,n} = 1\}$ is a sequence of partitions of the interval [0,1] such that $\sup(t_{k+1,n} - t_{k,n}) \to 0$ as $n \to \infty$. For the case when u_t is not adapted, a statement of this kind is given in (Fölmer, 1981).

Later, in (Pardoux and Protter, 1987) and (Nualart and Pardoux, 1988), the above statement was proved for the case when u_t and v_t are both anticipative processes (see, e.g., Appendix C and Theorem 5.4 in (Nualart and Pardoux, 1988)).

To prove the Itô–Ventsel' formula we need additionally the following statement: if $h(s,x,\omega)$ is a continuous (in x) bounded function and

$$U_t = U_0 + \int_0^t u_s\, dW_s + \int_0^t u_s\, ds$$

then, as $n \to \infty$,

$$\sum_{k=0}^{n-1} \int_{t_{k,n}}^{t_{k+1,n}} h(s, U_{t_{k,n}})\, dW_s \int_{t_{k,n}}^{t_{k+1,n}} u_s dW_s \longrightarrow \int_0^1 h(s, U_s)\, u_s ds$$

in probability.

We shall prove this statement in the anticipative case.

On the other hand, deriving the Itô formula in the anticipative case, we use the rule of stochastic differentiation of a composite function (see (Nualart and Pardoux, 1988, Proposition 2.3)): if $\varphi \in C_b^1(\mathbf{R}^m, \mathbf{R}^1)$ and $\xi^i \in \mathbf{D}_{2,1}$, $i = 1, \dots, m$, $\xi = (\xi^1, \dots, \xi^m)$ then $\varphi(\xi) \in \mathbf{D}_{2,1}$ and

$$D\varphi(\xi) = \sum_{i=1}^{m} \varphi'_{x^i}(\xi) D\xi^i.$$

However, for proving the Itô–Ventsel' formula we must to differentiate stochastically mixed–type functions $\varphi(t,\xi,\omega)$: argument x in the function $\varphi(t,x,\omega)$: $[0,1] \times \mathbf{R}^m \times \Omega \to \mathbf{R}^1$, belonging to $\mathbf{D}_{2,1}$ for each (t,x), and to $C_b^1(\mathbf{R}^m, \mathbf{R}^1)$ for any (t,ω), is replaced by a random vector $\xi = (\xi^1, \dots, \xi^m)$ with the components from $\mathbf{D}_{2,1}$. We shall describe a class of functions $\varphi(t,x,\omega)$ for which

$$D\varphi(t,\xi,\omega) = \sum_{i=1}^{m} \varphi'_{x^i}(\xi) D\xi^i + D\varphi(t,x,\omega)\big|_{x=\xi}.$$

Furthermore, one of the stages in the proof of the Itô–Ventsel' formula is connected with substituting some random value in the parameter–dependent stochastic integral. As it is well known, in the adapted case the statement

$$\int_{t_{k,n}}^{t_{k+1,n}} h(s,x,\omega)\,dW_s\Big|_{x=U_{t_{k,n}}} = \int_{t_{k,n}}^{t_{k+1,n}} h(s,U_{t_{k,n}},\omega)\,dW_s$$

is valid for some h. We shall see below that in the anticipative case

$$\int_{t_{k,n}}^{t_{k+1,n}} h(s,x,\omega)\,dW_s\Big|_{x=U_{t_{k,n}}} = \int_{t_{k,n}}^{t_{k+1,n}} h(s,U_{t_{k,n}},\omega)\,dW_s +$$

$$+\int_{t_{k,n}}^{t_{k+1,n}} h'_x(s,U_{t_{k,n}},\omega)D_sU_{t_{k,n}}\,ds.$$

It should be said that in the case when for each x from the bounded domain B the function h: $[0,1]\times B\times\Omega\to\mathbf{R}^1$ is $\mathcal{F}_t^W$–progressively measurable, this relation is proved in (Nualart and Pardoux, 1988, Proposition 4.2).
Finally note that the Itô–Ventsel' formula is derived in two stages. First we prove the formula under the Ventsel' condition (Ventsel', 1965), i.e., we consider the case when the process $F'_x(t,x,\omega)$, as well as $F(t,x,\omega)$, for each x can be represented as a sum of Lebesgue and Skorokhod integrals and this representation is obtained from a similar representation for $F(t,x,\omega)$ by differentiating with respect to x under the integral sign. Next, using Sobolev's averaging operator T_ε, we proceed to the general case following the method proposed in (Rozovskii, 1973).

1. SOME PROPERTIES OF A STOCHASTIC DERIVATIVE AND OF THE SKOROKHOD INTEGRAL

Let $(\Omega,\mathcal{F},\mathbf{P})$ be a full probability space with the given standard Wiener process W_t, $t\in[0,1]$. It is assumed that $\mathcal{F}=\mathcal{F}_1=\sigma\{W_s,\ s\in[0,1]\}$. Let $\{\Pi^n,\ n\in\mathbf{N}\}$ be a sequence of partitions of the segment [0,1] of the form $\Pi^n=\{0=t_{0,n}<t_{1,n}<<\ldots<t_{n,n}=1\}$ such that $|\Pi^n|=\sup_k(t_{k+1,n}-t_{k,n})\to 0$ as $n\to\infty$. We shall write t_k instead of $t_{k,n}$; $\mathbf{P}$–lim will denote the limit in probability as $n\to\infty$. In what follows we shall preserve the notation from (Nualart and Pardoux, 1988; Nualart, 1989); in particular, the Skorokhod integral is denoted by the same symbol as the Itô integral.
For the sake of simplicity we shall consider the one–dimensional case.

Definition 1.1. We shall define the functional

$$\|\varphi\|_{\mathrm{L}^{2,1}\times C_b^1} = \left(\mathbf{E}\int_0^1 \sup_x|\varphi(t,x)|^2dt\right)^{1/2} + \left(\mathbf{E}\int_0^1 \sup_x|\varphi'_x(t,x)|^4dt\right)^{1/4} +$$

$$+\left(\mathbf{E}\int_0^1\int_0^1 \sup_x|D_s\varphi(t,x)|^2dsdt\right)^{1/2}, \tag{1.1}$$

for the $\mathcal{B}([0,1]\times\mathbf{R}^1)\otimes\mathcal{F}$-measurable function $\varphi(t,x,\omega)$, that maps $[0,1]\times\mathbf{R}^1\times\Omega$ to $\mathbf{R}^1$, belongs to $\mathbf{L}^{2,1}$ for each x and to $C_b^1(\mathbf{R}^1,\mathbf{R}^1)$ for almost all (t,ω).
It is obvious that this functional is the norm. Besides, since the ordinary differentiation operator and the stochastic differentiation operator are both closable, it can be easily verified that this norm is complete.

Definition 1.2. Denote by $\mathbf{L}^{2,1}\otimes C_b^1$ the Banach space obtained by completing linear combinations of the function $f(x)\eta_t(\omega)$, where $f\in C_b^1(\mathbf{R}^1,\mathbf{R}^1)$ and $\eta\in\mathbf{L}^{2,1}$, with respect to the norm (1.1).

THEOREM 1.1. *Let $F(t,x,\omega)\in\mathbf{L}^{2,1}\otimes C_b^1$ and $\xi\in\mathbf{D}_{4,1}$. Then $F(t,\xi,\omega)\in\mathbf{L}^{2,1}$ and*

$$D_sF(t,\xi,\omega)=F_x'(t,\xi,\omega)\mathbf{D}_s\xi+\mathbf{D}_sF(t,x,\omega)\big|_{x=\xi}. \tag{1.2}$$

Proof. By Definition 1.2, there exist sequences of numbers $\{a^k\}_{k\geqslant 1}$, functions $\{f^k\}_{k\geqslant 1}$, $f^k\in C_b^1(\mathbf{R}^1,\mathbf{R}^1)$ and processes $\{\eta^k\}_{k\geqslant 1}$, $\eta^k\in\mathbf{L}^{2,1}$ such that

$$\lim_{n\to\infty}\left\|\sum_{k=1}^n a^kf^k\eta^k-F\right\|_{\mathbf{L}^{2,1}\times C_b^1}=0.$$

Assume $F^n(t,x,\omega)=\sum_{k=1}^n a^kf^k(x)\eta_t^k(\omega)$.
Now it is easy to observe that $F^n(t,\xi,\omega)\in\mathbf{L}^{2,1}$ and the relation

$$D_sF^n(t,\xi,\omega)=F_x^{n'}(t,\xi,\omega)D_s\xi+D_sF^n(t,x,\omega)\big|_{x=\xi} \tag{1.3}$$

is valid.
It remains to justify the passage to the limit in (1.3) as $n\to\infty$. First we have to verify that the sequence $\{F^n(t,\xi,\omega)\}_{n\geqslant 1}$ is fundamental in $\mathbf{L}^{2,1}$. By the definition of the norm in $\mathbf{L}^{2,1}$, using the Cauchy–Bunyakovski inequality and the elementary inequalities $(a+b)^2\leqslant 2a^2+2b^2$, $\sqrt{|a+b|}\leqslant\sqrt{|a|}+\sqrt{|b|}$, we have, for $m>k$,

$$\|F^m(t,\xi,\omega)-F^k(t,\xi,\omega)\|_{\mathbf{L}^{2,1}}=\left(\mathbf{E}\int_0^1\left|\sum_{i=k+1}^m a^if^i(\xi)\eta_t^i\right|^2dt\right)^{1/2}+$$

$$+\left(\mathbf{E}\int_0^1\int_0^1\left|D_s\left(\sum_{i=k+1}^m a^if^i(\xi)\eta_t^i\right)\right|^2dsdt\right)^{1/2}=\left(\mathbf{E}\int_0^1\left|\sum_{i=k+1}^m a^if^i(\xi)\eta_t^i\right|^2dt\right)^{1/2}+$$

$$+\left(\mathbf{E}\int_0^1\int_0^1\left|\left(\sum_{i=k+1}^m a^if_x^{i'}(\xi)\eta_t^i\right)D_s\xi+\sum_{i=k+1}^m a^if^i(\xi)D_s\eta_t^i\right|^2dsdt\right)^{1/2}\leqslant$$

$$\leqslant\left(\mathbf{E}\int_0^1\sup_x\left|\sum_{i=k+1}^m a^if^i(x)\eta_t^i\right|^2dt\right)^{1/2}+\sqrt{2}\left(\mathbf{E}\int_0^1\sup_x\left|\sum_{i=k+1}^m a^if_x^{i'}(x)\eta_t^i\right|^4dt\right)^{1/4}\times$$

$$\times\left(\mathbf{E}\int_0^1 |D_s\xi|^4 ds\right)^{1/4} + \sqrt{2}\left(\mathbf{E}\int_0^1\int_0^1 \sup_x \left|\sum_{i=k+1}^m a^i f^i(x) D_s\eta_t^i\right|^2 ds dt\right)^{1/2}. \quad (1.4)$$

Since $F^n(t,x,\omega)$ convergence to $F(t,x,\omega)$ in $\mathbf{L}^{2,1}\otimes C_b^1$ as $n\to\infty$, the sequence $\{F^n(t,x,\omega)\}_{n\geqslant 1}$ is fundamental in $\mathbf{L}^{2,1}\otimes C_b^1$. Therefore, it follows from (1.4) that $\{F^n(t,\xi,\omega)\}_{n\geqslant 1}$ is fundamental in $\mathbf{L}^{2,1}$.

But $\mathbf{L}^{2,1}$ is a complete Hilbert space. Thus, there exists $\widetilde{F}(t,\omega)\in\mathbf{L}^{2,1}$ such that

$$\lim_{n\to\infty}\|F^n(t,\xi,\omega)-\tilde{F}(t,\omega)\|_{\mathbf{L}^{2,1}}=0;$$

on the other hand,

$$\lim_{n\to\infty}\|F^n(t,\xi,\omega)-F(t,\xi,\omega)\|_{\mathbf{L}^2([0,1]\times\Omega)} =$$

$$= \lim_{n\to\infty}\left(\mathbf{E}\int_0^1 \left|F^n(t,\xi,\omega)-F(t,\xi,\omega)\right|^2 dt\right)^{1/2} \leqslant$$

$$\leqslant \lim_{n\to\infty}\left(\mathbf{E}\int_0^1 \sup_x\left|F^n(t,\xi,\omega)-F(t,\xi,\omega)\right|^2 dt\right)^{1/2} = 0.$$

Therefore, by definition of the norm in $\mathbf{L}^{2,1}$, it is obvious that $F(t,\xi,\omega) = \widetilde{F}(t,\omega)$ for almost all (t,ω), and

$$\lim_{n\to\infty}\|F^n(t,\xi,\omega)-F(t,\xi,\omega)\|_{\mathbf{L}^{2,1}}=0.$$

In particular, we obtain

$$D_sF^n(t,\xi,\omega)\longrightarrow D_sF(t,\xi,\omega) \quad (1.5)$$

in $\mathbf{L}^2([0,1]^2\times\Omega)$ as $n\to\infty$.

Next we have

$$\left(\mathbf{E}\int_0^1\int_0^1\left[F_x^{n'}(t,\xi,\omega)D_s\xi - F_x'(t,\xi,\omega)D_s\xi\right]^2 ds\,dt\right)^{1/2} \leqslant$$

$$\leqslant \left(\mathbf{E}\int_0^1 |D_s\xi|^4 ds\right)^{1/4}\left(\mathbf{E}\int_0^1 \sup_x |F_x^{n'}(t,x,\omega)-F_x'(t,x,\omega)|^4 dt\right)^{1/4} \longrightarrow 0 \quad (1.6)$$

and

$$\mathbf{E}\int_0^1\int_0^1\left[D_sF^n(t,x,\omega)\big|_{x=\xi} - D_sF(t,x,\omega)\big|_{x=\xi}\right]^2 ds\,dt \leqslant$$

$$\leqslant \mathbf{E}\int_0^1\int_0^1 \sup_x |D_sF^n(t,x,\omega)-D_sF(t,x,\omega)|^2 ds\,dt \longrightarrow 0 \quad (1.7)$$

as $n\to\infty$. Taking now into account (1.5)–(1.7), we obtain (1.2) from (1.3).

Remark 1.1. As follows from the proof of Theorem 1.1, for this theorem to be valid it is sufficient to require that a random variable ξ be such that $D_s\xi \in$ $\in \mathbf{L}^4([0,1]) \times \Omega)$.
Using the arguments of the proof of Theorem 1.1 and relation (1.2), we ascertain that the proposition below is valid.

PROPOSITION 1.1. *If the sequence of functions* $\{F^n(t,x,\omega)\}_{n\geqslant 1}$ *from* $\mathbf{L}^{2,1}\otimes C_b^1$ *converges to* $F(t,x,\omega) \in \mathbf{L}^{2,1}\otimes C_b^1$ *in* $\mathbf{L}^{2,1}\otimes C_b^1$, *then, for any random variable* ξ *with* $D_s\xi \in L^4([0,1]\times\Omega)$, *we have*

$$F^n(t,\xi,\omega) \longrightarrow F(t,\xi,\omega)$$

in $\mathbf{L}^{2,1}$ *as* $n\to\infty$. *In particular,*

$$\int_0^1 F^n(s,\xi,\omega)\,dW_s \longrightarrow \int_0^1 F(s,\xi,\omega)\,dW_s$$

in $\mathbf{L}^2(\Omega)$ *as* $n\to\infty$.

PROPOSITION 1.2. *If the function* $F(t,x,\omega)\in\mathbf{L}^{2,1}\otimes C_b^1$, *then the mapping* $\theta\mapsto D_sF(t,x,\omega)|_{x=\xi_\theta}$ *is continuous in* $\mathbf{L}^2([0,1]^2\times\Omega)$ *for any continuous process* ξ_θ.

Proof. Let $F^n(t,x,\omega) = \sum_{k=1}^n a^k f^k(x)\eta_t^k(\omega)$ be a sequence of elementary functions from $\mathbf{L}^{2,1}\otimes C_b^1$ which approximates $F(t,x,\omega)$. Then, on account of the continuity of $D_sF^n(t,x,\omega) = \sum_{k=1}^n a^k f^k(x)D_s\eta_t^k(\omega)$ with respect to x, and, by virtue of the Lebesgue dominated convergence theorem (LDCT), we ascertain that the mapping $\theta\mapsto D_sF^n(t,x,\omega)|_{x=\xi_\theta}$ is continuous in $\mathbf{L}^2([0,1]^2\times\Omega)$ for any fixed $n\geqslant 1$.
Furthermore, since the convergence in the metric $\mathbf{L}^{2,1}\times C_b^1$ is uniform with respect to x, it is not difficult to verify that the proposition is also valid for $F(t,x,\omega)$.
By virtue of Theorem 5.4 from (Nualart and Pardoux, 1988), we have

PROPOSITION 1.3. *If* $u,v\in\mathbf{L}^{2,1}_{\mathrm{loc}}$, *then*

$$\mathbf{P}\text{-}\lim\sum_{k=0}^{n-1}\int_{t_k}^{t_{k+1}} u_s\,dW_s\int_{t_k}^{t_{k+1}} v_s\,dW_s = \int_0^1 u_sv_s\,ds.$$

If, however, $u,v\in\mathbf{L}^{2,1}$, *the convergence takes place in* $\mathbf{L}^1(\Omega)$.

THEOREM 1.2. *Let* $u\in\mathbf{L}^{2,2}$; $v,g\in\mathbf{L}^{2,1}$; $h\in\mathbf{L}^{2,1}\otimes C_b^1$. *Let also*

$$\mathbf{E}\left(\int_0^1 |u_s|^4ds + \int_0^1\int_0^1 |D_ru_s|^4dr\,ds + \int_0^1\int_0^1\int_0^1 |D_tD_ru_s|^4dt\,dr\,ds\right) < \infty;$$

$$\mathbf{E}\int_0^1\int_0^1 |D_rv_s|^4dr\,ds<\infty;\quad \mathbf{E}\int_0^1 |D_rU_0|^4dr<\infty,$$

and

$$U_t = U_0 + \int_0^1 u_s dW_s + \int_0^1 v_s ds.$$

Then

$$\sum_{k=0}^{n-1} \int_{t_k}^{t_{k+1}} h(s, U_{t_k})\, dW_s \int_{t_k}^{t_{k+1}} g_s dW_s \longrightarrow \int_0^1 h(s, U_s) g_s ds$$

in $\mathbf{L}^1(\Omega)$ as $n \to \infty$.

Proof. We have

$$\sum_{k=0}^{n-1} \int_{t_k}^{t_{k+1}} h(s, U_{t_k})\, dW_s \int_{t_k}^{t_{k+1}} g_s dW_s = I_1^n + I_2^n, \tag{1.8}$$

where

$$I_1^n = \sum_{k=0}^{n-1} \int_{t_k}^{t_{k+1}} \left[h(s, U_{t_k}) - h(s, U_s)\right] dW_s \int_{t_k}^{t_{k+1}} g_s dW_s,$$

$$I_2^n = \sum_{k=0}^{n-1} \int_{t_k}^{t_{k+1}} h(s, U_s)\, dW_s \int_{t_k}^{t_{k+1}} g_s dW_s.$$

By virtue of Proposition 1.3 and Theorem 1.1, we have

$$I_2^n = \sum_{k=0}^{n-1} \int_{t_k}^{t_{k+1}} h(s, U_s)\, dW_s \int_{t_k}^{t_{k+1}} g_s dW_s \longrightarrow \int_0^1 h(s, U_s) g_s ds \tag{1.9}$$

in $\mathbf{L}^1(\Omega)$ as $n \to \infty$.
Next, by virtue of the Cauchy–Bunyakovski inequality, we can write

$$\mathbf{E} I_1^n \leqslant \left\{ \mathbf{E} \sum_{k=0}^{n-1} \Big(\int_{t_k}^{t_{k+1}} [h(s, U_{t_k}) - h(s, U_s)] dW_s \Big)^2 \right\}^{1/2} \left[\mathbf{E} \sum_{k=0}^{n-1} \Big(\int_{t_k}^{t_{k+1}} g_s dW_s \Big)^2 \right]^{1/2}.$$

On the other hand, by virtue of the $\mathbf{L}^2$–estimate of the Skorokhod integral (see (Nualart and Pardoux, 1988, Proposition 3.5)), we obtain

$$\mathbf{E} \sum_{k=0}^{n-1} \Big(\int_{t_k}^{t_{k+1}} \left[h(s, U_{t_k}) - h(s, U_s)\right] dW_s \Big)^2 \leqslant c_1 \mathbf{E} \sum_{k=0}^{n-1} \int_{t_k}^{t_{k+1}} \left[h(s, U_{t_k}) - h(s, U_s)\right]^2 ds +$$

$$+c_2\mathbf{E}\sum_{k=0}^{n-1}\int_{t_k}^{t_{k+1}}\int_{t_k}^{t_{k+1}}\left|D_r\left[h(s,U_{t_k})-h(s,U_s)\right]\right|^2 dr ds.$$

It is obvious, that the first term in the right–hand side vanishes. The second term is transformed, by means of Theorem 1.1, to

$$\mathbf{E}\sum_{k=0}^{n-1}\int_{t_k}^{t_{k+1}}\int_{t_k}^{t_{k+1}}\left|D_r\left[h(s,U_{t_k})-h(s,U_s)\right]\right|^2 dr ds =$$

$$=\mathbf{E}\sum_{k=0}^{n-1}\int_{t_k}^{t_{k+1}}\int_{t_k}^{t_{k+1}}\left[h'_x(s,U_{t_k})D_rU_{t_k}+D_rh(s,x)\Big|_{x=U_{t_k}}-\right.$$

$$\left.-h'_x(s,U_s)D_rU_s-D_rh(s,x)\Big|_{x=U_s}\right]^2 dr\, ds =$$

$$=\mathbf{E}\sum_{k=0}^{n-1}\int_{t_k}^{t_{k+1}}\int_{t_k}^{t_{k+1}}\left\{\left[h'_x(s,U_{t_k})-h'_x(s,U_s)\right]D_rU_0+\right.$$

$$+\left[h'_x(s,U_{t_k})-h'_x(s,U_s)\right]\int_0^{t_k}D_ru_\theta\, dW_\theta-h'_x(s,U_s)\int_{t_k}^{s}D_ru_\theta\, dW_\theta-$$

$$-h'_x(s,U_s)u_r\mathbf{1}\{[0,s]\}(r)+\left[h'_x(s,U_{t_k})-h'_x(s,U_s)\right]\int_0^{t_k}D_rv_\theta\, d\theta-$$

$$\left.-h'_x(s,U_s)\int_{t_k}^{s}D_rv_\theta\, d\theta+\left[D_rh(s,x)\Big|_{x=U_{t_k}}-D_rh(s,x)\Big|_{x=U_s}\right]\right\}^2 dr\, ds. \tag{1.10}$$

It should be noted that the second equality in (1.10) is based on Proposition 5.5 from (Nualart and Pardoux, 1988) and on the next statement

$$D_r\left(\int_0^s v_\theta\, d\theta\right)=\int_0^s D_rv_\theta\, d\theta.$$

Next, using the elementary inequality

$$\left(\sum_{i=1}^m a_i\right)^2\leqslant m\sum_{i=1}^m a_i^2,$$

the right–hand side of (1.10) is dominated by the expression $I_{11}^n+I_{12}^n+I_{13}^n$, where

$$I_{11}^n=7\mathbf{E}\sum_{k=0}^{n-1}\int_{t_k}^{t_{k+1}}\int_{t_k}^{t_{k+1}}\left[h'_x(s,U_{t_k})-h'_x(s,U_s)\right]^2\times$$

$$\times\left\{(D_rU_0)^2 + \left(\int_0^{t_k} D_r u_\theta\, dW_\theta\right)^2 + \left(\int_0^{t_k} D_r v_\theta\, d\theta\right)^2\right\} dr\, ds,$$

$$I_{12}^n = 7\mathbf{E}\sum_{k=0}^{n-1}\int_{t_k}^{t_{k+1}}\int_{t_k}^{t_{k+1}}\left[h_x'(s,U_s)\right]^2\left\{\left(\int_{t_k}^{s} D_r u_\theta\, dW_\theta\right)^2 + \right.$$

$$\left. + \left(\int_{t_k}^{s} D_r v_\theta\, d\theta\right)^2 + u_r^2 \mathbf{1}\{[0,s]\}(r)\right\} dr\, ds,$$

$$I_{13}^n = 7\mathbf{E}\sum_{k=0}^{n-1}\int_{t_k}^{t_{k+1}}\int_{t_k}^{t_{k+1}}\left\{D_r h(s,x)\Big|_{x=U_{t_k}} - D_r h(s,x)\Big|_{x=U_s}\right\}^2 dr\, ds.$$

Applying Proposition 1.2 to I_{13}^n, and the Cauchy–Bunyakovski inequality to I_{11}^n, by virtue of the continuity of $h_x'(s,x,\omega)$ (with respect to x), we ascertain that

$$\lim_{n\to\infty} I_{13}^n = \lim_{n\to\infty} I_{11}^n = 0.$$

After that, by virtue of the Cauchy–Bunyakovski inequality and the above-given elementary inequality, we obtain

$$I_{12}^n \leqslant 21\left(\mathbf{E}\sum_{k=0}^{n-1}\int_{t_k}^{t_{k+1}}\int_{t_k}^{t_{k+1}}\left[h_x'(s,U_s)\right]^4 dr\, ds\right)^{1/2}\left\{\mathbf{E}\sum_{k=0}^{n-1}\int_{t_k}^{t_{k+1}}\int_{t_k}^{t_{k+1}}\left[\left(\int_{t_k}^{s} D_r u_\theta\, dW_\theta\right)^4 + \right.\right.$$

$$\left.\left. + \left(\int_{t_k}^{s} D_r v_\theta\, d\theta\right)^4 + u_r^4\mathbf{1}\{[0,s]\}(r)\right] dr\, ds\right\}^{1/2}. \tag{1.11}$$

On the other hand,

$$\mathbf{E}\int_0^1 \left[h_x'(s,U_s)\right]^4 ds \leqslant \mathbf{E}\int_0^1 \sup_x \left|h_x'(s,x)\right|^4 ds < \infty.$$

Thus, the first multiplier in the right-hand side of (1.11) vanishes and the second one is finite (really, under assumptions the second multiplier also tends to zero). Therefore, $\lim_{n\to\infty} I_{12}^n = 0$. Summing up the above-said, we obtain

$$\lim_{n\to\infty} \mathbf{E} I_1^n = 0.$$

This together with (1.8) and (1.9) completes the proof of the theorem.
The following result will also be given in the form to be used in the discussion below.

Theorem 1.3. *Let $H \in \mathbf{L}^{2,1} \otimes C_b^1$, $D_s U_{t_k} \in \mathbf{L}^4([t_k, t_{k+1}] \times \Omega)$ and*

$$\int_{t_k}^{t_{k+1}} H(s,x,\omega)\,dW_s$$

have a modification continuous with respect to x. Then $\mathbf{P}$–a.s.

$$\int_{t_k}^{t_{k+1}} H(s,x,\omega)\,dW_s\Big|_{x=U_{t_k}} = \int_{t_k}^{t_{k+1}} H(s,U_{t_k},\omega)\,dW_s + \int_{t_k}^{t_{k+1}} H'_x(s,U_{t_k},\omega)D_s U_{t_k}\,ds. \qquad (1.12)$$

Proof. Let

$$H^n(s,x,\omega) = \sum_{i=1}^{n} a^i f^i(x)\eta_s^i(\omega), \quad n \geqslant 1,$$

be a sequence of elementary functions from $\mathbf{L}^{2,1} \otimes C_b^1$ converging to $H(s,x,\omega)$ in $\mathbf{L}^{2,1} \otimes C_b^1$. Then, by virtue of Theorem 3.2 and Proposition 2.3 from (Nualart and Pardoux, 1988), we have

$$\int_{t_k}^{t_{k+1}} H^n(s,x,\omega)\,dW_s\Big|_{x=U_{t_k}} =$$

$$= \sum_{i=1}^{n} a^i f^i(x) \int_{t_k}^{t_{k+1}} \eta_s^i(\omega)\,dW_s\Big|_{x=U_{t_k}} = \sum_{i=1}^{n} a^i f^i(U_{t_k}) \int_{t_k}^{t_{k+1}} \eta_s^i(\omega)\,dW_s =$$

$$= \sum_{i=1}^{n} \int_{t_k}^{t_{k+1}} a^i f^i(U_{t_k})\eta_s^i(\omega)\,dW_s + \sum_{i=1}^{n} \int_{t_k}^{t_{k+1}} a^i \Big[D_s f^i(U_{t_k})\Big]\eta_s^i(\omega)\,ds =$$

$$= \int_{t_k}^{t_{k+1}} H^n(s,U_{t_k},\omega)\,dW_s + \int_{t_k}^{t_{k+1}} H_x^{n\prime}(s,U_{t_k},\omega)D_s U_{t_k}\,ds. \qquad (1.13)$$

By virtue of Proposition 1.1 (also by the definition of the norm $\mathbf{L}^{2,1} \otimes C_b^1$), we have that

$$\int_{t_k}^{t_{k+1}} H^n(s,U_{t_k},\omega)\,dW_s \longrightarrow \int_{t_k}^{t_{k+1}} H(s,U_{t_k},\omega)\,dW_s \qquad (1.14)$$

and

$$\int_{t_k}^{t_{k+1}} H_x^{n\prime}(s,U_{t_k},\omega)D_s U_{t_k}\,ds \longrightarrow \int_{t_k}^{t_{k+1}} H'_x(s,U_{t_k},\omega)D_s U_{t_k}\,ds \qquad (1.15)$$

in $\mathbf{L}^2(\Omega)$ as $n \to \infty$.

Applying Proposition 1.1 for any x again, we have

$$\int_{t_k}^{t_{k+1}} H^n(s,x,\omega)\,dW_s \longrightarrow \int_{t_k}^{t_{k+1}} H(s,x,\omega)\,dW_s$$

in $\mathbf{L}^2(\Omega)$ as $n \to \infty$, and therefore in probability too. Hence, since the integrals

$$\int_{t_k}^{t_{k+1}} H^n(s,x,\omega)\,dW_s \quad \text{and} \quad \int_{t_k}^{t_{k+1}} H(s,x,\omega)\,dW_s$$

are continuous with respect to x, it is not difficult to verify that for some subsequence $\{n_m\}_{m\geqslant 1}$ of the sequence $\{n\}_{n\geqslant 1}$ ($n_m \to \infty$ as $m \to \infty$), for any x,

$$\int_{t_k}^{t_{k+1}} H^{n_m}(s,x,\omega)\,dW_s \longrightarrow \int_{t_k}^{t_{k+1}} H(s,x,\omega)\,dW_s$$

with probability 1 as $m \to \infty$. Therefore,

$$\int_{t_k}^{t_{k+1}} H^{n_m}(s,x,\omega)\,dW_s\Big|_{x=U_{t_k}} \longrightarrow \int_{t_k}^{t_{k+1}} H(s,x,\omega)\,dW_s\Big|_{x=U_{t_k}} \tag{1.16}$$

with probability 1 as $m \to \infty$.

Passing to the limit in (1.13) along the subsequence $\{n_m\}$ as $m \to \infty$ and taking into account (1.14)–(1.16), we complete the proof of the Theorem.

Before formulating our next result, we shall recall some properties of Sobolev's averaging operator T_ε.

Let

$$\zeta(x) = \begin{cases} \exp\left\{|x|^2\left(|x|^2-1\right)^{-1}\right\} & \text{for } |x| \leqslant 1, \\ 0 & \text{for } |x| > 1. \end{cases}$$

Then, by definition, for any locally integrable function $f\colon \mathbf{R}^1 \to \mathbf{R}^1$, we have

$$T_\varepsilon f(x) = \varkappa \int_{-1}^{1} f(x+\varepsilon y)\zeta(y)\,dy = \varkappa\varepsilon^{-1}\int_{\mathbf{R}^1} \zeta\Big(\frac{y-x}{\varepsilon}\Big) f(y)\,dy,$$

where

$$\varkappa = \Big(\int_{-1}^{1} \zeta(x)dx\Big)^{-1}.$$

PROPOSITION 1.4 (Rozovskii, 1983). *Let $f \in C^{0,0}(\mathbf{R}_+ \times \mathbf{R}^1)$. Then $T_\varepsilon f(t,x) \in C^{0,\infty}(\mathbf{R}_+ \times \mathbf{R}^1)$, and for any $N \in \mathbf{R}_+$, we have*

$$\lim_{\varepsilon\to 0} \sup_{|x|<N} \left|f(t,x) - T_\varepsilon f(t,x)\right| = 0.$$

If $f \in C^{0,1}(\mathbf{R}_+ \times \mathbf{R}^1)$, then $\left[T_\varepsilon f(t,x)\right]'_x = T_\varepsilon f'_x(t,x)$.

PROPOSITION 1.5. **a)** *We have*

$$\sup_x |T_\varepsilon f(x)|^n \leqslant K(n) \sup_x |f(x)|^n, \tag{1.17}$$

for any $n \geqslant 1$ and some constant $K(n)$;

b) *If $H \in \mathbf{L}^{2,1} \otimes C_b^1$, then $D_s[T_\varepsilon H(t,x,\omega)] = T_\varepsilon[D_s H(t,x,\omega)]$ ($\mathbf{P}$–a.s.) and $T_\varepsilon H \in \mathbf{L}^{2,1} \otimes C_b^1$.*

Proof. a) Using Holder's inequality, for $n > 1$, we obtain

$$\sup_x |T_\varepsilon f(x)|^n = \varkappa^n \sup_x \left| \int_{-1}^{1} f(x+\varepsilon y)\zeta(y)\,dy \right|^n \leqslant$$

$$\leqslant \varkappa^n \sup_x \left(\int_{-1}^{1} \left|f(x+\varepsilon y)\right|^n |\zeta(y)|^n dy \right) \left(\int_{-1}^{1} 1^{n/(n-1)}\,dy \right)^{n-1} \leqslant$$

$$\leqslant 2^{n-1}\varkappa^n \int_{-1}^{1} \sup_x \left|f(x+\varepsilon y)\right|^n |\zeta(y)|^n dy \leqslant$$

$$\leqslant 2^{n-1}\varkappa^n \sup_x |f(x)|^n \int_{-1}^{1} |\zeta(y)|^n dy = K(n) \sup_x |f(x)|^n.$$

Estimation (1.17) is also obvious for $n = 1$.
b) From Theorem 1.1, we have

$$D_s\left[T_\varepsilon H(t,x,\omega)\right] = D_s\left[\varkappa \int_{-1}^{1} H(t,x+\varepsilon y,\omega)\zeta(y)\,dy\right] =$$

$$= \varkappa \int_{-1}^{1} D_s\left[H(t,x+\varepsilon y,\omega)\right]\zeta(y)\,dy = \varkappa \int_{-1}^{1} D_s H(t,z,\omega)\Big|_{z=x+\varepsilon y} \zeta(y)\,dy =$$

$$= T_\varepsilon[D_s H(t,x,\omega)] \quad (\mathbf{P}\text{–a.s.}).$$

Hence, by virtue of (1.17) and Proposition 1.4, it is not difficult to see that

$$\|T_\varepsilon H\|_{\mathbf{L}^{2,1}\times C_b^1} = \left(\mathbf{E}\int_0^1 \sup_x |T_\varepsilon H(t,x)|^2 dt\right)^{1/2} +$$

$$+\left(\mathbf{E}\int_0^1 \sup_x |(T_\varepsilon H)'_x(t,x)|^4 dt\right)^{1/4} + \left(\mathbf{E}\int_0^1\int_0^1 \sup_x \Big|D_s\big[T_\varepsilon H(t,x)\big]\Big|^2 ds\,dt\right)^{1/2} \leqslant$$

$$\leqslant \left(K(2)\mathbf{E}\int_0^1 \sup_x |H(t,x)|^2 dt\right)^{1/2} + \left(K(4)\mathbf{E}\int_0^1 \sup_x |H'_x(t,x)|^4 dt\right)^{1/4} +$$

$$+\left(K(2)\mathbf{E}\int_0^1\int_0^1 \sup_x |D_s H(t,x)|^2 ds\,dt\right)^{1/2} < \infty.$$

On the other hand, if $H^n(t,x,\omega) = \sum_{k=1}^n a^k f^k(x)\eta_t^k(\omega)$, $n \geqslant 1$, is a sequence approximating $H(t,x,\omega)$, then it is not difficult to verify that the sequence

$$T_\varepsilon H^n(t,x,\omega) = \sum_{k=1}^n a^k T_\varepsilon f^k(x)\eta_t^k(\omega)$$

will approximate $T_\varepsilon H(t,x,\omega)$. Thus part **b)** and hence the whole of the proposition are proved.

COROLLARY 1.1. *Let $H^n(t,x,\omega) \to H(t,x,\omega)$ in $\mathbf{L}^{2,1} \otimes C_b^1$ as $n \to \infty$. Then $T_\varepsilon H^n(t,x,\omega) \to T_\varepsilon H(t,x,\omega)$ in $\mathbf{L}^{2,1} \otimes C_b^1$ as $n \to \infty$.*

PROPOSITION 1.6. *Let $H(s,x,\omega) \in \mathbf{L}^{2,1} \otimes C_b^1$ and $\int_0^1 H(s,x,\omega)\,dW_s$ have a modification continuous with respect to x. Then*

$$T_\varepsilon\left[\int_0^1 H(s,x,\omega)\,dW_s\right] = \int_0^1 T_\varepsilon H(s,x,\omega)\,dW_s \quad (\mathbf{P}\text{–a.s.}) \tag{1.18}$$

for any x.

Proof. Let $H^n(s,x,\omega) = \sum_{k=1}^n a^k f^k(x)\eta_s^k(\omega)$, $n \geqslant 1$, be a sequence of elementary functions from $\mathbf{L}^{2,1} \otimes C_b^1$ approximating $H(s,x,\omega)$. We have

$$T_\varepsilon\left[\int_0^1 H^n(s,x,\omega)\,dW_s\right] = T_\varepsilon\left[\sum_{k=1}^n a^k f^k(x)\int_0^1 \eta_s^k(\omega)\,dW_s\right] =$$

$$= \sum_{k=1}^n a^k T_\varepsilon f^k(x)\int_0^1 \eta_s^k(\omega)\,dW_s = \int_0^1 T_\varepsilon H^n(s,x,\omega)\,dW_s \tag{1.19}$$

for each $n \geqslant 1$.

By virtue of Corollary 1.1, $T_\varepsilon H^n$ converges to $T_\varepsilon H$ in $\mathbf{L}^{2,1}\otimes C_b^1$ as $n\to\infty$. Therefore, by Proposition 1.1, we have

$$\int_0^1 T_\varepsilon H^n(s,x,\omega)\,dW_s \longrightarrow \int_0^1 T_\varepsilon H(s,x,\omega)\,dW_s \tag{1.20}$$

for each x in $\mathbf{L}^2(\Omega)$ as $n\to\infty$.
On the other hand, by virtue of Proposition 3.5 from (Nulart and Pardoux, 1988), it is not difficult to see that

$$\sup_x \mathbf{E}\Big(\int_0^1 \big[H^n(s,x)-H(s,x)\big]\,dW_s\Big)^2 \leqslant c(2)\Big[\mathbf{E}\int_0^1 \sup_x \big|H^n(s,x)-H(s,x)\big|^2 ds +$$

$$+\mathbf{E}\int_0^1\int_0^1 \sup_x \big|D_tH^n(s,x)-D_tH(s,x)\big|^2 dt\,ds\Big] \longrightarrow 0$$

as $n\to\infty$. Hence, by virtue of Fubini's theorem and estimate (1.17), we obtain

$$\sup_n \mathbf{E}T_\varepsilon\Big|\int_0^1 [H^n(s,x)-H(s,x)]\,dW_s\Big|^2 = \sup_n T_\varepsilon\mathbf{E}\Big|\int_0^1 [H^n(s,x)-H(s,x)]\,dW_s\Big|^2 \leqslant$$

$$\leqslant c\sup_n\sup_x \mathbf{E}\Big|\int_0^1 [H^n(s,x)-H(s,x)]\,dW_s\Big|^2 < \infty$$

for any fixed $\varepsilon>0$ and x.
Therefore, the sequence $\{|\int_0^1 [H^n(s,x,\omega)-H(s,x,\omega)]\,dW_s|\}_{n\geqslant 1}$ is uniformly integrable with respect to the measure $d\mathbf{P}\times\zeta(y)\,dy$. So we have

$$\mathbf{E}T_\varepsilon\Big|\int_0^1 [H^n(s,x,\omega)-H(s,x,\omega)]\,dW_s\Big| \longrightarrow 0$$

as $n\to\infty$. Hence, we can conclude that, for some subsequence $\{n_i\}$ of the sequence $\{n\}$ ($n_i\to\infty$ as $i\to\infty$) and for any x, we have

$$T_\varepsilon\Big|\int_0^1 [H^{n_i}(s,x,\omega)-H(s,x,\omega)]\,dW_s\Big| \longrightarrow 0 \tag{1.21}$$

with probability 1 as $i\to\infty$. Passing now to the limit in (1.19) along the subsequence $\{n_i\}$ as $i\to\infty$ and taking into account (1.20) and (1.21), we obtain (1.18).

2. ITÔ–VENTSEL' FORMULA

THEOREM 2.1. *Let the assumptions below be fulfilled:*

a) $U_t = U_0 + \int_0^t u_s dW_s + \int_0^t v_s ds$,
where:

(i) $U_0 \in \mathbf{D}_{4,1}$;

(ii) u *is an element of* $\mathbf{L}^{2,2}$ *such that*

$$\mathbf{E}\Big(\int_0^1 |u_s|^4\, ds + \int_0^1\int_0^1 |D_t u_s|^4\, dt\, ds + \int_0^1\int_0^1\int_0^1 |D_r D_t u_s|^4\, dr\, dt\, ds\Big) < \infty;$$

(iii) v *is an element of* $\mathbf{L}^{2,1}$ *such that*

$$\mathbf{E}\int_0^1\int_0^1 |D_t v_s|^4\, dt\, ds < \infty;$$

b) $F\colon [0,1]\times\mathbf{R}^1\times\Omega \to \mathbf{R}^1$ *is* $\mathcal{B}([0,1]\times\mathbf{R}^1)\otimes\mathcal{F}$*-measurable function;* $\mathcal{F}, \mathcal{F}'_x \in \mathbf{L}^{2,1}\otimes C_b^1$; $F(\cdot,\cdot,\omega) \in C^{0,2}([0,1]\times\mathbf{R}^1)$ ($\mathbf{P}$-a.s)*; for each* $x \in \mathbf{R}^1$ *the function* $.F$ *can be represented in the form*

$$F(t,x,\omega) = F(0,x,\omega) + \int_0^t H(s,x,\omega)\, dW_s + \int_0^t I(s,x,\omega)\, ds,$$

where $H, I\colon [0,1]\times\mathbf{R}^1\times\Omega \to \mathbf{R}^1$ *are* $\mathcal{B}([0,1])\times\mathbf{R}^1)\otimes\mathcal{F}$*-measurable functions,* $H, H'_x \in \mathbf{L}^{2,1}\otimes C_b^1$ *and* $I(t,\cdot,\omega) \in C^0(\mathbf{R}^1)$ *for almost all* (t,ω);

c) $D_r F'_x(s,y,\omega)|_{y=U_s} \in \mathbf{L}^4([0,1]^2\times\Omega)$ *and the mapping* $s \mapsto D_r F'_x(s,y,\omega)|_{y=U_s}$ *is continuous in* $\mathbf{L}^4(\Omega)$ *uniformly with respect to* r.

Then we have the relation

$$F(t,U_t) - F(0,U_0) = \int_0^t F'_x(s,U_s)v_s\, ds + \int_0^t F'_x(s,U_s)u_s\, dW_s +$$

$$+\frac{1}{2}\int_0^t F''_{xx}(s,U_s)u_s\, ds + \int_0^t u_s F''_{xx}(s,U_s) D_s U_0\, ds +$$

$$+\int_0^t u_s F''_{xx}(s,U_s)\Big(\int_0^s D_s u_\theta\, dW_\theta\Big)\, ds + \int_0^t u_s F''_{xx}(s,U_s)\Big(\int_0^s D_s v_\theta\, d\theta\Big)\, ds +$$

$$+\int_0^t u_s D_s F'_x(s,y)|_{y=U_s}\, ds + \int_0^t I(s,U_s)\, ds + \int_0^t H(s,U_s)\, dW_s +$$

$$+\int_0^t H'_x(s,U_s)u_s\,ds+\int_0^t H'_x(s,U_s)D_sU_0\,ds+\int_0^t H'_x(s,U_s)\Big(\int_0^s D_su_\theta\,dW_\theta\Big)\,ds+$$

$$+\int_0^t H'_x(s,U_s)\Big(\int_0^s D_sv_\theta\,d\theta\Big)\,ds. \tag{2.1}$$

We shall first prove

THEOREM 2.2. *Let the conditions of Theorem 2.1 be fulfilled. Besides, it is assumed that $I(t,\cdot,\omega)\in C^1(\mathbf{R}^1)$, for almost all (t,ω), and the function $F'_x(t,x,\omega)$ admits the representation*

$$F'_x(t,x,\omega)=F'_x(0,x,\omega)+\int_0^t H'_x(s,x,\omega)\,dW_s+\int_0^t I'_x(s,x,\omega)\,ds$$

for each x. Then relation (2.1) is valid.

Proof. Without loss of generality the functions F, F'_x and F''_{xx} can be assumed to be bounded.

Let $\{\Pi^n,\ n\in\mathbf{N}\}$ be a sequence of partitions of the segment $[0,t]$ of the form $\Pi^n=\{0=t_{0,n}<t_{1,n}<\ldots<t_{n,n}=t\}$ such that $|\Pi^n|=\sup(t_{k+1,n}-t_{k,n})\to 0$ as $n\to\infty$. It is obvious that

$$F(t,U_t)-F(0,U_0)=\sum_{k=0}^{n-1}\Big[F(t_{k+1},U_{t_{k+1}})-F(t_k,U_{t_k})\Big]=$$

$$=\sum_{k=0}^{n-1}\Big[F(t_{k+1},U_{t_{k+1}})-F(t_k,U_{t_{k+1}})\Big]+\sum_{k=0}^{n-1}\Big[F(t_k,U_{t_{k+1}})-F(t_k,U_{t_k})\Big].$$

By Taylor's formula,

$$\sum_{k=0}^{n-1}\Big[F(t_k,U_{t_{k+1}})-F(t_k,U_{t_k})\Big]=I_1^n+I_2^n,$$

where

$$I_1^n=\sum_{k=0}^{n-1}F'_x(t_k,U_{t_k})(U_{t_{k+1}}-U_{t_k}),\quad I_2^n=\frac{1}{2}\sum_{k=0}^{n-1}F''_{xx}(t_k,\overline{U}_{t_k})(U_{t_{k+1}}-U_{t_k})^2,$$

and $\overline{U}_{t_k}$ is a random point between U_{t_k} and $U_{t_{k+1}}$.

In its turn,

$$I_2^n=\frac{1}{2}\sum_{k=0}^{n-1}F''_{xx}(t_k,\overline{U}_{t_k})(U_{t_{k+1}}-U_{t_k})^2=I_{21}^n+I_{22}^n+I_{23}^n,$$

where

$$I_{21}^n = \frac{1}{2}\sum_{k=0}^{n-1} F''_{xx}(t_k, \overline{U}_{t_k})\Big(\int_{t_k}^{t_{k+1}} u_s\, dW_s\Big)^2,$$

$$I_{22}^n = \sum_{k=0}^{n-1} F''_{xx}(t_k, \overline{U}_{t_k})\Big(\int_{t_k}^{t_{k+1}} u_s\, dW_s\Big)\Big(\int_{t_k}^{t_{k+1}} v_s\, ds\Big),$$

$$I_{23}^n = \frac{1}{2}\sum_{k=0}^{n-1} F''_{xx}(t_k, \overline{U}_{t_k})\Big(\int_{t_k}^{t_{k+1}} v_s\, ds\Big)^2.$$

Since $F''_{xx}(t,x)$ is continuous, using Theorem 5.4 and Lemma C2 from (Nualart and Pardoux, 1988), we obtain

$$\mathbf{P}\text{–}\lim I_{21}^n = \frac{1}{2}\int_0^t F''_{xx}(s, U_s)u_s^2\, ds.$$

Furthermore, since F''_{xx} is bounded, it is obvious that $\mathbf{P}\text{–}\lim I_{23}^n = \mathbf{P}\text{–}\lim I_{22}^n = 0$. On the other hand,

$$I_1^n = \sum_{k=0}^{n-1} F'_x(t_k, U_{t_k})(U_{t_{k+1}} - U_{t_k}) = I_{11}^n + I_{12}^n,$$

where

$$I_{11}^n = \sum_{k=0}^{n-1} F'_x(t_k, U_{t_k})\Big(\int_{t_k}^{t_{k+1}} u_s\, dW_s\Big),$$

$$I_{12}^n = \sum_{k=0}^{n-1} F'_x(t_k, U_{t_k})\Big(\int_{t_k}^{t_{k+1}} v_s\, ds\Big).$$

It is easy to verify that

$$\mathbf{P}\text{–}\lim I_{12}^n = \int_0^t F'_x(s, U_s)v_s\, ds.$$

Next, by virtue of Theorem 3.2 and Proposition 5.5 from (Nualart and Pardoux, 1988) and Theorem 1.1, we can write

$$I_{11}^n = \sum_{k=0}^{n-1} F'_x(t_k, U_{t_k})\Big(\int_{t_k}^{t_{k+1}} u_s\, dW_s\Big) = \sum_{k=0}^{n-1}\int_{t_k}^{t_{k+1}} F'_x(t_k, U_{t_k})u_s\, dW_s +$$

$$+\sum_{k=0}^{n-1}\int_{t_k}^{t_{k+1}} D_s\Big[F'_x(t_k, U_{t_k})\Big]u_s\, ds = \sum_{k=0}^{n-1}\int_{t_k}^{t_{k+1}} F'_x(t_k, U_{t_k})u_s\, dW_s +$$

$$+\sum_{k=0}^{n-1}\int_{t_k}^{t_{k+1}} F''_{xx}(t_k,U_{t_k})(D_sU_{t_k})u_s\,ds+\sum_{k=0}^{n-1}\int_{t_k}^{t_{k+1}} D_sF'_x(t_k,y)\Big|_{y=U_{t_k}}u_s\,ds=$$

$$=\sum_{k=0}^{n-1}\int_{t_k}^{t_{k+1}} F'_x(t_k,U_{t_k})u_s\,dW_s+\sum_{k=0}^{n-1}\int_{t_k}^{t_{k+1}} u_sF''_{xx}(t_k,U_{t_k})D_sU_0\,ds+$$

$$+\sum_{k=0}^{n-1}\int_{t_k}^{t_{k+1}} u_sF''_{xx}(t_k,U_{t_k})\Big(\int_0^{t_k}D_su_\theta\,dW_\theta\Big)\,ds+$$

$$+\sum_{k=0}^{n-1}\int_{t_k}^{t_{k+1}} u_sF''_{xx}(t_k,U_{t_k})\Big(\int_0^{t_k}D_sv_\theta\,d\theta\Big)\,ds+$$

$$+\sum_{k=0}^{n-1}\int_{t_k}^{t_{k+1}} u_sD_sF'_x(t_k,y)\Big|_{y=U_{t_k}}ds. \tag{2.2}$$

Let us prove that

$$\sum_{k=0}^{n-1}\int_{t_k}^{t_{k+1}} F'_x(t_k,U_{t_k})u_s\,dW_s\longrightarrow\int_0^t F'_x(s,U_s)u_s\,dW_s \tag{2.3}$$

in $\mathbf{L}^2(\Omega)$ as $n\to\infty$. To this end we have to show that

$$u_s\sum_{k=0}^{n-1}F'_x(t_k,U_{t_k})\mathbf{1}\{]t_k,t_{k+1}]\}(s)\longrightarrow u_sF'_x(s,U_s)$$

in $\mathbf{L}^{2,1}$ as $n\to\infty$. Since, due to the continuity of F'_x and by the LDCT, the convergence in $\mathbf{L}^2([0,1])\times\Omega)$ is obvious, it is sufficient to show, by virtue of Theorem 1.1, that we have the following relations in $\mathbf{L}^2([0,1]^2\times\Omega)$ as $n\to\infty$:

$$(D_ru_s)\sum_{k=0}^{n-1}F'_x(t_k,U_{t_k})\mathbf{1}\{]t_k,t_{k+1}]\}(s)\longrightarrow(D_ru_s)F'_x(s,U_s), \tag{2.4}$$

$$u_s\sum_{k=0}^{n-1}F''_{xx}(t_k,U_{t_k})D_rU_0\mathbf{1}\{]t_k,t_{k+1}]\}(s)\longrightarrow u_sF''_{xx}(s,U_s)D_rU_0, \tag{2.5}$$

$$u_s\sum_{k=0}^{n-1}F''_{xx}(t_k,U_{t_k})\Big(\int_0^{t_k}D_ru_\theta\,dW_\theta\Big)\mathbf{1}\{]t_k,t_{k+1}]\}(s)\longrightarrow$$

$$\longrightarrow u_sF''_{xx}(s,U_s)\Big(\int_0^s D_ru_\theta\,dW_\theta\Big), \tag{2.6}$$

$$u_s \sum_{k=0}^{n-1} F''_{xx}(t_k, U_{t_k}) u_r \mathbf{1}\{[0,t_k]\}(r) \mathbf{1}\{]t_k, t_{k+1}]\}(s) \longrightarrow u_s F''_{xx}(s, U_s) u_r \mathbf{1}\{r \leqslant s\}, \tag{2.7}$$

$$u_s \sum_{k=0}^{n-1} F''_{xx}(t_k, U_{t_k}) \Big(\int_0^{t_k} D_r v_\theta \, d\theta \Big) \mathbf{1}\{]t_k, t_{k+1}]\}(s) \longrightarrow u_s F''_{xx}(s, U_s) \Big(\int_0^{s} D_r v_\theta \, d\theta \Big), \tag{2.8}$$

$$u_s \sum_{k=0}^{n-1} D_r F'_x(t_k, y)\Big|_{y=U_{t_k}} \mathbf{1}\{]t_k, t_{k+1}]\}(s) \longrightarrow u_s D_r F'_x(s, y)\Big|_{y=U_s}. \tag{2.9}$$

Relation (2.4) is valid because u belongs to $\mathbf{L}^{2,2}$, since F'_x is continuous and bounded. Furthermore, it is obvious that $u_s D_r U_0$ and $u_r u_s$ belongs to $\mathbf{L}^2([0,1]^2 \times \times \Omega)$. Therefore, relations (2.5) and (2.7) follow from the continuity of F''_{xx}.
Before proving (2.6), note that, by virtue of the Cauchy–Bunyakovski inequality and Proposition 3.5 from (Nualart and Pardoux, 1988), $u_s(\int_0^s D_r u_\theta \, dW_\theta) \in \in \mathbf{L}^2([0,1]^2 \times \Omega)$. On the other hand,

$$u_s F''_{xx}(s, U_s) \Big(\int_0^s D_r u_\theta \, dW_\theta \Big) - u_s \sum_{k=0}^{n-1} F''_{xx}(t_k, U_{t_k}) \Big(\int_0^{t_k} D_r u_\theta \, dW_\theta \Big) \mathbf{1}\{]t_k, t_{k+1}]\}(s) =$$

$$= u_s \Big[F''_{xx}(s, U_s) - \sum_{k=0}^{n-1} F''_{xx}(t_k, U_{t_k}) \mathbf{1}\{]t_k, t_{k+1}]\}(s) \Big] \Big(\int_0^s D_r u_\theta \, dW_\theta \Big) +$$

$$+ u_s \sum_{k=0}^{n-1} F''_{xx}(t_k, U_{t_k}) \Big(\int_{t_k}^{s} D_r u_\theta \, dW_\theta \Big) \mathbf{1}\{]t_k, t_{k+1}]\}(s). \tag{2.10}$$

Since F''_{xx} is continuous and bounded, by the LDCT the first term in the right-hand side of (2.10) tends to zero in $L^2([0,1]^2 \times \Omega)$ for $n \to \infty$. Furthermore, by virtue of the Cauchy–Bunyakovski inequality and Proposition 3.5 from (Nualart and Pardoux, 1988), we have

$$\mathbf{E} \Big[\sum_{k=0}^{n-1} \int_{t_k}^{t_{k+1}} \int_0^1 u_s^2 F''_{xx}(t_k, U_{t_k})^2 \Big(\int_0^s D_r u_\theta \, dW_\theta \Big)^2 dr\, ds \Big] \leqslant$$

$$\leqslant \|(F''_{xx})^2\|_\infty \Big(\mathbf{E} \int_0^1 |u_s|^4 ds \Big)^{1/2} c_4 \Big[\mathbf{E} \sum_{k=0}^{n-1} \int_{t_k}^{t_{k+1}} \int_0^1 \int_{t_k}^{s} |D_r u_\theta|^4 d\theta\, dr\, ds +$$

$$+\mathbf{E}\sum_{k=0}^{n-1}\int_{t_k}^{t_{k+1}}\int_0^1\int_{t_k}^{s}\int_{t_k}^{s}|D_tD_ru_\theta|^4\,d\theta\,dt\,dr\,ds\Bigg]^{1/2} \tag{2.11}$$

The last term in (2.11) vanishes due to assumption (ii) of Theorem 2.1, part a). Therefore, the second term in the right-hand side of (2.10) tends to zero in $L^2([0,1]^2\times\Omega)$ as $n\to\infty$. This proves (2.6).

Relation (2.8) is proved similarly.

Finally, by virtue of condition **c)** of Theorem 2.1 and the Cauchy–Bunyakovski inequality, we have $u_sD_rF'_x(s,y)|_{y=U_s}\in L^2([0,1]^2\times\Omega)$ and

$$\mathbf{E}\int_0^1\int_0^1\Big[u_sD_rF'_x(s,y)\Big|_{y=U_s}-u_s\sum_{k=0}^{n-1}D_rF'_x(t_k,y)\Big|_{y=U_{t_k}}\mathbf{1}\{]t_k,t_{k+1}]\}(s)\Big]^2dr\,ds\leqslant$$

$$\leqslant\Big(\mathbf{E}\int_0^1|u_s|^4ds\Big)^{1/2}\Big(\sum_{k=0}^{n-1}\int_{t_k}^{t_{k+1}}\int_0^1\mathbf{E}\Big[D_rF'_x(s,y)\Big|_{y=U_s}-$$

$$-D_rF'_x(t_k,y)\Big|_{y=U_{t_k}}\Big]^4dr\,ds\Big)^{1/2}\longrightarrow 0$$

as $n\to\infty$.

Summing up the above results we obtain (2.3).

Similarly to relations (2.5), (2.6), (2.8) and (2.9), using the continuity (with respect to (t,x)) and boundedness of F''_{xx}, as well as the conditions **a)** and **c)** of Theorem 2.1, it is not difficult to verify that the following relations are valid:

$$\mathbf{P}\text{-lim}\sum_{k=0}^{n-1}\int_{t_k}^{t_{k+1}}u_sF''_{xx}(t_k,U_{t_k})D_sU_0\,ds=\int_0^t u_sF''_{xx}(s,U_s)D_sU_0\,ds,$$

$$\mathbf{P}\text{-lim}\sum_{k=0}^{n-1}\int_{t_k}^{t_{k+1}}u_sF''_{xx}(t_k,U_{t_k})\Big(\int_0^{t_k}D_su_\theta\,dW_\theta\Big)ds=$$

$$=\int_0^t u_sF''_{xx}(s,U_s)\Big(\int_0^s D_su_\theta\,dW_\theta\Big)ds,$$

$$\mathbf{P}\text{-lim}\sum_{k=0}^{n-1}\int_{t_k}^{t_{k+1}}u_sF''_{xx}(t_k,U_{t_k})\Big(\int_0^{t_k}D_sv_\theta\,d\theta\Big)ds=$$

$$=\int_0^t u_sF''_{xx}(s,U_s)\Big(\int_0^s D_sv_\theta\,d\theta\Big)ds,$$

$$\text{P-lim}\sum_{k=0}^{n-1}\int_{t_k}^{t_{k+1}} u_s D_s F'_x(t_k,y)\big|_{y=U_{t_k}}\,ds = \int_0^t u_s D_s F'_x(s,y)\big|_{y=U_s}\,ds.$$

Thus we have performed all passages to the limit in the right–hand side of (2.2). Therefore, we have obtained the limit expression for the second term in the representation of $F(t,U_t)-F(0,U_0)$. Let us now turn to the first term. By Taylor's formula, we have

$$\sum_{k=0}^{n-1}\Big[F(t_{k+1},U_{t_{k+1}})-F(t_k,U_{t_{k+1}})\Big] = \sum_{k=0}^{n-1}\Big[F(t_{k+1},U_{t_{k+1}})-F(t_{k+1},U_{t_k})\Big]+$$

$$+\sum_{k=0}^{n-1}\Big[F(t_{k+1},U_{t_k})-F(t_k,U_{t_k})\Big]+\sum_{k=0}^{n-1}\Big[F(t_k,U_{t_k})-F(t_k,U_{t_{k+1}})\Big]=$$

$$=\sum_{k=0}^{n-1}\Big[\sum_{m=0}^{1}(-1)^{m+1}F'_x\Big(t_{k+m},U_{t_k}+\theta_m(U_{t_{k+1}}-U_{t_k})\Big)(U_{t_{k+1}}-U_{t_k})\Big]+$$

$$+\sum_{k=0}^{n-1}\Big[F(t_{k+1},U_{t_k})-F(t_k,U_{t_k})\Big], \tag{2.12}$$

where $\theta_m=\theta_m(\omega)\in\,]0,1[,\ m=0,1$. By virtue of Theorem 1.3,

$$\sum_{k=0}^{n-1}\Big[F(t_{k+1},U_{t_k})-F(t_k,U_{t_k})\Big]=$$

$$=\sum_{k=0}^{n-1}\Big[\int_{t_k}^{t_{k+1}} H(s,x)\,dW_s+\int_{t_k}^{t_{k+1}} I(s,x)\,ds\Big]\Big|_{x=U_{t_k}}=\sum_{k=0}^{n-1}\int_{t_k}^{t_{k+1}} H(s,U_{t_k})\,dW_s+$$

$$+\sum_{k=0}^{n-1}\int_{t_k}^{t_{k+1}} H'_x(s,U_{t_k})D_sU_{t_k}\,ds+\sum_{k=0}^{n-1}\int_{t_k}^{t_{k+1}} I(s,U_{t_k})\,ds. \tag{2.13}$$

Let us prove that

$$\sum_{k=0}^{n-1}\int_{t_k}^{t_{k+1}} H(s,U_{t_k})\,dW_s \longrightarrow \int_0^t H(s,U_s)\,dW_s \tag{2.14}$$

in $\mathbf{L}^2(\Omega)$ as $n\to\infty$. To this end we have to prove that

$$\sum_{k=0}^{n-1} H(s,U_{t_k},\omega)\mathbf{1}\{]t_k,t_{k+1}]\}(s)\longrightarrow H(s,U_s)$$

in $\mathbf{L}^{2,1}$ as $n \to \infty$. Since, by virtue of the continuity of $H(s,x,\omega)$ (with respect to x), the convergence in $\mathbf{L}^2([0,1]\times\Omega)$ is obvious, due to Theorem 1.1 it is sufficient to prove that the following relations are valid in $\mathbf{L}^2([0,1]^2\times\Omega)$ as $n\to\infty$:

$$\sum_{k=0}^{n-1} H'_x(s,U_{t_k})D_rU_0\mathbf{1}\{]t_k,t_{k+1}]\}(s) \longrightarrow H'_x(s,U_s)D_rU_0, \tag{2.15}$$

$$\sum_{k=0}^{n-1} H'_x(s,U_{t_k})\Big(\int_0^{t_k} D_r u_\theta\, dW_\theta\Big)\mathbf{1}\{]t_k,t_{k+1}]\}(s) \longrightarrow H'_x(s,U_s)\Big(\int_0^{s} D_r u_\theta\, dW_\theta\Big), \tag{2.16}$$

$$\sum_{k=0}^{n-1} H'_x(s,U_{t_k})u_r\mathbf{1}\{[0,t_k]\}(r)\mathbf{1}\{]t_k,t_{k+1}]\}(s) \longrightarrow H'_x(s,U_s)u_r\mathbf{1}\{r\leqslant s\}, \tag{2.17}$$

$$\sum_{k=0}^{n-1} H'_x(s,U_{t_k})\Big(\int_0^{t_k} D_r v_\theta\, d\theta\Big)\mathbf{1}\{]t_k,t_{k+1}]\}(s) \longrightarrow H'_x(s,U_s)\Big(\int_0^{s} D_r v_\theta\, d\theta\Big), \tag{2.18}$$

$$\sum_{k=0}^{n-1} D_rH(s,y)\big|_{y=U_{t_k}}\mathbf{1}\{]t_k,t_{k+1}]\}(s) \longrightarrow D_rH(s,y)|_{y=U_s}. \tag{2.19}$$

By virtue of Definition 1.2, it is clear that all terms in representations (2.15)–(2.19) belong to the class $\mathbf{L}^2([0,1]^2\times\Omega)$. Furthermore, relations (2.15)–(2.18) are obvious corollaries of the LDCT, since $H'_x(s,x,\omega)$ is continuous with respect to x and relation (2.19) follows from Proposition 1.2. This completes the proof of (2.14). In a similar way it is not difficult to find the limit of the second term on the right–hand side of (2.13)

$$\mathbf{P}\text{–}\lim\sum_{k=0}^{n-1}\int_{t_k}^{t_{k+1}} H'_x(s,U_{t_k})D_sU_{t_k}\,ds = \int_0^t H'_x(s,U_s)D_sU_0\,ds +$$

$$+\int_0^t H'_x(s,U_s)\Big(\int_0^s D_s u_\theta\,dW_\theta\Big)\,ds + \int_0^t H'_x(s,U_s)\Big(\int_0^s D_s v_\theta\,d\theta\Big)\,ds.$$

Because of the continuity of $I(s,x,\omega)$ (with respect to x) it is also obvious that

$$\mathbf{P}\text{–}\lim\sum_{k=0}^{n-1}\int_{t_k}^{t_{k+1}} I(s,U_{t_k})\,ds = \int_0^t I(s,U_s)\,ds.$$

Next, we transform the first term in (2.12) as follows:

$$\sum_{k=0}^{n-1}\sum_{m=0}^{1}(-1)^{m+1}F'_x\Big(t_{k+m},U_{t_k}+\theta_m(U_{t_{k+1}}-U_{t_k})\Big)(U_{t_{k+1}}-U_{t_k})=K_1^n+K_2^n,$$

where

$$K_1^n=\sum_{k=0}^{n-1}\Bigg\{\sum_{m=0}^{1}(-1)^{m+1}\Big[F'_x\Big(t_{k+m},U_{t_k}+\theta_m(U_{t_{k+1}}-U_{t_k})\Big)-$$
$$-F'_x(t_{k+m},U_{t_k})\Big](U_{t_{k+1}}-U_{t_k})\Bigg\},$$

$$K_2^n=\sum_{k=0}^{n-1}\Big[F'_x(t_{k+1},U_{t_k})-F'_x(t_k,U_{t_k})\Big](U_{t_{k+1}}-U_{t_k}).$$

Since $F'_x(s,x,\omega)$ is continuous (with respect to x), we have $\mathbf{P}$–lim $K_1^n=0$. Then, from the representation of F'_x and Theorem 1.3, we obtain $K_2^n=\sum_{j=1}^{4}K_{2j}^n$, where

$$K_{21}^n=\sum_{k=0}^{n-1}\Big(\int_{t_k}^{t_{k+1}}H'_x(s,U_{t_k})\,dW_s\Big)\Big(\int_{t_k}^{t_{k+1}}u_s\,dW_s\Big),$$

$$K_{22}^n=\sum_{k=0}^{n-1}\Big(\int_{t_k}^{t_{k+1}}H'_x(s,U_{t_k})\,dW_s\Big)\Big(\int_{t_k}^{t_{k+1}}v_s\,ds\Big),$$

$$K_{23}^n=\sum_{k=0}^{n-1}\Big(\int_{t_k}^{t_{k+1}}H''_{xx}(s,U_{t_k})D_sU_{t_k}\,ds\Big)(U_{t_{k+1}}-U_{t_k}),$$

$$K_{24}^n=\sum_{k=0}^{n-1}\Big(\int_{t_k}^{t_{k+1}}I'_x(s,U_{t_k})\,ds\Big)(U_{t_{k+1}}-U_{t_k}).$$

It is easy to verify that

$$\mathbf{P}\text{–lim } K_{22}^n=\mathbf{P}\text{–lim } K_{23}^n=\mathbf{P}\text{–lim } K_{24}^n=0.$$

On the other hand, from Theorem 1.2 we have

$$\mathbf{P}\text{–lim } K_{21}^n=\int_0^t H'_x(s,U_s)u_s\,ds.$$

Summing up the above–obtained limit expressions, we ascertain that relation (2.1) is valid.

Proof of Theorem 1.2. Fix $\varepsilon > 0$ and assume that $E_\varepsilon(t,x,\omega) = T_\varepsilon F(t,x,\omega)$, $H_\varepsilon(t,x,\omega) = T_\varepsilon H(t,x,\omega)$, $I_\varepsilon(t,x,\omega) = T_\varepsilon I(t,x,\omega)$. Check whether the functions $F_\varepsilon, H_\varepsilon$ and I_ε satisfy the conditions of Theorem 2.2. The continuity of $F(t,x,\omega)$ and $\int_0^t I(s,x,\omega)\,ds$ (with respect to (t,x)) and the connection of the integral $\int_0^t H(s,x,\omega)\,dW_s$ with these functions in the representation of $F(t,x,\omega)$ show that the Skorokhod integral $\int_0^t H(s,x,\omega)\,dW_s$ has a **P**–a.s. continuous modification with respect to (t,x). Therefore, it can be assumed that $\int_0^t H(s,x,\omega)\,dW_s$ is a **P**–a.s. locally integrable function with respect to x (for each t), on which the operator T_ε is defined. Therefore, from Proposition 1.6 and Fubini's theorem, we conclude that F_ε admits the representation

$$F_\varepsilon(t,x,\omega) = F_\varepsilon(0,x,\omega) + \int_0^t H_\varepsilon(s,x,\omega)\,dW_s + \int_0^t I_\varepsilon(s,x,\omega)\,ds.$$

By virtue of Proposition 1.4, $F_\varepsilon(\cdot,\cdot,\omega) \in C^{0,\infty}([0,1]\times\mathbf{R}^1)$ (**P**–a.s.) and $I_\varepsilon(t,\cdot,\omega) \in$ $\in C^\infty(\mathbf{R}^1)$ (**P**–a.s.) for all t. Moreover, Proposition 1.5 leads us to $F_\varepsilon, (F_\varepsilon)'_x, H_\varepsilon,$ $(H_\varepsilon)'_x \in \mathbf{L}^{2,1} \otimes C_b^1$. It is also clear that

$$\left(\int_0^t I_\varepsilon(s,x,\omega)\,ds\right)'_x = \int_0^t (I_\varepsilon(s,x,\omega))'_x\,ds \quad (\mathbf{P}\text{–a.s.}).$$

On the other hand, from Propositions 1.4 and 1.6, we have

$$\left(\int_0^t H_\varepsilon(s,x,\omega)\,dW_s\right)'_x =$$

$$= \left[T_\varepsilon\left(\int_0^t H(s,\cdot,\omega)\,dW_s\right)(x)\right]'_x = \varepsilon^{-1}\varkappa \int_{\mathbf{R}^1} \zeta'_x\left(\frac{x-y}{\varepsilon}\right) \int_0^t H(s,y,\omega)\,dW_s\,dy =$$

$$= \varepsilon^{-1}\varkappa \int_0^t \int_{\mathbf{R}^1} \zeta'_x\left(\frac{x-y}{\varepsilon}\right) H(s,y,\omega)\,dy\,dW_s = \int_0^t \Big[H_\varepsilon(s,x,\omega)\Big]'_x\,dW_s \quad (\mathbf{P}\text{–a.s.}),$$

where the last equality is proved similarly to (1.18).
Thus $(F_\varepsilon)'_x$ admits the representation

$$[F_\varepsilon(t,x,\omega)]'_x = [F_\varepsilon(0,x,\omega)]'_x + \int_0^t \Big[H_\varepsilon(s,x,\omega)\Big]'_x\,dW_s + \int_0^t \Big[I_\varepsilon(s,x,\omega)\Big]'_x\,ds.$$

Therefore, with Proposition 1.5 taken into account, we see that F_ε satisfies the assumptions of Theorem 2.2. Thus the relation (2.1) when F, H and I replaced by $F_\varepsilon, H_\varepsilon,$ and I_ε, respectively, is valid.

Passing in this equality to the limit as $\varepsilon \to 0$, and taking into account Propositions 1.4, 1.5, Corollary 1.1 and using the reasoning similar to that used above in justifying passages to the limit, we have that Theorem 2.1 is valid.

REFERENCES

Follmer, H. (1981). Calcul d'Ito sans probabilites. *Lect. Notes Math.* **850**, 143–150.

Hitsuda, H. (1979). *Formula for Brownian partial derivatives.* Faculty of Integrated Arts and Sciences, Hiroshima University, **3**, 1–15.

Martias, C. (1988). Une formula d'Ito–Ventsel' pour les processues anticipatifes. *C. r. Acad. Sci., Ser.* **1**, **307**, 675–678.

Nualart, D. (1989). The Maliavin Calculus and some of its Applications. *Preprint.*

Nualart, D. and Pardoux, E. (1988). Stochastic calculus with anticipating integrands. *Probab. Theory Rel. Fields,* **78**, 535–581.

Pardoux, E. and Protter, Ph. (1987). Two–sided stochastic integral and calculus. *Probab. Theory Rel. Fields,* **76**, 15–50.

Rozovskii, B. (1973). On the Itô–Ventsel' formula. *Vestnik MGU,* 26–32. (in Russian).

Rozovskii, B. (1983). *Evolution Stochastic Systems.* Nauka, Moscow (in Russian).

Ustunel, A. (1988). The Ito formula for anticipative processes with nonmonotonous time scale via the Malliavin calculus. *Probab. Theory Rel. Fields,* **79**, 249–269.

Ventsel, A.D. (1965). On equations of the theory of conditional Markov processes. *Theory Probab. Appl.* **10**, 390–393.

New Trends in Probab. and Statist., pp. 528 – 539
V. Sazonov and T. Shervashidze (Eds.)
1991 VSP/Mokslas

CONNECTIVITY PROPERTY AND OPTIMALITY ALMOST SURELY AND IN PROBABILITY

V.I. ROTAR'
Central Economic Mathematical Institute
of the USSR Academy of Sciences,
Krasikova str. 32, 117418 Moscow, USSR

ABSTRACT

This paper concerns problems of stochastic dynamic control. We are interested in controls which maximize a reward function for all realizations from a set the probability of which asymptotically, for large time horizons, is close to one (controls asymptotically optimal almost surely). We show that for a rather general controlled random process optimal in the mean controls are also optimal almost surely under rather mild conditions. These conditions concern the possibility of transition from any state to a trajectory which is close, in some sense, to a trajectory generated by a control optimal in the mean.

INTRODUCTION

Normally, the optimal control theory concerns strategies maximizing the mathematical expectation of a reward function (strategies optimal in the mean) for a dynamic, as a rule, system. For different models some authors (see, e.g., (Lippman, 1970; Mandl, 1982; Rotar', 1986 b; Mandl and Romera Ayllon, 1987; Chen and Guo, 1987; Konyuhova and Rotar', 1987, 1988; Leizarowitz, 1988; Asriev and Rotar', 1990)) considered strategies having a much stronger property, namely those maximizing a reward function itself for all realizations of a controlled random process from a set with probability close (asymptotically for large time horizons) to one. The investigation of a general scheme was undertaken in this direction in (Rotar', 1986b; Asriev and Rotar', 1990). The latter paper includes the results of (Rotar', 1986b).

In (Rotar', 1986b; Asriev and Rotar', 1990) it was shown that, under rather mild conditions, optimal in the mean strategies possess the property mentioned above. The main conditions were formulated in terms of a value function for the usual problem of maximizing the mathematical expectation of a reward. In the present paper we consider essentially another type of conditions concerning the possibility of transition from any state to a trajectory close, in some sense, to a trajectory generated by a control which is optimal in the mean. Conditions of such type were also considered in (Asriev and Rotar', 1990) but only in the frame of a special model of economic growth. The task of this paper consists in investigating the general case in the same direction.

We use a rather general but not very customary framework from (Asriev and Rotar', 1990). From the author's point of view, this framework is convenient and allows to simplify formulations and proofs. On the other hand, the basic scheme is rather abstract, and we elucidate it and the results in Section 1 by an example of a homogeneous controlled Markov chain. The corresponding proposition plays the role of an illustration; we shall not yearn for generality. In Sections 2, 3 we construct a general scheme, in Section 4 the main conditions and a general theorem are formulated. Sections 5, 6 are devoted to proofs.

1. HOMOGENEOUS CONTROLLED MARKOV CHAIN

Any such chain may be specified by the recursive formula

$$x_t = h(\xi_t, x_{t-1}, d_t) \tag{1.1}$$

where x_t is the state of the chain at time t; $\xi_1, \xi_2, \ldots$ are identically distributed independent random variables (r.v.), d_t is a decision or an action at time t and $h(\cdot)$ is a function (which is measurable with respect to the corresponding σ-algebra, we omit the obvious formalism).

Remark. Often, one specifies a chain by a measure $Q(\cdot \mid x, d)$ which is the conditional distribution of x_t given $x_{t-1} = x$, $d_t = x$. For every measure Q there exists such a function h that the distribution of the r.v. $h(\xi_t, x, d)$ coincides with $Q(\cdot \mid x, d)$. Moreover, without loss of generality we may assume the r.v. ξ_t to be uniformly distributed on [0,1]; see, e.g., (Gihman and Skorohod, 1979). The representation (1.1) seems, at least for our purposes, more convenient because of the existence of a single measure on the space of elementary events; see also (Asriev and Rotar', 1990).

Below for any sequence of arbitrary elements $(a_1, a_2, \ldots)$ we will set $a^t = (a_1, \ldots, a_t)$.

Let us denote the time horizon by n and consider the reward function

$$f_n(\xi^n, d^n) = (1/n) \sum_{t=1}^{n} q(x_t), \tag{1.2}$$

where x_t on the right-hand side are subjected to (1.1), $q(\cdot)$ is the single-period reward.

Setting $d_1 = \psi_1(x_0)$ and $d_t = \psi_t(x^{t-1}, d^{t-1})$ for $t > 1$, where ψ_t are some functions, we specify the control at time t. A strategy is a finite or infinite sequence $(\psi_1, \psi_2, \ldots)$ of such functions. If $\psi_t(x^{t-1}, d^{t-1}) = \psi_t(x_{t-1})$ for all t, the strategy is said to be Markov.

Let $X_{t\psi z}$ $(t = 0, 1, \ldots)$ be the random process generated by a strategy $\psi = (\psi_1, \psi_2, \ldots)$ and an initial state z, that is,

$$X_{0\psi z} = z, \qquad X_{t\psi z} = h(\xi_t, X_{(t-1)\psi z}, u_{tz}) \quad \text{for } t > 0,$$

and the r.v. u_{tz} are specified by the relations

$$u_{1z} = \psi_1(z), \qquad u_t = \psi_{tz}(X_{\psi z}^{t-1}, u_z^{t-1}) \quad \text{for } t > 1,$$

where $X_{\psi z}^t = (X_{0\psi z}, \ldots, X_{t\psi z})$, $u_z^t = (u_{1z}, \ldots, u_{tz})$.
We shall sometimes omit the index ψ in $X_{t\psi z}$. Thereafter X_{tz} denotes a random process generated by some strategy and an initial state z. We call any such process X_{tz} a *trajectory*.
Any strategy ψ and an initial state z generate the sequence (for different n) of random rewards

$$f_{n\psi z} = (1/n) \sum_{t=1}^{n} q(X_{t\psi z}).$$

Let us call a strategy Ψ *asymptotically optimal almost surely* (as.opt.a.s.) for a fixed initial state z if, for all ψ,

$$\limsup_n [f_{n\psi z} - f_{n\Psi z}] \leqslant 0 \quad \text{a.s.} \tag{1.3}$$

Note that the definition (1.3) deals with the difference $[f_{n\psi} - f_{n\Psi}]$ (we left out the index z) and does not assume finiteness of $\limsup_n f_{n\psi}$, that is r.v. $f_{n\psi}$ may tend to infinity as $n \to \infty$. If it is not so, (1.3) implies

$$\limsup_n f_{n\psi} \leqslant \limsup_n f_{n\Psi} \quad \text{a.s.}$$

or

$$\liminf_n f_{n\psi} \leqslant \liminf_n f_{n\Psi} \quad \text{a.s.},$$

but these weaker than (1.3) properties do not seem very natural because it is not clear why one should consider only *lim sup* or *lim inf*. Certainly, it would be very good if a strategy Ψ existed such that

$$\limsup_n f_{n\psi} \leqslant \liminf_n f_{n\Psi} \quad \text{a.s.},$$

but in the general case we cannot hope for this. On the other hand, if $\lim_n f_{n\Psi}$ a.s. exists, (1.3) coincides with

$$\limsup_n f_{n\psi} \leqslant \lim_n f_{n\Psi} \quad \text{a.s.}$$

Further, assume that, for all z and n, there exist a strategy $\widehat{\psi} = \widehat{\psi}^n$ which maximizes $\mathbf{E} f_{n\psi z}$. Of course, $\widehat{\psi}$ may depend on n and for $t > n$ the functions $\widehat{\psi}_t$ are arbitrary. Let $\hat{f}_{nz} = f_{n\hat{\psi}z}$ and ${}_n\widehat{X}_{tz} = X_{t\hat{\psi}^n z}$. For $t > n$ the process ${}_n\widehat{X}_{tz}$ is arbitrary.

PROPOSITION. *Assume that the function $q(\cdot)$ is bounded and there exist a number $p > 2$ and a constant C_p with the following property:*

for all n and any pair of initial states z and z' there exist such a trajectory X_{tz} (possibly depending on n and z') and such a r.v. N (which may depend on n, z and z') that $\mathbf{E}N^p \leqslant C_p$ and

$$N \leqslant t \leqslant n \Rightarrow q(X_{tz}) = q({}_n\widehat{X}_{tz'}). \tag{1.4}$$

Then, for all strategies ψ and for every initial state z,

$$\limsup_n [f_{n\psi z} - \hat{f}_{nz}] \leqslant 0 \quad a.s.$$

Remarks. 1) Since components of the strategy $\hat{\psi}^n$ may depend on n, the proposition does not mean the existence of a strategy as.opt.a.s. (In fact, we are to consider the case of triangular arrays; it will be done in next sections.) If there exists such a strategy Ψ (in the homogeneous case it is possible, as a rule, to choose a stationary one) that

$$f_{n\Psi z} - \hat{f}_{nz} \to 0 \quad \text{as } n \to \infty \quad \text{a.s.,}$$

then, as it is easy to see, under the conditions of Proposition the strategy Ψ is as.opt.a.s.

2) Certainly for any fixed n one may choose a r.v. N such that $N > n$ almost surely, so the former event in (1.4) is empty and (1.4) holds automatically. But we cannot do this for all n in view of the condition $\mathbf{E}N^p \leqslant C_p$. So the latter condition of Proposition concerns the behavior of the controlled process (and properties of the optimal in the mean trajectory) for large horizon.

It is worth also to emphasize the following two circumstances.

Firstly, all said above is valid if we consider arbitrary sets of possible states and decisions and any, possibly "narrow", class of admissible strategies.

Secondly, one can always extend sets of admissible states and decisions and the class of admissible strategies in such a way that for the new problem, arising after this operation, the main condition of Proposition holds. The matter is whether the optimal in the mean strategy for the new problem will be the same as for the parent one. If so, this strategy will be as.opt.a.s. for both the new and the old problems.

We illustrate this by the concrete example from (Asriev and Rotar', 1990). Let $\mathbf{P}(\xi_t = 1) = \mathbf{P}(\xi_t = -1) = 1/2$; variables x_t and d_t take values from sets $\{-1, 0, +1\}$ and $\{\alpha, \beta\}$, respectively. Let $h(\xi, +1, d) = +1$ and $h(\xi, -1, d) = -1$ for all ξ and d; $h(\xi, 0, \alpha) = 0$ for all ξ; $h(+1, 0, \beta) = +1$ and $h(-1, 0, \beta) = -1$. (The states ± 1 are absorbing for all strategies, the state 0 is absorbing for decision α. Under decision β the system transfers to the states ± 1 from the state 0 with equal probabilities.) Let also $q(x) \equiv x$.

It is easy to understand (see also (Asriev and Rotar', 1990)) that for this chain any strategy is optimal in the mean (and $\mathbf{E}f_n \equiv 0$), but an optimal a.s. strategy does not exist.

On the other hand, we may add to the set of possible decisions such a decision γ that, for all ξ and x,

$$h(\xi, x, \gamma) = +1.$$

Then conditions of Proposition hold for the new problem and the strategy $(\gamma, \gamma, \ldots)$ is, of course, optimal both in the mean and almost surely. But this strategy is not admissible for the old problem.

Now we turn to the general scheme. For completeness of exposition we shall repeat below some definitions from (Asriev and Rotar', 1990).

2. THE GENERAL SCHEME

Let $\{\Omega, \mathcal{F}, \mathcal{F}^t, \mathbf{P}\}$ be a probability space with a flow of σ-algebras $\{\mathcal{F}^t\}$, $t = 0, 1, \ldots$. For convenience and, in fact, without loss of generality, we may assume that any elementary event

$$\omega = (\omega_0, \omega_1, \ldots), \tag{2.1}$$

where ω_t are elements of an arbitrary nature and $\mathcal{F}^t$ is the σ-algebra of all measurable cylindric sets specified by the vectors $\omega^t = (\omega_0, \ldots, \omega_t)$. Formally, however, representation (2.1) is not necessary for the following constructions.

Let $(D_t, \mathcal{D}_t)$, where $t = 1, 2, \ldots$, be a sequence of measurable spaces. We regard $(D_t, \mathcal{D}_t)$ as a space of decisions or actions taken at $t = 1, 2, \ldots$. Let $D^t = D_1 \times \ldots \times D_t$, $\mathcal{D}^t = \mathcal{D}_1 \times \ldots \times \mathcal{D}_t$, $d^t = (d_1, \ldots, d_t)$, where $d_j \in D_j$.

The next element of the scheme is a sequence of reward functions $\{f_n(\omega, d^n) = f_n(\omega^n, d^n)\}_{n=1}^{\infty}$, that is, $f_n(\cdot)$ is measurable with respect to $\mathcal{F}^n \times \mathcal{D}^n$. Below we consider only the reward functions

$$f_n(\omega, d^n) = (1/n) \sum_{t=1}^{n} g_t(\omega, d^t), \tag{2.2}$$

where $g_t(\omega, d^t) = g_t(\omega^t, d^t)$ are $\mathcal{F}^t \times \mathcal{D}^t$- measurable.

For a fixed n, we define a *control* as a vector $u^n = (u_1, \ldots, u_n)$, where the map u_t: $\Omega \to D_t$ is $\mathcal{F}^{t-1}$-measurable, i.e., r.v. $u_t = u_t(\omega^{t-1})$ and is predictable. Assume that all controls under consideration belong to a class U^n.

Generally speaking, a control determines not resourceful behavior because it does not take into account previous decisions. Given n, we define a strategy as a vector $v^n = \{v_1, \ldots, v_n\}$, where the map v_t: $\Omega \times D^{t-1} \to D_t$ is $\mathcal{F}^{t-1} \times \mathcal{D}^{t-1}$-measurable, that is, $v_t = v_t(\omega^{t-1}, d^{t-1})$.

For any strategy v^n we define the set of operators ${}_t\mathbf{V}^n$ $(t = 1, 2, \ldots, n)$ transforming the space of controls into itself and such that for all u^n

$${}_t\mathbf{V}^n u^n = \tilde{u}^n = (\tilde{u}_1, \ldots, \tilde{u}_n),$$

where

$$\tilde{u}_j = \begin{cases} u_j & \text{if } j < t; \\ v_j(\omega, \tilde{u}_1(\omega), \ldots, \tilde{u}_{j-1}(\omega)) & \text{if } j \geqslant t. \end{cases}$$

In other words, the map ${}_t\mathbf{V}^n$ demands to use the control u^n before time t and after that time to control the system in accordance with the strategy v^n. It is clear that the control ${}_1\mathbf{V}^n u^n$ does not depend on u^n. We shall call ${}_1\mathbf{V}^n u^n$ the control generated by the strategy v^n. Clearly, any class of strategies generates the corresponding class of controls.

Let us illustrate this scheme by the example of a Markov chain from Section 1. Let independent r.v. ξ_t be defined on a space $\{\Omega, \mathcal{F}, \mathbf{P}\}$, $\mathcal{F}^t = \sigma(\xi^t)$ and d_t be an element of a space $(D_t, \mathcal{D}_t)$.

In view of (1.1), there exist $\mathcal{F}^t \times \mathcal{D}^t$-measurable functions $g_t(\omega, d^t) = q(x_t)$. For any function $\psi_t(\cdot)$ from Section 1, there exist $\mathcal{F}^{t-1} \times \mathcal{D}^{t-1}$-measurable functions

$$v_t(\omega, d^{t-1}) = \psi_t(x^{t-1}, d^{t-1}).$$

Thus, we are in the limits of the general scheme and can work with (2.2). Certainly, the form of the functions $g_t(\cdot)$ and $v_t(\cdot)$ depends on the structure of the function $h(\cdot)$ and on fixed x_0.

For some concrete purposes, the knowledge of a concrete structure of a controlled process, for example, of relations similar to (1.1) and (1.2), may be even necessary. But for other general problems including one under discussion, this information is useless; moreover, dealing with the general scheme, we in fact simplify considerations. Note also that, as a rule, the necessary information on a process structure can be given in terms of the functions $g_t(\cdot)$ in (2.2), and the structure of the latter can be described in terms of the σ-algebras with respect to which $g_t(\cdot)$ are measurable. Certainly, these σ-algebras usually are more "rough" than $\mathcal{F}^t \times \mathcal{D}^t$ (see (Rotar', 1986 a)).

3. DEFINITIONS OF OPTIMALITY

Assume that the mathematical expectations of all considered r.v. are finite.
A control $\hat{u}^n$ from U^n is *optimal in the mean* if, for all $u^n \in U^n$,

$$\mathbf{E}f_n(\omega, u^n) \leqslant \mathbf{E}f_n(\omega, \hat{u}^n). \tag{3.1}$$

Let $\mathbf{E}_t(\cdot) = \mathbf{E}(\cdot \mid \mathcal{F}^{t-1})$, $\mathbf{E}_0 = \mathbf{E}$. We call a strategy v^n *persistently optimal* (see also, e.g., (Kertz, 1982)) if, for all controls $u^n \in U^n$ and for all $t = 1, \ldots, n$,

$$\mathbf{E}_t f_n(\omega, {}_{t+1}\mathbf{V}^n u^n) \leqslant \mathbf{E}_t f_n(\omega, {}_t\mathbf{V}^n u^n) \quad \text{a.s.}$$

(We assume that ${}_t\mathbf{V}^n u^n = u^n$ for $t > n$.)
Obviously, a control generated by a persistently optimal strategy is optimal in the mean.
Let us turn to the asymptotic analysis. Note that if, for a fixed n, an optimal, in some sense, control exists, then its components may depend also on n, that is, $u_t = u_{tn}$. Thereby, varying n, we deal, in the general case, with a *sequence of controls* (seq.c.) or triangular arrays

$$u = \{u^n\} = \{u_{1n}, \ldots, u_{nn}\}_{n=1}^{\infty}.$$

This framework is more general than the usual one for which at any time t a control may depend on t but not on the time horizon n. If the situation is such that it is reasonable to restrict ourselves to the latter case, i.e., to consider for different values of n just the same sequence $\{u_1, u_2, \ldots\}$, then it means that we consider such a class of seq.c. $u = \{u^n\}$ that $u_{tn} = u_{tm}$ for all t, n, m.
Let U be the class of all seq.c. $u = \{u^n\}$ such that $u^n \in U^n$ for all n.
We call a seq.c. $\hat{u} = \{\hat{u}^n\} \in U$ *asymptotically optimal* (as.opt.) *in probability* if, for any $\varepsilon > 0$ and all $u = \{u^n\} \in U$,

$$\mathbf{P}\big(f_n(\omega, u^n) - f_n(\omega, \hat{u}^n) > \varepsilon\big) \to 0 \quad \text{as } n \to \infty. \tag{3.2}$$

We call a seq.c. $\hat{u} = \{\hat{u}^n\} \in U$ *as.opt.almost surely* if, for all $u = \{u^n\} \in U$,

$$\mathbf{P}\Big(\limsup_n \left[f_n(\omega, u^n) - f_n(\omega, \hat{u}^n)\right] \leqslant 0\Big) = 1. \tag{3.3}$$

4. A GENERAL THEOREM

Introduce some conditions on the class of seq.c. U.
Assume that for all n there exists a persistently optimal strategy v^n. Let $\hat{u}^n$ and ${}_t\mathbf{V}^n$ be the control and the sequence of operators, respectively, which are generated by v^n; $\hat{u} = \{\hat{u}^n\}$. Assume that $\hat{u} \in U$.
We shall sometimes omit the index n in $u_{tn}, v_{tn}, {}_t\mathbf{V}^n$ and so on and often write $g_t(\omega, u^n)$ instead of $g_t(\omega, u^t)$ remembering that, in fact, $g_t(\cdot)$ depends only on the first t coordinates of the vector $(u_1, \dots, u_n)$. Such notation will be convenient and, above all else, emphasize that we deal with the n-th term of the seq.c. $u = \{u^n\}$.
Let $\|\cdot\|_p$ be the norm in L_p, i.e., $\|\xi\|_p = \left(\mathbf{E}|\xi|^p\right)^{1/p}$ for a r.v. ξ.

Condition **A**. There exist $p > 2$ and a constant A_p with the following property: for all n, all pairs of controls $u^n, \bar{u}^n \in U^n$, and all $t \leqslant n$, there exist such a control $z^n \in U^n$ that

$$z^n = (\bar{u}_1, \dots, \bar{u}_t, z_{t+1}, \dots, z_n), \tag{4.1}$$

and

$$\left\| \sum_{j=t+1}^{n} \{g_j(\omega, z^n) - g_j(\omega, {}_{t+1}\mathbf{V}^n u^n)\} \right\|_p \leqslant A_p. \tag{4.2}$$

Note that the number of summands in (4.2) is equal to $(n-t)$, so (4.2) holds only if the difference $\{g_j(\omega, z^n) - g_j(\omega, {}_{t+1}\mathbf{V}^n u^n)\}$ is "small" for large j, that is, the control z^n is such that the corresponding reward $g_j(\cdot)$ is approaching, beginning from the time t, the reward generated by a persistently optimal strategy.
For example, in the case of a nonhomogeneous Markov chain, in order to verify Condition **A** we should consider an arbitrary time t as initial, consider two arbitrary attainable states, consider an optimal in the mean trajectory beginning (at time t) at the first state and try to construct a trajectory which begins at the second state and goes close to the optimal trajectory in the sense of (4.2).
Dealing with optimality in probability, we shall use a weaker

Condition **A'**. There exists a function $\varepsilon(n) \to 0$ as $n \to \infty$, with the following property: for all n, all pairs of controls $u^n, \bar{u}^n \in U^n$, and all $t \leqslant n$, there exists a control $z^n \in U^n$ such that (4.1) holds and

$$\mathbf{E}\left[\sum_{j=t+1}^{n} \{g_j(\omega, z^n) - g_j(\omega, {}_{t+1}\mathbf{V}^n u^n)\} \right]^2 \leqslant \varepsilon(n) n. \tag{4.3}$$

THEOREM. 1) *Let condition* **A** *hold and, for all* $u = \{u^n\} \in U$,

$$(1/n) \sum_{t=1}^{n} \{g_t(\omega, u^n) - \mathbf{E}_t g_t(\omega, u^n)\} \to 0 \quad \text{as } n \to \infty \tag{4.4}$$

almost surely. Then the sequence of controls $\hat{u}$ *is as.opt.a.s.*
2) *If condition* **A'** *holds and, for all* $u = \{u^n\} \in U$, (4.4) *is valid in probability, then the sequence of controls* $\hat{u}$ *is as.opt. in probability.*

Remark. Since the summands in (4.4) are martingale differences with respect to the flow $\{\mathcal{F}^t\}$, it is obvious that (4.4) holds in probability if, for example,

$$\frac{1}{n^2}\sum_{t=1}^{n}\mathbf{E}\left[g_t(\omega,u^n)-\mathbf{E}_t g_t(\omega,u^n)\right]^2 \to 0 \quad \text{as } n\to\infty. \tag{4.5}$$

Certainly, (4.5) holds if $g_t(\cdot)$ is uniformly bounded, that is,

$$\sup_{t,\omega,d^t}\left|g_t(\omega,d^t)\right| < \infty. \tag{4.6}$$

It is easy also to verify (see also (6.7) below and the remark after it) that (4.4) holds a.s. if, for some $l \geqslant 2$ and all $u \in U$,

$$\sum_{n=1}^{\infty}\frac{1}{n^l}\left[\sum_{t=1}^{n}\left\|g_t(\omega,u^n)-\mathbf{E}_t g_t(\omega,u^n)\right\|_l^2\right]^{l/2} < \infty. \tag{4.7}$$

Clearly, (4.7) holds, for example, if there exists such a constant C and a number $l > 2$ that, for all controls $u^n \in U^n$,

$$\left\|g_t(\omega,u^t)-\mathbf{E}_t g_t(\omega,u^t)\right\|_l \leqslant C \tag{4.8}$$

(because in this case the left-hand side of (4.7) is smaller than $C^l\sum_{n=1}^{\infty} n^{-l/2}$). Obviously, (4.6) yields (4.8).
We show first that Proposition is a corollary of the above Theorem.

5. PROOF OF PROPOSITION

It is easy to understand that, since the chain is homogeneous and $\hat{\psi}^n$ exists for any initial state, this strategy is persistently optimal. Since the function $q(\cdot)$ is bounded, condition (4.4) holds a.s. (see remark above). Thus, it remains to show that Condition **A** holds for any initial state.
Since the chain is homogeneous, in order to verify Condition **A**, it suffices to set $t = 0$ in (4.1) and (4.2). Namely, in terms of Introduction, we should show that there exist $p > 2$ and a constant A_p such that, for all n and any pair of initial states z and z', one can choose a trajectory X_{tz} such that

$$\left\|\sum_{j=1}^{n}\left\{q_j(X_{tz})-q_j({}_n\hat{X}_{tz'})\right\}\right\|_p \leqslant A_p. \tag{5.1}$$

Let X_{tz} and r.v. N be the same as in the proposition. Let $\sum_k^m = 0$ if $k > m$, and $\sum_1^N = \sum_1^n$ if $N > n$. Then, in view of (1.4),

$$\left|\sum_{j=1}^{n}\left\{q_j(X_{tz})-q_j({}_n\hat{X}_{tz'})\right\}\right| = \left|\sum_{j=1}^{N}\left\{q_j(X_{tz})-q_j({}_n\hat{X}_{tz'})\right\}\right| \leqslant 2AN,$$

where $A = \sup_x q(x)$. Thus, (5.1) holds with $A_p = 2AC_p^{1/p}$, where C_p is the same as in Introduction. The proof is completed.
Note also that in a similar way it is not difficult to obtain the corresponding corollary of Theorem related to a general controlled process using a random time N of transition from an arbitrary state to the "trajectory" generated by a persistently optimal strategy.

6. PROOF OF THEOREM

The first step is the same as in the proof of Theorem 1 from (Asriev and Rotar', 1990). Let the strategy v^n, the control $\hat{u}^n$ and the maps ${}_t\mathbf{V}^n$ be the same as in Section 3, and let $\hat{u} = \{\hat{u}^n\} \in U$. Let us fix some other seq.c. $u = \{u^n\}$. We omit below index n in $u_{tn}, \hat{u}_{tn}, v_{tn}$ and ${}_t\mathbf{V}^n$.
Let r.v.

$$Y_t = Y_{tn} = \mathbf{E}_t^* f_n(\omega,\ {}_{t+1}\mathbf{V}u^n) - \mathbf{E}_t f_n(\omega,\ {}_t\mathbf{V}u^n),$$

$$W_t = W_{tn} = \sum_{j=t}^{n} [Z_{tj} - \mathbf{E}_t Z_{tj}],$$

where

$$Z_{tt} = Z_{tt}(n) = g_t(\omega, u^n) - g_t(\omega, \hat{u}^n),$$

and, for $j > t$,

$$Z_{tj} = Z_{tj}(n) = \mathbf{E}_{t+1}\{g_j(\omega,\ {}_{t+1}\mathbf{V}u^n) - g_j(\omega, \hat{u}^n)\}.$$

We use the approach of (Asriev and Rotar', 1990; Rotar', 1986b) based on the identity

$$f_n(\omega, u^n) - f_n(\omega, \hat{u}^n) = \sum_{t=1}^{n} Y_t + (1/n) \sum_{t=1}^{n} W_t. \tag{6.1}$$

Since v^n is persistently optimal, $Y_t \leqslant 0$ a.s. for all $t \leqslant n$. Thus, it remains to show that

$$(1/n) \sum_{t=1}^{n} W_t \to 0 \quad \text{as } n \to \infty, \tag{6.2}$$

either a.s. or in probability (depending on the choice of conditions).
Let

$$V_t = V_{tn} = \sum_{j=t+1}^{n} \mathbf{E}_{t+1} g_j(\omega,\ {}_{t+1}\mathbf{V}u^n), \quad \widehat{V}_t = \widehat{V}_{tn} = \sum_{j=t+1}^{n} \mathbf{E}_{t+1} g_j(\omega, \hat{u}^n),$$

$$R_t = R_{tn} = V_{tn} - \widehat{V}_{tn}.$$

Then

$$W_t = Z_{tt} - \mathbf{E}_t Z_{tt} + R_t - \mathbf{E}_t R_t. \tag{6.3}$$

Since (4.4) holds for all $u \in U$, under conditions of Theorem, we have

$$(1/n) \sum_{t=1}^{n} [Z_{tt} - \mathbf{E}_t Z_{tt}] \to 0 \quad \text{as } n \to \infty \tag{6.4}$$

(in probability or a.s.). Therefore, it suffices to show that

$$(1/n)\sum_{t=1}^{n}\{R_t - \mathbf{E}_t R_t\} \to 0 \quad \text{as } n \to \infty, \tag{6.5}$$

in probability or a.s.
Fix $p > 2$ and set $U_t = R_t - \mathbf{E}_t R_t$. Summands in (6.5) are martingale-differences with respect to the flow $\{\mathcal{F}^t\}$, and, because of Burkholder's inequality, there exist a constant C_p depending only on p, such that

$$\mathbf{E}\Big|\sum U_t\Big|^p \leqslant C_p \mathbf{E}\Big|\sum U_t^2\Big|^{p/2} = C_p \Big\|\sum U_t^2\Big\|_{p/2}^{p/2} \leqslant$$

$$\leqslant C_p\Big[\sum \|U_t^2\|_{p/2}\Big]^{p/2} = C_p\Big[\sum \|U_t\|_p^2\Big]^{p/2}. \tag{6.6}$$

Using (6.6), the Markov inequality and the Borel–Cantelli lemma, we conclude that the convergence in (6.5) takes place a.s. if for some $p > 2$

$$\sum_{n=1}^{\infty} \frac{1}{n^p}\left[\sum_{t=1}^{n-1} \|U_t\|_p^2\right]^{p/2} < \infty. \tag{6.7}$$

Condition (4.7) (see Remark in Section 3) can be obtained in exactly the same way.
Consider (6.7). Clearly, it holds if

$$\sum_{n=1}^{\infty} \frac{1}{n^p}\left[\sum_{t=1}^{n-1} \|R_t\|_p^2\right]^{p/2} < \infty. \tag{6.8}$$

Let us fix n and $t \leqslant n$ and turn to Condition **A**. Set $\bar{u} = \hat{u}$. Then the control (4.1) from the condition has the form

$$z^n = (\hat{u}_1, \ldots, \hat{u}_t,\ z_{t+1}, \ldots, z_n). \tag{6.9}$$

Since the strategy v^n is persistently optimal,

$$\widehat{V}_t \geqslant \mathbf{E}_{t+1}\sum_{j=t+1}^{n} g_j(\omega, z^n) = V_t + \mathbf{E}_{t+1}\sum_{j=t+1}^{n} g_j(\omega, z^n) - \mathbf{E}_{t+1}\sum_{j=t+1}^{n} g_j(\omega,\ {}_{t+1}\mathbf{V}^n u^n)$$

and

$$\widehat{V}_t - V_t \geqslant -\left|\mathbf{E}_{t+1}\sum_{j=t+1}^{n} g_j(\omega, z^n) - \mathbf{E}_{t+1}\sum_{j=t+1}^{n} g_j(\omega,\ {}_{t+1}\mathbf{V}^n u^n)\right|. \tag{6.10}$$

Now let us replace the pair of controls $(u^n, \bar{u}^n)$ by the pair $(\hat{u},\ {}_{t+1}\mathbf{V}^n u^n)$. Control (4.1) in this case, certainly, differs from control (6.9); denote it, say, by k^n. Since in our case $\bar{u}^n = {}_{t+1}\mathbf{V}^n u^n$, and $\bar{u}_j = u_j$ for $j \leqslant t$, control (4.1) has the form

$$k^n = \{u_1, \ldots, u_t,\ k_{t+1}, \ldots, k_n\}.$$

Recall also that ${}_{t+1}\mathbf{V}^n\hat{u}^n = \hat{u}^n$ for all t. Thus, since the strategy v^n is persistently optimal,

$$V_t \geqslant \mathbf{E}_{t+1} \sum_{j=t+1}^{n} g_j(\omega, k^n) = \mathbf{E}_{t+1} \sum_{j=t+1}^{n} g_j(\omega, {}_{t+1}\mathbf{V}^n\hat{u}^n)+$$

$$+\mathbf{E}_{t+1} \sum_{j=t+1}^{n} g_j(\omega, k^n) - \mathbf{E}_{t+1} \sum_{j=t+1}^{n} g_j(\omega, {}_{t+1}\mathbf{V}^n\hat{u}^n) =$$

$$= \mathbf{E}_{t+1} \sum_{j=t+1}^{n} g_j(\omega, \hat{u}^n) + \mathbf{E}_{t+1} \sum_{j=t+1}^{n} g_j(\omega, k^n) - \mathbf{E}_{t+1} \sum_{j=t+1}^{n} g_j(\omega, {}_{t+1}\mathbf{V}^n\hat{u}^n) =$$

$$= \widehat{V}_t + \mathbf{E}_{t+1} \sum_{j=t+1}^{n} g_j(\omega, k^n) - \mathbf{E}_{t+1} \sum_{j=t+1}^{n} g_j(\omega, {}_{t+1}\mathbf{V}^n\hat{u}^n)$$

and

$$V_t - \widehat{V}_t \geqslant -\left| \mathbf{E}_{t+1} \sum_{j=t+1}^{n} g_j(\omega, k^n) - \mathbf{E}_{t+1} \sum_{j=t+1}^{n} g_j(\omega, {}_{t+1}\mathbf{V}^n\hat{u}^n) \right|. \tag{6.11}$$

Clearly, Condition **A**, (6.10), and (6.11) imply that

$$\|\widehat{V}_t - V_t\|_p \leqslant 2A_p \tag{6.12}$$

and

$$\sum_{n=1}^{\infty} \frac{1}{n^p} \left[\sum_{t=1}^{n-1} \|R_t\|_p^2 \right]^{p/2} \leqslant 2^p A_p^p \sum_{n=1}^{\infty} \frac{1}{n^{p/2}} < \infty$$

if $p > 2$. The first assertion of Theorem is proved.

Consider now the convergence in (6.5) in probability. Similarly to the proof of (6.12), one may obtain that, under Condition **A**′,

$$\mathbf{E}|\widehat{V}_t - V_t|^2 \leqslant 2\varepsilon(n)n.$$

Then the variance of r.v. on the left-hand side of (6.5) is smaller than

$$\frac{1}{n^2} \sum_{t=1}^{n} 2\mathbf{E}|\widehat{V}_t - V_t|^2 \leqslant 4\varepsilon(n) \to 0.$$

The proof is completed.

REFERENCES

Asriev, A.V. and Rotar', V.I. (1990). On asymptotic optimality in probability and almost surely in dynamic control. *Stochastics and Stoch. Rep.* (to appear).

Chen, H.F. and Guo, L. (1987). Adaptive control and consistent parameter estimates for ARMAX Model with quadratic costs. *SIAM J. Control Optim.* **25**, 845–867.

Gihman, I.I. and Skorohod, A.V. (1979). *Controlled Stochastic Processes.* Springer, Berlin, etc.

Kertz, R.P. (1982). Renewal plans and persistent optimality in countably additive gambling. *Math. Oper. Res.* **7**, 361–382.

Konyuhova, T.A. and Rotar', V.I. (1987). On optimal in probability solutions for a linear stochastic model. In: *Probability Problems of Discrete Mathematics.* Moscow Institute of Electronic Engineering, Moscow, pp. 131–135.

Konyuhova, T.A. and Rotar', V.I. (1988). Optimal in probability stochastic control in a linear model with quadratic cost. In: *Probability Problems of Discrete Mathematics.* Moscow Institute of Electronic Engineering, Moscow, pp. 111–114.

Leizarowitz, A. (1988). On almost sure optimization for stochastic control systems. *Stochastics* **23**, 85–110.

Lippman, S. (1971). Maximal average-reward policies for semi-Markov decision processes with arbitrary space and action space. *Ann. Math. Statist.* **42**, 1717–1726.

Mandl, P. (1982). Martingale methods in discrete state random processes. *Kybernetica.* **18**, supplement.

Mandl, P. and Romera Ayllon, M.R. (1987). On controlled Markov processes with average cost criterion. *Kybernetica.* **23**, 433–442.

Nazin, A.V. and Pozdniak, A.S. (1983). *Adaptive Choice of Variants.* Nauka, Moscow (in Russian).

Rotar', V.I. (1986 a). On sufficient control in the dynamical problems of the stochastic optimization. *Math. Notes.* **40**, 542–551.

Rotar', V.I. (1986 b). Some remarks about asymptotical optimality. In: *Investigation of Stochastic Control Problems of the Economical Processes,* CEMI AS USSR, Moscow, pp. 93–116.

New Trends in Probab. and Statist., pp. 540 – 549
V. Sazonov and T. Shervashidze (Eds.)
1991 VSP/Mokslas

ABSOLUTE CONTINUITY OF SMOOTH MEASURES GENERATED BY NONLINEAR EQUATIONS

G. SOKHADZE
Kutaisi N.Muskhelishvili Polytechnical Institute,
384014 Kutaisi, Georgia, USSR

ABSTRACT

Some classes of nonlinear differential equations with random perturbations are considered. It is proved that the measures generated by the solutions of such equations are absolutely continuous with respect to more simple ones and the Radon–Nikodym densities are calculated.

1. INTRODUCTION

First we remind one theorem (Sokhadze, 1989) and then we discuss applications of this result in different situations. Separately we will consider the case of analogues of stochastic differential equations. Here the right sides of the equations are random elements with smooth distributions (see (Daletskii and Belopolskaya, 1989)). In particular, the Gaussian measure and the measures that are obtained from it by smooth transformations are of this kind.

Let us agree upon some notation.

Let E be a nuclear operator in Hilbert space H and a_k, $k = 1,2,\dots$, be its eigenvalues. Then

$$\det(I+E) = \prod_{k=1}^{\infty}(1+a_k).$$

For Hilbert–Schmidt operators one defines a regularized determinant (see (Gokhberg and Krein, 1965)):

$$\widetilde{\det}(I+E) = \prod_{k=1}^{\infty}(1+a_k)e^{-a_k}.$$

Let $(x,y)_H$ and $\|x\|_H$ denote the scalar product and the norm in H.
Consider an equipped Hilbert space with Hilbert–Schmidt inclusions

$$H_+ \subset H \subset H_-$$

and a measure μ that has the logarithmic derivative $\lambda(x)$ in H_- along constant directions in H_+. Consider the expression

$$\beta_\mu^H(t, F, y) = \langle \lambda(y + tF(y)),\ F(y) \rangle \ +\ \operatorname{tr} F'(y), \tag{1}$$

where $F(y)$ is a smooth nonlinear function such that $F'(y)$ is a nuclear operator and $\langle x, y \rangle$ denotes the pairing of elements of H_- and H_+. Usually we will write below the pairing in the form of scalar product in H. It is known that (1) makes sense also in the case when F: $H_- \to H$ and $F'(y)$ is a Hilbert–Schmidt operator. In what follows, we will consider a class $A^{(1)}$ of bounded open domains $G \subset \mathbf{R}^n$ such that the boundary ∂G can be locally represented in an analytical form $t_n = \varphi(t_1, \ldots, t_{n-1})$, where $\varphi \in C^1$ (see (Miranda, 1955)).
The usual and generalized derivatives are denoted as follows:

$$D_j = \frac{\partial}{\partial x_j} \quad (j = 1, \ldots, n), \quad D^\alpha = D_1^{\alpha_1} \ldots D_n^{\alpha_n}$$

$$\alpha = (\alpha_1, \ldots, \alpha_n), \quad |\alpha| = \alpha_1 + \ldots + \alpha_n.$$

We denote $\mathcal{L}_2(G)$ the space of square integrable with respect to the Lebesgue measure real valued functions defined on G with the inner product

$$(u, v)_{\mathcal{L}_2(G)} = \int_G u(x) v(x)\, dx, \qquad u, v \in \mathcal{L}_2(G);$$

$W_2^p(G)$ is the Sobolev space with the scalar product

$$(u, v)_{W_2^p(G)} = (u, v)_{\mathcal{L}_2(G)} + \sum_{|\alpha| = p} (D^\alpha u, D^\alpha v)_{\mathcal{L}_2(G)};$$

$C_0^p(G)$ is the space of all finite on G functions from $C^p(G)$, $p = 0, 1, \ldots, \infty$.
Let $\overset{\circ}{W}{}_2^p(G)$ be a subspace of $W_2^p(G)$ obtained by taking closure in $W_2^p(G)$ of the linear set $C_0^\infty(G) \subset W_2^p(G)$.
The terminology and notation agree with (Berezanskii, 1965).
Below $\{\Omega, \mathcal{F}, \mathbf{P}\}$ denotes a fixed probability space.
In H_- we consider equations

$$A\eta + B(\eta) = \xi \tag{2}$$

and

$$A\zeta = \xi, \tag{3}$$

where A is a linear and B is a nonlinear operator in H_-; ξ is a random element in H_-.
If solutions of (2) and (3) exist, then we denote their probability distributions by μ_η and μ_ζ, respectively.

THEOREM 1 (Sokhadze, 1989). *Let in an equipped Hilbert space $H_+ \subset H \subset$ $\subset H_-$ the following conditions for equations (2) and (3) be satisfied:*

A) *A is a linear, closed, possibly unbounded operator defined on a domain $D(A)$ which is dense in H_- and there exists a bounded inverse A^{-1};*

B) *B is a nonlinear operator B: $H_- \to H$, $B(x)|_{D(A)}$: $D(A) \to D(A)$, there exists $B'(x)$ and $\|B'(x)A^{-1}\| < 1$ for any $x \in D(A)$;*

C) *ξ is a random element in H_- for which there exists a continuous function $\lambda(x)$: $H_- \to H_-$ such that, for any $\Phi \in C^1(H_-)$, the condition*

$$\mathbf{E}(\Phi'(\xi), h)_H = \mathbf{E}\Phi(\xi)(\lambda(\xi), h)_H$$

holds for any $h \in H_+$.

Then the measures μ_η and μ_ζ are equivalent and the Radon–Nikodym density can be written as

$$\frac{d\mu_\eta}{d\mu_\zeta}(u) = \widetilde{\det}(I + A^{-1}B'(u))\exp\int_0^1 \beta^H_{\mu_\xi}(t, B \circ A^{-1}, Au)\,dt \tag{4}$$

with $u \in D(A)$. If, in addition, it is known that $B(x)$: $H_- \to H_+$, then

$$\frac{d\mu_\eta}{d\mu_\zeta}(u) = \det(I + A^{-1}B'(u))\exp\Big\{\Big(\int_0^1 \lambda(Au + tB(u))dt, B(u)\Big)_H\Big\}. \tag{5}$$

For simplicity of the exposition we will use below formula (5) although its generalization (4) holds true.

2. SUPPLEMENTS TO THE THEOREM

To Theorem 1 one can reduce some problems on absolute continuity connected with nonlinear equations. Here we will give two such applications.
In a separable real Hilbert space H consider the equation

$$A\eta + B(\eta) = \xi \tag{2'}$$

and suppose that the following conditions are satisfied:

a) A is a linear unbounded operator with domain $D(A)$ which is dense in H, A is invertible and A^{-1} is a Hilbert–Schmidt operator;

b) B: $H \to D(A^2)$, there exists $B'(x)$ and $\|B'(x)A^{-1}\| < 1$;

c) ξ is a random element in H with probability distribution μ_ξ, which has logarithmic derivative along $A^{-2}H$ of the form $\lambda(x)$: $H \to H$.

Under these conditions there exists a solution η of equation (2′) with the distribution μ_η. Besides (2′) we also consider a linear equation

$$A\zeta = \xi \tag{3'}$$

with the solution $\zeta = A^{-1}\xi$ having the distribution μ_ζ. In the spaces $D(A)$ and $D(A^2)$ introduce the scalar product

$$(x,y)_{D(A)} = (Ax, Ay)_H, \quad (x,y)_{D(A^2)} = (Ax, Ay)_{D(A)}.$$

Then we will get the equipped Hilbert space

$$D(A^2) = H_+ \subset D(A) = H_0 \subset H = H_-.$$

Now we can apply Theorem 1. We get $\mu_\zeta \sim \mu_\eta$ and

$$\frac{d\mu_\eta}{d\mu_\zeta}(u) = \det(I + A^{-1}B'(u))\exp\left\{\left(\int_0^1 \lambda(Au + tB(u))dt, A^*AB(u)\right)_H\right\}.$$

The case when in equations (2′) and (3′) the right-hand side is a generalized random variable is interesting. This is the case, e.g., when we deal with stochastic differential equations.

Let in the equipped Hilbert space

$$H_+ \subset H \subset H_- \tag{6}$$

we have the equation

$$A\eta + g(\eta) = \xi \tag{7}$$

where A, g and ξ satisfy the following conditions:

α) A is a linear unbounded operator with a domain $D(A)$ which is dense in H and there exists A^{-1} which is a Hilbert–Schmidt operator;

β) $g(x)$ is a nonlinear function in H with values in $D(A)$, $g'(x)$ exists and is a Hilbert–Schmidt operator;

γ) ξ is a random element in H_- such that

$$\mathbf{E}(\Phi'(\xi), h)_H = \mathbf{E}\Phi(\xi)(\lambda(\xi), h)_H,$$

for any $\Phi \in C^1(H_-)$, $h \in H_+$ and some λ: $H_- \to H_-$.

We must explain in what sense we understand equation (7). Under condition α) the operator A^{-1} can be extended to H_-. We denote the extension by $\widehat{A}^{-1}$. In this case A can be extended to H_-. Denote this extension by $\widehat{A}$. Then we will understand (7) as an equivalent form of the equation

$$\eta + A^{-1}g(\eta) = \widehat{A}^{-1}\xi, \tag{8}$$

which is correctly defined under our conditions. From the general theory it follows that equation (8) has a solution η with distribution μ_η in H. In an analogous sense we understand the equation

$$A\zeta = \xi. \tag{9}$$

From (8) and (9) we obtain the relation

$$\zeta = \eta + A^{-1}g(\eta), \tag{10}$$

which defines a transformation f in the space H with inverse $f^{(-1)}$. It is clear that $\mu_\zeta = \mu_\eta \circ f^{-1}$. Now we can apply the results of (Daletskii and Sokhadze, 1988). We get

THEOREM 2. *Let conditions α), β), γ) be fulfilled for equations* (7) *and* (8) *with a generalized random right side in an equipped Hilbert space* (6). *Then* $\mu_\zeta \sim \mu_\eta$ and

$$\frac{d\mu_\eta}{d\mu_\zeta}(u) = \det\big(I + A^{-1}g'(u)\big)\exp\Big(\int_0^1 \lambda\big(\widehat{A}u + tg(u)\big)dt, g(u)\Big)_H. \tag{11}$$

For example, if ξ is a Gaussian random element, then $\lambda(x) = -x$ and under conditions α) and β) we have $\mu_\zeta \sim \mu_\eta$ and

$$\frac{d\mu_\eta}{d\mu_\zeta}(u) = \det\big(I + A^{-1}g'(u)\big)\exp\Big\{ - \big(\widehat{A}u, g(u)\big)_H - \frac{1}{2}\|g(u)\|_H^2 \Big\};$$

in particular for the classical equation

$$d\xi_t = a(t, \xi_t)\, dt + dw_t, \quad \xi_0 = 0,$$

in the Hilbert space $\mathcal{L}_2([0, a])$ this gives

$$\frac{d\mu_\xi}{d\mu_w}(w) = \exp\Big\{ - \int_0^a a(\tau, w_\tau)\, dw_\tau - \frac{1}{2}\int_0^a a^2(\tau, w_\tau)\, d\tau \Big\}.$$

3. APPLICATIONS TO DIFFERENTIAL EQUATIONS

Let $D \subset \mathbf{R}^n$ be a bounded open domain of class $A^{(1)}$. Consider the differential expression

$$Lu = \sum_{|\alpha| \leqslant p} a_\alpha(x) D^\alpha u, \tag{12}$$

where $a_\alpha(x) \in C^{|\alpha|}(D \cup \partial D)$. Symbol L' denotes the adjoint expression

$$L'u = \sum_{|\alpha| \leqslant p} (-1)^{|\alpha|} D^\alpha\big(a_\alpha(x)u\big).$$

In order to include the most general situation we introduce the boundary conditions in the following way. We will call every subspace of $W_2^p(G)$ containing

$\overset{\circ}{W}{}_2^p(G)$ the subspace of functions which satisfy specific boundary conditions and we will denote it by $\overline{W}{}_2^p(\partial G)$.
Consider the triple of spaces

$$W_2^{2p}(G) \subset W_2^p(G) \subset \mathcal{L}_2(G) \tag{13}$$

and let $\xi = \xi(x)$, $x \in G$, be a random field integrable with the second moment

$$\int\limits_G \mathbf{E}\xi^2(x)dx < \infty. \tag{14}$$

Suppose also that the distribution μ_ξ in $\mathcal{L}_2(G)$ has logarithmic derivative along $W_2^{2p}(G)$. So, for any $\Phi \in C^1(\mathcal{L}_2(G))$, we have

$$\mathbf{E}\big(\Phi(\xi), h\big)_{W_2^p(G)} = \mathbf{E}\Phi(\xi)\big(\lambda(\xi), h\big)_{W_2^p(G)}, \tag{15}$$

where λ: $\mathcal{L}_2(G) \to \mathcal{L}_2(G)$ and $h \in W_2^{2p}(G)$.
With the differential expression L from (12) we connect the boundary value problem

$$L\zeta = \xi, \quad \zeta \in \overline{W}{}_2^p(\partial G) \tag{16}$$

and suppose that L and L' satisfy the energetic inequalities

$$\|Lu\|_{\mathcal{L}_2(G)} \geqslant c\|u\|_{\mathcal{L}_2(G)}, \quad \|L'v\|_{\mathcal{L}_2(G)} \geqslant c\|v\|_{\mathcal{L}_2(G)}, \tag{17}$$

where $c > 0$, u, $v \in C_0^\infty(G)$.
We understand the solvability of problem (16) in the following sense: it is known that under condition (17) there exists a solvable extension L, which has a continuous inverse in all $\mathcal{L}_2(G)$. We take $\zeta = L^{-1}\xi$ to be a solution of (16). In an analogous sense we understand the solvability of

$$L\eta + g\big(x, \eta(x)\big) = \xi(x), \qquad \eta \in \overline{W}{}_2^p(\partial G), \tag{18}$$

where $g(x, u)$ is a smooth function in $\mathcal{L}_2(G)$.

THEOREM 3. *Let D be an open bounded domain of the class $A^{(1)}$ with boundary ∂D. Consider on D a partial differential equation with general boundary conditions (16) and (18), where the operator L is defined by equality (12) with $a_\alpha(x) \in C^{|\alpha|}(D \cup \partial D)$, ξ is a random field satisfying (14) and (15) in the equipped space (13); $g(x, u)$: $G \times \mathcal{L}_2(G) \to \mathcal{L}_2(G)$ and has generalized derivatives of order $2p$ in the Sobolev sense; $F = \partial g/\partial u$ and $\|F\| < 1/\|L^{-1}\|$ for every u.*
Then, if energetic inequality (17) is satisfied, we have $\mu_\zeta \sim \mu_\eta$ and for $u \in W_2^p(G)$ we have

$$\frac{d\mu_\eta}{d\mu_\zeta}(u) = \det\big(I + L^{-1}F(u)\big) \times$$

$$\times \exp\Bigg\{ \int\limits_0^1 \int\limits_G \lambda\Big(\sum_{|\alpha| \leqslant p} a_\alpha(x) D^\alpha u + t\, g(x, u)\Big)\, g(x, u)\, dx\; dt +$$

$$+(-1)^p \int\limits_0^1 \int\limits_G \lambda\Big(\sum_{|\alpha| \leqslant p} a_\alpha(x) D^\alpha u + t g(x, u)\Big) \sum_{|\alpha| = p} (D^\alpha)^2 g(x, u)\, dx\; dt \Bigg\}. \tag{19}$$

Remark. In fact, in this theorem we deal with a strong solution of the equation. In reality it holds in the case when we consider a semistrong solution since energetic inequalities — the basic solvability apparatus — hold in this case too and have the following form

$$\|Lu\|_{\mathcal{L}_2(G)} \geqslant C\|u\|_{\mathcal{L}_2(G)}, \quad \|L'v\|_{\mathcal{L}_2(G)} \geqslant C\|v\|_{\mathcal{L}_2(G)}$$

for some $C > 0$ and any $u \in \overline{W}_2^p(\partial G)$ and $v \in \overline{W}_2^p(\partial G)^+$, " + " means the adjoint boundary condition. In the case when $g(x,u)$ is linear weak solutions of the equations can also be considered.
Now let us extend these considerations to the case of generalized solutions. Consider the triple of spaces

$$W_2^p(G) \subset \mathcal{L}_2(G) \subset W_2^{-p}(G) \tag{20}$$

and let the random field $\xi = \xi(x)$, $x \in G$, belong to $W_2^{-p}(G)$ (a random linear functional on $W_2^p(G)$) and satisfies the equation

$$\mathbf{E}\int_G [\Phi'(\xi)](x)h(x)\,dx = \mathbf{E}\Phi(\xi)\int_G [\lambda(\xi)](x)h(x)\,dx \tag{21}$$

for any functional $\Phi:\ W_2^{-p}(G) \to \mathbf{R}$ and $h \in W_2^p(G)$, $\lambda:\ W_2^{-p}(G) \to W_2^{-p}(G)$. Consider the boundary value problem

$$Lu = \xi, \quad u \in \overline{W}_2^p(\partial G) \tag{22}$$

with the solution

$$u = (\Lambda(\partial G)^{-1})(\xi), \tag{23}$$

where the operator Λ is constructed in the following way: suppose that closures of the spaces $\overline{W}_2^p(\partial G)$ and $\overline{W}_2^p(\partial G)^+$ coincide in the space $W_2^m(G)$. Denote it by $\overline{W}_2^m(G)$. Define the m-strong operator $\Lambda_{-m}(\partial G)$: $\overline{W}_2^m(G) \to \overline{W}_2^{-m}(G)$:

$$\Lambda_{-m}(\partial G)u = Lu \quad \text{for} \quad u \in \overline{W}_2^p(\partial G)$$

and denote its closure by $\Lambda(\partial G)$.
It is known that if the order of differential expression L is even ($p = 2k$) and the estimate

$$|\,(Lu,v)_{\mathcal{L}_2(G)}\,| \leqslant c\|u\|_{W_2^m(G)}|u|_{W_2^m(G)} \tag{24}$$

holds for $u \in \overline{W}_2^p(\partial G)$, $v \in \overline{W}_2^p(\partial G)^+$, then there exists the closure, i.e., $\Lambda(\partial G)$ is a homeomorphism between $\overline{W}_2^p(G)$ and $\overline{W}_2^{-p}(G)$. Under these conditions the solution of equation (22) has the form (23).
For the nonlinear problem

$$Lv + g(x, v(x)) = \xi(x), \quad v \in \overline{W}_2^p(\partial G), \tag{25}$$

with the above–mentioned conditions, the solvability of equation (25) is equivalent to the solvability of the equation

$$v(x) + \Lambda(\partial G)^{-1} g(x, v(x)) = \Lambda(\partial G)^{-1} \xi(x),$$

which has a solution if we assume that $g(x,v)$ is smooth. So, we have

THEOREM 4. *Let in the triple of spaces* (20) *a random field* $\xi = \xi(x)$, $x \in G$, *be considered. Suppose that it satisfies condition* (21), *where* G *is a domain of the class* $A^{(1)}$. *Suppose also that for the differential expression* $L = \sum_{|\alpha| \leqslant 2k} a_\alpha(x) D^\alpha$ *inequality* (24) *holds*, $g(\cdot, u)$: $W_2^{-p}(G) \to W_2^p(G)$ *and there exists* $F = \partial g/\partial u$ *such that* $\|F\| < \|\Lambda^{-1}(\partial G)\|^{-1}$.
Then the distributions of the solutions of (22) *and* (25) *are equivalent and*

$$\frac{d\mu_u}{d\mu_v}(z) = \det\left(I + \Lambda(\partial G)^{-1} F\right) \times$$

$$\times \exp\left\{ \int_0^1 \int_G \lambda\Big(\sum_{|\alpha| \leqslant 2k} a_\alpha(x) D^\alpha z \; + \; t g(x,z) \Big) \, g(x,z) \, dx \; dt \right\}.$$

Remark. The main condition in this theorem is that the energetic inequalities should hold. This question is well investigated in the theory of differential operators. It is known, e.g., that the energetic inequalities are always satisfied for a differential operator with constant coefficients (see (Berezanskii, 1965)). If we want to get an analogous result for semistrong solutions then it is necessary to impose additional boundary conditions and additional conditions on configuration of the domain D. Leaving the details aside, we remark only that the Dirichlet problem

$$Lu = \xi, \qquad u\,|_{\partial G} = 0$$

for the operator of the second order with constant coefficients

$$L\,u = \sum_{j=1}^{n} c_j D_j^2 u + \sum_{j=1}^{n} b_j D_j u + b\,u \tag{26}$$

satisfies all the conditions for domains of the class $A^{(1)}$ which do not contain parts of characteristics of (26).

4. MODELS WITH GENERALIZED NOISE

Consider the triple of Hilbert spaces

$$W_2^p(G) \subset \mathcal{L}_2(G) \subset W_2^{-p}(G),$$

where G is a domain of the class $A^{(1)}$. Consider the boundary value problem

$$\sum_{|\alpha| \leqslant p} a_\alpha(x) D^\alpha u + g(x, u(x)) = \xi(x), \qquad u \in \overline{W}_2^p(\partial G), \tag{27}$$

and

$$\sum_{|\alpha|\leqslant p} a_\alpha(x)D^\alpha v = \xi(x), \qquad v \in \overline{W}_2^p(\partial G). \tag{28}$$

Let the following conditions be realized:

i) for the differential expression

$$Lu = \sum_{|\alpha|\leqslant p} a_\alpha(x)D^\alpha u, \qquad a_\alpha(x) \in C^{|\alpha|}(D \cup \partial D),$$

the energetic inequalities hold:

$$\|Lu\|_{\mathcal{L}_2(G)} \geqslant c\|u\|_{\mathcal{L}_2(G)}, \quad \|L'v\|_{\mathcal{L}_2(G)} \geqslant c\|v\|_{\mathcal{L}_2(G)},$$

where $c > 0$, $u, v \in C_0^\infty(G)$ and the minimal and maximal operators Λ and $\overline{\Lambda}$ have inverses which are Hilbert–Schmidt operators in $\mathcal{L}_2(G)$;

ii) $g(x,u)$: $G\times\mathcal{L}_2(G) \to \mathcal{L}_2(G)$ and, for any u, there exist generalized in Sobolev sense derivatives of order $2m$, and $F(u) = \partial g/\partial u$, the contraction of which leaves $W_2^p(G)$-invariant;

iii) $\xi = \xi(x)$ is a random element in $W_2^{-p}(G)$, satisfying for any smooth function $\Phi \in C'(W_2^{-p}(G))$ and any $h \in W_2^p(G)$, the equality

$$\mathbf{E}\int_G [\Phi'(\xi)](x)h(x)\,dx = \mathbf{E}\Phi(\xi)\int_G [\lambda(\xi)](x)h(x)\,dx,$$

where $\lambda: \ W_2^{-p}(G) \to W_2^{-p}(G)$.

If condition i) holds for the operator L, there exists a solvable extension Λ_L for which

$$\Lambda \subseteq \Lambda_L \subseteq \overline{\Lambda},$$

and thus there exists Λ_L^{-1} which is a Hilbert–Schmidt operator. In this case Λ_L^{-1} can be extended to $W_2^{-p}(G)$ and so Λ_L can also be extended to $\mathcal{L}_2(G)$. Denote the extension by $\widehat{L}$, so that

$$\widehat{A}: \ \mathcal{L}_2(G) \to W_2^{-p}(G), \qquad \widehat{A}^{-1}: \ W_2^{-p}(G) \to \mathcal{L}_2(G).$$

Equations (27) and (28) are called stochastic differential equations and we consider them as an equivalent form for the following "integral" equations

$$u(x) + \widehat{A}^{-1}g(x,u(x)) = \widehat{A}^{-1}\xi(x),$$

$$v(x) = \widehat{A}^{-1}\xi(x).$$

THEOREM 5. *Let, for the boundary value problems* (27) *and* (28), *conditions* i), ii) *and* iii) *hold. Then the measures* μ_u *and* μ_v *are equivalent in* $\mathcal{L}_2(G)$ *and*

$$\frac{d\mu_u}{d\mu_v}(z) = \det\left(I + L^{-1}F(z)\right)\times$$

$$\times \exp\left\{ \int_G \int_0^1 \lambda(\widehat{A}z + tg(x, z(x)))g(x, z(x))dt\, dx \right\};$$

if, moreover, $\xi(x)$ *is the Gaussian "white noise", then*

$$\frac{d\mu_u}{d\mu_v}(z) = \det\left(I + L^{-1}F(z)\right)\times$$

$$\times \exp\left\{ -\int_G g\big(x, z(x)\big)\widehat{A}\, z(x)\, dx - \frac{1}{2}\int_G g^2\big(x, z(x)\big)\, dx \right\}.$$

REFERENCES

Berezanskii, Yu.M. (1965). *Expansion of Self-Adjoint Operators with Respect to Eigenfunctions.* Naukova Dumka, Kiev (in Russian).

Gohberg, I.C. and Krein, M.G. (1969). *Introduction to the Theory of Linear Nonselfadjoint Operators.* Amer. Math. Soc., Providence.

Daletsskii, Yu.L. and Belopolskaya, Ya.I. (1989). *Stochastic Equations and Differential Geometry.* Visha Shkola, Kiev (in Russian).

Daletskii, Yu.L. and Sokhadze, G.A. (1988). Absolute continuity of smooth measures. *Funct. Anal.Appl.* **22**, 77–78 (in Russian).

Miranda, C. (1955). *Equazioni alle derivate parziali di tipo ellittico.* Springer, Berlin etc.

Sokhadze, G.A. (1989). *Absolute continuity of distributions of solutions of equations with random perturbation.* In: *Statistics and Control of Random Processes,* Nauka, Moscow, pp. 195–198 (in Russian).

Part Six

MATHEMATICAL STATISTICS

New Trends in Probab. and Statist., pp. 553 - 566
V. Sazonov and T. Shervashidze (Eds.)
1991 VSP/Mokslas

A NEW METHOD OF ASYMPTOTIC ANALYSIS OF SIMPLE LINEAR RANK STATISTICS

D.M. CHIBISOV
Steklov Mathematical Institute, Vavilova 42, 117966 Moscow, GSP-1, USSR

ABSTRACT

A new proof of Hájek's theorem on asymptotic normality of simple linear rank statistics (SLRS's) under H_0 is presented. Its main step is a conditioning which makes the rank statistic a sum of conditionally independent SLRS's so that the technique of characteristic functions is applicable. The purpose of the publication is to demonstrate in this setup the method of dealing with SLRS's which was motivated by a problem of asymptotic optimality of the chi-square test within the class of symmetric tests when the number of outcomes grows to infinity. We also discuss this problem and its relation to SLRS's.

1. ASYMPTOTIC OPTIMALITY OF THE CHI-SQUARE TEST

In this section we discuss the problem of asymptotic optimality of the chi-square test within the class of tests symmetric in frequencies for testing the equiprobability of the outcomes in a multinomial distribution as the number of outcomes grows to infinity. We show how the SLRS's arise in this problem and state a result we expect to prove.

The method of dealing with SLRS's employed in the proof is demonstrated by the proof of the theorem on asymptotic normality of SLRS's given in the subsequent sections.

Consider the problem of testing the equiprobability hypothesis

$$H_0\colon p_i \doteq p_{0i} = N^{-1}, \quad i = 1, \dots, N, \tag{1.1}$$

given the vector $\nu = (\nu_1, \dots, \nu_N)$, $\sum \nu_i = n$, of frequencies observed in n independent trials with N outcomes having probabilities $p_1, \dots, p_N$. A common test for this problem is the chi-square test which rejects H_0 for large values of the statistic

$$X^2_{N,n} = \sum \frac{(\nu_i - n/N)^2}{n/N}.$$

(Throughout the paper, i runs over $1, \dots, N$ unless otherwise stated.) The question is whether there are better tests and if so how much better they can be. Of course, there are more powerful tests against specific alternatives. Suppose, however, that the set of alternatives is symmetric with respect to the orderings of the outcomes. Then it is natural to restrict oneself to the class of tests which are symmetric (in frequencies), i.e., tests whose critical functions $\varphi(\nu)$ satisfy

$$\varphi(\nu) = \varphi(r\nu), \quad r\nu = (\nu_{r_1}, \dots, \nu_{r_N}), \tag{1.2}$$

for any permutation $r = (r_1, \dots, r_N)$ of $(1, \dots, N)$.
The problem we are interested in is to prove that the sequence of the chi-square tests is asymptotically optimal within this class as both n and N tend to infinity (perhaps with some limitations on the growth of N relative to n). More precisely, consider a sequence of alternatives

$$H_1\colon p_i = p_{1i} = N^{-1}\left(1 + tn^{-1/2}a_i\right), \ t > 0, \quad i = 1, \dots, N, \tag{1.3}$$

where a_i depend on n and N and are normalized by

$$\sum a_i = 0, \quad \sum a_i^2 = N. \tag{1.4}$$

Denote the power of the chi-square test of size $\alpha > 0$ (to be fixed for varying n, N) by $\beta_{N,n}(t) = \beta_{N,n}(t; a_1, \dots, a_N)$.
We are going to prove that for any sequence of level α symmetric tests and for any sequence of a's satisfying some uniform asymptotic negligibility condition, the chi-square test is no less powerful, i.e.,

$$\limsup_{n,N\to\infty} \sup_{t>0} \left[\beta'_{N,n}(t) - \beta_{N,n}(t)\right] \leqslant 0,$$

where $\beta'_{N,n}(t)$ is the power of the competing test.
Note that there are tests (based on the knowledge of H_1) in the unrestricted class of tests whose power is nontrivial (bounded off from α and 1) for bounded $t > 0$. Unlike this, the chi-square test requires t to grow as $N^{1/4}$ in order to have a nontrivial power. Namely,

$$\beta_{N,n}\left(sN^{1/4}\right) \longrightarrow \Phi\left(2^{-1/2}s^2 - u_\alpha\right), \quad s > 0, \tag{1.5}$$

where Φ denotes the $\mathcal{N}(0,1)$ (standard normal) distribution function (d.f.) and $u_\alpha = \Phi^{-1}(1-\alpha)$ (see, e.g., (Tumanyan, 1982; Ivchenko and Medvedev, 1978)).
One can "obtain" (1.5) by first taking the limit as $n \to \infty$ so that $X^2_{N,n}$ becomes distributed as χ^2_{N-1} under H_0 and as $\chi^2_{N-1}(t^2)$ (t^2 is the noncentrality parameter) under H_1, and then letting $N \to \infty$ to use the asymptotic normality of χ^2_{N-1}.
This argument, though incorrect, provides the correct formula (1.5) which explains the order of magnitude of deviations from H_0 distinguishable by χ^2-tests.
There are results on asymptotic optimality of χ^2 within some specific classes of tests. For example, let $\beta_{N,n}(t; f)$ denote the power against H_1 of a size α test based on $\sum f(\nu_i)$. Then, under some conditions on f and $\{a_i\}$ in (1.3),

$$\beta_{N,n}\left(sN^{1/4}; f\right) \longrightarrow \Phi\left(2^{-1/2}c(f)s^2 - u_\alpha\right)$$

and $c(f)$ is maximized by $f(x) = x^2$ which corresponds to the chi-square test (see (Ivchenko and Medvedev, 1978)) for the case $n \asymp N$).

Here the assumed structure of the test allows one to explicitly describe its performance and the problem thus reduces to the maximization of certain functionals. Unlike this we have to consider the likelihood ratio (LR) of the distributions under H_1 and H_0 restricted to the field of permutation invariant sets and to show that asymptotically it is a function of $X^2_{N,n}$. The LR of the distributions themselves is, obviously,

$$h(\nu) = \prod \left(p_{1i}/p_{0i}\right)^{\nu_i}, \tag{1.6}$$

where p_{1i} are given by (1.3) and p_{0i} by (1.1) (in fact, $\prod p_{0i}^{\nu_i} = N^{-n}$). Of course, $h(\nu)$ depends on n, N, t and $\{a_i\}$. The LR we need, $\bar{h}(\nu)$, say, is obtained by symmetrization of $h(\nu)$, i.e.,

$$\bar{h}(\nu) = (N!)^{-1} \sum_r h(r\nu). \tag{1.7}$$

Another way to obtain (1.7) is to consider the power, $\beta(\varphi)$, say, of a test φ satisfying (1.2),

$$\beta(\varphi) = N^{-n} \sum_\nu \varphi(\nu) h(\nu).$$

Since the domain of ν is permutation invariant, one has

$$\beta(\varphi) = N^{-n} \sum_\nu \varphi(r\nu) h(r\nu) = N^{-n} \sum_\nu \varphi(\nu) h(r\nu),$$

for any r, hence,

$$\beta(\varphi) = N^{-n} \sum_\nu \varphi(\nu) \bar{h}(\nu).$$

This shows that the most powerful symmetric test is obtained by the Neyman–Pearson lemma with $\bar{h}(\nu)$ as the LR.
Let $R = (R_1, \ldots, R_N)$ be a random permutation taking on any of the $N!$ permutations with equal probabilities. Then $\bar{h}(\nu) = \mathbf{E}_R h(R\nu)$, where $\mathbf{E}_R$ denotes the corresponding expectation.
By (1.6), (1.1) and (1.3), this becomes

$$\bar{h}(\nu) = \mathbf{E}_R \prod \left(1 + tn^{-1/2} a_i\right)^{\nu_{R_i}}.$$

The product under $\mathbf{E}_R$ may be treated by taking its logarithm and expanding it by the Taylor formula. Taking the first two terms of the expansion one obtains

$$\bar{h}(\nu) = \mathbf{E}_R \exp\left(S_N\right) = \mathbf{E}_R \exp\left[tS_{N1} - \frac{1}{2}t^2 S_{N2} + \ldots\right], \tag{1.8}$$

where

$$S_N = \sum \nu_{R_i} \log\left(1 + tn^{-1/2} a_i\right),$$

$$S_{N1} = n^{-1/2} \sum \nu_{R_i} a_i, \quad S_{N2} = n^{-1} \sum \nu_{R_i} a_i^2.$$

One obtains from the asymptotic theory of SLRS's (see, e.g., (Hájek and Šidak, 1967)) that under some uniform asymptotic negligibility conditions on $\{a_i\}$

$$S_{N1} \xrightarrow{d} \mathcal{N}(0,1) \quad \text{and} \quad S_{N2} \xrightarrow{\mathbf{P}} 1 \tag{1.9}$$

(*with respect to the permutation distribution) for the "largest part" of ν's, i.e.,* on a set of probability tending to 1 with respect to the distribution of ν under H_0. One can use (1.9) to show that symmetric tests are asymptotically insensitive to alternatives (1.3) with bounded t. Indeed, for the largest part of ν's, $S_N \xrightarrow{d}$ $\xrightarrow{d} tZ - t^2/2$ with Z a normal $\mathcal{N}(0,1)$ random variable, hence

$$\bar{h}(\nu) \longrightarrow \mathbf{E} \exp\left[tZ - t^2/2\right] = 1.$$

A similar argument was used in (Gvanceladze and Chibisov, 1979), where it was shown that goodness-of-fit tests symmetric in frequencies of N cell intervals equiprobable under H_0 cannot asymptotically distinguish contiguous alternatives if $N \to \infty$ as $n \to \infty$.

However the passage to the limit of S_N under the $\mathbf{E}_R$ sign cannot be used when t grows to infinity. To overcome this difficulty we introduce a conditioning which makes S_N a sum of conditionally independent SLRS's so that the expectation in (1.8) becomes conditionally a product of the corresponding expectations and may be treated by the technique commonly used for characteristic functions (ch.f.). The result which we expect to obtain by this method can be formulated in the following way:

Assume that $N^{-1} \sum a_i^4 \leqslant A < \infty$. Then, under H_1 as in (1.3) with $t = sN^{1/4}$ and $\{a_i\}$ satisfying (1.4), the convergence

$$\mathbf{E}\left|\bar{h}(\nu) - \exp\left[\frac{s^2(X_{N,n}^2 - N)}{2N^{1/2}} - \frac{s^4}{4}\right]\right| \to 0$$

holds as $n, N \to \infty$ (subject possibly to some limitation on the growth of the ratio N/n).

Although some essential parts of the proof have been carried out before it has not been accomplished is full by now. We shall demonstrate the method of conditioning in a simpler setup of asymptotic normality of SLRS's.

2. THEOREM ON ASYMPTOTIC NORMALITY OF SLRS

Let be given two sequences

$$\mathbf{a}_N = (a_{N1}, \ldots, a_{NN}), \quad \mathbf{c}_N = (c_{N1}, \ldots, c_{NN}) \tag{2.1}$$

of real vectors and let $\mathbf{R}_N = (R_{N1}, \ldots, R_{NN})$ be a random (equiprobable) permutation of $1, \ldots, N$. For notational simplicity, the additional subscript N will often be suppressed. Consider the statistic

$$S_N = \sum a_{R_i} c_i. \tag{2.2}$$

Recall that (see (Hájek and Šidak, 1967, Chapter 2, Section 3.1))

$$\mathbf{E}S_N = N^{-1}\sum a_i \sum c_i = N\bar{a}\bar{c}, \tag{2.3}$$

$$\operatorname{var} S_N = (N-1)^{-1}\sum(a_i-\bar{a})^2\sum(c_i-\bar{c})^2 =$$

$$= \left[N^2/(N-1)\right](A_2-\bar{a}^2)(C_2-\bar{c}^2), \tag{2.4}$$

where $\bar{a} = A_1$, $\bar{c} = C_1$, and

$$A_m = N^{-1}\sum a_i^m, \quad C_m = N^{-1}\sum c_i^m, \qquad m = 1, 2, \dots \tag{2.5}$$

Note that

$$\frac{S_N - \mathbf{E}S_N}{(\operatorname{var} S_N)^{1/2}} = \frac{(N-1)^{1/2}}{N}\sum a'_{R_i} c'_i \tag{2.6}$$

with constants

$$a'_i = \frac{a_i - \bar{a}}{(A_2 - \bar{a}^2)^{1/2}}, \quad c'_i = \frac{c_i - \bar{c}}{(C_2 - \bar{c}^2)^{1/2}}$$

which satisfy $\sum a'_i = \sum c'_i = 0$, $\sum(a'_i)^2 = \sum(c'_i)^2 = N$. Therefore, in order to prove the asymptotic normality of S_N normalized as in (2.6), we may prove that

$$N^{-1/2} S_N \xrightarrow{d} \mathcal{N}(0,1), \tag{2.7}$$

assuming from the outset that

$$\bar{a} = \bar{c} = 0, \quad A_2 = C_2 = 1. \tag{2.8}$$

Moreover, we shall impose on a's and c's the following conditions. Let

$$\rho_N^2(u) = N^{-1}\sum_{|a_i|>u} a_i^2. \tag{2.9}$$

Condition (i). For any sequence $u_N \to \infty$,

$$\rho_N^2(u_N) \longrightarrow 0 \quad \text{as} \quad N \to \infty.$$

Condition (ii). Noether's condition on $\{\mathbf{c}_N\}$, which in the case of (2.8) is

$$\delta_N := N^{-1/2}\max_{1\leqslant i\leqslant N}|c_i| \longrightarrow 0 \quad \text{as} \quad N \to \infty. \tag{2.10}$$

Condition (i) is equivalent to condition (4.10) in (Hájek, 1961) where it is shown that this is a necessary and sufficient condition on $\{\mathbf{a}_N\}$ for (2.7) to hold for any sequence $\{\mathbf{c}_N\}$ satisfying condition (ii).
Thus we shall prove the following

THEOREM. *Let sequences* (2.1) *satisfy* (2.8) *and conditions* (i) *and* (ii). *Then for S_N as in* (2.2), *the asymptotic normality* (2.7) *holds.*

Note that no scores generating function will be introduced in the proof.

3. PROOF

We start with the following truncation argument. Take $\{u_N\}$ in the condition (i) such that

$$u_N \longrightarrow \infty \quad \text{and} \quad u_N \delta_N \longrightarrow 0 \quad \text{as} \quad N \to \infty \tag{3.1}$$

(see (2.10)). We shall write $\rho_N^2 = \rho_N^2(u_N)$ (see (2.9)) for this particular sequence. Let

$$a_{i1} = a_i \mathbf{1}(|a_i| \leqslant u_N), \quad a_{i2} = a_i \mathbf{1}(|a_i| > u_N),$$

where $\mathbf{1}(\ldots)$ denotes the indicator function. Then $S_N = S_{N1} + S_{N2}$ with

$$S_{Nj} = \sum a_{R_i j} c_i, \quad j = 1, 2.$$

Among the two summands, S_{N2} is negligible because $\mathbf{E} S_{N2} = 0$ (since $\bar{c} = 0$) and, by condition (i),

$$\operatorname{var}(N^{-1/2} S_{N2}) \leqslant (N-1)^{-1} \sum a_{i2}^2 \leqslant (N-1)^{-1} N \rho_N^2 \longrightarrow 0.$$

Thus we have to prove (2.7) for S_{N1}.
One has

$$|\bar{a}_1| = N^{-1} \Big| \sum a_{i2} \Big| \leqslant \rho_N \longrightarrow 0, \quad N^{-1} \sum a_{i1}^2 = 1 - \rho_N^2 \longrightarrow 1.$$

Hence after renormalizing $\{a_{i1}\}$ to satisfy (2.8), the problem reduces to the proof of Theorem with condition (i) replaced by

Condition (i'). There exists a sequence $\{u_N\}$ satisfying (3.1) such that

$$\max_i |a_i| \leqslant u_N. \tag{3.2}$$

Consider the ch.f. of $N^{-1/2} S_N$:

$$f_N(s) = \mathbf{E} \exp\left[isN^{-1/2} S_N\right]. \tag{3.3}$$

We are to prove that, for any $s \in \mathbf{R}^1$,

$$f_N(s) \longrightarrow \exp\left[-s^2/2\right]. \tag{3.4}$$

It will be convenient for us to have $\{c_i\}$ in (2.2) in a random order too. Thus we introduce another random permutation $\mathbf{Q}_N = (Q_1, \ldots, Q_N)$ of $1, \ldots, N$ independent of $\mathbf{R}_N$ and redefine S_N as

$$S_N = \sum a_{R_i} c_{Q_i},$$

which obviously does not change its distribution.
Take a sequence $\{k_N\}$ of integers such that $k_N \to \infty$ as $k_N/N \to 0$, and define integers l_N and d_N by $N = k_N l_N + d_N$, $0 \leqslant d_N < l_N$. Then $l_N \to \infty$, $l_N/N \to 0$ and $k_N l_N/N \to 1$. In the sequel, the subscript N at k, l, d will be omitted.
We split the set of indices $\{1, \ldots, N\}$ into the consecutive "blocks" $I_1, \ldots, I_k$ of size l and (possibly empty) I_0 of size d, i.e.,

$$I_j = \{(j-1)l + 1, \ldots, jl\}, \quad j = 1, \ldots, k,$$
$$I_0 = \{kl + 1, \ldots, N\}.$$

For the random permutation $\mathbf{R}$ (respectively, $\mathbf{Q}$), denote by $\mathcal{R}_1, \ldots, \mathcal{R}_k, \mathcal{R}_0$ (respectively, $\mathcal{Q}_1, \ldots, \mathcal{Q}_k, \mathcal{Q}_0$) the "contents" of the respective blocks, i.e., the (unordered) sets

$$\mathcal{R}_j = \{R_i,\ i \in I_j\}, \quad \mathcal{Q}_j = \{Q_i,\ i \in I_j\}.$$

The ordered collection of the sets $(\mathcal{R}_j)_{j=0}^k$ will be denoted by $\mathcal{R}$ (respectively, $\mathcal{Q} = (\mathcal{Q}_j)_{j=0}^k$). We shall treat the ch.f. (3.3) by first taking the conditional expectation given $\mathcal{R}$ and $\mathcal{Q}$ which we shall denote, for brevity, by $\mathbf{E}_1$. Then conditionally the randomness of S_N is only generated by the random permutations of a's and c's within each block. Write $S_N = \sum S_{Nj}$, where

$$S_{Nj} = \sum_{i \in I_j} a_{R_i} c_{Q_i}, \quad j = 0, 1, \ldots, k \tag{3.5}$$

($S_{N0} = 0$ whenever $d = 0$). Then these S_{Nj} are conditionally independent SLRS's. Therefore, denoting

$$f_N\big(s \big| \mathcal{R}, \mathcal{Q}\big) = \mathbf{E}_1 \exp\left[isN^{-1/2} S_N\right]$$

and, for $j = 0, 1, \ldots, k$,

$$f_{Nj}\big(s \big| \mathcal{R}, \mathcal{Q}\big) = \mathbf{E}_1 \exp\left[isN^{-1/2} S_{Nj}\right], \tag{3.6}$$

we have

$$f_N\big(s \big| \mathcal{R}, \mathcal{Q}\big) = \prod_{j=0}^{k} f_{Nj}\big(s \big| \mathcal{R}, \mathcal{Q}\big). \tag{3.7}$$

Since $f_N(s) = \mathbf{E} f_N(s \big| \mathcal{R}, \mathcal{Q})$ and $\big|f_N(s \big| \mathcal{R}, \mathcal{Q})\big| \leqslant 1$, it is sufficient for (3.4) to prove that, for any $s \in \mathbf{R}^1$,

$$f_N\big(s \big| \mathcal{R}, \mathcal{Q}\big) \xrightarrow{\mathbf{P}} \exp\left[-s^2/2\right], \tag{3.8}$$

where $\xrightarrow{\mathbf{P}}$ denotes the convergence in probability.
Let, for $j = 0, 1, \ldots, k$,

$$\alpha_j = \mathbf{E}_1 S_{Nj}, \quad \beta_{rj} = \mathbf{E}_1 \widetilde{S}_{Nj}^r, \quad r = 2, 3, \ldots, \tag{3.9}$$

where $\widetilde{S}_{Nj} = S_{Nj} - \alpha_j$. Let also

$$\tilde{f}_{Nj}(s) = \mathbf{E}_1 \exp\left[isN^{-1/2}\widetilde{S}_{Nj}\right], \quad j = 0, 1, \ldots, k.$$

Note that α_j, β_{rj} and $\tilde{f}_{Nj}(s)$ depend on $\mathcal{R}, \mathcal{Q}$ which dependence we suppress for notational simplicity. We shall also suppress the dependence of f_{Nj} as in (3.6) on $\mathcal{R}, \mathcal{Q}$. Thus, in particular,

$$f_{Nj}(s) = \exp\left[isN^{-1/2}\alpha_j\right]\tilde{f}_{Nj}(s). \tag{3.10}$$

The summations with respect to j within this section will be over $0, \ldots, k$. First we shall show that, in order to obtain (3.8), it suffices to prove that

$$N^{-1/2}\sum \alpha_j \xrightarrow{\mathbf{P}} 0, \tag{3.11}$$

$$N^{-1} \max \beta_{2j} \xrightarrow{\mathbf{P}} 0, \tag{3.12}$$

$$\sum \zeta_{N_j}(s) \xrightarrow{\mathbf{P}} s^2/2 \tag{3.13}$$

and

$$\sum \left|\zeta_{N_j}(s)\right|^2 \xrightarrow{\mathbf{P}} 0, \tag{3.14}$$

where $\zeta_{N_j}(s) = 1 - \tilde{f}_{Nj}(s)$. Indeed, by (3.7), (3.10) and (3.11), one has to prove that

$$\sum \log \tilde{f}_{Nj}(s) \xrightarrow{\mathbf{P}} -s^2/2. \tag{3.15}$$

Now we use the fact that

$$\left|\zeta_{N_j}(s)\right| \leqslant s^2 N^{-1}\beta_{2j}/2, \tag{3.16}$$

which follows from the definition of ζ_{N_j} and the inequality

$$\left|e^{ix} - (1 + ix)\right| \leqslant x^2/2, \quad x \in \mathbf{R}^1$$

(see, e.g., (Feller, 1971, Chapter 15, § 4, Lemma 1)). Then (3.12) implies that, for an arbitrary $0 < \varepsilon < 1$ and a fixed $s \in \mathbf{R}^1$,

$$\mathbf{P}\left\{\max_j \left|\zeta_{N_j}(s)\right| \leqslant \varepsilon\right\} \longrightarrow 1. \tag{3.17}$$

By the Taylor formula, $\left|\log(1-\zeta)+\zeta\right| \leqslant \left|\zeta\right|^2/2(1-\varepsilon)^2$ if $|\zeta| \leqslant \varepsilon$. Hence, whenever the event in (3.17) occurs, one has

$$\left|\sum \log \tilde{f}_{Nj}(s) + \sum \zeta_{N_j}(s)\right| \leqslant \sum \left|\zeta_{N_j}(s)\right|^2/2(1-\varepsilon)^2,$$

thus, (3.15) follows from (3.13) and (3.14).

Next we state sufficient conditions for (3.13) and (3.14) in terms of moments (3.9). By the inequality

$$\left|e^{ix} - \left(1 + ix - \frac{x^2}{2} + \frac{(ix)^3}{6}\right)\right| \leqslant \frac{x^4}{24}, \quad x \in \mathbf{R}^1,$$

(see the same lemma in (Feller, 1971)), one has

$$\left|\zeta_{Nj}(s) - \frac{s^2}{2N}\beta_{2j} + \frac{(is)^3}{6N^{3/2}}\beta_{3j}\right| \leqslant \frac{s^4}{24N^2}\beta_{4j}.$$

Therefore, it suffices to prove that

$$N^{-1}\sum \beta_{2j} \xrightarrow{\mathbf{P}} 1, \tag{3.18}$$

$$N^{-3/2}\sum \beta_{3j} \xrightarrow{\mathbf{P}} 0 \tag{3.19}$$

and

$$N^{-2}\sum \beta_{4j} \xrightarrow{\mathbf{P}} 0, \tag{3.20}$$

to obtain (3.13). Moreover, it follows from (3.18) and (3.12) that $N^{-2}\sum \beta_{2j}^2 \xrightarrow{\mathbf{P}} 0$, which in view of (3.16) implies (3.14).
Thus, in order to obtain the Theorem, we are to prove (3.11), (3.12), (3.18), (3.19) and (3.20).

4. PROOF OF (3.11)

Let, for $i = 1, \ldots, N$ and $j = 0, 1, \ldots, k$,

$$\delta_{ij} = \begin{cases} 1, & i \in I_j, \\ 0, & \text{otherwise.} \end{cases} \tag{4.1}$$

Let, for $r = 1, 2, \ldots$,

$$A_{rj} = l^{-1}\sum a_{R_i}^r \delta_{ij}, \quad j = 1, \ldots, k, \tag{4.2}$$

$$A_{r0} = \begin{cases} d^{-1}\sum a_{R_i}^r \delta_{i0} & \text{if } d \geqslant 1, \\ 0 & \text{if } d = 0, \end{cases} \tag{4.3}$$

define C_{rj} in terms of $c_{Q_i}^r$ in a similar way. Note that A_{rj} are independent of C_{rj} since they are determined by independent random permutations $\mathbf{R}$ and $\mathbf{Q}$. By (2.3) applied to S_{Nj}, one has (see (3.9))

$$\alpha_0 = dA_{10}C_{10}, \quad \alpha_j = lA_{1j}C_{1j}, \quad j = 1, \ldots, k. \tag{4.4}$$

It is seen from (4.2), (4.3) that A_{rj} and C_{rj} are SLRS's of two–sample form. By (2.3) applied to (4.2), (4.3), for $r = 1$,

$$\mathbf{E}A_{1j} = \mathbf{E}C_{1j} = 0, \quad j = 0, 1, \ldots, k,$$

hence, by the independence of A_{1j} and C_{1j}, $\mathbf{E}\alpha_j = 0$, $j = 0, 1, \dots, k$. Consider separately the term α_0 to show that

$$N^{-1/2}\alpha_0 \xrightarrow{\mathbf{P}} 0. \tag{4.5}$$

By (2.4), one has from (2.8) that (for $d \geqslant 1$)

$$\mathbf{E}A_{10}^2 = \mathbf{E}C_{10}^2 \leqslant \frac{1}{d^2(N-1)} \sum a_i^2 \sum \delta_{i0}^2 = \frac{N}{d(N-1)}. \tag{4.6}$$

Hence (see (4.4)), $N^{-1}\mathbf{E}\alpha_0^2 \leqslant N/(N-1)^2 \to 0$, which implies (4.5).
In what follows the summations with respect to j will be over $1, \dots, k$.
In view of (4.4), we are to prove that

$$N^{-1/2} l \sum A_{1j} C_{1j} \xrightarrow{\mathbf{P}} 0. \tag{4.7}$$

We shall estimate the second moment of the left–hand side of (4.7) by first conditioning on the unordered collections $\{\mathcal{R}_1, \dots, \mathcal{R}_k\}$ and $\{\mathcal{Q}_1, \dots, \mathcal{Q}_k\}$. Under this condition, the sets of values $\{A_{11}, \dots, A_{1k}\}$ and $\{C_{11}, \dots, C_{1k}\}$ are fixed, but these A_{1j} and C_{1j} occur in a random order. Thus,

$$W_1 = \sum A_{1j} C_{1j}$$

becomes again a SLRS. Denote by $\mathbf{E}_2$ the conditional expectation as described. Then, by (2.3) and (2.4),

$$\mathbf{E}_2 W_1 = k^{-1} \sum A_{1j} \sum C_{1j},$$

$$\mathrm{var}_2 W_1 \leqslant (k-1)^{-1} \sum A_{1j}^2 \sum C_{1j}^2.$$

Now (4.7) will follow if we prove that

$$(l^2/Nk^2)\mathbf{E}\Big(\sum A_{1j}\Big)^2 \mathbf{E}\Big(\sum C_{1j}\Big)^2 \longrightarrow 0 \tag{4.8}$$

and

$$(l^2/N(k-1))\mathbf{E}\sum A_{1j}^2\, \mathbf{E}\sum C_{1j}^2 \longrightarrow 0. \tag{4.9}$$

By (2.8), (4.2) and (4.3), $\sum A_{1j} = -(d/l)A_{10}$; the same holds for C_{1j} and (4.8) follows from (4.6).
Similarly as in (4.6), we have, for $j = 1, \dots, k$,

$$\mathbf{E}A_{1j}^2 = \mathbf{E}C_{1j}^2 \leqslant N/l(N-1), \tag{4.10}$$

hence, the left–hand side of (4.9) is $O(k/N) \to 0$. This completes the proof of (3.11).

5. PROOF OF (3.12)

To prove (3.12) we shall show that

$$\mathbf{E}\beta_{2j}^2 = O(A_4C_4) + O(l^2), \quad j = 0, 1, \ldots, k \tag{5.1}$$

(see (2.5)). One has, by (2.8), (3.2) and (2.10),

$$A_4 \leqslant u_N^2, \quad C_4 \leqslant N\delta_N^2, \tag{5.2}$$

whence, in view of (3.1),

$$A_4C_4 = o(N). \tag{5.3}$$

Hence, using (5.1) one has, by the Chebyshev inequality,

$$\mathbf{P}\Big\{N^{-1} \max_{0\leqslant j\leqslant k} \beta_{2j} > \varepsilon\Big\} \leqslant \sum_{j=0}^{k} \mathbf{P}\Big\{\beta_{2j} > N\varepsilon\Big\} \to 0,$$

which proves (3.12).

We shall prove (5.1) for $j = 1, \ldots, k$; the case $j = 0$ is treated similarly. By (2.4) applied to $\widetilde{S}_{Nj}$,

$$\beta_{2j} = \frac{l^2}{l-1}(A_{2j} - A_{1j}^2)(C_{2j} - C_{1j}^2) \leqslant \frac{l^2 A_{2j}C_{2j}}{l-1} \tag{5.4}$$

(see (4.2)). By (2.3), (2.8) and (4.1),

$$\mathbf{E}A_{2j} = (lN)^{-1} \sum a_i^2 \sum_i \delta_{ij} = 1,$$

and, by (2.4), (2.5) and (4.1),

$$\operatorname{var} A_{2j} \leqslant l^{-2}(N-1)^{-1} \sum a_i^4 \sum \delta_{ij}^2 = O(l^{-1}A_4).$$

Similar relations hold for C_{2j}. Hence,

$$\mathbf{E}A_{2j}^2 = O(1 + l^{-1}A_4), \quad \mathbf{E}C_{2j}^2 = O(1 + l^{-1}C_4), \tag{5.5}$$

which, in view of (5.4), implies (5.1).

6. PROOF OF (3.18)

In view of (3.12) and (5.4), (3.18) is equivalent to

$$k^{-1}W_2 \xrightarrow{\mathbf{P}} 1, \tag{6.1}$$

where

$$W_2 = \sum (A_{2j} - A_{1j}^2)(C_{2j} - C_{1j}^2).$$

We shall show that

$$k^{-1}\mathbf{E}_2W_2 \xrightarrow{\mathbf{P}} 1 \quad \text{and} \quad k^{-2}\text{var}_2W_2 \xrightarrow{\mathbf{P}} 0, \tag{6.2}$$

where $\mathbf{E}_2$ and var_2 refer to the conditioning described in Section 4. It is easily seen that (6.2) implies (6.1) and, hence, (3.18).

By (2.3),

$$\mathbf{E}_2W_2 = k^{-1}\sum (A_{2j} - A_{1j})^2 \sum (C_{2j} - C_{1j})^2. \tag{6.3}$$

Note that, by (4.1)–(4.3) and (2.8), $\sum A_{2j} = l^{-1}(N - dA_{20})$ and similarly for $\sum C_{2j}$. Now A_{20} and C_{20} satisfy (5.5) with l replaced by d. Hence, in view of (5.2),

$$k^{-1}\sum A_{2j} = 1 + o_{\mathbf{P}}(1), \quad k^{-1}\sum C_{2j} = 1 + o_{\mathbf{P}}(1). \tag{6.4}$$

Next, (4.10) implies that

$$k^{-1}\sum A_{1j}^2 \xrightarrow{\mathbf{P}} 0, \quad k^{-1}\sum C_{1j}^2 \xrightarrow{\mathbf{P}} 0. \tag{6.5}$$

The first relation in (6.2), thus, follows from (6.3)–(6.5).

By virtue of (2.4),

$$\text{var}_2W_2 \leqslant (k-1)^{-1}\sum (A_{2j} - A_{1j}^2)^2 \sum (C_{2j} - C_{1j}^2)^2 \leqslant$$
$$\leqslant (k-1)^{-1}\sum A_{2j}^2 \sum C_{2j}^2.$$

Making use of (5.5) and (5.2), we obtain

$$k^{-2}\mathbf{E}\text{var}_2W_2 = O(k^{-1}) + o(l^{-1}),$$

which implies the second relation in (6.2). The proof of (3.18) is completed.

7. PROOF OF (3.19)

Using the formula for the third moment of a SLRS (see (Hájek and Šidak, 1967, Problems and Complements to Chapter II, No. 25)) and definitions (3.5), (3.9), (4.1)–(4.3), we obtain that, for $j = 1, \ldots, k$,

$$\beta_{3j} = \frac{l^3}{(l-1)(l-2)}(A_{3j} - 3A_{2j}A_{1j} + 2A_{1j}^3)(C_{3j} - 3C_{2j}C_{1j} + 2C_{1j}^3).$$

We shall consider only $\sum \beta_{3j}$ in (3.19), leaving to the reader to check that β_{30} is negligible.

Let (cf. (4.2))

$$A'_{3j} = l^{-1}\sum |a_{R_i}|^3 \delta_{ij}, \quad j = 1, \ldots, k.$$

Then, obviously, $|A_{3j}| \leqslant A'_{3j}$ and

$$|A_{1j}| \leqslant (A'_{3j})^{1/3}, \quad A_{2j} \leqslant (A'_{3j})^{2/3}, \quad j = 1, \ldots, k.$$

With a similar notation for C'_{3j}, we see that

$$|\beta_{3j}| \leqslant 36 l A'_{3j} C'_{3j} (1 + o(1)), \tag{7.1}$$

and (3.19) follows if we show that

$$k^{-1} N^{-1/2} W_3 \xrightarrow{\mathbf{P}} 0, \tag{7.2}$$

where $W_3 = \sum A'_{3j} C'_{3j}$.
In order to prove (7.2), it suffices to show that

$$k^{-1} N^{-1/2} \mathbf{E} W_3 \longrightarrow 0. \tag{7.3}$$

Applying (2.3) to A'_{3j} and C'_{3j}, $j = 1, \ldots, k$, one obtains

$$\mathbf{E} A'_{3j} = N^{-1} \sum |a_i|^3, \quad \mathbf{E} C'_{3j} = N^{-1} \sum |c_i|^3.$$

Hence, in view of (2.8), (2.10) and (3.2),

$$\mathbf{E} A'_{3j} \leqslant u_N, \quad \mathbf{E} C'_{3j} \leqslant N^{1/2} \delta_N.$$

Therefore, (7.3) follows from (3.1) which completes the proof of (3.19).

8. PROOF OF (3.20)

Using the formula for the fourth moment of a SLRS (see the same problem in (Hájek and Šidak, 1967)) and proceeding in a similar way as in the proof of (7.1), we arrive after some algebraic transformations at the inequality, for $j = 1, \ldots, k$,

$$\beta_{4j} \leqslant \left(121\, l A_{4j} C_{4j} + 3 l^2 A_{2j}^2 C_{2j}^2\right)\left(1 + o(1)\right)$$

(we again omit the case $j = 0$). Thus, we have to prove that

$$k^{-1} N^{-1} W_4 \xrightarrow{\mathbf{P}} 0 \quad \text{and} \quad k^{-2} W_{22} \xrightarrow{\mathbf{P}} 0, \tag{8.1}$$

where

$$W_4 = \sum A_{4j} C_{4j} \quad \text{and} \quad W_{22} = \sum A_{2j}^2 C_{2j}^2.$$

By means of (2.3) applied to A_{4j} and C_{4j} (see (4.2)), we obtain that $\mathbf{E} A_{4j} = A_4$ and $\mathbf{E} C_{4j} = C_4$. Hence, $\mathbf{E} W_4 = k A_4 C_4$ and, by (5.3), $k^{-1} N^{-1} \mathbf{E} W_4 \to 0$. Also, by (5.5) and (5.3), $k^{-2} \mathbf{E} W_{22} \to 0$. These relations imply (8.1) which completes the proof.

REFERENCES

Feller, W. (1971). *An Introduction to Probability Theory and Its Applications.* **2**, 2nd Ed., Wiley, New York.

Gvanceladze, L.G. and Chibisov, D.M. (1979). On tests of fit based on grouped data. In: *Contributions to Statistics. J. Hájek Memorial Volume.* Ed. by J. Jurečkova. Academia, Prague.

Hájek, J. (1961). Some extensions of the Wald–Wolfowitz–Noether theorem. *Ann. Math. Statist.* **32**, 506–523.

Hájek, J. and Šidak Z. (1967). *Theory of Rank Tests.* Academia, Prague.

Ivchenko, G.I. and Medvedev, Yu.I. (1978). Separable statistics and hypotheses testing. The case of small samples. *Theory Probab. Appl.* **23**, 764–775.

Tumanyan, S.H. (1982). On the power of the chi–square test for "close" alternatives. In: *Matematika,* Erevan Univ., Erevan, pp. 6–18 (in Russian).

New Trends in Probability and Statistics, pp. 567–596
V.V. Sazonov and T. Shervashidze (Eds.)
1991 VSP/Mokslas

ON STABLE M-ESTIMATORS IN THE PARTIAL LIKELIHOOD SCHEME

N.L. LAZRIEVA and T.A. TORONJADZE
Razmadze Mathematical Institute of the Georgian Academy of Sciences, Rukhadze str.1, 380093 Tbilisi, Georgia, USSR

ABSTRACT

The generalization of Cox' method of partial likelihood is obtained for a filtered statistical experiment. The concept of stability of estimators with respect to a wide class of alternatives is introduced. Stable M-estimators are constructed and it is shown that among stable M-estimators the partial likelihood estimator is the optimal one. Several examples are given.

1. INTRODUCTION

In this paper we generalize Cox' method of partial likelihood (Cox, 1975) to the case of a general statistical experiment with a filtration. Our generalization of Cox' method consists in representing the full likelihood in the form of product of two Dolean exponents, one of which can be considered as a partial likelihood, whereas by varying the second exponent (cofactor) we introduce a wide class of alternatives into consideration (see (3.1) or (3.1′)).
Our principal tasks here are to construct M-estimators which are stable (in the sense of Definition 2.1) with respect to the introduced class of alternatives, to study their asymptotic behaviour and to prove that among stable estimators the partial likelihood estimator (PLE) is the optimal one.
In Section 2 we prove basic Lemma 2.1 and Theorem 2.1 on the asymptotic behaviour of solutions of the estimational equation and M-estimators. The stability concept of estimators is also introduced in this Section (Definition 2.1).
Section 3 deals, properly speaking, with describing the partial likelihood scheme. Main attention is given to describing the class of alternatives (see (3.1) or (3.1′)) and to characterizing processes which are martingales simultaneously with respect to all alternatives (Lemmas 3.1 and 3.2).
The latter property is important in constructing stable estimators and makes it possible to consider M-estimators as stable when they are defined as solutions of estimational equations given by L-transformations (see (2.2)) of special form (see (3.4′)).
Section 4 contains the examples illustrating the partial likelihood scheme. Section 5 deals with the asymptotic efficiency of PLE. The partial likelihood method based on full likelihood factorization is introduced in (Cox, 1985). The partial likelihood

theory in the discrete time case is developed in (Wong, 1986), where the reader can also find a detailed list of relevant literature. The generalization of Cox' model for point processes is presented in (Dzhaparidze, 1986). Cox' method is extended in (Jacod, 1989), where the partial likelihood process (PLP) associated with the (nonunique) solution of some martingale problem is defined. Jacod also defines partial regularity and a partial information process, obtains the Rao-Cramer inequality and shows that the PLP is optimal for construction of estimators (in the asymptotic sense) among all quasi-likelihood processes admitting an integral representation.

In (Greenwood and Wefelmeyer, 1989a, 1989b) various schemes of realization of partial likelihood method are considered. In particular the scheme of partial likelihood, related to the component of two-dimensional semimartingale of general form is studied. Conditions, allowing the factorization of full likelihood are investigated and the concept of regularity of estimators is introduced, the LAN property, convolution theorem and the asymptotic variance bound are established for partial likelihood scheme (parameter to be estimated is functional one).

2. BASIC CONCEPTS. AUXILIARY RESULTS

2.1. *Specification of the model*

Let

$$\mathcal{E} = (\Omega,\ \mathcal{F},\ F,\ \mathbf{P},\ \{\mathbf{P}_\theta, \in \Theta\}), \quad \Theta \subset \mathbf{R}^d,\ d \geqslant 1,$$

be a general statistical experiment with filtration, where $(\Omega, \mathcal{F}, F, \mathbf{P})$ is a probability space with a filtration $F = (\mathcal{F}_t)_{t\geqslant 0}$ satisfying the usual conditions; $\{\mathbf{P}_\theta,\ \theta \in \Theta\}$ is a family of probability measures parametrized by $\theta \in \Theta$, Θ is an open subset of the Euclidean space $\mathbf{R}^d$. It is assumed for all $\theta \in \Theta$ that $\mathbf{P}_\theta \overset{\text{loc}}{\sim} \mathbf{P}$.

Let $\mathbf{P}(t) = \mathbf{P} \mid \mathcal{F}_t$, $\mathbf{P}_\theta(t) = \mathbf{P}_\theta \mid \mathcal{F}_t$ be the restrictions of the measures $\mathbf{P}$ and $\mathbf{P}_\theta$, respectively, on the σ-algebra $\mathcal{F}_t$ and let $\rho_\theta = (\rho_\theta(t))_{\geqslant 0}$ be the likelihood ratio process (we shall consider the right-continuous modification with the left limits)

$$\rho_\theta(t) = d\mathbf{P}_\theta(t)/d\mathbf{P}(t).$$

For simplicity we set $\rho_\theta(0) = 1$.

As it is known (Jacod, 1979), there exists a local $\mathbf{P}$-martingale $M_\theta = (M_\theta(t))_{t\geqslant 0}$ such that the process ρ_θ can be written as Dolean's exponent

$$\rho_\theta = \mathcal{E}(M_\theta) := \exp\left\{M_\theta - \frac{1}{2}\langle M_\theta^c\rangle\right\}\Pi(1+\Delta M_\theta)e^{-\Delta M_\theta},$$

where M^c is the continuous part, $\langle M\rangle$ is the square characteristic and ΔM is the jump of the martingale M.

2.2. *L-transformation*

Let Q be some other probability measure on $(\Omega,\ \mathcal{F},\ F)$ such that $Q \overset{\text{loc}}{\ll} P$ and $dQ/dP = \mathcal{E}(M)$. If m is a local $\mathbf{P}$-martingale, then by the general Girsanow theorem (Jacod, 1979) the process L (L-transformation)

$$L(m, M) := m - \langle m^c, M^c \rangle - \sum \frac{\Delta m \, \Delta M}{1 + \Delta M}$$

is a local Q-martingale.

2.3. *Regularity*

The experiment $\mathcal{E}$ is called regular if the conditions below are satisfied.

(1) For each $t \geqslant 0$ **P**-a.s. the function $\theta \mapsto M_\theta$ is differentiable and the derivative $\dot{M}_\theta := (\partial M_\theta / \partial \theta_1, \ldots, \partial M_\theta / \partial \theta_d)'$ is a local **P**-martingale for all $\theta \in \Theta$ (here and in what follows the sign "'" denotes transposition).

(2) For all $t \geqslant 0$ **P**-a.s.:

a) there exists $\partial \ln \rho_\theta / \partial \theta$;

b) $\partial \ln \rho_\theta / \partial \theta = L(\dot{M}_\theta, M)$, where

$$L(\dot{M}_\theta, M_\theta) = \left(L\left(\frac{\partial}{\partial \theta_1} M_\theta, M_\theta\right), \ldots, L\left(\frac{\partial}{\partial \theta_d} M_\theta, M_\theta\right)\right)' ;$$

c) $L(\dot{M}_\theta, M_\theta) \in \mathcal{M}^2(\mathbf{P}_\theta)$, where $\mathcal{M}^2(\mathbf{P}_\theta)$ is a class of square integrable $\mathbf{P}_\theta$-martingales.

Let $\widehat{I}_\theta = (\widehat{I}_\theta(t))_{t \geqslant 0}$ denote a Fisher information process, i.e.,

$$\widehat{I}_\theta = \langle L(\dot{M}_\theta, M_\theta) \rangle :=$$

$$:= \left(\left\langle L\left(\frac{\partial}{\partial \theta_i} M_\theta, M_\theta\right), L\left(\frac{\partial}{\partial \theta_j} M_\theta, M_\theta\right) \right\rangle \right)_{i,j=1}^{d} .$$

(3) The Fisher information matrix

$$I_\theta := \mathbf{E}_\theta \widehat{I}_\theta$$

is finite and positively defined.

2.4. MLE *and M-estimators*

Under the regularity conditions the likelihood equation takes the from

$$L(\dot{M}_\theta, M_\theta) = 0.$$

This is certainly a special case of the family of equations

$$L(m_\theta, M_\theta) = 0$$

with certain d-dimensional **P**-martingales m_θ, $\theta \in \Theta$. They are estimational equations in the sense that their solutions are viewed as estimators of the unknown parameter θ (M-estimators).

2.5. *Ergodicity*
In what follows it is convenient to consider a sequence of regular experiments $\mathsf{E} = (\mathcal{E}_n)_{n\geqslant 1}$,

$$\mathcal{E}_n = \left(\Omega^n,\ \mathcal{F}^n,\ F^n = (\mathcal{F}^n_t)_{0\leqslant t\leqslant T},\ \mathbf{P}^n,\ \{\mathbf{P}^n_\theta,\ \theta\in\Theta\}\right),$$

where $\Theta\subset\mathbf{R}^d$, $T>0$ is a number. For each $n\geqslant 1$, $\theta\in\Theta$,

$$\rho^n_\theta = \frac{d\mathbf{P}^n_\theta}{d\mathbf{P}^n} = \mathcal{E}(M^n_\theta), \quad M^n_\theta\in\mathcal{M}_{\rm loc}(F^n,\ \mathbf{P}^n).$$

The sequence $\mathsf{E} = (\mathcal{E}_n)_{n\leqslant 1}$ will be called ergodic if the condition below is satisfied. There exists a numerical sequence $(C_n)_{n\leqslant 1}$ such that $C_n>0$, $\lim_{n\to\infty} C_n = 0$ and, for each $\theta\in\Theta$, we have

$$C^2_n I^n_\theta(T) \longrightarrow I_\theta, \quad n\to\infty,$$

$$C^2_n \widehat{I}^n_\theta(T) \xrightarrow{\mathbf{P}^n_\theta} I_\theta, \quad n\to\infty,$$

where $I^n_\theta(T)$, $\widehat{I}^n_\theta(T)$ are defined in 2.3, I_θ is a finite positively defined matrix, the sign "$\xrightarrow{\mathbf{P}^n_\theta}$" denotes the convergence in probability $\mathbf{P}^n_\theta$.

2.6. *Asymptotic properties of solutions of estimational equations*
Asymptotic properties of M-estimators will be obtained from Lemma 2.1. which states the asymptotic properties of solutions of the equation

$$L_n(\theta,\omega) = 0$$

provided the behaviour of the sequence $(L_n(\theta,\omega))_{n\geqslant 1}$ is known.
Let a sequence of probability measures $\{Q^n_\theta\}_{n\geqslant 1}$ ($Q^n_\theta\sim\mathbf{P}^n$) and d-dimensional random vectors $L_n(\theta,\omega)$ be given on the measurable space $(\Omega^n,\ \mathcal{F}^n,\ \mathbf{P}^n)$, for every $\theta\in\Theta$. Let $(C_n)_{n\geqslant 1}$ be a sequence of positive numbers.

LEMMA 2.1 (Chitashvili et al., 1990). *Let the conditions below be fulfilled:*

a) $\lim_{n\to\infty} C_n = 0$;

b) *the mapping* $\theta\mapsto L_n(\theta)$ *is continuously* θ*-differentiable for each* $n\geqslant 1$ $\mathbf{P}$-a.s.;

c) *for each* $\theta\in\Theta$ *there exists a function* $\Delta_Q(\theta,y)$, $\theta,y\in\Theta$, *such that*

$$Q^n_\theta - \lim_{n\to\infty} C^2_n L_n(y) = \Delta_Q(\theta,y)$$

and the equation (with respect to y*)*

$$\Delta^n_Q(\theta,y) = 0$$

has an unique solution $\theta^* = b(\theta)$, $\theta^*\in\Theta$;

d)

$$Q_\theta^n - \lim_{n\to\infty} C_n^2 \dot{L}_n(\theta^*) = -\gamma_Q(\theta),$$

where $\gamma_Q(\theta)$ is a positively defined matrix for every $\theta \in \Theta$;

e)

$$\lim_{r\to 0} \overline{\lim_{n\to\infty}} Q_\theta^n \Big\{ \sup_{y:\|y-\theta\|\leqslant r} C_n^2 \|\dot{L}_n(y) - \dot{L}_n(\theta)\| > \rho \Big\} = 0 \quad (\forall_\rho > 0).$$

Then for each $\theta \in \Theta$ there exists a sequence of random vectors $\hat{\theta} = (\hat{\theta}_n)_{n\geqslant 1}$ with values in Θ such that

i) $\lim_{n\to\infty} Q_\theta^n \{L_n(\hat{\theta}_n) = 0\} = 1$;

ii) $Q_\theta^n\text{-}\lim_{n\to\infty} \hat{\theta}_n = b(\theta)$;

iii) *if $\tilde{\theta} = (\tilde{\theta}_n)_{n\geqslant 1}$ is another sequence with properties* i), ii), *then*

$$\lim_{n\to\infty} Q_\theta^n \{\hat{\theta}_n = \tilde{\theta}_n\} = 1;$$

iv) *if $\mathcal{L}(C_n L_n(\theta^*) \mid Q_\theta^n) \Rightarrow \Phi$ ($\mathcal{L}(\xi_n \mid \mathbf{P}^n) \Rightarrow \Phi$ denotes the weak convergence of distribution of r.v. ξ_n calculated with respect to $\mathbf{P}^n$, Φ is some distribution), then*

$$\mathcal{L}\Big(\gamma_Q(\theta) C_n^{-1} (\hat{\theta}_n - \theta^*) \mid Q_\theta^n\Big) \Rightarrow \Phi, \quad n \to \infty.$$

2.7. *Asymptotic properties of M-estimators*

Let the family of martingales $\{m_\theta^n,\ \theta \in \Theta\}_{n\geqslant 1}$ be regularly related to the experiment (see Subsection 2.5) in the sense as follows. For each $n \geqslant 1$:

(a) the components of $m_\theta^n = (m_\theta^{n,1}, \dots, m_\theta^{n,d})'$ satisfy condition (1) from Subsection 2.3;

(b) $L(m_\theta^n, M_\theta^n), L(\dot{m}_\theta^n, M_\theta^n) \in \mathcal{M}^2(\mathbf{P}_\theta^n)$;

(c)

$$\dot{L}(m_\theta^n, M_\theta^n) := \left(\frac{\partial}{\partial\theta_i} L\left(m_\theta^{n,j}, M_\theta^n\right)\right)_{i,j=1}^d =$$

$$= L(\dot{m}_\theta^n, M_\theta^n) - \left[L(m_\theta^n, M_\theta^n),\ L\left(\dot{M}_\theta^n, M_\theta^n\right)\right],$$

where $[M, N] = [M^i, N^j]_{i,j=1}^d$ is the matrix of mutually square variations;

(d) elements of the vector $L(m_\theta^n, M_\theta^n)$ are θ-continuous for all $t \in [0, T]$, $\mathbf{P}^n$-a.s.

Using the central limit theorem for martingales (Liptser and Shiryaev, 1986) and Lemma 2.1, we shall obtain the asymptotic properties of M-estimators defined by

$$L(m_\theta^n, M_\theta^n) = 0. \tag{2.1}$$

To this end, we introduce the following conditions.

For all $\theta \in \Theta$:

($\mathbf{c}_1$) $\mathbf{P}_\theta^n\text{-}\lim_{n\to\infty} C_n^2\langle L(m_\theta^n, M_\theta^n)\rangle_T = \Gamma(\theta)$, where $\Gamma(\theta) = (\Gamma_{i,j}(\theta))_{i,j=1}^d$ is a positively defined symmetric matrix;

($\mathbf{c}_2$)

$$\mathbf{P}_\theta^n - \lim_{n\to\infty} \int_0^T \int_{\|x\|\geqslant\varepsilon} \|x\|^2 \nu_\theta^n(ds, dx) = 0, \quad \forall\varepsilon \in (0,1),$$

where ν_θ^n is a compensator of the jump measure of the process $C_n L(m_\theta^n, M_\theta^n)$. It follows from conditions ($\mathbf{c}_1$) and ($\mathbf{c}_2$) that $\theta^* = \theta$ is the solution of

$$\Delta(\theta, y) := \mathbf{P}_\theta^n - \lim_{n\to\infty} C_n^2 L(m_y^n, M_y^n) = 0. \tag{2.2}$$

($\mathbf{c}_3$) $\theta^* = \theta$ is the unique solution of (2.2).

THEOREM 2.1 (Chitashvili et al., 1990). *Let conditions* ($\mathbf{c}_1$)–($\mathbf{c}_3$), *and conditions* **d**), **e**) *of Lemma* 2.1 *with* $\mathbf{P}_\theta^n$ *instead of* Q_θ^n, $L(m_\theta^n, M_\theta^n)$ *instead of* $L_n(\theta)$, θ *instead of* $b(\theta)$, $\gamma(\theta)$ *instead of* $\gamma_Q(\theta)$ *be satisfied. Then statements* i)–iii) *of Lemma* 2.1 *(with the above substitutions) hold and*

$$\mathcal{L}\left(C_n^{-1}(\hat{\theta}_n - \theta) \mid \mathbf{P}_\theta^n\right) \Longrightarrow \mathcal{N}\left(0, \gamma^{-1}(\theta)\Gamma(\theta)(\gamma^{-1}(\theta))'\right).$$

2.8. *Definition of estimator stability*

From Lemma 2.1 it easily follows (Chitashvili et al., 1990) that in studying the asymptotic behaviour of M-estimators as solutions of (2.1) when some alterative sequence $\{\widetilde{\mathbf{P}}_\theta^n,\ \theta \in \Theta\}_{n\geqslant 1}$ of measures (forming a regular ergodic experiment) is given, there arises an asymptotic bias, i.e.,

$$\widetilde{\mathbf{P}}_\theta^n - \lim_{n\to\infty} \hat{\theta}_n = b(\theta),$$

where $b(\theta)$ is the solution (with respect to y) of

$$\Delta(\theta, y) = 0$$

with

$$\Delta(\theta, y) = \widetilde{\mathbf{P}}_\theta^n - \lim_{n\to\infty} C_n^2 L(m_y^n, M_y^n).$$

In this connection it is natural to introduce the following definition of stability. Consider a set of experiments

$$E = \left\{\mathsf{E} = (\mathcal{E}_n)_{n\geqslant 1} = (\Omega^n,\ \mathcal{F}^n,\ F^n,\ \{Q_\theta^n,\ \theta \in \Theta\}, \mathbf{P}^n)_{n\geqslant 1}\right\}$$

everyone of which is regular and ergodic.

Definition 2.1. An estimator $\hat{\theta} = (\hat{\theta}_n)_{n\geqslant 1}$ is stable (in the sense of consistency) with respect to the class E if $\hat{\theta}$ is a consistent estimator of θ, for any family of measures $\{Q_\theta^n,\ \theta \in \Theta\}_{n\geqslant 1}$ forming the experiment $\mathsf{E} \in E$, i.e.,

$$Q_\theta^n - \lim_{n\to\infty} \hat{\theta}_n = \theta.$$

From the foregoing reasoning and Theorem 2.1 it follows that the main requirement for the stability of the estimator defined by (2.1) is

$$Q_\theta^n - \lim_{n\to\infty} C_n^2 L(m_\theta^n, M_\theta^n) = 0,$$

for all $\{Q_\theta^n,\ \theta \in \Theta\}_{n\geqslant 1}$, which naturally leads to a suggestion that the process $L(m_\theta^n, M_\theta^n)$ is a martingale with respect to all $\{Q_\theta^n,\ \theta \in \Theta\}_{n\geqslant 1}$. The method of partial likelihood to be discussed below is a special method for such a situation.

3. PARTIAL LIKELIHOOD ESTIMATORS

Let $(\Omega,\ \mathcal{F},\ F = (\mathcal{F}_t)_{0\leqslant t\leqslant T},\ \mathbf{P})$ be a probability space with filtration, satisfying the usual conditions. Let, besides, $\{N_\theta,\ \theta \in \Theta\}$ and H be families of martingales from

$$H^1(F,\mathbf{P}) = \left\{ M \in \mathcal{M}_{\mathrm{loc}}(F,\mathbf{P}) :\ \mathbf{E} \sup_{0\leqslant t\leqslant T} \|M_t\| < \infty \right\}$$

such that the conditions below are satisfied:

(I_1) $N_\theta \perp X$ for all $\theta \in \Theta$ and $X \in H$*;

(I_2) $\Delta X > -1,\ X(0) = 0$ for any $X \in H$;

(I_3) for any $\theta \in \Theta$ and $X \in H$ Dolean's exponent

$$\mathcal{E}_T(M_\theta^X) := \mathcal{E}_T(X)\mathcal{E}_T(N_\theta) \tag{3.1}$$

defines the density of some measure Q_θ^X with respect to the measure $\mathbf{P}$, i.e., $\mathbf{E}\mathcal{E}_T(M_\theta^X) = 1$, where $M_\theta^X = X + N_\theta + [X, N_\theta]$;

(I_4) for any $\theta \in \Theta$

$$\mathbf{E}\mathcal{E}_T(N_\theta) = 1, \quad \Delta N_\theta > -1, \quad N_\theta(0) = 0.$$

Furthermore, let $\mathcal{L}^1(H)$ be a stable subspace of martingales from $H^1(F,\mathbf{P})$ generated by H, i.e., $\mathcal{L}^1(H)$ is a minimal subspace of $H^1(F,\mathbf{P})$ containing, together with each $M \in H$, all stochastic integrals $h \cdot M$ with $h \in L^1(M)$, where

$$L^1(M) = \left\{ h :\ \mathbf{E}\Bigl(\int_0^T h^2 d[M] \Bigr)^{1/2} < \infty \right\}.$$

*The local martingales M and N are called orthogonal if their product MN is a local martingale and $M_0 N_0 = 0$.

Denote by H_0 the largest subset from $\mathcal{L}^1(H)$ satisfying, together with $\{N_\theta,\ \theta \in \Theta\}$, conditions (I_1)–(I_4). Evidently, $\mathcal{L}^1(H_0) = \mathcal{L}^1(H)$.
Consider a set $\mathbf{X}$ of various mappings $X\colon \Theta \to H_0$ and a set of experiments $\overline{\mathcal{E}} = (\mathcal{E}^X,\ X \in \mathbf{X})$, where

$$\mathcal{E}^X = (\Omega,\ \mathcal{F},\ F,\ \mathbf{P},\ Q^X = \{Q_\theta^X,\ \theta \in \Theta\})$$

and the family Q^X is such that for all $\theta \in \Theta$ we have $Q_\theta^X \sim \mathbf{P}$, $Q_\theta^X \mid \mathcal{F}_0 = \mathbf{P} \mid \mathcal{F}_0$, and

$$\frac{dQ_\theta^X}{d\mathbf{P}} = \mathcal{E}(X_\theta)\mathcal{E}(N_\theta). \tag{3.1'}$$

For convenience, let us fix some basic experiment $\mathcal{E}^{X_0}$. All the remaining experiments from $\overline{\mathcal{E}}$ will be called alternatives.
Therefore, the family of martingales $\mathbf{X}$ will be called the family defining alternatives.
Following (Cox, 1975), the second product in the right-hand side of (3.1′) will be called the partial likelihood; accordingly, the estimator defined by the relation

$$\hat{\theta}_t = \arg\max_\theta \mathcal{E}_t(N_\theta)$$

or by the equation

$$L_t(\dot{N}_\theta, N_\theta) = 0 \tag{3.2}$$

will be called the partial likelihood estimator (PLE).
At the end of this Section there are examples making it clear that the introduced PLE is the generalization of the PLE introduced by Cox for models in discrete time in the case of a general statistical experiment.
Before studying of the asymptotic behaviour of the PLE, we shall prove two lemmas which will play the key role in our further discussion.
Let the martingales N and m from $\mathcal{M}_{\text{loc}}(F, \mathbf{P})$ be such that $N = N_{\theta_0}$ for some $\theta_0 \in \Theta$ and $m \perp H_0$.

LEMMA 3.1. *The process $L(m, N)$ is a local $\mathbf{P}^X$-martingale with respect to any measure $\mathbf{P}^X$ ($\mathbf{P}^X \sim \mathbf{P}$) with the density*

$$\frac{d\mathbf{P}^X}{d\mathbf{P}} = \mathcal{E}(X)\mathcal{E}(N) = \mathcal{E}(M^X), \tag{3.3}$$

where $X \in H_0$, $M^X = X + N + [X, N]$.

Proof. By virtue of Girsanov's generalized theorem, it is sufficient to prove that, for any $X \in H_0$, there exists a $\mathbf{P}$-martingale m^X such that

$$L(m, N) = L(m^X, M^X).$$

Since $m \perp H_0$ and hence $[m, X] = \langle m^c, X^c\rangle + \sum \Delta M \Delta X \in \mathcal{M}_{\text{loc}}(F, \mathbf{P})$ we also have $m^X := m + [m, X] \in \mathcal{M}_{\text{loc}}(F, \mathbf{P})$. We shall show that

$$L(m^X, M^X) = L(m, N).$$

By the definition of the L-transformation, we have

$$L(m^X, M^X) = m^X - \langle m^{X,c}, M^{X,c}\rangle - \sum \frac{\Delta m^X \Delta M^X}{1 + \Delta M^X}.$$

Furthermore, since $[m, X], [X, N] \in \mathcal{M}^d_{\text{loc}}(F, \mathbf{P})$, we have $m^{X,c} = m^c$, $M^{X,c} = = X^c + N^c$, $\Delta m^X = \Delta m(1 + \Delta X)$, $\Delta M^X = \Delta X(1 + \Delta N) + \Delta N$. Therefore,

$$L(m^X, M^X) = m + \langle m^c, X^c\rangle + \sum \Delta m \Delta X -$$

$$-\langle m^c, X^c + N^c\rangle - \sum \frac{\Delta m(1 + \Delta X)(\Delta X(1 + \Delta N) + \Delta N)}{(1 + \Delta X)(1 + \Delta N)} =$$

$$= m - \langle m^c, N^c\rangle - \sum \frac{\Delta m \Delta M}{1 + \Delta N} = L(m, N).$$

It follows from Lemma 3.1 that each process $L(m, N)$ with some local $\mathbf{P}$-martingale $m \perp H_0$ is a local martingale with respect to all measures $\mathbf{P}^X$, $X \in H_0$, simultaneously.
The lemma below claims that the processes $L(m, N)$, $m \perp H_0$, exhaust the class of processes which are local martingales with respect to all measures $\mathbf{P}^X$, $X \in H_0$ (with property (3.3)), simultaneously.

LEMMA 3.2. *Let $l = (l_t,\ 0 \leqslant t \leqslant T)$ be an F-adapted process with regular trajectories such that $l \in \mathcal{M}_{\text{loc}}(F, \mathbf{P}^X)$ for all measures $\mathbf{P}^X$, $X \in H_0$, with property (3.3). Then there exists a local $\mathbf{P}$-martingale m such that $m \perp H_0$ and $l = l(m, N)$.*

Proof. Since for any $X \in H_0$ we have $l \in \mathcal{M}_{\text{loc}}(F, \mathbf{P}^X)$ and

$$\frac{d\mathbf{P}}{d\mathbf{P}^X} = \mathcal{E}(L(-M^X, M^X)),$$

where $M^X = X + N + [X, N]$, by Girsanov's generalized theorem we have

$$L(l, L(-M^X, M^X)) \in \mathcal{M}_{\text{loc}}(F, \mathbf{P})$$

for any $X \in H_0$.
It can be easily calculated that

$$L(l, L(-M^X, M^X)) = l + [l, M^X].$$

Indeed, by the definition of the L-transformation we have

$$L(l, L(-M^X, M^X)) = l - \langle l^c, L^c(-M^X, M^X)\rangle -$$

$$- \sum \frac{\Delta l \Delta L(-M^X, M^X)}{1 + \Delta L(-M^X, M^X)}.$$

On the other hand, by virtue of the well-known properties of the L-transformation, we have

$$\langle l^c, L^c(-M^X, M^X)\rangle = -\langle l^c, M^{X,c}\rangle,$$

$$\Delta L(-M^X, M^X) = -\Delta M^X/(1+\Delta M^X).$$

Substituting these equalities into the previous ones, we obtain the desired result. Thus, the process $m^X = l + [l, M^X]$ is a local martingale for any $X \in H_0$, Since $X \equiv 0 \in H_0$, we have

$$l + [l, N] := m \in \mathcal{M}_{\rm loc}(F, \mathbf{P}).$$

It can be easily verified that

$$l = L(m, N).$$

Now it remains to prove that $m \perp H_0$. We have $m^X - m \in \mathcal{M}_{\rm loc}(F, \mathbf{P})$ for all $X \in H_0$. But

$$m^X - m = \langle l^c, X^c\rangle + \sum \Delta l(1+\Delta N)\Delta X = [m, X],$$

since $\langle l^c, X^c\rangle = \langle m^c, X^c\rangle, \Delta l(1+\Delta N) = \Delta m$. Hence, $[m, X] \in \mathcal{M}_{\rm loc}(F, \mathbf{P})$ for any $X \in H_0$.

Note that it can be easily verified that the process l can be represented as

$$l = L(m^X, M^X)$$

for any $X \in H_0$.

To study the asymptotic behaviour of the PLE, we shall consider as usual the scheme of series.

Let all the objects introduced previously be defined for each $n \geqslant 1$. In particular, among these objects we have families $\{N^n_\theta,\ \theta \in \Theta\}, H^n,\ H^n_0 \subset H^1(F^n, \mathbf{P}^n)$, satisfying conditions $(I_1) - (I_4)$, and the set $\mathbf{X}^n$ of all possible mappings $X^n : \Theta \to H^n_0$. Consider the set $\mathbf{X}$ of sequences $X = (X^n)_{n\geqslant 1}$ with $X^n \in \mathbf{X}^n$ and associate to each $X \in \mathbf{X}$ the family of measures

$$Q^X = \{Q^{X,n}_\theta,\ \theta \in \Theta\}_{n\geqslant 1},$$

where, for every $n \geqslant 1$ and all $\theta \in \Theta$, we have $Q^{X,n}_\theta \sim \mathbf{P}^n$,

$$\frac{dQ^{X,n}_\theta}{d\mathbf{P}^n} = \mathcal{E}(X^n_\theta)\mathcal{E}(N^n_\theta) = \mathcal{E}(M^{X,n}_\theta),$$

$M^{X,n}_\theta = X^n_\theta + N^n_\theta + [X^n_\theta, N^n_\theta]$ and the corresponding experiment

$$\mathsf{E}^X = (\mathcal{E}^X_n)_{n\geqslant 1} = \left(\Omega^n,\ \mathcal{F}^n,\ F^n,\ \mathbf{P}^N,\ \{Q^{X,n}_\theta;\ \theta \in \Theta\}\right)_{n\geqslant 1}.$$

Denote by E the set of all regular and ergodic experiments $\mathcal{E}^X,\ X \in \mathbf{X}$.

By the previous condition, choose now some basic experiment E^{X_0} from the entire family of experiments $E = (\mathcal{E}^X,\ X \in \mathbf{X})$. The corresponding measure $Q^{X_0,n}_\theta$ will be denoted by $\mathbf{P}^n_\theta$ and $M^{X_0,n}_\theta$ by M^n_θ.

A question arises: under what choice of $\mathbf{P}^n$-martingales $\{m^n_\theta\}$ can we obtain the stability (in the sense of Definition 2.1) of M-estimators defined by

$$L(m^n_\theta, M^n_\theta) = 0$$

with respect to the class E ?

It is evident from the arguments of Subsection 2.8 and Lemmas 3.1 and 3.2 that only M-estimators constructed by the equation

$$L(m^n_\theta, N^n_\theta) = 0 \tag{3.4}$$

with some family of martingales $\{m^n_\theta,\ \theta \in \Theta\}_{n\geqslant 1}$, where $m^n_\theta \perp H_0$ for all $n \geqslant 1$ and might be stable.

Remark 3.1. Note that these estimators are usual M-estimators. Indeed, it can easily be verified that if we define martingales $m_\theta^{X,n}, M_\theta^{X,n}$ as

$$m_\theta^{X,n} = m_\theta^n + [m_\theta^n, X_\theta^n], \quad M_\theta^{n,X} = X_\theta^n + N_\theta^n + [X_\theta^n, N_\theta^n],$$

then $L(m_\theta^{n,X}, M_\theta^{n,X}) = L(m_\theta^n, N_\theta^n)$ for any $X \in \mathbf{X}$. The asymptotic behaviour with respect to all alternatives $\{Q^X,\ X \in \mathbf{X}\}$ of the above estimators can be investigated are usual by Lemma 2.1. For simplicity we shall consider the one-dimensional case. All the statements of Theorem 3.1 are valid in the multidimensional case, too. Assume that in (3.4) the families $\{N_\theta^n, \theta \in \Theta\}_{n\geqslant 1}$ and $\{m_\theta^n,\ \theta \in \Theta\}_{n\geqslant 1}$ satisfy the conditions below:

(1) the mapping $\theta \mapsto N_\theta^n(t,\omega)$, $\theta \mapsto m_\theta^n(t,\omega)$ are differentiable for every $n \geqslant 1$, $0 \leqslant t \leqslant T$, $\mathbf{P}^n$-a.s., $\dot{N}_\theta^n, \dot{m}_\theta^n \in \mathcal{M}_{\mathrm{loc}}(F^n, \mathbf{P}^n)$, $m_\theta^n, \dot{m}_\theta^n, \dot{N}_\theta^n \perp H_0^n$ for all $\theta \in \Theta$;

(2) for every $n \geqslant 1$

$$\frac{\partial}{\partial\theta} L(m_\theta^n, N_\theta^n) = L(\dot{m}_\theta^n, N_\theta^n) - \left[L(m_\theta^n, N_\theta^n), L(\dot{N}_\theta^n, N_\theta^n)\right],$$

$$\frac{\partial}{\partial\theta} L(\dot{N}_\theta^n, N_\theta^n) = L(\ddot{N}_\theta^n, N_\theta^n) - \left[L(\dot{N}_\theta^n, N_\theta^n)\right];$$

(3) processes $L(\dot{N}_\theta^n, N_\theta^n)\ L(m_\theta^n, N_\theta^n)$, $L(\dot{m}_\theta^n, N_\theta^n)$, $L(\ddot{N}_\theta^n, N_\theta^n) \in \mathcal{M}^2(F^n, Q_\theta^{X,n})$ for all Q^X, $X \in \mathbf{X}$, and are θ-continuous for all $t \in [0,T]$, $\mathbf{P}^n$-a.s., $n \geqslant 1$.

THEOREM 3.1. *Let C_n, $n \geqslant 1$, be a sequence of real numbers with $C_n > 0$, $\lim_{n\to\infty} C_n = 0$, such that, for all $Q^X = \{Q_\theta^X,\ \theta \in \Theta\}_{n\geqslant 1}$ and $X \in \mathbf{X}$, the conditions below are satisfied:*

(a') $Q_\theta^{X,n}\text{-}\lim_{n\to\infty} C_n^2 \langle L(m_\theta^n, N_\theta^n)\rangle_T = \Gamma_X(\theta),\ \Gamma_X(\theta) > 0$;

(b') $Q_\theta^{X,n}\text{-}\lim_{n\to\infty} C_n^2 [L(m_\theta^n, N_\theta^n),\ L(\dot{N}_\theta^n, N_\theta^n)]_T = \gamma_X(\theta),\ \gamma_X(\theta) > 0$;

(c') $Q_\theta^{X,n}\text{-}\lim_{n\to\infty} C_n^2 L_T(\dot{m}_\theta^n, N_\theta^n) = \theta$;

(d') $Q_\theta^{X,n}\text{-}\lim_{n\to\infty} \int_0^T \int_{|u|>\varepsilon} u^2 \nu_\theta^{n,X}(ds, du) = 0$, $\forall \varepsilon \in (0,1]$, *where* $\nu_\theta^{X,n}$ *is a* $Q_\theta^{X,n}$*-compensator of the process* $C_n L(m_\theta^n, N_\theta^n)$;

(e') $\lim_{r\to 0} \overline{\lim}_{n\to\infty} Q_\theta^{X,n}\{\sup_{y:|y-\theta|\leqslant r} C_n^2 |\dot{L}(m_y^n, N_y^n) - \dot{L}(m_\theta^n, N_\theta^n)| > \rho\} = 0$, $\forall \rho > 0$.

Then for any $\theta \in \Theta$ and Q^X, $X \in \mathbf{X}$, there exists an M-estimator $\tilde{\theta} = (\tilde{\theta}_n)_{n\geqslant 1}$ (depending in general on both Q^X and θ) such that

i) $\lim_{n\to\infty} Q_\theta^{X,n}\{L_T(m_{\tilde{\theta}_n}^n, N_{\tilde{\theta}_n}^n) = 0\} = 1$;

ii) $Q_\theta^{X,n}\text{-}\lim_{n\to\infty} \tilde{\theta}_n = \theta$;

iii) $\tilde{\theta}$ *is asymptotically unique (as in Lemma 2.1)*;

iv) $\mathcal{L}\{C_n^{-1}(\tilde{\theta}_n - \theta) \mid Q_\theta^{X,n}\} \Rightarrow \mathcal{N}(0, \gamma_X^{-2}(\theta)\Gamma(\theta))$.

Proof. Let us verify the conditions of Lemma 2.1. Condition c) of Lemma 2.1 follows from condition (a′); condition d) follows from conditions (b′) and (c′); conditions (e′) and e) coincide and, finally, applying the central limit theorem for martingales, from (a′) and (b′) we obtain

$$\mathcal{L}\{C_n L(m_\theta^n, N_\theta^n) \mid Q_\theta^{X,n}\} \Rightarrow \mathcal{N}(0, \Gamma_X(\theta)).$$

Remark 3.2. From statements i and ii of Theorem 3.1 we conclude that if for each $n \geqslant 1$ and $\omega \in \Omega^n$ (or for an arbitrary large n) the equation (3.4) has the unique solution $\tilde{\theta}_n(\omega) = \tilde{\theta}_n \in \mathcal{F}^n_T$, then the estimator $\tilde{\theta} = (\tilde{\theta}_n)_{n\geqslant 1}$ is stable (in the sense of Definition 2.1) with respect to the class $E = (\mathcal{E}^X,\ X \in \mathbf{X})$. Moreover, this estimator has an additional property of stability (see statement iii): under all alternative measures Q^X, $X \in \mathbf{X}$, the asymptotic distribution of $C_n^{-1}(\tilde{\theta}_n - \theta)$ is Gaussian with the zero mean (i.e., the second-order bias does not appear either). When equation (3.4) has many solutions, we can construct a stable estimator which will be an asymptotic solution of equation (3.4), using a technique similar to the one from (Chitashvili et al., 1990, § 6, Corollary 6.1.1).
The partial likelihood estimator (PLE) is an important particular case with $m^n_\theta = = \dot{N}^n_\theta$. In addition to being stable, it also possesses a certain optimality property. Namely, PLE $\hat{\theta} = (\hat{\theta}_n)_{n\geqslant 1}$ has a minimal asymptotic variance among all estimators $\tilde{\theta} = (\tilde{\theta}_n)_{n\geqslant 1}$ defined by (3.4) with some $\{m^n_\theta,\ \theta \in \Theta\}_{n\geqslant 1}$ $(m^n_\theta \perp H^n_0)$.
Indeed, since $[M,N] \leqslant [M]^{1/2}[N]^{1/2}$, $M,N \in \mathcal{M}_{\text{loc}}$, we have, for each $X \in \mathbf{X}$,

$$\frac{(\sigma^X_{\hat{\theta}})^2}{(\sigma^X_{\tilde{\theta}})^2} = Q^{X,n}_\theta - \lim_{n\to\infty} \frac{[L(m^n_\theta, N^n_\theta),\ L(\dot{N}^n_\theta, N^n_\theta)]^2_T}{[L(m^n_\theta, N^n_\theta)]_T [L(\dot{N}^n_\theta, N^n_\theta)]_T} \leqslant 1.$$

4. EXAMPLES

First we shall describe a rather general scheme including as it will be shown below, a number of cases studied by various authors, e.g., (Cox, 1975; Wong, 1986; Dzhaparidze, 1985).
Let $(\Omega^n,\ \mathcal{F}^n,\ F^n,\ \mathbf{P}^n)$, for any $n \geqslant 1$, be a probability space with filtration F^n satisfying the usual conditions, E be a Lusin space, $\widetilde{\Omega}^n = \Omega^n \times \mathbf{R}_+ \times E$, $\widetilde{\mathcal{F}}^n = = \mathcal{F}^n \times \mathcal{B}(\mathbf{R}_+) \otimes \mathcal{B}(E)$, μ^n be an integer-valued measure, $\mu^n \in \widetilde{\mathcal{A}}_\sigma(F^n, \mathbf{P}^n)$ and ν^n be its $\mathbf{P}^n$-compensator, $D^n = \{\omega, t : \mu^n(\omega, \{t\}, E) = 1\}$.
Following (Jacod, 1979), we introduce the classes

$$\mathcal{K}(\mu^n) = \{W:\ W \in \widetilde{\mathcal{O}}(F^n, \mathbf{P}^n),\quad W * \mu^n \in \widetilde{\mathcal{A}}_\sigma(\mathbf{P}^n)\},$$

$$G^q(\mu^n) = \Big\{W:\ W \in \widetilde{\mathcal{P}}(F^n, \mathbf{P}^n),\quad \big(\sum (\widetilde{W}^n)^2\big)^{q/2} \in \mathcal{A}(\mathbf{P}^n)\Big\},\quad q = 1,2,$$

$$H^q(\mu^n) = \Big\{V:\ V \in \mathcal{K}(\mu^n),\quad M^{\mathbf{P}^n}_{\mu^n}(VI_{]0,\infty[} \mid \widetilde{\mathcal{P}}(F^n, \mathbf{P}^n)) = 0,$$
$$(V^2 * \mu^n)^{q/2} \in \mathcal{A}(\mathbf{P}^n)\Big\},$$

where $\widetilde{\mathcal{O}}(F^n, \mathbf{P}^n) = \mathcal{O}(F^n, \mathbf{P}^n) \otimes \mathcal{B}(E)$, $\widetilde{\mathcal{P}}(F^n, \mathbf{P}^n) = \mathcal{P}(F^N, \mathbf{P}^N) \otimes \mathcal{B}(E)$, $\mathcal{O}(F^n, \mathbf{P}^N)$, $\mathcal{P}(F^n, \mathbf{P}^n)$ are optional and predictable σ-algebras, respectively; $\mathcal{A}(\mathbf{P}^n)$ is a class of processes with an integrable (with respect to the measure $\mathbf{P}^n$) variation; $\mu^n \in \widetilde{\mathcal{A}}_\sigma(\mathbf{P}^n)$ if there exists a sequence of subsets $\widetilde{\Omega}^{n,k} \in \mathcal{P}(F^n, \mathbf{P}^n)$ such that

$$\widetilde{\Omega}^{n,k} \uparrow \widetilde{\Omega}^n,\quad k \to \infty, \text{ and } M^{\mathbf{P}^n}_{\mu^n}(I_{\widetilde{\Omega}^{n,k}}) < \infty,$$

$M^{\mathbf{P}^n}_{\mu^n}$ is Dolean's measure associated with μ^n, i.e.,

$$M^{\mathbf{P}^n}_{\mu^n}(d\omega, dt, dx) = \mathbf{P}^n(d\omega)\mu^n(\omega, dt, dx);$$

$$\widetilde{W}^n_t = \int_E W(\omega, t, x)\mu^n(\{t\}, dx) - \widehat{W}^n_t, \quad W^n_t = \int_E W(\omega, t, x)\nu^n(\{t\}, dx).$$

Let now

$$\mathcal{K}^{q,1}(\mu^n) = \{W * (\mu^n - \nu^n)\colon\ W \in G^q(\mu^n)\},$$

$$\mathcal{K}^{q,2}(\mu^n) = \{V * \mu^n\colon\ V \in H^q(\mu^n)\},$$

$$\mathcal{K}^{q,3}(\mu^n) = \{M \in H^q\colon\ \Delta M = 0 \text{ on } D^n\}, \quad q = 1, 2.$$

Assume that, for each $n \geqslant 1$, we have a family $\{N^n_\theta,\ \theta \in \Theta\}$ such that

1) $N^n_\theta \in \mathcal{K}^{1,2}(\mu^n)$ for any $\theta \in \Theta$, i.e.,

$$N^n_\theta = V^n_\theta * \mu^n;$$

2) $\mathbf{E}\mathcal{E}_T(N^n_\theta) = 1$ for all $\theta \in \Theta$;

3) $\Delta N^n_\theta = \int_E V^n_\theta \mu^n(\{t\}, dx) > -1$.

For every $n \geqslant 1$ introduce the set H^n:

$$H^n = \left\{ W * (\mu^n - \nu^n) : \begin{array}{l} \text{a) } W \in G^1(\mu^n); \\ \text{b) } (W - \widehat{W})V^n_\theta \in H^1_{\text{loc}}(\mu^n) \text{ for all } \theta \in \Theta\ ; \\ \\ \text{c) } \widetilde{W}^n_t > -1 \text{ for all } t \in [0, T]; \\ \text{d) } \mathbf{E}\mathcal{E}_T(M^{W,n}_\theta) = 1 \text{ for each } \theta \in \Theta. \end{array} \right.$$

where $M^{W,n}_\theta = W * (\mu^n - \nu^n) + (W - \widehat{W} + 1)V^n_\theta * \mu^n$.

Obviously, for every $n \geqslant 1$ the families $\{N^n_\theta,\ \theta \in \Theta\}_{n\geqslant 1}$ and H^n satisfy conditions (I_1)–(I_4). Indeed, condition b) implies $N^n_\theta \perp H^n$ for each $\theta \in \Theta$, since for any $W * (\mu^n - \nu^n) \in H^n$ and N^n_θ we have

$$[W * (\mu^n - \nu^n), N^n_\theta] = (W - \widetilde{W})V^n_\theta * \mu^n \in H^1(F^n, \mathbf{P}^n),$$

condition c) means that, for any $X \in H^n$, we have $\Delta X = \widetilde{W}^n > -1$, condition d) coincides with (I_3) and, finally, conditions 1) and 2) in the definition of the class $\{N^n_\theta\}$ are equivalent to (I_4). Since $H^n \subset \mathcal{K}^{1,1}(\mu^n)$ and the class $\mathcal{K}^{1,1}(\mu^n)$ is a stable subspace of $H^1(F^n, \mathbf{P}^n)$, then $\mathcal{L}^1(H^n) \subset \mathcal{K}^{1,1}(\mu^n)$ and so $H^n_0 = H^n$.

The class of alternatives is introduced in a usual manner, i.e., the set of all classes of measures $Q^X = \{Q^{X,n}_\theta,\ \theta \in \Theta\}_{n\geqslant 1}$

$$dQ^{X,n}_\theta / d\mathbf{P}^n = \mathcal{E}(X^n_\theta)\mathcal{E}(N^n_\theta)$$

with $X^n_\theta = W^n_\theta * (\mu^n - H^n) \in H^n$ is considered and stable estimators as above are constructed from the equation $L(m^n_\theta, N^n_\theta) = 0$, where $m^n_\theta \perp H^n$, for all $n \geqslant 1$, and $\theta \in \Theta$.

For example, for m_θ^n we can take $m_\theta^n = V_\theta^{m,n} * \mu^n$ with $V_\theta^{m,n}$ such that, for any W with $W * (\mu^n - \nu^n) \in H^n$, we have $V_\theta^{m,n}(W - \widehat{W}^n) \in H^1_{\text{loc}}(\mu^n)$.
If $N_\theta^n \in \mathcal{K}^{2,2}(\mu^n)$ for all $\theta \in \Theta$, then for H^n we can take a subset of $\mathcal{K}^{2,1}(\mu^n)$ (or $\mathcal{K}^{2,1} \cup \mathcal{K}^{2,3}$), satisfying conditions (I_2), (I_3) only, since in this case condition (I_1) is satisfied automatically (recall that classes $\mathcal{K}^{2,i}$, $i = 1,2$, are mutually orthogonal).
The examples below will illustrate the scheme under consideration.

Example 4.1. *I.i.d. case.*
Let $(Y_1, Z_1), \dots, (Y_n, Z_n)$ be independent identically distributed (i.i.d.) observations with the distribution $\mathbf{P}_\theta$ equivalent to some probability measure $\mathbf{P}$, $\mathbf{P}(dy, dz) = \mathbf{P}_1(dy)\mathbf{P}_2(dz)$ with the density

$$f_0(y, z; \theta) = \varphi_0(y, \theta)\Psi(z; \theta \mid y) > 0.$$

Then the likelihood ratio process $\rho_n(\theta) = \Pi_{i=1}^n f_0(y_i, Z_i; \theta)$ has the multiplicative decomposition

$$\rho_n(\theta) = \prod_{i=1}^n \varphi_0(Y_i; \theta)\Psi(Z_i; \theta \mid Y_i), \tag{4.1}$$

where, according to (Cox, 1975), the second cofactor is called the partial likelihood based on Z, and the estimator defined by the relation

$$\hat{\theta}_n = \arg\max_\theta \prod_{i=1}^n \Psi(Z_i; \theta \mid Y_i)$$

or the equation

$$\sum_{i=1}^n \dot{\Psi}(Z_i; \theta \mid Y_i)/\Psi(Z_i; \theta \mid Y_i) = 0$$

is called the partial likelihood estimator (PLE).
Let $\mathbf{P}_\theta^n, \mathbf{P}^n$ be the Cartesian products of the measures $\mathbf{P}_\theta, \mathbf{P}$ on $(\Omega^n, \mathcal{F}^n) = ((\mathbf{R}^2)^n, \mathcal{B}((\mathbf{R}^2)^n))$, respectively; $\mathcal{F}_t^n = \sigma(X_i, Y_i\colon\ i \leqslant [nt])_{0 \leqslant t \leqslant 1}$, $\mathbf{P}_\theta^n(t)$ and $\mathbf{P}^n(t)$ be the restrictions of the measures $\mathbf{P}_\theta^n$ and $\mathbf{P}^n$ on $\mathcal{F}_t^n$. Then

$$\rho_\theta^n(t) = \frac{d\mathbf{P}_\theta^n(t)}{d\mathbf{P}^n(t)} = \mathcal{E}_t(M_\theta^n) = \mathcal{E}_t(X_{0,\theta}^n)\mathcal{E}_t(N_\theta^n), \tag{4.2}$$

where the processes $M_\theta^n, X_{0,\theta}^n, N_\theta^n$ defined by the relations

$$M_\theta^n(t) = \sum_{i=1}^{[nt]} \big(f_0(Y_i, Z_i; \theta) - 1\big),$$

$$X_{0,\theta}^n(t) = \sum_{i=1}^{[nt]} \big(\varphi_0(Y_i; \theta) - 1\big),$$

$$N_\theta^n(t) = \sum_{i=1}^{[nt]} \left(\Psi(Z_i;\theta \mid Y_i) - 1\right)$$

are $\mathbf{P}^n$-martingales with martingales $X_{0,\theta}^n$ and N_θ^n being orthogonal for all $\theta \in \Theta$ and $n \geqslant 1$.

After comparing decompositions (4.1) and (3.1) it becomes obvious that the partial likelihood of Cox is the same as $\mathcal{E}(N_\theta^n)$. Thus, the concept of partial likelihood we introduced here for the case of a general statistical experiment coincides with Cox' definition.

Assume further that μ^n is a jump measure of the process $\sum_{i=1}^{[nt]} Y_i$, i.e.,

$$\mu^n\big([0,t] \times B\big) = \sum_{i=1}^{[nt]} I_{\{Y_i \in B\}}, \quad B \in \mathcal{B}(R), \quad \mu^n\big(\{k/n\}, B\big) = I_{\{Y_k \in B\}},$$

and ν^n is a $\mathbf{P}^n$-compensator of the measure μ^n, i.e.,

$$\nu^n\left([0,t] \times B\right) = [nt]\mathbf{P}_1(B), \quad \nu^n\left(\{k/n\}, B\right) = \mathbf{P}_1(B).$$

Then the martingales $X_{0,\theta}^n$ and N_θ^n can be written as

$$X_{0,\theta}^n = W_\theta^n * (\mu^n - \nu^n), \; N_\theta^n = V_\theta^n * \mu^n,$$

where $W_\theta^n(\omega, t, y) = \varphi_0(y,\theta) - 1$, $V_\theta^n(\omega,t,y) = \Psi(Z_{[nt]};\theta \mid y)$, $\omega = (Y_i, Z_i)_{i \geqslant 1}$.
It is easy to verify that $W_\theta^n \in G^1(\mu^n)$ and $V_\theta^n \in H^1(\mu^n)$.
Indeed, the former relation is obvious while the latter follows from

$$\mathbf{E}\big(\Psi(Z_i;\theta \mid Y_i) \mid \mathcal{F}_{i-1} \vee \sigma(Y_i)\big) = 0$$

and from the obvious fact $\mathbf{E}(V_\theta^2 * \mu^n)^{1/2} < \infty$.

In common with the general case we could consider all alternative measures $Q^X = \{Q_\theta^{X,n}, \; \theta \in \Theta\}_{n \geqslant 1}$ corresponding to any $X = \{X^n\}_{n \geqslant 1}$, $X^n \colon \theta \to H^n$, where H^n are defined in the usual manner but our consideration will be restricted to natural alternatives, namely, in the set H^n we shall choose only martingales of the from

$$X^n(t) = \sum_{i=1}^{[nt]} (\varphi(Y_i) - 1),$$

i.e., $X^n = \varphi * (\mu^n - \nu^n)$ with all possible one-dimensional densities $\varphi (\int \varphi \mathbf{P}_1(dy) = 1, \; \varphi > 0)$. This set will also be denoted by H^n .

Thus, in capacity of alternative experiments we shall consider i.i.d. observations $(Y_i, Z_i)_{i \geqslant 1}$ with densities $f(y,z,;\theta) = \varphi(y;\theta)\psi(z;\theta \mid y)$ with all possible $\varphi(y;\theta)$ such that $\int \varphi(y;\theta)\mathbf{P}_1(dy) = 1$, $\varphi(y;\theta) > 0$. Estimators that are stable with respect to such alternatives will naturally be constructed by the equation

$$L(m_\theta^n, N_\theta^n) = 0,$$

where $m^n_\theta \perp H^n$. In particular, for m^n_θ we can take

$$m^n_\theta(t) = \sum_{i=1}^{[nt]} g(Y_i, Z_i; \theta),$$

i.e., $m^n_\theta = V_\theta * \mu^n$, where $V_\theta(\omega, t, y) = g(y, Z_{[nt]}; \theta), g(y, z; \theta)$ is a functions such that, for all $\theta \in \Theta$, the condition below are fulfilled:

1) $\int g(y, z; \theta) \mathbf{P}_z(dz) = 0$ for all $y \in \mathbf{R}^1$;

2) $\int \int \varphi(y)|g(y, z; \theta)| \mathbf{P}(dy, dz) < \infty$ for any positive density φ (including $\varphi \equiv 1$);

3) $\int \dot{g}(y, z; \theta) \mathbf{P}_2(dz) = 0$ $(\dot{g} = \partial g / \partial \theta)$, $\forall y \in \mathbf{R}^1$;

4) $\int \int \varphi(y)|\dot{g}(y, z; \theta)| \mathbf{P}(dy, dz) < \infty$ for all $\varphi > 0, \int \varphi \mathbf{P}_1(dy) = 1$.

It is clear that if $\Psi(z; \theta \mid y)$ is differentiable in θ and $\Psi, \dot{\Psi}$ satisfy conditions 1) – 4), then in the conditions of Theorem 4.1 the PLE will be stable with respect to the considered class of alternatives.

The examples below will deal only with the decomposition of the local density of the measures μ^n and ν^n, the class of natural alternatives and, in some cases, with the class of martingales defining stable estimators.

Example 4.2. *Censored i.i.d. observations.*
Let $X_1, X_2, \ldots, X_n, n \geqslant 1$, be i.i.d. observations with a density $f(x; \theta)$, $\theta \in \Theta \subset \mathbf{R}^1, f(x; \theta) > 0$ (with respect to some measure μ). The density $f(x; \theta)$ is assumed to be known only on the set $\{x : |x| \leqslant c\}$ and θ is to be estimated.
Observations $\{X_i\}_{i \geqslant 1}$ can be transformed into a two-dimensional sequence $\{Y_i, Z_i\}_{i \geqslant 1}$, where $Y_i = X_i I_{\{|X_i| > c\}}$, $Z_i = X_i I_{\{|X_i| \leqslant c\}}$. Denote by $\mathbf{P}^{Y,Z}_\theta$ the distribution of a couple (Y_1, Z_1) on $(\mathbf{R}^2, \mathcal{B}(\mathbf{R}^2))$ calculated with respect to the distribution of X_1 (i.e., with respect to the $\mathbf{P}^X_\theta$ with $d\mathbf{P}^X_\theta / d\mu = f(x; \theta)$) and by $P^{Y,Z}$ the distribution of the couple (Y, Z) calculated with respect to μ. It is now obvious that, for every $\theta \in \Theta$, we have $\mathbf{P}^{Y,Z}_\theta \sim \mathbf{P}^{Y,Z}$. Let

$$d\mathbf{P}^{Y,Z}_\theta / d\mathbf{P}^{Y,Z} = f(y, z; \theta).$$

Evidently, the density $f(y, z; \theta)$ can be factorized by the following way:

$$f(y, z; \theta) = \varphi_0(y; \theta) \Psi(z; \theta \mid y),$$

where

$$\varphi_0(y; \theta) = \begin{cases} f(y; \theta), & |y| > c, \\ \int\limits_{|u| \leqslant c} f(u; \theta) \mu(du) / \int\limits_{|u| \leqslant c} \mu(du), & |y| \leqslant c, \end{cases}$$

$$\Psi(z; \theta \mid y) = \begin{cases} f(z; \theta) \int\limits_{|u| \leqslant c} \mu(du) / \int\limits_{|u| \leqslant c} f(u; \theta) \mu(du), \quad y = 0, & |z| \leqslant c, \\ 1, & \text{otherwise}, \end{cases}$$

where

$$\varphi_0(y;\theta) = d\mathbf{P}_\theta^Y / d\mathbf{P}^Y,$$

$\mathbf{P}_\theta^Y, \mathbf{P}^Y$ are martingales (first) of the distribution $\mathbf{P}_\theta^{Y,Z}$ and $\mathbf{P}^{Y,Z}$, respectively. Recall that

$$\mathbf{P}^Y(dy) = \begin{cases} \mu(dy), & |y| > c, \\ \sigma_0(dy) \int\limits_{|u|\leqslant c} \mu(du), & |y| \leqslant c, \end{cases}$$

where $\sigma_0(dy)$ is Dirac measure.

Thus the initial problem is reduced to the one of Example 1 dealing with the estimation of the parameter θ with respect to two-dimensional observations $(Y_i, Z_i)_{i\geqslant 1}$ with a factorized density $f(y,z;\theta) = \varphi_0(y;\theta)\Psi(z;\theta \mid y)$. Here we have, as above,

$$\rho_\theta^n(t) = \mathcal{E}_t(X_{0,\theta}^n)\mathcal{E}_t(N_\theta^n),$$

where $X_{0,\theta}^n = W * (\mu^n - \nu^n)$; $N_\theta^n = V_\theta * \mu^n$, μ^n is a jump measure of the process

$$\sum_{i=1}^{[nt]} Y_i = \sum_{i=1}^{[nt]} X_i I_{|X_i|>c},$$

ν^n is its compensator. The natural class of alternative experiments is the class of experiments corresponding to two-dimensional i.i.d. observations $(Y_i, Z_i)_{i\geqslant 1}$ with a fixed conditional density $\Psi(z;\theta \mid y)$ and an arbitrary marginal density $\varphi(y;\theta)$ (with respect to the measure μ) which in the initial problem corresponds to the alternatives under which the density value is fixed on the set $\{x\colon\ |x| \leqslant c\}$ and arbitrary on the set $\{x\colon\ |x| > c\}$.

Example 4.3. *Discrete time* (Cox, 1975; Wong, 1976).

Let $\{X_1 = (Y_1, Z_1), \dots, X_n = (Y_n, Z_n)\}_{n\geqslant 1}$ be a sequence of two-dimensional observations with a local density $f_n(x_1,\dots,x_n;\theta)$, $\theta \in \Theta \subset \mathbf{R}^d$ (with respect to some probability measure $\mathbf{P}$ on $((\mathbf{R}^2)^\infty), \mathcal{B}((\mathbf{R}^2)^\infty))$ $(f_n(x_1,\dots,x_n;\theta)$ will also be called the full likelihood as in (Cox, 1975)).

The full likelihood is assumed to be factorized as follows:

$$f(X_1,\dots,X_n;\theta) = \prod_{i=1}^n \varphi_0(Y_i;\theta) \mid d_i) \prod_{i=1}^n \Psi(Z_i;\theta \mid C_i),$$

where φ_0, Ψ are conditional densities,

$$d_i = (Y_1, Z_1, \dots, Y_{i-1}, Z_{i-1}), \quad C_i = (Y_1, Z_1, \dots, Y_{i-1}, Z_{i-1}, Y_i).$$

In Cox' approach the conditional density may turn out either unknown or, in addition to the parameter θ, containing some nuisance parameter ν. In this case Cox suggests to replace the full likelihood by the partial likelihood

$$\prod_{i=1}^n \Psi(Z_i;\theta \mid C_i).$$

The local density can be written as

$$\rho_\theta^n(t) = \mathcal{E}_t(M_\theta^n) = \mathcal{E}_t(X_{0,\theta}^n)\mathcal{E}_t(N_\theta^n),$$

where the processes $M_\theta^n, X_{0,\theta}^n, N_\theta^n$ defined by the relations

$$M_\theta^n(t) = \sum_{i=1}^{[nt]} \big(\varphi_0(Y_i;\theta \mid d_i)\Psi(Z_i;\theta \mid C_i) - 1\big),$$

$$X_{0,\theta}^n(t) = \sum_{i=1}^{[nt]} \varphi(Y_i;\theta \mid d_i) - 1),$$

$$N_\theta^n(t) = \sum_{i=1}^{[nt]} \big(\Psi(Z_i;\theta \mid C_i) - 1\big)$$

are $(\mathbf{P}^n, F^n)$-martingales, where $F^n = (\mathcal{F}^n)_{0\leqslant t\leqslant 1}$, $\mathcal{F}_t^n = \mathcal{F}_{[nt]}$, $\mathcal{F}_n = \sigma(Y_i, Z_i, i \leqslant n)$, $\mathbf{P}^n$ is the restriction of the measure $\mathbf{P}$ on the σ-algebra $\mathcal{F}_n$. For all $n \geqslant 1$ and $\theta \in \Theta$ we have $X_{0,\theta}^n \perp N_\theta^n$. The partial likelihood has the form

$$\prod_{i=1}^{[nt]} \Psi(Z_i;\theta \mid C_i) = \mathcal{E}_t(N_\theta^n).$$

Let μ^n be a jump measure of the process $\sum_{i=1}^{[nt]} Y_i$, i.e.,

$$\mu^n((0,t] \times B) = \sum_{i=1}^{[nt]} I_{\{Y_i \in B\}},$$

and ν^n its $\mathbf{P}^n$-compensator, i.e., $\nu^n((0,t] \times B) = \sum_{i=1}^{[nt]} \mathbf{P}\{Y_i \in B \mid \mathcal{F}_{i-1}\}$. Then we have

$$X_{0,\theta}^n = W_\theta^n * (\mu^n - \nu^n),$$
$$N_\theta^n = V_\theta^n * \mu^n,$$

where

$$W_\theta^n(\omega,t,y) = \varphi_0(y;\theta \mid d_{[nt]}), \quad V_\theta^n(\omega,t,y) = \Psi(Z_{[nt]};\theta \mid d_{[nt]}, y) - 1.$$

Hence, $X_{0,\theta}^n \in \mathcal{K}^{1,1}(\mu^n)$ and $N_\theta^n \in \mathcal{K}^{1,2}(\mu^n)$. The class H^n of martingales determining the alternatives can be defined in the usual manner, but as above we shall confine ourselves by natural alternatives and consider only the subclass H_1^n of H^n which consists of martingales of the form

$$X_\theta^n(t) = W_\theta^n * (\mu^n - \nu^n)$$

with $W^n(\omega,t,y) = \varphi(y \mid d_{[nt]})$, where $\varphi(y \mid d_i)$ is some conditional density, i.e.,

$$\int \varphi(y \mid d_i)\, \mathbf{P}(dy \mid d_i) = 1, \quad \varphi > 0.$$

This means that alternatives are specified as measure Q_θ^n on $((R^2)^\infty, \mathcal{B}((R^2)^\infty))$ for which the densities of conditional distributions of Z_i with respect to C_i are fixed and coincide with $\Psi(Z_i; \theta \mid C_i)$ and conditional densities of Y_i with respect to d_i are allowed to be arbitrary. Stable M-estimators will be defined by martingales of the form

$$m_\theta^n = V_\theta^{m,n} * \mu^n$$

with $V_\theta^{m,n}(\omega, t; y) = g_n(Y_1, Z_1, \ldots, Y_{[nt]-1}, Z_{[nt]-1}, Y, Z_{[nt]}, \theta)$, where the functions $g_i(Y_1, Z_1, \ldots, Y_i, Z_i; \theta)$ are chosen so that $m_\theta^n \perp H_1^n$.
In particular, we can write out some sufficient conditions

$$\int g_i(Y_1, Z_1, \ldots, Y_i, z; \theta)\mathbf{P}(dz \mid C_i) = 0$$

$$\int\int \varphi(y \mid d_i) \mid g(C_i, z; \theta) \mid \mathbf{P}(dy \mid d_i)\mathbf{P}(dz \mid C_i) < \infty$$

for all $C_i, d_i, \varphi, \theta$ etc.

Example 4. Multivariate point process.
Let $(\Omega, \mathcal{F}, \widetilde{F}, \mathbf{P})$ be a probability space with filtration satisfying the usual conditions.
A multivariate point process on $(\mathbf{R}_+ \times E, \mathcal{B}(\mathbf{R}_+) \otimes \mathcal{B}(E))$ (E is a Luzin space) is assumed to be a sequence $(T_n, X_n)_{n \geqslant 1}$, where T_n are Markov moments such that $T_1 > 0, T_n < T_{n+1}$ on $\{T_n < \infty\}$, while X_n are random elements with the property: $X_n \in E$ on $\{T_n < \infty\}$, $X_n = \delta$ on $[T_n = \infty]$, where δ is a "marginal point", $\delta \notin E$ and $\{X_n \in C\} \in \mathcal{F}_{T_n}$, $\forall C \in \mathcal{B}(E)$, $n \geqslant 1$.
A complete description of the multivariate point process is given by the integer-valued random measure

$$\mu(\omega; dt, dx) = \sum_{n \geqslant 1} I_{\{T_n < \infty\}} \varepsilon_{T_n, X_n}(dt, dx),$$

where $\mathcal{E}_{(T_n, X_n)}(dt, dx)$ is Dirac measure. As it is well-known, the random measure μ has a compensator $\widetilde{\nu}$ with respect to $(\mathbf{P}, \widetilde{F})$ (a version can be chosen such that $\widetilde{\nu}(\omega, \{t\}, E) \leqslant 1$). Let $G = (G_t)_{t \geqslant 0}$ be a natural flow of σ-algebras corresponding to the measure μ with

$$G_t = \sigma\Big\{\mu\big(\omega; (0, s] \times B\big), \ s \leqslant t, \quad B \in \mathcal{B}(E)\Big\}.$$

Let $F = \{\mathcal{F}_0 \vee G_t\}_{t \geqslant 0}$ and assume that F satisfy the usual conditions.
Consider the stochastic basis $(\Omega, \mathcal{F}, F, \mathbf{P})$. In the case the compensator ν of the random measure μ with respect to $(\mathbf{P}, F)$ can be explicitly defined, and it will uniquely define the measure $\mathbf{P}$ in the sense that if $\overline{\mathbf{P}}$ is some other measure and $\overline{\nu}$ is the $(\overline{\mathbf{P}}, F)$-compensator, then

$$\overline{\mathbf{P}}_0 = \mathbf{P}_0, \quad \overline{\nu} = \nu \Rightarrow \overline{\mathbf{P}} = \mathbf{P}.$$

Let for every $\theta \in \Theta$ on $(\Omega, \mathcal{F}, F)$ the measure $\mathbf{P}_\theta$ be defined and let ν_θ be a $\mathbf{P}_\theta$-compensator of μ.

We will assume below that

1) $\{T_n < \infty\} = \{\lim_{n\to\infty} T_n = \infty\} = \Omega$,

2) $\nu(\{t\}, E) = \nu_\theta(\{t\}, E) = 0$.

Assumption 1) implies that, for any $\theta \in \Theta$, we have

$$\mu(t) = \mu(\omega; (0,t] \times E) = \int_0^t \int_E \mu(\omega, ds, dx) < \infty$$

which, for all $\theta \in \Theta$ and $t > 0$, implies

$$\nu_\theta(t) = \nu_\theta(\omega; (0,t] \times E) < \infty,$$

$$\nu(t) = \nu(\omega; (0,t] \times E) < \infty, \quad \forall \omega \in \Omega.$$

Note that $\mu(\omega; dt)$ is a counting process

$$\mu(\omega; (0,t] \times E) = \sum_n I_{\{T_n \leqslant t\}}$$

and $\nu_\theta(t)$ and $\nu(t)$ are its compensators with respect to $\mathbf{P}_\theta$ and $\mathbf{P}$, respectively. From the local equivalence of the measures $\mathbf{P}_\theta$ and $\mathbf{P}$ it follows that $\nu_\theta \sim \nu$ with

$$\frac{\nu_\theta(ds, dx)}{\nu(ds, dx)} = \lambda(s, x, \theta) > 0$$

(the argument ω is omitted), and

$$\rho_\theta(t) = \frac{d\mathbf{P}_\theta(t)}{d\mathbf{P}(t)} = \mathcal{E}_t(M_\theta),$$

where M_θ is a $\mathbf{P}$-local martingale ($\forall \theta \in \Theta$). By assumption 2), we have

$$M_\theta(t) = \int_0^t \int_E (\lambda(s, x, \theta) - 1)(\mu - \nu)(ds, dx).$$

In this example we give only the multiplicative decomposition of the exponential (or, what is the same, the decomposition of the martingale M_θ) and specify the measure μ (with respect to which the classes $\mathcal{K}^{1,q}(\mu)$, $q = 1, 2$, have been introduces). Thus, Example 4 fits into the general scheme described above.
Let us now decompose the martingale M_θ.
The measures $\nu_\theta(ds, dx)$ and $\nu(ds, dx)$ can be written in the factorized from

$$\nu_\theta(ds, dx) = \nu_\theta(ds)\, q_t(ds, \theta),$$

$$\nu(ds, dx) = \nu(ds)\, q_t(dx),$$

where, roughly speaking, $q_t(dx)$ is a conditional distribution of the jump under the condition that the jump occurs at the moment t.
Denote

$$\lambda_\theta^0(t) = \frac{\nu_\theta(dt)}{\nu(dt)}, \quad g_t(x,\theta) = \frac{q_t(dx,\theta)}{q_t(dx)}.$$

Then

$$\lambda(t,x,\theta) = \lambda_\theta^0(t)\, g_t(x,\theta)$$

and, as a result, M_θ can be decomposed as follows:

$$M_\theta(t) = \int_0^t \int_E (\lambda(s,x,\theta) - 1)(\mu - \nu)(ds,dx) =$$

$$= \int_0^t \int_E (\lambda_\theta^0(s) - 1)(\mu - \nu)(ds,dx) +$$

$$+ \int_0^t \int_E (g_s(x,\theta) - 1)(\mu - \nu)(ds,dx) +$$

$$+ \int_0^t \int_E (\lambda_\theta^0(s) - 1)\big(g_s(x,\theta) - 1\big)(\mu - \nu)(ds,dx).$$

Here the first term can be rewritten as

$$X_{0,\theta}(t) = \int_0^t \int_E (\lambda_\theta^0(s) - 1)(\mu - \nu)(ds,dx) =$$

$$= \int_0^t (\lambda_\theta^0(s) - 1)\big(\mu(ds) - \nu(ds)\big).$$

As to the second term, with the equality

$$\int_E \big(g_t(x,\theta) - 1\big)\, q_t(dx) = 0 \tag{4.3}$$

in mind, it can be rewritten as

$$N_\theta(t) = \int_0^t \int_E \big(g_s(x,\theta) - 1\big)\, \mu(ds,dx).$$

The martingale N_θ can be written in a different way, namely, as a stochastic integral with respect to the measure $\mu(\omega, ds)$

$$N_\theta(t) = \int_0^t \big(g_s(\beta_s, \theta) - 1\big)\, \mu(\omega, ds),$$

where $\beta_s(\omega) \in E$ is such that

$$\mu(\omega, \{s\}, dx) = \mathcal{E}_{\beta_s(\omega)}(dx).$$

Indeed,

$$\Delta N_\theta(t) = \big(g_t(\beta_t, \theta) - 1\big) I_D$$

and, besides,

$$\Delta_t \int_0^{\cdot} \big(g_s(\beta_s, \theta) - 1\big)\, \mu(\omega, ds) =$$

$$= \big(g_t(\beta_t, \theta) - 1\big)\, \mu(\omega, \{t\}, E) = \big(g_t(\beta_t) - 1\big) I_D.$$

The martingales N_θ and $X_{0,\theta}$ are orthogonal, since

$$[X_{0,\theta}, N_\theta] = \int_0^t \int_E \big(\lambda_\theta^0(s) - 1\big)\big(g_s(x, \theta) - 1\big)\, \mu(ds, dx) =$$

$$= \int_0^t \int_E \big(\lambda_\theta^0(s) - 1\big)\big(g_s(x, \theta) - 1\big)\big(\mu - \nu\big)(ds, dx)$$

is a $\mathbf{P}$-martingale. (The last equality is true by virtue of (4.3).)
Hence, we obtain the decomposition

$$M_\theta = X_{0,\theta} + N_\theta + [X_{0,\theta}, N_\theta]$$

with orthogonal $N_\theta, X_{0,\theta}$ and the martingales can be rewritten in two ways:

$$X_{0,\theta} = \int_0^t \int_E \big(\lambda_\theta^0(s) - 1\big)\big(\mu - \nu\big)(ds, dx),$$

$$N_\theta = \int_0^t \int_E \big(g_s(x, \theta) - 1\big)\, \mu(ds, dx)$$

and, alternatively,

$$X_{0,\theta} = \int_0^t \big(\lambda_\theta^0(s) - 1\big)\big(\mu(\omega, ds) - \nu(\omega, ds)\big),$$

$$N_\theta = \int_0^t (g_s(\beta_s, \theta) - 1)\mu(\omega, ds).$$

It is evident now how the classes $\mathcal{K}^{1,q}(\mu)$ (with two different $\mu(dt)$, $\mu(dt, dx)$, H_0, H etc.), can be introduced.
Observe that the classes of natural alternatives are obtained by choosing all possible $\lambda_\theta(s)$ with fixed $g_s(x, \theta)$ (certainly of such $\lambda_\theta(s)$) which define a martingale

$$X_\theta(t) \doteq \int_0^t (\lambda_\theta(s) - 1)(\mu - \nu)(ds, dx)$$

such that $\mathbf{E}_\theta(M_\theta^X) = 1$, $M_\theta^X = X_\theta + N_\theta + [X_\theta, N_\theta]$.
In (Dzhaparidze, 1986) the case $E = \{1, 2, \ldots, r\}$, $N_t = \{N_t^1, \ldots, N_t^r\}$ with $N_t^i = \mu(\omega; (0, t], E)$ is considered (N_t is called the r-variate counting process) with the $\mathbf{P}$-compensator of the process N_t

$$A_t = \{A_t^1, \ldots, A_t^r\}, \quad A_t^i = \nu((0, t], i)$$

and the $\mathbf{P}_\theta$-compensator is $A_t(\theta) = \{A_t^1(\theta), \ldots, A_t^r(\theta)\}$, where $A_t^i(\theta) = \nu_\theta((0, t], i)$.
In this case

$$\lambda_s(\theta) = \nu_\theta(ds, E)/\nu(ds, E) = d\,\overline{A}_t(\theta)/d\,\overline{A}_t,$$

where

$$\overline{A}_t(\theta) = \sum_{i=1}^r A_t^i(\theta), \quad \overline{A}_t = \sum_{i=1}^r A_t^i$$

(in the notation of (Dzhaparidze, 1986)

$$g_s(i, \theta) = \Psi_t^i(\theta)/\Psi_t^i, \quad \Psi_t^i(\theta) = dA_t^i(\theta)/d\,\overline{A}_t^i(\theta), \quad \Psi_t^i = dA_t^i/d\,\overline{A}_t).$$

The martingale N_θ then has the form

$$N_\theta = \int_0^t (g_s(x, \theta) - 1)\,\mu(ds, dx) =$$

$$= \sum_{i=1}^r \int_0^t (g_s(i, \theta) - 1) dN_t^i.$$

Example 5.
Let $\xi = (\eta, \zeta)$ be a two-dimensional diffusion type process satisfying the stochastic differential equation

$$d\eta_t = \Phi^0(t, \eta, \zeta, \theta)dt + dW_t^1,$$

$$d\zeta_t = \Psi(t, \eta, \zeta, \theta)dt + dW_t^2, \tag{4.4}$$

where W^1 and W^2 are independent Wiener processes and the coefficients Φ^0 and Ψ are such that there exists the unique weak solution of equation (4.4) and $\mathbf{P}_\theta^{\eta,\zeta} \overset{\text{loc}}{\sim} \mathbf{P}^{W^1,W^2}$, where $\mathbf{P}_\theta^{\eta,\zeta}$ is a distribution of the couple (η,ζ) with a given θ, $\mathbf{P}^{W_1,W_2}$ is a distribution of the couple (W^1,W^2). Then

$$d\mathbf{P}_\theta^{\eta,\zeta}/d\mathbf{P}^{W_1,W_2} = \mathcal{E}(M_\theta),$$

where

$$M_\theta(t) = \int_0^t \Phi^0(s,y,z,\theta)dy_s + \int_0^t \Psi(s,y,z,\theta)dz_s.$$

Note that the measures $\mathbf{P}_\theta^{\eta,\zeta}$ and $\mathbf{P}^{w_1,W_2}$are defined on $(C^2_{[0,T]},\mathcal{B}(C^2_{[0,T]}))$ with the filtration $\mathcal{F}_t = (y,z:\ Y_s,Z_s,s\leqslant t)$, $Y=(Y_s)_{s\geqslant 0}$, $Z=(Z_s)_{s\geqslant 0}$ are coordinate processes.

Let

$$X_{0,\theta}(t) = \int_0^t \Phi^0(s,Y,Z,\theta)dY_s, \quad N_\theta(t) = \int_0^t \Psi(s,Y,Z,\theta)dZ_s.$$

The processes $X_{0,\theta}$ and N_θ are $\mathbf{P}^{W_1,W_2}$-martingales and $\langle X_{0,\theta},N_\theta\rangle = 0$.

Thus,

$$d\mathbf{P}_\theta^{\eta,\zeta}/d\mathbf{P}^{W^1,W^2} = \mathcal{E}(M_\theta) = \mathcal{E}(X_{0,\theta})\mathcal{E}(N_\theta).$$

In this case a class of natural alternatives is given by the variation of Φ with a given Ψ, and stable estimators with respect to this class should be constructed by means of $\mathbf{P}_\theta$-martingales m_θ of the form

$$m_\theta(t) = \int_0^t g(s,Y,Z,\theta)dZ_s.$$

(Note that $m_\theta \perp X_\theta$ and $X_\theta(t) = \int_0^t \Phi(s,Y,Z,\theta)dY_s$.)

Example 6. *Censored diffusion.*

Let ξ be a diffusion process with the differential

$$d\xi(t) = I_{\{|\xi_t|>C\}}\Phi_0(t,\xi,\theta)dt + I_{\{|\xi_t|\leqslant C\}}\Psi(t,\xi,\theta)dt + dW_t, \tag{4.5}$$

where $\Phi_\theta(t,\xi,\theta)$ and $\Psi(t,\xi,\theta)$ are functionals such that, for every $\theta\in\Theta$, there exists the unique weak solution of the equation (4.5). Let $\mathbf{P}_\theta^\xi$ be the distribution of ξ with a given θ.

It is assumed that $\mathbf{P}_\theta^\xi \overset{\text{loc}}{\sim} \mathbf{P}^W$.

Then

$$d\mathbf{P}_\theta^\xi/d\mathbf{P} = \mathcal{E}(M_\theta) = \mathcal{E}(X_{0,\theta})\mathcal{E}(N_\theta),$$

where

$$M_\theta = X_{0,\theta} + N_\theta,$$

$$X_{0,\theta} = \int_0^t I_{\{|X_s|>C\}}\Phi_0(s,X_s,\theta)dx_s,$$

$$N_\theta(t) = \int_0^t I_{\{|X_s|\leqslant C\}} \Psi(s, X_s, \theta) dX_s$$

and $\langle X_{0,\theta}, N_\theta \rangle = 0$.
It is evident that the class of martingales defining alternatives consists of martingales of the form

$$X(t) = \int_0^t I_{\{|X_s|>C\}} \Phi(s, X_s, \theta) dX_s$$

and stable estimators will be obtained by means of martingales m_θ of the form

$$m_\theta(t) = \int_0^t I_{\{|\xi_s|\leqslant C\}} g(s, X_s, \theta) dX_s.$$

5. ASYMPTOTIC EFICIENCY OF PLE

Consider the set of experiments $E_0 = (\mathcal{E}^X,\ X \in \mathbf{X}_0)$

$$\mathsf{E}^X = (\mathcal{E}_n^X)_{n\geqslant 1} = \big(\Omega^n,\ \mathcal{F}^n,\ F^n = (\mathcal{F}_t^n)_{0\leqslant t\leqslant T},\ \mathbf{P}^n, \{Q_\theta^{X,n},\ \theta \in \Theta\}\big)_{n\geqslant 1}$$

with

$$\frac{dQ_\theta^{X,n}}{d\mathbf{P}^n} = \mathcal{E}(X_\theta^n)\mathcal{E}(N_\theta^n) = \mathcal{E}(M_\theta^{X,n}),$$

$$M_\theta^{X,n} = X_\theta^n + N_\theta^n + [X_\theta^n, N_\theta^n];\ X = (X_\theta^n,\ \theta \in \Theta)_{n\geqslant 1}.$$

Assume that the set $\mathbf{X}_0 \subset \mathbf{X}$ and $\{N_\theta^n,\ \theta \in \Theta\}_{n\geqslant 1}$ are such that the experiment $\mathcal{E}^X$ satisfies the conditions of Theorem 3.1 (Chitashvili et al., 1990) and also the conditions of Theorem 3.1 of Section 3, for all $X \in \mathbf{X}_0$. Assume, besides, the families $X = (X_\theta^n,\ \theta \in \Theta)_{n\geqslant 1}$ such that $X_\theta^n \equiv X^n \in H_0^n$ (i.e., X_θ^n does not depend on θ) belong to this set $\mathbf{X}_0$. The family of measures $\{Q_\theta^{X,n},\ \theta \in \Theta\}_{n\geqslant 1}$ satisfies the LAN condition for each $X \in \mathbf{X}_0$. This means, of course, that

$$\log \frac{dQ_{\theta+C_n(X,\theta)u_n}^{X,n}}{dQ_\theta^{X,n}} = \Delta_\theta^{X,n} u - \frac{1}{2}u^2 + \Psi_\theta^{X,n}(u_n), \tag{5.1}$$

where

$$C_n^2(X,\theta) = \mathbf{E}_\theta^{X,n}\langle L(\dot{M}_\theta^{X,n}, M_\theta^{X,n})\rangle,\ \Delta_\theta^{X,n} = C_n(X,\theta)\ L(\dot{M}_\theta^{X,n}, M_\theta^{X,n}),$$

$$\mathcal{L}\{\Delta_\theta^{X,n} \mid Q_\theta^{X,n}\} \Rightarrow \mathcal{N}(0,1),\ Q_\theta^{X,n} - \lim_{n\to\infty} \Psi_\theta^{X,n}(u_n) = 0$$

for every sequence $(u_n)_{n\geqslant 1}$ with $\lim_{n\to\infty} u_n = u$,

$$\theta + C_n(X,\theta)u_n \in \Theta.$$

Note that

$$L(\dot{M}_\theta^{X,n}, M_\theta^{X,n}) = L(\dot{N}_\theta^n, N_\theta^n) + L(\dot{X}_\theta^n, X_\theta^n).$$

From this and in accordance with the specification of the model (in particular, owing to the orthogonality of martingales $\dot{X}_\theta^n$ and $\dot{N}_\theta^n$) it immediately follows that

$$\langle L(\dot{m}_\theta^{X,n}, M_\theta^{X,n})\rangle = \langle L(\dot{N}_\theta^n, N_\theta^n)\rangle + \langle L(\dot{X}_\theta^n, X_\theta^n)\rangle. \tag{5.2}$$

Under conditions of Theorem 3.1 of Section 3 (with the normalizing sequence $C_n(X,\theta)$), we have for PLE $\hat{\theta} = (\hat{\theta}_n)_{n\geqslant 1}$ (see assertion iv) of Theorem 3.1)

$$\mathcal{L}\{C_n^{-1}(X,\theta)(\hat{\theta}_n - \theta) \mid Q_\theta^{X,n}\} \Longrightarrow \mathcal{N}(0, \sigma_X(\theta)), \tag{5.3}$$

where

$$\sigma_X(\theta) = Q_\theta^{X,n} - \lim_{n\to\infty} \frac{\langle L(\dot{M}_\theta^{X,n}, M_\theta^{X,n})\rangle}{\langle(\dot{N}_\theta^n, N_\theta^n)\rangle} \geqslant 1$$

Let $\widetilde{C}_n^{-2}(X,\theta) = \mathbf{E}_\theta^{X,n}\langle L(\dot{N}_\theta^n, N_\theta^n)\rangle$. Then from (5.3) it directly follows that

$$\mathcal{L}\{\widetilde{C}_n^{-1}(X,\theta)(\hat{\theta}_n - \theta) \mid Q_\theta^{X,n}\} \Rightarrow \mathcal{N}(0,1). \tag{5.4}$$

From last relation and LAN property of family $\{Q_\theta^{X,n},\ \theta \in \Theta\}_{n\geqslant 1}$ it is easy to obtain the regularity property of PLE $\hat{\theta}$ (see for this the proof of Theorem 7.1 in (Chitashvili et al., 1990))

$$\mathcal{L}\{\widetilde{C}_n^{-1}(X,\theta)(\hat{\theta}_n - \theta) - u_n \mid Q_{\theta+\widetilde{C}_n(X,\theta)u_n}\} \Rightarrow \mathcal{N}(0,1) \tag{5.5}$$

for all $(u_n)_{n\geqslant 1}$ such that $\lim_{n\to\infty} u_n = u,\ u \in \mathbf{R}^1$.
Let now $W(z),\ z \in \mathbf{R}^1$, be usual loss function with properties $\mathbf{E}W(\xi) < \infty$, $\xi \in \mathcal{N}(0,1)$. Denote $W_a(z) = \min(W(z), a)$.

LEMMA 5.1. *For each estimator $\tilde{\theta} = (\tilde{\theta}_n)_{n\geqslant 1}$ the following inequality is true:*

$$\sup_{X\in\mathbf{X}_0} \sup_a \lim_{k\to\infty} \underline{\lim}_{n\to\infty} \sup_{|\widetilde{C}_n^{-1}(X,\theta)(\theta-t)|\leqslant k} \mathbf{E}_t^{X,n}\Big\{W_a(X,\theta)(\widetilde{C}_n^{-1}(X,\theta)(\tilde{\theta}_n - t))\Big\} \geqslant$$

$$\geqslant \mathbf{E}W(\xi), \tag{5.6}$$

where $\xi \in \mathcal{N}(0,1)$.

Proof. Note that $\widetilde{C}_n(X,\theta) = C_n(X,\theta)$ for every $X = (X_\theta^n,\ \theta \in \Theta)_{n\geqslant 1}$ such that $X_\theta^n \equiv X^n$, $X^n \in H_0^n$, $n \geqslant 1$ (i.e., for X independent of θ).
For such X from Hajek's inequality we have

$$\sup_a \lim_{k\to\infty} \underline{\lim}_{n\to\infty} \sup_{|\widetilde{C}_n^{-1}(X,\theta)(\theta-t)|\leqslant k} \mathbf{E}_t^{X,n}\Big\{W_a(\widetilde{C}_n^{-1}(X,\theta)(\tilde{\theta}_n - t))\Big\} \geqslant \mathbf{E}W(\xi).$$

Now (5.6) directly follows from the last relation.

We say that estimator $\theta^* = (\theta^*_n)_{n\geqslant 1}$ is efficient (locally asymptotically minimax) at θ with respect to loss function W if equality in (5.6) holds for θ^*.
From (5.5) it immediately follows that PLE $\hat\theta = (\hat\theta_n)_{n\geqslant 1}$ is efficient with respect to loss functions W. Moreover, we have

$$\sup_a \lim_{k\to\infty} \varliminf_{n\to\infty} \sup_{|\widetilde{C}_n^{-1}(X,\theta)(\theta-t)|\leqslant k} \mathbf{E}_t^{X,n}\Big\{W_a\big(\widetilde{C}_n^{-1}(X,\theta)(\hat\theta_n - t)\big)\Big\} = \mathbf{E}W(\xi) \quad (5.7)$$

for all $X \in \mathbf{X}_0$.
Denote now by $\mathbf{X}_0^c$ the set of families $X = \{X_\theta^n,\ \theta\in\Theta\}_{n\geqslant 1}$ such that $X_\theta^n \equiv X^n$, for some $X^n \in H_0^n$ (i.e., X does not depend on θ).
Consider the set of alternatives $\{Q_\theta^{X,n},\ \theta\in\Theta\}_{n\geqslant 1}$ for which

$$\frac{dQ_\theta^{X,n}}{d\mathbf{P}^n} = \mathcal{E}(X^n)\mathcal{E}(N_\theta^n),$$

where $X \in \mathbf{X}_0^c$. Then PLE coincides with MLE for such alternatives and PLE is unimproved in the following sense: there is no estimator $\tilde\theta = (\tilde\theta_n)_{n\geqslant 1}$ such that

$$\sup_a \lim_{k\to\infty} \varliminf_{n\to\infty} \sup_{|\widetilde{C}_n^{-1}(\overline{X},\theta)(\theta-t)|\leqslant k} \mathbf{E}_t^{\overline{X},n}\Big\{W_a\big(\widetilde{C}_n^{-1}(\overline{X},\theta)\tilde\theta - t)\big)\Big\} <$$

$$< \sup_a \lim_{k\to\infty} \varliminf_{n\to\infty} \sup_{|\widetilde{C}_n^{-1}(\overline{X},\theta)(\theta-t)|\leqslant k} \mathbf{E}_t^{\overline{X},n}\Big\{W_a\big(\widetilde{C}_n^{-1}(\overline{X},\theta)(\hat\theta_n - t)\big)\Big\} \quad (5.8)$$

for some $\overline{X} \in \mathbf{X}_0^c$ and in (5.8) the equality holds for other $X \in \mathbf{X}_0^c$.
The situation is quite different if we consider all alternatives $\{Q_\theta^{X,n},\ \theta\in\Theta\}_{n\geqslant 1}$, $X \in \mathbf{X}_0$ (i.e., alternatives for which X_θ^n essentially depends on θ). For simplicity let us consider the case $W(z) = z^2$.
Here estimators "better" than PLE may arise, i.e., estimators for which (5.8) takes place, for some $X \in \mathbf{X}_0$, and the equality is true for other one. In particular, if the family $\mathbf{X}_0$ allows the regular adaptive estimators at point θ, i.e., estimators $\overline\theta = (\overline\theta_n)_{n\geqslant 1}$ for which the relation

$$\mathcal{L}\{C_n^{-1}(X,\theta)(\overline\theta_n - \theta) - u_n \mid Q^{X,n}_{\theta+C_n(X,\theta)u_n}\} \Rightarrow \mathcal{N}(0,1) \quad (5.9)$$

holds for every $(u_n)_{n\geqslant 1}$ such that $\lim_{n\to\infty} u_n = u$ and $X \in \mathbf{X}_0$, then strong inequality (5.8) is true for all $X = \{X_\theta^n,\ \theta\in\Theta\}_{n\geqslant 1}$ such that X_θ^n essentially depends on θ (of course with $\overline\theta$ instead of $\tilde\theta$).
Indeed, from (5.9) we have

$$\mathcal{L}\left\{\widetilde{C}_n^{-1}(X,\theta)(\overline\theta_n - \theta) - u_n \mid Q^{X,n}_{\theta+\widetilde{C}_n(X,\theta)u_n}\right\} \Rightarrow \mathcal{N}\left(0,\ \lim_{n\to\infty}\frac{\widetilde{C}_n^{-1}(X,\theta)}{C_n^{-1}(X,\theta)}\right)$$

which leads to inequality

$$\sup_a \lim_{k\to\infty} \varliminf_{n\to\infty} \sup_{|\widetilde{C}_n^{-1}(X,\theta)(\theta-t)|\leqslant k} \mathbf{E}_t^{X,n}\Big\{W_a\big(\widetilde{C}_n^{-1}(X,\theta)(\hat\theta - t)\big)\Big\} =$$

$$= \lim_{n\to\infty} \frac{\widetilde{C}_n^2(X,\theta)}{C_n^{-1}(X,\theta)} < 1.$$

This is evident from (5.2).

As important example we consider the scheme of partial likelihood in presence of nuisance parameter. Namely, let

$$\frac{d\mathbf{P}^n_{\theta,\eta}}{d\mathbf{P}^n} = \mathcal{E}(M^n_{\theta,\eta}) = \mathcal{E}(N^n_\theta)\mathcal{E}(X^n_{\theta,\eta}), \quad \theta,\eta \in \mathbf{R}^1,$$

where θ is a parameter of interest, η is a nuisance one, $X^n_{\theta,\eta}$ is a $\mathbf{P}^n$-martingale depending on θ,η while the martingale X^n_θ depends only on θ and $X^n_{\theta,\eta}\perp N^n_\theta$ for all $\theta,\eta \in \mathbf{R}^1$. Suppose also that N^n_θ is θ-differentiable and $\overset{*}{N}\perp \overset{*}{X}{}^n_{\theta,\eta}$ for all $\theta,\eta \in \mathbf{R}^1$, $n \geqslant 1$, where

$$\overset{*}{N}{}^n_\theta = \partial N^n_\theta/\partial\theta, \quad \overset{*}{X}{}^n_{\theta,\eta} = \partial X_{\theta,\eta}/\partial\theta.$$

The PLE $\hat\theta = (\hat\theta_n)_{n\geqslant 1}$ of the parameter θ is defined as usual by the equation $L(\overset{*}{N}{}^n_\theta, N^n_\theta) = 0$. Put

$$l^n_1(\theta,\eta) = L(\frac{\partial}{\partial\theta}M^n_{\theta,\eta}, M^n_{\theta,\eta}) = L(\overset{*}{M}{}^n_{\theta,\eta}, M^n_{\theta,\eta}),$$

$$l^n_2(\theta,\eta) = L(\frac{\partial}{\partial\eta}M^n_{\theta,\eta}, M^n_{\theta,\eta}) = L(\dot{M}^n_{\theta,\eta}, M^n_{\theta,\eta}),$$

$$I_{ij}(\theta,\eta) = \mathbf{P}^n_{\theta,\eta}-\lim_{n\to\infty} C^2_n\langle l^n_i(\theta,\eta),\ l^n_i(\theta,\eta)\rangle, \quad i,j=1,2,$$

$$I^*(\theta,\eta) = I_{11}(\theta,\eta) - I^2_{12}(\theta,\eta)\, I^{-1}_{22}(\theta,\eta),$$

where $(C_n)_{n\geqslant 1}$ is some normalizing sequence.

Now applying Theorem 3.1 from Section 3 to this case, we get

$$\mathcal{L}\{C^{-1}_n(\hat\theta_n - \theta) \mid \mathbf{P}^n_{\theta,\eta}\} \Rightarrow \mathcal{N}\big(0, \sigma_1(\theta,\eta)\big),$$

where

$$\sigma^{-1}_1(\theta,\eta) = \mathbf{P}^n_{\theta,\eta}-\lim_{n\to\infty} C^2_n\langle L(\overset{*}{N}{}^n_\theta, N^n_\theta\rangle.$$

If $\tilde\theta = (\tilde\theta_n)_{n\geqslant 1}$ is the estimator, constructed by means of classical projection technique (see (Chitashvili et al., 1990)), i.e., by the equation

$$\widetilde{l}^n_1(\theta,\overline{\eta}_n) = 0,$$

where $\widetilde{l}^n_1(\theta,\eta)^n_1$ is a modified informant, i.e.,

$$\widetilde{l}^n_1(\theta,\eta) = l^n_1(\theta,\eta) - I_{12}(\theta,\eta)I^{-1}_{22}(\theta,\eta)l^n_2(\theta,\eta),$$

$\overline{\eta} = (\overline{\eta}_n)_{n\geqslant 1}$ is the C_n-consistent estimator of η, then

$$\mathcal{L}\{C_n^{-1}(\tilde{\theta}_n - \theta) \mid \mathbf{P}^n_{\theta,\eta}\} \Rightarrow \mathcal{N}(0, \sigma_2(\theta, \eta)),$$

where $\sigma_2^{-1}(\theta,\eta) = I^*_{11}(\theta,\eta)$ (see Theorem 9.2 in (Chitashvili et al., 1990)).
Under LAN condition at point (θ,η) the PLE $\hat{\theta} = (\hat{\theta}_n)_{n\geqslant 1}$ and $\tilde{\theta} = (\tilde{\theta}_n)_{n\geqslant 1}$ are regular in usual sense (see, e.g., Theorem 9.3 in (Chitashvili et al., 1990)).
If in addition the condition for adaptation is satisfied, i.e., $I_{12}(\theta,\eta) = 0$, then the estimator $\tilde{\theta} = (\tilde{\theta}_n)_{n\geqslant 1}$ is adaptive regular estimator and so is "better" that the PLE (by virtue of arguments mentioned above).
Moreover, even though the condition for adaptation is not fulfilled the inequality

$$\sigma_2(\theta,\eta) \leqslant \sigma_1(\theta,\eta)$$

is true for all $\theta,\eta \in \mathbf{R}^1$ (see (Chitashvili et al., 1990)).
The relation between σ_1 and σ_2 is given ibidem. Here for simplicity we give this relation for the case of two-dimensional i.i.d. observations $(Y_i, Z_i)_{i\geqslant 1}$ with a factorized joint density $f(y,z,\theta,\eta) = \phi(y;\theta,\eta)\psi(z;\theta,\eta \mid y)$:

$$\sigma_1^{-1} = \int \left(\frac{\overset{*}{\psi}}{\psi}\right)^2 \phi\psi \, d\mu,$$

$$\sigma_2^{-1} = \sigma_1^{-1} + \int \left(\frac{\overset{*}{\phi}}{\phi}\right)^2 \phi\psi d\mu \left[1 - \frac{\left(\int \left(\overset{*}{\phi}/\phi\right)\left(\dot{\phi}/\phi\right)\phi\psi d\mu\right)^2}{\left(\int \left(\overset{*}{\phi}/\phi\right)\psi\phi d\mu\right)\left(\int \left(\dot{\phi}/\phi\right)\psi\phi d\mu\right)}\right].$$

If the density ϕ does not depend on θ (in general case it means that $X^n_{\theta,\eta} \equiv X^n_\eta$), then $\sigma_2(\theta,\eta) = \sigma_1(\theta,\eta)$ for all $\eta \in \mathbf{R}^1$.
In (Greenwood and Wefelmeyer, 1989b) the analogous questions are considered for partially specified semimartingale models with functional parameter θ. Namely, the concepts of regularity and efficiency are given and the efficiency of PLE is investigated.
In our notation their concepts of regularity at point θ_0 coincides with usual one in models with $X^n_\theta \equiv X^n_{\theta_0}$, $\theta \in \mathbf{R}^1(\phi(y,z;\theta,\eta) \equiv \phi(y,z;\theta_0,y)$ in i.i.d. case). For such alternatives PLE coincides with MLE and so is efficient.

REFERENCES

Chitashvili, R.J., Lazrieva, N.L. and Toronjadze, T.A. (1990). Asymptotic theory of M-estimators in general statistical models. Parts I, II. *Reports* BS-R9019; 9020. Centre for Mathematics and Computer Science, Amsterdam.

Cox, D.H. (1975). Partial likelihood. *Biometrika* **64**, 269-276.

Dzhaparidze, K. (1986). On asimptotic inference about intensity parameters of a counting process. *Report* MS-R8614. Centre for Mathematics and Computer Science, Amsterdam.

Greenwood, P.E. and Wefelmeyer, W. (1989a). Efficiency of estimators for partially specified filtered models. *Preprint in Statistics* **119**, University of Cologne, Cologne.

Greenwood, P.E. and Wefelmeyer, W. (1989b). Partially and fully specified semimartingale models and efficiency. *Preprint in Statistics* **124**, University of Cologne, Cologne.

Jacod, J. (1979). *Calcul stochastique et problemes de martingales.* Springer, New York etc.

Jacod, J. (1987). *Partial Likelihood Process and Asymptotic Normality.* Laboratoire de Probabilites Universite P. et M.Curie, Paris.

Jacod, J. (1989). *Regularity, Partial Regularity, Partial Information Process for a Filtered Statistical Model.* Laboratoire de Probabilites University P. et M.Curie, Paris.

Liptser, R.Sh. and Shiryaev, A.N. (1989). *Theory of Martingales.* Kluwer, Dordrecht.

Wong, W.H. (1976). Theory of partial likelihood. *Ann. Statist.* **14**, 88-124.

New Trends in Probab. and Statist., pp. 597 – 611
V. Sazonov and T. Shervashidze (Eds.)
1991 VSP/Mokslas

ASYMPTOTIC ESTIMATION OF INTEGRALS OF CERTAIN RANDOM FIELDS

A.V. NAGAEV
Institute of Mathematics of the Uzbek Academy of Sciences,
F. Khodzaeva str. 29, 700143 Tashkent, USSR

ABSTRACT

A generalization of the likelihood function as a realization of a random field is considered. The Laplace technique is applied to establish asymptotic properties of integrals of such realizations.

1. INTRODUCTION

Let $\{\mathbf{P}_u\}$, $u \in U$, be a family of measures defined on the σ-algebra of Borel sets of the Euclidean space $\mathbf{R}^d$, $d \geqslant 1$. In what follows we assume that for every $u \in U$ the measure $\mathbf{P}_u$ is dominated by some σ-finite measure μ.
Consider a function $l(x, v)$ defined over $\mathbf{R}^d \times \mathbf{R}^k$ and measurable with respect to the σ-algebra generated by $\mathcal{B}^{(d)} \times \mathcal{B}^{(k)}$.
Let x_j, $j = 1, \ldots, n$, be independent and

$$\mathbf{P}(x_j \in A) = \mathbf{P}_u(x_j \in A) = \mathbf{P}_u(A).$$

Put

$$L(X, v) = \sum_{j=1}^{n} l(x_j, v) \tag{1}$$

and consider the statistic

$$T(X) = \int_V r(v)\, e^{L(X,v)} dv, \tag{2}$$

where $X = (x_1, x_2, \ldots, x_n)$ and the function $r(v)$ is Lebesgue measurable.
The general purpose of this paper is to investigate asymptotic properties of the statistic $T(X)$ which is an integral of a realization of a random field having very specific nature. Under rather mild regularity conditions the estimation of (2) can be realized with the aid of some modification of the Laplace method which is one of the basic methods of asymptotic analysis. Similar statistics were considered in (Borovkov, 1984; Ibragimov and Khas'minsky, 1981).
Consider three important examples of statistics $T(X)$.

Example 1. Put

$$p(x,u) = d\,\mathbf{P}_u/d\mu.$$

Assume that $U = V$, μ is the Lebesgue measure,

$$l(x;u) = \log p(x,u).$$

It is clear that under some rather mild regularity conditions the principal contribution into the asymptotic of (2) is given by a neighborhood of the value of the maximum likelihood estimator. Here the technique based on a systematical usage of the Hellinger distance proved to be very useful as shown in (Borovkov, 1984; Ibragimov and Khas'minskii, 1981).

Example 2. Let $U = V$, but

$$l(x,u) \neq \log p(x,u).$$

Under rather general conditions imposed on $l(\cdot\,,\cdot)$ and $p(\cdot\,,\cdot)$ the principal role is played by a neighborhood of the value of the minimal contrast estimator. Note that the Hellinger distance technique mentioned above is not applicable here.

Example 3. Let U and V have different nature and

$$l(x,v) = \log q(x,v),$$

where $q(x,v) = d\,Q_v/d\nu$, $\{Q_v\}$ is a family of measures, ν is a σ-finite measure. Here the principal contribution into the asymptotic of (2) for a fixed u is given by a neighborhood of the value which minimizes the Kullback–Leibler distance between $\mathbf{P}_u$ and $\{Q_v\}$, i.e.,

$$v(u) = \arg\min_v \int_{\mathbf{R}^d} \log \frac{p(x,u)}{q(x,v)}\,\mathbf{P}_u(dx).$$

In the general case the asymptotic of (2) is determined by the behavior of the above considered random field near the point

$$v(u) = \arg\min_v \int_{\mathbf{R}^d} l(x,v)\,\mathbf{P}_u(dx).$$

It is worth noting that investigation of stochastic integrals (2) was stimulated by author's interest in the most powerful tests destined to distinguish location-scale families. Let

$$U = V = (u\colon\ u_2 > 0) \quad \text{and} \quad \{\mathbf{P}_u^{(i)}\}, \quad i = 1,2,$$

be two such families of measures defined on Borel sets of $\mathbf{R}^1$. Put $u^{(0)} = (0,1)$ and

$$p_i(x) = d\,\mathbf{P}_{u^{(0)}}^{(i)}/dx, \quad i = 1,2.$$

Consider the statistics

$$t_i(X) = \int_0^{\infty} v_2^{n-2}\left(\int_{-\infty}^{\infty} \prod_{j=1}^{n} p_i(v_2 x_j + v_1)\, dv_1\right) dv_2, \quad i = 1,2, \tag{3}$$

and

$$d(X) = t_1(X)/t_2(X).$$

As it was shown in (Hájek and Šidak, 1967), $d(X)$ represents the likelihood ratio corresponding to the measures on the σ-algebra of location-scale invariant sets generated by $p_1(x)$ and $p_2(x)$.
Thus $d(X)$ can be used to construct the most powerful test for the above composite hypotheses.
It is easily seen that the statistics $t_i(X)$ are of the form (1), where $r(v) = v_2^{-2}$,

$$l(x,v) = \log p_i(x,v), \quad i = 1,2.$$

If the distribution of x_1 is characterized by the parameter u, then $t_1(X)$ satisfies the conditions of Example 1 and $t_2(X)$ satisfies those of both Examples 2 and 3.

2. STATEMENT OF THE RESULTS

Let us agree about some notation. The letter c denotes any positive constant while $\omega(t)$ denotes any function such that $\lim_{t\to 0}\omega(t) = 0$. The symbols $\mathbf{E}_u$ and var_u are used for the expectation and the variance corresponding to $\mathbf{P}_u$.
We denote

$$a(u,v) = \mathbf{E}_u l(x,v), \qquad u \in U, \quad v \in V.$$

Let us introduce the contrast condition for $a(u,v)$.

Condition C. For any $u \in U$ the function $a(u,v)$ attains its least upper bound at a unique point $v = v(u)$. Moreover, for any $\delta > 0$,

$$\sup_{|v-v(u)|\geqslant\delta} a(u,v) < a\big(u,v(u)\big)$$

and, for some $\gamma > 0$,

$$\inf_{u\in U}\ \inf_{|v-v(u)|\leqslant\delta} \frac{a\big(u,v(u)\big) - a(u,v)}{|v-v(u)|_\gamma} > 0. \tag{4}$$

The following assertion holds true.

THEOREM 1. *Let, in addition to the condition* C,

1° *the sets U and V be compact;*

2° *for some $m \geqslant r \geqslant 2k > 2$ or $m \geqslant r > 2k = 2$ and any $v, v', v'' \in V$,*

$$\mathbf{E}_u|l(x_1,v') - l(x_1,v'')|^m \leqslant c|v'-v''|^r, \qquad \mathbf{E}_u|l(x_1,v)|^m \leqslant c,$$

and

$$|a(u,v') - a(u,v'')| \leqslant c|v' - v''|^{r/m}$$

uniformly in $u \in U$;

3° $0 < \inf_{v\in V} r(v) \leqslant \sup_{v\in V} r(v) < \infty$.

Then

a) *if* $\gamma > r/m$ *in* (4), *then*

$$\log T(X) = L\big(X, v(u)\big) + \Delta_n(X,u),$$

where

$$\overline{\lim_{n\to\infty}} \sup_{u\in U} \mathbf{P}_u\big(|\Delta_n(X,u)| \geqslant t\log n\big) \leqslant \omega(1/t);$$

b) *if* $\gamma = r/m$, *then*

$$\log T(X) = L\big(X, v(u)\big) - (r/m)\log n + \Delta_n(X,u),$$

where

$$\overline{\lim_{n\to\infty}} \sup_{u\in U} \mathbf{P}_u\big(|\Delta_n(X,u)| \geqslant t\big) \leqslant \omega(1/t).$$

This theorem has simple but very important corollaries.

COROLLARY 1. *Under conditions of Theorem* 1 *we have, as* $n \to \infty$,

$$\mathbf{P}_u\Big(\log T(X) - na\big(u, v(u)\big) < z\sigma(u)\sqrt{n}\Big) = \Phi(z) + o(1)$$

uniformly in $u \in U$ *and* z, $-\infty < z < \infty$, *where*

$$\sigma^2(u) = \operatorname{var}_u l\big(x_1,\ v(u)\big).$$

Here and in what follows $\Phi(z)$ is the standard normal distribution function.

COROLLARY 2. *Let* $\hat{v} = \hat{v}_n(X)$ *be a uniformly consistent estimator of* $v(u)$, *i.e., for any* $\varepsilon > 0$,

$$\lim_{n\to\infty} \sup_{u\in U} \mathbf{P}_u\big(|\hat{v} - v(u)| > \varepsilon\big) = 0.$$

Under the conditions of Theorem 1,

$$L(X,\hat{v}) - na(u,\hat{v}) = L\big(X, v(u)\big) - na\big(u, v(u)\big) + \sqrt{n}\,\delta_n(X,u),$$

where, for any $\varepsilon > 0$,

$$\lim_{n\to\infty} \sup_{u\in U} \mathbf{P}_u\big(|\delta_n(X,u)| > \varepsilon\big) = 0.$$

Under more rigid restrictions on $l(x,v)$ one can derive some analogs of Theorems 2.24.2 and 2.28.5 in (Borovkov, 1984).

Let $V_0 \subset V$ be the image of U under the mapping $v = v(u)$. Set, for $i, s = 1, \ldots, k$,

$$l_i(x,v) = \frac{\partial l}{\partial v_i}, \qquad l_{is}(x,v) = \frac{\partial^2 l}{\partial v_i \partial v_s}, \qquad v = (v_1, v_2, \ldots, v_k).$$

Condition R

1° The function $l(x,v)$ is twice differentiable in v for all $v \in V$ and for almost all x.

2° For some $\varepsilon > 0$ and all i and s

$$\sup_{v \in V_0^\varepsilon} |l_{is}(x,v)| \leqslant h(x),$$

where $V_0^\varepsilon = (v\colon \inf_{v' \in V_0} |v - v'| \leqslant \varepsilon)$ and the integral $\mathbf{E}_u h(x_1)$ converges uniformly in $u \in U$.

3° The integral $\mathbf{E}_u(\sum_{i=1}^k l_i(x_1, v(u)))^2$ also converges uniformly in $u \in U$.

4° The matrix

$$\Lambda = \Lambda_u = \|\lambda_{is}(u)\| = -\|\mathbf{E}_u l_{is}(x_1, v(u))\|, \quad i, s = 1, \ldots, k,$$

is strictly positive definite in U, i.e.,

$$\inf_{u \in U} \det \Lambda_u > 0.$$

The following statement is true.

THEOREM 2. *Let conditions* C *and* R *hold and* $r(v)$ *be continuous in* V_0. *Then*

$$T(X) = \exp\Big(-L(X,\, v(u))\Big)(2\pi/n)^{k/2}(\det \Lambda)^{-1/2} \exp\Big(-\frac{1}{2}(\Lambda\xi, \xi)(1+\delta_n(X,u))\Big),$$

where

$$\xi = \xi_n(u) = n^{-1/2} \sum_{j=1}^n \Big(l_1(x_j, v(u)), \ldots, l_k(x_j, v(u))\Big).$$

Note that the assertion of this theorem is somewhat weaker than that of Theorem 2.28.5 in (Borovkov, 1984). This is due to the fact that, generally, in our case L is not the logarithm of a likelihood function. So, in order to obtain in Theorem 2 $\mathbf{P}_u$-a.s. convergence to zero of the remainder term, one has to impose very restrictive conditions on $l(x,v)$.

Note also that Theorem 2 is not directly applicable to statistics $t(X)$ because the domain of integration in (3) is not compact. That is why we devote a special theorem to this case.

Let $U = V = (u\colon\ u = (u_1, u_2),\ u_2 > 0)$, $d = 1$, and let μ be the Lebesgue measure. Set

$$l(x,u) = \log \frac{u_2 q(u_2 x + u_1)}{p(x)}, \tag{5}$$

where $p(x)$ and $q(x)$ are density functions and $\{\mathbf{P}_u\}$ is a location-scale family generated by $p(x)$.

It is easily seen that in our case

$$a(u,v) = a(u_1, u_2;\ v_1, v_2) = \mathbf{E}_u l(x_1, v) = a(0, 1;\ v_1 - u_1 v_2/v_1,\ v_2/u_2).$$

That is why instead of $a(u,v)$ we can use the function

$$\alpha(u) = \int p(x)\, l(x,u)\, dx = \mathbf{E}\, l(x,u), \tag{6}$$

where $\mathbf{E}$ corresponds to the measure generated by $p(x)$.
Set $u^{(0)} = (0,1)$. Due to condition C, $\alpha(u)$ has its maximum at some point $u^{(1)}$. There is no loss of generality in assuming that $u^{(0)} = u^{(1)}$. It is worth reminding that $-\alpha(u)$ is the Kullback–Leibler distance between $p(x)$ and $u_2 q(u_2 x + u_1)$. Consider the statistic

$$t(X) = \int_0^\infty v_2^{-2} \left(\int_{-\infty}^\infty \exp\left(L(X,v)\right) dv_1 \right) dv_2, \tag{7}$$

where L is determined by (1) and (5). It is readily seen that the distribution of $t(X)\, u_2^{-(n-1)}$ does not depend on u.

THEOREM 3. *Let the following conditions hold:*

1° *for any compact* $V_1 = (v\colon\ |v_1| < 1/\delta_1,\ \delta_2 < v_2 < 1/\delta_3)$ *and for some* r *and* m, $m \geqslant r \geqslant 4$, *uniformly in* $v, v', v'' \in V_1$,

$$\mathbf{E}\,|l(x_1,v') - l(x_1,v'')|^m \leqslant c\,|v' - v''|^r, \qquad \mathbf{E}\,|l(x_1,v)|^m \leqslant c,$$

and

$$|\alpha(v') - \alpha(v'')| \leqslant c\,|v' - v''|^{r/m};$$

2° *in* (4) $\gamma \geqslant r/m$, *i.e., in some neighborhood of* $u^{(0)}$

$$\alpha(u^{(0)}) - \alpha(u) \geqslant c\,|u - u^{(0)}|^\gamma;$$

3° $q(x) \leqslant q_0(1+|x|)^{-\beta}, \quad \beta > 1.$

Then

a) *if* $\gamma > r/m$, *then*

$$\log\left(t(X)\, u_2^{-(n-1)}\right) = L(X, u^{(0)}) + \Delta_n(X),$$

where

$$\varlimsup_{n\to\infty} \mathbf{P}\Big(|\Delta_n(X)| > t\log n\Big) \leqslant \omega(1/t);$$

b) *if* $\gamma = r/m$, *then*

$$\log\left(t(X)\, u_2^{-(n-1)}\right) = L(X, u^{(0)}) - (r/m)\log n + \Delta_n(X),$$

where

$$\varlimsup_{n\to\infty} \mathbf{P}(|\Delta_n(X)| > t) \leqslant \omega(1/t).$$

Here $\mathbf{P}$ *is a measure on the* σ*-algebra of location-scale invariant Borel sets of* $\mathbf{R}^n$ *generated by* $p(x)$.

Note that in Theorem 3 noncompactness of $U = V$ is compensated by rather mild condition 3°.
Let $T_1(X) = t(X)$ if, in (1) and (7),

$$l(x,v) = \log \frac{v_2 p_1(v_2 x + v_1)}{p_1(x)}$$

and $T_2(X) = t(X)$ if

$$l(x,v) = \log \frac{v_2 p_2(v_2 x + v_1)}{p_1(x)}.$$

Further, let $\alpha_i(u)$, $i = 1,2$, correspond to the above $l(x,v)$ according to (6). It is clear that $\alpha_1(u^{(0)}) = 0$ and $\alpha_2(u^{(0)}) < 0$. Note also that, in view of (3) and (7),

$$d(X) = t_1(X)/t_2(X) = T_1(X)/T_2(X).$$

From Theorem 3 we obtain two simple corollaries.

COROLLARY 3. *Suppose that the conditions of Theorem 3 hold both for* $p(x) = p_1(x) = q(x)$ *and for* $p(x) = p_1(x)$, *and* $q(x) = p_2(x) \neq p_1(x)$. *Then, as* $n \to \infty$,

$$\mathbf{P}\Big(\log\big(t_1(X)/t_2(X)\big) - n\alpha_2(u^{(0)}) < z\sigma\sqrt{n}\Big) = \Phi(z) + o(1)$$

uniformly in z, $-\infty < z < \infty$, *where*

$$\sigma^2 = \operatorname{var}\log\frac{p_1(x_1)}{p_2(x_1)}.$$

Here var *corresponds to* $\mathbf{P}_{u^{(0)}}$, *and* $\mathbf{P}$ *has the same meaning as in Theorem* 3.

COROLLARY 4. *Let* u_i^*, $i = 1,2$, *be statistics such that, for any compact* V_1 *in Condition* 1° *of Theorem* 3,

$$\varlimsup_{n\to\infty} \sup_{u\in V_1} \mathbf{P}_u(\sqrt{n}|u_i^* - u| \geqslant t) \leqslant \omega(1/t).$$

If, in addition, in some neighborhood of $u^{(0)}$,

$$\alpha_i(u^{(0)}) - \alpha_i(u) \leqslant c\,|u - u^{(0)}|^{\gamma_i}, \qquad \gamma_i > 1, \quad i = 1,2,$$

then as $n \to \infty$

$$\mathbf{P}_u\big(L_2(X,u_2^*) - L_1(X,u_1^*) - n\alpha_2(u^{(0)}) < z\sigma\sqrt{n}\big) = \Phi(z) + o(1)$$

uniformly in z, $-\infty < z < \infty$, $u \in V_1$.

We conclude this section with some remarks about the above corollaries.

It is evident that with the aid of Corollary 3 we derive the asymptotically best invariant procedure for choosing one of the two given location-scale families. When

$$u_i^* = \arg\max L_i(X,u),$$

the statistic $L_2(X,u_2^*) - L_1(X,u_1^*)$ is location-scale invariant and in accordance with Corollary 4 can be used for similar purpose. Note that the calculation of statistics $L_i(X,u_i^*)$ is sometimes much easier than that of $t_i(X)$. Unfortunately, it is very difficult to compare the power of these two tests.

3. PROOF OF THEOREM 1

Consider the random function

$$\eta_n(v) = n^{-1/2}\big(L(X,v) - na(u,v)\big). \tag{8}$$

In accordance with Lemma 3.3.1 in (Ibragimov and Khas'minskii, 1981),

$$\begin{aligned} \mathbf{E}_u|\eta_n(v') - \eta_n(v'')|^m &\leqslant c\,\mathbf{E}_u|l(x_1,v') - l(x_1,v'')- \\ -a(u,v') + a(u,v'')|^m &\leqslant c\big(\mathbf{E}_u|l(x_1,v') - l(x_1,v'')|^m + \\ +|a(u,v') - a(u,v'')|^m\big) &\leqslant c|v' - v''|^r. \end{aligned}$$

Similarly,

$$\mathbf{E}_u|\eta_n(v)|^m \leqslant c.$$

Since the convergence of finite dimensional distributions is obvious, in view of the Theorem 20 in Appendix I (ibidem), we have

$$\sup_{v\in V} |\eta_n(v)| \xrightarrow{\text{d}} \eta \tag{9}$$

uniformly in $u \in U$, where η is a non-negative variable. Consider the following representation

$$T(X)\exp\Big(-L(X,v(u))\Big) = \int_V r(v)\exp\Big(L(X,v)-L(X,v(u))\Big)dv = I_1+I_2, \tag{10}$$

where

$$I_1 = \int_{V\cap\big(v:\ |v-v(u)|\leqslant\varepsilon\big)} r(v)\exp\Big(L(X,v) - L(X,v(u))\Big)\,dv,$$

$$I_2 = \int_{V\cap\big(v:\ |v-v(u)|>\varepsilon\big)} r(v)\exp\Big(L(X,v) - L(X,v(u))\Big)\,dv.$$

Further,

$$L(X,v) - L(X,v(u)) = \sqrt{n}\Big(\eta_n(v) - \eta_n(v(u))\Big) + n\Big(a(u,v) - a(u,v(u))\Big).$$

From condition C and from hypotheses of the theorem we have

$$-cn\,|v-v(u)|^{r/m}+\sqrt{n}\Big(\eta_n(v)-\eta_n\big(v(u)\big)\Big)<L(X,v)-L\big(X,v(u)\big)<$$

$$<-cn\,|v-v(u)|^{\gamma}+\sqrt{n}\Big(\eta_n(v)-\eta_n\big(v(u)\big)\Big). \tag{11}$$

In accordance with Theorem 19 from Appendix I in (Ibragimov and Khas'minskii, 1981), we obtain

$$\mathbf{E}_u \sup_{|v-v(u)|<h} \big|\eta_n(v)-\eta_n\big(v(u)\big)\big| \leqslant ch^{(r-k)/m}. \tag{12}$$

For $h \leqslant \varepsilon$ put

$$\mathcal{X}=\Big(X\colon \sup_{|v-v(u)|\leqslant h} \big|\eta_n(v)-\eta_n\big(v(u)\big)\big| \leqslant Nh^{(r-k)/m}\Big). \tag{13}$$

From (12) and (13) it follows that

$$\mathbf{P}_u(\overline{\mathcal{X}}) \leqslant cN^{-1}. \tag{14}$$

By virtue of (11) and (13), for $X \in \mathcal{X}$, we obtain

$$-nc\,|v-v(u)|^{r/m}-N\sqrt{n}\,|v-v(u)|^{(r-k)/m} \leqslant$$

$$\leqslant L(X,v)-L\big(X,v(u)\big) \leqslant -nc\,|v-v(u)|^{\gamma}+N\sqrt{n}\,|v-v(u)|^{(r-k)/m}. \tag{15}$$

From (10), (11), and (15)

$$c\int_0^{\varepsilon} \exp\big(-nct^{r/m}-N\sqrt{n}\,t^{(r-k)/m}\big)\,dt \leqslant I_1 \leqslant$$

$$\leqslant c\int_0^{\varepsilon} \exp\big(-nct^{\gamma}+N\sqrt{n}\,t^{(r-k)/m}\big)\,dt. \tag{16}$$

Since $r \geqslant 2k$ and $\gamma \geqslant r/m$, we have, for all sufficiently large n,

$$\int_0^{\varepsilon} \exp\big(-nct^{r/m}-N\sqrt{n}\,t^{(r-k)/m}\big)\,dt \geqslant c_N n^{-m/r} \tag{17}$$

and

$$\int_0^{\varepsilon} \exp\big(-nct^{\gamma}+N\sqrt{n}\,t^{(r-k)/m}\big)\,dt \leqslant c_N n^{-1/\gamma}. \tag{18}$$

Thus, due to (16)–(18),

$$c_N n^{-m/r} \leqslant I_1 \leqslant c_N n^{-1/\gamma}. \tag{19}$$

Let us estimate I_2. From condition 3° and (10) we have

$$I_2 \leqslant c \int\limits_{(v:\ |v-v(u)|\geqslant\varepsilon)\cap V} \exp\Big(L(X,v) - L\big(X,v(u)\big)\Big)\,dv.$$

For $X \in \mathcal{X}_1 = (X\colon\ \sup_{v\in V} |\eta_n(v)| \leqslant N_1)$, in view of (9) and condition C,

$$L(X,v) - L\big(X,v(u)\big) \leqslant -nc + \sqrt{n}\sup_{v\in V}\eta_n(v) - \sqrt{n}\,\eta_n\big(v(u)\big) \leqslant -nc + 2\sqrt{n}N_1.$$

Thus, due to the compactness of V, for all sufficiently large n, we obtain

$$I_2 \leqslant ce^{-nc}. \tag{20}$$

If $\gamma > r/m$, we put

$$\Delta_n(X,u) = \log T(X) - L(X,u).$$

For $X \in \mathcal{X} \cap \mathcal{X}_1$, from (10), (19), and (20) we have

$$-(m/r)\log n - c_N \leqslant \Delta_n(X,u) \leqslant -(1/\gamma)\log n + c_N. \tag{21}$$

If $\gamma = r/m$, then we put

$$\Delta_n(X,u) = \log T(X) - L(X,u) - (r/m)\log n$$

and, instead of (21), we obtain

$$|\Delta_n(X,u)| \leqslant c_N. \tag{22}$$

Since N and N_1 are arbitrary, both assertions follow immediately from (21) and (22). Theorem 1 is proved.

Corollary 1 is quite evident.

Proof of Corollary 2. Put $\mathcal{X}_2 = (X\colon\ |\hat{v} - v(u)| \leqslant \varepsilon)$ and, in accordance with (8), $\delta_n = \delta_n(X,u) = \eta_n(\hat{v}) - \eta_n\big(v(u)\big)$. For $X \in \mathcal{X}_2$ we have

$$|\delta_n| \leqslant \sup_{|v-v(u)|\leqslant\varepsilon} |\eta_n(v) - \eta_n\big(v(u)\big)|. \tag{23}$$

For any $\delta > 0$,

$$\mathbf{P}_u(|\delta_n| \geqslant \delta) \leqslant \mathbf{P}_u(|\delta_n| \geqslant \delta;\ \mathcal{X}_2) + \mathbf{P}_u(\overline{\mathcal{X}}_2). \tag{24}$$

According to Theorem 19 in (Ibragimov and Khas'minskii, 1981, Appendix I), from (23) it follows that

$$\mathbf{P}_u(|\delta_n| \geqslant \delta; \mathcal{X}_2) \leqslant \mathbf{P}_u\left(\sup_{|v-v(u)|\leqslant\varepsilon} \big|\eta_n(v) - \eta_n\big(v(u)\big)\big| \geqslant \delta\right) \leqslant$$

$$\leqslant \delta^{-1}\mathbf{E}_u \sup_{|v-v(u)|\leqslant\varepsilon} \big|\eta_n(v) - \eta_n\big(v(u)\big)\big| \leqslant c\delta^{-1}\varepsilon^{(r-k)/m}. \tag{25}$$

Since $\hat{v}$ is uniformly consistent, $\sup_{u\in U}\mathbf{P}_u(\overline{\mathcal{X}}_2) \to 0$ as $n \to \infty$. From this and from (24), (25), taking into account that ε is arbitrary, we deduce that $\delta_n \to 0$ in $\mathbf{P}_u$.

4. PROOF OF THEOREM 2

Consider the following expansion of $L(X,v)$ in a neighborhood of the point $v = v(u)$:

$$L(X,v) = L(X,v(u)) + (L_1, v - v(u)) + \frac{1}{2}\Big(\tilde{L}_2(v - v(u)), v - v(u)\Big), \tag{26}$$

where

$$L_1 = L_1(X,v(u)) = \sum_{j=1}^{n}\Big(l_1(x_j,v(u)),\dots,l_k(x_j,v(u))\Big),$$

$$\tilde{L}_2 = L_2(X,\tilde{v}) = \left\|\sum_{j=1}^{n} l_{is}(x_j,\tilde{v})\right\|, \quad i,s = 1,\dots,k,$$

$$\tilde{v} = v(u) + \theta(v - v(u)), \quad |\theta| \leqslant 1.$$

Put

$$L_2 = L_2(X,v(u)). \tag{27}$$

If $|v - v(u)| \leqslant \varepsilon$, then

$$n^{-1}\left|\sum_{j=1}^{n}\Big(l_{is}(x_j,\tilde{v}) - l_{is}(x_j,v(u))\Big)\right| \leqslant$$

$$\leqslant n^{-1}\sum_{j=1}^{n}\sup_{|v-v(u)|<\varepsilon}\big|l_{is}(x_j,v) - l_{is}(x_j,v(u))\big|. \tag{28}$$

From condition R it follows that

$$\mathbf{E}_u \sup_{|v-v(u)|\leqslant\varepsilon}\big|l_{is}(x_1,v) - l_{is}(x_1,v(u))\big| \tag{29}$$

(see Lemmas 2.24.1 and 2.28.1 in (Borovkov, 1984)).
For any matrix $B = \|b_{is}\|$ define

$$\langle B\rangle = \max_{i,s}|b_{is}| \tag{30}$$

and

$$\mathcal{X}_3 = \Big(X\colon \langle n^{-1}L_2 - \|\mathbf{E}_u l_{is}(x_1,v(u))\|\rangle < \delta,\ n^{-1}\langle L_2 - \tilde{L}_2\rangle < \delta\Big). \tag{31}$$

From (27)–(31) and Theorem 10 in (Ibragimov and Khas'minskii, 1981, Appendix I) it follows that

$$\mathbf{P}_u(\overline{\mathcal{X}}_3) \to 0$$

as $n \to \infty$, if ε is sufficiently small with respect to δ.

Put $\mathcal{X}_4 = (X\colon |n^{-1/2}L_1| \leqslant N)$. In view of conditions C and R, we have

$$\mathbf{E}_u L_1(X, v(u)) = 0$$

and, therefore, $\xi = \xi_n(u) = n^{-1/2}L_1(X, v(u))$ converges in distribution to a normal vector having zero expectation and the covariance matrix

$$B(u) = \|\mathbf{E}_u l_i(x_1, v(u))\, l_s(x_1, v(u))\|, \quad i, s = 1, \dots, k.$$

Thus,

$$\mathbf{P}_u(\mathcal{X}_3) > 1 - \omega(\delta), \quad \mathbf{P}_u(\mathcal{X}_4) > 1 - \omega(1/N)$$

and, in view of (26),

$$-\frac{1}{2}\Big((\Lambda + k\delta I)(v - v(u)), v - v(u)\Big) \leqslant L(X, v) - L(X, v(u)) - \sqrt{n}(\xi_n, v - v(u)) \leqslant$$

$$\leqslant -\frac{1}{2}\Big((\Lambda - k\delta I)(v - v(u)), v - v(u)\Big),$$

where I is the unit matrix. Hence, due to (26),

$$I_1 \geqslant n^{-k/2}\Big(r(v(u)) + \omega(\varepsilon)\Big) \int\limits_{|v| < \varepsilon\sqrt{n}} \exp\Big((\xi_n, v) - \frac{1}{2}((\Lambda + k\delta I)v, v)\Big)\, dv,$$

whence, for $X \in \mathcal{X}_3 \cap \mathcal{X}_4$ and sufficiently large n,

$$I_1 \geqslant n^{-k/2}\Big(r(v(u)) + \omega(\varepsilon)\Big)\left(\frac{\exp\Big(((\Lambda + k\delta I)\xi_n, \xi_n)/2\Big)}{(2\pi)^{-k/2}\big(\det(\Lambda + k\delta I)\big)^{1/2}} - \right.$$

$$\left. -e^{cN^2} \int\limits_{|v| > \varepsilon\sqrt{n}} \exp\Big(-\frac{1}{2}((\Lambda + k\delta I)v, v)\Big)\, dv\right) \geqslant$$

$$\geqslant \left(\frac{2\pi}{n}\right)^{k/2} \frac{r(v(u))}{(\det \Lambda)^{1/2}}\, e^{(\Lambda\xi_n, \xi_n)/2}\big(1 + \omega(\delta) + \omega(\varepsilon)\big) \tag{32}$$

uniformly in $u \in U$.

Similarly,

$$I_1 \leqslant \left(\frac{2\pi}{n}\right)^{k} \frac{r(v(u))}{(\det \Lambda)^{1/2}}\, e^{(\Lambda\xi_n, \xi_n)/2}\big(1 + \omega(\delta) + \omega(\varepsilon)\big). \tag{33}$$

Combining (10), (20), (32) and (33), we obtain the desired result. Theorem 2 is proved.

5. PROOF OF THEOREM 3

Instead of (10) we put

$$t(X)\, u_2^{-(n-1)} = \int\limits_{-\infty}^{\infty} \int\limits_{0}^{\infty} \lambda^{n-2} \prod_{j=1}^{n} q(\lambda x_j + \nu)\, d\nu\, d\lambda = \sum_{i=1}^{4} I_i, \tag{34}$$

where

$$I_1 = \int\limits_{|v-v^{(0)}|\leqslant\varepsilon} G(v)\,dv, \qquad I_2 = \int\limits_{\substack{\varepsilon<|v-v^{(0)}|<1/\varepsilon_1 \\ v\in V_1}} G(v)\,dv,$$

$$I_3 = \int\limits_{\substack{0<\lambda\leqslant\varepsilon_2 \\ -\infty<\nu<\infty}} G(v)\,dv, \qquad I_4 = \int\limits_{\substack{|v-v^{(0)}|\geqslant 1/\varepsilon_1 \\ \lambda>\varepsilon_2}} G(v)\,dv,$$

$$G(v) = \lambda^{n-2}\prod_{j=1}^{n} q(\lambda x_j + \nu), \qquad v = (\nu,\lambda), \quad v^{(0)} = (0,1).$$

Note that

$$\mathbf{P}_u\big(t(X)\,u_2^{-(n-1)} \in A\big) = \mathbf{P}_{u^{(0)}}\big(t(X)\,u_2^{-(n-1)} \in A\big).$$

In view of Theorem 1, it remains only to estimate I_3 and I_4. From 3° it follows that

$$I_3 \leqslant \varepsilon_2^{n-2} q_0^{n-1} \int\limits_0^{\varepsilon_2} d\lambda \int\limits_{-\infty}^{\infty} q(\lambda x_1 + \nu)\,d\nu \leqslant (\varepsilon_2 q_0)^{n-1}. \tag{35}$$

Put $\varepsilon_3 = 2\varepsilon_1$. Then

$$I_4 \leqslant \int\limits_{1/\varepsilon_3}^{\infty} \int\limits_{-\pi/2}^{\pi/2} r^{n-1}(\cos\varphi)^{n-2} \prod_{j=1}^{n} q\big(r(x_j\cos\varphi + \sin\varphi)\big)\,dr d\varphi. \tag{36}$$

Let $M > 0$ be an even number. Put

$$\varphi_1 = -\pi/2 + \pi/2M, \qquad \varphi_i = \varphi_1 + (i-1)\,\pi/M, \qquad i = 1,\dots,M,$$

$$\varphi_M = \pi/2 - \pi/2M \tag{37}$$

and

$$p_{iM} = \mathbf{P}\big(|\mathrm{arctg}(1/x_j) - \psi_i| \leqslant \pi/M\big),$$

where

$$\psi_i = \begin{cases} \varphi_i + \pi/2, & 1 \leqslant i \leqslant M/2, \\ \varphi_i - \pi/2, & M/2 < i \leqslant M, \end{cases} \tag{38}$$

and $I_0 = (i\colon\ p_{iM} = 0)$. Furthermore, put

$$J_i' = \big(j\colon\ |\mathrm{arctg}(1/x_j) - \psi_i| \leqslant \pi/M\big),$$

$$J_i'' = \big(j\colon\ |\mathrm{arctg}(1/x_j) - \psi_i| > \pi/M\big) \tag{39}$$

and

$$I_{4i} = \int\limits_{1/\varepsilon_3}^{\infty} \int\limits_{|\varphi-\varphi_i|<\pi/(2M)} r^{n-1}(\cos\varphi)^{n-2}\Pi_i'\Pi_i''\,dr d\varphi, \qquad i = 1,\dots,M, \tag{40}$$

where

$$\Pi_i' = \prod_{j\in J_i'} q\big(r(x_j\cos\varphi + \sin\varphi)\big), \qquad \Pi_i'' = \prod_{j\in J_i''} q\big(r(x_j\cos\varphi + \sin\varphi)\big). \tag{41}$$

Note that $J_i' = \varnothing$ a.s. for $i \in I_0$. By (37), (38), and (39), for $1 \leqslant i \leqslant M/2$ and $j \in J_i''$, we have

$$|x_j\cos\varphi + \sin\varphi| = \sqrt{1+x_j^2}\,\Big|\cos\big(\varphi - \operatorname{arctg}(1/x_j)\big)\Big| =$$

$$= \Big|\sin\big(\varphi - \varphi_i + \psi_i - \operatorname{arctg}(1/x_j)\big)\Big| \geqslant \sin\big(\pi/(2M)\big). \tag{42}$$

Quite similarly, for $M/2 < i \leqslant M$ and $j \in J_i''$, we obtain

$$|x_j\cos\varphi + \sin\varphi| \geqslant \sin\big(\pi/(2M)\big). \tag{43}$$

Put, for $i \notin I_0$,

$$\chi_i(x_j) = \begin{cases} 1 & \text{if } |\operatorname{arctg}(1/x_j) - \psi_i| \leqslant \pi/M, \\ 0 & \text{otherwise} \end{cases}$$

and let $\xi_i = \sum\limits_{j=1}^{n} \chi_i(x_j)$.

From condition 3°, (41)–(42), we obtain

$$\Pi_i' \leqslant q_0^{\zeta_i},$$

$$\Pi_i'' \leqslant \left(\frac{q_0}{\big(r\sin(\pi/2M)\big)^{\beta}}\right)^{n-\zeta_i} \leqslant q^{n-\zeta_i}\Big(\sin\frac{\pi}{2M}\Big)^{-n\beta} r^{-(n-\zeta_i)\beta},$$

hence, taking into account (40), we have

$$I_{4i} \leqslant q_0^n(\pi/M)\Big(\sin\frac{\pi}{2M}\Big)^{-n\beta} \int\limits_{1/\varepsilon_3}^{\infty} r^{n-1-\beta(n-\zeta_i)}\,dr. \tag{44}$$

Put

$$\mathcal{X}_5 = \big(X\colon \max_{1\leqslant i\leqslant M} |\xi_i - np_{iM}| \leqslant N_2\sqrt{n}\big).$$

It is readily seen that

$$\mathbf{P}(\mathcal{X}_5) = \mathbf{P}_{u^{(0)}}(\mathcal{X}_5) > 1 - MN_2^{-2}/4. \tag{45}$$

Hence, for all sufficiently large n and $X \in \mathcal{X}_5$, we have

$$\zeta_i \leqslant 2np_+, \tag{46}$$

where $p_+ = \max\limits_{1\leqslant i\leqslant M} p_{iM}$.

From (44) and (46), for $X \in \mathcal{X}_5$ and $i \in I_0$, it follows that

$$I_{4i} \leqslant \frac{\pi}{M}\left(q_0\left(\sin\frac{\pi}{2M}\right)^{-\beta}\right)^n \int\limits_{1/\varepsilon_3}^{\infty} r^{n-1-n\beta(1-p_+)}dr.$$

Since $\beta > 1$ and p_+ can be made arbitrarily small by a proper choice of M, say, $p_+ < (1-1/\beta)/2$, we have, for $i \in I_0$,

$$I_{4i} \leqslant \frac{\pi}{nM}\left(q_0\left(\sin\frac{\pi}{2M}\right)^{-\beta}\right)^n \varepsilon_3^{n(\beta_0-1)}/(\beta_0-1), \quad \beta_0 = \beta(1-2p_+). \tag{47}$$

For $i \notin I_0$,

$$I_{4i} \leqslant \frac{\pi}{nM}\left(q_0\left(\sin\frac{\pi}{2M}\right)^{-\beta}\right)^n \varepsilon_3^{n(\beta-1)}/(\beta-1). \tag{48}$$

From (47), (48), (36) and (40) we have, for $X \in \mathcal{X}_5$,

$$I_4 \leqslant \left(q_0\left(\sin\frac{\pi}{2M}\right)^{-\beta}\right)^n \varepsilon_3^{n(\beta_0-1)}/((\beta_0-1)n). \tag{49}$$

Note that M has already been chosen. Now we should choose ε_1 (or ε_3) so small as to have

$$\max(I_3, I_4) \leqslant ce^{-nc} \tag{50}$$

on $\mathcal{X}_5$, which is possible in view of (35) and (49).

Thus, for $X \in \mathcal{X} \cap \mathcal{X}_1 \cap \mathcal{X}_5$, relations (19), (20) and (50) are valid. Since N, N_1 and N_2 are arbitrary, by virtue of (34), (19), (20), and (50), we are led to the desired result. Theorem 3 is proved.

Corollary 3 is evident. As to Corollary 4, under the conditions of Theorem 3, we have

$$a_i(u, u_i^*) = \alpha_i(u_{i1}^* - u_{i2}^* u_1/u_2, u_{i2}^*/u_2), \qquad u_i^* = (u_{i1}^*, u_{i2}^*)$$

and

$$\alpha(u^{(0)}) - \alpha_i(u, u_i^*) \leqslant c\left(|u_{i1}^* - u_1|^2 + |u_{i2}^* - u_2|^2\right)^{\gamma/2}$$

uniformly in $u \in U_0$. It remains only to refer to Corollary 3.

REFERENCES

Borovkov, A.A. (1984). *Mathematical Statistcs.* Nauka, Moscow (in Russian).

Ibragimov, I.A. and Khas'minskii, R.Z. (1981). *Statistical Estimation: Asymptotic Theory.* Springer, Berlin etc.

Hájek, J. and Šidak, Zb. (1967). *Theory of Rank Tests.* Academia, Prague.

New Trends in Probab. and Statist., pp. 612 – 616
V. Sazonov and T. Shervashidze (Eds.)
1991 VSP/Mokslas

ON MINIMAX BOUND FOR PARAMETER ESTIMATION IN BALL (BIAS ACCOUNTING)

A.V. NAZIN
Institute of Control Sciences, Profsoyuznaya str. 65, 117806 Moscow, USSR

ABSTRACT

A minimax bound for the parameter estimation in a ball of radius $r > 0$ is obtained under quadratic loss function. An example of its implementation for the problem of a function value asymptotic estimation at a given point by its noised observations at random points is considered for a nonparametric class of functions with a known smoothness.

INTRODUCTION

The obtaining of precise informational bounds is an important problem of statistical estimation and decision theory. It is considered systematically in a series of monographs (see, e.g., (Chentsov, 1982; Ibragimov and Khas'minskii, 1981)). This matter is especially interesting for nonparametric estimation problems for which the known lower bounds are usually attainable by the order only. It is connected particularly with asymptotical biasness of estimations optimal by the order. A possible approach consists in considering the problem for a parametric family that is a part of original nonparametric class of uncertainty. In this case the problem is reduced to obtaining an attainable minimax nonasymptotical bound of finite–dimensional parameter estimation when this parameter belongs to some compact set. An attempt of obtaining such inequality is presented in this paper for the case when this compact set is a ball. It may be considered as an extension of a result of Nemirovskii (Nemirovskii, 1981). The inequality obtained here is used for the problem of a function value asymptotical estimation at a given point by its noised observations at random points. A nonparametric class of functions with known smoothness is considered.

1. STATEMENT OF THE PROBLEM. MINIMAX BOUND

Let z be an observation with values in a Euclidian space $\mathcal{Z}$ and let $p(z\,|\,\theta)$ be its density with respect to Lebesgue measure with an unknown parameter $\theta \in W_r$, where W_r is a ball of radius $r > 0$ in an Euclidian space Θ. Assume that the following conditions are satisfied:

a) the function $p(z|\theta)$ is continuously differentiable with respect to θ on W_r for almost all z (with respect to Lebesgue measure);

b) the family of functions $\{\nabla_\theta p(z|\theta)\colon \theta \in W_r\}$ is uniformly integrable on any ball $|z| \leqslant R < \infty$;

c) for any $\theta \in W_r$, the Fisher information matrix

$$J(\theta) = \mathbf{E}_\theta\{\nabla_\theta \log p(z|\theta)\ \nabla_\theta^\tau \log p(z|\theta)\} = \int \nabla_\theta p(z|\theta) \nabla_\theta^\tau p(z|\theta)\ p^{-1}(z|\theta)\, dz$$

exists, is nondegenerate and continuous on W_r.

THEOREM 1. *Let conditions* **a)–c)** *be satisfied and v be an arbitrary nonzero vector. Then, for any measurable function $\widehat{\theta}\colon \mathcal{Z} \to \Theta$, the following inequality holds:*

$$\sup_{\theta \in W_r} \mathbf{E}_\theta(\widehat{\theta} - \theta, v)^2 \geqslant \varkappa(r/\sqrt{\nu}) \inf_{\theta \in W_r} (J^{-1}(\theta)v, v). \tag{1}$$

Here

$$\nu = \sup_{\theta \in W_r} \|J^{-1}(\theta)\|,$$

$\varkappa$: $[0, \infty) \to [0, 1)$ *is the increasing function with inverse*

$$\frac{\arccos(-\sqrt{\varkappa})}{\sqrt{1-\varkappa}} - \frac{\pi}{2}.$$

Proof (in many aspects) is similar to the proof of a previous result (Nemirovskii, 1981). It is based on the use of a "more precise" version of the Rao–Cramér inequality which takes into account the value of estimation bias. It is sufficient to consider the estimate for which the left-hand side of (1), say α, is finite. Denote the right-hand side of (1) by β. In addition, let θ^0 be the center of the ball W_r, $|v| = 1$, $\theta_v = (\theta, v)$, $\widehat{\theta}_v = \widehat{\theta}_v(z)$ be the nearest element to $(\widehat{\theta}, v)$ from the segment $[(\theta^0, v) - r,\ (\theta^0, v) + r]$. It is evident that

$$\sup_{\theta \in W_r} \int (\widehat{\theta}_v - \theta_v)^2 p(z|\theta)\, dz \leqslant \alpha.$$

Since conditions **a)–c)** hold, for any $\varepsilon > 0$, there exists $R = R(\varepsilon)$ such that, for

$$\widetilde{\theta}_v(z) = \begin{cases} \widehat{\theta}_v(z), & |z| \leqslant R, \\ \theta^0, & |z| > R, \end{cases}$$

the inequality

$$\sup_{\theta \in W_r} \int (\widetilde{\theta}_v(z) - \theta_v)^2 p(z|\theta)\, dz \leqslant \alpha + \varepsilon \tag{2}$$

is true. Denote $\alpha_\varepsilon = \alpha + \varepsilon$. The bias

$$b(\theta) = \int (\widetilde{\theta}_v(z) - \theta_v) p(z \mid \theta)\, dz \tag{3}$$

is continuously differentiable in W_r, and

$$\nabla b(\theta) + v = \int (\widetilde{\theta}_v(z) - \theta_v) \nabla_\theta p(z \mid \theta) dz.$$

Now we may use the Rao–Cramér inequality

$$\mathbf{E}_\theta (\widetilde{\theta}_v(z) - \theta_v)^2 \geqslant b^2(\theta) + \Big(v + \nabla b(\theta),\ J^{-1}(\theta)(v + \nabla b(\theta))\Big)$$

on the trajectory of the differential equation

$$\theta'(\tau) = -J^{-1}(\theta(\tau))v, \quad \theta(0) = \theta^0. \tag{4}$$

This equation has the continuously differentiable solution $\theta(\tau)$ on the finite segment $[-\tau_1, \tau_2]$ with $\tau_1, \tau_2 > 0$, $\theta(\tau) \in W_r$, $\theta(-\tau_1) \in \partial W_r$, $\theta(\tau_2) \in \partial W_r$. Furthermore, taking into account (2) and denoting

$$g(\tau) = (J^{-1}(\theta(\tau))v, v), \quad \varphi(\tau) = b(\theta(\tau)), \tag{5}$$

we obtain

$$\alpha_\varepsilon \geqslant \varphi^2(\tau) + g(\tau) - 2\varphi'(\tau) + \big(J^{-1}(\theta)\nabla b(\theta), \nabla b(\theta)\big)_{\theta = \theta(\tau)}. \tag{6}$$

From (6) and the Schwartz inequality,

$$(J^{-1}(\theta)v, \nabla b(\theta))^2 \leqslant (J^{-1}(\theta)v, v)\big(J^{-1}(\theta)\nabla b(\theta), \nabla b(\theta)\big),$$

it follows that

$$\varphi'(\tau) \geqslant g(\tau) - \sqrt{g(\tau)\big(\alpha_\varepsilon - \varphi^2(\tau)\big)}, \tag{7}$$

for any $\tau \in [-\tau_1, \tau_2]$. Denote

$$g_0 = \inf_{\theta \in W_r} (J^{-1}(\theta)v, v)$$

and observe that if $g_0 \leqslant \alpha_\varepsilon$ then $\alpha_\varepsilon > \beta$, because $\varkappa(\cdot)$ is bounded by 1. Let us show that $\alpha_\varepsilon \geqslant \beta$ also in the case when $\alpha_\varepsilon < g_0$, and, consequently,

$$g(\tau) > \alpha_\varepsilon, \quad \forall \tau \in [-\tau_1, \tau_2]. \tag{8}$$

From (7) we have

$$\varphi'(\tau) \geqslant \sqrt{g(\tau)}\Big(\sqrt{g_0} - \sqrt{\alpha_\varepsilon - \varphi^2(\tau)}\Big), \tag{9}$$

for any $\tau \in [-\tau_1, \tau_2]$. Consider the case $\varphi(0) \geqslant 0$. Then, from (8), (9), we have

$$\varphi(\tau) \geqslant \varphi_+(\tau) \geqslant 0, \quad \forall \tau \in [0, \tau_2],$$

where $\varphi_+(\tau)$ is the solution of differential equation

$$\varphi'(\tau) = \sqrt{g(\tau)}\Big(\sqrt{g_0} - \sqrt{\alpha_\varepsilon - \varphi^2(\tau)}\Big), \quad \varphi(0) = 0, \tag{10}$$

when $\tau \in [0, \tau_2]$. Taking into account (2), (3), (5) and the Schwartz inequality, we obtain from (10)

$$\sqrt{\alpha_\varepsilon} \geqslant \varphi_+(\tau_2). \tag{11}$$

Now integrate (10) and put $\tau = \tau_2$. Then it follows, from (11) and from the monotony of $\varphi_+(\tau)$ (if $\alpha_\varepsilon < g_0$), that

$$\int_0^{\tau_2} \sqrt{g(\tau)}d\tau \leqslant \frac{2\sqrt{g_0}}{\sqrt{g_0 - \alpha_\varepsilon}} \operatorname{arctg} \frac{\sqrt{g_0} + \sqrt{\alpha_\varepsilon}}{\sqrt{g_0 - \alpha_\varepsilon}} - \frac{\pi}{2}. \tag{12}$$

Let us estimate from below the integral in the left-hand side of (12).
From (4) we obtain

$$r = \left|\int_0^{\tau_2} J^{-1}(\theta(\tau))v d\tau\right| \leqslant \left(\int_0^{\tau_2} \sqrt{g(\tau)}d\tau\right) \sup_{\theta \in W_r} \left\|J^{-1/2}(\theta)\right\|$$

and thus (12) leads to the inequality

$$\frac{r}{\sqrt{\nu}} \leqslant \frac{\arccos\left(-\sqrt{\alpha_\varepsilon/g_0}\right)}{\sqrt{1 - \alpha_\varepsilon/g_0}} - \frac{\pi}{2}$$

or, by the monotony of $\varkappa(\cdot)$,

$$\alpha_\varepsilon g_0^{-1} \geqslant \varkappa(r/\sqrt{\nu}).$$

The case $\varphi(0) \leqslant 0$ is treated in a similar way with the only difference that the solution of the differential equation (10) is used on the segment $[-\tau_1, 0]$. Since $\varepsilon > 0$ is arbitrary, Theorem 1 is proved.

2. AN EXAMPLE OF NONPARAMETRIC ESTIMATION

Consider the following problem. Let the measurements $y_n = f(x_n) + \xi_n$ of the values of a function $f\colon \mathbf{R}^1 \to \mathbf{R}^1$, having random noise ξ_n, are carried out at random points x_n, $n = 1, 2, \ldots$ We have to estimate the value $f(\bar{x})$ at a given point $\bar{x}$ by these measurements. We shall assume that the function f has the derivative of the $l-1$ order satisfying the Lipschitz condition with a constant L. Denote by $\mathcal{F}_l(L)$ the class of such functions. Let the sequences $\{\xi_n\}$ and $\{x_n\}$ be independent and have the densities q and p, respectively. We assume that p is continuous and positive at $\bar{x}$, and q has finite Fisher information $I(q)$. Denote

$$\gamma = 2(2l + 1)^{-1}.$$

THEOREM 2. *Let the assumptions made in Section 2 be true. Then the following inequality holds for any sequence of estimations* $\hat{f}_n = \hat{f}_n(x_1, y_1, \ldots, x_n, y_n)$*:*

$$\liminf_{n\to\infty} n^{\gamma l} \sup_{f\in\mathcal{F}_l(L)} [p(\bar{x})]^{\gamma l}\mathbf{E}_f(\hat{f}_n - f(\bar{x}))^2 \geqslant \rho_l L^{\gamma}[I(q)]^{-\gamma l},$$

where the constant ρ_l depends on l only.

Proof. We may assume $\bar{x} = 0$ without loss of generality. Consider (for any fixed $n > 1$) a parametric family of functions

$$f(x|\,\theta) = \begin{cases} 0, & |x| \geqslant h_n, \\ \theta\eta(xh_n^{-1}), & |x| < h_n, \end{cases}$$

where θ is the parameter of the family, $h_n = hn^{-\gamma/2}$, $h > 0$, η: $(-1,1) \to [0,+\infty)$ is an even, bounded function from $\mathcal{F}_l(K)$ satisfying the following conditions:

$$\eta(0) = 1, \quad \eta_-^{(k)}(1) = 0 \quad (k = 0,\ldots,l-1), \tag{13}$$

$K > 0$ is sufficiently large, $\eta_-^{(k)}$ is the left-hand derivative of the k order. It is obvious that

$$f(\cdot\,|\,\theta) \in \mathcal{F}_l(L), \quad \forall\,|\theta| \leqslant r_n = LK^{-1}h_n^l.$$

In addition, $f(\bar{x}|\,\theta) \equiv \theta$. Hence, for the considered family of functions the original problem is transformed to the problem of a scalar parameter θ estimation, $|\theta| \leqslant r_n$, given the observations x_t, $y_t = f(x_t|\,\theta) + \xi_t$, $t = 1,\ldots,n$. Using Theorem 1, maximizing the obtained minimax bound in η and letting $n \to \infty$, we get the desired lower bound. We obtain also the following expression for the constant factor:

$$\rho_l = \left(\sup_{\tau>0} \tau^{-\gamma}\varkappa(\tau)\right)\left[\inf K\left(\int_{-1}^{1}\eta^2(u)\,du\right)^l\right]^{-\gamma},$$

where inf is calculated with respect to $\eta \in \mathcal{F}_l(K)$ satisfying conditions (13) and then with respect to K. Theorem 2 is proved.

Note that if $l = 1$, we obtain $\rho_l \approx 0.45157$, and a similar constant factor from the Nemirovskii inequality is equal approximately to 0.36668. The upper bound for the considered problem has been investigated for $l = 1$ in (Tsybakov, 1982). It coincides with the lower bound obtained in Theorem 2 in its dependence on n, L, $I(q)$, $p(\bar{x})$, l, and differs only by an absolute constant factor which approximately is equal to 0.6933.

REFERENCES

Chentsov, N.N. (1982). *Statistical Decision Rules and Optimal Inference.* AMS, Providence.

Ibragimov, I.A. and Khas'minskii, R.Z. (1981). *Statistical Estimation: Asymptotic Theory.* Springer, Berlin etc.

Nemirovskii, A.S. (1981). Optimization of recursive algorithms of estimation of parameters of linear plants. *Autom. Remote Contr.* **42**, 775–783.

Tsybakov, A.B. (1982). Robust estimates of a function. *Probl. Inf. Transm.* **18**, 190–201.

New Trends in Probab. and Statist., pp. 617 - 628
V. Sazonov and T. Shervashidze (Eds.)
1991 VSP/Mokslas

LIMIT THEOREMS FOR SOME RANDOM PROCESSES BASED ON FREQUENCIES

R.M. MNATSAKANOV
A. Razmadze Mathematical Institute of the Georgian Academy of Sciences,
Z. Rukhadze str. 1, 380093 Tbilisi, Georgia, USSR

ABSTRACT

The paper deals with random processes which are sums of functions of a multinomial frequency vector $(\nu_{1,nt}, \dots, \nu_{N,nt})$ with probabilities $(p_{1n}, \dots, p_{Nn})$, the number of trials being equal to $[nt]$, $t \in [0,1]$. Three types of limit theorems are obtained under various behaviour of the ratio $\alpha = n/N$ as $n, N \to \infty$.

INTRODUCTION

Let for each $n = 1, 2, \dots$ and $t \in [0,1]$ the random vector $(\nu_{1,nt}, \dots, \nu_{N,nt})$ have the multinomial distribution with probabilities $(p_{1n}, \dots, p_{Nn})$ and the number of trials equal to $[nt]$ ($[a]$ denotes the integer part of a number a). The weak convergence of the finite–dimensional distributions of processes of the type

$$\sum_{i \leqslant Nt} g\left(\nu_{in}, \frac{i}{N}\right), \qquad t \in [0,1], \quad (g\colon \mathbf{N} \times [0,1] \to \mathbf{R}, \quad \mathbf{N} = \{0, 1, \dots\})$$

is proved in (Ivchenko and Levin, 1976; Kolchin et al., 1978) whereas the functional limit theorems in (Khmaladze, 1983; Mnatsakanov, 1984, 1985, 1987a,b; Savchuk, 1983). More complete bibliography can be found in (Ivanov et al., 1984; Ivchenko et al., 1986).
As distinct from the object which has been introduced above and which is a partial sum process, here we shall investigate the limit behaviour of processes

$$\widetilde{X}^{n,N}(t) = \sum_{i=1}^{N} g\left(\nu_{i,nt}, \frac{i}{N}\right), \quad t \in [0,1].$$

The following three cases of the behaviour of the ratio $\alpha = n/N$ are considered as $n, N \to \infty$:

a) $\alpha \to 1$;

b) $n\alpha^{r-1} \to \lambda$, $0 < \lambda < \infty$, for some integer $r \geqslant 2$;

c) $\alpha\sqrt{n} \to 0$.

Statements on the convergence of finite-dimensional distributions of $\widetilde{X}^{n,N}$-type processes (appropriately normalized) are proved, for example, in (Ivchenko and Levin, 1976) for the case **a)**, and in (Kolchin et al., 1976) they are proved with functions $g(x,t) = I\{x = 0\}$ using various relations between n and N.

The approach proposed below allows us to obtain quite easily functional limit theorems for $\widetilde{X}^{n,N}$ in the cases **a)** - **c)**.

It is assumed for the sake of simplicity that in cases **a)** and **b)** the function $g(x,t) = I\{x \geqslant r\}$, i.e., processes of the form

$$\xi_r^{n,N}(t) = \sum_{i \leqslant N} I\{\nu_{i,nt} \geqslant r\}, \qquad t \in [0,1], \quad r \geqslant 1, \tag{1}$$

are studied ($I\{A\}$ is the indicator of a set A).

The main theorems are formulated in Section 1.

Since some parts of their proofs are long and similar, in Section 2 we shall give the main common steps of the proofs.

1. RESULTS

Prior to the formulation of the main results we shall consider the following scheme. Let $X_1, \dots, X_n$ be independent identically distributed random variables (r.v.'s) with a common absolutely continuous distribution function (d.f.) F having probability density f. In what follows it will be assumed that $\|f\| = \sup_t | f(t) | < \infty$. Without loss of generality it will also be assumed that f vanishes outside the interval [0,1]. Divide the interval [0,1] into subintervals of length $1/N$. For each $t \in [0,1]$ denote by $\nu_{i,nt}$ the number of elements of the sample among the first $[nt]$ observations $X_i, \dots, X_{[nt]}$ that have taken values from the i-th interval $[(i-1)/N,\ i/N]$, $i = 1, \dots, N$. Introduce also the flow of σ-algebras

$$\mathcal{F}_{[nt]}^n = \sigma\{\nu_{i,ns}, \quad s \leqslant t, \quad i = 1, \dots, N\}.$$

It is convenient to consider the processes $\xi_r^{n,N}$ as semimartingales with respect to the flow $\mathcal{F}_{[nt]}^n$. The convenience consists in that the conditional expectation of the increment

$$\Delta\xi_r^{n,N}\left(\frac{m}{n}\right) = \xi_r^{n,N}\left(\frac{m}{n}\right) - \xi_r^{n,N}\left(\frac{m-1}{n}\right)$$

under the condition $\mathcal{F}_{m-1}^n$ is of a rather simple form. Indeed, introduce the Bernoulli r.v.

$$\xi_{i,k} = I\{X_k \in [(i-1)/N,\ i/N]\}.$$

Note that the equality

$$I\{\nu_{i,n} \geqslant r\} = \sum_{k \leqslant n} I\{\nu_{i,k-1} = r-1\}\xi_{i,k}$$

is valid. Having in mind the above discussion it is easy to write the conditional expectation for $\Delta\xi_r^{n,N}(\frac{m}{n})$:

$$\mathbf{E}\left[\Delta\xi_r^{n,N}\left(\frac{m}{n}\right)\middle|\mathcal{F}_{m-1}^n\right] = \sum_{i \leqslant N} I\{\nu_{i,m-1} = r-1\}p_{in}, \tag{2}$$

whete $p_{in} = \Delta F(i/N)$.
Thus, for the semimartingale $\{X_r^{n,N}(t),\ \mathcal{F}_{[nt]}^n\}$, $t \in [0,1]$, where

$$X_r^{n,N} = (\xi_r^{n,N} - \mathbf{E}\xi_r^{n,N})/\sqrt{n},$$

Doob's decomposition takes the form

$$X_r^{n,N} = W_r^{n,N} + K_r^{n,N};$$

here

$$W_r^{n,N}(t) = \frac{1}{\sqrt{n}} \sum_{k \leqslant nt} C_{n,k-1}^{*}\, \eta_k, \quad t \in [0,1], \tag{3}$$

$C_{n,k}$ is the column vector with coordinates $I\{\nu_{i,k} = r-1\}$, $i = 1, \dots, N$, and η_k is the column vector with coordinates $\xi_{i,k} - p_{in}$, $i = 1, \dots, N$ (the symbol * denotes the transposition).
Denote by W_r the Wiener process with respect to the time

$$\rho_r(t) = \int_0^t \varphi_r(s)(1 - \varphi_r(s))ds, \quad t \in [0,1],$$

where

$$\varphi_r(s) = \int_0^1 \pi(r-1,\ sf(y))dF(y),$$

$$\pi(k,\lambda) = \frac{\lambda^k}{k!}e^{-\lambda}, \quad k = 0, 1, \dots$$

The convergence in distribution in the Skorohod space $\mathcal{D}[0,1]$ will be denoted by the symbol $\xrightarrow{\mathcal{D}}$.
For the martingale part $W_r^{n,N}$ of the process $X_r^{n,N}$ we have

THEOREM 1. *If condition* a) *is fulfilled, then*

$$W_r^{n,N} \xrightarrow{\mathcal{D}} W_r.$$

Remark 1. Denote by $M^{n,N}$ the martingale part of the semimartingale $\{U^{n,N}(t),\ \mathcal{F}_{[nt]}^n\}$, $t \in [0,1]$, where

$$U^{n,N} = (\widetilde{X}^{n,N} - \mathbf{E}\widetilde{X}^{n,N})/\sqrt{n}.$$

When certain conditions are imposed on the function g, we can prove that in the case a) the processes $M^{n,N} \xrightarrow{\mathcal{D}} M$, where M is the Wiener process with respect to the time

$$\gamma(t) = \int_0^t \sigma^2(s)ds, \quad t \in [0,1].$$

Here

$$\sigma^2(s) = \int_0^1 \mathbf{E}\left[\Delta_1 g(Z_s, y)\right]^2 dF(y) - \left\{\int_0^1 \mathbf{E}\left[\Delta_1 g(Z_s, y)\right] dF(y)\right\}^2$$

and $\Delta_1 g(x,t) = g(x+1,t) - g(x,t)$; Z_s is the Poisson r.v. with parameter $sf(y)$. The proof of this statement will be given elsewhere.
Define

$$Y^{n,N}(t) = \frac{1}{\sqrt{n}} \sum_{i \leqslant N} \left[I\{\nu_{i,nt} = 0\} - \mathbf{E}I\{\nu_{i,nt} = 0\}\right], \quad t \in [0,1].$$

The trajectories of the processes $Y^{n,N}$ and $X_1^{n,N}$ coincide up to the sign. Recall also that the statistic $Y^{n,N}(1)$ is widely used in the nonparametric statistics and is known as the empty cell test statistic.
The theorem given below enables us to construct a process which is the limit one for $Y^{n,N}$ in the case **a**).

THEOREM 2. *If $f \equiv 1$ and condition* **a**) *is fulfilled, then $X_1^{n,N} \xrightarrow{\mathcal{D}} X_1$, where X_1 is the Gaussian process with the mean 0, satisfying the stochastic differential equation*

$$dX_1(t) = -X_1(t)dt + dW_1(t).$$

Here W_1 is the Wiener process with respect to the time $\tau(t) = (1-e^{-t})^2/2$.

Remark 2. The covariance function of the process X_1 is equal to

$$e^{-s}\left[1 + (1+t)e^{-t}\right], \quad t \leqslant s. \tag{4}$$

The statement on convergence of the finite–dimensional distributions of the processes $Y^{n,N}$ to the corresponding finite–dimensional distributions of the Gaussian process with the covariance function (4) was proved in (Kolchin et al., 1978).
Let us consider now the case **b**). Denote by ξ_r a Poisson process with mean function

$$A_r(t) = \frac{\lambda t^r}{r!} \int_0^1 f^r(s)ds, \quad r \geqslant 2, \ t \in [0,1].$$

THEOREM 3. *If condition* **b**) *is fulfilled, then*

$$\xi_r^{n,N} \xrightarrow{\mathcal{D}} \xi_r.$$

And finally we shall consider the case **c**). Assume

$$X_0^{n,N}(t) = \frac{1}{\sqrt{n}} \sum_{i \leqslant N} \left[g_N\left(\nu_{i,nt}, \frac{1}{N}\right) - \mathbf{E}g_N\left(\nu_{i,nt}, \frac{i}{N}\right)\right], \quad t \in [0,1].$$

Consider the conditions

($\mathbf{g}_1$) $g_N(0,\cdot) \equiv 0$ and there exists a measurable function l: $[0,1] \to \mathbf{R}$ such that

$$\int_0^1 l^2(t)dF(t) < \infty, \quad \int_0^1 \left[g_N\left(1, \frac{[Nt]}{N}\right) - l(t)\right]^2 dF(t) \to 0, \quad N \to \infty;$$

($\mathbf{g}_2$) $| g_N(x,t) | \leqslant be^{ax}k_N(t), \quad a,b > 0, \ x \in \mathbf{N}, \ t \in [0,1]$, where k_N: $[0,1] \to \mathbf{R}_+$ and

$$\int_0^1 k_N^2\left(\frac{[Nt]}{N}\right)dF(t) = O(1), \quad N \to \infty.$$

Denote $X_0(t) = \int_0^1 l(y)u(t,dy), \ t \in [0,1]$, where u is a Gaussian field with mean 0 and covariance function

$$(t_1 \wedge t_2)\big[F(s_1 \wedge s_2) - F(s_1)F(s_2)\big].$$

THEOREM 4. *If conditions* ($\mathbf{g}_1$), ($\mathbf{g}_2$), *and* **c**) *are fulfilled, then* $X_0^{n,N} \xrightarrow{\mathcal{D}} X_0$.

Remark 3. X_0 is a Wiener process with respect to the time $\rho_0(t) = t \operatorname{var} l(X_1)$.

2. PROOF OF THE THEOREMS

To prove Theorems 1 to 3 we shall need an auxiliary statement.
Put

$$b(k,n,p) = \frac{n!}{k!\,(n-k)!}\, p^k\,(1-p)^{n-k}, \quad k = 0,1,\ldots,n, \quad 0 \leqslant p \leqslant 1.$$

LEMMA. *Let* $0 < \tau < \sigma \leqslant 1$. *Then*

$$\mathbf{P}\{\nu_{i,n\tau} = M, \ \nu_{i,n\sigma} = L\} =$$

$$= \begin{cases} b(M, [n\tau], p_{in})\, b\left(L-M, [n\sigma]-[n\tau], p_{in}\right) & \text{if } L \geqslant M, \\ 0 & \text{if } L < M; \end{cases} \tag{5}$$

$$\mathbf{P}\{\nu_{i,n\tau} = M, \ \nu_{j,n\sigma} = L\} = \frac{p_{in}^M\, p_{jn}^L}{M!} \sum_{k=0}^{L} \left\{ \frac{[n\tau]!}{k!\,([n\tau]-k-M)!} \times \right.$$
$$\times \frac{([n\sigma]-[n\tau])!}{(L-k)!\,([n\sigma]-[n\tau]-L+k)!}\,(1-p_{jn})^{[n\sigma]-[n\tau]-L+k} \times$$
$$\left. \times (1-p_{in}-p_{jn})^{[n\tau]-M-k} \right\}, \quad i \neq j; \tag{6}$$

$$\mathbf{P}\{\nu_{i,n\tau} = M,\ \nu_{j,n\sigma} = L,\ \nu_{l,n\sigma} = N\} =$$

$$= (M!)^{-1}\, p_{in}^{M}\, p_{jn}^{L}\, p_{ln}^{N} \sum_{y=0}^{L}\sum_{z=0}^{N}\left\{ \frac{[n\tau]!\,([n\sigma]-[n\tau])!}{y!\,z!\,(L-y)!\,(N-z)!}\times\right.$$

$$\times\Big(([n\sigma]-[n\tau]-L+y-N+z)!\,([n\tau]-M-y-z)!\Big)^{-1}\times$$

$$\left.\times(1-p_{in}-p_{jn}-p_{ln})^{[n\tau]-M-y-z}(1-p_{jn}-p_{ln})^{[n\sigma]-[n\tau]-L+y-N+z}\right\},\quad i\neq j\neq l; \quad (7)$$

$$\mathbf{P}\{\nu_{i,n\tau} = M,\ \nu_{j,n\tau} = L,\ \nu_{l,n\sigma} = N,\ \nu_{\gamma,n\sigma} = K\} =$$

$$= \frac{p_{in}^{M}\, p_{jn}^{L}\, p_{ln}^{N} p_{\gamma n}^{K}}{M!\,L!} \sum_{y=0}^{N}\sum_{z=0}^{K}\left\{ \frac{[n\tau]!\,([n\sigma]-[n\tau])!}{y!\,z!\,(K-z)!\,([n\tau]-M-L-y-z)!}\times\right.$$

$$\times\Big(([n\sigma]-[n\tau]-N+y-K+z)!\Big)^{-1}(1-p_{in}-p_{jn}-p_{ln}-p_{\gamma n})^{[n\tau]-M-L-y-z}\times$$

$$\left.\times(1-p_{ln}-p_{\gamma n})^{[n\sigma]-[n\tau]-N+y-K+z}\right\},\quad i\neq j\neq l\neq j. \quad (8)$$

The proof of the lemma follows from the total probability formula and the independence of the r.v.'s $X_1,\dots,X_{[nt]},\dots,X_n$.

Proof of Theorem 1. By virtue of (3) and Corollary 6 from (Liptser and Shiryaev, 1980) the statement of Theorem 1 will follow from the relations

(A) for each $t\in[0,1]$,

$$\langle W_r^{n,N}\rangle_t = \frac{1}{n}\sum_{k\leqslant nt}\mathbf{E}\big[C^*_{n,k-1}\,\eta_k\,\eta_k^*\,C_{n,k-1}\mid \mathcal{F}^n_{k-1}\big]\xrightarrow{\mathbf{P}}\rho_r(t),$$

(B) for any $\varepsilon>0$,

$$\alpha_n = \frac{1}{n}\sum_{k\leqslant n}\mathbf{E}\Big[C^*_{n,k-1}\,\eta_k\,\eta_k^*\,C_{n,k-1}\,I\{\,|\,C^*_{n,k-1}\,\eta_k\,|>\varepsilon\sqrt{n}\}\mid \mathcal{F}^n_{k-1}\Big]\xrightarrow{\mathbf{P}}0,$$

$$n,N\to\infty,\quad n/N\to 1.$$

Let us prove (A). Since

$$\psi_{n,N}\left(\frac{k-1}{n}\right) := \mathbf{E}\big[C^*_{n,k-1}\eta_k\eta_k^*\,C_{n,k-1}\mid\mathcal{F}^n_{k-1}\big] =$$

$$= C^*_{n,k-1}p_n - C^*_{n,k-1}p_n p_n^* C_{n,k-1}, \quad (9)$$

where $p_n^* = (p_{1n},\dots,p_{Nn})$, we have

$$\langle W_r^{n,N}\rangle_t = \int_0^t \psi_{n,N}\left(\frac{[ns]}{n}\right)ds. \quad (10)$$

To prove (A) it is sufficient to show that

$$\mathbf{E}\langle W_r^{n,N}\rangle_t \to \rho_r(t), \tag{11}$$

$$\operatorname{var}\langle W_r^{n,N}\rangle_t \to 0, \qquad n, N \to \infty, \quad n/N \to 1. \tag{12}$$

Note that for each $s \in [0,1]$ and for almost all $y \in [0,1]$ we have $[ns]\,\Delta F\left(\frac{[ny]}{N}\right) \to$ $\to sf(y)$ as $n/N \to 1$, $n, N \to \infty$. Thus, applying formulas (5), (6) from Lemma, we obtain

$$\mathbf{E}\psi_{n,N}\left(\frac{[ns]}{n}\right) = \sum_{i\leqslant N} b(r-1, [ns], p_{in})\,\Delta F\left(\frac{i}{N}\right) -$$

$$-\sum_{i\leqslant N}\sum_{j\leqslant N} \mathbf{P}\{\nu_{i,ns} = r-1,\ \nu_{j,ns} = r-1\}\,\Delta F\left(\frac{i}{N}\right) \to$$

$$\to \int_0^1 \pi\big(r-1,\ sf(y)\big)dF(y) - \left[\int_0^1 \pi\big(r-1,\ sf(y)\big)dF(y)\right]^2.$$

Hence, applying Lebesque's majorized convergence theorem (LMCT), we obtain (11).
To prove (12), by virtue of (11) it suffices to show that

$$\mathbf{E}\langle W_r^{n,N}\rangle_t^2 \to \rho_r^2(t), \quad n, N \to \infty, \quad n/N \to 1. \tag{13}$$

By (10) we have

$$\mathbf{E}\langle W_r^{n,N}\rangle_t^2 = \mathbf{E}\int_0^t\int_0^t \psi_{n,N}\left(\frac{[n\tau]}{n}\right)\psi_{n,N}\left(\frac{[n\sigma]}{n}\right)d\tau d\sigma,$$

where by virtue of (9) we can write the integrand in the form

$$\psi_{n,N}\left(\frac{[n\tau]}{n}\right)\psi_{n,N}\left(\frac{[n\sigma]}{n}\right) = C^*_{n,[n\tau]}\,p_n\,p^*_n\,C_{n,[n\sigma]} - \tag{14}$$

$$-C^*_{n,[n\tau]}\,p_n\,(C_{n,[n\sigma]}\,p^*_n\,p_n\,C^*_{n,[n\sigma]}) -$$

$$-(C^*_{n,[n\tau]}\,p_n\,p^*_n\,C_{n,[n\tau]})\,p^*_n C_{n,[n\sigma]} + (C^*_{n,[n\tau]}\,p_n\,p^*_n\,C_{n,[n\tau]})\times$$

$$\times(C_{n,[n\sigma]}\,p^*_n\,p_n\,C^*_{n,[n\sigma]}).$$

Let us now prove that the first term in (14) satisfies the relation

$$\mathbf{E}\int_0^t\int_0^t C^*_{n,[nt]}\,p_n\,p^*_n\,C_{n,[n\sigma]}d\tau d\sigma \longrightarrow$$

$$\longrightarrow \int_0^t\int_0^t d\tau d\sigma \left[\int_0^1\int_0^1 \pi(r-1,\tau f(y))\pi(r-1,\sigma f(z))dF(y)dF(z)\right] =$$

$$= \int_0^t \varphi_r(\tau)d\tau \int_0^t \varphi_r(\sigma)d\sigma, \quad n,N\to\infty, \quad n/N\to 1. \tag{15}$$

We divide the domain of integration in the second integral on the left-hand side of (15) into two parts: $0\leqslant\sigma\leqslant\tau$ and $\tau<\sigma\leqslant t$.
To prove (15), by virtue of the symmetry property it suffices to show that

$$\int_0^t d\tau \int_\tau^t \mathbf{E}[C^*_{n,[n\tau]}\, p_n\, p^*_n\, C_{n,[n\sigma]}]d\sigma \longrightarrow \int_0^t \varphi_r(\tau)d\tau \int_\tau^t \varphi_r(\sigma)d\sigma,$$

$$n,N\to\infty, \quad n/N\to 1. \tag{16}$$

However, again applying formula (6) from Lemma, it is not difficult to verify that for $\sigma>\tau$, as $n/N\to 1$, $n,N\to\infty$, we have

$$\mathbf{E}C^*_{n,[n\tau]}\, p_n\, p^*_n\, C_{n,[n\sigma]} \longrightarrow \int_0^1\int_0^1 \left\{\frac{[\tau f(y)]^{r-1}}{(r-1)!} f^{r-1}(z)\times\right.$$

$$\left.\times\sum_{k=0}^{r-1}\frac{\tau^k(\sigma-\tau)^{r-1-k}}{k!\,(r-1-k)!}\exp\{-\tau[f(y)+f(z)]-(\sigma-\tau)f(z)\}\right\}dF(y)\,dF(z) =$$

$$= \int_0^1\int_0^1 \pi(r-1,\,\tau f(y))\pi(r-1,\sigma f(z))dF(y)\,dF(z) = \varphi_r(\tau)\,\varphi_r(\sigma).$$

In the latter equality we used the fact that

$$\sum_{k=0}^{r-1}\frac{\tau^k(\sigma-\tau)^{r-1-k}}{k!\,(r-1-k)!} = \frac{\sigma^{r-1}}{(r-1)!} \quad (\sigma>\tau).$$

Thus, applying the LMCT, we obtain (16).
For the rest of the terms in (14) relations similar to (15) are proved by the same procedure, using (6) to (8). Thus (13) is proved and so condition (A) is verified. Condition (B) is verified by a quite similar procedure, taking into account the properties of the r.v.'s $\xi_{i,k}$. We obtain

$$\max_{k\leqslant n}\mathbf{E}\left[C^*_{n,k-1}\,\eta_k\,\eta^*_k\,C_{n,k-1}\,I\{|\,C^*_{n,k-1}\,\eta_k\,|>\varepsilon\sqrt{n}\}\mid\mathcal{F}^n_{k-1}\right] = o_{\mathbf{p}}(1), \quad n,N\to\infty.$$

Theorem 1 is proved.

Proof of Theorem 2. It is not difficult to show that when $f\equiv 1$ for $\{X_1^{n,N}(t),\ \mathcal{F}^n_{[nt]}\}$,

$t \in [0,1]$, Doob's decomposition takes a rather simple form

$$X_1^{n,N}(t) = -\alpha \int_0^t X_1^{n,N}(s)ds + W_1^{n,N}(t).$$

The solution of this stochastic integral equation can be written in the form

$$X_1^{n,N}(t) = W_1^{n,N}(t) - \alpha e^{-\alpha t} \int_0^t e^{\alpha\tau} W_1^{n,N}(\tau)d\tau.$$

By Theorem 1 the processes $W_1^{n,N} \xrightarrow{\mathcal{D}} W_1$, where W_1 is a Wiener process with respect to the time

$$\tau(t) = \rho_1(t) = \int_0^t [e^{-s} - e^{-2s}]ds = (1-e^{-t})^2/2.$$

Let us now consider the sequence of mappings from $\mathcal{D}[0,1]$ into $\mathcal{D}[0,1]$ defined by

$$h_n x(t) = x(t) - \alpha e^{-\alpha t} \int_0^t e^{\alpha\tau} x(\tau)d\tau, \quad x \in \mathcal{D}[0,1]. \tag{17}$$

Since the trajectories of the process W_1 are continuous almost everywhere, the mappings given by (17) satisfy the conditions of Theorem 5.5 from (Billingsley, 1968). Therefore as $\alpha = n/N \to 1$, $n, N \to \infty$, the convergence $W_1^{n,N} \xrightarrow{\mathcal{D}} W_1$ yields the convergence $X_1^{n,N} \xrightarrow{\mathcal{D}} X_1$, where

$$X_1(t) = W_1(t) - e^{-t} \int_0^t W_1(\tau)e^{\tau} d\tau.$$

Theorem 2 is proved.

Proof of Theorem 3. It is convenient to consider the sequence of processes $\xi_r^{n,N}(t)$, $t \in [0,1]$, defined by (1) as a sequence of point processes with respect to the flow of σ-algebras $\mathcal{F}_{[nt]}^n$.

Therefore to prove Theorem 3 it suffices to show that for the compensator $A_r^{n,N}$ of the process $\xi_r^{n,N}$ the relation

$$A_r^{n,N}(t) \xrightarrow{\mathbf{P}} A_r \tag{18}$$

is fulfilled for each $t \in [0,1]$ (see, e.g., (Brown, 1981)). By virtue of (2), $A_r^{n,N}$ can be written in the form

$$A_r^{n,N}(t) = \sum_{k \leqslant nt} C_{n,k-1} P_n^* = \int_0^t \varphi_{n,N}(s)ds,$$

where

$$\varphi_{n,N}(s) = n\sum_{i=1}^{N} I\{\nu_{i,ns} = r-1\}\, p_{in}, \quad p_{in} = \Delta F\left(\frac{i}{n}\right).$$

Relation (18) follows from relations

$$\mathbf{E}A_r^{n,N}(t) \to A_r(t), \tag{19}$$

$$\operatorname{var} A_r^{n,N}(t) \to 0, \tag{20}$$

as $n\alpha^{r-1} \to \lambda$, $n, N \to \infty$.
Indeed, since for each $s \in [0,1]$ and for almost all $y \in [0,1]$ we have $[ns]\Delta F([Ny]/N) \to 0$ and $N\Delta F([Ny]/N) \to f(y)$ as $\alpha = n/N \to 0$, $n, N \to \infty$, we find

$$\mathbf{E}\varphi_{n,N}(s) = n\sum_{i\leqslant N} b(r-1, [ns], p_{in})\, p_{in} \to$$

$$\to \frac{\lambda s^{r-1}}{(r-1)!}\int_0^1 f^{r-1}(y)\, dF(y).$$

Therefore, applying the LMCT, we obtain (19).
Relation (20) is verified in a similar manner, using (5), (6). Theorem 3 is proved.

Proof of Theorem 4. Denote $g_N(1,t) = l_N(t)$, $\tilde{g}_N(x,t) = g_N(x,t) - x l_N(t)$ and introduce the following random fields and the process:

$$z_n(t,s) = \sum_{k\leqslant nt} I\{X_k \leqslant s\},$$

$$u_n(t,s) = \left(z_n(t,s) - [nt]F(s)\right)/\sqrt{n},$$

$$Z^{n,N}(t) = \int_0^1 l_N\left(\frac{[Ny]}{N}\right) u_n(t, dy).$$

Since $g_N(0,\cdot) \equiv 0$, it is not difficult to obtain the representation

$$X_0^{n,N} = Z^{n,N} + \widetilde{Y}^{n,N}, \tag{21}$$

where

$$\widetilde{Y}^{n,N}(t) = \frac{1}{\sqrt{n}}\sum_{i=1}^{N}\Big[\tilde{g}_N\left(\nu_{i,nt}, \frac{i}{N}\right) I\{\nu_{i,nt} \geqslant 2\} -$$

$$-\mathbf{E}\left(\tilde{g}_N\left(\nu_{i,nt}, \frac{i}{N}\right) I\{\nu_{i,nt} \geqslant 2\}\right)\Big].$$

The proof of Theorem 4 repeats the proof of Theorem 2 in (Mnatsakanov, 1985). To be more precise the processes $Z^{n,N}$ will be approximated by the processes

$$\widetilde{Z}^n(t) = \int_0^1 l(y)\, u_n(t, dy) = \frac{1}{\sqrt{n}}\sum_{k\leqslant nt}\left[l(X_k) - \mathbf{E}l(X_k)\right].$$

Let $\varphi_N(x) = l_N\left(\frac{[Nx]}{N}\right) - l(x)$. Then

$$\zeta^{n,N}(t) = Z^{n,N}(t) - \widetilde{Z}^n(t) = \frac{1}{\sqrt{n}} \sum_{k \leqslant nt} \left[\varphi_N(X_k) - \mathbf{E}\varphi_N(X_k)\right].$$

From condition $(\mathbf{g}_1)$ and Donsker's theorem from (Billingsley, 1968) it follows that

$$\widetilde{Z}^n \xrightarrow{\mathcal{D}} X_0 \text{ and } \zeta^{n,N} \xrightarrow{\mathcal{D}} 0.$$

Hence, by virtue of (21), Theorem 4 will be proved if we show that

$$\|\widetilde{Y}^{n,N}\| \xrightarrow{\mathbf{P}} 0. \tag{22}$$

Let us prove (22). Note that the functions $\tilde{g}_N$ satisfy condition $(\mathbf{g}_2)$ with $\tilde{a} = \max(a, 1)$, $\tilde{b} = 2be^a$. Therefore,

$$\|\widetilde{Y}^{n,N}\| \leqslant I_1^{n,N} + \mathbf{E}I_1^{n,N},$$

where

$$I_1^{n,N} = \frac{1}{\sqrt{n}} \tilde{b} \sum_{i \leqslant N} e^{a\nu_{i,n}} k_N\left(\frac{i}{N}\right) I\{\nu_{i,n} \geqslant 2\}.$$

We shall show that $\mathbf{E}I_1^{n,N} \to 0$ and $\operatorname{var} I_1^{n,N} \to 0$ as $\alpha\sqrt{n} \to 0$, $n, N \to \infty$. Indeed, it is easy to check that

$$\mathbf{E}I_1^{n,N} \leqslant \frac{\tilde{b}}{\sqrt{n}} \sum_{i \leqslant N} k_N\left(\frac{i}{N}\right) \sum_{x \geqslant 2} e^{\tilde{a}} \left\{\frac{n\,\Delta F(i/N)}{x\,(x-1)!} \left[e^{\tilde{a}}\, n\,\Delta F(i/N)\right]^{x-1}\right\} \leqslant$$

$$\leqslant \frac{\tilde{b}\, e^{\tilde{a}} \sqrt{n}}{2} \sum_{i \leqslant N} k_N\left(\frac{i}{N}\right) \Delta F\left(\frac{i}{N}\right) \sum_{x \geqslant 1} \frac{(e^{\tilde{a}}\, \alpha\, \|f\|)^x}{x!} \leqslant$$

$$\leqslant \frac{1}{2} \alpha \sqrt{n}\, \tilde{b}\, e^{\tilde{a}}\, \|f\|\, e^{\alpha \|f\| e^{\tilde{a}}} \sum_{i \leqslant N} k_N\left(\frac{i}{N}\right) \Delta F\left(\frac{i}{N}\right).$$

Hence by virtue of condition $(\mathbf{g}_2)$ it follows that $\mathbf{E}I_1^{n,N} \to 0$. Similarly, it is not difficult to show that $\mathbf{E}(I_1^{n,N})^2 \to 0$. Theorem 4 is proved.

REFERENCES

Billingsley, P. (1968). *Convergence of Probability. Measures.* Wiley, New York.

Brown, T.C. (1981). Compensators and Cox convergence. *Math. Proc. Cambridge Phil. Soc.* **9**, 305–310.

Ivanov, V.A., Ivchenko, G.I. and Medvedev, Yu.I. (1984). Discrete problems in probability theory. In: *J. Soviet Mathematics* (*Theory Probab., Math. Statist., Theor. Cybern.*). **22**, pp. 3–60.

Ivchenko, G.I. and Levin, V.V. (1976). Asymptotic normality of one class of statistics in the multinomial scheme. *Theory Probab. Appl.* **21**, 188–192.

Ivchenko, G.I., Medvedev, Yu.I. and Ronzhin, A.F. (1986). Separable statistics and goodness-of-fit tests for polynomial samples. In: *Proc. V.A. Steklov Math. Inst.,* **177**, pp. 60–74.(in Russian).

Khmaladze, E.V. (1983). Martingale limit theorems for separable statistics. *Theory Probab. Appl.* **28**, 530–548.

Kolchin, V.F., Sevastyanov, B.A. and Chistyakov, V.P. (1978). *Random Allocations.* Wiley, New York.

Liptser, R.Sh., Shiryaev, A.N. (1980). A functional central limit theorem for semimartingales. *Theory Probab. Appl.* **25**, 667–688.

Mnatsakanov, R.M. (1984). Martingale limit theorems for divisible statistics in the case of very rare events. In: *Proc. of 3rd Prague Symposium on Asymptotical Statist.,* Akademia, Prague, pp. 355–362.

Mnatsakanov, R.M. (1985). A functional limit theorem for additively separable statistics in the case of very rare events. *Theory Probab. Appl.* **30**, 622–625.

Mnatsakanov, R.M. (1987a). On the convergence of separable statistics to Wiener process. *Theory Probab. Appl.* **32**, 152–157.

Mnatsakanov, R.M. (1987b). On the convergence of divisible statistics to the Poisson process. *Theory Probab. Appl.* **32**, 361–365.

Savchuk, M.N. (1983). Some limit theorems in the scheme of equiprobable location of particles by complete sets. *Theory Probab. Math. Statist.* **28**, 530–548.

New Trends in Probab. and Statist., pp. 629 - 635
V. Sazonov and T. Shervashidze (Eds.)
1991 VSP/Mokslas

FROM ASYMPTOTIC EFFICIENCY IN MINIMAX SENSE TO BAHADUR EFFICIENCY

M. RADAVIČIUS
Institute of Mathematics and Cybernetics, Lithuania Academy Sci.,
Akademijos str. 4, 232600 Vilnius, Lithuania

ABSTRACT

A generalization of Hajek's inequality for the minimax risk is proposed. The new inequality takes into account the large deviations of estimators. It is proved in some special cases including the Bahadur inequality.

1. INTRODUCTION

Let $X_1, X_2, \ldots, X_n$ be i.i.d. random variables with a common distribution density $f(\cdot, \theta) = d\mathbf{P}_\theta / d\mu$ in the measurable space $(\mathfrak{X}, \mathfrak{A}, \mu)$, where μ is some σ-finite measure. The unknown parameter θ belongs to a bounded interval $\Theta = (\alpha, \beta)$. Let us suppose that the Fisher information $I(\theta)$ exists and is positive. Denote by Φ the distribution function of a standard normal variable ξ. Let W be the class of all loss functions, i.e., non-negative, symmetric, nonidentically equal to zero, and nondecreasing on $[0, \infty)$ functions w with $w(0) = 0$ such that the integral $\mathbf{E}w(\xi)$ exists. The theory of asymptotic efficiency in the minimax sense of estimators $\widehat{\theta}$ of θ is based on the following Hajek's inequality (Hajek, 1972)

$$\liminf_{C \to \infty} \liminf_{n \to \infty} \inf_{\hat{\theta}} \sup_{|u-\theta| \leqslant Cn^{-1/2}} \mathbf{E}_u w\Big(\big(nI(\theta)\big)^{1/2} (\widehat{\theta} - u) \Big) \geqslant \mathbf{E}w(\xi) \tag{1}$$

which under the assumption of local asymptotic normality is valid for each $w \in W$. An estimator θ^* is said to be asymptotically efficient (in the minimax sense) for a loss function $w \in W$ if the equality

$$\lim_{C \to \infty} \limsup_{n \to \infty} \sup_{|u-\theta| \leqslant Cn^{-1/2}} \mathbf{E}_u w\Big(\big(nI(\theta)\big)^{1/2} (\theta^* - u) \Big) = \mathbf{E}w(\xi) \tag{2}$$

holds. So, for the asymptotic efficiency of an estimator θ^* it is sufficient (leaving apart the question of uniform convergence of the corresponding integrals) that

$$T_u^* = \big(nI(\theta)\big)^{1/2} (\theta^* - u) \xrightarrow{d} \xi.$$

But in practice one always deals with a finite number of observations and it is not enough to know that an estimator behaves well at infinity. Thus, the rate of convergence of the distribution of T_u^* to the corresponding normal distribution must be also taken into account. It is natural to consider an estimator to be better if the rate of convergence for it is faster or the probabilities of large deviations are smaller. This idea was realized in the notions of second order asymptotic efficiency (Levit, 1982) and asymptotic efficiency in the Bahadur sense (Bahadur, 1971).

In the present paper we consider a generalization of the inequality (1) that takes into account large deviations of the estimators. The generalization is that a loss function may depend on n in a rather arbitrary way. Moreover, a class of functions that define the exactness (or the level) of comparison of minimax risks of estimators is introduced. To give an explicit formulation we need some notation.

Let $\psi = \{\psi_n\}_{n=1}^{\infty}$ be some nondecreasing sequence of positive numbers tending to infinity as $n \to \infty$ slower than $n^{1/2}$, so $\delta_n = \big(nI(\theta)\big)^{1/2}\psi_n \to 0$. Let us denote by Ψ the class of all such sequences and by W_ψ the class of sequences $w_\psi = \{w_n\}_{n=1}^{\infty}$ of loss functions $w_n \in W$ such that

$$\big(\mathbf{E}w_n(\xi)\big)^{-1} \int\limits_{|y|>\psi_n} w_n(y)d\Phi(y) \longrightarrow 0, \quad n \to \infty.$$

Further, denote by H the class of strictly increasing continuous functions h: $\mathbf{R}_+ \to \mathbf{R}_+$ $(\mathbf{R}_+ = [0,\infty))$, $h(0) = 0$, which are asymptotically equivalent at zero and at infinity, i.e., $h(s)h(1/s) \to 1$ as $s \to 0$. For any estimator $\widehat{\theta}$, define

$$\widehat{T}_u = \big(nI(\theta)\big)^{1/2}(\widehat{\theta} - u).$$

The problem is given $\psi \in \Psi$, $h \in H$, and subclass $W_\psi^0 \subset W_\psi$ to prove that the inequality

$$\liminf_{C\to\infty} \liminf_{n\to\infty} \inf_{\hat{\theta}} \sup_{|u-\theta|\leqslant C\delta_n} \frac{h\big(\mathbf{E}_u w_n(\widehat{T}_u)\big)}{h\big(\mathbf{E}w_n(\xi)\big)} \geqslant 1 \tag{3}$$

holds for each $w_\psi = \{w_n\}_{n=1}^{\infty} \in W_\psi^0$.

Remark 1. Let us suppose that W_ψ^0 consists only of those w_ψ for which the equality $w_n(y) = w(y)$ holds for some $w \in W$ and for all $y \in \mathbf{R}$ and $n = 1, 2, \ldots$. Then we have $0 < c \leqslant \mathbf{E}w_n(\xi) \leqslant C < \infty$, and the choice of h is of no importance (so we may assume $h(y) \equiv y$). Thus, in this case the inequality (3) is equivalent to (1).

Remark 2. Let us denote by W_ψ^1 the class of sequences $w_\psi = \{w_n\}_{n=1}^{\infty} \in W_\psi$ of indicator functions $w_n(y) = \chi(|y| \geqslant t)$, where $t = \{t_n\} \in \Psi$ and $\psi_n^2 - t_n^2 \to \infty$ as $n \to \infty$. If $h \in H$ and $h(s) \sim \log s$, $s \to \infty$, and $W_\psi^0 = W_\psi^1$, the inequality (3) concides with the minimax version of the Bahadur inequality (cf. (Bahadur, 1971; Radavičius, 1983))

$$\liminf_{\tau\to 0} \liminf_{n\to\infty} \inf_{\hat{\theta}} \sup_{|u-\theta|\leqslant\tau} \frac{2}{t^2} \log \mathbf{P}_u\big\{|\widehat{T}_u| \geqslant t\big\} \geqslant -1. \tag{4}$$

Taking into consideration (2) and Remarks 1 and 2, we say that an estimator θ^* is asymptotically efficient with an exactness $h \in H$ in the region $|t| \leqslant \psi_n$, $\psi \in \Psi$, for a loss function class $W_\psi^0 \subset W_\psi$, if the equality

$$\lim_{C\to\infty} \limsup_{n\to\infty} \sup_{|u-\theta|\leqslant C\delta_n} \frac{h(\mathbf{E}_u w_n(\widehat{T}_u))}{h(\mathbf{E} w_n(\xi))} = 1 \tag{5}$$

holds for each $w_\psi \in W_\psi^0$. It follows from (Radavičius, 1983) that under some additional conditions the maximum likelihood estimator satisfies (5) with $h(y) \equiv y$, $W_\psi^0 = W_\psi$, and $\psi \in \Psi$, $\psi_n = o(n^{1/6})$ (see Remark 4 and Corollary below).

We prove inequality (3) for some interesting special cases including the Bahadur inequality (4). The results are formulated in the next section and the proofs are given in the last one.

2. RESULTS

Everywhere below the following regularity condition is supposed to be satisfied (see (Ibragimov and Khas'minskii, 1981, Chapter 1, § 7)).

(R) The function $g(x,\theta) = f^{1/2}(x,\theta)$ is continuous on Θ for almost all x with respect to the measure μ and continuously differentiable in $\mathbf{L}_2(\mu)$, moreover, $0 < c \leqslant I(\theta) \leqslant C < \infty$, where

$$I(\theta) = 4\int \left|\frac{\partial}{\partial\theta} g(x,\theta)\right|^2 \mu(dx) = 4\left\|\frac{\partial}{\partial\theta} g(\cdot\,,\theta)\right\|_2^2$$

is the Fisher information.

Denote

$$l_j(\tau) = \log\big(f(X_j,\theta+\tau)/f(X_j,\theta)\big), \quad j = 1,2,\dots,n,$$

and for a given non-negative $\varepsilon = \varepsilon(\tau) \to 0$, $\tau \to 0$, define

$$\bar{l}_j(\tau) = l_j(\tau)\chi\big(|l_j(\tau)| \leqslant \varepsilon(\tau)\big), \quad j = 1,2,\dots,n$$

(here $\chi(A)$ is the indicator function of an event A).

It follows from condition (R) that there exist non-negative, symmetric, and monotone functions $\varepsilon = \varepsilon(\tau)$ and $\lambda = \lambda(\tau)$ tending to zero as $\tau \to 0$ such that the following relations hold

$$p(\tau) = \mathbf{P}_\theta\Big\{|l_1(\tau)| \leqslant \varepsilon(\tau)\Big\} \geqslant 1 - \tau^2\lambda(\tau), \tag{6}$$

$$a(\tau) = \mathbf{E}_\theta \bar{l}_1(\tau) = -I(\theta)\frac{\tau^2}{2}\Big(1 + O(\lambda(\tau))\Big), \tag{7}$$

$$\sigma^2(\tau) = \mathbf{E}_\theta \bar{l}_1^2(\tau) = I(\theta)\tau^2\Big(1 + O(\varepsilon(\tau))\Big) \tag{8}$$

(see (Ibragimov and Khas'minskii, 1981, Chapter 3, § 1)).

The Bahadur inequality (4) is an immediate consequence of the following theorem.

THEOREM 1. *Let us assume that relations* (6)–(8) *are fulfilled and* $\varphi, \psi \in \Psi$, $\varphi_n < \psi_n$. *Then, for* $\delta_n = \big(nI(\theta)\big)^{-1/2}\psi_n$, *the asymptotic* (*as* $n \to \infty$) *inequality*

$$\sup_{|u-\theta|\leqslant\delta_n} \mathbf{P}_u\{|\widehat{T}_u| \geqslant t\} \geqslant \frac{1}{2e}\mathbf{P}\{|\xi| \geqslant t\} \exp\{O(t^2\lambda(\delta_n))\}(1+o(1)) \qquad (9)$$

is valid for any estimator $\widehat{\theta}$ *of* θ *uniformly in* $t \in [\varphi_n, \psi_n]$.

Remark 3. It is easy to check following the proof of the Hajek inequality (see (Hajek, 1972; Ibragimov and Khas'minskii, 1981, Theorem 2.12.1)) that there exists $b \in \Psi$ such that

$$\sup_{|u-\theta|\leqslant b_n n^{-1/2}} \mathbf{P}_u\{|\widehat{T}_u \geqslant t\} \geqslant \mathbf{P}\{|\xi| \geqslant t\}(1+o(1)) \qquad (10)$$

uniformly in $t \in [0, \psi_n]$ provided $\psi \in \Psi$, $\psi_n = o(b_n)$, $n \to \infty$. If $\psi \in \Psi$ increases sufficiently slowly so that $\psi_n^2\lambda(\delta_n) \to 0$, $n \to \infty$, we obtain from (9) the same inequality except that the constant is $2e$ times worse and the region over which the supremum is taken is much smaller.

Let us introduce a new class of loss function sequences W_ψ^2. It consists of all sequences $w_\psi = \{w_n\}_{n=1}^\infty$ of the type

$$w_n(y) = \cosh(t_n y) - 1,$$

where $\cosh(y) = 0.5(e^{-y} + e^y)$, $\{t_n\} \in \Psi$ and $\psi_n - t_n \to \infty$ as $n \to \infty$.

THEOREM 2. *Let us assume that relations* (6) *and* (7) *are fulfilled and* $\psi \in \Psi$. *Then, for each* $w_\psi \in W_\psi^2$,

$$\sup_{|u-\theta|\leqslant\delta_n} \mathbf{E}_u w_n(\widehat{T}_u) \geqslant \frac{1}{2}\mathbf{E}w_n(\xi) \exp\{t_n^2 O(\lambda(\delta_n))\}(1+o(1)).$$

COROLLARY. *Let relations* (6)–(8) (*or* (6) *and* (7)) *be valid with* $\lambda(u) \leqslant C|u|^\nu$, $0 < \nu \leqslant 1$, *and let* $h_\rho \in H$ *be such that* $h_\rho(s) \sim \exp\{\rho(s)\log s\}$, $s \to \infty$, *where* $\rho(s) \to 0$ *as* $s \to \infty$. *Then*

$$\liminf_{n\to\infty} \inf_{\widehat{\theta}} \sup_{|u-\theta|\leqslant\delta_n} \frac{h_\rho(\mathbf{E}_u w_n(\widehat{T}_u))}{h_\rho(\mathbf{E}w_n(\xi))} \geqslant 1 \qquad (11)$$

for every $\psi \in \Psi$, $\psi_n = o(n^{\nu/(4+2\nu)})$, *and every* $w_\psi \in W_\psi^1$ (*or* $w_\psi \in W_\psi^2$).

Remark 4. Let us suppose that relations (6)–(8) are fulfilled with $\lambda(u) \leqslant C|u|$. Then for $t \leqslant \psi_n$, $\psi_n = o(n^{1/6})$, $\psi \in \Psi$, Theorem 1 and (10) yield

$$\sup_{|u-\theta|\leqslant\delta_n} \mathbf{P}_u\{|\widehat{T}_u| \geqslant t\} \geqslant e^{-1}\Phi(-t)(1+o(1)).$$

On the other hand, under some additional conditions (see (Radavičius, 1983)) the maximum likelihood estimator θ^* satisfies the asymptotic ($n \to \infty$) relation

$$\mathbf{P}_u\{|T_u^*| \geqslant t\} = 2\Phi(-t)(1+o(1)) \qquad (12)$$

uniformly in $t \in [0, \psi_n]$, $\psi_n = o(n^{1/6})$. Consequently, the expression in the right-hand side of (9) is accurate up to the constant $(2e)^{-1}$. In view of (12) and Remark 1, it is very likely that this constant (at least for $\psi_n = o(n^{1/6})$) could be removed if the supremum in (9) were taken over the same region as in (3). The same remark applies to Theorem 2.

Remark 5. Let us suppose that the conditions of Corollary are fulfilled and $h(s) \sim \exp\{\log^{\beta} s\}$ as $s \to \infty$, $0 < \beta < 1$. Then it is easy to verify that (11) holds for each $w_\psi \in W_\psi^1$ ($w_\psi \in W_\psi^2$) with $\psi_n = o\big(n^{\nu/(2\nu+4\beta)}\big)$, $n \to \infty$.

3. PROOFS

Proof of Theorem 1. Let us introduce some notation. Denote

$$A^{\pm} = \Big\{|\theta \pm \tau - \widehat{\theta}| \geqslant \tau\Big\}, \quad B^{\pm} = \Big\{\sum_{j=1}^{n} l_j(\pm\tau) \geqslant z\Big\},$$

$$B = \Big\{\max\big(|l_j(\tau)|, |l_j(-\tau)|\big) \leqslant \varepsilon(\tau),\ j = 1, 2, \ldots, n\Big\},$$

where z will be made precise later. Given $\varphi, \psi \in \Psi$, $\varphi < \psi$, we have, for any $t \in [\varphi_n, \psi_n]$ and $\tau = \big(nI(\theta)\big)^{-1/2} t$,

$$\sup_{|u-\theta| \leqslant \delta_n} \mathbf{P}_u\{|\widehat{T}_u| \geqslant t\} \geqslant \frac{1}{2}\Big(\mathbf{P}_{\theta+\tau}(A^+) + \mathbf{P}_{\theta-\tau}(A^-)\Big). \tag{13}$$

Let us estimate the probability $\mathbf{P}_{\theta+\tau}(A^+)$ from below. By changing the probability measure, we obtain

$$\mathbf{P}_{\theta+\tau}(A^+) \geqslant \mathbf{P}_{\theta+\tau}(A^+ \cap B^+ \cap B) \geqslant e^{z}\big(\mathbf{P}_\theta(B^+ \cap B) - \mathbf{P}_\theta(B \setminus A^+)\big). \tag{14}$$

Consider

$$\mathbf{P}_\theta(B^+ \cap B) = \mathbf{P}_\theta(B)\mathbf{P}_\theta\Big\{\sum_{j=1}^{n} Y_j \geqslant z\Big\},$$

where $Y_1, Y_2, \ldots, Y_n$ are i.i.d. random variables with the common distribution function

$$F_\theta(y) = \mathbf{P}_\theta\Big\{l_1(\tau) \leqslant y \,\Big|\, \max\Big(|l_1(\tau)|, |l_1(-\tau)|\Big) \leqslant \varepsilon(\tau)\Big\}.$$

To apply the Berry–Esseen inequality to i.i.d. random variables $Y_1, Y_2, \ldots, Y_n$, an estimate for their third moments is needed (see, for example, (Petrov, 1975)). Due to (6) and (8),

$$\mathbf{E}_\theta|Y_1|^3 \leqslant \big(p(\tau) + p(-\tau) - 1\big)^{-1}\mathbf{E}_\theta\big|\bar{l}_1(\tau)\big|^3 \leqslant C\varepsilon(\tau)\tau^2. \tag{15}$$

Further, by definition,

$$\bar{a}(\tau) = \mathbf{E}_\theta Y_1 = -I(\theta)\frac{\tau^2}{2}\Big(1 + O\big(\lambda(\tau)\big)\Big), \tag{16}$$

$$\bar{\sigma}^2(\tau) = \operatorname{var}_\theta Y_1 = I(\theta)\tau^2\big(1 + o(1)\big). \tag{17}$$

Therefore, taking $z = n\bar{a}(\tau) - 1$, we obtain

$$\mathbf{P}_\theta(B^+ \cap B) \geqslant \mathbf{P}_\theta(B)\bigg(\Phi(-n^{-1/2}\bar{\sigma}^{-1}(\tau)) - \frac{C\varepsilon(\tau)\tau^2}{\sqrt{n}\bar{\sigma}^{3/2}(\tau)}\bigg),$$

and the substitution of (17) into this inequality gives

$$\mathbf{P}_\theta(B^+ \cap B) \geqslant \mathbf{P}_\theta(B)\Big(\frac{1}{2} + \frac{1}{t\sqrt{2\pi}}(1+o(1))\Big). \tag{18}$$

It is easy to verify by the same arguments that the estimates similar to that in (14) and (18) for probabilities.$\mathbf{P}_{\theta+\tau}(A^+)$ and $\mathbf{P}_{\theta+\tau}(B^+ \cap B)$, respectively, arc also valid for $\mathbf{P}_{\theta-\tau}(A^-)$ and $\mathbf{P}_{\theta-\tau}(B^- \cap B)$. Thus, from (6), (13), (14), (16) and (18), we conclude that

$$\sup_{|u-\theta|\leqslant\delta_n} \mathbf{P}_u\{|\widehat{T}_u| \geqslant t\} \geqslant \frac{1}{2e}\exp\{n\min(\bar{a}(\tau), \bar{a}(-\tau))\}\times$$

$$\times\mathbf{P}_\theta(B)\Big[\frac{1+o(1)}{\sqrt{2\pi}t} + 1 - \frac{\mathbf{P}_\theta(B\backslash A^+) + \mathbf{P}_\theta(B\backslash A^-)}{\mathbf{P}_\theta(B)}\Big] \geqslant$$

$$\geqslant \frac{1+o(1)}{e\sqrt{2\pi}t}\exp\Big\{-\frac{t^2}{2}(1+O(\lambda(\tau)))\Big\},$$

because

$$\mathbf{P}_\theta(B) - \mathbf{P}_\theta(B\backslash A^+) - \mathbf{P}_\theta(B\backslash A^-) = \mathbf{P}_\theta(B \cap A^+ \cap A^-) \geqslant 0.$$

Proof of Theorem 2. We will use the same notation as in the proof of Theorem 1. Given $w_\psi \in W_\psi$, $w_n(y) = \cosh(t_n y) - 1$, we have

$$\sup_{|u-\theta|\leqslant\delta_n} \mathbf{E}_u w_n(\widehat{T}_u) \geqslant \frac{1}{2}\Big(\mathbf{E}_{\theta+\tau} w_n(\widehat{T}_{\theta+\tau})) + (\mathbf{E}_{\theta-\tau} w_n(\widehat{T}_{\theta-\tau})\Big) \geqslant$$

$$\geqslant \frac{1}{2}\Big(\mathbf{E}_{\theta+\tau}\cosh(t\widehat{T}_{\theta+\tau})\chi(B) + \mathbf{E}_{\theta-\tau}\cosh(t\widehat{T}_{\theta-\tau})\chi(B)\Big) - 1. \tag{19}$$

Since

$$\mathbf{E}_\theta\big[l_1(\tau)|\, B\big] = \mathbf{E}_\theta Y_1 = \bar{a}(\tau) = -\frac{1}{2}I(\theta)\tau^2(1+O(\lambda(\tau)))$$

(see (16)), by changing the probability measure and applying Jensen's inequality to the convex function $\varphi(y_1, y_2) = \exp(y_1)\cosh(y_2)$, we find that

$$\mathbf{E}_{\theta+\tau}\cosh(t\widehat{T}_{\theta+\tau})\chi(B) \geqslant \mathbf{P}_\theta(B)\exp\{n\bar{a}(\tau)\}\cosh\Big(t\mathbf{E}_\theta\big[\widehat{T}_{\theta+\tau}|\, B\big]\Big) =$$

$$= \mathbf{P}_\theta(B)\exp\Big\{-\frac{t^2}{2}\Big(1+O(\lambda(\tau))\Big)\Big\}\cosh\Big(t\mathbf{E}_\theta\big[\widehat{T}_\theta|\, B\big] - t^2\Big). \tag{20}$$

Similarly,

$$\mathbf{E}_{\theta-\tau}\cosh(t\widehat{T}_{\theta-\tau})\chi(B) \geqslant \mathbf{P}_\theta(B)\Big\{-\frac{t^2}{2}\Big(1+O(\lambda(\tau))\Big)\Big\}\cosh\Big(t\mathbf{E}_\theta\big[\widehat{T}_\theta|\, B\big] + t^2\Big). \tag{21}$$

The elementary inequality $\cosh(y+s)+\cosh(y-s) \geqslant 2\cosh(s)$, $y, s \in \mathbf{R}$, together with (19)–(21) and (6) shows that

$$\sup_{|u-\theta|\leqslant\delta_n} \mathbf{E}_u w_n(T_u) \geqslant \mathbf{P}_\theta(B)\exp\left\{-\frac{t^2}{2}\Big(1+O(\lambda(\tau))\Big)\right\}\cosh(t^2) - 1 =$$

$$= \frac{1}{2}\exp\left\{\frac{t^2}{2}\Big(1+O(\lambda(\delta_n))\Big)\right\}(1+o(1)), \quad n\to\infty.$$

The proof is completed.

REFERENCES

Bahadur, R.R. (1971). *Some Limit Theorems in Statistics*. SIAM, Philadelphya.

Hajek, J. (1972). Local asymptotic minimax and admissiability in estimation. In: *Proc. Sixth Berkeley Symp. on Math. Statist. and Probab.* 1, California Univ. Press, Los Angelos, pp. 175–194.

Ibragimov, I.A. and Khas'minskii, R.Z. (1981). *Statistical Estimation: Asymptotic Theory*. Springer, Berlin etc.

Levit, B.Ya. (1982). Minimax estimation and positive solutions of elliptic equations. *Theory Probab. Appl.* **27**, 563–586.

Petrov, V.V. (1975). *Sums of Independent Random Variables*. Springer, Berlin etc.

Radavičius, M. (1983). On the probability of large deviations of maximum likelihood estimators. *Soviet Math. Dokl.* **27**, 127–131.

New Trends in Probab. and Statist., pp. 636 – 643
V. Sazonov and T. Shervashidze (Eds.)
1991 VSP/Mokslas

ON ESTIMATING PARAMETERS OF A DISTRIBUTION CONCENTRATED IN A CIRCLE

A.E. SHEMYAKIN
Novosibirsk State University, Mathematical Department, Pirogova str. 2, 630090 Novosibirsk, USSR

ABSTRACT

An applied physical problem is discussed. It concerns the estimation of coordinates of a point emitter radiating through a circular hole in an impermeable screen.
Consider a related problem of asymptotical parametric estimation: based on a sample from the distribution of a given class concentrated in a circle, estimate the location of its center.
Two schemes of construction of such estimates are suggested. One of them is based on the sufficient statistic introduced in (Ermakov, 1980). The other deals with construction of the circumcircle of the sample. Their comparative characteristics are studied for the case of the uniform distribution.

1. INTRODUCTION

Let us begin with formulation of an applied problem related to computer tomography. Assume that a point θ of a fixed plane A in $\mathbf{R}^3$ emits Poisson flow of particles at a random angle to A distributed according to some specific law. Let these particles be recorded on a parallel plane B, the distance between A and B being equal to 1. The problem is to estimate the location of the emitter θ knowing the coordinates of the recorded points in B, when the period of observation is sufficiently long.
The accuracy of estimation may increase essentially due to an admissible alteration of experimental design: invest between A and B (say, on the same distance from the both planes) an impermeable screen S with a circular hole of diameter δ. In this case the particles are recorded on B inside some circle of diameter 2δ, and estimation of the location of its center with respect to the center of the hole is equivalent to estimation of θ.
Therefore, we face the problem of estimating the unknown center of a circle based on a sample from a distribution concentrated in this circle. This problem is complicated by the randomness (Poisson) of sample size and the form of the distribution of observations which may be rather awkward to handle (in applications frequently two-dimensional Cauchy). Prior assumptions on θ (localization on a subset of A or

the Bayesian approach, where, θ is a point distributed on A with a preset density) are also of some importance. From the point of view of experimental design it might be also useful to determine the optimal size of δ.
This problem was considered in (Shemyakin, 1985) for the planar case only (A, B and S are parallel lines on the plane).
In the next section, the simplest mathematical model derived for the spatial case (A, B and S are planes in $\mathbf{R}^3$) is used as an example illustrating the prime difficulties one encounters when trying to estimate, based on a sample from a distribution concentrated in a circle, the unknown center of the circle. In Sections 3 and 4 the description and comparative characteristics of two types of such estimates are given. Iterative algorithms for obtaining these estimates are introduced and discussed in Section 5. Section 6 contains proofs of the results formulated in Sections 3 and 4.

2. MATHEMATICAL MODEL

Let $X = (X_1, ..., X_n)$ be i.i.d. random vectors in $\mathbf{R}^2$, $X_i = (X_i^1,\ X_i^2)$, distributed in a circle $C_\rho(\theta)$ of radius ρ with center at θ. Assuming that ρ is known, estimate the unknown parameter θ. Let us restrict ourselves, unless otherwise specified, to the case of a uniform distribution.
Considering this problem as a two-dimensional extension of the problem of estimating the location on the real line of the uniform distribution on an interval, one might expect the extremal sample values to play the most important role. The most natural analogue for the sample extremes in the two-dimensional case is the convex hull of X in $\mathbf{R}^2$, conv(X). In (Renyi and Sulanke, 1963) it was demonstrated that, for the uniform distribution,

$$\mathbf{E}\Big[\mathrm{Card}\,(\mathrm{conv}\,(X))\Big] \sim n^{1/3}, \quad n \to \infty.$$

It makes us believe that conv (X), as a base for constructing reasonable estimates of θ, is too numerous a statistic which contains a lot of superfluous information. In the next two sections two types of statistics are introduced that enable one to obtain some simple and rather efficient estimates.

3. ERMAKOV'S SUFFICIENT STATISTIC

Denote

$$T = T(X) = \bigcap_{i=1}^{n} C_\rho(X_i) = \{t\colon \max_i |t - X_i| \leqslant \rho\}$$

the set of all admissible estimates of θ (all points $\widehat{\theta}$ that have nonzero likelihood). In (Ermakov, 1980) it was demonstrated that, as $n \to \infty$, there exist a bounded in probability number ν (i.e., $\forall \varepsilon > 0\ \exists K : \mathbf{P}\{\nu \geqslant K\} < \varepsilon$) and $i_1, ..., i_\nu$, $1 \leqslant i_j \leqslant n$, such that

$$T = \bigcap_{j=1}^{\nu} C_\rho(X_{i_j}).$$

Thus, the observations $X_{i_1}, ..., X_{i_\nu}$ form a sufficient statistic of random but low enough cardinality. Numerical simulation performed for the uniform distribution with $n = 10000$ showed that $E\nu \simeq 5$ and $\mathbf{P}\{\nu \geqslant 8\} \simeq 0,02$ (Ermakov, 1980).
Denote the vertices of arc polygon T by Y_{i_j}. Consider linear statistics of Y_{i_j} having the form

$$\widehat{\theta}_\alpha = \sum_{j=1}^{\nu} \alpha_j Y_{i_j}, \quad \sum_{j=1}^{\nu} \alpha_j = 1, \ \alpha_j \geqslant 0. \tag{1}$$

Since $\widehat{\theta}_\alpha \in T$, we may take any $\alpha = (\alpha_1, ..., \alpha_\nu)$, α_j satisfying (1), and consider $\widehat{\theta}_\alpha$ as an estimate of θ. According to Pittman's principle (Strasser, 1982), the optimal choice of α for the uniform distribution of X is $\alpha_1 = ... = \alpha_\nu = \nu^{-1}$.
So the algorithm of obtaining such estimates includes the following:

1. Stepwise elimination of "superfluous" observations in order to obtain the sufficient statistic $X_{i_1}, ..., X_{i_\nu}$.
2. Determination of $Y_{i_1}, ..., Y_{i_\nu}$.
3. Appropriate choice of $\alpha = (\alpha_1, ..., \alpha_\nu)$ and direct calculation of $\widehat{\theta}_\alpha$.

The realization of this algorithm, especially of its first stage, requires rather cumbersome calculations. Furthermore, some additional complications in analysis of the constructed estimates are related to the random nature of ν. Thus, after the formulation of a result concerning this estimation scheme, we pass to a simpler one.

THEOREM. *If X_i are distributed uniformly in $C_\rho(\theta)$, then*

$$\lim_{n\to\infty} \mathbf{P}\Big\{n \sup_{t\in T} |t - \theta| < u\Big\} = 1 - \exp\Big(-\frac{2u}{\pi\rho}\Big).$$

Proof of Theorem and some additional notes concerning efficiency are given in Section 6.

4. THE CIRCUMCIRCLE OF X

Another approach to the estimation of θ is related to the construction of the circumcircle of X, i.e., of the circle of minimal diameter containing all the points of X.
Denote the circumcircle of a set A in $\mathbf{R}^2$ by $C[A]$. According to the corollary of Helly's theorem (Berger, 1977), there always exist points X_{i_j} in X such that $C[X] = C\big[\{X_{i_1}, X_{i_2}\}\big]$ or $C[X] = C\big[\{X_{i_1}, X_{i_2}, X_{i_3}\}\big]$. We call these points the *supporting points* of the circumcircle.
If $C[X] = C_{\rho^*}(\theta^*)$, then $\rho^* \xrightarrow{\mathbf{P}} \rho$, $\theta^* \xrightarrow{\mathbf{P}} \theta$ as n tends to infinity. It means that at least θ^* is a consistent estimate of θ. The reasoning given above guarantees that θ^* may be represented as

$$\theta^* = \sum_{j=1}^{r} \alpha_j X_{ij}, \qquad \sum \alpha_j = 1, \quad \alpha_j \geqslant 0, \quad r = 2, 3. \tag{2}$$

An algorithm providing the construction of θ^* should include the following items:

1. Determination of supporting points X_{i_j}.
2. Estimation of ρ by ρ^*.
3. Calculation of the coefficients α_j and direct calculation of θ^*.

An iterative algorithm implementing simultaneously all these three items has been kindly suggested to the author by professor G.S. Rubinstein.
Some asymptotic properties of the estimates θ^* and ρ^* are stated below.

LEMMA 1. *If X_i are distributed uniformly in a circle $C_\rho(\theta)$ then*

$$\lim_{n\to\infty} \mathbf{P}\Big\{|\theta^* - \theta| < u\Big\} \geqslant 1 - \exp\Big(-\frac{2u}{\pi\rho}\Big).$$

Denote $\Delta_\rho = \rho^* - \rho$, $\Delta_\theta = |\theta^* - \theta|$. The following statement concerns the relationship between Δ_θ and Δ_ρ.

LEMMA 2. *If X_i are distributed uniformly in $C_\rho(\theta)$, then for any $u, v \in (0, \rho)$*

$$\mathbf{P}\Big\{n\Delta_\rho \geqslant v \mid n\Delta_\theta = u\Big\} \sim \exp\Big\{-\frac{2v}{\rho}h(u,v)\Big\}$$

as $n \to \infty$, where

$$h(u,v) = \begin{cases} 1 & v \geqslant u, \\ 1 - \pi^{-1}\Big(\arccos(v/u) - (v/u)\sqrt{1-(v/u)^2}\Big) & v < u. \end{cases}$$

5. CALCULATION OF $\widehat{\theta}_\alpha$ AND θ^*

When dealing with applications, one must make sure that there exist convenient realizations of the algorithms leading to construction of the estimates. Two basic features of any practicable algorithm are a simple computational scheme and a reasonable amount of calculation required for the implementation. From this point of view, estimate (2) seems to be much easier to handle than (1). Note also that simultaneous estimation of θ and ρ might provide some additional information even in the case of known ρ (see Section 4, Lemma 2).
Consider first the class of estimates $\widehat{\theta}_\alpha$ defined in (1). According to the scheme introduced above, the first and the most important step in the construction of $\widehat{\theta}_\alpha$ is the determination of sufficient statistic $X_{i_1}, \dots, X_{i_\nu}$. For this purpose consider all the pairs (X_i, X_j), $i, j \leqslant n$, determine the points Z_{ij}, Z'_{ij} such that $|X_i - Z_{ij}| = |X_j - Z_{ij}| = \rho = |X_i - Z'_{ij}| = |X_j - Z'_{ij}|$, and eliminate all the X_k $(k \neq i, j)$ for which $|X_k - Z_{ij}| \leqslant \rho$, $|X_k - Z'_{ij}| \leqslant \rho$. This procedure ends up with only $X_{i_1}, \dots, X_{i_\nu}$ remained. However, it requires approximately n^2 steps, which makes it very inconvenient for large samples.

The other steps of the algorithm are simpler. Determine Y_{i_j} as cross-sections of circles of radius ρ centered at X_{i_j} (although here we also need a kind of selection). Then take any $\alpha = (\alpha_1, ..., \alpha_\nu)$ satisfying (1) and obtain an estimate of the class $\widehat{\theta}_\alpha$.

Pass now to the construction of θ^*. Let us introduce the detailed description of the algorithm enabling us to find the set K_0 of supporting points, the radius and the center of the circumcircle $C[X]$. It can be started from any two- or three-point subset of X and requires no more than n steps, each containing a finite number of arithmetic operations each.

Consider an arbitrary affinely independent subset of X of the form

$$K = \{X_i\}_{i=1}^r, \quad r = 2 \quad \text{or} \quad r = 3. \tag{3}$$

Denote scalar product $\delta_{ij} = (X_i - X_1,\ X_j - X_1)$ and find the barycentric coordinates h_i of $\theta^*(K)$, the center of $C[K]$, from the system of equations

$$\begin{cases} \sum\limits_{j=2}^{r} \delta_{ij} h_j = \delta_{ii}/2, \quad i = 2, ..., r; \\ h_1 = 1 - \sum\limits_{j=2}^{r} h_j, \end{cases}$$

so that $\theta^*(K) = \sum h_i X_i$ and the radius of $C[K]$ may be expressed as $\rho^*(K) = |X_j - \theta^*(K)|$.

Then take a point $x \in X \backslash K$ and check whether

$$|x - \theta^*(K)| \geqslant \rho^*(K). \tag{4}$$

If the chosen point does not satisfy (4), then take another one until (4) holds or $X \backslash K$ is empty. The points having not satisfied (4) once, are totally eliminated from the consideration. If x satisfies (4), denote $X_{r+1} = x$ and from the system of equations

$$\begin{cases} \sum\limits_{j=2}^{r} \delta_{ij} g_j = \delta_{i,j+1}; \\ g_1 = 1 - \sum\limits_{j=2}^{r} g_j \end{cases}$$

determine the barycentric coordinates g_i of the orthogonal projection $\overline{X}_{r+1}$ of the point X_{r+1} on

$$\text{aff}\,(K) = \{x = \sum \alpha_i X_i\colon \sum \alpha_i = 1\}.$$

Calculate

$$\bar{\varepsilon} = \max\ \{\varepsilon\colon\ h_j - \varepsilon g_j \geqslant 0, \quad j = 1, ..., r\}$$

and denote

$$K' = \{X_i \in K\colon\ h_i - \bar{\varepsilon} g_i > 0\}.$$

Furthermore, denote by $\eta = X_{r+1} - \bar{X}_{r+1}$ the component of X_{r+1} orthogonal to aff (K) and consider two alternatives:

1) $\eta = 0$ (X_{r+1} belongs to aff (K), $r = 2, 3$). In this case two situations are possible:

   1a) If Card $K' = 1$, then denote $K^* = \{K' \bigcup X_{r+1}\}$.

   1b) If Card $K' = 2$, then return to (3), replace K by K', and replicate the procedure with the same X_{r+1}.

2) $\eta \neq 0$ (X_{r+1} does not belong to aff (K), $r = 2$). Let ε_0 satisfy the equation $\rho^*(K \bigcup X_{r+1}) = \rho^*(K) + \varepsilon_0 \eta$. It may be expressed in the form

$$\varepsilon_0 = \frac{1}{2|\eta|^2} \left(|X_{r+1} - \rho^*(K)|^2 - \rho^*(K)^2 \right).$$

   Here also two situations may occur:

   2a) If $\varepsilon_0 \leqslant \bar{\varepsilon}$, then denote $K^* = \{K' \bigcup X_{r+1}\}$.

   2b) If $\varepsilon_0 > \bar{\varepsilon}$, then denote $K^* = \{K \bigcup X_{r+1}\}$.

Each step of the algorithm ends up with the determination of two- or three-point subset of X, $K^* = K^*(K)$. Replacing K by K^* in (3) and taking a new point of $X \backslash K^*$ in (3), replicate the entire procedure. Evidently, $\rho^*(K^*) \geqslant \rho^*(K)$, so each transition from K to K^* increases the radius of $C[K]$. It takes no more than n steps to attain $K^* = K_0$ and consequently $C[K^*] = C[X]$, $\theta^*(K^*) = \theta^*(X)$, $\rho^*(K^*) = \rho^*(X)$. In fact, the algorithm usually contains only several full steps, since a substantial number of observations is eliminated while testing condition (4).

This algorithm admits very simple computer implementations. Furthermore, it can be easily extended to the case of $\mathbf{R}^m$, $m \geqslant 2$.

6. DEMONSTRATION OF THE RESULTS

Proof of Theorem. Denote by $F(u)$ the distribution function of the random variable $\xi = n \sup_{t \in T} |t - \theta|$. It is evident that for any spherosymmetrically distributed X_i and any z, $|z| = 1$, we have

$$\mathbf{P}\{\xi \geqslant u\} = 1 - F(u) = \mathbf{P}\left\{T \subset C_{u/n}(\theta)\right\} =$$

$$= \mathbf{P}\left\{X \subset C_\rho(\theta + un^{-1}z)\right\}.$$

Since X_i are independent and identically distributed, we can express $F(u)$ as

$$F(u) = 1 - \mathbf{P}^n\left\{X_1 \in C_\rho(\theta + un^{-1}z)\right\}.$$

Since X_i are distributed uniformly,

$$\mathbf{P}\{X_1 \in C_\rho(\theta + un^{-1}z)\} =$$

$$= S\Big(C_\rho(\theta + un^{-1}z) \cap C_\rho(\theta)\Big)\Big[S(C_\rho(\theta))\Big]^{-1},$$

where $S(\cdot)$ is the area in $\mathbf{R}^2$. The right-hand side of the latter formula does not depend on z and admits the following asymptotic representation

$$S(C_\rho(\theta + un^{-1}z) \cap C_\rho(\theta)) = \rho^2\Big(\pi - \frac{2u}{\rho n} + o\left(n^{-1}\right)\Big).$$

Hence,

$$\mathbf{P}\{X_1 \in C_\rho(\theta + un^{-1}z)\} = 1 - \frac{2u}{\pi\rho n} + o\left(n^{-1}\right)$$

and $F(u) \sim 1 - \exp\left(-2u/\pi\rho\right)$, $n \to \infty$.
Thus we see that the random variable ξ is asymptotically exponentially distributed with the parameter $2/\pi\rho$. Hence the proof is completed.

Proof of Lemma 1. It is evident that

$$\max_i |\theta^* - X_i| \leqslant \rho^* \leqslant \rho.$$

Recall that

$$T = \{t\colon \max_i |t - X_i| \leqslant \rho\},$$

so $\theta^* \in T$. Now the statement of Lemma 1 follows directly from Theorem 1.

Proof of Lemma 2. Denote

$$H(u,v) = \mathbf{P}\{n\Delta_\rho \geqslant v \mid n\Delta_\theta = u\}.$$

For any spherosymmetrical distribution of X_i independently of z, $|z| = 1$, we have

$$H(u,v) = \mathbf{P}\Big\{X \subset C_\rho(\theta) \cap C_{\rho - v/n}(\theta + un^{-1}z)\Big\}.$$

Similarly to the proof of Theorem 1, by virtue of i.i.d. property of X_i, we obtain

$$H(u,v) = 1 - \mathbf{P}^n\Big\{X_1 \in C_\rho(\theta) \cap C_{\rho - v/n}(\theta + un^{-1}z)\Big\}.$$

Consider first the case $v \geqslant u$. We have

$$C_{\rho - v/n}(\theta + un^{-1}z) \subseteq C_\rho(\theta).$$

and, for uniformly distributed, X_i

$$\mathbf{P}\Big\{X_1 \in C_{\rho - v/n}(\theta + un^{-1}z)\Big\} =$$

$$= \pi(\rho - v/n)^2/\pi\rho^2 = 1 - 2v/\rho n + o\left(n^{-1}\right), \quad n \to \infty.$$

Therefore, we obtain

$$H(u,v) \sim 1 - \exp\left\{-\frac{2v}{\rho}\right\}, \quad n \to \infty.$$

Pass now to the case $v < u$. A purely geometrical reasoning leads to the representation:

$$\mathbf{P}\Big\{X_1 \in C_\rho(\theta) \cap C_{\rho-v/n}(\theta + un^{-1}z)\Big\} =$$
$$= (1 - v/\rho n)^2 + \pi^{-1}\Big[(\varphi - \sin\varphi) - (1 - v/\rho n)^2(\psi - \sin\psi)\Big], \tag{5}$$

where

$$\varphi = 2\arccos\left(\frac{v}{u} + \frac{u^2 - v^2}{2\rho un}\right),$$
$$\psi = 2\arccos\left(\frac{v}{u} - \frac{u^2 - v^2}{2\rho un(1 - v/\rho n)}\right).$$

An accurate asymptotic analysis of (5) shows that, as $n \to \infty$,

$$\mathbf{P}\Big\{X_1 \in C_\rho(\theta) \cap C_{\rho-v/n}(\theta + un^{-1}z)\Big\} =$$
$$= 1 - \frac{2v}{\rho n}h(u,n) + o\left(n^{-1}\right).$$

Therefore we obtain the asymptotical representation

$$H(u,v) \sim 1 - \exp\Big\{-\frac{2v}{\rho}h(u,v)\Big\}, \quad n \to \infty.$$

which completes the proof.

Remark. The above results demonstrate that the estimate θ^* is at least not worse than the worst of $\hat{\theta}_\alpha$, if we consider the case of uniform distribution. Nevertheless, an appropriate choice of α can substantially increase the accuracy of the estimation. It would be helpful to construct a nontrivial lower bound for the risk of $\hat{\theta}_\alpha$ in order to evaluate the optimal attainable accuracy. But this problem lies beyond the scope of this paper.

Acknowledgements

The author would like to thank the organizers of the colloquium in Bakuriani for the opportunity to present a paper. Special thanks are also due to M.S. Ermakov for helpful discussions and to G.S. Rubinstein for the idea of the algorithm introduced in Section 5.

REFERENCES

Berger, M. (1977). *Géométrie.* CEDIC/Nathan, Paris.

Ermakov, M.S. (1980). On sufficient statistics for distribution families with variable density support. *Notes Sci. Seminars Leningrad Branch of Math. Inst.* **97**, 62–82 (in Russian).

Renyi, A. and Sulanke, R. (1963). Über die konvexe Hülle von n zufälling gewählten Punkten. *Z. Wahrsch. verw. Geb.* **2**, 75–84.

Shemyakin, A.E. (1985). A problem of estimating the coordinates of a radiation source. *Theory Probab. Appl.* **30**, 128–140.

Strasser, H. (1982). Local asymptotic minimax properties of Pitman estimators. *Z. Warsch. verw. Geb.* **60**, 223–247.

New Trends in Probab. and Statist., pp. 644 - 651
V. Sazonov and T. Shervashidze (Eds.)
1991 VSP/Mokslas

ON THE APPROXIMATION OF THE LIKELIHOOD FUNCTION BY A VECTOR ANALOGUE OF THE WHITTLE STATISTICS

K.A. TSERTSVADZE
Gori Economic Institute, Chavchavadze str. 53,
383500 Gori, Georgia, USSR

ABSTRACT

Conditions are found on the spectral density function $f(\lambda)$ of a Gaussian stationary vector sequence, under which the quadratic form in the likelihood function is approximated by a vector analogue of the Whittle statistics.

Let the Gaussian stationary vector sequence

$$X(k) = \big(x^1(k), x^2(k), \ldots, x^m(k)\big), \quad k \in \mathbf{Z},$$

have spectral density

$$f(\lambda) = \big\|f_{ij}(\lambda)\big\|, \quad i,j = 1,2,\ldots,m,$$

and covariance function

$$\beta(\tau) = \mathbf{E}X(t)X^*(t+\tau),$$

where $X^*(t)$ is the transposed vector. Then

$$\beta(\tau) = \int_{-\pi}^{\pi} e^{i\lambda\tau} f(\lambda)\, d\lambda.$$

Consider the Toeplitz matrix $B_n^f = \big\|\beta(\tau - s)\big\|$, $\tau, s = 0,1,\ldots,n$, connected with the spectral density $f(\lambda)$.
Let $Y(n)$ denote a vector of observations on $X(k)$, whose components are $X(1), X(2), \ldots, X(n)$. From the probability density expression of the $n \times m$-dimensional vector $Y(n)$ we obtain the log-likelihood function

$$\mathcal{L}_n = \log P_n\big(Y(n)\big) = -\frac{1}{2}\Big\{m\; n \log 2\pi + \log\det B_n^f + Y^*(n)\big[B_n^f\big]^{-1} Y(n)\Big\}.$$

Solving statistical problems of Gaussian random processes, we must know the behaviour of $\mathcal{L}_n$ for large n, which in turn leads us to study the asymptotic behaviour of the Toeplitz determinant $D_n(f) = \det B_n^f$ and the quadratic form

$$T_1 = Y_n^*\big[B_n^f\big]^{-1} Y(n).$$

It was shown in (Tsertsvadze, 1988) that under suitable conditions on the spectral density $f(\lambda)$ of the sequence $X(k)$, $k \in Z$, one has the relation

$$\log D_n(f) = \frac{n}{2\pi} \int_{-\pi}^{\pi} \log \det f(z) d\lambda + o(\sqrt{n}), \quad n \to \infty.$$

In this paper we consider approximation of the statistics T_1 by a simpler statistics

$$T_2 = \int_{-\pi}^{\pi} \Gamma_n^*(\lambda) f^{-1}(\lambda) \Gamma_n(\lambda),$$

where

$$\Gamma_n(\lambda) = \frac{1}{2\pi} \sum_{k=1}^{n} X(k) e^{-i\lambda k},$$

$*$ is the symbol of Hermitian conjugation. The statistics T_2 is a natural vector analogue of the well-known Whittle statistics.
We will find the conditions under which

$$\mathbf{E}(T_1 - T_2) = o(\sqrt{n}), \quad \operatorname{var}(T_1 - T_2) = o(\sqrt{n}), \quad n \to \infty.$$

Let H be the subspace of the space $L^2(d\mathbf{P})$, where $\mathbf{P}$ is a probability measure, and let $\{x_j\}$ be a basis in H.
If $\|a_{ij}\|$, $i, j \in Z$, is a symmetric matrix, we define the operator A spanned by the relation

$$A\varphi = \sum_i \sum_j a_{ij} x_i (\varphi, x_j), \tag{1}$$

where $(\cdot\,,\cdot)$ is the scalar product in H. Thus A is the integral operator in $L^2(d\mathbf{P})$ with the kernel

$$\Phi(w, w') = \sum_i \sum_j a_{ij} x_i(w) x_j(w')$$

(it is assumed here that the matrix $\|a_{ij}\|$ and the basis $\{x_j\}$ are chosen so that formula (1) determines nuclear operator). It is obvious that A is self-adjoint.
Then it is know that

$$\mathbf{E} \sum_i \sum_j a_{ij} x_i x_j = \operatorname{tr} A, \quad \operatorname{var} \sum_i \sum_j a_{ij} x_i x_j = \alpha \|A\|_2^2, \tag{2}$$

where $\|\cdot\|_2^2$ is the Hilbert–Schmidt norm in H and α is a constant. Let $x_1, x_2, \ldots$ be some Gaussian sequence, $\mathcal{H}$ is a linear space generated by values x_i. In $\mathcal{H}$ we introduce the scalar product

$$(h_1, h_2) = \mathbf{E} h_1 h_2.$$

The Hilbert space constructed by the pre–Hilbert space $\mathcal{H}$ will be denotes by $H_.$.
It will be assumed that the system $\{x_j\}$ forms a basis in H. Let $\{x_j^*\}$ by the conjugate system of $\{x_j\}$: $(x_i, x_j^*) = \delta_{ij}$.

Let another scalar product $(\cdot\,,\cdot)_*$ be given in $\mathcal{H}$. Define in H the positive operator A which is given by the relation $(\cdot\,,\cdot) = (A\cdot\,,\cdot)_*$. It is easy to verify that $x_j^* = A^{-1}x_j$.
Let H_* be a Hilbert space with the scalar product $(\cdot\,,\cdot)_*$ constructed by H_*. Assume

$$A_n = P_n(*)AP_n(*),$$

where $P_n(*)$ is the orthoprojector in H_* on the subspace H_n generated by $(x_1, x_2, \ldots, x_n)$. Let $\|a_{ij}\|$ be the matrix of the operator A_n^{-1},

$$A_n^{-1}A_n = A_nA_n^{-1} = P_n(*),$$

in the basis $(x_1, x_2, \ldots, x_n)$ and $\|b_{ij}\|$ be the matrix of the operator $P_n(*)A^{-1}P_n(*)$ in the same basis.
Consider two quadratic forms

$$T_1^n = \sum_{i=-n}^{n}\sum_{j=-n}^{n} a_{ij}x_ix_j, \quad T_2^n = \sum_{i=-n}^{n}\sum_{j=-n}^{n} b_{ij}x_ix_j.$$

We set $T_n = T_1^n - T_2^n$. Let P_n, Q_n be the orthoprojectors in H on the subspaces H_n, $A^{-1}H_n$, respectively. Let also $S_n = P_nQ_nP_n$.

LEMMA 1. *The formulas*

$$\mathbf{E}(T_n) = \operatorname{tr}(P_n - S_n^{-1}), \quad \operatorname{var}(T_n) = \alpha\big\|P_n - S_n^{-1}\big\|_2^2$$

are valid.

Proof. Let G_n be an operator in $L^2(d\mathbf{P})$ with the kernel

$$\sum_{i=-n}^{n}\sum_{j=-n}^{n} a_{ij}x_i(w)x_j(w')$$

and Φ_n be an operator in $L^2(d\mathbf{P})$ with the kernel

$$\sum_{i=-n}^{n}\sum_{j=-n}^{n} b_{ij}x_i(w)x_j(w').$$

Then, by virtue of (2), we have

$$\mathbf{E}(T_n) = \operatorname{tr}(G_n - \Phi_n), \quad \operatorname{var}(T_n) = \alpha\big\|G_n - \Phi_n\big\|_2^2. \tag{3}$$

Note that

$$G_nx_j^* = \sum_{i=-n}^{n}\sum_{j=-n}^{n} a_{ij}x_i = A_n^{-1}x_j.$$

Since

$$(A_n^{-1}x_i, x_j) = (AA_n^{-1}x_i, x_j)_* = (A_nA_n^{-1}x_i, x_j)_* = \delta_{ij},$$

we have $A_n^{-1}x_j = P_n x_j^*$, from which we obtain $G_n x_j^* = P_n x_j^*$ and hence, $G_n = P_n$. Let us now consider the operator

$$\Phi_n x_j^* = \sum_{i=-n}^{n} \sum_{j=-n}^{n} b_{ij} x_i = P_n(*)A^{-1}P_n(*)x_j = P_n(*)A^{-1}P_n(*)Ax_j^*.$$

Thus, we obtain (because the system $\{x_j\}$ is complete in $\mathcal{H}$)

$$\Phi_n = P_n(*)A^{-1}P_n(*)A.$$

Let

$$\varphi \in A^{-1}H_n, \quad \varphi = A^{-1}\varphi_., \quad \varphi_. \in H_n,$$

then

$$\Phi_n\varphi = \Phi_n A^{-1}\varphi_. = P_n(*)A^{-1}\varphi_. = P_n(*)\varphi. \tag{4}$$

Choose systems $\{\varphi_j\}$, $\{\psi_j\}$ such that

1) the system $\{\psi_j\}$ is the orthonormalized basis in the space H_n;
2) the system $\{\varphi_j\}$ is the orthonormalized basis in the space $A^{-1}H_n$;
3) $(\psi_i, \varphi_j) = \delta_{ij}$.

It is easy to verify that

$$Q_n\psi_j = \lambda_j\varphi_j, \quad P_n\varphi_j = \lambda_j\psi_j.$$

Therefore ψ_j are eigenvectors of the operator $S_n = P_nQ_nP_n$ and λ_j^2 are eigenvalues of the same operator.
Assume that

$$\xi_j = \varphi_j - \lambda_j\psi_j, \quad \eta_j = \psi_j - \lambda_j\varphi_j.$$

Obviously,

$$\xi_j \in A^{-1}H^n \quad (A^{-1}H^n = H \ominus H_n),$$
$$\eta_j \in H^n \quad (H^n = H \ominus A^{-1}H_n).$$

Hence

$$P_n(*)\eta_j = P_n(*)\psi_j - \lambda_j P_n(*)\varphi_j = \psi_j - \lambda_j\Phi_n\varphi_j = 0,$$
$$P_n(*)\xi_j = \Phi_n\varphi_j - \lambda_j\Phi_n\psi_j = 0.$$

Therefore, taking into account (4), we obtain

$$\Phi_n\psi_j = \frac{1}{\lambda_j^2}\psi_j.$$

Since the operator Φ_n vanishes on elements φ, $\varphi \in H \ominus H_n$, we have $\Phi_n = S_n^{-1}$, which together with relation (3) proves the lemma. The angle $\varphi = (H_-, H_+)$, $0 \leqslant \varphi \leqslant \pi/2$, determined by

$$\rho = \cos(H_-, H_+) = \sup_{\substack{h_- \in H_-, h_+ \in H_+ \\ \|h_-\| = \|h_+\| = 1}} |(h_-, h_+)|,$$

(where $(\cdot,\cdot)$, $\|\cdot\|$ are the scalar product and norm in H, respectively,) is called the minimal angle between the subspaces H_- and H_+ of the Hilbert space H. By definition,

$$\sin^2(H_-, H_+) = 1 - \cos^2(H_-, H_+^{\perp}).$$

LEMMA 2. *Let*

1) $\inf_n \sin(H_n, H_n^*) > 0$,

2) $\log\det P_n Q_n P_n = o(\sqrt{n})$.

Then

$$\mathbf{E}(T_n) = o(\sqrt{n}), \quad \operatorname{var}(T_n) = o(\sqrt{n}), \quad n \to \infty.$$

Proof. Lemma 1 implies that it is sufficient to show that the relations

$$\operatorname{tr}(P_n - S_n^{-1}) = o(\sqrt{n}), \quad \|P_n - S_n^{-1}\|_2^2 = o(\sqrt{n}), \quad n \to \infty$$

are valid. Since, by condition 1), $\sup\|S_n^{-1}\| < \infty$ (here $\|A\|$ is the uniform norm of the operator A), it is sufficient to show that as $n \to \infty$

$$\operatorname{tr}(P_n - S_n) = o(\sqrt{n}), \quad \|P_n - S_n\|_2^2 = o(\sqrt{n}).$$

We have

$$P_n - S_n = P_n - P_n Q_n P_n = P_n Q_n^{\perp} P_n,$$
$$P_n Q_n P_n = I - P_n Q_n P_n.$$

Since 1) implies $\sup\|P_n Q_n P_n\| < 1$, we have

$$-\log\det P_n Q_n P_n = -\log\det(I - P_n Q_n^{\perp} P_n) > k \operatorname{tr} P_n Q_n P_n.$$

Therefore 2) yields the relation

$$\operatorname{tr} P_n Q_n P_n = \operatorname{tr}(P_n - S_n) = o(\sqrt{n}), \quad n \to \infty.$$

Since $O < (P_n - S_n) < I$ (here O and I denote the zero and the unit matrix, respectively, and the inequality $M_1 < M_2$ between the matrices M_1 and M_2 means that $M_2 - M_1$ is positive definite), we get

$$\|P_n - S_n\| \leqslant \operatorname{tr}(P_n - S_n) = o(\sqrt{n}), \quad n \to \infty.$$

Lemma 2 is proved.
Let the system $\{x_j\}$ be the basis of the Hilbert space H and $\{x_j^*\}$ is the conjugate system of $\{x_j\}$. We introduce the notation

$$H_n = \operatorname{sp}\{x_j,\ |j| \leqslant n\}, \quad H_n^* = \operatorname{sp}\{x_j^*,\ |j| \leqslant n\}$$

(here and in what follows $\operatorname{sp}\{M\}$ is the linear span of M).

LEMMA 3. *If $\{x_j\}$ is the basis in H, then*

$$\inf_n \sin(H_n, H_n^*) > 0. \tag{5}$$

Proof. Let the element $h \in H_n$ be expanded as $h = \sum_{k=-\infty}^{\infty} \alpha_k x_k$. It is assumed that

$$P_n h = \sum_{k=-n}^{n} \alpha_k x_k.$$

As it is known (Kashin and Saakyan, 1984; Theorem 6, p. 19), the uniform norm of operators P_n is bounded uniformly with respect to n: $\|P_n\| < M < \infty$. Hence, we obtain

$$\sup_n \cos(H_n, H^n) < 1, \tag{6}$$

where $H^n = \mathrm{sp}\,\{x_j,\ |j| > n\}$. Since $H^n = H \ominus H_n$, we have

$$\sin^2(H_n, H_n^*) = 1 - \cos^2(H_n, H^n).$$

From the last formula and (6) we obtain (5).

We introduce the notation

$$f_m = \|f_{ij}\|, \quad i, j = 1, 2, \ldots, m;$$

$$\widehat{f}_m = \frac{\det f_{m+1}}{\det f_m}, \quad \alpha_j^m = \frac{\det f_{m+1}^j}{\det f_m},$$

$$f_m^j = \begin{pmatrix} f_{11} & f_{12} & \cdots & f_{1j-1} & f_{1m} & f_{1j+1} & \cdots & f_{1m-1} \\ f_{21} & f_{22} & \cdots & f_{2j-1} & f_{2m} & f_{2j+1} & \cdots & f_{2m-1} \\ \cdots & \cdots & \cdots & \cdots & \cdots & \cdots & \cdots & \cdots \\ f_{m-11} & f_{m-12} & \cdots & f_{m-1j-1} & f_{m-1m} & f_{m-1j+1} & \cdots & f_{m-1m-1} \end{pmatrix}.$$

Determine the operator α_j^k by

$$\alpha_j^k \varphi = \int \alpha_j^k \widehat{\varphi}(z)\, dz_j,$$

where $\varphi = \int \widehat{\varphi}(z)\, dz_j$ and dz_j is the orthogonal measure involved in the spectral representation of the sequence $X(n)$: $x^j(n) = \int e^{in\lambda} dz_j$. Let

$$f^{ij} = \begin{pmatrix} f_{11}^{ij} & f_{12}^{ij} \\ f_{21}^{ij} & f_{22}^{ij} \end{pmatrix},$$

where

$$f_{11}^{ij} = |\alpha_i^{j-1}|^2 f_{ii} + \widehat{f}_j, \quad f_{12}^{ij} = \alpha_i^{j-1} f_{ii},$$

$$f_{21}^{ij} = p^{ij} \sqrt{f_{11}^{ij} f_{22}^{ij}}, \quad f_{22}^{ij} = f_{ii}.$$

Let further,

$$R_-^{lj}(x) = \sum_{k=-\infty}^{0} a_k^{lj} e^{ikx}, \quad I_-^{lj} = \sum_{k=-\infty}^{0} b_k^{lj} e^{ikx},$$

where a_k^{lj}, b_k^{lj} are the Fourier coefficients of the functions

$$R^{lj} = \operatorname{Re}\frac{p^{lj}\gamma^{lj}}{g^{lj}}, \quad I^{lj} = \operatorname{Im}\frac{p^{lj}\gamma^{lj}}{g^{lj}},$$

respectively, and the function γ^{lj}, g^{lj} are determined from relations

$$\gamma^{lj} = \sqrt{f_{11}^{lj}/f_{22}^{lj}}, \quad g^{lj} = f_{22}^{lj}(1 - |p^{lj}|^2).$$

Denote also

$b_{11} = \det(f_{ij})$, $i,j = 1,\dots,m-1$, $b_{12} = \det(f_{ij})$, $1 = 2,\dots,m$; $j = 1,\dots,m-1$,
$b_{21} = \det(f_{ij})$, $i = 1,\dots,m-1$; $j = 2,\dots,m$, $b_{22} = \det(f_{ij})$, $i,j = 2,\dots,m$,

$$p = \frac{b_{12}}{\sqrt{b_{11}b_{22}}}, \quad \rho = \frac{b_{12}}{\tau_{b_{11}}\tau_{b_{22}}\tau_{(1-|p|^2)}},$$

where

$$\tau_s = \exp\left[\frac{1}{2\pi}\int_{-\pi}^{\pi}\frac{e^{i\lambda}+z}{e^{i\lambda}-z}\log S(e^{i\lambda})\,d\lambda\right].$$

We impose the condition on the decrease velocity of coefficients of the function h

$$\sum_{|k|>n} |\widehat{h}_k|^2 = o(\sqrt{n}), \quad n \to \infty. \tag{7}$$

Before formulating the theorem, we would like to remind the reader of the following two definitions. The non–negative function f is said satisfy the Muckenhoupt condition if

$$\sup_I\left\{\frac{1}{|I|}\int_I f(\lambda)\,d\lambda\,\frac{1}{|I|}\int_I f^{-1}(\lambda)\,d\lambda\right\} < \infty,$$

where sup is taken over all segments I of length $|I| \leqslant 2\pi$.
The set of all functions $f \in L^1[-\pi,\pi]$, satisfying the condition

$$\sup_I\left\{\frac{1}{|I|}\int_I |f(t) - f_I|\,dt\right\} < \infty,$$

where

$$f_I = \frac{1}{|I|}\int_I f(t)\,dt$$

and sup is taken over all sets $I \subset [-\pi,\pi]$ such that $\Theta(I) = \{z\colon\ z = e^{ix},\ x \in I\}$ is an arc of the unit circle, is called the BMO space.

THEOREM. *Let the spectral density $f = \|f_{ij}\|$, $i,j = 1,\dots,m$, of the sequence $X(n) = \big(x^1(n), x^2(n), \dots, x^m(n)\big)$ is such that*

1) *functions b_{11}, b_{22}, $(1-|p|^2)$, f_{11}^{ij}, f_{22}^{ij}, $f_{11}^{ij}(1-|p^{ij}|^2)$ satisfy the Muckenhoupt condition when $i < j$;*
2) *functions $\log b_{11}$, $\log b_{22}$, $\log(1-|p|^2)$, $\rho \in L^1$ and satisfy condition (7);*
3) *functions R_-^{ij}, I_-^{ij} are contained in the BMO space.*

Then

$$\mathbf{E}(T_1 - T_2) = o(\sqrt{n}), \quad \operatorname{var}(T_1 - T_2) = o(\sqrt{n}), \quad n \to \infty.$$

Proof. Note that

$$\log\det P_n Q_n P_n = -\log\det B_n^f - \log\det B_n^{f^{-1}}.$$

By the conditions of the theorem from (Tsertsvadze, 1988) we have

$$\log\det B_n^f = \frac{n}{2\pi}\int_{-\pi}^{\pi}\log\det f(\lambda)d\lambda + o(\sqrt{n}), \quad n \to \infty,$$

$$\log\det B_n^{f^{-1}} = -\frac{n}{2\pi}\int_{-\pi}^{\pi}\log\det f(\lambda)d\lambda + o(\sqrt{n}), \quad n \to \infty.$$

Therefore,

$$\log\det P_n Q_n P_n = o(\sqrt{n}), \quad \text{as} \quad n \to \infty.$$

This formula and Lemma 2 yield the proof of the theorem.

REFERENCES

Kashin, B.S. and Saakyan, A.A. (1984). *Orthogonal Series,* Nauka, Moscow (in Russian).

Tsertsvadze, K.A. (1988). Information operator and asymptotic behaviour of the toeplitz determinant, *Zap. Nauchn. Semin. LOMI,* **166**, 179–186 (in Russian).

Part Seven

TOWARDS NATURAL SCIENCES

New Trends in Probab. and Statist., pp. 655 – 667
V. Sazonov and T. Shervashidze (Eds.)
1991 VSP/Mokslas

ON CONDENSABLE POINT PROCESSES

R.V. AMBARTZUMIAN
Institute of Mathematics of Armenian Academy of Sciences,
Marshal Bagramian pr. 24b, 380019 Erevan, Armenia, USSR

ABSTRACT

The operation of α–condensation $(\alpha > 1)$ is defined as the converse to the operation of independent $1/\alpha$–thinning of a point process. The paper presents sufficient conditions under which a point process is 2–condensable. This permits to establish 2–condensability of a class Gibbs processes. A remark by Sukiasian relates the result for the renewal case in one dimension to a classical result of Schürer dating to 1912.

1. INTRODUCTION

Definition. The point process P will be called a α*–condensable*, $\alpha > 1$, if P can be obtained by applying the operation of independent α^{-1}–thinning to some other point process P_1 which we call *parent*.
The operation of independent ρ–thinning, $\rho < 1$, we define as usual (see (Daley and Vere–Jones, 1988)) by means of a stochastic transformation of point process realizations

$$\mathfrak{m} \mapsto \mathfrak{m}^*$$

under which

$$\mathfrak{m}^* \subseteq \mathfrak{m}$$

and, for each point x of a realization $\mathfrak{m}$, the event

$$x \in \mathfrak{m}^*$$

has probability ρ; for different points these events are assumed to be independent.
We also consider a stochastic transformation which maps $\mathfrak{m}$ into a sequence $\mathfrak{m}_1, \ldots, \mathfrak{m}_r$ of realizations which we term as *fragments.* Under this map each point of $\mathfrak{m}$ is assigned to one of the fragments independently with probability $1/r$. If $\mathfrak{m}$ is random and has distribution $\mathbf{P}$ then the fragments are "doubly" random and determine r jointly distributed point processes. Their joint distribution $\mathfrak{P}$ we term as obtained from $\mathbf{P}$ by means of *independent* $1/r$*–splitting.*
Under such symmetrical splittings the fragments are always identically distributed: each fragment is actually obtained from the parent process $\mathfrak{m}$ by means of independent $1/r$–thinning.

In the paper two general results on this topic (Propositions 2 and 3) are derived within the framework of density functions approach to point processes first exposed in (Ambartzumian and Sukiasian, 1990). Proposition 2 presents conditions which guarantee that a sequence of point processes can be obtained as fragments of some parent point process $\mathfrak{m}$ by means of independent $1/r$–splitting.

In Proposition 3 a criterion of 2–condensability is given in terms of the Radon–Nycodim density of the Palm distribution with respect to the distribution of the point process. (It seems that α–condensability for other values of α requires special treatment.)

Our proof of Propositions 2 and 3 is partly based on Proposition 1 where conditions (having combinatorial origin) are given which ensure that a system of functions generate several jointly distributed point processes in which case these functions become the joint densities for the processes.

The main application concerns Gibbs processes governed by two–point interactions and non–negative potentials (subject to a usual convergence condition). It follows from Proposition 3 in conjunction with the results of the earlier papers (see (Ambartzumian, 1990a; Ambartzumian and Sukiasian, 1990)), that for smaller values of the activity parameter each process from the mentioned Gibbs class is 2–condensable.

This conclusion has an important corollary concerning hard core unit diameter balls Gibbs processes from our class: their intensities are at least two times less then the highest possible.

At the end we consider an example of renewal Gibbs process in one dimension where we are able to describe completely the parent process. It is rather immediate that any renewal point process produces after independent $1/2$–splitting a pair of fragment processes each of which is again renewal. First we show that this property is characteristic for renewal point processes (at least in the framework of our density approach). In other words, the fact that each of the fragment processes $\mathfrak{m}_1$ and $\mathfrak{m}_2$ is renewal already implies that the parent process $\mathfrak{m}$ was renewal. This enables us to interpret our main result in the renewal case as a conclusion concerning a functional series.

Sukiasian noticed that the series in question with its remarkable properties has been discovered by Schürer as early as 1912. The remark provides a valuable illustration for Proposition 3.

2. JOINT DENSITIES

We begin by a brief outline of the result of (Ambartzumian and Sukiasian, 1990) concerning the generation of a point process in $\mathbf{R}^d$ by means of a system of density functions. By doing this we aim at providing an illustration to Proposition 1 that follows where we formulate an extension of this result to the case of several jointly distributed point processes within the same space $\mathbf{R}^d$. A complete proof of Proposition 1 is too long to be included here and will be published elsewhere. We mention only that it follows (without substantial extra difficulties) the general idea of the explanation we give for the case where one point process is in question. The main features of the problem and its solution in the case of a single point process as presented in (Ambartzumian and Sukiasian, 1990) are as follows. We consider non–negative functions $f(X)$ defined on finite point sets $X \subset \mathbf{R}^d$. The specification of such an f is equivalent to the specification of an infinite system of functions

$$f(x_1,\ldots,x_n), \quad n = 1,2,\ldots,$$

which depend symmetrically on the increasing number of arguments $x_i \in \mathbf{R}^d$. Where necessary, we will indicate the number of points in the set X using the upper index (n), i.e.,

$$X = X^{(n)} = \{x_1,\ldots,x_n\}.$$

One of the main features of the theory is the requirement:

$$f(X^{(n)}) = f(x_1,\ldots,x_n) < C^n \tag{1}$$

for some constant C. We consider the expressions

$$F_D(X) = \sum_{m=0}^{\infty} \frac{(-1)^m}{m!} \int\limits_{D^m} f\Big(X \cup Y^{(m)}\Big) dY^{(m)}, \tag{2}$$

where, naturally, $X \cup Y^{(m)} = \{x_1,\ldots,x_n,y_1,\ldots,y_m\}$, $X \subset D$, $D \subset \mathbf{R}^d$ is a d–dimensional rectangle, while

$$dY^{(m)} = dy_1 \ldots dy_m$$

is the product of the Lebesgue measures dy in $\mathbf{R}^d$. A slight abuse of notation under which a set $Y^{(m)}$ is considered as a point in D^m is settled by the recourse to symmetry. Alternatively we could extend the integrals in (2) over the corresponding spaces of subsets of D and remove the coefficients $(m!)^{-1}$. It was proved in (Ambartzumian and Sukiasian, 1990) that under (1), if the expressions (2) remain non–negative for every choice of D and for every finite X then a (unique) point process P exists for which $f(X)$ constitutes its system of density functions. This means that at the points where f is continuous

$$\mathbf{P}\big\{N(dx_1) = 1,\ldots,N(dx_n) = 1\big\} =$$

$$= f(x_1,\ldots,x_n)\, dx_1 \ldots dx_n = f(X^{(n)})\, dX^{(n)}, \tag{3}$$

where $N(D)$ denotes the number of points of process realization within the domain D.

The complete proof of this remarkable result consists of a series of steps governed by the following idea. Assume that the functions $f(X)$ are actual density functions of a point process P. Then the functions $F_D(X)$ as given by (2) have the probabilistic interpretation of *relative densities* of P, i.e.,

$$F_D\big(X^{(n)}\big)dX^{(n)} = \mathbf{P}\big\{N(dx_i) = 1 \text{ for } i = 1,\ldots,n,\ N(D \setminus \cup\, dx_i) = 0\big\}. \tag{4}$$

Proof. Split the rectangle D into small rectangles dz_i, $i = 1, \ldots, s$, so that $dx_i = dz_i$, for $i = 1, \ldots, n$, and rewrite the probability

$$\mathbf{P}\{N(dz_i) = 1 \text{ for } i = 1, \ldots, n, \text{ and } N(dz_i) = 0 \text{ for } i = n+1, \ldots, s\}$$

in terms of probabilities

$$\mathbf{P}\{N(dz_i) = 1 \text{ for } i = 1, \ldots, n, \text{ and } N(dz_i) = 1 \text{ for } i \in J\},$$

where $J \subseteq \{n+1, \ldots, s\}$. This we do using standard inclusion–exclusion formula. Then using (3), the limit of the corresponding expression (as diameter of each dz_i tends to zero and s tends to infinity) can be written as (2). The legitimacy of the latter action is guaranteed by (1).

The next observation is that the so-called finite–dimensional distributions of P, i.e., the probabilities of the events

$$A = \{N(D_i) = k_i, \ i = 1, \ldots, s\}$$

can be directly expressed in terms of relative densities by applying an integration operator which we denote by $\mathfrak{I}$. We write thjs symbolically as a relation between the families of quantities

$$\{\mathbf{P}(A)\} = \mathfrak{I}\{F_D\}.$$

For instance, in the case of a single rectangle D,

$$\mathbf{P}\{N(D) = k\} = (k!)^{-1} \int\limits_{D^k} F_D(X^{(k)})\, dX^{(k)}. \tag{5}$$

A system of conditions which should be imposed on a set of functions

$$W(D_1, \ldots, D_s; \ k_1, \ldots, k_s) = W(\overline{D}; \bar{k})$$

to guarantee the existence of a (unique) point process P for which

$$W(\overline{D}; \bar{k}) = \mathbf{P}(A)$$

can be found for instance in (Daley and Vere–Jones, 1988). Now, given an arbitrary set of functions f satisfying (1), we use the mentioned system of conditions to check that the necessary condition

$$\{F_D(X) \geqslant 0 \quad \text{for all rectangles } D \text{ and all finite } X\}$$

is also sufficient. This means that the expressions

$$W(\overline{D}; \bar{k}) = \mathfrak{I} F_D$$

(which we construct by formal substitution of our functions F_D in the expressions like (5)) satisfy the mentioned conditions on the functions W. Finally, to check

that our functions f are actual densities of P in the sense of (3) is a rather simple matter (follows from the nature of the operator $\mathfrak{I}$).
We turn now to the question of generation of several jointly distributed point processes in $\mathbf{R}^d$,

$$\mathfrak{m}_1, \dots, \mathfrak{m}_r,$$

by means of *joint densities.*
Let $\mathbf{P}$ be a joint probability distribution of $\mathfrak{m}_1, \dots, \mathfrak{m}_r$. The joint densities $f(\overline{X})$, where $\overline{X} = (X_1; \dots; X_r)$ (note the use of semicolon) and

$$X_i = \{x_{i1}, \dots, x_{in_i}\}, \quad x_{ik} \in \mathbf{R}^d$$

of the two processes are defined by the relations

$$\mathbf{P}\{N_i(dx_{is}) = 1 \text{ for } s = 1, \dots, n_i,\ i = 1, \dots, r\} = f(\overline{X})\, dX_1^{(n_1)} \dots dX_r^{(n_r)}$$

where N_i is the number of points in the corresponding domain from the point process $\mathfrak{m}_i, i = 1, \dots, r$.
We also have *joint relative densities* $F_{\overline{D}}(\overline{X})$ which depend on r rectangles $\overline{D} = (D_1, \dots, D_r)$ and have a probabilistic meaning similar to (4). Under the additional condition

$$f(\overline{X}) < C^M; \quad M = n_1 + \dots + n_r, \quad n_i = \operatorname{card} X_i, \tag{6}$$

we have the identity

$$F_{\overline{D}}(\overline{X}) = \sum_{n_i=0;\ i=1,\dots,r}^{\infty} \frac{(-1)^{\Sigma n_i}}{n_1! \dots n_r!} \int_{\widehat{D}} f(Z_1; \dots; Z_r)\, dY_1^{(n_1)} \dots dY_r^{(n_r)}, \tag{7}$$

where $\widehat{D} = D_1^{n_1} \times \dots \times D_r^{n_r}$, and

$$Z_i = X_i \cup Y_i^{(n_i)}.$$

This again is a limiting form of an inclusion–exclusion formula.

PROPOSITION 1. *Let the functions $f(X_1, \dots, X_r)$ which depend on r finite sets from $\mathbf{R}^d$ satisfy condition (6) for some constant C.*
If the results of substitution of these functions in the right–hand side of (7) remain non–negative for all sequences of rectangles $D_1, \dots, D_r$ and all values of $\overline{X}$ then there exist r jointly distributed point processes for which f turn to be the joint density functions. The joint distribution is unique.

3. COMPLETE SYMMETRY AND CONDENSABILITY

Let $f_1(X)$ be the density functions of a point process which we subject to independent 1/2–splitting and let $f(X;Y)$ be the joint density function of the fragments. Then, clearly,

$$f(X^{(n)}; Y^{(m)}) = (1/2)^{n+m} f_1(X^{(n)} \cup Y^{(m)})$$

and we observe that the function $f(X;Y)$ necessarily possesses the *complete symmetry* property, i.e., it depends only on the union of the set arguments. We have a similar symmetry effect if we perform the independent $1/r$–splitting.

PROPOSITION 2. *Let the functions $f(X_1;\ldots;X_r)$ be the joint density functions of r point processes in $\mathbf{R}^d$,*

$$\mathfrak{m}_1,\ldots,\mathfrak{m}_r$$

and satisfy (6).
Then the condition of complete symmetry

$$f(X_1;\ldots;X_r)=f_2(Z), \tag{8}$$

where $Z=X_1\cup\ldots\cup X_r$ is sufficient in order that $\mathfrak{m}_1,\ldots,\mathfrak{m}_r$ be the fragments obtained from

$$\mathfrak{m}_1\cup\ldots\cup\mathfrak{m}_r \tag{9}$$

by means of independent $1/r$–splitting. In such a case the functions

$$f_1(Z^{(k)})=r^k f_2(Z^{(k)})$$

are the density functions of the point process (9).

Proof. The problem is to show that f_1 is a system of density functions for some point process. We consider the case $r=2$. For a rectangle $D\subset\mathbf{R}^d$ and a set $X\in D^n$ we calculate the value of the joint relative density, i.e., we write a version of (7) (for $r=2$)

$$0\leqslant F_{D,D}(X;\varnothing)=$$

$$=\sum_{k,m=0}^{\infty}\frac{(-1)^{k+m}}{k!\,m!}\iint\limits_{D^k\times D^m} f(X\cup Y_1^{(k)};Y_2^{(m)})\,dY_1^{(k)}dY_2^{(m)}=$$

$$=\sum_{l=0}^{\infty}\sum_{k=0}^{l}\frac{(-1)^l}{k!(l-k)!}\int\limits_{D^l} f_2(X\cup Y^{(l)})\,dY^{(l)}=$$

$$=\sum_{l=0}^{\infty}\frac{(-1)^l}{l!}\int\limits_{D^l} f_1(X\cup Y^{(l)})\,dY^{(l)}.$$

By Proposition 1 (the case $r=1$), this completes the proof.
If we have one point process (rather than a number) we ask the question: when this process is 2–condensable?
Below we give an answer to this question in terms of the Radon–Nycodim density

$$\frac{Q_x(d\mathfrak{m})}{\mathbf{P}(d\mathfrak{m})}=V_x(\mathfrak{m}), \tag{10}$$

where $\mathbf{P}$ is the distribution of our point process and Q_x is its Palm distribution. Here $\mathfrak{m}$ is a *realization*, i.e., a point in the realization space and thus V is a function defined on the latter space (which depends also on a parameter $x\in\mathbf{R}^n$).

Roughly speaking Q_x is the conditional distribution of the point process P under the condition that $\mathfrak{m}$ has a point $x \in \mathbf{R}^n$.
We consider the products

$$U_X(\mathfrak{m}) = U_{x_1,\ldots,x_n}(\mathfrak{m}) = \prod_{i=1}^{n} f(x_i) V_{x_i}(\{x_{i+1},\ldots,x_n\} \cup \mathfrak{m}), \tag{11}$$

where $f(x)$ is the first order density. They have the following interpretation (see (Ambartzumian, 1990a))

$$U_X(\mathfrak{m}) = \lim \frac{F_D(X \cup Y_D)}{F_D(Y_D)} = \frac{\mathbf{P}(d(X \cup \mathfrak{m}))}{\mathbf{P}(d\mathfrak{m})dX}, \tag{12}$$

where the limit is calculated as D expands to fill the whole of $\mathbf{R}^d$ while $\mathfrak{m}$ remains fixed and

$$Y_D = \mathfrak{m} \cap D.$$

The quantities (10) and (12) have been considered in a number of papers including (Nguyen and Zessin, 1979; Matthes et al., 1979) as well as (Daley and Vere–Jones, 1988). We will be interested in a new situation which we term as situation

$\mathcal{S}$: with probability 1 in $\mathfrak{m}$ the quantities $U_X(\mathfrak{m})$ exist and as functions of X they constitute a system of density functions of a point process (whose distribution depends on $\mathfrak{m}$).

We briefly describe a class of point processes for which $\mathcal{S}$ is satisfied.
Let $\mathbf{P}$ be the distribution of a point process which was obtained by independent 1/2–thinning from a parent process. Let $\mathfrak{m}_1, \mathfrak{m}_2$ be the two point processes obtained by means of independent splitting of the parent process. By $\mathbf{P}(d\mathfrak{m}_1|\mathfrak{m}_2)$ we denote the conditional distribution of $\mathfrak{m}_2$ conditional upon $\mathfrak{m}_1$. Assume that this conditional distribution possesses densities

$$f(X|\mathfrak{m}_2) = \mathfrak{P}\Big(\big\{N_1(dx_i) = 1 \text{ for } x_i \in X\big\} \times d\mathfrak{m}_2\Big)\Big[dX\ \mathbf{P}(d\mathfrak{m}_2)\Big]^{-1},$$

where $\mathfrak{P}$ is the joint distribution of $\mathfrak{m}_1, \mathfrak{m}_2$ and we consider a product event. Because of the symmetry, we have

$$\mathfrak{P}\Big(\big\{N_1(dx_i) = 1 \text{ for } x_i \in X\big\} \times d\mathfrak{m}_2\Big) = \mathbf{P}\Big(d(X \cup \mathfrak{m}_2)\Big).$$

Thus we obtain the existence of ratios (12) which possess the probabilistic significance required by $\mathcal{S}$.

PROPOSITION 3. *Let P be a point process in $\mathbf{R}^d$. If $\mathcal{S}$ is satisfied and if the functions $U_X(\mathfrak{m})$ satisfy* (1) *uniformly in $\mathfrak{m}$ then P is 2–condensable.*
In this case functions (11) *represent the conditional density functions of a fragment of the parent process of P under independent 1/2–splitting, conditional upon the realization of the other fragment.*

Proof. Let $K_{\mathfrak{m}}(d\mathfrak{m}_1)$ be the (distribution of) point process defined by the functions $U_X(\mathfrak{m})$. According to Proposition 2 it is enough to show the complete symmetry of the joint density function $f(X;Y)$ of the two point processes which have

$$K_{\mathfrak{m}}(d\mathfrak{m}_1)\mathbf{P}(d\mathfrak{m}) \tag{13}$$

for their joint distribution.

Clearly, we have

$$f(X;Y)dY = \int_{B(Y)} U_X(\mathfrak{m})\mathbf{P}(d\mathfrak{m}), \tag{14}$$

where $B(Y) = \{N(dy_i) = 1 \text{ for every } y_i \in Y\}$.

By a change of integration variable $\mathfrak{m} = Y \cup \mathfrak{m}_2$, we transform (14) to

$$f(X,Y)dY = \int U_X(Y \cup \mathfrak{m})\mathbf{P}\big(d(Y \cup \mathfrak{m})\big)$$

(integration is over the whole space of realizations) and then, by virtue of (12), to

$$f(X,Y) = \int U_X(Y \cup \mathfrak{m})U_Y(\mathfrak{m})\mathbf{P}(d\mathfrak{m}).$$

We have a functional relation which follows from (11):

$$U_X(Y \cup \mathfrak{m})U_Y(\mathfrak{m}) = U_{X\cup Y}(\mathfrak{m}),$$

from which we conclude that

$$f(X,Y) = \int U_{X\cup Y}(\mathfrak{m})\mathbf{P}(d\mathfrak{m}),$$

hence the required symmetry and (6) follow. The concluding part of the proposition also follows from the above formulae.

4. GIBBS PROCESSES

The existence of the density (10) for general Gibbs processes has been shown in (Nguyen and Zessin, 1979) and in the stationary case with two–point interaction and non–negative potential this can be represented in the form

$$V_x(\mathfrak{m}) = a^{-1}b \prod_{x_j \in \mathfrak{m}} h(x - x_j),$$

where a is the intensity of the point process in question, b is a parameter called the *activity* and

$$0 \leqslant h(x) \leqslant 1$$

is a function defined on $\mathbf{R}^d$.

In this case $f(x)$ identically equals a and we have

$$U_{X^{(n)}}(m) = b^n \prod_{x_i, x_j \in X^{(n)};\ i>j} h(x_i - x_j) \prod_{x_i \in X^{(n)}} \prod_{z_i \in m} h(x_i - z_j). \tag{15}$$

In (Ambartzumian and Sukiasian, 1990) were found sufficient conditions which guarantee that (15) is a system of density functions of a point process (i.e., $V_x(m)$ satisfies S). They are: the convergence condition

$$I = \int [1 - h(x)]\,dx < \infty \tag{16}$$

and the following bound for the activity parameter

$$b < (I\ e)^{-1}, \tag{17}$$

where e is the base of natural logarithm. The conditions of Proposition 3 concerning (1) are also satisfied (for the properties of the density functions of our Gibbs processes (see (Ambartzumian, 1990a)).
Let us summarize the above observations. Let P be a stationary Gibbs point process governed by pair interaction and non-negative potential. If (16) holds then, for the values of activity parameter satisfying (17), P is 2-condensable.
The joint distribution of the fragments of the corresponding parent process is given by (13).

Example 1. Let us consider the case

$$h(x) = \begin{cases} 1 \text{ for } |x| < 1, \\ 0 \text{ otherwise.} \end{cases} \tag{18}$$

If we center unit diameter balls at the points of the corresponding Gibbs process then the balls will be pairwise disjoint. In the range of activity (17) the parent process will share similar property. (If we assume that the parent process possesses pairs of intersecting balls then the result of independent 1/2-thinning will share the same property, but this was not the case.) Thus, loosely speaking, in the range of activity (17) the intensity of our Gibbs process is at least two times less then the intensity of the centers under densest packing of disjoint balls of unit diameter.

Example 2. In the one dimensional case stationary renewal point processes with interpoint distance probability density function given by

$$r(u) = \begin{cases} 0 & \text{if} \quad u < 1, \\ c\exp\left(-c(u-1)\right) & \text{if} \quad u > 1 \end{cases} \tag{19}$$

(where $c > 0$ is a parameter) belong to the class of Gibbs processes of Example 1 with (see (Ambartzumian, 1990b))

$$b = a\left[\mathbf{P}\left\{N((-1,1)) = 0\right\}\right]^{-1}.$$

Here a is the intensity of the renewal process, $a = (1 + 1/c)^{-1}$.
Applying a Palm formula (ibidem), we easily get

$$\mathbf{P}\{N((-1,1)) = 0\} = 1 - a \int_0^2 dx \int_x^\infty r(u)\,du = \left[(c+1)e^c\right]^{-1}.$$

This yields

$$b = c\,e^c. \tag{20}$$

According to (17), for values

$$b < 1/2e \tag{21}$$

our renewal process can be obtained from some parent process by means of symmetrical 1/2–thinning. We obtain a complete description of the parent process using the following

STATEMENT. *Under the condition of existence of bounded renewal density, any parent process (in the sense of independent thinning with constant probability) of a renewal process can be renewal only.*

Proof. The density functions of a renewal process have the form: $f(x) = a$, the intensity: $f(x_1, x_2) = av(x_2 - x_1)$, v is the "renewal density";

$$f(x_1, \ldots, x_n) = a \prod_{i=1}^{n-1} v(x_{i+1} - x_i),$$

where we assume that $x_1 < x_2 < \ldots < x_n$.
The system of density functions f^* of the parent process has the form

$$f^*(X^{(n)}) = q^n f(X^{(n)}), \quad q > 1,$$

where $1/q$ is the corresponding thinning probability.
If we take x_1, x_2 to be the endpoints of an interval D and calculate $F_D^*(x_1, x_2)$ by substituting f^* into (2) we get

$$F_D^*(x_1, x_2) = qar^*\left(|x_1 - x_2|\right),$$

where r^* is the interpoint distance probability density function for the parent process. It remains to check that the substitution of f^* in (2) (where we choose X so that x_1, x_n coincide with the endpoints of D and $x_1 < x_2 < \ldots < x_n$) results in

$$F_D^*(x_1, \ldots, x_n) = qa \prod_{i=1}^{n-1} r^*\left(|x_i, x_{i+1}|\right).$$

But this is the definition of a renewal process.
We return to our special case.

Since the parent process is renewal, we have

$$r(u) = \sum_{k=1}^{\infty}(1/2)^k r^{*(k)}(u),$$

where the upper (k) denqtes the convolution power. In terms of the Laplace transforms

$$L(w) = \int \exp(-wu) r(u)\, du, \quad L^*(w) = \int \exp(-wu) r^*(u)\, du,$$

this corresponds to

$$L = \sum_{k=1}^{\infty}(1/2)^k (L^*)^k = L^*(2 - L^*)^{-1},$$

which can be rewritten as

$$L^* = 2L(1+L)^{-1} = 2\sum_{k=0}^{\infty}(-1)^k L^{k+1}.$$

This corresponds to

$$r^* = 2\sum_{k=0}^{\infty}(-1)^k r^{(k+1)} \tag{22}$$

and we deduce from (19) that $r^{(k+1)}(u)$ is the shifted gamma density:

$$r^{(k)}(u) = \begin{cases} 0 & \text{if} \quad u < k, \\ (u-k)^{k-1} c^k \big((k-1)!\big)^{-1} \exp\big(-c(u-k)\big) & \text{if} \quad u > k. \end{cases}$$

We conclude from (20), (21) that, for the values of c less then h which is the root of the equation

$$te^t = \frac{1}{2e}, \tag{23}$$

the series (22) represents a probability density function. We point out that the bound given by (23) comes from our general theory and a priory is smaller than the exact bound in each special case.

Remark (Sukiasian). The purpose of this remark is to relate the result of the above example to some classical work done as early as 1912.
It was shown in (Ambartzumian and Sukiasian, 1990) that if inequalities (16) and (17) hold then there exist a point process P_0 in $\mathbf{R}^d$ with the following densities (in the sense of (3), compare with (15) where the same expression is present as a factor)

$$f_0(X^{(n)}) = b^n \prod_{x_i, x_j \in X^{(n)};\ i<j} h(x_i - x_j).$$

The processes are necessarily translation–invariant and of finite intensity.

In the particular case of $d = 1$ and h given by (18), P_0 is a renewal process because the above reduces to

$$f_0(x_1, \ldots, x_n) = b^n \prod_{i=1}^{n-1} h(x_{i+1} - x_1), \quad x_1 < x_2 < \ldots < x_n.$$

This renewal process has the following properties. We denote by Q_t the corresponding Palm probability of the event

$$A_t = \{N([0,t]) = 0\}$$

(equivalently, $1 - Q(t)$ is the distribution function of the distance between the neighboring points in the process).
In (Ambartzumian and Sukiasian, 1990) it is shown that

$$P_0(A_t) = Q(t+1). \tag{24}$$

From (24), using a Palm formula (see [5]), we get an equation which was studied in (Schürer, 1912)

$$\frac{dQ(t)}{dt} = -bQ(t-1), \quad t > 1,$$

$$Q(t) = 1, \quad 0 \leqslant t \leqslant 1.$$

This equation has the following solution

$$Q(t) = \sum_{n=0}^{[t]} \frac{(-b)^n}{n!} (t-n)^n, \tag{25}$$

where $[t]$ denotes the integer part of t.
The interpoint distance probability density function $q(u) = -dQ(u)/du$ is found to be

$$q(u) = \begin{cases} bQ(u-1), & u > 1, \\ 0, & u < 1. \end{cases} \tag{26}$$

From (22), (25) and (26), we get

$$r^*(u) = 2e^{-cu} q(u), \tag{27}$$

where the parameters b and c are connected via (20).
We have the fundamental Schürer result: the alternating sum (25) is non-negative for all $t > 0$ if and only if $b < 1/e$. Hence, the bound given by (23) is indeed less then the precise one.
It can be instructive to derive (27) in a different way, using the conditional distribution K_m (see Proposition 3), which in this special case can be explicitly described. It follows from (15) that the process K_m can be obtained from P_0 by independent thinning with the "survival" law

$$\prod_{z \in m} h(x - z),$$

i.e., a realization of the process $K_{\mathfrak{m}}$ consists of those points x of a realization of P_0 for which there is no point from $\mathfrak{m}$ in the segment $[x-1, x+1]$.
We consider the Palm distribution Q^* of the parent process. Roughly speaking Q^* is the conditional distribution of the parent process under the condition that the realization $\mathfrak{m} \cup \mathfrak{m}_1$ contains the origin 0. We can suppose that $0 \in \mathfrak{m}$ because of the symmetry. We divide the event A_t in two subevents:

a) $\mathfrak{m} \cap [t, t+1] = \emptyset$ and

b) there is a point $u = \mathfrak{m} \cap [t, t+1]$

(the probability of the event that there are more points in this segment is zero because the parent process is a process of nonintersecting unit diameter balls). We have

$$Q^*(t) = Q^*(A_t) = 1, \quad 0 \leqslant t \leqslant 1,$$

$$Q^*(t) = P_0(A_{t-1} \int_{t+1}^{\infty} r(u)\, du + \int_{t}^{t+1} r(u) P_0(A_{u-2})\, du, \quad t > 1.$$

From (19), (20), (24) and (26) we get

$$Q^*(t) = e^{-ct} + \int_{t}^{t+1} e^{-cu} q(u)\, du.$$

Differentiating we find

$$r^*(t) = -\frac{dQ^*(t)}{dt} = ce^{-ct}Q(t) + be^{-ct}Q(t-1) + e^{-ct}q(t) - e^{-ct-c}q(t+1).$$

Now, from (20) and (26), we get (27).

REFERENCES

Ambartzumian, R.V. (1990a). Random graph approach to Gibbs processes with pair interaction. *Acta Applic. Math.* (to appear).

Ambartzumian, R.V. (1990b). *Factorization Calculus and Geometric Probability.* Cambridge University Press, Cambridge.

Ambartzumian, R.V. and Sukiasian, H.S. (1990). Inclusion–exclusion and point processes. *Acta Applic. Math.* (to appear).

Daley, D.J. and Vere–Jones, D. (1988). *An Introduction to the Theory of Point Processes.* Springer, Berlin etc.

Matthes, K., Warmuth, W. and Mecke, J. (1979). Bemerkungen zu einer Arbeit vom Nguen Xuan Xanh and Hans Zessin. *Math. Nachr.* **88**, 117–127.

Nguyen, X.X. and Zessin, H. (1979). Integral and differential characterizations of the Gibbs process. *Math. Nachr.* **88**, 105–115.

Schürer, F. (1912). Ueber die Funktional–Differentialgleichung $f(x+1)=\alpha f(x)$. *Berichte Verhandlungen kgl. sachs. Ges. Wiss. zu Leipzig, Math.-Phys. Kl.* **64**, 167–236.

New Trends in Probab. and Statist., pp. 668 – 682
V. Sazonov and T. Shervashidze (Eds.)
1991 VSP/Mokslas

THE GROUND STATE OF A RANDOM STATIONARY MEDIUM IN THE MEAN FIELD APPROXIMATION

A. ASTRAUSKAS AND S.A. MOLCHANOV
Institute of Mathematics and Informatics, Lithuanian Academy Sci., Akademijos str. 4, 232600 Vilnius, Lithuania
Moscow State University, Department of Mathematics and Mechanics, Len. Gory, 119899 Moscow, USSR

ABSTRACT

In the present article we have obtained limit theorems for the maximal eigenvalue $\lambda_{|V|}$ of the operator $H_V = \kappa\overline{\Delta} + \xi(x)$, $x \in V \subset \mathbf{Z}^d$, as $|V| \to \infty$. Here $\overline{\Delta}$ is the mean field diffusion operator in V and $\xi(x)$, $x \in \mathbf{Z}^d$, an i.i.d. random potential. The result depends strongly both on the distribution function of ξ and on the parameter $\kappa > 0$. As a consequence of these limit theorems, some statements on the particle localization corresponding to the ground state $\lambda_{|V|}$ are obtained.

INTRODUCTION

In the paper we analyze spectral properties of a quantum mechanical model of random medium, described by the equation

$$\kappa\overline{\Delta}\psi + \xi(x)\psi(x) = \lambda\psi(x), \quad x \in V \subset \mathbf{Z}^d, \tag{1}$$

where the operator $\overline{\Delta}$ is given by

$$\overline{\Delta}\psi = |V|^{-1} \sum_{x' \in V} \psi(x'), \tag{2}$$

and $|V|$ is the number of sites in V. Here $\xi(x)$, $x \in \mathbf{Z}^d$, are independent identically distributed random variables and $\kappa > 0$ is a physical parameter.
The mean field diffusion model represents a simplified modification of the physically more real Anderson model with the local second order operator Δ (Anderson, 1958). The model was first discussed in the papers (Bogachov and Molchanov, 1990; Zel'dovich et al., 1987) related to the investigation of intermittence phenomena for evolutionary problems with the mean field diffusion Hamiltonian. On the other hand, the operator $\overline{\Delta}$ is widely used in statistical physics; rather, e.g., the Curie—Weiss model (Kac and Thompson, 1966).
Consider the spectral problem (1). From (1) it follows that

$$\psi(x) = k\overline{\Delta}\psi(\lambda - \xi(x))^{-1}. \tag{3}$$

Averaging (3) with respect to $x \in V$ and taking into account that $\overline{\Delta}\psi \neq 0$, we arrive at the following equation for eigenvalues λ:

$$1/\kappa = |V|^{-1} \sum_{x \in V} (\lambda - \xi(x))^{-1}. \tag{4}$$

We denote by $\xi_{(1)}, \ldots, \xi_{(|V|)}$ the order statistics of $\xi(x)$, $x \in V$. The inequalities $\xi_{(1)} < \ldots < \xi_{(|V|)}$ are strict with probability 1 provided the distribution function of $\xi(0)$ is continuous. Consequently, with probability 1 the equation (4) has exactly $|V|$ real roots $\lambda_{(1)} < \ldots < \lambda_{(|V|)}$ such that $\xi_{(i)} < \lambda_{(i)} < \xi_{(i+1)}$, $i = 1, \ldots, |V| - 1$, and $\lambda_{|V|} > \xi_{(V)}$. By (3), the eigenfunction $\psi(x) = \psi(x;\lambda) = \psi_V(x;\lambda)$, corresponding to an eigenvalue λ and normed by the condition $\sum_{x \in V} \psi^2(x;\lambda) = 1$, can be written in the form

$$\psi(x;\lambda) = \psi_V(x;\lambda) = (\lambda - \xi(x))^{-1} \Big/ \Big(\sum_{x' \in V} (\lambda - \xi(x'))^{-2} \Big)^{1/2} \tag{5}$$

We remark that the maximal eigenvalue $\lambda_{\max} = \lambda_{|V|}$ plays a particular role for the asymptotics of the solution of the equation

$$\partial C / \partial t = \kappa \overline{\Delta} C + \xi C$$

(see (Bogachov and Molchanov, 1990; Zel'dovich et al., 1987)).
In the present article we prove limit theorems for $\lambda_{(|V|)}$ as $|V| \to \infty$. As a consequence of limit theorems, some statements on the particle localization for $\lambda_{\max}$ are obtained.
We shall illustrate a phase transition in κ for a special case, assuming that $\xi(0)$ is distributed in [0,1] with the distribution function $F(t)$ such that $1 - F(t) \sim$ $\sim \text{const}\,(1-t)^{\alpha}$ as $t \uparrow 1$, where $\alpha \in (1,2)$. Namely,

a) if $1/\kappa > \langle (1-\xi)^{-1} \rangle$ ("low temperature"), then a single-site localization takes place:

$$\max_{x \in V} \psi_V^2(x;\lambda_{|V|}) \to 1 \quad (|V| \to \infty)$$

in probability;

b) in the case $1/\kappa < \langle (1-\xi)^{-1} \rangle$ ("high temperature") a complete delocalization is observed: $\psi_V^2(x;\lambda_{|V|})$ is concentrated (with $\varepsilon > 0$ accuracy) on a set $V' \subset V$ such that $\mathbf{P}\text{-}\lim_{|V| \to \infty} |V'|/|V|$ is positive;

c) if $1/\kappa = \langle (1-\xi)^{-1} \rangle$ ("critical point"), then a partial localization takes place: $\psi_V^2(x;\lambda_{|V|})$ is concentrated (with $\varepsilon > 0$ accuracy) on a finite number of sites which does not increase with $|V|$.

Let us make a short comment on other sections. In Section 1 auxiliary statements on the asymptotical behavior of extremal values $\xi_{(|V|)} > \ldots > \xi_{(1)}$ and some nonlinear function of them are proved. Here Lemma 3 plays a central role. In

Section 2 limit theorems for $\lambda_{|V|}$ as $|V| \to \infty$ are presented (Theorems 1–3). Section 3 is given to the investigation of the localization problem for $\lambda_{\max} = \lambda_{|V|}$.

1. AUXILIARY LEMMAS

Let ξ_k, $k = 1, 2, \ldots$, be a sequence of i.i.d. random variables on a probability space $(\Omega, \mathcal{F}, \mathbf{P})$ having a continuous distribution function $F(t)$. We denote by $\xi_{(1)} < \, < \ldots < \xi_{(N)}$ the order statistics of $\xi_1, \ldots, \xi_N$, $N = 1, 2, \ldots$
Let ω_F stand for the essential supremum of ξ_k, i.e.,

$$\omega_F = \sup\{t\colon F(t) < 1\}. \tag{6}$$

For each $s \in (0,1)$ we set

$$\underline{\Phi}(s) = \inf\{t \leqslant \omega_F\colon F(t) \geqslant s\}, \quad \overline{\Phi}(s) = \sup\{t \leqslant \omega_F\colon F(t) \leqslant s\}.$$

LEMMA 1. (Galambos, 1978).

(i) $\mathbf{P}\text{–}\lim\limits_N \xi_{(N)} = \omega_F$;

(ii) *For each* $\delta \in (0,1)$

$$\underline{\Phi}(1-\delta) \leqslant \mathbf{P}\text{–}\varliminf_N \xi_{(N-[N\delta])} \leqslant \mathbf{P}\text{–}\varlimsup_N \xi_{(N-[N\delta])} \leqslant \overline{\Phi}(1-\delta) < \omega_F,$$

here [ ] *denotes the integer part.*

LEMMA 2. *Let* $\omega_F < \infty$. *If*

$$\langle(\omega_F - \xi_0)^{-q}\rangle < \infty \tag{7}$$

for some $q \in (0, \infty)$, *then*

$$\mathbf{P}\text{–}\lim_N N^{-1/q}(\omega_F - \xi_{(N)}) = \infty.$$

Proof. From (7) it follows that

$$(\omega_F - t)^{-q}(1 - F(t)) \to 0 \quad (t \uparrow \omega_F).$$

Thus, for any $L > 0$, $N \to \infty$,

$$\ln \mathbf{P}(N^{-1/q}(\omega_F - \xi_{(N)}) > L) = -N\big(1 - F(\omega_F - LN^{-1/q})\big)(1 + o(1)) \to 0.$$

For an arbitrary $\omega_F \in (-\infty, \infty]$, let us define the function $E(t) = E(t; \omega_F)$ by

$$E(t) = \langle(t - \xi_0)^{-1}\rangle = \int_{-\infty}^{\omega_F} (t-s)^{-1} dF(s), \quad t \in [\omega_F, \infty) \tag{8}$$

and $E(\infty;\infty) = 0$. Further on, let

$$Q_N^{(1)} = \sum_{k=1}^{N-1} (\xi_{(N)} - \xi_{(N-k)})^{-1}, \tag{9}$$

$$Q_N^{(2)} = \sum_{k=1}^{N-1} (\xi_{(N)} - \xi_{(N-k)})^{-2}. \tag{10}$$

LEMMA 3. (**A**) *Let in a neighbourhood of* ω_F *there exists the derivative* $p(t) = F'(t)$ *such that:*

$$\lim_{\delta \downarrow 0} \int_0^1 dt\, t^{-1} |p(\delta^{-1} - t) - p(\delta^{-1} + t)| = 0 \quad \textit{if} \quad \omega_F = \infty,$$

$$\lim_{\delta \downarrow 0} \int_0^\delta dt\, t^{-1} |p(\omega_F - \delta - t) - p(\omega_F - \delta + t)| = 0 \quad \textit{if} \quad \omega_F < \infty. \tag{11}$$

If, moreover, $E(\omega_F) < \infty$, *then*

$$\mathbf{P}\text{–}\lim_N N^{-1} Q_N^{(1)} = E(\omega_F) \tag{12}$$

and

$$\mathbf{P}\text{–}\lim_N N^{-2} Q_N^{(2)} = 0. \tag{13}$$

(**B**) *Suppose that* $\omega_F < \infty$ *and that* $1 - F(t) \sim \text{const}\,(\omega_F - t)^\alpha$ *as* $t \uparrow \omega_F$, *where* $\alpha \in (1, 2)$. *Then the relations* (12) *and*

$$N^{-2/\alpha} Q_N^{(2)} \stackrel{\mathbf{P}}{=} O(1) \quad (N \to \infty) \tag{14}$$

hold.

Proof. (**A**) Because of the law of large numbers for $(\omega_F - \xi_k)^{-1}$, $k = 1, 2, \ldots$, and Lemma 2 with $q = 1$, (12) can be rewritten as

$$\mathbf{P}\text{–}\lim_N N^{-1} \sum_{k=1}^{N-1} \left(\xi_{(N)} - \xi_{(N-k)}\right)^{-1} - \left(\omega_F - \xi_{(N-k)}\right)^{-1} = 0. \tag{15}$$

The k–th summand in (15) is non–negative and does not exceed $g(\xi_{(N)}, \xi_{(N-k)}) + h(\xi_{(N)}, \xi_{(N-k)})$, where

$$g(t,s) = (\omega_F - t)(t-s)^{-1}(\omega_F - s)^{-1} \mathbf{1}(t - s > \omega_F - t),$$
$$h(t,s) = (t-s)^{-1} \mathbf{1}(0 < t - s < \omega_F - t), \tag{16}$$

$\mathbf{1}(\cdot)$ is the indicator function.
Hence (15) follows from

$$I_N := N^{-1}\Big\langle \sum_{k=1}^{N} g(\xi_{(N)},\xi_k)\Big\rangle \to 0 \tag{17}$$

and

$$\mathbf{P}\Big(N^{-1}\sum_{k=1}^{N-1} h\big(\xi_{(N)},\xi_{(N-k)}\big) > \varepsilon\Big) \to 0 \quad (\forall\varepsilon > 0). \tag{18}$$

Let us prove (17). As $g(\xi_{(N)},\xi_k) \leqslant (\omega_F - \xi_k)^{-1} \in L^1(\Omega)$ and $g(\xi_{(N)},\xi_k) \to 0$ $(N \to \infty)$ for each $k = 1,2,\dots$, so $I_{N,k} := \langle g(\xi_{(N)},\xi_k)\rangle \to 0$ for each $k = 1,2,\dots$ and

$$I_N = \frac{1}{N}\sum_{k=1}^{N} I_{N,k} \to 0 \quad (N \to \infty).$$

Now we shall prove (18) by using condition (11). Choose $\varepsilon_N > 0$, $N = 1,2,\dots$, in such a way that $\varepsilon_N \to 0$, $N(1 - F(\omega_F - \varepsilon_N)) \to \infty$ $(N \to \infty)$. Then

$$p_N := \mathbf{P}(\omega_F - \xi_{(N)} > \xi_N) \leqslant \exp\{N(F(\omega_F - \varepsilon_N) - 1)\} \to 0 \quad (N \to \infty). \tag{19}$$

On the other hand, the left–hand side of (18) does not exceed

$$p_N + r_N + 2\varepsilon_N^{-1} q_N, \tag{20}$$

where

$$r_N := \mathbf{P}\Big(N^{-1}\big(\xi_{(N)} - \xi_{(N-1)}\big)^{-1} > \varepsilon_N/2,\ \xi_{(N-1)} > \omega_F - 2\varepsilon_N\Big),$$

$$q_N := \Big\langle N^{-1}\sum_{k=2}^{N-1} h\big(\xi_{(N)},\xi_{(N-k)}\big)\mathbf{1}\big(\xi_{(N)} > \omega_F - \varepsilon_N\big)\Big\rangle.$$

According to the formula

$$\mathbf{P}\Big(\xi_{(N)} < t,\ \xi_{(N-k)} < s\Big) = N!\Big((N-k-1)!(k-1)!\Big)^{-1} \times$$
$$\times \int_{-\infty}^{t}\int_{-\infty}^{s} dF(t')\,dF(s')\,F(s')^{N-k-1}\big(F(t') - F(s')\big)^{k-1},$$

for all $t > s$ and all $k = 1,\dots,N-1$, we have

$$r_N = N\int_{-\infty}^{\infty} dF^{N-1}(s)\Big(F(s + 2(\varepsilon_N)^{-1}) - F(s)\Big)\mathbf{1}(s > \omega_F - 2\varepsilon_N),$$

$$q_N = (N-1)\int_{-\infty}^{\infty} dF(t)\int_{-\infty}^{\infty} dF(s)\,\Big(F(t)^{N-2} - F(s)^{N-2}\Big)h(t,s)\mathbf{1}(t > \omega_F - \varepsilon_N).$$

Since $p(t) \to 0$ $(t \uparrow \omega_F)$,

$$r_n \to 0 \quad (N \to \infty). \tag{21}$$

Now we estimate q_N. After the change of the variables $t - s = v$, $t = t$ and the order of integration, we obtain

$$q_N \leqslant \left\langle \int_{-\infty}^{\infty} dv\, v^{-1} \mathbf{1}\big(0 < v < \omega_F - \tilde{\xi}_{(N-1)}\big) \times \right.$$

$$\left. \times \Big| p\big(\tilde{\xi}_{(N-1)} - v\big) - p\big(\tilde{\xi}_{(N-1)} + v\big) \Big| \mathbf{1}\big(\tilde{\xi}_{N-1} \geqslant \omega_F - 2\varepsilon_N\big) \right\rangle +$$

$$+ \left\langle \int_{-\infty}^{\infty} dv\, v^{-1} \mathbf{1}\Big(\big(\omega_F - \tilde{\xi}_{(N-1)}\big)/2 < v < \omega_F - \tilde{\xi}_{(N-1)} \Big) p\big(\tilde{\xi}_{(N-1)} + v\big) \right\rangle, \tag{22}$$

where $\tilde{\xi}_{(N-1)} = \max(\xi_1, \ldots, \xi_{N-1})$ and N is sufficiently large. Note that the last term in (22) does not exceed

$$2\left\langle \big(\omega_F - \tilde{\xi}_{(N-1)}\big)^{-1} \Big(1 - F\big((\tilde{\xi}_{(N-1)} + \omega_F)/2\big)\Big) \right\rangle,$$

which tends to zero as $N \to \infty$. This and (11) yield $q_N \to 0$ $(N \to \infty)$, which together with (19), (20) and (21) implies (18), and consequently complete the proof of (12) for $\omega_F < \infty$.

To establish (13) in the case $\omega_F < \infty$, let us represent $Q_N^{(2)}$ in the form of two sums with the summands $h^2(\xi_{(N)}, \xi_{(N-k)})$ and $((\xi_{(N)} - \xi_{(N-k)})^{-1} - h(\xi_{(N)}, \xi_{(N-k)}))^2$, respectively. Applying to the first sum the inequality: $a^2 + b^2 \leqslant (a+b)^2$ for all $a > 0,\ b > 0$, we get

$$N^{-2} Q_N^{(2)} \leqslant \left(N^{-1} \sum_{k=1}^{N-1} h\big(\xi_{(N)}, \xi_{(N-k)}\big) \right)^2 + N^{-2} \big(\omega_F - \xi_{(N)}\big)^{-1} Q_N^{(1)}.$$

From the just proved relations (17), (12) and Lemma 2 with $q = 1$, we obtain (13). The proof of (12) and (13) for $\omega_F = \infty$ is similar to that in the case $\omega_F < \infty$ and is omitted.

(B) In this case we use the fact that the random variables $\eta_k := -\log(1 - F(\xi_k))$ are independent exponentially distributed with mean 1 and, thus, by the following formula

$$\xi_k = \omega_F - c(\xi_{(k)}) \exp\{-\alpha^{-1} \eta_{(k)}\} \quad (k = 1, \ldots, N), \tag{23}$$

where $c(t) \to \text{const} > 0$ $(t \uparrow \omega_F)$, the investigation of asymptotic behavior of $\xi_{(N)}, \xi_{(N-1)}, \ldots$ can be reduced to the investigation of asymptotic behavior of $\eta_{(N)}, \eta_{(N-1)}, \ldots$ We shall need the following two lemmas about $\eta_{(k)}$.

LEMMA 4 (Galambos, 1978). *We have*

$$\eta_{(N)} = T_1 + T_2/2 + \cdots + T_N/N,$$

$$\cdots\cdots\cdots\cdots\cdots\cdots$$

$$\eta_{(N-k)} = T_{k+1}/k+1 + \cdots + T_N/N,$$

$$\cdots\cdots\cdots\cdots\cdots\cdots$$

$$\eta_{(1)} = T_N/N,$$

where $T_1, \ldots, T_N$ are independent exponentially distributed random variables with mean 1, $N = 1, 2, \ldots$

LEMMA 5 (ibidem). *For $T_1, T_2, \ldots$ as above,*

$$\sum_{k=1}^{N} T_k/k - \log N \overset{\mathbf{P}}{=} \gamma + o(1) \quad (N \to \infty),$$

where γ is some random variable.

Let us prove (14). By Lemma 1

$$\mathbf{P}\text{–}\lim_N N^{-2/\alpha} \sum_{k=[N\delta]}^{N-1} \left(\xi_{(N)} - \xi_{(N-k)}\right)^{-2} = 0 \quad \forall \delta \in (0,1). \tag{24}$$

On the other hand, using (23) and Lemma 4, one can represent the variables $\xi_{(N)} - \xi_{(N-k)}$, $k = 1, \ldots, [N\delta]$, as

$$\xi_{(N)} - \xi_{(N-k)} = \exp\left\{-\alpha^{-1}\eta_{(N)}\right\} \times$$

$$\times \left(c\big(\xi_{(N-k)}\big) \exp\left\{\alpha^{-1} \sum_{l=1}^{k} T_l/l\right\} - c\big(\xi_{(N)}\big) \right). \tag{25}$$

Lemma 1 shows that for any $\delta \in (0,1)$ there exists (nonrandom) $\rho(\delta) > 0$ such that

$$\underline{\lim}_N \inf_{k<N\delta} c\big(\xi_{(N-k)}\big)/c\big(\xi_{(N)}\big) > 1 - \rho(\delta)$$

and $\rho(\delta) \to 0$ ($\delta \downarrow 0$). From here, using Lemma 5, we get that for any $L > 0$,

$$\overline{\lim}_N \mathbf{P}\left(N^{-2/\alpha} \sum_{k=1}^{[N\delta]} \left(\xi_{(N)} - \xi_{(N-k)}\right)^{-2} > L,\, T_1 > \sqrt{\rho(\delta)} \right) \leqslant$$

$$\leqslant \mathbf{P}\left(\gamma' \sum_{k\geqslant 1} \left((1-\rho(\delta)) \exp\left\{\alpha^{-1} \sum_{l=1}^{k} T_l/l\right\} - 1 \right)^{-2} > L,\, T_1 > \sqrt{\rho(\delta)} \right), \tag{26}$$

provided δ is small enough, where $\gamma' > 0$ is some random variable. Again turning to Lemma 5 we see that the last series in (26) converges in probability. Together with (24), this implies (14) as δ is arbitrary small. Lemma 3 is proved.

2. LIMIT THEOREMS FOR $\lambda_{(N)} - \xi_{(N)}$ AND $\lambda_{(N)}$

As before, let us suppose that the variables ξ_k, $k = 1, 2, \ldots$, are independent having the same continuous distribution function F and $\lambda_{(N)}$ is the maximal solution of the equation

$$1/\kappa = N^{-1} \sum_{k=1}^{N} (\lambda - \xi_k)^{-1} = N^{-1} \sum_{k=0}^{N-1} (\lambda - \xi_{(N-k)})^{-1}, \quad \lambda > \xi_{(N)}, \tag{27}$$

$N = 1, 2, \ldots$ Below we formulate three limit theorems for $\lambda_{(N)} - \xi_{(N)}$ and $\lambda_{(N)}$ in the situations when, respectively: (**a**) $1/\kappa > E(\omega_F)$; (**b**) $1/\kappa < E(\omega_F)$ and (**c**) $1/\kappa = E(\omega_F)$, where the function $E(t)$ is defined in (8) and $\omega_F \in (-\infty, \infty]$.

THEOREM 1. *Let $1/\kappa > E(\omega_F)$, and let F satisfy* (11). *Then*

(i) $\mathbf{P}\text{–}\lim_N N(\lambda_{(N)} - \xi_{(N)}) = (\kappa^{-1} - E(\omega_F))^{-1}$;

(ii) *If for some constants b_N and $a_N > 0$, $a_N = o(1)$, the weak limit: $w\text{–}\lim_N F^N(a_N^{-1} t + b_N) = H(t)$ exists, then*

$$w\text{–}\lim_N \mathbf{P}\Big(a_N(\lambda_{(N)} - b_N) < t\Big) = H(t).$$

It is well known (see, e.g., (Galambos, 1978)) that the class of limiting nondegenerate distribution functions for $\mathbf{P}(a_N(\xi_{(N)} - b_N) < t) = F^N(b_N + t a_N^{-1})$, where $a_N > 0$ and b_N are correspondingly chosen constants, contains only the laws of the three following types:

$$H_{1,\alpha}(t) = \begin{cases} \exp\{-t^{\alpha}\}, & t \in (0, \infty); \\ 0, & t \in (-\infty, 0], \end{cases}$$

$$H_{2,\alpha}(t) = \begin{cases} 1, & t \in (0, \infty); \\ \exp\{-(-t)^{\alpha}\}, & t \in (-\infty, 0], \end{cases}$$

$$H_{3,0}(t) = \exp\{-e^{-t}\}, \; t \in (-\infty, \infty),$$

where $\alpha > 0$.

THEOREM 2. *Let $1/\kappa < E(\omega_F)$. Suppose that the constant $\lambda_0 > \omega_F$ is given by the equation*

$$1/\kappa = E(\lambda_0).$$

Then

$$\mathbf{P}\text{–}\lim_N (\lambda_{(N)} - \xi_N) = \lambda_0 - \omega_F;$$

$$w\text{–}\lim_N \mathbf{P}\Big(\sqrt{N}(\lambda_{(N)} - \lambda_0) < t\Big) = \mathbf{P}(Z_\sigma < t), \tag{28}$$

where Z_σ denotes the Gaussian variable with zero mean and covariance

$$\sigma = 1 - \big\langle(\lambda_0 - \xi_0)^{-1}\big\rangle^2 \big\langle(\lambda_0 - \xi_0)^{-2}\big\rangle^{-1}.$$

THEOREM 3. *Let $1/\kappa = E(\omega_F)$. If $\langle(\omega_F - \xi_0)^{-2}\rangle < \infty$, then*

(i) $\mathbf{P}\text{–}\lim_N (\lambda_{(N)} - \xi_{(N)})(\omega_F - \xi_{(N)})^{-1} = 1$ *and*

(ii) $w\text{–}\lim_N \mathbf{P}(\sqrt{N}(\lambda_{(N)} - \omega_F) < t) = \mathbf{P}(Z_{\sigma_0} < t)$, *where* $\sigma_0 = 1 - \langle(\omega_F - \xi_0)^{-1}\rangle^2 \langle(\omega_F - \xi_0)^{-2}\rangle^{-1}$.

If

$$\mathbf{P}(\xi_0 > t) \sim \text{const}\,(\omega_F - t)^\alpha \tag{29}$$

as $t \uparrow \omega_F$, and $\alpha \in (1,2)$, then there exist an increasing subsequence N_k and random variables Z' and Z'', $\mathbf{P}(Z' \neq 0,\ Z'' > 0) = 1$, such that

$$w\text{–}\lim_k \mathbf{P}\left(N_k^{1/\alpha}(\lambda_{(N_k)} - \omega_F) < t\right) = \mathbf{P}(Z' < t)$$

and

$$w\text{–}\lim_k \mathbf{P}\left(N_k^{1/\alpha}(\lambda_{(N_k)} - \xi_{(N_k)}) < t\right) = \mathbf{P}(Z'' < t).$$

Proof of Theorem 1. (i) Let $\omega_F < \infty$. We rewrite equation (27) as

$$1/\kappa - N^{-1}\sum_{k=1}^{N}(\omega_F - \xi_k)^{-1} =$$

$$= (\omega_F - \lambda_{(N)})N^{-1}\sum_{k=1}^{N}\int_0^1 d\theta\left(\omega_F - \xi_k - \theta(\omega_F \lambda_{(N)})\right)^{-2}. \tag{30}$$

Since the limit in probability of the left-hand side of (30) equals $1/\kappa - E(\omega_F) > 0$, it follows that

$$\lim_N \mathbf{P}\left(\lambda_{(N)} > \omega_F\right) = 0. \tag{31}$$

Turning to (27), we see that $\lambda_{(N)} \leqslant \omega_F$ implies the following inequality

$$1/\kappa - N^{-1}(\lambda_{(N)} - \xi_{(N)})^{-1} \geqslant N^{-1}\sum_{k=1}^{N-1}\left(\omega_F - \xi_{(N-k)}\right)^{-1} = E(\omega_F) + o(1), \tag{32}$$

as $N \to \infty$, where we have used the law of large numbers for $\{(\omega_F - \xi_k)^{-1}\}$ and Lemma 2 with $q = 1$. From (27) and Lemma 3 (**A**), we also obtain:

$$1/\kappa - N^{-1}\left(\lambda_{(N)} - \xi_{(N)}\right)^{-1} \leqslant N^{-1}Q_N^{(1)} = E(\omega_F) + o(1),$$

as $N \to \infty$. This together with (31) and (32) implies (i) with $\omega_F < \infty$.
For $\omega_F = \infty$, the proof of (i) does not differ from the reasoning above and can be omitted.
Finally, applying (i) to the equality

$$a_N\left(\lambda_{(N)} - b_N\right) = a_N\left(\xi_{(N)} - b_N\right) + a_N\left(\lambda_{(N)} - \xi_{(N)}\right),$$

we obtain (ii).

Proof of Theorem 2 is similar to that of Theorem 3 in the case $\langle(\omega_F-\xi_0)^{-2}\rangle <$ $< \infty$ (see bellow).

Proof of Theorem 3. Let $\langle(\omega_F-\xi_0)^{-2}\rangle < \infty$. Let us show first that as $N \to \infty$,

$$\sqrt{N}\Big|\lambda_{(N)} - \omega_F\Big| = O(1) \tag{33}$$

in probability. For this purpose we represent (27) as

$$E(\omega_F) - N^{-1}\sum_{k=1}^{N}(\omega_F - \xi_k)^{-1} =$$

$$= (\omega_F - \lambda_{(N)})N^{-1}\sum_{k=1}^{N}\int_0^1 d\theta\,\big(\omega_F - \xi_k - \theta(\omega_F - \lambda_{(N)})\big)^{-2}. \tag{34}$$

Denote by R_N the left–hand side of (34). For any $N = 1,2,\ldots$ and any $L \in$ $\in (0,\infty)$, we conclude from (34) and the trivial inequality $\lambda_{(N)} \leqslant \omega_F + \kappa$ that $\omega_F - \lambda_{(N)} \leqslant -L/\sqrt{N}$ implies

$$R_N < -L\sqrt{N}\sum_{k=1}^{N}(\omega_F + \kappa - \xi_k)^{-2},$$

while $\omega_F - \lambda_{(N)} > L/\sqrt{N}$ implies

$$R_N > L\sqrt{N}\sum_{k=1}^{N}(\omega_F - \xi_k)^{-2}.$$

Combining this with CLT for R_N and the law of large numbers for $(\omega_F - \xi_k)^{-2}$, we immediately obtain (33).

Using (33) it is easy to prove Theorem 3. In particular, (i) follows from Lemma 2 with $q = 2$ and (33); to prove (ii), let us multiply both sides of (27) by $\sqrt{N}$ and rewrite the obtained equality as

$$\sqrt{N}R_N - \sqrt{N}(\omega_F - \lambda_{(N)})N^{-1}\sum_{k=1}^{N}(\omega_F - \xi_k)^{-2} =$$

$$= \sqrt{N}(\omega_F - \lambda_{(N)})^2 N^{-1}\sum_{k=1}^{N}\int_0^1 d\theta\,(1-\theta)\Big(\omega_F - \xi_k - \theta(\omega_F - \lambda_{(N)})\Big)^{-3}. \tag{35}$$

Applying (33) and Lemma 6 below to (35), we obtain (ii).

LEMMA 6. *If* $1/\kappa = E(\omega_F)$ *and* $\langle(\omega_F - \xi_0)^{-2}\rangle < \infty$, *then*

$$\mathbf{P}\text{–}\lim_N N^{-3/2}\sum_{k=1}^{N}\Big(\omega_F - \xi_k - \big|\omega_F - \lambda_{(N)}\big|\Big)^{-3} = 0. \tag{36}$$

Proof. Let us note first that for any $N = 1, 2, \ldots$, the event

$$\left\{ |\omega_F - \lambda_{(N)}| < 1/2(\omega_F - \xi_{(N)}) \right\}$$

implies that the sum in (36) is dominated by $8\sum_{k=1}^{N}(\omega_F - \xi_k)^{-3}$. Thus, it is sufficient to show that

$$\mathbf{P}\text{--}\lim_N |\omega_F - \lambda_{(N)}|(\omega_F - \xi_{(N)})^{-1} = 0 \tag{37}$$

and

$$\mathbf{P}\text{--}\lim_N N^{-3/2} \sum_{k=1}^{N} (\omega_F - \xi_k)^{-3} = 0. \tag{38}$$

But (37) follows immediately from (33) and Lemma 2 with $q = 2$. The proof of (38) is standard. Lemma 6 is proved.
Let us suppose now that condition (29) is fulfilled. We must show that

$$N^{-1/\alpha}\left|\omega_F - \lambda_{(N)}\right|^{-1} = O(1), \tag{39}$$

$$N^{1/\alpha}\left|\omega_F - \lambda_{(N)}\right| = O(1) \tag{40}$$

and

$$N^{1/\alpha}\left(\lambda_{(N)} - \xi_{(N)}\right) = O(1), \tag{41}$$

$$N^{-1/\alpha}\left(\lambda_{(N)} - \xi_{(N)}\right)^{-1} = O(1) \tag{42}$$

in probability.
In order to prove (39), let us represent (27) in the form (34). Set

$$\begin{aligned} R_N^{(1)} &= \sum_{k=1}^{N} \left(\omega_F - \xi_k\right)^{-1} - NE(\omega_F), \\ R_N^{(2)} &= \sum_{k=1}^{N} \left(\omega_F - \xi_k\right)^{-2}. \end{aligned} \tag{43}$$

In these notations we conclude from (34) that, for any $N = 1, 2, \ldots$ and any $\varepsilon > 0$, under the occurrence of the event

$$\left\{ \left|\omega_F - \lambda_{(N)}\right| < \varepsilon N^{-1/\alpha} \right\} \cap \left\{ \omega_F - \xi_{(N)} > 2\varepsilon N^{-1/\alpha} \right\},$$

the following estimates hold

$$|R_N^{(1)}| \leqslant |\omega_F - \lambda_{(N)}| \sum_{k=1}^{N} \left(\omega_F - \xi_k - \left|\omega_F - \lambda_{(N)}\right| \right)^{-2} \leqslant 4\varepsilon N^{-1/\alpha} R_N^{(2)}.$$

Consequently,

$$\mathbf{P}\left(N^{1/\alpha}\left|\omega_F - \lambda_{(N)}\right| < \varepsilon\right) \leqslant$$

$$\leqslant \mathbf{P}(N^{1/\alpha}|R_{N.}^{(1)}|/R_N^{(2)} < 4\varepsilon) + \mathbf{P}(N^{1/\alpha}(\omega_F - \xi_{(N)}) < 2\varepsilon).$$

Since the sequences $N^{1/\alpha}|R_N^{(1)}|^{-1}$, $N^{-2/\alpha}R_N^{(2)}$ and $N^{-1/\alpha}(\omega_F - \xi_{(N)})^{-1}$ are bounded from above (see (29)), it follows (39).
Since $\omega_F - \lambda_{(N)} < \omega_F - \xi_{(N)} = O(N^{-1/\alpha})$, it suffices for (40) to show that

$$\lim_{L\to\infty} \overline{\lim_N} \mathbf{P}\Big(N^{1/\alpha}(\omega_F - \lambda_{(N)}) < -L\Big) = 0. \tag{44}$$

By (27) we have that the event $\{\omega_F - \lambda_{(N)} < -LN^{-1/\alpha}\}$ implies

$$1/\kappa \leqslant N^{-1} \sum_{k=1}^{N} \Big(\omega_F - \xi_k + LN^{-1/\alpha}\Big)^{-1},$$

which together with the Taylor formula yields $R_N^{(1)} \geqslant LN^{-1/\alpha}D_N$, where

$$D_N = \sum_{k=0}^{N-1} \Big(\omega_F - \xi_{N-k} + LN^{-1/\alpha}\Big)^{-2}.$$

From this we get

$$\mathbf{P}\Big(N^{-1/\alpha}(\omega_F - \lambda_{(N)}) < -L\Big) \leqslant \mathbf{P}\Big(R_N^{(1)} > LN^{-1/\alpha}D_N\Big) \leqslant$$
$$\leqslant \mathbf{P}\Big(N^{-2/\alpha}D_N < L^{-1+\delta}\Big) + \mathbf{P}\Big(N^{-1/\alpha}R_N^{(1)} > L^{\delta}\Big) \tag{45}$$

for any $\delta > 0$. By arguments similar to those used to prove Lemma 3 (**B**), we obtain the following estimate for the first term on the right–hand side of (45)

$$\overline{\lim_N} \mathbf{P}(N^{-2/\alpha}D_N < L^{-1+\delta}) \leqslant$$
$$\leqslant \mathbf{P}\left(\sum_{k\geqslant 1}\Big(\tilde{\gamma}\exp\Big\{\alpha^{-1}\sum_{l=1}^{k} T_l/l\Big\} + L\Big)^{-2} < L^{-1+\delta}\right), \tag{46}$$

where T_l are defined in Lemma 4 and $\tilde{\gamma} > 0$ is some random constant. By Lemma 5, the sum over k in (46) equals

$$L^{\alpha-2}\int_1^{\infty} dk(\gamma k^{1/\alpha} + 1)^{-2}(1 + o(1))$$

as $L \to \infty$. For $\delta \in (0, \alpha - 1)$, we conclude from this that

$$\lim_{L\to\infty} \overline{\lim_N} \mathbf{P}(N^{-2/\alpha}D_N < L^{-1+\delta}) = 0.$$

This limit and the same one for $R_N^{(1)}$ (see (29)) yield (44).
The estimate (41) follows immediately from (40) and the boundedness from above of $N^{1/\alpha}(\omega_F - \xi_{(N)})$.
To show (42) let us represent (27) as

$$1/\kappa = N^{-1}\big(\lambda_{(N)} - \xi_{(N)}\big)^{-1} + N^{-1}\sum_{k=1}^{N-1}\big(\omega_F - \xi_{(N-k)}\big)^{-1} +$$

$$+\big(\omega_F - \lambda_{(N)}\big)N^{-1}\sum_{k=1}^{N-1}\int_0^1 d\theta\left(\omega_F - \xi_{(N-k)} - \theta\big(\omega_F - \lambda_{(N)}\big)\right)^{-2}. \tag{47}$$

Having multiplied both sides of (47) by $N^{1-1/\alpha}$, we get:

$$N^{-1/\alpha}\big(\lambda_{(N)} - \xi(N)\big)^{-1} \leqslant N^{-1/\alpha}|R_N^{(1)}| + N^{-1/\alpha}\big(\omega_F - \xi_{(N)}\big)^{-1} +$$
$$+N^{1/\alpha}|\omega_F - \lambda_{(N)}|\left\{N^{-2/\alpha}R_N^{(2)} + N^{-2/\alpha}Q_N^{(2)}\right\}, \tag{48}$$

where $R_N^{(2)}$ and $Q_N^{(2)}$ are defined in (43) and (10), respectively. Since all the sequences on the right-hand side of (48) are bounded from above (see, e.g., Lemma 3 (**B**) for $Q_N^{(2)}$), it follows (42). Theorem 3 is proved.

3. THE STATEMENTS ON THE LOCALIZATION FOR $\lambda_{(|V|)}$

Let λ be an eigenvalue of operator (2) and $\psi_V(x;\lambda)$, $x \in V$, be the eigenfunction corresponding to λ and defined by (5).
Let us put the variables $\psi(x;\lambda)$, $x \in V$, in the increasing order and denote by $\psi_{(1)}(\lambda) < \ldots < \psi_{(|V|)}(\lambda)$ their order statistics.

Definition. Write $N_V(\varepsilon,\lambda) = \#\{i : \Sigma\psi_{(|V|-i)}(\lambda) > 1-\varepsilon\}$, for any $\varepsilon > 0$. We say that it takes place *complete localization* if

$$\mathbf{P}(N_V(\varepsilon,\lambda) = 1) \to 1 \quad \text{as} \quad N = |V| \to \infty;$$

complete delocalization if

$$\mathbf{P}\text{–}\lim_N N_V(\varepsilon,\lambda)/N > 0;$$

partial localization if

$$\lim_{\varepsilon\downarrow 0}\overline{\lim_N}\,\mathbf{P}(N_V(\varepsilon,\lambda) = 1) = 0$$

and there are nonrandom numbers $r(N)$ such that $r(N) = o(N)$ and

$$\mathbf{P}\text{–}\lim_N N_V(\varepsilon,\lambda)/r(N) > 0.$$

Let us consider the problem of localization for $\lambda_{\max} = \lambda_{|V|}$ from the point of view of limit Theorems 1–3 above.

THEOREM 4. (i) *Suppose that $1/\kappa > E(\omega_F)$ and F satisfies condition* (11). *Then the complete localization takes place.*

(ii) *Let either $1/\kappa < E(\omega_F)$ or $1/\kappa = E(\omega_F)$ and $\langle(\omega_F - \xi_0)^{-2}\rangle < \infty$. Then the complete delocalization is observed.*

(iii) *Let us suppose that $1/\kappa = E(\omega_F)$ and F satisfies condition* (29). *Then one has a partial localization with $\sup r(N) < \infty$.*

Proof. (i) Since

$$\psi^2_{(N)}(\lambda_{(N)}) \geqslant N^{-2}\left(\lambda_{(N)} - \xi_{(N)}\right)^{-2}\left(N^{-2}\left(\lambda_{(N)} - \xi_{(N)}\right)^{-2} + N^{-2}Q_N^{(2)}\right)^{-1}$$

($Q_N^{(2)}$ is defined in (10)), so (i) follows from Theorem 1 (i) and Lemma 3 (**A**).
(ii) Complete delocalization is equivalent to the following relation: for any sufficiently small $\varepsilon > 0$ there is a number $\delta = \delta(\varepsilon) \in (0,1)$ such that

$$\overline{\lim_N} \mathbf{P}\left(\sum_{k>N\delta} \psi^2_{(N-k)}(\lambda_{(N)}) < \varepsilon \right) = 0. \tag{49}$$

We shall prove (49) in the case $1/\kappa = E(\omega_F)$. The case $1/\kappa < E(\omega_F)$ is similar. Clearly,

$$N^{-1}\sum_{k=1}^{N} (\lambda_{(N)} - \xi_k)^{-2} - (\omega_F - \xi_k)^{-2} \leqslant$$

$$\leqslant \sqrt{N}|\omega_F - \lambda_{(N)}|N^{-3/2}\sum_{k=1}^{N}\left(\omega_F - \xi_k - |\omega_F - \lambda_{(N)}|\right)^{-3} = o(1)$$

as $N \to \infty$, where we have used Theorem 3 (ii) and Lemma 6. Thus

$$\sum_{k>[N\delta]}^{N-1} \psi^2_{(N-k)}(\lambda_{(N)}) \geqslant$$

$$\geqslant \sum_{k\geqslant[N\delta]}^{N-1} (\lambda_{(N)} - \xi_k)^{-2} \Big/ \sum_{k=1}^{N}(\lambda_{(N)} - \xi_k)^{-2} = (1-\delta)(1+o(1))$$

as $N \to \infty$. This implies (49) with $\delta = 1 - \sqrt{\varepsilon}$.
(iii) It is sufficient to prove that

$$\mathbf{P}\text{-}\overline{\lim_N}\, \psi^2_{(N)}(\lambda_{(N)}) < 1 \tag{50}$$

and

$$\mathbf{P}\text{-}\lim_{L\to\infty} \overline{\lim_N} \sum_{k=L}^{N-1} \psi^2_{(N-k)}(\lambda_{(N)}) = 0. \tag{51}$$

Let us show (51). Since

$$\sum_{k=L}^{N-1} \psi^2_{(N-k)}(\lambda_{(N)}) \leqslant \sum_{k=L}^{N-1} \left(\lambda_{(N)} - \xi_{(N-k)}\right)^{-2}\left(\lambda_{(N)} - \xi_{(N)}\right)^2,$$

(51) follows now from $N^{1/\alpha}(\lambda_{(N)} - \xi_{(N)}) = O(1)$ and

$$\mathbf{P}\text{-}\lim_{L\to\infty} \overline{\lim_N} N^{-2/\alpha} \sum_{k=L}^{N-1} \left(\xi_{(N)} - \xi_{(N-k)}\right)^{-2} = 0. \tag{52}$$

The proof of (52) is quite similar to that of Lemma 3 (**B**) and can be omitted. Finally, since

$$\psi^2_{(N)}(\lambda_{(N)}) \leqslant N^{-2/\alpha}\big(\lambda_{(N)} - \xi_{(N)}\big)^{-2} \Big/ \Big\{ \big(N^{1/\alpha}(\lambda_{(N)} - \xi_{(N)})\big)^{-2} + \\ +1/2\big(N^{2/\alpha}(\lambda_{(N)} - \omega_F)^2 + N^{2/\alpha}(\omega_F - \xi_{(N-1)})^2\big)^{-1}\Big\}, \tag{53}$$

and all the sequences on the right-hand side of (53) are bounded from above and from zero, we have (50). This completes the proof of Theorem 4.

REFERENCES

Anderson, P.W. (1958). Absence of diffusion in certain random lattices. *Phys. Rev.* **109**, 1492–1505.

Bogachov, L.I. and Molchanov, S.A. (1990). Mean field models in the theory of random media *Teor. Mat. Fiz.* **81**, 281–290, **82**, 143–154.

Galambos, J. (1978). *The Asymptotic Theory of Extreme Order Statistics.* Wiley, New York.

Kac, M. and Thompson, C.J. (1966). On the mathematical mechanism of phase transition. *Proc. Nat. Acad. Sci.* USA, **55**, 676–689.

Zel'dovich, Ya.B., Molchanov, S.A., Ruzmaikin, A.A. and Sokoloff, D.D. (1987). Intermittency in random medium. *Uspekhi Phys. Nauk.* **152**, 3–32.

New Trends in Probab. and Statist., pp. 683 - 693
V. Sazonov and T. Shervashidze (Eds.)
1991 VSP/Mokslas

SCALING BEHAVIOR OF RANDOM WALKS WITH TOPOLOGICAL CONSTRAINTS

S.K. NECHAEV, Ya.G. SINAI
Institute of Chemical Physics of the USSR Academy of Sciences,
117334 Moscow, USSR
Landau Institute for Theoretical Physics of the USSR Academy of Sciences,
117334 Moscow, USSR

ABSTRACT

We study the limiting joint probability distribution for two-dimensional random walks with topological constraints.

1. FORMULATION OF THE PROBLEM AND DESCRIPTION OF THE RESULT

A random path ω on $\mathbf{Z}^2$ going out of $0 \in \mathbf{Z}^2$ can be represented by a continuous function $\omega(t)$ such that $\omega(0) = 0$, $\omega(k) \in \mathbf{Z}^2$ for $k \in \mathbf{Z}^+$, and $\omega(t)$ belongs to a bond of $\mathbf{Z}^2$ for other t. We consider only ω which are closed paths of length $2n$, i.e., $0 \leqslant t \leqslant 2n$ and $\omega(2n) = 0$.
Denote by $G(\mathbf{Z}^2)$ the one-dimensional complex consisting of the unit bonds of $\mathbf{Z}^2$. It is clear that any path ω is a union of such bonds. A deformation of ω is a continuous family $\omega^{(s)}$, $0 \leqslant s \leqslant 1$, such that, for any $s \in [0,1]$, $\omega^{(s)}$ is a continuous mapping of $[0, 2n]$ to $G(\mathbf{Z}^2)$ and $\omega^{(s)}(0) = \omega^{(s)}(2n) = 0$ for all s, $\omega^{(0)} = \omega$.

Definition 1. A path ω is called topologically trivial if there exists a deformation for which $\omega^{(1)}(t) \equiv 0$, $0 \leqslant t \leqslant 2n$.
Fig. 1 gives examples of topologically trivial and non-trivial paths. Take the lattice $\tilde{\mathbf{Z}}^2$ of points $(m + 1/2,\ n + 1/2)$, where $(m, n) \in \mathbf{Z}^2$. For each ω and each $(m + 1/2, n + 1/2)$ one can consider the winding number ρ of ω around this point. A necessary condition for a path to be topologically trivial is the equality $\rho = 0$ for all $(m+1/2,\ n+1/2)$. However it is not sufficient (see Fig. 1c). Denote by Ω_{2n} the set of all topologically trivial paths. Introduce the probability distribution on Ω_{2n} by putting $\mathbf{P}(\omega) = |\Omega_{2n}|^{-1}$. We will call Ω_{2n} with probability distribution $\mathbf{P}(\omega)$ *the random walk with topological constraints* (RWTC).
In (Khokhlov and Nechaev, 1985; Nechaev et al., 1987) it was in fact shown that for any s, $0 \leqslant s \leqslant 1$, there exists a limiting probability distribution for $\omega(2sn)/n^{1/4}$. The appearance of the normalization $n^{-1/4}$ is quite surprising. We explain it in Sections 2 and 3 reproducing the result of (ibidem). Various statistical problems concerning the ensemble Ω_{2n} arise in quantum field theory and physics of polymer

chains (Nechaev, 1988, 1990). It is noteworthy that the $n^{1/4}$-law appears also for randomly branching ideal polymers without any topological constraints (see, for example, (Grosberg and Khokhlov, 1989)).

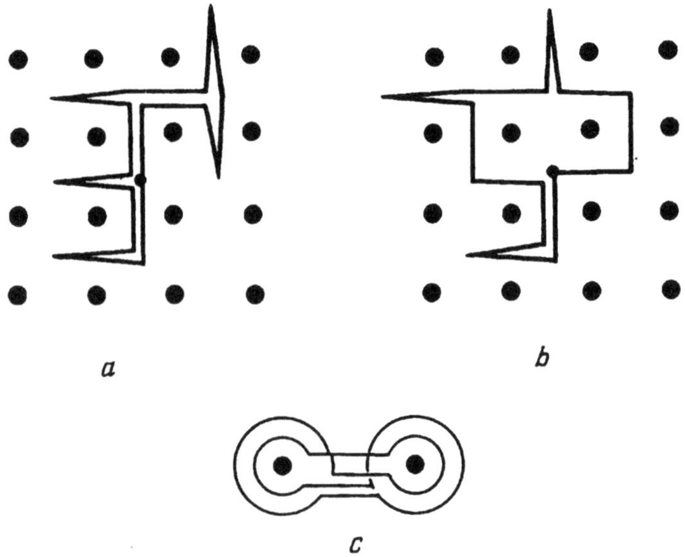

Figure 1. *Closed paths on* $\mathbf{Z}^2$ *with different topological states with respect to the points of dual lattice:* a) *trivial configuration;* b) *non-trivial one;* c) *non-trivial unentangled path (Pochgammer contour)*

The main purpose of this paper is to study the limiting behavior of the process $\omega_n(s) = \omega(2sn)/n^{1/4}$, $0 \leqslant s \leqslant 1$. More precisely, we want to investigate the limiting joint probability distribution of $\omega_n(s_1), \ldots, \omega_n(s_r)$ for any $0 < s_1 < s_2 < \ldots$ $\ldots < s_r < 1$ as $n \to \infty$.

To describe the result, we need to introduce the notion of a *skeleton*. It will be constructed by induction on r. For $r = 1$ a skeleton is an interval of some length $h^{(1)}$ (Fig. 2a). For $r = 2$ it consists of two intervals having a common point as it is shown in Fig. 2b. (The initial point is denoted by 0 in all cases.) The length between the initial and the common points is h_1, the union of this interval and the next one is $h^{(2)}$. Assume that we have defined skeletons for $r - 1$ as unions of intervals. The isolated end points of these intervals are denoted by $h^{(1)}, h^{(2)}, \ldots, h^{(r-1)}$. The end points which belong to two intervals are denoted by $h_1, h_2, \ldots, h_{r-2}$. We assume that the $(r-1)$-skeleton is defined completely by its lengths $h^{(1)}, \ldots, h^{(r-1)}$; $h_1, \ldots, h_{r-2}$. The angles between intervals are unimportant and only linear sizes are essential. The r-skeleton is constructed by adding an interval passing through an inner point of the $(r-1)$-skeleton. Points of a skeleton can be isolated end points $h^{(1)}, \ldots, h^{(r)}$ or branching points $h_1, \ldots, h_{r-1}$. The notations $h^{(r)}$ and h_r mean also the distances from the initial point to these ones (along the intervals). A fragment of a skeleton is an interval whose end points are the points of the skeleton and which does not contain any branches. Note that the skeleton obtained does not contain cycles.

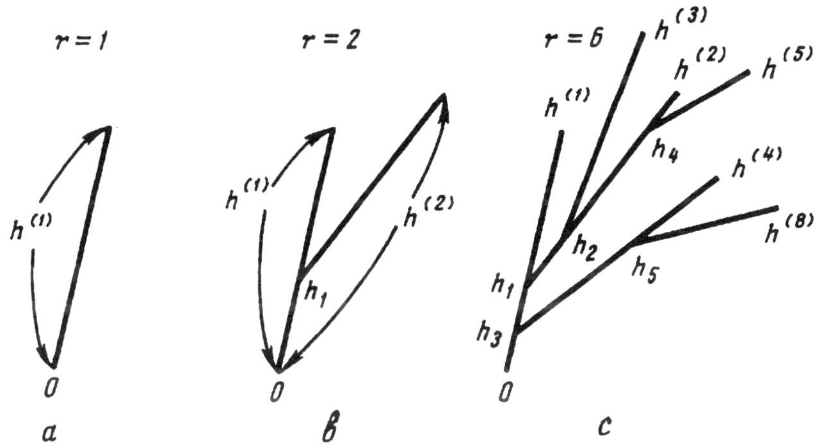

Figure 2. *Illustration to recurrent construction of the skeletons (see the text)*

Let H_r be an r-skeleton and $D_j = (d'_j, d''_j)$ be its fragments, $j = 1, \ldots, J$. A branching Wiener process with time H_r is the space of continuous mappings $w : H_r \to \mathbf{R}^2$ such that:

1. $w(0) = 0$;
2. For any fragment D_j the induced measure on the space of functions $w(d)- -w(d'_j)$, $d \in D_j$, is the Wiener measure with parameter σ not depending on j;
3. For different D_j the processes $w(h) - w(d'_j)$, $h \in D_j$, are mutually independent.
4. $w(h'_{j_1}) = w(h''_{j_2})$ if $d'_{j_1} = d''_{j_2}$.

In fact, 4 follows from the continuity of w. The meaning of this definition is the following. First we take a Brownian path on the interval $[0, h^{(1)}]$. Then we add an independent path which bifurcates from the first one at time h_1 and goes for the interval of time $[h_1, h^{(2)}]$ and so on. A typical curve is presented in Fig. 3, where the marked points correspond to bifurcating moments.

Return to our problem. Our main result (Section 4) shows that the joint probability distribution of $\omega_n(s_1), \ldots, \omega_n(s_r)$ can be described in the following way. We consider the space $\mathcal{H}^{(r)}$ of all r-skeletons H_r and define some probability measure on this space. Then the limiting conditional distribution of $\omega_n(s_1), \ldots, \omega_n(s_r)$ as $n \to \infty$ is the distribution of the end points $w(h^{(1)}), \ldots, w(h^{(r)})$ of the branching Wiener process for which H_r is also random and is distributed according to the measure mentioned above.

The derivation of the main result is based upon the reduction of our problem to the investigation of some statistical properties of closed random walks on the Cayley tree. This reduction is described in Section 2. For the first time it appeared in

(Khokhlov and Nechaev, 1985). In Section 3 we study some statistical properties of closed random walks on the Cayley tree. All results we derive for the general p-adic Cayley tree taking into account the fact that our case of $\mathbf{Z}^2$-lattice corresponds to the value $p = 3$. Apparently the statements presented here are particular cases of much more general ones. We hope to discuss them in future. As was already mentioned, the main result is derived in Section 4.

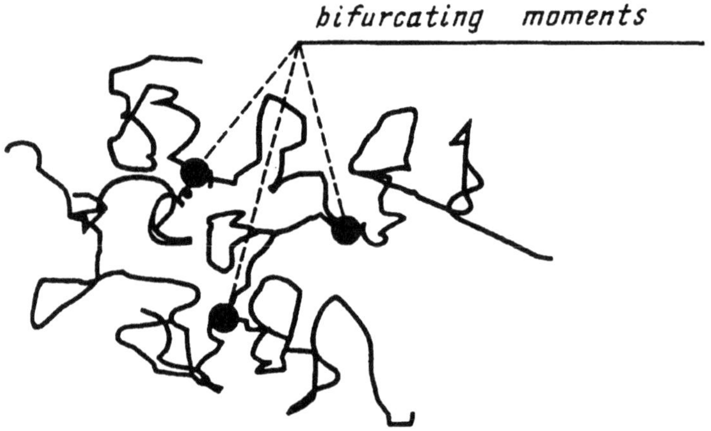

Figure 3. *Branching Wiener path on* $\mathbf{Z}^2$

It is our pleasure to dedicate this paper to Yu.V. Prokhorov.

2. CONNECTION OF RANDOM WALKS ON CAYLEY TREE WITH RWTC

The connection of random walks with topological constraints on $\mathbf{Z}^2$ with random walks on the Cayley tree (Fig. 4) for the first time was considered in (ibidem, see also Grosberg and Khokhlov, 1989). Now this connection will be described more precisely.

The Cayley tree which we will consider is a universal covering of $G(\mathbf{Z}^2)$. Let us describe its construction.

First of all fix the starting point α_0 for four bonds $e^{(1)}, \dots, e^{(4)}$. We denote by $\alpha_j(1)$ $(1 \leqslant j \leqslant 4)$ the vertices e_j connected by the bonds $e^{(1)}, \dots, e^{(4)}$ with the point α_0. They will be called the "first floor vertices". Every first floor vertex is the starting point for three new bonds which belong to the second floor, their ends being denoted by $\alpha_j(2)$ $(1 \leqslant j \leqslant 3)$. The total number of vertices built on both floors is 12. Using the induction, we construct from the k-th floor vertices the vertices of the $(k+1)$-th floor by taking three new vertices going out from each vertex of the k-th floor. The whole graph G is a union of the vertices of each floor, i.e., $G = \bigcup_{k=1}^{\infty} G_k$. Let $G^{(1)}$ be the set of connecting bonds involved in the construction of G. For each point $\alpha \in G$ we denote its floor by $h(\alpha)$. For any two points α_0 and $\alpha \neq \alpha_0$ there exists the minimal path connecting α_0 and α. By analogy with the papers (Khokhlov and Nechaev, 1985; Nechaev et al., 1987), we will call it "the primitive path" and denote by γ_α. This path plays the role of geodesics on our tree and has the length $h(\alpha)$. A natural parameterization of this primitive path is $\gamma_\alpha(s)$ $(0 \leqslant s \leqslant h)$.

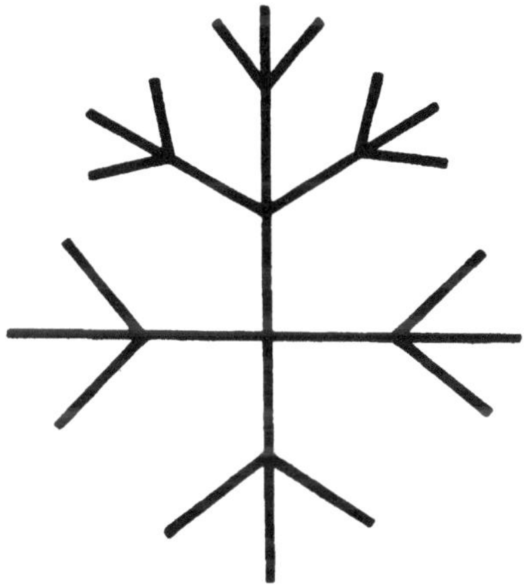

Figure 4. *Cayley tree with* $p = 3$

Now construct the canonical mapping π of the graph G to $\mathbf{Z}^2$ which generates also the mapping of $G^{(1)}$ to $G(\mathbf{Z}^2)$. Any path γ_α will be mapped to a path Φ_α on $G(\mathbf{Z}^2)$. If the first bond of γ_α is $e^{(j)}$ then the first bond of Φ_α is also $e^{(j)}$. We apply the induction again. Let k bonds of the path Φ_α which are the images of k steps of the path γ_α be constructed and let the last bond of γ_α be $e^{(j)}$. Then the path Φ_α can be continued in three possible ways by means of addition of one of three new oriented links for which the direction $e^{(j)}$ is forbidden. Our construction is finished as soon as the point α is reached. By definition, $\pi(\alpha)$ is the last point of Φ_α. It is clear that π can be extended to the mapping of the one-dimensional complex $G^{(1)}$ to $G(\mathbf{Z}^2)$ because for any two points α' and α'' which are the ends of one bond on $G^{(1)}$ their images are the end of one bond on $G(\mathbf{Z}^2)$.
The following topological properties of π are important for our purposes.

COROLLARY A. *If ω is an arbitrary path on $G(\mathbf{Z}^2)$ starting at the point 0, then there exists a lifted path $\Omega = \pi^{-1}(\omega)$ which corresponds uniquely to ω.*

In topology the corresponding general statement is known as the theorem about covering homotopy.

COROLLARY B. *The transition to the Cayley tree is important because a path ω beginning and ending at the point 0 on $\mathbf{Z}^2$ is topologically trivial if and only if $\Omega = \pi^{-1}(\omega)$ is a closed path on $G^{(1)}$. Moreover the connection between RWTC and closed paths on the Cayley tree is one-to-one.*

In this section we describe briefly the basic statistical properties of π. Let $\alpha \in G_m$ and $\gamma_\alpha(s)$ be the primitive path,

$$\gamma_\alpha^{(m)}(t) = \frac{\pi(\gamma_\alpha(tm))}{\sqrt{m}}, \quad 0 \leqslant t \leqslant 1, \quad t = \frac{q}{m},$$

where q is an integer.
Connecting the neighboring points $\gamma_\alpha^{(m)}(\frac{q}{m})$ and $\gamma_\alpha^{(m)}(\frac{q+1}{m})$ by straight segments, we get a continuous curve on $\mathbf{R}^2$ starting at the point 0. The space of such paths is denoted by $\Omega_m(G^{(m)})$. It is clear that $\gamma_\alpha^{(m)} \subset C([0,1])$, where $C([0,1])$ is the space of continuous mappings of the segment $[0,1]$ to $\mathbf{R}^2$. Supposing that for any m the values of α and $h(\alpha)$ are uniformly distributed, we obtain a probability measure on the space $\mathbf{C}[0,1]$. This distribution is denoted by $\mathfrak{P}_m$.

THEOREM 1. *The distributions $\mathfrak{P}_m$ converge weakly to the Wiener measure on the plane as $m \to \infty$.*

Note that, for this Wiener measure, the increments $\gamma(t_1)-\gamma(t_0),\ \gamma(t_2)-\gamma(t_1),\dots$ $\dots,\gamma(t_r)-\gamma(t_{r-1})$ are mutually independent for any $t_0 < t_1 < \dots < t_r$, and every increment $\gamma(t_k)-\gamma(t_{k-1})$ has a symmetric Gaussian distribution with $\mathbf{E}(\gamma(t_k)-$ $-\gamma(t_{k-1}))=0$, $\mathbf{E}\|\gamma(t_k)-\gamma(t_{k-1})\|^2 = \sigma(t_k - t_{k-1})$ for some $\sigma = \text{const} > 0$.

Proof of Theorem 1 is quite simple. Consider the following Markov chain with four different states $e^{(1)},\dots,e^{(4)}$ which are the unit vectors of the initial $\mathbf{Z}^2$-lattice. From any state the moving point passes to any of the three remaining states with equal transition probabilities 1/3. The invariant distribution of the Markov chain is uniform. Let us consider the space M_m of realizations of the Markov chain of length m. The points $u \in U_m$ have the form $u = (e_1, e_2, \dots, e_m)$, where $e_k \in E = \{e^{(1)},\dots,e^{(4)}\}$. The space M_m is in one-to-one correspondence with U_m. This correspondence arises if e_j is considered as the increment $\gamma(j)-\gamma(j-1)$. The vector $\gamma_\alpha^{(m)}\left(\frac{q}{m}\right)$ has the form $\left(\frac{\nu_1(q)-\nu_3(q)}{\sqrt{m}}, \frac{\nu_2(q)-\nu_4(q)}{\sqrt{m}}\right)$, where $\nu_j(q)$ is the number of appearances of the state e_j in the time interval $[0,m]$. Now the statement of Theorem 1 follows from the central limit theorem for our Markov chain.

3. SOME STATISTICAL PROPERTIES OF RANDOM WALKS ON CAYLEY TREE

In this section we study the probability distributions of closed random walks starting at α_0 and returning to the initial point on the Cayley tree at $t = 2n$. Denote this random walk by $\Gamma = \{\gamma(t),\ 0 \leqslant t \leqslant 2n\}$. Fix s, $0 \leqslant s \leqslant 1$, and consider the random variable $h_n(s) = h(\gamma[2ns])$.

THEOREM 2. *Let z belong to a fixed interval $a_1 \leqslant z \leqslant a_2$ $(0 < a_1 < a_2 <$ $< \infty)$, $z = h/\sqrt{n}$, where h is an integer and a_1, a_2 are constants; let also $m \sim 2ns$. Then*

$$\mathbf{P}\{h_n(s) = h\} \sim \frac{1}{24\pi s^{3/2}(1-s)^{3/2}}(3/4)^n z^2 n^{-2} \exp\left\{-\frac{z^2}{4s(1-s)}\right\}$$

as $n \to \infty$.

Proof. Let $Z(m,h)$ be equal to the number of arbitrary m-step paths starting at α_0 and ending at an arbitrary point α of height h on the Cayley tree at the moment m. Then $Z(m,h)$ satisfies the following recurrent equations (boundary and initial conditions are included):

$$\begin{aligned} Z(m+1,h) &= p\,Z(m,h-1) + Z(m,h+1) \quad \text{for} \quad h \geqslant 2, \\ Z(m+1,1) &= (p+1)\,Z(m,0) + Z(m,2), \\ Z(m+1,0) &= Z(m,1), \\ Z(0,h) &= \delta(h), \end{aligned} \tag{1}$$

where p is the number of branches going out of one point of the Cayley tree to the next floor. In our case $p = 3$, but our arguments work for any $p \geqslant 2$, i.e., for any p-adic simple tree.

The substitutions

$$\begin{aligned} Z(m,h) &= 2^m\ p^{m/2+h/2}\ Q(m,h), \quad h \geqslant 1, \\ Z(m,0) &= 2^m\ p^{m/2}\frac{p}{p+1}Q(m,0) \end{aligned} \tag{2}$$

transform equations (1) to the following symmetric form:

$$\begin{aligned} Q(m+1,h) &= \frac{1}{2}Q(m,h-1) + \frac{1}{2}\,Q(m,h+1), \quad h \geqslant 1, \\ Q(m+1,0) &= \frac{1}{2}\frac{p+1}{p}\,Q(m,1), \\ Q(0,h) &= \frac{p+1}{p}\delta(h). \end{aligned} \tag{3}$$

As $m \to \infty$, the solution of equations (3) has the following asymptotics (Nechaev et al., 1987)

$$Q(m,h) \sim 2\frac{p-1}{p}\ \frac{2^{-m}m!}{(m/2+h/2)!(m/2-h/2)!}\ \frac{h+(p+1)/(p-1)}{m+h}. \tag{4}$$

Here it is assumed that $(m-h)/2$ is an integer.

For $1 \ll h \ll m$ we easily derive

$$Q(m,h) \sim 2\,\frac{p+1}{p}\sqrt{\frac{2}{\pi}}\frac{h}{m^{3/2}}\exp\left(-\frac{h^2}{2m}\right). \tag{5}$$

Now define $Z(2n,m,h)$ to be equal to the total number of trajectories consisting of $2n$ steps starting and ending at the point 0 with the condition that at the m-th step $(0 \leqslant m \leqslant 2n)$ the moving point belongs to the h-th floor of the tree. Then

$$Z(2n,m,h) = \frac{Z(m,h)Z(2n-m,h)}{(p+1)p^{h-1}}. \tag{6}$$

This is true because every such closed path consists of two parts: the part going from α_0 to α and having m steps and the path going back and having $2n-m$ steps. In the denominator we have the number of primitive paths going from α_0 to the h-th floor. Naturally, for the first step of the primitive path there are $p+1$ different ways to go to the second floor and for each of the remaining $h-1$ steps there are only p possibilities. Thus, for $Z(2n,m,h)$ we obtain the asymptotic expression

$$Z(2n,m,h) \sim \frac{2}{\pi}\frac{p-1}{p(p+1)}(4p)^n \frac{h^2}{m^{3/2}(2n-m)^{3/2}} \exp\left\{-\frac{h^2 n}{m(2n-m)}\right\} \qquad (7)$$

which for $p=3$ and $h=\sqrt{n}\,z$, $m \sim 2ns$, $n \to \infty$, gives the expression in the formulation of the theorem.

4. SCALING BEHAVIOR OF RWTC

Take s, $0 \leqslant s \leqslant 1$, fix $\gamma([2ns]) = \gamma_s$ and $h([2ns]) = h_s$. Consider the primitive path $\eta_s(\Gamma) = \eta_s$ from α_0 to γ_s. The intersection of Γ with η_s can be considered as a random path $\zeta(u)$ along η_s. Denote by $2\tau_s(\Gamma) = 2\tau_s$ the step of this path $\zeta(u)$. In other words, $2\tau_s$ is the number of t, $0 \leqslant t \leqslant 2n$, such that $\gamma(t) \in \tau_s$. It is clear that $2\tau_s \geqslant 2\eta_s$. We will prove the following surprising

THEOREM 3. *Let $n \to \infty$. Assume that $a_1\sqrt{n} \leqslant h_s \leqslant a_2\sqrt{n}$ for some positive $a_1 < a_2$. Consider the conditional distribution $\mathbf{P}$ of $2\tau_s^{(1)} = 2\varkappa\tau_s - 2h_s$, where $\varkappa = (1-(\log 2)/\log(4p))$ and γ_s is fixed. Then*

$$\mathbf{P}(\tau_s^{(1)} \geqslant k) \leqslant \text{const}\cdot\rho^k$$

for some $\rho < 1$.

Proof. The total number N of random paths on G, for which γ_s, h_s, and τ_s are fixed, is equal to

$$N(\tau_s) = \sum_\zeta \sum\nolimits'_{k_0,\dots,k_{2\tau_s-1}} \prod_{j=0}^{2\tau_s-1} N_{2k_j}(\zeta(j)),$$

where $\sum'$ denotes the summation under the condition

$$\sum_{j=0}^{2\tau_s-1} 2k_j = 2n - 2\tau_s,$$

$N_{2k}(\alpha)$, $\alpha \in \eta_s$, is the number of closed paths of length $2k$ going out of α, lying outside $\eta_s - \alpha$ and returning back to α after $2k$ steps. It is clear that, if $h(\alpha) \geqslant 1$, then this number does not depend on α. For $\alpha = \alpha_0$ it is slightly different. Write now N_{2k} in the form $N_{2k} = (4p)^k\, n_{2k}(\alpha)$.

LEMMA. *The following equalities are valid:*

$$\sum n_{2k}(\alpha) = 2p/(p+1) \quad \text{for } \alpha \neq \alpha_0$$

and

$$\sum n_{2k}(\alpha_0) = 2 \quad \text{for } \alpha = \alpha_0;$$

also

$$n_{2k} \sim \text{const} \cdot k^{-3/2} \quad \text{as } k \to \infty \text{ for all } \alpha.$$

Proof. For the initial point of a primitive path the number of possible branches, going out of α_0 and lying outside of the primitive path, is p, whereas for intrinsic points it is $p-1$ (in our case of $\mathbf{Z}^2$-lattice, $p=3$). Let $U_{2k}(\alpha)$ be the number of paths lying outside of the primitive path, beginning at a point α of this path and returning the first time to α after $2k$ steps. Let us write $U_{2k}(\alpha)$ in the following form:

$$U_{2k}(\alpha) = \begin{cases} (1/2)p^k u(2k) & \text{for } \alpha = \alpha_0, \\ (1/2)(p-1)p^{k-1}u(2k) & \text{for } \alpha \neq \alpha_0,\ \alpha \in h, \end{cases} \tag{8}$$

where $u(2k)$ is the number of trajectories of a simple one-dimensional random walk returning to the initial point for the first time at the $2k$-th step. The factor p^k in (8) is connected with the fact that there are p different possibilities per one step to go away from the initial point. The function $u(2k)$ can be written as $u(2k) = 2^{2k}\, v(2k)$, $\sum_k v(2k) = 1$ and $v(2k) \sim \text{const}\, k^{-3/2}$ as $k \to \infty$.
The numbers $N_{2k}(\alpha)$ and $U_{2k}(\alpha)$ are connected by the usual renewal equation

$$N_{2k}(\alpha) = \sum_{i=0}^{k} U_{2k}(\alpha) N_{2(k-i)}(\alpha) \tag{9}$$

with the boundary conditions $N_0(\alpha) = 1$ and $U_0(\alpha) = 0$. From equations (8) and (9) we obtain immediately:

$$\sum_{k=0}^{\infty} n_{2k}(\alpha) = \sum_{k=0}^{\infty} N_{2k}(\alpha)(4p)^{-k} \begin{cases} 2 & \text{for } \alpha = \alpha_0 \\ 2p/p+1 & \text{for } \alpha \neq \alpha_0, \quad \alpha \in h. \end{cases} \tag{10}$$

Equation (10) completes the proof of Lemma 1. Now let us write $N(\tau_s)$ as follows:

$$N(\tau_s) = 2^{2\tau_s}(4p)^{2n-2\tau_s} \sum_{b=0}^{2\tau_s-2h_s} 2^b \left(\frac{2p}{p+1}\right)^{2\tau_s-b} \times$$

$$\times \sum_{k_0+\ldots+k_{2\tau_s}-1=n-\tau_s} p_{2k_0}(x_0)\ldots p_{2k_{2\tau_s}-1}(x)$$

Here $p_{2k}(\alpha_0) = n_{2k}(\alpha_0)/2$, $p_{2k}(\alpha) = n_{2k}(\alpha)(p+1)/2p$, $\alpha \neq \alpha_0$, and b is the number of visits to α_0. It is clear that $0 \leqslant b \leqslant 2\tau_s - 2h_s$. The sum

$$\sum_{k_0+\ldots+k_{2\tau_s-1}=n-\tau_s} p_{2k_0}(x_0)\ldots p_{2k_{2\tau_s-1}}(x)$$

is the convolution of b distributions $\{p_{2k}(\alpha_0)\}$ and $2\tau_s - b$ distributions $\{p_{2k}(\alpha)\}$. Both distributions belong to the domain of attraction of a stable law with exponent

1/2. Thus, from the local limit theorem for this distribution (see, for example (Feller, 1962)), it follows easily that

$$\sum_{k_0+\ldots+k_{2\tau_s-1}=n-\tau_s} p_{2k_0}(x_0)\ldots p_{2k_{2\tau_s-1}}(x) \sim F\Big(\frac{n-\tau_s}{\tau_s^2}\Big)\frac{1}{\tau_s^2}, \quad n\to\infty, \tag{11}$$

where F is the density of this stable distribution, $F(x) = \text{const}\cdot x^{-3/2}\exp\left\{-\frac{\text{const}}{x}\right\}$. This yields

$$N(\tau_s) \sim (4p)^{2n-2\tau_s}\sum_{b\geqslant 0}(4p)^{-b}(4p)^{-(2\varkappa\tau_s-2h_s-b)}2^b\Big(\frac{2p}{p+1}\Big)^{2\tau_s-b}F\Big(\frac{n-\tau_s}{\tau_s^2}\Big)\frac{1}{\tau_s^2}.$$

We see that the last expression takes its maximal value when $2\tau_s - 2h_s = 0$. Therefore,

$$\frac{N(\tau_s)}{N(2h_s)} \leqslant \text{const}\cdot\rho^{2\varkappa\tau_s-2h_s}, \qquad \rho=\frac{1}{4p}, \qquad \varkappa = 1-(\log 2)/\log(4p).$$

Theorem 1 is proved.
Take two points α' and $\alpha'' \in G$; $\gamma(\alpha',\alpha'')$ is the shortest primitive path on G connecting α' and α''. Let η be its length. Consider the ensemble of random walks on G going out of α' and coming to α'' at $t = O(h^2)$, let τ be the random variable equal to the number of steps which our random walk spends in $\gamma(\alpha',\alpha'')$. The same arguments as in the proof of Theorem 3 give

THEOREM 3$'$. *There exists $\rho \leqslant 1$ such that*

$$\mathbf{P}\{\tau - h \geqslant k\} \leqslant \text{const}\cdot\rho^k.$$

Return to our situation. Take $0 < s_1 < s_2 < \ldots < s_r < 1$ and fix $\gamma_{s_j} = \gamma([2ns_j])$, $1 \leqslant j \leqslant r$. Connect the points γ_{s_j} by primitive paths $\gamma(\eta_{s_j},\eta_{s_{j+1}})$. There arise branching points γ_1,γ_j which are the last intersection of $\gamma(\eta_{s_j})$ and $\gamma(\eta_{s_{j+1}})$. Considering the heights of all points $\gamma(s_j)$, we get a skeleton H_r (see Section 1 and Fig. 2). For any path Γ, for which $\gamma([2ns_j]) = \gamma s_j$, its intersection with H_r is a random walk along H_r. Each fragment D_j of H_r is covered by this walk even number of times (equal times in the opposite directions). Divide [0,1] to the intervals $s^{(0)} = 0 < s^{(1)} < s^{(2)} < \ldots < s^{(m)} < s^{(m+1)} = 1$, where $[s^{(j)}, s^{(j+1)}]$ is a time interval which the moving point spends passing along one of the fragments of H_r, i.e., the terminal points $h^{(1)}, h^{(2)},\ldots,h^{(r-1)}$ of the skeleton are reached at the moments $s^{(1)}, s^{(2)},\ldots,s^{(r-1)}$ and $s^{(j)} \leqslant s_j \leqslant s^{(j+1)}$.
Consider now the ensemble $E(H_r, s^{(1)}, s^{(2)},\ldots,s^{(m)})$ of all paths Γ generating an r-skeleton (i.e., inducing a measure on the ensemble of skeletons $\mathcal{H}^{(r)}$). Denote $E_r = \bigcup_{j=1}^r \Gamma$ for each skeleton H_r. It is easy to show that the conditional probability distribution on $\pi(E_r)$ converges to a branching Wiener process with "time" H_r. A small strengthening of theorems 3 and 3$'$ gives

THEOREM 4. *The following equality is valid*

$$\left|E(H_r, s^{(1)}, s^{(2)}, \dots, s^{(m)})\right| =$$

$$= F\left(\frac{2ns_1}{h^2(\gamma^{(1)})}\right)\frac{1}{h^2(\gamma^{(1)})} \cdots F\left(\frac{2n(s_r - s_{r-1})}{\left(h(\gamma^{(r)}) - h(\gamma_r)\right)^2}\right)\frac{1}{\left(h(\gamma^{(r)}) - h(\gamma_r)\right)^2}.$$

This theorem shows that if we introduce the statistical weight

$$w\left(H_r, s^{(1)}, s^{(2)}, \dots, s^{(m)}\right) = \left|E(H_r, s^{(1)}, s^{(2)}, \dots, s^{(m)})\right|,$$

then it converges, as $n \to \infty$, to some probability measure on the space of framed r-skeletons $\left(H_r, s^{(1)}, s^{(2)}, \dots, s^{(m)}\right)$ which is the marginal distribution for our conditional branching Wiener process. It can be directly proved that if we sum $w(H_r, s^{(1)}, s^{(2)}, \dots, s^{(m)})$ over all variables s_j and $\Delta h_j = h(\gamma^{(j)}) - h(\gamma_j)$ $(1 \leqslant j \leqslant r)$, we obtain

$$\sum_{\{s_j\}} \sum_{\{\Delta h_j\}} w\left(H_r, s^{(1)}, s^{(2)}, \dots s^{(m)}\right) = 1 + o(1), \quad n \to \infty.$$

Thus we can define our RWTC as a Wiener process with a "randomly branching time".

REFERENCES

Khokhlov, A.R. and Nechaev, S.K. (1985). Polymer chain in an array of obstacles. *Phys. Lett.* **112** A, 156.

Nechaev, S.K., Semenov, A.N., and Koleva, M.K. (1987). Dynamics of a polymer chain in an array of obstacles. *Physica,* **140** A, 506.

Nechaev, S.K. (1988). Topological proporties of a two-dimensional polymer chain in the lattice of obstacles. *J. Phys.* **21** A, 3659.

Nechaev, S.K. (1990). Overview in polymer topology. *Int. J. Mod. Phys. (to appear).*

Grosberg, A.Yu. and Khokhlov, A.R. (1989). *Statistical Physics of Macromolecules.* Nauka, Moscow (in Russian).

Feller, W. (1962). *An Introduction to Probability Theory and its Applications.* **1**, 2-nd ed. Wiley, New York.

New Trends in Probab. and Statist., pp. 694 - 699
V. Sazonov and T. Shervashidze (Eds.)
1991 VSP/Mokslas

CONVERGENCE OF THE STOCHASTIC QUANTIZATION METHOD FOR LATTICE R–GAUGE THEORIES

V. SIDORAVIČIUS
Institute of Mathematics and Informatics of the Lithuanian Academy Sci., Akademijos st. 4, 232600 Vilnius, Lithuania

ABSTRACT

In the present paper the evolution of Langevin's stochastic differential equations for **R**–gauge fields is considered. We prove convergence of the method of stochastic quatization for small perturbations of Yang–Mills–Wilson potential in infinite volume.

INTRODUCTION

Lattice gauge theories were first studied by Wegner under a different name for the gauge group $\mathbf{Z}_2$. He was interested in generalizations of the Ising model which possess phase transitions without a local order parameter. A few years later Wilson introduced a rather general class of lattice gauge theories in order to understand the permanent confinement of quarks.
There are two review articles on the subject, (see (Drouffe and Itzykson, 1974; Kogut, 1979)), and the monograph (Seiler, 1982). Both articles contain a lot of useful information and are extremely valuable introductions into the subject. The book of Seiler is based on the lattice approximation method and contains rigorously established results of physical relevance. This method provides a tool for the construction of continuum models (see (Seiler, 1982)).
But at this time there are a large number of mathematically rigorous work on the various aspects of this subject.
In the present paper we consider a noncompact $\mathbf{Z}^d$ lattice gauge theory with Abelian gauge group **R** (see (Zinovjev, 1981)). We study the behaviour of the limiting Gibbsian measures on the gauge fields and use the stochastic quantization method (see (Damgaard and Hüffel, 1987; Parizi and Wu, 1981; Migdal, 1986)).
This method was first introduced by physicists (see (Parisi and Wu, 1981)), and its main idea is to view Euclidean quantum field theory as the equilibrium limit of a statistical system coupled to a thermal reservoir. This system evolves in a new fictitious time direction t until it reaches the equilibrium limit as $t \to \infty$. The coupling to a heat reservoir is simulated by means of a stochastic noise field which forces the original Euclidean field to wander randomly on its manifold.

The main result of the present paper is the proof of convergence for infinite system of Langevin's equations associated with Yang–Mills–Wilson action on the cubic lattice $\mathbf{Z}^d$, $d \geqslant 3$, with additional α–term, $\alpha > 0$ (see (Borgs and Nill, 1986)), moreover, we show that the expectation of gauge–invariant observables with respect to Gibbsian measure does not depend on additional α–term.

1. THE BASIC DEFINITIONS OF LATTICE R–GAUGE THEORIES

Let $\Lambda_n \subset \mathbf{Z}^d$, $d \geqslant 3$, be the cube centered around the origin with edges of length $2n+1$.

The function $\xi = \{\xi_x\}$ on Λ_n with values in $\mathbf{R}^d$ is called a "field configuration" of $\mathbf{R}$–gauge field on the Λ_n (see (Malyshev and Minlos, 1985; Brydges, Frölich and Seiler, 1978; Osterwalder and Seiler, 1978)), i.e., the variable at the point $x \in \Lambda_n$ is a vector $(\xi_{x,1}, \dots, \xi_{x,d})$, $\xi_{x,j} \in \mathbf{R}$, and we can associate each component variable $\xi_{x,j}$ with directed lattice bound $(x, x+\vec{e}_j)$ in volume Λ_n, where $\vec{e}_j$ is the unit vector on j–axis.

With any subset $\Lambda_n \subset \mathbf{Z}^d$ we associate a polynomial S^{Λ}_{YMW} (the Yang–Mills–Wilson action on the lattice), depending on variables $\xi_{x,j} \in \mathbf{R}$, $x \in \Lambda_n$,

$$S^{\Lambda}_{YMW}(\xi) = \sum_{x\in\Lambda,\ i,j=1,\dots,d} (\xi_{x,j} + \xi_{x+\vec{e}_j,i} - \xi_{x+\vec{e}_j,j} - \xi_{x,i})^2 .$$

In this paper only periodical border conditions are considered.

Unfortunately, the statistical sum $\mathbf{Z}^{\Lambda}_{YMW}$ of this model does not satisfy the stability condition:

$$\mathbf{Z}^{\Lambda}_{YMW} = \int_{\mathbf{R}^{d\cdot|\Lambda|}} \exp\{-S^{\Lambda}_{YMW}\{\xi\}\} \prod_{x\in\Lambda} \prod_{j=1}^{d} d\xi_{x,j} = \infty,$$

where $\prod_{x\in\Lambda} \prod_{j=1}^{d} d\xi_{x,j}$ is the product of Lebesgue measures $d\xi_{x,j}$.

In this situation we shall introduce a new term in the action S^{Λ}_{YMW} and consider new action

$$S^{\Lambda}_{\alpha} = S^{\Lambda}_{YMW} + \frac{1}{\alpha} S^{\Lambda}_{N},$$

where

$$S^{\Lambda_n}_{N} = \sum_{x\in\Lambda_n} \left(\sum_{j=1}^{d} (\xi_{x,j} - \xi_{x-\vec{e}_j,j}) \right)^2 ,$$

and $\alpha > 0$ is fixed (see (Borgs and Nill, 1986)).

With action $S^{\Lambda_n}_{\alpha}$ we define a Gibbs measure

$$d\mu^{\Lambda}_{\alpha} = \frac{1}{\mathbf{Z}^{\Lambda}_{\alpha}} \exp\{-S^{\Lambda_n}_{\alpha}\} \prod_{x\in\Lambda} \prod_{j=1}^{d} d\xi_{x,j}$$

on $\mathbf{R}^{d\cdot|\Lambda|}$, where

$$\mathbf{Z}_{\alpha}^{\Lambda} = \int_{\mathbf{R}^{d\cdot|\Lambda|}} \exp\left\{ - S_{YMW}^{\Lambda_n} - \frac{1}{\alpha} S_{N}^{\Lambda_n} \right\} \prod_{x\in\Lambda} \prod_{j=1}^{d} d\xi_{x,j} < \infty.$$

We define a local observable on Λ_n to be a function F in $L_0(\mathbf{R}^{d\cdot|\Lambda|})$ which depends only on finitely many variables ξ_x. The expectation of F with respect to the measure $\mu_{\alpha}^{\Lambda_n}$ is defined by

$$\langle F\rangle_{\Lambda,\alpha} = \int F d\mu_{\alpha}^{\Lambda_n}.$$

A gauge transformation is a map h from lattice points $x \in \mathbf{Z}^d$ into the group $\mathbf{R}$. Associated with h there is a transformation of the field variables:

$$\xi_{x,j} \rightarrow h\xi_{x,j} = h_x + \xi_{x,j} - h_{x+\vec{e}_j}.$$

We denote by T_g the set of all gauge transformations.
An observable F is gauge invariant if, for all gauge transformations $h \in T_g$, we have

$$F(h(\xi)) = F(\xi).$$

By $\mathcal{A}_{\Lambda_n}$ we denote the algebra of local observables F, which depends on variables $\xi_{x,j}$, $x \in \Lambda_n$, and denote by $\mathcal{A}^0$ all local observables, i.e.,

$$\mathcal{A}^0 = \bigcup_{\Lambda_n} A_{\Lambda_n}.$$

Finally the norm closure of $\mathcal{A}^0$ is a quasi-local algebra denoted by

$$\mathcal{A} = \overline{\bigcup_{\Lambda_n} \mathcal{A}_{\Lambda_n}}.$$

By $\mathcal{A}_{\Lambda_n}^{INV}$ we denote the algebra of gauge invariant observables in volume Λ_n and denote

$$\mathcal{A}^{INV} = \overline{\bigcup_{\Lambda_n} \mathcal{A}_{\Lambda_n}^{INV}},$$

respectively.

2. THE MAIN RESULT

Definition 1 (see (Malyshev, Ignatiuk and Sidoravičius, 1990)). By the system of Langevin's equations for the measure $\mu_{\alpha}^{\Lambda_n}$ we call the system of stochastic differential equations:

$$d\xi_{x,j}(t) = -\frac{1}{2}\frac{\partial S_{\alpha}^{\Lambda}}{\partial \xi_{x,j}}(t)\, dt + dw_{x,j}(t), \quad x \in \Lambda_n, \ \alpha > 0, \tag{1}$$

where $w_{x,j}(t)$ are standard Wiener processes independent for different $x \in \Lambda_n$ and $j = 1,\ldots,d$.
The measure $\mu_{\alpha}^{\Lambda_n}$ is invariant for the corresponding system (1) of Langevin's equations and it is easy to see that $\lim_{\Lambda_n\uparrow\mathbf{Z}^d} \mu_{\alpha}^{\Lambda_n} = \mu_{\alpha}$ exists in the sense of weak convergence of finite dimensional distributions. Moreover, for any $x \in \mathbf{Z}^d$, $j = 1,\ldots,d$, $F_{x,j}^{\alpha} = \partial S_{\alpha}^{\Lambda}/\partial\xi_{x,j}$ does not depend on Λ_n if Λ_n is sufficiently large.

Definition 2 (see (Malyshev, Ignatiuk and Sidoravičius, 1991)). By the system of Langevin's equations for the limiting measure μ_α we call the infinite system of stochastic differential equations:

$$d\xi_{x,j}(t) = -\frac{1}{2}F^{\alpha}_{x,j}(t)\,dt + dw_{x,j}(t); \quad x \in \mathbf{Z}^d, \ \alpha > 0, \tag{2}$$

where $w_{x,j}(t)$ are standard Wiener processes independent for different $x \in \mathbf{Z}^d$ and $j = 1, \ldots, d$.
We rewrite system (2) in a form convenient for us:

$$d\vec{\xi}_x(t) = -\frac{1}{2}\sum_{q\in Q} A_\alpha(q)\vec{\xi}_{x+q}(t)\,dt + d\vec{w}_x(t), \quad x \in \mathbf{Z}^d, \tag{3}$$

where $\xi_x(t) = \{\xi_{x,j};\ j = 1, \ldots, d\}$; $Q = \{0;\ \vec{e}_i; -\vec{e}_i;\ \vec{e}_k - \vec{e}_l;\ i, k, l = 1, \ldots, d\}$ and matrixes $A_\alpha(q)$ have the following form:

a) $A_\alpha(0)_{ii} = 4(d-1) + 4\alpha$;
$A_\alpha(0)_{ij} = -2 + 2\alpha; \quad i, j = 1, \ldots, d; \quad i \neq j$;

b) $A_\alpha(\vec{e}_i)_{ii} = -2\alpha$; $A_\alpha(\vec{e}_i)_{jj} = -2; \quad i, j = 1, \ldots, d, \quad i \neq j$;
$A_\alpha(\vec{e}_i)_{ij} = 2 - 2\alpha$; $j \neq i$, $A_\alpha(\vec{e}_i)_{kl} = 0; \quad k \neq i, \quad l \neq j$;
$A_\alpha(\vec{e}_i) = A_\alpha(-\vec{e}_i)^T$;

c) $A_\alpha(\vec{e}_i - \vec{e}_j)_{ij} = -2 + 2\alpha$; $A_\alpha(\vec{e}_i - \vec{e}_j)_{kl} = 0; \quad k \neq i, \quad l \neq j$;
$A_\alpha(\vec{e}_i - \vec{e}_j) = A_\alpha(\vec{e}_j - \vec{e}_i)^T$.

Let us consider translation–invariant Gaussian random field $\{\xi_{x,j},\ x \in \mathbf{Z}^d\}$ on the lattice $\mathbf{Z}^d$ with zero mean $\langle \xi_{x,j} \rangle = 0$, $x \in \mathbf{Z}^d$, and the spectral density

$$f_2(\lambda) = \left[2\sum_{y\in\mathbf{Z}^d} A_\alpha(y)e^{-i\langle\lambda,y\rangle}\right]^{-1}, \quad \lambda \in [-\pi, \pi]^d.$$

By $\mu_{\alpha,0}$ we denote the measure corresponding to this Gaussian field.
Consider now a Gibbsian modification $\mu^{\Lambda_n}_{\alpha,\varepsilon}$, $\Lambda_n \subset \mathbf{Z}^d$, of the Gaussian measure $\mu_{\alpha,0}$. Let

$$\frac{d\mu^{\Lambda}_{\alpha,\varepsilon}}{d\mu_{\alpha,0}} = \frac{1}{\mathbf{Z}^{\Lambda}_{\alpha}}\exp\left\{-\varepsilon\sum_{x\in\Lambda} P(\xi_{x,j})\right\},$$

where $P(\xi)$ is a polynomial bounded from below.
It is known (see (Malyshev and Minlos, 1985)) that in this case, for sufficiently small $\varepsilon > 0$, there exists $\lim_{\Lambda_n \uparrow \mathbf{Z}^d} \mu^{\Lambda}_{\alpha,\varepsilon} = \mu_{\alpha,\varepsilon}$ in the sense of weak convergence of finite dimensional distributions.
For the measure $\mu_{\alpha,\varepsilon}$, we write the system of Langevin's equations

$$d\xi_{x,j}(t) = -\frac{1}{2}F^{\alpha}_{x,j}(t)\,dt - \varepsilon\frac{\partial P(\xi)}{\partial \xi_{\xi,j}}(t)\,dt + dw_{x,j}(t), \tag{4}$$

where $x \in \mathbf{Z}^d$, $j = 1, \ldots, d$, $\alpha > 0$.

The system (4) has a solution in the real Hilbert space $H = \ell_2(\mathbf{Z}^d; e^{-|x|})$ with the scalar product

$$\langle f, g \rangle_H = \sum_{x \in \mathbf{Z}^d} f(x) g(x) e^{-|x|}$$

(see (Rozovsky, 1981)).
Let $\{\xi_{x,j}(t),\ x \in \mathbf{Z}^d,\ j = 1, \ldots, d\}$, $t \in \mathbf{R}_+$, be the solution of the system (4). By $\mu_{\alpha,\varepsilon}(t)$ we denote the probability measure on $\mathbf{R}^{d \cdot \mathbf{Z}^d}$, which corresponds to the solution $\xi_{x,j}(t)$, $x \in \mathbf{Z}^d$, $j = 1 \ldots d$, at the time moment $t \in \mathbf{R}_+$. The following theorem is our main result for the $\mathbf{R}$-gauge fields.

THEOREM 1. *Let $\varepsilon > 0$ be sufficiently small, $\alpha > 0$. Then the measure $\mu_{\alpha,\varepsilon}$ is invariant for the system (4). Moreover, if the measure corresponding to the initial field $\xi_{x,j}(0)$, $x \in \mathbf{Z}^d$, $j = 1, \ldots, d$, is the Gibbsian modification of the Gaussian measure $\mu_{\alpha,0}(0)$ with potential $\varepsilon \sum P_0(\xi_{x,j}(0))$, where $P_0(\xi)$ is polynomial bounded from below such that*

$$P_0(\xi) - \frac{1}{2} P(\xi) \geqslant c,$$

for all $\xi \in \mathbf{R}$, then

$$\mu_{\alpha,\varepsilon}(t) \longrightarrow \mu_{\alpha,\varepsilon} \quad (t \to \infty)$$

in the sense of weak convergence of finite dimensional distributions.

The proof of this theorem is not complicated modification of the proof of Theorem 1 in (Malyshev, Ignatiuk and Sidoravičius, 1991) and can be found in (Sidoravičius, 1991).
Further we shall consider the quasi-states $\langle \cdot \rangle_{\alpha,\varepsilon,t}$, $t \in \mathbf{R}_+$, on the algebra $\mathcal{A}$, i.e., linear functional on $\mathcal{A}$ corresponding to the measure $\mu_{\alpha,\varepsilon}(t)$ (see (Malyshev and Minlos, 1985)).

THEOREM 2. *The restriction of quasi-states $\langle \cdot \rangle_{\alpha,\varepsilon,t}$, $t \in \mathbf{R}_+$, on subalgebra $\mathcal{A}^{INV} \subset \mathcal{A}$ does not depend on parameter α, $\alpha > 0$, and, therefore, it is true for the limiting quasi-state $\langle \cdot \rangle_{\alpha,\varepsilon}$.*

Proof. From Stone–Weierstrass theorem follows that the plaquette-variables ξ_x^{ij}, where $\xi_x^{ij} = \xi_{x,i} + \xi_{x+e_i,j} - \xi_{j+e_j,i} - \xi_{x,j}$, $x \in \mathbf{Z}^d$, $i, j = 1, \ldots, d$, $i < j$, generate the algebra $\mathcal{A}^{INV}$ and the application of Itô's formula shows that the system of Langevin's stochastic differential equations for plaquette-variables ξ_x^{ij} does not depend on parameter α, $\alpha > 0$.
Generally, we consider the gauge field on $\mathbf{Z}^d$ with values on real q-dimensional manifold M. We assume that M is connected and σ-compact. For complete definitions and the necessary informations about stochastic differential equations on manifolds see the monograph (Ikeda and Watanabe, 1981).
We remark that the solution of corresponding system of Langevin's stochastic differential equations can be constructed directly by appealing to Whitney's imbedding theorem (see (Whitney, 1957)). The manifold M is imbedded into $\mathbf{R}^{2q+1}$ as a closed submanifold of $\mathbf{R}^{2q+1}$. The stochastic differential equations is defined globally in the Euclidean coordinate system and solution is constructed in standard way. If the initial value is on M, then it is easy to see that the solution remains

on M (see (Ikeda and Watanabe, 1981)). Thus this solution actually defines the solution of the system of Langevin's equations on M.
Finally, the application of cluster expansion technique allows to prove the complete analogues of Theorem 1 and Theorem 2 for M-values fields.

REFERENCES

Brydges, D., Frolich, J. and Seiler, E. (1978). *On the Construction of Quantized Gauge Fields.* I. Preprint of IHES, Bures-sur-Yvette.

Borgs, C. and Nill, F. (1986). *The Phase Diagram of the Abelian Lattice Higgs Model — a Review of Rigorous Results.* Preprint of ETH, Zürich.

Damgaard, P. and Hüffel, H. (1987). Stochastic quantization. *Phys. Reports* **152**, No. 5, 6, 227–398.

Malyshev, V. and Minlos, R. (1985). *Gibbsian Random Fields.* Nauka, Moscow (in Russian).

Malyshev, V.A., Ignatiuk, I.A. and Sidoravičius, V. (1991). Convergence of the stochastic quantization method. *Theory Probab. Appl.* (to appear).

Migdal, A.A. (1986). The stochastic quantization in the field theory. *Uspekhi Fiz. Nauk* **149**, 3–44.

Parisi, G. and Wu, Y. (1981). Perturbation theory without gauge fixing. *Scientia Sin.* **24**, 483–496.

Rozovsky, B.R. (1981). *Evolutional Stochastic Systems.* Nauka, Moscow (in Russian).

Osterwalder, K. and Seiler, E. (1978). Gauge field theories on a lattice. *Ann. Phys.* **110**, 440–471.

Sidoravičius, V. (1991). Convergence of stochastic quantization method for lattice gauge fields. *Lithuanian Math. J.* (to appear).

Zinovjev, Yu.M. (1981). $\mathbb{R}$-gauge theories on the lattice. *Theor. Math. Phys.* **49**, 156–163.

Ikeda, N. and Watanabe, S. (1981). *Stochastic Differential Equations and Diffusion Processes.* North-Holland/Kodansha, Amsterdam, Tokyo etc.

Kogut, J. (1979). An Introduction to lattice gauge theory and spin systems. *Rev. Mod. Phys.* **51**, 659–713.

Drouffe, J.M. and Itzykson, C. (1974). Gauge fields on a lattice. I. General autlook. *Phys. Rev. D.* **10**, 3376–3395.

Seiler, E. (1982). *Gauge Theories as a Problem of Constructive Quantum Field Theory and Statistical Mechanics.* Springer, Berlin etc.

Whitney, H. (1957). *Geometric Integration Theory.* Princeton University Press, Princeton, New Jersey.

LIST OF CONTRIBUTORS

Ambartzumian, R.V.	655 - 667
Astrauskas, A.	668 - 682
Bendikov, A.D.	221 - 232
Bloznelis, M.	73 - 77
Buldygin, V.	78 - 92
	289 - 300
Chibisov, D.M.	553 - 566
Chitashvili, R.J.	301 - 350
Chobanyan, S.A.	93 - 105
Chuprunov, A.N.	106 - 110
Dorogovtsev, A.Ya.	375 - 390
Ebralidze, S.S.	3 - 10
Evstigneev, I.V.	391 - 412
Gafurov, M.U.	11 - 29
Giorgobiani, G.J.	93 - 105
Glonti, O.A.	351 - 374
Gudynas, P.	111 - 122
Hamdamov, I.M.	11 - 29
Jamburia, L.G.	351 - 374
Kabanov, Yu.M.	413 - 435
Katyshev, P.K.	436 - 444
Khokhlov, V.I.	445 - 456
Klebanov, L.B.	30 - 37
Klesov, O.I.	38 - 42
Kolchin, V.F.	445 - 456
Kolchinskii, V.I.	167 - 182
Koval, V.	289 - 300
Lifshits, M.A.	233 - 240
Liptser, R.Sh.	183 - 189
Lugavov, V.S.	457 - 472
Makhno, S.	474 - 484
Mania, M.G.	301 - 350
Melamed, J.A.	30 - 37
Mishura, Yu.S.	485 - 502
Mnatsakanov, R.M.	617 - 628
Molchanov, S.A.	668 - 682
Mushtari, D.H.	241 - 245
Nagaev, A.V.	597 - 611
Nagaev, S.V.	190 - 197
Nazin, A.V.	612 - 616
Nechaev, S.K.	683 - 693
Ostrovskii, E.I.	246 - 254
Pergamenshchikov, S.M.	413 - 435
Puhalskii, A.	198 - 218
Purtukhia, O.	503 - 527
Radavičius, M.	629 - 635
Rogozin, B.A.	457 - 472
Rotar', V.I.	528 - 539
Saulis, L.	43 - 56
Sazonov, V.V.	123 - 136
Shemyakin, A.E.	636 - 643
Sidoravičius, V.	694 - 699
Sinai, Ya.G.	683 - 693
Sokhadze, G.	540 - 549
Solntsev, S.	57 - 69
Stoyanov, J.M.	413 - 435
Tsertsvadze, K.A.	644 - 651
Ulyanov, V.V.	123 - 136
	255 - 263
Vladimirskii, Yu.N.	264 - 275
Zerakidze, Z.S.	276 - 284
Zitikis, R.	137 - 164

Organizing Committee of the XXIII Bakuriani Colloquium on Probability Theory and Mathematical Statistics

I. Bokuchava, Tbilisi I. Javakhishvili State University, Tbilisi
D.M. Chibisov, Steklov Mathematical Institute of the USSR Academy of Sciences, Moscow
S.A. Chobanyan, Muskhelishvili Institute of Computational Mathematics of the Georgian Academy of Sciences, Tbilisi
V.I. Khokhlov, Steklov Mathematical Institute of the USSR Academy of Sciences, Moscow
A. Nikabadze, Razmadze Mathematical Institute of the Georgian Academy of Sciences, Tbilisi
A.A. Novikov, Steklov Mathematical Institute of the USSR Academy of Sciences, Moscow
V.V. Sazonov, Steklov Mathematical Institute of the USSR Academy of Sciences, Moscow
T. Shervashidze, Razmadze Mathematical Institute of the Georgian Academy of Sciences, Tbilisi
R. Tkebuchava, Tbilisi I. Javakhishvili State University, Tbilisi
N. Vakhania, Muskhelishvili Institute of Computational Mathematics of the Georgian Academy of Sciences, Tbilisi

Co-chairmen: V.V. Sazonov, T. Shervashidze and N. Vakhania.
Scientific Secretaries: V.I. Khokhlov and A. Nikabadze.

Acknowledgement

It is our please to express our gratitude to the editorial group that has prepared these proceedings for publication. Of those people whose contribution to this work was of particular importance we would like to mention V.I. Khokhlov, V.V. Statulevičius, T. Szejbak, N.V. Tret'yakov and E. Žalys.

The Editors

Bei Fragen zur Produktsicherheit wenden Sie sich bitte an:
If you have any questions regarding product safety,
please contact:

Walter de Gruyter GmbH
Genthiner Straße 13
10785 Berlin
productsafety@degruyterbrill.com